WORLD RAILWAY ACCIDENTS 1811-1880

Francis Voisey

ISBN: 978-1-9162011-0-1

DEDICATION

This book is dedicated to signalman Brian Stone, who first awakened my lifelong interest in railways way back in 1966.

CONTENTS

ACKNOWLEDGMENTS

Thanks to the staff of the Exeter Reference Library and the University of Exeter Library for their assistance while I was researching this book, to my friend June Nightingale for encouraging me to go ahead and get it finished when my enthusiasm was waning a little, and to the staff of Kindle Book Publishing for getting it published, especially Yanina Goldenberg, without whose guidance and assistance it is doubtful if I would have been able to navigate the mysteries of getting my first book into print.

INTRODUCTION

Although prototype railways have been known since ancient times and waggonways existed in Europe from the late Middle Ages the railway as we know it today only came into being coincident with the development of steam locomotion. The first railway in the modern sense was the Stockton and Darlington Railway, opened on 27th September 1825. The first true main line railway, the Liverpool and Manchester, followed in 1830 and the railway network in Great Britain subsequently rapidly expanded. Outside of Great Britain, the United States had opened its first railway in 1827 and France in 1828. Germany and Belgium opened their first lines in 1835, Canada in 1836, Austria and Cuba in 1837, Russia in 1838 and Italy and The Netherlands in 1839. In the 1840s Denmark, Switzerland, Spain, Hungary and British Guiana acquired railways, followed by Norway, Sweden, Egypt, Chile, Brazil, Argentina, India and Australia (the colonies of New South Wales, Victoria and South Australia) in the 1850s. By 1870 Portugal, Romania, Turkey, Mexico, Peru and Uruguay and South Africa (Cape Colony) had railways and the first railway in China opened in 1883.[1] By 1901 only a relatively few countries in the world lacked a railway system of some sort.

With the bringing into use of this new means of transportation, there came a new class of accident to add to road accidents and shipwrecks, while due to their much greater carrying capacity than road vehicles the new railways had the potential for much more serious land accidents. At the same time the running of flanged wheels on rails proved to be much safer than the operation of horse-drawn and later motorised vehicles upon roads, and serious railway accidents, horrible though they might have been when they did occur, proved to be much rarer than road accidents, even in the early days when the railways were considerably more dangerous than they are today in the early 21st century. To highlight the relative safety of road and rail transport, in the year 2017 alone, 18 people, not including pedestrians or cyclists, were killed in the county of Devon, England, in road traffic accidents.[2] By comparison, no one was killed in a train accident in Devon in 2017. In fact, since the opening of the first railway in that county, the Bristol and Exeter Railway to Exeter in 1844, 175 years previously, only one passenger and about half a dozen railwaymen had died in train accidents in Devon. (Train accidents may be defined as accidents happening to the trains themselves, such as derailments, collisions, fires and

[1] The Railway Magazine 1925, pg. 216 and 279

[2] Webpage - Devon Live These are people killed on Devon's roads in 2017

so on, as opposed to movement accidents, which are accidents to people involving the movement of trains excluding those classified as train accidents - mostly people being run down by or falling from trains, and non-movement accidents which are accidents on railway property not involving the movement of trains, such as falling down a flight of stairs or suffering electrocution. {These definitions are derived from Report to the Secretary of State for the Environment on the Safety Record of the Railways of Great Britain during the year 1970 pg. 33}.) It is no doubt the very rarity of railway accidents which makes them so newsworthy and gives them a (perhaps slightly morbid) interest.

A good many books have been written over the years on the subject of such accidents. Some have concentrated on single disasters, others on accidents in a given country or geographical area. Railway accidents in countries such as Great Britain, Germany, The Netherlands, the United States, Australia, New Zealand and so on have been documented comprehensively in one or more books but no attempt has been made previously to detail every significant railway accident that has occurred worldwide. This is clearly a very ambitious project, but this volume will attempt to, at any rate, detail all accidents involving five or more fatalities to train passengers which occurred up until the end of the year 1880, with further volumes hopefully to follow dealing with nineteenth century accidents between 1881 and 1900 and covering the twentieth and early twenty-first centuries. The original intention had been to cover the whole of the nineteenth century in one volume, but it was found that this would have made for an overly large work, so the decision was taken to limit the first volume to accidents up to the end of 1880 instead. A good number of less serious accidents which have been considered of interest have also been included. The accidents have been grouped into chapters according to their primary cause and are dealt with almost entirely chronologically within each chapter.

The gathering of information about early railway accidents in distant parts of the world has not been easy, though much easier nowadays than when this project was begun in the days before the internet and the World Wide Web. Outside of Great Britain and Ireland, few countries seem to have had a formal body charged with investigating all serious railway accidents during this period, though one or two major disasters such as, for example, the 1854 Baptiste Creek collision in Canada were the subject of full formal investigations by specially appointed bodies. Otherwise, reliance has largely had to be placed on contemporary newspaper reports and previously published works on the subject, though the author has tried as much as possible to keep to primary sources. If the book appears to be biased towards accidents which occurred in Great Britain and the United States this is partly because these are the countries about which information on accidents has been easiest to obtain, though these were also in any case the countries where many of the most serious accidents occurred.

Contemporary spellings of place names have been used where appropriate and for the many accidents in the United States, American terms such as railroad, engineer, caboose and soon have been used in place of the English terms of railway, driver, brake van and so on. The 24-hour clock was not used anywhere in the world up to 1881 (its use first began in Italy from 1893) and all times are thus quoted using the 12-hour clock.

The names and boundaries of countries have undergone many changes over the years. Thus, for example, German and Italian unification was not completed until 1871, prior to which each country consisted of a collection of independent states. On the other hand, modern Eastern European countries such as the Czech Republic, Slovenia, Slovakia, Estonia, Lithuania and so on did not exist as independent entities before 1919 or later. Australia was a collection of separate colonies until 1901, likewise South Africa until 1910. India, on the other hand, included the present (2019) independent countries of Pakistan and Bangladesh until 1947. The world atlas of say 1840 looked quite different from that of 2019.

CHAPTER 1

DERAILMENTS: TRACK DEFECTS

The tracks over which trains run are obviously a vital part of the railway and any problems with the track are likely to lead to the train being derailed. The earliest rails used on waggonways were made of timber. These tended to wear our quickly and an advance was the laying of cast iron strips upon them to make them more durable. Such strap rails were quick and cheap to build but they required a lot of maintenance and tended to work loose (see Plainfield accident below). An alternative, developed in Great Britain, was the use of tracks with 'L' shaped rails or plates, which came to be known as plateways. The plates were laid on stone blocks, sometimes without sleepers, though this rendered the track likely to spread and the stone blocks tended to shift out of position. The cast iron plates were about three feet long. Edge rails, with flanged waggon wheels, proved, however, to be the way forward and there was a breakthrough in 1820 with the development of wrought iron rails in 15 ft lengths, which were strong enough to bear the weight of a steam locomotive and its train and were used on the Stockton and Darlington Railway. With improved manufacturing techniques, the lengths of rails gradually increased to 39 ft in Great Britain by the late 1840s, 39 ft in the United States by the 1850s and 60 ft in England in 1895. Wrought iron rails were also gradually replaced by steel ones, initially used in Germany in 1861.

The use of individual stone blocks to support the rails soon gave way to the employment of wooden sleepers, which were easier to maintain. The wrought iron rails of the 1820s also required the use of chairs in which the rails sat, keys to hold the rails in place and spikes or bolts to secure the chairs to the sleepers. In 1830 Robert L. Stevens invented the flat-bottomed rail which could be spiked directly to the sleepers without the need for chairs, and these were soon adopted in most of the world, though not in Great Britain.[3]

The earliest tracks were laid directly on the ground, but this proved to be unsatisfactory when steam engines came into use, as the ground would rarely be strong enough to bear the weight of relatively heavy locomotives without settlement. Ballast, consisting of stones or ash, thus came into use to support the track and to help spread the

[3] Wikipedia – Permanent Way (history)

load, while it also improved drainage and inhibited the growth of vegetation.

The earliest train derailment, clearly due to track defects, of which records appear to have survived, was one on the Liverpool and Manchester Railway in England on 24th September 1830. The accident occurred on the high approach embankment to the Sankey Viaduct and involved Messrs Braithwaite and Ericsson's engine, *The William the Fourth*, on its way from Liverpool to Manchester. The accident ended luckily, for having left the rails the locomotive proceeded to embed itself in the soft earth and sand on the top of the embankment rather than falling to the bottom of the slope with much more serious consequences. Only one person was slightly hurt, and the engine was little damaged.

Workmen had been engaged in repairing the line at the point of the accident and the driver and others on the engine initially attributed the accident to some tool which had been carelessly left on the track by the workmen, which in the darkness (it was 7 p.m. at the time of the accident) they had not seen. A check next morning showed, however, that all the workmen's tools had been carefully laid out on the side of the embankment at the end of their day's labours and none of the tools showed any signs of damage.

Measurements with a bricklayer's level from amongst the tools disclosed the real cause of the derailment; ten yards from the point of derailment the left-hand rail was a quarter inch higher than the opposite rail, at nine yards they were level and at eight yards and closer the left-hand rail was an inch and a half lower than its counterpart. 'This sudden deflection at once accounted for the whole affair. The flange of the off-wheel having lost its hold of the rail, the engine, as a matter of course, slipped off, the tendency to do which was probably increased by the circumstances of the wheels of the tender (which was in advance of the engine) not being fixed on the axles, as is the case with Mr Stephenson's tenders, and ought, we think, to be the case with all carriages which have to travel on such unprotected embankments as those of the Liverpool and Manchester Railway.'[4]

On Thursday 10th November 1836 (date uncertain, possibly Thursday 3rd) a Baltimore and Ohio Railroad train was in trouble near Point of Rocks, Maryland, United States. Its engine and several passenger cars ran off the line, two cars were 'shattered to pieces' and seven passengers and its conductor were injured. The conductor was the most seriously injured while Mrs James L. Hawkins of Baltimore suffered a dangerous head injury. One of the rails had come loose and was out of place. 'Snow was falling at the time.'[5]

Problems had also occurred from time to time with the track of the United States Portsmouth and Roanoke Railroad, a newspaper reporter writing that, 'the iron rails having from time to time become loose and projected over the timber to which they had been secured, timely precautions were taken against accident from this cause. Whenever a rail had started from its place, measures of security were ordered to be taken.' On 11th November 1837 these precautions proved insufficient, for as a passenger train hauled by a new Norris engine was approaching Rochelle's or Newsom's, Virginia, United States, its engineer and a passenger riding on the engine simultaneously noticed a problem with the track a hundred yards ahead and jumped from the engine. On reaching the defective track, the engine was derailed, and its wheels and lower framework broken. The leading

[4] Webpage – Newton Heritage: Sankey Viaduct Train Derailment

[5] Huron Reflector, Norwalk, Ohio, United States, 15th November 1836

car was crushed and the second was lifted up by the third running under it and thrown fifteen feet to one side, but it was the passengers in the first and third cars who were injured. There were 25 or 30 passengers in these two cars, including six or eight ladies travelling just for the novelty of the ride. About fourteen of them were more or less hurt, and two of them, a Mrs Rochelle and a Miss Blow, shortly died of their injures, while the condition of a Miss Simmons and a Mr Crocker was giving rise to concern.[6]

The failure of a railwaymen to shift the points after a Birmingham train had passed caused the derailment of a heavily loaded goods train from Liverpool to Manchester at Newton Junction, England, on the Liverpool and Manchester Railway, between 8 and 9 p.m. on the evening of Thursday 18th October 1838. A drover named Hannon, who was riding in a waggon with a load of pigs, was killed on the spot and the brakeman so badly injured that his life was despaired of. Thirty or forty of the pigs were also killed but the driver and fireman escaped injury despite the engine being thrown off the line.[7]

A serious derailment occurred on the North Midland Railway between South Wingfield and Ambergate, England, on 13th September 1840. A passenger train was involved, and the first six of its fourteen carriages left the rails, though the engine and tender remained on the line. The third carriage, a London and Birmingham Railway second-class one, was the most badly damaged, being twisted round and overturned and its upper part knocked to pieces. A M. Deneux, described by *The Times* as a foreigner, 'supposed to have been shooting in the north,' was thrown out of it and killed when it then fell onto him. Another passenger, a Mr Wright, in the same carriage, was very seriously injured and died a few days later. A number of other passengers and Guard Shaw were more or less badly hurt.[8]

The *Times* report attributed the accident to a broken axle. Lt-Col Frederic Smith, who investigated the accident for the Board of Trade, was not so sure. There was no doubt that both of the axles of the leading carriage had been fractured in the accident, but he was of the opinion that this was a result of the derailment rather than its cause. There was a temporary siding at the site of the derailment leading to a spoil bank, with a facing connection with the up line. The inspecting officer decided that despite his assertion that he had done so, the flagman, Benjamin Smith, had failed to reset the spoil bank points properly before the arrival of the passenger train. It was true that the engine and tender had passed over the points without leaving the track, but he thought that in doing so they have shifted the points such that they fouled the left-hand wheel of the leading carriage. As contributory causes of the accident, Lt-Col Smith listed the fact that the ballasting was insufficient, that the rails were laid on stone blocks, that the axle-trees had imperfections and that the speed of the train was possibly too high. The ballasting of the line, he stated, should have been completed before it was used for passenger trains, not afterwards. As a rule, no temporary connections should be permitted after the line had been opened, and under no circumstances any such connections involving facing points. He considered transverse sleepers to be more suitable than blocks on unsettled embankments, and

[6] Washington Globe, Washington, D.C., United States,14th December 1837

[7] The Times, London, England, 23rd October 1838

[8] Ibid, 16th September 1840

longitudinal sleepers with sufficient cross-ties to be even more secure. The use of the best quality iron for axle-trees and wheels was essential, and officers of the company should have the right to visit contractor's works at any time to check on the quality of their material. He considered that the safety of passengers would be increased by the substitution of six-wheeled for four-wheeled carriages. Drivers and guards should be supplied with tables detailing the maximum permissible speeds over different portions of the line and issued with watches. Guards should submit reports at the end of each journey stating if the maximum speed had anywhere been exceeded.[9]

Some early American railroads were laid to begin with with strap rails, longitudinal timbers with a thin iron bar surmounting them. As noted above, these could eventually become loose with the passage of trains and spring up and penetrate the floor of a wooden coach. Passengers became very wary of the possibility of such accidents though they were in truth rather rare. One did, however, cause a fatality on the New Jersey Central Railroad, then still under construction, near Plainfield, New Jersey, on 12th August 1843, when it penetrated the floor of a car and pierced the throat of a mechanic, who was killed outright[10] (see also Miscellaneous Accidents chapter).

Just before 11 a.m. on 31st March 1845, a passenger and mail train from Altenburg to Leipzig, carrying 300 people, became derailed at Gerstenberg, Saxony, probably because the engine and leading vehicles were too heavy for the track. The leading postal van and following well-laden passenger carriage went down the embankment, the breaking of the coupling behind this vehicle preventing the remainder of the train from following. Seven seriously injured passengers were conveyed by horse-drawn vehicles to hospital in Altenburg. It would seem that one of them later died from his or her injuries.[11]

Derailments could sometimes occur due to an interaction of several causes but that between Luddenden Foot (contemporary spelling) and Sowerby Bridge, England, on the Manchester and Leeds Railway on 16th September 1847 was clearly due to a single cause, namely a broken rail. The last (first-class) carriage of the 1 p.m. down passenger train from Manchester to Leeds left the track at a broken rail a few miles beyond Luddenden Foot station and ran off the rails for two miles and through a tunnel to Sowerby Bridge station. There it was noticed, and the Sowerby Bridge stationmaster, observing that the last carriage was out of position, 'hoisted his red signal' and the driver was signalled to stop, but before he could do so the carriage ran over an ash-pit and lost its wheels. The passengers in the front coupé were thrown out at this point. One, Mr Gillard, superintendent of the telegraph, was killed outright and a second, Mr Robert Weston, Surveyor General of Customs, was fatally injured but the third, Mr Thomas Moon, secretary to Mr Weston, was not injured.[12]

Mr Moon described the accident, 'Immediately after passing Luddenden Foot station,

---

[9] Board of Trade report

[10] A History of Railroad Accidents, Safety Precautions and Operating Practices by Robert B. Shaw, pgs. 20-1

[11] Eisenbahnunfälle in Sachsen by Claus Schlegel pg. 41

[12] The Times, London, England, 18th September 1847

the oscillation and uneasy motion of the carriage convinced us that we were off the line. The rapid motion. The noise of the wind, and pattering of the rain, added to the fact that ours was the last carriage, convinced us also that we were too removed from the guard to enable us to draw attention to our position, and we therefore sat still, waiting tremulously the result. After the lapse of about three minutes, during which time the carriage was tossing to and fro in a frightful manner and dragged along at a rate of a full 30 miles an hour, we found we were entering the Sowerby Bridge tunnel, which is about a quarter of a mile in length, and which terminates just before reaching the station. In the tunnel, lights appeared to be dancing about, and I fancied I could perceive the faces of men, whom I took to be porters holding up lights, and our hopes were that they had seen the predicament our carriage was in. The train, however, proceeded, and on approaching the Sowerby-Bridge station, the deceased, Mr Gillard, remarked that he thought the carriage had righted itself upon the rails again; they became conscious that the oscillation was considerably less. I sought this opportunity to seize the straps which passed along the roof of the coupé, and, raising my feet out of the window, I thought it the most secure way to escape danger. I advised Mr Weston and Mr Gillard to do the same, but whether they did so or not I cannot say, as immediately after a terrible shock entirely rendered me insensible to what was passing around me. I remembered nothing further until after I was picked up and conveyed to the Royal Hotel.'[13]

Captain J.L.A. Simmons, who investigated the accident, stated that at the point where the rail broke the track consisted of T-shaped rails, weighing originally 58 lbs per yard, seated in chairs fastened to two ft square sandstone blocks. Seven years of heavy traffic meant that the weight of the rails had diminished to about 56 lbs per yard and the sandstone blocks were much worn down, requiring constant adjustment of the bolts of the chairs. As there were no lateral ties between the two rails, the track needed constant watching, especially in wet weather, lest it be forced out of gauge. The track was in the process of being re-laid with superior rails weighing 84 lbs per yard and the broken rail had occurred between the second and third blocks short of the new track. Captain Simmons examined a quarter mile of old track short of the point of derailment and found it wide to gauge, at one point by as much as an inch, the sandstone blocks much worn, the bolts not driven home so that the chairs sunk half an inch under the wheels of a train and the ballast in need of renewal. Further back again the track consisted of transverse wooden sleepers and was in better condition, but the sleepers were some of them decayed and there was the same need of ballast.[14]

The jury at the inquest had referred to the need for a guard's van at the end of the train,[15] and the inspecting officer agreed, pointing out that it had been recommended in 1840 that a brake carriage with a guard should be provided at least for every fourth carriage of a train. In this case the driver, on a very windy day, had failed to notice the extra drag caused by the rearmost of the seven carriages of the train being off the track, and hence a guard in the rear carriage would have been of little use unless the guard could have communicated with the driver. Numerous suggestions for doing this had been made but none, as yet, had been of practical use. The necessity for a guard at the rear of the train had, however, been reinforced by this accident, where the unlit rear carriage

---

[13] Ibid, 20th September 1847

[14] Board of Trade Report

[15] The Times, London, England, 21st and 22nd September 1847

could easily have come to a stand in the tunnel, with a goods train close behind.

Captain Simmons stated that the road was badly maintained and certainly not suitable for the running of express trains at the average speed of 43 miles an hour demanded by the timetable, especially with such heavy outside cylindered engines as that which had pulled the train involved. Nor was he impressed with the permanent way inspector's view that a quarter inch of vertical play in the rail was of no importance, stressing that the slightest defect in the road should be strictly avoided as causing undue movements in trains and engines and jars and wear and tear in their parts might lead to eventual accidents. If the inspector's statement that the five men employed maintaining two miles of track were fully occupied was true, then clearly more staff were needed so that the problems with the old track could be speedily corrected pending its renewal.[16]

An accident near the station at Newton Road, England, on Saturday 2nd September 1848, was another of those which was due to a combination of factors, one of which was the type of track used.

An up London and North Western Railway express from Liverpool left Wolverhampton 25 minutes late and its passengers described rough riding from there, with much oscillation before the final derailment seven miles along the line. As a result of the derailment, the engine and tender were thrown across the up line, coming to rest in the ballast on the west side of it, the leading luggage van ended up foul of the up line and the three passenger carriages, which made up the remainder of the train, were thrown in the opposite direction, one remaining on the up line and two clear of it. Shortly afterwards, a down pilot engine, which had been sent out from Birmingham because of the non-arrival of the Liverpool train, collided with the derailed luggage van and was then itself thrown off the line, the track of the down line having been damaged by the derailed engine of the up train. Two passengers in the train were badly injured, a Lieutenant Colonel Baird, aged 71, of Stirling, Scotland, who was badly shaken up; and a Mr H Shuard, an architect, of Paddington, London, who suffered a broken thigh. Colonel Baird died on the following Tuesday and Mr Shuard over a month later, on 11th October.[17]

Captain J.L.A. Simmons was again the Board of Trade inspecting officer. He had some difficulty in deciding how the derailment had occurred. The officers of the London and North Western Railway attributed it entirely to the defective construction of the last carriage of the train, one belonging to the Scottish Central Railway Company, the riding of which the passengers had complained of. Captain Simmons found that the bad riding was due to the fact that there was insufficient space between the carriage wheels and its floor, so that the wheels were scraping the bottom of the carriage. He was not convinced, however, that this was the cause of the derailment, the bottom of the carriage being too slight to cause a sudden resistance to the wheels sufficient to throw it off the line, as suggested by the coroner at the inquest into the death of Colonel Baird.

The inspecting officer noted that the train had been running at high speed, probably at least 50 miles an hour as it was a light train running on a falling gradient. The engine driver, not having been cautioned not to run at high speed, could not be blamed for doing

[16] Board of Trade Report

[17] The Times, London, England, 6th September and 12th October 1848

so, but Captain Simmons considered that the speed was too high for the type of track in use, which consisted of rails laid on stone blocks, with cast-iron chairs and wooden keys. There were no ties to maintain the gauge, and the superintendents did not consider the track to be in bad order if it were an eighth or a quarter inch out of gauge, despite this being enough to produce severe oscillation in engines or carriages passing at high speed. He considered that the accident was probably due to a high speed with an unstable engine over poor track.

As for the involvement of the pilot engine, it was apparently the practice to send them out when an express was 15 minutes late in arriving, and when other trains were 20 or 25 minutes late. He considered that if they were sent out, they should travel at very moderate speed, so as to be able to stop short of any possible obstruction, and with the extension of the electric telegraph they could hopefully be dispensed with anyway.[18]

The London and North Western Railway was again in trouble at Bicester, England, on Saturday 6th September 1851. The accident occurred at about 7 p.m., to a return excursion train from London Euston to Oxford. The branch from Bletchley to Oxford was single beyond Winslow, and Bicester was a crossing place with a loop and two platforms. Up trains took the main line through the station, down trains taking the facing points at the east end of the station onto the loop. Problems had arisen, however, with the non-stop Saturday return excursion trains passing through the connections at high speed, the drivers of them contending that they had to do so to maintain their schedules, and the Bicester stationmaster, Charles Bruin, therefore decided to let them run through on the straight up line. On the one occasion, previous to the day of the accident, when this was not possible because there was an up train from Oxford in the loop, the down excursion train had been specially cautioned at Winslow. At the end of August, the excursion train arrangements were altered, excursionists travelling up in the week then being permitted to return on the Saturday, and Stationmaster Bruin, finding that half of the 100 excursion tickets issued on Tuesday 2nd were still outstanding, expected the return excursion train to stop at his station, rather than running through non-stop as it had done on previous Saturdays. He therefore instructed Policeman John Wilmot at the Winslow end of the station to route the train into the down platform, and assuming that the driver and guard of the train were well aware that they were to stop at Bicester, did not take any special measures to have the train cautioned at Winslow as before.

The eastern side of the station was protected by a station semaphore signal and an auxiliary signal. When Policeman Wilmot was operating the signals for through excursion trains, he set the auxiliary signal at "Caution" but the station signal at "All right" whereas when Stationmaster Bruin or one of the porters operated them both were set at "Caution". The drivers of Saturday excursion trains thus became accustomed to running through the station on the wrong line at high speed past the "Caution" signals, which had little significance for them. On the day of the accident, Policeman Wilmot, for additional safety, equipped himself with two red flags, giving one to another policeman, Henry Harris, at the level crossing at the east (Winslow) end of the station for him to stop the train at the platform.

Wyld's self-acting points were provided at the east end of the crossing loop, set for

---

[18] Board of Trade report

the down line, and when opened, by hand or by an up train, would automatically return to the down line if they were kept well-greased and free of obstructions. When Policeman Wilmot reached them, he moved them to and fro to check that they were working properly, and before he knew it the down excursion was upon him. Leaving the points lever he went a few yards down the line to flag the train, failing from inexperience to realise how quickly the train was running, and before he could get back to the points, which were half open, the train was upon them, and the engine took the straight road. It was here that the policeman, in his vain effort to correct his mistake, made matters a lot worse. Had he just left the points alone the whole train would probably have just gone down the straight road and the harm would have been limited to a bit of damage to the point work and the wheels of the train. Instead, he moved the points lever with full force, diverting the tender and carriages to the left-hand line and wrecking the train. The engine was thrown off the line, ploughed through the ballast and across the turnpike road at the level crossing to stop close to the stationmaster's house; the tender and guard's van stopped on the ballast, but the leading carriage, striking a gate post at the crossing, was run into by the carriage behind and demolished, and the four or five following carriages were crushed and broken and piled up on one another. Five passengers were killed outright, a sixth died of his injuries, 20 were seriously injured and a larger number less badly hurt.[19]

The engine and driver of the excursion train had been changed at Bletchley, where the locomotive superintendent was unaware that the train was meant to stop at Bicester. Driver William Carrier, whose 14-year-old son was one of those killed, took the excursion from Bletchley. He knew nothing of having to stop at Bicester. When he reached Winslow, the stationmaster waved him through, but he pulled up anyway as he needed to take water, but he only stopped briefly to take enough to get him to Oxford. He had then run at high speed to keep time, and reaching Bicester saw the auxiliary and station signals at "Caution," but that was just as they normally stood when he was clear to run through on the straight line and he did not apply the brakes until he saw Policeman Wilmot's red flag, when there was only time to shut off steam, and not to reverse his engine, before he reached the points.

Reviewing the circumstances of the accident, Captain R.M. Laffan, the inspecting officer, concluded that Stationmaster Bruin was most to blame for disregarding the company's regulations as to the signals and speed at junctions, and then having done so having failed to have the excursion train driver specially cautioned at Winslow. Next to blame were the staff at Euston who had failed to inform the driver or guard of the Bicester stop. Policeman Wilmot was clearly directly the cause of the accident, but his actions were the result of lack of experience rather than neglect. Driver Carrier was to blame for disregarding caution signals, but great excuse could be made for him in the desire of all parties to run the excursion trains at the greatest possible speed.

Captain Laffan recommended that the branch be doubled, and meanwhile the speed of trains running over it be reduced. The jury at the inquest into the victims of the accident found a verdict of "Accidental death" and likewise recommended that the line be doubled.[20] [21]

---

[19] Ibid.

[20] Ibid.

An express train from New York to Boston was thrown off the track on Saturday afternoon, 30th October 1852, by a broken rail at Windsor Locks, Connecticut, United States. The passenger cars were thrown from the line and the rearmost was thrown fifteen feet into the basin of a canal, in which the water was twelve feet deep. There were about 20 passengers in the car, which immediately began to fill with water as it entered the canal. The male passengers, by great effort, managed to force its doors open, and the occupants were pulled out one by one. As soon as possible, the roof of the car was broken open with axes and the bodies of two brothers named Parker, of Pawtucket, Rhode Island, who had just returned from California, were then extracted. It was feared that there were other victims still in the car, but it was later found that the two siblings were the only fatalities, though all of the other passengers were more or less injured. The brothers who lost their lives had been away in the west for three years, having been drawn to the 1849 gold rush, and had spoken repeatedly on their eastward journey of how they were looking forward to being reunited with their wives.[22]

Broken rails were basically an unavoidable hazard throughout the nineteenth century, no means having been developed to detect them before they failed.

On 4th March 1853 a Lancashire and Yorkshire Railway express train, consisting of portions from Edinburgh and Liverpool, left Bolton for Manchester at 5.48 p.m., three minutes late, consisting of an engine and tender, six passenger carriages and two guard's vans. Having called at Halshaw Moor station, it passed through Stoneclough and Dixonfold and was passing through a deep cutting beyond the latter station, travelling at between 40 and 50 miles an hour, when the engine and tender left the rails and became separated from carriages, the first three of which were hurled from the line and smashed to pieces. Driver Thomas Croston was killed, his fireman, Henry Wood, was seriously injured and a third man riding on the engine, Richard Simmons, a night cokeman on his way to work in Manchester, was also killed. Among the passengers George Barbour, a five-year-old boy, was killed outright and a number were injured, of whom two more died later.[23]

An initial theory by the railway officials stated that the derailment had been caused by the breakage of a driving axle of the engine, yet this was dismissed by the inspecting officer, Captain George Wynne, after an examination of the engine in the workshops. He concluded instead that it was the track which was at fault. The section of line concerned had been laid 15 years previously with longitudinal timbers resting at nine feet intervals with cross sleepers dovetailed into them. The original rails had been replaced in 1846 by new rails weighing 68 lbs per yard, secured to the sleepers at joints by chairs of an obsolete pattern and elsewhere by spikes. From having new spikes being driven into them and the effects of the weather, the longitudinal timbers were greatly deteriorated and their bearing strength very materially impaired. The use of longitudinal timbers was one no longer practised and had the disadvantage of producing points of unequal bearing underneath the timber, making the line more difficult to maintain. The majority of chairs

---

[21] The Times, London, England, 8th, 9th,10th, 11th, 12th and 16th September 1851

[22] Gettysburg Adams Sentinel, Gettysburg, Pennsylvania, United States, 8th November 1852

[23] The Times, London, England, 7th, 8th, 9th, 12th, 13th and 16th March 1853

were loose in their beds from want of packing, and from being insufficiently screwed down, and the wooden keys generally required renewing. He continued, 'I should say that, from one or another of the above causes, full seventy-five per cent of the chairs were in an inefficient state, and the whole line required beating up under the timbers. In fact, to sum up, I would say, that the maintenance of the line was most carelessly performed; there were but two men to a mile employed on the part that I went over, a number hardly sufficient to screw up and pack the chairs and place them in an efficient state. The season has, undoubtedly, been very unfavourable to railways; but on this line there is less excuse than on many others; the ballast is in abundance, near at hand, and of the best description, and the line having a continued fall of 1 in 200 for several miles, there is every facility for drainage, and a line well drained and well ballasted can be kept in good order in any weather.'[24]

Turning to the question of speed, Captain Wynne stated that the motion of a train on an uneven road would consist of a series of bounds over bad joints, the magnitude and frequency of which would increase with increasing speed until they became highly dangerous He did not consider a speed of more than 20 or 25 miles an hour safe over the poor track between Bolton and a point short of Clifton Junction, and added that poor track likely resulted in an increased deterioration of rolling stock as well, leading to an increased number of broken axles and broken tyres. He was also critical of the use, by the Lancashire and Yorkshire Company, of contractors to perform track maintenance rather than using their own staff.

Finally, the inspecting officer commented on the positions, fourth and sixth in the train, of the two brake vans of the express and on the apparent running of some of the engines on the line tender first.[25]

On Sunday 27th March 1853 a Baltimore and Ohio Railroad passenger train consisting of a baggage car and three passenger coaches hauled by two engines was descending an incline of 116 feet to the mile (about 1 in 45) on a curve to Cheat River, West Virginia, United States, when it was derailed. The 'large engine started the nails' in the words of the newspaper report, presumably meaning that the spikes were ripped out of the track. The engines, baggage car and leading passenger coach kept to the line but owing to the spreading of the track the last two coaches were thrown 100 feet down an embankment, performing four somersaults in their descent. Of the 74 passengers in them nine were killed and almost all the others more or less badly injured. Only 12 of them were well enough to continue to Baltimore. Those killed were David Holt, a Baltimore merchant; A. Sallie, supposedly from South Carolina; Lewis Deline, returning from California (from the gold rush?); Richard Clayton, of Wellsville, Virginia; Miss Isaacs, a Jewess, travelling from Indiana to Pennsylvania; Mr Wooten, residence unknown; a stepson of Mr Murray, a railroad supervisor; and two children of Mr Giese, of St Louis. Mrs Giese was herself, badly cut and her face covered in blood, bathed and washed two of her children in the creek to revive them, though one died later. The other child was killed outright, and two others survived. The scene of the accident was described as a 'most

---

[24] Board of Trade report

[25] Ibid.

agonising one.'[26]

A New York Central Railroad train was running at full speed as it approached Nine Mile Creek, between Warner's and Syracuse, New York, United States, on 15th October 1857, when its engine suddenly left the track and plunged fifteen or twenty feet down an embankment to the creek below. The baggage car, mail car and three passenger cars left the rails in turn and piled up on top of the engine. The cars were well filled but the accident happened so suddenly that only one scream was heard. The last car remained on the track and its occupants hurried to the rescue of their less fortunate fellow travellers. The mail car was turned upside down and set alight by spilt coals from the stove.[27] The fire spread to involve one side of the passenger car, part of which was submerged in the creek, which was running high after heavy rain.

After most of the passengers had been extricated from the first passenger car, a shawl was discovered, which when pulled was found to contain the body of a Miss Catharine Brown of Toronto, 'daughter of Mr Brown, late of Toronto Globe,' who had evidently been stunned in the initial accident and had subsequently drowned in three feet of water. Another passenger, a Mr Clinton Bronson, of New Britain, Connecticut, agent for Townsend and Co Agricultural Implement Factory, Westfield, New York, had his abdomen pierced by mahogany splinters from a seat, and although he was said to be of strong build and constitution and was still alive at the time of the report his survival was thought impossible. The recovery of Mrs Lizzie Franklin, of Warren, Rhode Island, of Baggagemaster Peter Bettinger and of Baggageman Patrick Nolan was also thought doubtful and a number of other passengers were less badly hurt. Some had lucky escapes; Mr Ogsburg, of Jefferson, was in the second car and had the woodwork splinter a foot above his head, the car having been telescoped by the third car, fortunately at a point just below its roof. D.A. Levien had left the front car for the rear one only five minutes before the accident.

The derailment was attributed to the swollen creek washing away the embankment and causing the track to spread.[28]

The colony of New South Wales suffered a serious railway accident on 10th July 1858. A Great Southern Railway passenger train which had left Campbelltown at 7.35 a.m. for Sydney, and which consisted of an engine and tender, two horse-boxes and six second and third-class carriages, became derailed at Haslam's Creek, between Parramatta and Homebush. The horse-boxes and some of the carriages left the track and the coupling between the front horse-box and the tender giving way the engine and tender shot forward, while the horse-boxes and carriages remained more or less on the track for three hundred yards, when on reaching a curve they were precipitated down an embankment. Two passengers; a Mr George Want of Parramatta, of the firm of Want, Roxburgh and Want, Sydney attorneys; and a Mrs Hackett, of Pennant Hills Road, who owned a large orchard and was a dealer at the George-street fruit market in Sydney, were thrown down the embankment and killed by the descending carriages falling upon them, and several

---

[26] Adams Sentinel, Gettysburg, Pennsylvania, United States, 4th April 1853

[27] Reedsburg Herald, Reedsburg, Wisconsin, United States, 22nd October 1857

[28] Burlington Daily Iowa State Gazette, Burlington, Iowa, United States, 21st October 1857

other passengers were badly hurt. Ephraim Kater, an elderly man, would probably also have been killed had not his wife snatched him out of the way of a falling carriage at the last moment.

Richard Boan, the driver of the train, gave evidence at the inquest held at Parramatta. He stated that his train had been travelling at about 22 miles an hour at the time of the derailment and he was unable to account for the horse-boxes leaving the line.

John Whitton, engineer-in-chief for the New South Wales railways, stated that the track was in good condition for the low speeds at which trains ran, that there was no danger in having the horse-boxes marshalled at the front of the train, and that he was likewise unable to account for the accident.

Guard Richard Darby, who had been in charge of the train, estimated its speed to have been 23 miles an hour and he too was unable to account for the derailment.

The inquest jury not surprisingly returned a verdict of "Accidental death," stating they could not account for the accident from the evidence that they had heard, and no one appeared to be to blame.[29]

If the inquest was unable to reach a conclusion as to how the accident had happened, then the government inquiry which followed did. The investigation committee found that the location of the accident, at the commencement of a curve and just beyond the foot of a falling gradient of 1 in 132, the position of the horse-boxes next to the tender and fastened by loose chains or centre-links instead of screw couplings, and the speed of the train, estimated at 40 miles an hour by independent witnesses, considerably more than the speed given by the train crew, had all been factors in the accident, but the principal cause appeared to be the condition of the track.

Despite Mr Whitton saying that he could find no cause for the accident, the accident investigation committee considered that the permanent way was badly laid out and badly constructed, while the bridges were faulty both in design and materials.

The committee's report continued, 'The rail laid down – Barlow's rail – is not adapted to the nature of the soil; the cross-sleepers, for which this description of rail was not designed, were found not to have originally laid down in sufficient numbers, and the ballast employed is not adapted for the purpose. The curves upon the line are not only unnecessarily numerous, but most objectionably severe, and in many instances, curves are found where a straight line would be easier, safer and more economical.'

It also appeared that Driver Boan had been reported for dangerous driving the day before the accident, but no steps were taken to stop him driving a train the next day.[30]

Five days after the New South Wales accident, the Erie Railroad in the United States experienced the worst accident to occur on its system up until that date. An express train which had left Jersey City at 5.30 p.m. on Thursday 15th July 1858 had stopped at Turner's Corner for its passengers to take tea and was proceeding westwards at about 35 miles an hour at about 9 p.m. when it was derailed on a forty-foot high embankment at Shin Hollow, New York, six miles east of Port Jervis. The engine struck a broken rail. It,

---

[29] Sydney Morning Herald, Sydney, New South Wales, 12th July 1858

[30] Ibid,13th August 1858

along with the 'crate' (presumably the baggage car) and first four passenger cars, negotiated the faulty piece of track, but the two rear cars left the rails and after being dragged over the sleepers for some distance the coupling between them and the front of the train snapped, and they plunged down the embankment. The penultimate car performed one and a half somersaults and the rear car two or three, and it was in the rear car that six passengers died, while a number of others in both cars were seriously injured.

The *New York Times* reported the scene after the accident, 'For over an hour a scene of the wildest confusion prevailed around the spot. All the lights having been extinguished, the surrounding darkness added to the horror of their situation. While some laboured industriously to extricate those who were buried beneath the wreck, others piled portions of the fragments together and made bonfires to give light, and others brought water from a spring nearby to revive those who were fainting from the pain and the loss of blood. As soon as possible, after the full extent of the disaster was known, the locomotive, with one car, was despatched to Port Jervis to bring assistance, and in about one hour (11½ p.m.), it returned, having from four to five physicians, with such medicines and restoratives as were at hand, and several citizens, who promptly came down to render their assistance. By this time, all the injured but living persons were placed in the remaining cars; those most seriously hurt reclining upon double seats, and the dead being laid in one of the cars together. The train reached Port Jervis about midnight, but the news of the disaster had been circulated among the citizens, who rose, lighted their houses, and made every arrangement to receive and care for the wounded passengers.'

Those killed were Harvey Wood, of Wellsboro', Pennsylvania; George. W. Robertson, of New York City; a child of Mr and Mrs Turner, of New York City; a child of Mrs Brown, of Waterbury, Connecticut; Mrs Sarah Ray (coloured), of Binghampton, New York, who had been visiting her daughter at Newark; and a German named John Schultz, of Buffalo, a journeyman tinsmith. The same report listed 53 passengers injured to a greater or lesser degree.[31]

The inquest into the accident seems to have been a very thorough, if not long-winded, affair, given that a broken rail was quickly apparent as the cause of the accident, but apparently the popular public sentiment was against the Erie Company. Conductor Phineas Thompson and Engineer William Maxon (name illegible in newspaper) of the train gave evidence before the Coroner, along with other railwaymen and passengers, and the proceedings were very fully detailed in the *New York Times* newspaper. The jury eventually brought in the obvious verdict of "Accidental death."[32]

Disaster struck the Michigan Southern and Northern Indiana Railroad night express from Chicago, Illinois to Toledo, Ohio, just before midnight on Monday 27th June 1859 as it was crossing an embankment incorporating a culvert, known as Springbrook Bridge, between Mishawaka and South Bend, Indiana, United States. Exceptionally heavy rain on the previous afternoon and evening had swollen the usually small stream into a raging torrent and driftwood carried downstream by the torrent had probably blocked the culvert so that a dam formed on the upstream side of the embankment. The accumulation of the

---

[31] New York Times, New York, New York, United States, 17th July 1858

[32] Ibid, 19th July 1858

floodwater, combined with the weight of the train, caused the embankment to give way beneath it and the whole train was precipitated into the flooded valley below. The engine ended up buried in quicksand and mud and its tender, baggage and express car and two second-class were piled on top of it and 'shattered almost into kindling wood.' Next came two first-class cars, which were torn to pieces and carried downstream, and finally a sleeping car, which was less badly damaged. There were about 150 passengers on the train and a report the next day stated that 60 bodies had been recovered, while 50 or 60 more were injured. Those killed included the engineer and fireman, both named Chulp, Baggagemaster Babbington and Express-man Hartwell. The conductor of the train, William Osgood, was amongst those injured. Mr Bliss, the President of the Company, and Mr Hiram Sibley, one of its directors, were in the sleeping car; they escaped unhurt. The train had only been travelling at 10 or 20 miles an hour, Engineer Chulp having been running cautiously from La Porte, but the night was dark, the site of the accident was around a curve and he had no warning of disaster. The westbound express had passed over the embankment safely at 8.30 p.m.[33]

The citizens of South Bend and Mishawaka turned out in large numbers to assist the injured and recover the bodies of the dead. One woman was carried downstream towards the St Joseph River but lodged in a tree top and was rescued. She said that while she was caught there, she heard several other victims being carried along past her, calling for help. A safe of the U.S. Express containing $30,000 was broken to pieces in the accident but the agent managed to recover most of the contents from the wreckage. By the next day one of the through mail bags had also been found near the river, 200 yards from the embankment. It was not known if the other bags were lost.

A newspaper account of the disaster included a detailed statement from a passenger on the express, a Mr J.J.H. Rice. 'He was in sleeping car at the time of the accident, which occurred about midnight. The first he knew of anything being the matter was when the water was rushing over the cars. The lights were out, and nothing but an occasional flash of lightning was then to see by. He jumped from the cars, but instead of landing on *terra firma*, he leaped into the torrent, but succeeded in making his way to a clump of bushes and got up the bank afterwards. He heard a great deal of screaming while in the current, and after landing the cries of distress were awful. There being no lights, everything was confusion, and survivors were rushing wildly about the banks of the ravine in vain search of their missing friends. He says no one was killed in the sleeping car, but one old gentleman jumped out and was drowned in the current. Mr Rice says Osgood (the conductor) seemed crazy. He ran to Mishawaka, and to South Bend and then returned, a distance of eight miles, without his hat. He took charge of him and brought him to the city {South Bend} at 3½ o'clock yesterday. One woman who was on the train with her husband and five children ran wildly about all night seeking her family, but without success until morning, when she found them all dead. She then went to a farmhouse a few yards off, where after sitting some minutes, the wretched wife and mother expired. Mr Rice says that a gentleman, with his wife and two children, were on the train on a pleasure trip to the east. But in the accident the parties became separated. The husband found one of his young ones dead in the morning; next he found his wife dead, but with her arms closely embracing the other child, who was saved.'[34]

---

[33] Davenport Daily Gazette, Davenport, Iowa, United States,30th June 1859

[34] Ibid.

Another newspaper listed various incidents relating to the disaster. A woman passenger should have alighted at South Bend, and had she done so would have avoided the disaster, but she fell asleep on the train, was carried past her destination, and died. Three young men had seats opposite one another; two sitting and the other lying down. The former were killed and the latter but slightly hurt. A woman from Illinois owed her life and that of her child to Mr Richards, the Sleeping Car Conductor, who caught and saved them as they were being swept away. Mr E.P. Gillot, of Jefferson County, New York, was seated with his wife and child; they were returning from a trip to the West. He was the only one of the three to survive; to continue his journey with their corpses.[35]

As is often the case in such disasters, the exact number killed is very unclear. A newspaper report of 9th July spoke of the recovery of 57 bodies, with private dispatches putting the figure as high as 73.[36]

In the opinion of the author, the accident was to all intents and purposes unavoidable, but after an inquest lasting five days the coroner's jury found someone to blame, namely the local permanent way foreman, whom they felt should have checked the condition of the embankment in view of the heavy rainfall. Whether, had he done so, he would have been able to detect any problem, seems open to doubt, given that it was very likely the extra weight of the train itself which caused the embankment to finally give way. They also found (being wise after the event?) that the dimensions of the culvert were insufficient. The jury's verdict read: "That they are of opinion that the embankment and culvert at the place were well and substantially built; but that the culvert was not of sufficient capacity and size to carry off the water safely in time of flood, and liable to be obstructed by driftwood, rails, trees, etc. The said culvert was four and a half feet wide and six and a half feet high and should have been as large again. The jury, further upon their oaths, say that the foreman, Dennis Kane, was guilty of wilful neglect of duty in not examining the track on and over the embankment, after the storm of the night, before the time of the arrival of the train, in accordance with the instructions of the managers of the road."[37]

A train on the Hudson River Railroad carrying the 94th New York Regiment was derailed near Tivoli, New York, United States by a broken rail on Saturday 15th March 1862. Five of the cars were thrown into the river and four privates and a civilian were killed and 15 injured. The names of those killed were Samuel and Barton Glazier (father and son), Nelson Fenton, John Sherman and Master Driscoll (the civilian). 'None of those injured suffered broken bones.'[38] The train had only just set off from Tivoli and hence was not travelling at full speed. The five cars which went into the water included the two baggage cars in which the Quartermaster and his sergeant were riding, and the two were lucky that the baggage floated and did not bounce about, injuring them. The car in which the band were travelling turned upside down in the water and was smashed by another car falling on top of it. To escape from it the men had first to climb to the car above them. The band's instruments had been purchased in New York only the week before, the

[35] Fort Wayne Daily Times, Fort Wayne, Indiana, United States, 30th June 1859

[36] Defiance Democrat, Defiance, Oho, United States, 9th July 1859

[37] New York Times, New York, New York, United States, 21st July 1859, from the South Bend Register

[38] Pittsfield Berkshire County Eagle, Pittsfield, Massachusetts, United States, 20th March 1862

accident had rendered them useless and the officers were not in a position to buy more, so the regiment would have to leave without them.

The civilian who died, Master Driscoll, had been put off the train two or three times by the Quartermaster but had insisted on catching it, saying that when they reached New York he would enlist.[39]

The London Chatham and Dover Railway 7.10 a.m. down mail express from London Victoria to Dover met with an accident at Ospringe, England, a mile short of Faversham, on 9th May 1862. The train consisted of an engine and tender, two second-class carriages, two first-class carriages and a brake van and most fortunately was carrying only six or seven passengers. As the train reached Ospringe it was completely derailed, though the engine subsequently re-railed itself. The last three carriages and the brake van separated from the leading carriage and ran down a 15 ft embankment. Two of the carriages knocked down the high brick wall of an allotment and were broken to pieces. Two of the passengers in these carriages were killed and three seriously injured. Those who lost their lives were a Mr James Plumb, a commercial traveller; and Mr William Harris, a storekeeper for the railway company. One of those injured, Mr W. Crockford, of Rochester, died several days afterwards.[40]

There was little doubt as to the cause of the accident. Driver Mathew Simpson told the inquest into the deaths of Messrs Plumb and Harris that he had been travelling at about 30 miles an hour as he approached Ospringe (his guard, Charles Bonham, said 30 to 35 miles an hour) when he had noticed that there was a problem with the track, but he had had no time to stop or reduce speed very much before the train was derailed.

Ganger Henry Ramsden said that he had walked that part of the line that morning before the arrival of the express and had seen nothing wrong with it.

Colonel W. Yolland, who investigated the accident, pointed out the striking discrepancy between the statements of the engine driver and the ganger as to the condition of the track. He could not pretend to know which statement was true but if the driver's statement was then the damage to the track must have been caused by the preceding goods train and should certainly have been noticed by the ganger.

As to the cause of the accident, there was not the least uncertainty; it was due to the breaking of the wooden trenails securing the chairs on the outside rails of the curve. An accident had occurred previously on the same line from the same cause and the London Chatham and Dover had admitted the insufficiency of this kind of fastening for the permanent way and had directed that iron spikes should be inserted in half-inch holes bored down the centre of the trenails, a considerable portion of the line having already been so strengthened and the remainder being in hand.

In conclusion, Colonel Yolland stated that the continuing recurrence of accidents, even with trenails of the best construction and form, nearly new, and supplied by the best manufacturers, would oblige the inspecting officers in future to recommend that the

[39] New York Times, New York, New York, United States, 17th March 1862

[40] The Times, London, England, 10th and 12th May 1862

opening of any new line exclusively using such trenails should be declined.[41]

The jury at the inquest believed the driver's statement and concluded that the ganger must have missed the damage to the track but returned verdicts of "Accidental death."[42]

The Victoria portion of the London Brighton and South Coast Railway's 5 p.m. up express from Brighton met with a serious accident just on the Croydon side of Streatham Common station, England, on 29th May 1863. Its engine suddenly left the rails and, after running another 200 yards or so, overturned and its steam dome striking a rail was knocked off and the boiler exploded. The sixteen vehicles of the train – thirteen carriages and three brake vans – also all left the rails, some of them being overturned and a number badly damaged. Driver Salmon was killed, and Fireman Edward Tribe was badly scalded, while among the passengers three lost their lives and 59 were injured. Among the passengers on the train were 150 Grenadier Guards, returning from a periodical rifle practice at Eastbourne, two of whom, Charles Stone and George Blundin, were killed and 39 others injured. The third fatality was Mrs Eliza Chilver, the wife of a London doctor.

Dr Chilver was also a passenger on the train and gave evidence at the inquest, which must have been a distressing experience: 'I am a medical practitioner, residing at 14, New Burlington-street. The deceased, Eliza Chilver, was my wife. We were on our return from St Leonards to London and travelling in a first-class carriage. After passing Haywards-heath there was more shaking of the carriages than usual. I did not speak of it, as I thought it might create alarm. My opinion was that the carriages were not well braced up. The motion afterwards suddenly became very uneasy, the carriages rocked very much; this increased to a great extent, and it seemed that we were tearing up the road, as if the carriages were off the line. The swinging increased very rapidly, and the carriages fell over to the left side. I had hold of the arm hand and kept my seat; my wife was thrown on the floor. The carriage parted in several places. I had not much difficulty in removing the deceased. She was fearfully wounded in the face and head. I do not know at what rate the train was travelling. I can only say that it was going fast.'[43]

The line on which the accident happened, the Croydon and Balham branch, had only been opened at the end of the previous year. Its track consisted of double-headed rails weighing 75 lbs per yard, fixed in cast-iron chairs by patent compressed keys. The chairs were fastened to the sleepers by two iron spikes and there was a good quantity of gravel ballast.

Several witnesses other than Dr Chilver had spoken of the oscillation of the train near the Streatham Common distant signal, including the head guard, who had tried to apply the brakes, and Colonel Burnaby, of the Grenadier Guards.

Colonel W. Yolland, the Board of Trade inspecting officer, stated that non-stop expresses were allowed 16 minutes for the 10¾ miles from East Croydon to London Victoria, implying an average speed of about 40·3 miles an hour, but as the section included three junctions and three relatively sharp curves, a considerably higher speed would be required on the more favourable stretches. To prove his assertion the colonel

---

41 Board of Trade report

42 The Times, London, England, 15th May 1862

43 Ibid, 1st June 1863

and several officers of the company boarded the 11 a.m. train from Brighton to Victoria at East Croydon, unbeknown to its driver, and the train, which took 17 minutes for the journey, was timed by the inspecting officer over the stretch where the accident occurred at 60 miles an hour.[44]

The train had been drawn by a six-wheeled tank engine, weighing, without coal and water, 29 tons 15 cwt. It had been running tender first. It was attached to the leading brake van by a double screw coupling.

Continuing his report, Colonel Yolland stated that he had decided that the accident had resulted a combination of circumstances. All engines, he stated, were liable to oscillate when running at high speed, with irregularities in the track being among the causes of such oscillation, but short wheel-base engines, such as that involved in the accident, were more likely to do so, especially when not properly coupled up to the vehicle behind.

Another factor was the lack of strength of the permanent way, the spikes having been insufficient to bear the lateral strain upon them and having bent or been pulled out, the chairs broken, and the track entirely destroyed. The track would have been much stronger had the chairs been secured by fang or through bolts instead of spikes. The brake power of the train was also insufficient, with only one brake for eight vehicles exclusive of the engine (there were three brake vans but only two guards to operate the brakes). Three times the retarding force could have been provided with the use of continuous brakes without increasing the number of guards.

In conclusion, Colonel Yolland recommended that the LBSCR consider abandoning the use of tank engines of 13 feet 6 inches wheel-base on express trains or of strengthening the permanent way with heavier chairs and stronger fastenings. A portion of the line in question, around sharp 15 chain curves, had he understood already been so strengthened, but the accident demonstrated that the whole line needed to be dealt with.[45]

Guided by the evidence of the inspecting officer, the inquest jury returned a verdict that: 'the deceased persons severally came to their deaths from accident, and we, the jury, are of opinion that it was attributable to the high rate of speed at which the express trains run over the line from Croydon to Victoria. The jury would urge on the directors of the London and Brighton Company the necessity of allowing more time for the journey, and that careful attention should be given to the coupling of the trains.'[46]

The Great Indian Peninsula Railway morning up passenger train from Kussara, which left there at 3 a.m. on 27th June 1864, met with an accident four miles on the Wassind side of Titwalla, India, at 4.30 a.m. The train consisted of an engine, a first-class, three second-class and eight third-class carriages and two brake vans. The whole of the train was wrecked except one third-class carriage and the rear brake van, the engine toppling over and being partially buried in mud. Fifteen persons, all male 'natives,' were killed and eighteen persons were seriously injured; one of them being the European fireman, named Kemp, who suffered a fractured thigh, and the others, Hindus. A 'native' woman

[44] Board of Trade report
[45] Ibid.
[46] The Times, London, England, 2nd and 4th June and 2nd July 1863

had both her arms broken. In addition, European guard Joseph Price lost one of his ears, Second Guard Coales received several severe bruises and a Portuguese inspector of water cranes suffered a seriously injured cheekbone. The engine driver was unhurt as were a number of convicts riding in one of the third-class carriages. The accident was due to heavy rains washing away the earthwork of a culvert. Immediately after the accident one of the passengers, Mr P. Grogan, a superintendent's clerk at Bhusawal, jumped from his carriage and ran to Titwalla to telegraph from there for assistance, incurring some personal risk in doing so as the water was covering the line.[47]

A hospital train of six cars, carrying 275 sick and wounded Union soldiers from the U.S. hospital in New Haven, Connecticut to Readville, Massachusetts, was derailed, probably by a broken rail, in the deep Rocky Ledge cutting, four miles east of the crossing of the Connecticut River and rather further east of Old Saybrook, Connecticut, United States on the Shore Line Railroad. The accident happened at about 11 a.m. on Saturday 15th October 1864. The entire train was derailed, and its cars were thrown against the rocky walls of the cutting; three of the cars being completely smashed. The bodies of nine soldiers were extricated from the wreckage and Brakemen Horace Beebe and Edgar Parsons were also killed instantly. A third brakeman, Samuel Chittenden, was so badly injured that it was thought that he could hardly survive. Ten or twelve soldiers were seriously and twenty slightly injured. The badly injured soldiers were returned to the hospital at New Haven, one of them dying on the journey thence and an eleventh in the hospital the next day. The uninjured and slightly injured men continued their journey to Readville on another train. Superintendent Calhoun visited the scene and ordered the damaged rolling stock to be burnt as it was beyond repair.[48]

England was unlucky enough to suffer two major disasters in two days in June 1865, both due to the failure to properly protect work on the track. The second, on 9th June at Staplehurst, involved a bridge failure and will be described in the chapter relating to those, the first, on 7th June, at Rednal, will be described here.

The train involved in the Rednal disaster, on the Shrewsbury and Chester main line of the Great Western Railway, was a return excursion from Manchester and Birkenhead (for Liverpool) to Birmingham and London. The Manchester and Birkenhead portions had been combined at Chester into one long train of 28 carriages and two brake vans, hauled by two engines, and it left there at 12.29 p.m., an hour and four minutes late. Water was taken at Ruabon, 17 miles from Chester, and four more carriages were attached at Gobowen, 24¼ miles from Chester. Departure from there was at 1.50 p.m., an hour and 25 minutes late. Whittington, 26 miles from Chester, was passed without stopping, and after a brief climb and a level stretch the train began to descend a gradient of 1 in 132 to Rednal. Close to the start of the descent, a green flag on a 12 ft pole had been placed at the foot of the cutting to warn the drivers of up trains of work being done on the track 1,100 yards further on. Driver William Anderton on the pilot engine failed to notice the flag and in his dying declaration blamed the platelayers for not having exhibited one. Driver David Evans on the train engine did see the flag, shut off steam and whistled to

---

[47] The Express, London, England, 4th August 1864

[48] The Boston Post, Boston, Massachusetts, United States, 17th October 1864

attract Anderton's attention, while his fireman applied the tender brake. Fireman David Griffiths on the train engine stated that Anderton first gave a shrill whistle for the platelayers to get out of the way and then when they did not do so the brake whistle and the tender brake was applied.

Steam had thus been shut off and the brakes applied on both engines and both tenders by the time the train reached the spot where the platelayers were working. Fireman Griffith said that their engine ran steadily up to that point where he felt its front-end drop and it "shook tremendously." The pilot engine was thrown off the line to the left, the train engine to the right, and the following carriages piled themselves up against the obstruction so formed, four of them being destroyed and 11 others badly damaged. Driver Anderton was fatally injured, the fireman of the train engine was killed, and 11 passengers lost their lives and 30 were injured, some very seriously.

Evidence at the inquest and the Board of Trade inquiry into the disaster disclosed that the platelayers had been ballasting and lifting the up line near Rednal for about a fortnight and the flag had been placed in the cutting about a mile distant from where they were working in the first instance, but they had gradually moved towards it without moving the flag further backwards. Captain H.A. Tyler, the Board of Trade inspecting officer, noted that a green flag was not very easy to make out from a distance against the background of a green cutting and a red flag in the hands of a platelayer would have been better, and was the least that should have been provided in the circumstances. As the train had approached, five platelayers and Ganger Thomas Davis had been at work and Ganger Davis told one of his men, John Crump, to run back and signal to the drivers of the excursion as they seemed to be coming rather fast, given the state of the track.

Captain Tyler remarked that the evidence of the platelayers was not satisfactory. They alleged that they had done no lifting of the track in the two and a half hours before the accident, that the track had been in a satisfactory condition, and that the engines had not shut off steam until they had passed them. However, two of the platelayers admitted that they had been beating up the ballast under the sleepers up to the last moment as the train approached (hence the platelayers being slow to clear the line when Driver Anderton whistled) and a third man stated that the ballast under the sleepers supporting two lengths of rail had not been beaten up when the train passed over them. This, Captain Tyler remarked, no doubt pointed to the cause of the accident. The leading wheels of the pilot engine, (No. 5) doubtless left the track at this point due to the defective track. The remainder of the train did not derail here but because of its short wheel-base and the weight overhanging its trailing wheels, No. 5 was more sensitive to imperfections in the track. Even then it would not have turned to the left so suddenly had it not met with a check rail and siding connections at the station.[49]

Another important factor in making the accident so severe, the inspecting officer added, was the lack of brake power on the train, with only two brake vans for 32 carriages, 'a proportion which is quite insufficient for a reasonable degree of safety on any gradient, and which was especially so on gradients which this train had to descend, the steepest having been 1 in 82½.' One of the guards had applied his brakes on hearing a whistle from the engine and the other when he found his van bumping over the track after it had been disturbed by the derailed engine. Frequently, in the case of an accident, the

---

[49] Board of Trade report

amount of brake power meant the difference between life and death for the passengers. In the present case there was a clear 600 yards between the initial derailment of the train engine and the final pile-up and had there been more brake power the carriages would have acted as a drag upon the engines instead of forcing them forward and so far reduced the speed of the train that the results of the accident would have been much less serious.

In conclusion, Captain Tyler stated that it was desirable that whenever it was necessary to warn a train driver, due to work on the permanent way or otherwise, that explosive fog signals should always be used in addition to hand signals. Although train drivers were required to keep a constant look out it was inevitable that they were more likely to miss a hand signal at a point where they did not expect to see one, than a fixed signal, the location of which they were familiar with, especially when travelling at speed, when a mere glance at a gauge or the machinery might be sufficient to miss an unexpected hand signal.[50]

The jury at the inquest in the victims of the disaster returned a verdict of "Accidental death" adding that: 'We find that there was great blame attached to the officials of the Great Western Railway in not providing sufficient break power before leaving Chester, and in not providing better carriages and engines in this excursion train. We consider that the enginemen were guilty of gross and culpable negligence in not attending to the signal put up by the platelayers and in going at too great a speed over the defective portion of the line between Babingswood-bridge and the scene of the accident. We think the platelayers were also guilty of gross and culpable negligence in not properly and sufficiently packing the sleepers so as to make them secure enough to bear so heavy a train as the excursion train.' The foreman afterwards added that the jury had carefully avoided the use of the word "wilful" in their verdict, because they did not wish to bring in a verdict of manslaughter.[51]

In early September 1866 the 17-mile-long line between Carnarvon (contemporary spelling) and Afonwen in North-West Wales had not yet been opened to the public or inspected by the Board of Trade, but the locals were already using the line to make journeys, including a party of Volunteers, and excursionists. On Thursday 6th September 1866 a number of people wished to travel from Portmadoc (contemporary spelling) and Pwllheli to a Calvinistic meeting at Carnarvon. The contractors in charge of the line were not happy about running the train but, in the end, gave in to pressure for them to do so. The operation of the train seems to have been arranged in a very casual manner, and there could potentially have been a collision with one or the other of two engines in its path on the single line, but one of them was laid up for repairs and the driver of the other heard from a navvy of the running of the excursion train, and so was careful to keep out of its way. In contrast to previous trips on the line, which had been free, this time the passengers were asked to pay fares. Four volunteer brakesmen assisted the guard and Mr John Savin, the brother of the contractor, rode on the engine.

The train travelled slowly to Carnarvon, whistling at intervals lest one of the other engines should be on the line, and reached their safely. The return journey to Portmadoc began at 7 p.m. on a dark and stormy night with heavy rain; the train consisting of an engine and tender and 20 open ballast waggons. Twelve miles from Carnarvon and 350

---

[50] Ibid.

[51] The Times, London, England, 8th, 9th, 10th and 12th June 1865

yards short of Brynkir station the train stopped to take water. Mr Savin was again riding on the engine with the driver and fireman, also Mr Pearce, a banker. Its engine having taken on sufficient water, the train was restarted but it had only progressed 250 yards when it became derailed on the facing points where the line became double-track at the north end of the station, the driver only having time to pull the reversing lever. The engine stopped upright, embedded in the soil, but the leading ballast waggons were thrown off the line and the passengers riding in them were thrown out, five being killed outright and a sixth fatally hurt. Other passengers were injured.

Captain Henry Tyler of the Board of Trade remarked that the accident was plainly caused by the facing points having been half open. The engine had passed safely through them on the west side, the tender and all the trucks on the east side. The tender had then dragged the engine off of the line. The points were the permanent ones and were secured by a spike against the left rail but were not locked and there was no indicator to show their position. Pointsman Alfred Harvey had held the points in the morning for the passing of the excursion train, the running of which he was unaware, nor did he know at what time the train would be coming back, and he did not mean the points for its return journey, not having been specially instructed to do so.[52]

The coroner's jury at the inquest on the six victims returned a verdict that "We find that the death of the six persons whose bodies we have viewed was caused by an accident on the Carnarvonshire Railway, near Brynkir station, on Thursday, the 6 inst. Such accident having been caused by the points having been opened, and a stone placed between the points and the stock rail, thereby keeping the points slightly open, but by whom the points were opened, and the stone so placed we have no evidence to show. We also find that due caution and vigilance were exercised by all of the officials connected with the railway in order to avoid the accident." Mr Breese, the Carnarvonshire Company solicitor, then announced that the Company was offering a reward of £100 for the conviction of the perpetrators.[53]

Captain Tyler stated that after examining the scene of the accident and various witnesses he was unable to be certain either if the points had been spiked, as Pointsman Harvey alleged, or that a stone had been placed in them. He agreed that there were traced of crushed stone in the points but believed that it might have been left there unnoticed rather than placed there maliciously, especially as Harvey admitted that he had never cleaned the points since they were first installed.[54]

A United States derailment on the Buffalo and Erie Railroad on 13th November 1866, near Wesleyville, Pennsylvania, was due, like those in England at Rednal and Staplehurst the previous year, to the inadequate protection of work on the track. Gang of labourers had been carrying out repairs and in the process of doing so had removed two rails. No protection had been provided and it was only when the New York express was seen approaching, when it was only a few rods away, that a labourer ran towards it waving his hat. The engineer probably did not see even this belated warning and only reversed his engine when it began to derail.

---

[52] Board of Trade Report

[53] The Express, London, England, 14th September 1866

[54] Board of Trade report

The train was thrown down the embankment ten or fifteen feet high and its cars smashed. Immediately after the accident the passengers who had escaped more or less unhurt seemed stunned, but they then began to assist the workmen in going to the assistance of their less fortunate fellow passengers. News of the wreck was sent to Erie and an engine with two cars was despatched and brought back the most seriously injured and the bodies of the four who had been killed, which were lodged in the superintendent's room at the depot. The engine and cars then returned to the accident site and picked up the slightly hurt and unhurt passengers.

Four passengers lost their lives; a Dr William Wheeler from Ellridge, New York and his little grandchild, John West, the four-year-old son of Mrs Anna West of Milwaukee; Mathew Harland, from Sheridan, Iowa; and Thomas Hunt, from Nunda, New York. Several other passengers were probably fatally injured, including the mother of the dead child, and 40 or 50 less badly hurt.

The accident was attributed in the newspaper report to the criminal negligence of the track gang in failing to send out a flag or provide other protection against a train they knew to be due, and it was stated that no blame attached to the engineer or conductor of that train.[55]

Another report stated that the gang was in charge of a man named Mahony. The two old rails had been removed and one of the new ones had been inserted. The men were hurriedly trying to replace the other when the train appeared in the distance, but the new rail was slightly too long. Mahony apparently admitted at the coroner's inquest that he alone was to blame for the accident. Why he had failed to protect the track work was not explained, perhaps he was short staffed and needed all the men to do the actual work, or perhaps his watch had stopped, and he thought that he had more time than was actually the case.

This report added that, 'Mrs West, of Milwaukee, was found completely wedged in between heavy timbers, and covered by the debris. In her grasp was the little boy dead. She clung to him without uttering a cry or groan, and all unconscious that his young life had been crushed out. She only released him from her hold when informed of his death, and afterwards with heroic patience and fortitude awaited the time when the energetic hands should rescue her from imprisonment. She was taken out, badly crushed, and at last accounts hopes were entertained of her recovery. Her conduct and general bearing under the terrible trial through which she passed is spoken of in the highest terms by all who saw her at the time.'[56]

Negligence on the part of men working on the track was of course not confined to England and the United States, as was shown by an accident to an excursion train from Marseilles to Paris at Saint-Albain, France on the Chemins de fer de Paris à Lyon et à la Méditerranée, on 1st August 1867. The accident occurred at 4.40 a.m., when there were probably less trains running, and as it involved an excursion train perhaps the men in charge of the work were unaware that it was coming. At any rate a man was sent out to protect the obstruction, but only to a distance of 300 metres, and although the driver of the excursion train saw the signal and at once shut off steam and whistled for the guard's

---

55 Philadelphia Age, Philadelphia, Pennsylvania, United States, 15th November 1866

56 The Philadelphia Inquirer, Philadelphia, Pennsylvania, United States, 19th November 1866

brakes the distance was too short to pull up in. The engine managed to cross the gap in the rails but about 80 metres further on the coupling between it and the tender gave way and the tender and carriages, filled with 800 excursionists, left the track. The tender ran into a pile of ballast and sleepers stacked at the lineside for the relaying work and was brought to an abrupt stop, whereby the following carriages were crushed up against it and some knocked down the wall of a lineside presbytery. The train consisted of seven second and eleven third-class carriages, and it was the second-class carriages which suffered, two being smashed to pieces.

After the accident, the railwaymen and the uninjured passengers reportedly worked zealously to extricate the victims while some country people stood idly by looking on with their hands in their pockets. 'The work lasted more than an hour, and the devotedness and activity of a country doctor and of the curé of St Albin were most conspicuous.'[57]

The accident claimed six dead and twenty more or less badly hurt, with an unknown number slightly injured.

A month after the accident the Mâcon Correctional Tribunal sentenced those responsible to prison terms ranging between 15 days and a year, plus fines.[58]

Carelessness on the part of a pointsman led to a major accident near Samnuggar, India, on the Eastern Bengal Railway on 7th May 1868. Part of a passenger train had gone into a siding when he prematurely reversed the points (or perhaps just let go of the points lever momentarily) whereby the middle coaches of the train collided with each other in the space between the siding and the main line and the rear coaches continued down the main line. The latter carriages met a train coming down that line and three of them were smashed to pieces and all their occupants more or less hurt. Three other carriages damaged to a greater or lesser extent. Twelve passengers were killed outright or died the same night, eight more were seriously injured and several slightly hurt. Mr Wilkinson, Officiating Administrator-General, had a very narrow escape. He was riding in the centre compartment of a first-class carriage and the adjoining compartment was smashed in by the buffers of the next carriage.[59 60]

Back in the United States, on Friday 23rd October 1868, it was according to the newspaper reports almost certainly a broken rail which led to the derailment of the 6.30 p.m. night express from New York as it approached Albany, New York, United States, just before midnight. The train consisted of two express cars, a baggage car and eight passenger coaches and was well-filled with passengers, drawn by the Hudson River Railroad engine *Poughkeepsie*. According to different reports, when he felt a problem beneath his locomotive the engineer either opened the throttle to get clear or, more likely, applied his brakes. At any rate the engine, baggage and express cars passed the spot safely but then a coupling broke and the following smoking car, containing fifteen or

---

57 The Times, London, England, 5th August 1867

58 Wikipedia – Liste des accidents ferroviaires en France au XIX siècle

59 The Times of India, 12th May 1868

60 Allen's Indian Mail, London, England, 18th June 1868

twenty men (no women were permitted in the male-only smoking car) left the track, went down the embankment and overturned in the soft clay at the bottom. The passenger coaches and one sleeping car followed suit, the second and third passenger cars somersaulted twice as they did so and were badly damaged. The second car crushed to death a Mrs Sarah A. Ross of Pennsylvania, a milliner returning from a shopping exhibition to New York. The four sleeping cars at the end of the train remained on the track.

After the accident, the engineer proceeded to Greenbush for assistance and the train for Troy, which was waiting to connect with the New York express, was backed down to the site of the accident and picked up the injured and uninjured passengers. About fifteen other passengers were more or less badly hurt and John Davidson, of Lyons, New York, died on the Saturday morning. It seems likely that a third passenger might have also died later. Most fortunately, the stove was very firmly secured and so did not overturn and start a fire to add to the calamity.[61]

Early on the morning of 12th January 1869 a night train on the Pittsburgh Cincinnati Chicago and St Louis (Panhandle) Railroad met with an accident at Cork's River Point, three miles from Pittsfield, Pennsylvania, United States. Heavy rain the previous day had caused a stream to overflow and the foundations had been washed away from about a hundred sleepers. When the train reached the spot, a sleeping car was derailed and thrown down a 60 ft high embankment. Three of the seven passengers in the car were killed outright and the others injured.[62]

A serious derailment on the Long Island Railroad near Willow station, a mile or so east of Jamaica, New York, United States, on 23rd April 1869, was another caused by a broken rail. The train, which had left Hunter's Point at 10 a.m., was only travelling at 20 miles an hour and only two of its cars were derailed but the train was well-filled, and the cars telescoped one another. Six people died including William C. Rushmore, President of the Atlantic Bank in Brooklyn and Dr Orestes Pray of Brooklyn and his mother Matilda of Milwaukee. The other passenger fatalities were P. Shanahan, a railroad contractor and three-year-old Jennie Adams. Brakeman George van Nostrand also lost his life and a number of passengers were injured, some very seriously.[63]

The rail which broke (in two places) was described as having been of the best English manufacture. It had only been laid six months earlier and had broken under the *Thurlow Weed*, the lightest engine in use on the line {Thurlow Weed was a prominent newspaper publisher, Whig and later Republican politician}.[64] The newspaper was moved to comment, 'An examination should be made today of the broken rail. If this rail is of the best English manufacture, as claimed, a decent regard for the lives of the passengers transported over the road should lead the company to remove them at once and substitute others. The fractured rail is unmistakably imperfect and teems with flaws. On each of the pieces there are two flaws on the bottom of the rail that are coated with hardened dirt. That the rail was broken by the smallest engine used by the road is additional evidence

---

[61] Philadelphia Inquirer, Philadelphia, Pennsylvania, United States, 27th October 1868

[62] Newport Daily News, Newport, Rhode Island, United States,14th January 1869

[63] Boston Post, Boston, Massachusetts, United States, 24th April 1869

[64] Wikipedia – Thurlow Weed

that it was of imperfect manufacture. This feature of the disaster will undoubtedly be thoroughly investigated by the jury on Monday. It is also charged that the tie on which the rail rested was perfectly rotten.'[65]

At the conclusion of a long inquest, the Coroner's jury found that the victims had died, 'in consequence of a car being thrown from the track by a defective rail and hold that the Long Island Railroad Company are responsible for said accident by neglecting their duty in keeping the track in proper order.'[66]

At about 4.15 p.m. on 17th June 1869 disaster stuck a passenger train near Otumba, Mexico. As a result, presumably, of a sudden downpour, a river suddenly became greatly swollen and the mass of water reached the point where the railway line crossed it just before the arrival of the train. The river bridge itself held firm but the force of the floodwaters undermined the approach embankment. The train driver apparently remained unaware of the danger until his train was derailed and fell into the river. The driver, two firemen, three guards, and three passengers, namely a Senora Zamacona, of Tiaxcala state, and her two children, lost their lives. The other passengers on the train were more or less injured but survived.[67]

The next accident to be considered, at Welwyn Junction, England, arose not from defective track as such but, like that at Bicester in 1851, from a set of facing points being shifted under the wheels of a train. The 8 p.m. Great Northern Railway down passenger train from London Kings Cross to Peterborough on 24th October 1869 had started on time and consisted of an engine and tender, two horse boxes, second and first-class carriages and a guard's van. There were no problems as far as Welwyn Junction and the engine and tender, two horse boxes and leading carriage passed through a set of facing points there safely, but the remainder of the train was derailed on these points with the exception of part of the last first-class carriage and the rear brake van, which were diverted onto the Hertford branch. The derailed carriages were thrown about the line and some of them much damaged. William and Jane Walden, husband and wife, were killed instantly and a young man named Charles Sell grievously injured. He lived long enough to have his left leg amputated just below the thigh but died a couple of hours later, probably from loss of blood. Several other passengers and the front guard were injured.[68]

The passenger train had run through the junction at about 25 miles an hour under clear signals. The Welwyn Junction signalman, George Williamson, stated that after the engine and leading two or three vehicles had passed the junction signals, he had replaced them. He swore repeatedly at the inquest that he had not touched the facing points, which would have become unlocked by the premature replacing of the signals. His wife had been in the box at the time, having brought him his supper, but had not touched anything. Lt-Col F.H. Rich, who inquired into the accident for the Board of Trade, was unconvinced by Williamson's denials, and sure that he had inadvertently moved the points and thus caused the accident.

It was apparently the custom on the Great Northern at this time for signalmen to

---

65 Philadelphia Inquirer, Philadelphia, Pennsylvania, United States, 26th April 1869

66 Indianapolis Journal, Indianapolis, Indiana, United States, 15th May 1869

67 Repocilacion de leyes, decretos y providencios de los poderes by Manuel Azpiroz, pg 615

68 The Times, London, England, 27th October 1869

replace their signals at junctions as soon as the engines and tenders of trains had past and the inspecting officer urged that this should be changed. He also desired the company to revise their regulations and strike out such archaic rules as No. 70, which stated that "the speed of trains when approaching any junction must be slackened to 10 or 12 miles per hour." At the inquest Driver John Moss, in charge of the next down train, was pressed hard by Mr Merriman, the solicitor representing the relatives of the deceased, on Rule 70, and could only reply that it was not understood to apply to main line trains. Lt-Col Rich observed that no express trains could keep to their booked times where they obliged to reduce speed so drastically at every junction.[69]

The Coroner's jury returned a verdict that the points had been shifted accidentally by the signalman as the train was passing over them and recommended the construction of an independent line from Hatfield to Hertford.[70]

Chemin de fer de Paris à Lyon et à la Méditerranée company mail train No 715 from Saint-Germain-des-Fosses to Clermont Ferrand left the track at 5.30 a.m. on 2nd July 1871, 300 metres before Port-Mort station, France. A first-class carriage mounted a van and was itself crushed by a postal van. It was in this coach that the victims were found; six died and a dozen were injured. The victims were mostly from the Issoire, Riom and Clermont-Ferrand communes of the Auvergne region. The accident was blamed on the track spreading due to its poor condition.

Victor Belin witnessed the accident as a child: "It was the hour when the express was due, and for a child living in Saint-Myon watching the express go by was an event! My father stopped his carriage for a moment. As the express neared us, I saw it suddenly stop and its carriages climbed into the air and crashed into one another."

A year later, two Paris Lyon et Méditerranée executives were committed to Riom Correctional Court to stand trial on a charge of permitting the normal running of trains over a stretch of track which was known to be in bad shape, but eventually the case was dropped.[71]

The North Missouri Railroad suffered an accident near Wellsville, Missouri, United States, on 10th November 1871. The usual regular train of seven cars had been strengthened by the addition of two extra cars at its rear for the use of 96 soldiers of the Fourteenth US infantry regiment, in charge of Lieut. S. McConihe and en route from their barracks at Newport, Kentucky, to Utah. Half a mile from Wellsville these two rear cars were precipitated down an embankment due to a broken rail which the remainder of the train had passed over or had fractured. The coupling between the two extra cars and the cars in front snapped so the latter escaped disaster. Brakeman Edward Colvin, son of Judge Colvin of St Louis, was killed outright and two soldiers, Butler Fletcher, of Tennessee and Joseph White, of Logansport, Indiana, were fatally injured. About thirty other soldiers were injured.[72]

---

[69] Board of Trade report

[70] The Times, London, England, 2nd and 9th November 1869

[71] Wikipedia – Liste des accidents ferroviaires en France au XIX siècle

[72] Liberty Tribune, Liberty, Missouri, United States, 17th November 1871

The thermometer stood at four degrees below zero (Fahrenheit) and the ground was frozen hard as an eastbound Lehigh Valley Railroad train was running at about 25 miles an hour along the hillside overlooking the Lehigh River when it struck a broken rail at Wild Run, near Rockport, Pennsylvania, United States on 1st February 1872. As was often the case the front of the train, consisting of the engine, baggage car and leading coach, passed over the faulty track safely, but the rail broke beneath the next car, which went over the embankment, dragging the rear car with it. Turning over and over, the cars fell down the embankment and the first ended up halfway across the river, but did not break through the ice, while the second ended up broken to pieces at the foot of the embankment. Railwaymen and passengers from the front car, which was dragged off the rails as well but did not go down the embankment, managed to rescue all the passengers from the car at the bottom of the slope before it caught fire from the stove and was completely burnt out. The passengers who died on the spot were listed as Mrs L. Praetorius, of Wilkes-Barre; Mr A. Smith, of the firm of Gordon and Smith, of Wilkes-Barre; Mrs M. Miller, of Tannery Station; Mrs Till and daughter, of Reading and Theodore Baxter, of 398, Arch Street, Philadelphia. In addition, it was stated that Mr Wolfe, of White Haven, who had a bolt through his throat, could not recover and H.H. Gillock, a lumber dealer, of Leslie's Run, had been thrown 100 feet and fatally injured. These were reportedly the first passenger fatalities on the line since it opened in 1855. About fifteen other passengers were less badly hurt.[73] [74]

The rail which broke had only been down three months. It was thought that the intense cold had made it brittle and the hardness of the frozen ground probably also contributed to the fracture.

In keeping with the style of contemporary newspapers, the Philadelphia Inquirer did not spare its readers the grisly details of the accident: 'Mrs Till when found was lying on the ice in a dying condition. A large iron bolt had passed entirely through her body. An attempt was made by a bystander to remove it, but the unfortunate woman suffered such excruciating pains that he desisted, and it was not removed until after she had breathed her last.

'Mr Baxter, of Philadelphia, was one of the numbers killed. His head was so badly smashed that in attempting to remove the hat, part of the head came off with it, presenting a most ghastly sight.

'An elderly man named Sampson, an old resident of Brooklyn, was taking a drink of water at the time the accident occurred and was violently pitched against the water tank. He grasped one of the racks on the side of the car, and maintained a firm hold on it until the car struck the ice in the river, when he was thrown under one of the seats, in which place he was found in an unconscious state. Restoratives were applied, and he soon regained consciousness, but the shock was so great that it is feared that her will not recover. He was sent to Easton by the evening train and his friends notified by telegraph.'[75]

The Oil City accommodation train on the Alleghany Valley Railroad, running from

---

[73] Philadelphia Inquirer, Philadelphia, Pennsylvania, United States, 3th February 1872

[74] Bloomington Pantagraph, Bloomington, Illinois, United States, 9th February 1872

[75] Philadelphia Inquirer, Philadelphia, Pennsylvania, United States, 3rd February 1872

Brady's Bend to Oil City on 10th February 1873 consisted of a passenger coach and five oil cars, which seems a very hazardous combination and one which would surely not have been permitted on a British railway, even at this early date; but this was the United States. The train was running on time when it neared Black's Siding, near Scrubgrass station, a few miles north of Emlenton, Pennsylvania, at about 4 p.m. Disaster struck at this point when it struck a broken rail, and the passenger coach and oil cars were thrown down an embankment. Fire broke out at once in the wreckage and although most of the passengers were able to escape, a newsboy named James Cordell, a William Casey 'said to be a sporting man' and a third unnamed man lost their lives.

'The track where the accident occurred was as sound as any on the road. The ties are all sound, and the iron is good, but the hard-freezing weather has caused the rails to break worse than ever before.'[76]

The inquest into the accident was in the opinion of a local newspaper a perfunctory affair. The jury did not bother to visit the scene of the accident and so had no idea of the condition of the road. The railwaymen stated that the train had been travelling at 17 to 20 miles an hour at the moment of the derailment, which they blamed on either a broken rail or a broken axle. The inquest jury accordingly passed a verdict to the effect that the deceased had died from an accident due to one or other of those causes. They did at least give their opinion that crude oil should not be carried on passenger trains. A local newspaper, the *Venango Spectator*, wrote: 'The loss of life in this case and the injury to those who were not killed is justly set down to the practice of carrying oil in connection with passenger coaches. It is a practice that should be abandoned at once and forever. There are enough unavoidable dangers without adding to them a supply of fuel to roast the victims. The frequent occurrence of disasters on the Alleghany Valley Road suggest grave doubts as to its careful management or its regard for the safety of passengers.'[77]

A disaster occurred in England early on the morning of Saturday 2nd August 1873 to the London and North Western Railway's down tourist special, 8 p.m. from London Euston to the North, consisting of two locomotives, 22 carriages and three vans. The train was not booked to stop at Wigan and was passing through at about 40 miles an hour when its eighteenth vehicle left the rails at the facing points leading from the down main to the down platform. The sixteenth and seventeenth vehicles in the train, family carriage No. 14 and a brake van, were dragged off in turn and ran through the station off the rails, the brake van striking a shunter's cabin, and re-railed themselves at the north end of the station. The eighteenth vehicle, a Caledonian Railway composite carriage, ended up on the down platform and the next five vehicles were thrown around in great confusion, one carriage breaking through a boundary wall and the roof of a wooden forge, a passenger falling from it to the foundry below. The rear guard's van and last carriage came to a stand on their wheels but off the rails. Thirteen passengers died in the accident and about 30 passengers and Conductor Alexander Harper were injured. The dead included Sir John Anson, Deputy-Lieutenant for Lancashire. A good proportion of those killed were servants, maids and nurses accompanying affluent passengers and their families.

The accident created something of a sensation, especially as its cause was not at all

---

[76] Titusville Herald, Titusville, Pennsylvania, United States,11th February 1873

[77] Ibid, 14th February 1873

obvious. *The Times* wrote in its editorial: 'Nothing is more unsatisfactory than to be obliged to conclude that an accident was "accidental." Such a phrase can only be a cloak for our ignorance. It implies that some cause was in operation of which we know nothing, and we are all left with the uncomfortable reflection that we remain exposed to an unknown danger.'[78]

Driver William Stawpert of the train engine told the Board of Trade inquiry into the accident that he had passed through the station under clear signals and looking back from his engine at the north end of the station he had observed sparks flying from the wheels about two-thirds of the way down the train, then had felt a jerk as the train broke loose. He had whistled for brakes and brought his train to a gradual stop lest the rear portion of the train should run into it a quarter mile north of Wigan North box. The pilot engine had then proceeded to Rylands Siding box to stop up trains and he had gone back to find what the problem was when he had discovered that he only had 17 vehicles attached to his engine. He had assisted Conductor Harper from his van and reaching the south end of the down platform was "so overpowered by the sight of the poor people in the wrecked carriages" that he had felt helpless and had retreated to the north box.

Driver Stawpert's evidence was followed by that of Conductor Alexander Harper and Head Guard Frederick May (who had been in the rear brake van), neither of whom could explain the accident, then by that of a good number of the passengers and two station porters. The latter had witnessed the luggage being thrown from the derailed guard's van occupied by Conductor Harper and were busy picking it up, unaware of the devastation at the south end of the station, until a passenger came up to them and asked, "What station is this?" and told them what had happened.

The next witness was Joshua Goodall, the Wigan South signalman who controlled the facing points at which the derailment had occurred. Not surprisingly, he was subject to a searching examination on the first day of the inquiry. He said that he had accepted the tourist train on the gong from Springs Branch, got it accepted from the north box and cleared his signals for it at about 1.17 or 1.18 a.m., he could not be exact as he was not required to keep a train register book. The facing points must have been locked for the main line for him to be able to clear his signals. He saw the train approaching, replaced his distant signal, was gonged regarding a train coming from Manchester and at that moment saw fire flying from the tourist train and its carriages jumping about. He replaced his home signal and, afraid that the train was going to hit his box, then sat down on a bench for several minutes and did nothing. After three or four minutes, with the Canal Bridge signalman repeatedly gonging him about the Stalybridge (Manchester) mail he finally sent the "Line blocked" signal. He stated most positively that he had not replaced the home signal, and thus released the locking on the facing points until after he had heard the noise of the accident. He admitted that the Stalybridge mail went into the platform line but said that he would have had to get permission from the north box before setting the road for the platform loop. He never replaced the home signal until the whole of the train had passed it. He had looked at the points after the accident and they had been undamaged and had worked perfectly since. He had never known anything to go wrong with them.

Various other witnesses followed, including London and North Western Locomotive

[78] The Times, London, England, 4th August 1873

Superintendent Francis Webb and Assistant Locomotive Superintendent George Whale; Midland Railway Outdoor Inspector Herbert Loveday and Great Northern Railway Chief Signal Inspector Amos Pigott, each trying to fathom out how the disaster could have been caused.[79]

Finally, Captain H.W. Tyler, the inspecting officer, stated his conclusions. He pointed out that the derailment could have arisen from a number of causes; the signalman might have moved the points beneath the train, there might have been a defect with the points, or the permanent way, or the rolling stock, and so on. The derailment must have been initiated by the near wheel of some vehicle of the train getting on the wrong side of the near tongue of the facing points. The most obvious means of causing the derailment would have been the signalman moving the points beneath the train, presumably having been under the impression that the whole train had passed and to reset the road for the Stalybridge mail train, but it would have very difficult to have shifted the points between the wheels of a train of six-wheeled carriages moving at 40 miles an hour, even deliberately, in daylight, let alone accidentally at night. Furthermore, the signalman stated that he had not replaced the home signal until after he had heard the crash, and his statement was verified by the observation of the Canal Bridge signalman. Until the home signal had been replaced the points would have been held by the interlocking. The signalman would also have needed to pull another lever, No. 9, to alter another set of points, or the latter would have been found to be strained after the accident, which they were not.

Having thus dismissed the possibility of Signalman Goodall having caused the derailment, Captain Tyler turned to other possibilities. There were various defects in the rolling stock after the accident such as bent axles and broken springs and axle boxes, but these were almost certainly the result of the accident rather than its cause. As regards the permanent way the situation was different. It was of a most substantial character but not in good adjustment with irregularities of gauge, line and level in the neighbourhood of the facing points which would tend to cause oscillation in passing vehicles. There was also a bulge inside of the off rail to the extent of $\frac{5}{16}$ of an inch between the off tongue of the facing points and the off tongue of the leading points south of them in consequence of the standard rail having been bent to receive the tongues of the points, as was the practice of a number of companies. Such a practice was objectionable at the site in question. It was true that trains had been running over the facing points for the previous two years without a problem and had done so at high speed since the accident, but the margin between safety and danger was small. The general condition of the track, combined perhaps with a little extra oscillation in a vehicle, might have been sufficient to cause an unfortunate lurch in one of them and precipitate the disaster.[80]

While admitting that the case presented great difficulty, Captain Tyler suggested that: 'A train of maximum length, drawn by two engines, after passing round a curve of 30 chains radius - on leaving which there would be a certain amount of oscillation in the carriages - ran at, say 40 miles an hour, along 60 yards of straight line, over a portion of the permanent way on which the traffic and shunting were almost incessant, and a high condition of maintenance was not preserved, through leading points and facing points

[79] Board of Trade report

[80] Ibid.

within 5 feet 9 inches of one another, with a bent rail between them to receive the tongues of the points, and past a signal cabin the locking apparatus in which in the course of six years of wear and tear become impaired in usefulness, and was in need of repair.

'Under these conditions, the flange of the leading wheel of the 18th vehicle slipped over the leading tongue of the facing points, fractured the heel chairs, and marked the chairs further forward, while the corresponding off wheel dropped off the off rail of the main line. The leading wheels of the carriage which thus left the main line were diverted from their course by the off rail of the loop-line, and further by their contact of the near leading wheel with a check-rail. The carriages in front of it were drawn off, and those behind it dragged off the main line as the train proceeded forward. Meeting with other obstacles, these various vehicles ran against the shunter's cabin, and up the platform, or against the platform, and finally came to rest in the positions in which they were found. And all this happened, apparently, without the points having been moved, either by the action of the pointsman, or by the effect of the wheels of the train, or from any other cause.'

Although a greater degree of safety could now be obtained than previously in the use of facing points through the employment of interlocking and locking bars, there was always the possibility of a problem such as the check rail being out of gauge or the heel of a point too low and stretches of line which included them could never be as safe as stretches without. The moral of the accident was obvious; the speed of trains should be reduced through stations where such points existed.

The inspecting officer concluded his report with a lengthy piece criticising the railway companies of the East and West Coast routes to Scotland for the competition between them which had involved them in trying to run their trains faster and faster at the expense of passenger safety.[81]

The inquest on the victims of the accident was a drawn-out affair but finally, on 9th September, the jury returned a verdict of "Accidental death," adding a rider repeating Captain Tyler's opinion that trains should pass through Wigan station at reduced speed.[82]

The train which had left Arlon at 10.25 a.m. on 1st September 1873 was derailed between Poix and Grupont stations on the Luxembourg Railway in the South-eastern corner of Belgium. The accident resulted from a broken rail. The engine escaped derailment but the leading carriage did not and was 'literally reduced to fragments,' while 'the passengers in the second {carriage} saw the roof of the first pass over their heads, and the flooring form across the line a sort of barrier, against which the succeeding carriages were overturned in confusion. The second suffered little damage, while those which followed were broken, with the exception of the rearmost and the baggage waggon.'

The accident occurred in a deep cutting near the Château of Mireswart, and the derailed carriages were badly damaged by being thrown against the walls of the rock cutting. There were no screw jacks available close at hand and many of the victims were trapped in the

---

[81] Ibid.

[82] The Times, London, England, 5th, 6th, 7th. 8th, 9th, 11th, 12th, 15th, 18th, 19th, 20th, 21th, 22nd and 29th August and 1st, 9th and 10th September 1873

wreckage for two hours until equipment could be obtained from Jemelle. Fortunately, there was no fire. A doctor named Delvaux chanced to be a passenger on the train and was unhurt so was able to begin treating the injured before help arrived from outside.

Four passengers died in the accident: Mr Graff, hotel-keeper at Wellin; M. Wala, a judge at the Liege Court of Appeal; Mlle Schong, of Metz and an unidentified young woman. Guard M. Pastiel of Brussels was also killed. Another sixteen passengers were injured including Madame Schanck, of Metz, who suffered two broken legs and was buried two hours in the wreckage and General Dens, who had a leg caught under a carriage for three-quarters of an hour.[83]

A Russian troop train carrying 420 recruits was involved in a major catastrophe near Odessa on 8th January 1876. It was derailed, apparently as a result of the inadequate protection of work being carried out on the track, and the whole of its 27 carriages fell down an embankment and caught fire before many of the recruits could be rescued. Sixty-eight of them were killed and another 54 injured.[84]

The *Times* correspondent in Vienna referred to the catastrophe almost a year later, referring to an article by the German Railway Association describing the inadequacies of the Russian railway system to cope with the traffic which would arise during a war with Turkey. His report concluded: 'Lastly, the utter want of a staff of well-trained and trustworthy officials is touched upon, and the case of the railway accident is cited which happened about a year ago at Filigut, where a whole military transport was almost destroyed, entailing a loss of life and limb such as has not occurred in the whole history of railways. The inquiry which was going on was stopped at the request of the Minister of Communication, because he could not spare the officials at this present time.'[85]

The up Special Scotch Express, already referred to unofficially and in the Times as the Flying Scotsman, though that name was not adopted officially until 1924,[86] was in 1877 an overnight express from Glasgow and Edinburgh to London, leaving Edinburgh at 10.30 p.m. At 1.25 a.m. on the morning of Sunday 27th March 1877 it met with a bad accident at Morpeth, south of the border in England and on the North Eastern Railway, becoming derailed on the facing points immediately south of the station platforms leading to the Wansbeck branch. The engine and the whole of the train, except the rear van, left the track, the carriages were badly damaged, and five passengers were killed and 17 injured, some very seriously. Driver Enoch Shipley was also very badly hurt.

The five passengers who died in the accident were R. Cottam, of Edinburgh; Guard John Burton, travelling as a passenger; Alexander Malcolm, of London; Henry Horn jun., of Kirkcaldy and John Donaln, a commercial traveller, of Edinburgh.[87]

An upholstery salesman from Glasgow named Thomas Diamond, who was a passenger on the train, gave a graphic account of the accident to a newspaper reporter: -

---

[83] Ibid, 9th September 1873

[84] Ibid, 10th January 1876

[85] Ibid, 13th December 1876

[86] Wikipedia - Flying Scotsman

[87] The Times, London, England, 25th March 1877

"I joined the train at Edinburgh. There was another gentleman and myself in the compartment, which formed part of a composite carriage, placed, I believe, second from the van next the engine. It was a third-class compartment. The train had hardly passed Morpeth Station-indeed, I almost think the hinder portion of it would be in the station-when I felt an oscillation of the carriage. Momentarily it turned over, but as gently as if it had been let down by persons provided with appliances for the purpose. It remained fixed at an angle of about 45 degrees. I was not in the least hurt, nor was my fellow passenger. I did not stay to think, but, though I have no definite recollection of all the incidents which were crowded into the moment-I found myself escaping through the window and across the roof of the carriage. I was unhurt. I had lost my hat and personal luggage, and was exposed, bareheaded, to the rain which began to descend. When I recovered a little, I saw what seemed to me to be the tender of the engine on fire, while the engine itself was further ahead, and was lying upon its side on the down line. The scene which presented itself to my view was of the most terrible description. The carriages were reduced to mere splinters and planks of wood, and it was impossible to distinguish where first and where a second-class carriage had been situated when the train was in full motion. All those who were immediately in front were reduced to a heap of useless rubbish. Strange to say, the carriage in which I had been seated, though near the engine, was not completely broken; but I believe that I and the gentleman who was with me had extricated ourselves with such speed that we avoided the crushing collision of the carriages which came after. One gentleman in the hindermost compartment of this composite carriage was greatly hurt. I missed my fellow passenger about five minutes after the accident. He had probably gone to assist the sufferers in other portions of the train. There were most harrowing screams for assistance, but it was difficult to discern where to begin our efforts first. I ran to the station to procure a lamp. It was very dark. The stationmaster was aroused, and the porters were soon in attendance. With one of the porters, I tore down an advertisement board to be used as a "stretcher." We first got a man on to the stretcher. I did not ascertain his name. His leg was broken and lacerated, from what I saw, and I think his foot was amputated. He was laid in the waiting room. We next brought from among the wreck a female. Her groans were piteous, and we conveyed her to the station as tenderly as we could. I could not say whether her leg was broken, but it was much cut and bruised. She was lying outside the carriage altogether when we came upon her. We next assisted a Scotchman and his wife. They appeared to have been in a third-class carriage. They were on top of the debris. With much difficulty we succeeded in getting them on to the stretchers and conveyed to the waiting room. The man's leg was broken, and his wife's leg was very much maimed, I believe broken. Immediately beside them we discerned a man in a most horrible position. Pressing with vice-like force against his abdomen was a plank of wood, while on his back, wedged against him with equal strength, was another piece of timber. At the same time his legs were firmly embedded in a mass of wreckage. At each attempt we made to free him he sent up piercing screams. His agony was extreme. We felt the want of hand-saws and crowbars and our duties were of the most heartrending nature. The unfortunate man was at length taken out, and he was the last recovered when I left. In close vicinity to him I saw the head and arms of a lifeless man. He was beyond need of help, and we confined our services to those who were beseeching aid. I came along with the rest of the passengers to Newcastle in a train sent on from that station. A considerable number had been shaken and received slight bruises. We arrived in Newcastle between 4 and 5 o'clock. Most of the passengers went South. Some got out at Newcastle. All the passengers worked with human zeal, but I thought that the people residing in the neighbourhood seemed paralyzed for a while. There appeared to be four or five carriages smashed to pieces. The remainder were standing upright, but I could not state whether or not they had left the ways. The sensation I felt when the carriage was heeling over was of the most curious character. At first, I thought we were going over an embankment, and I was anticipating a crash. But nothing could have been more gentle. I covered the man who got his leg taken off with my top-coat, and afterwards I made it into a pillow for him. I have not seen it since. I also lost my rug, portmanteau and hat."[88]

Driver Shipley was still too ill to give evidence either at the inquest or the Board of Trade inquiry. The first person to give evidence at the latter was Conductor Philip James, who said that the train had left Edinburgh 15 minutes late, awaiting the Glasgow portion of the train, and Berwick 24 minutes late. They had been 22 or 23 minutes late at Morpeth. His train had consisted of an engine and tender, East Coast brake van, Highland Railway fish truck, composite carriage from Edinburgh, third-class carriage from Edinburgh, composite carriage from Edinburgh, third-class brake composite carriage from Edinburgh, East Coast van from Aberdeen, composite carriage from Aberdeen, composite carriage from Perth, sleeping carriage from Glasgow, composite carriage from Glasgow, third-class carriage from Glasgow and Glasgow van; thirteen vehicles in all. Steam had been shut off and speed reduced as usual on approaching Morpeth on account of the sharp curve beyond the station, and the train had been travelling at 20 or 30 miles an hour.

Fireman Thomas Wylie (who unlike his driver had escaped injury) agreed that the speed had been 20 or 30 miles an hour and said the signals had been clear for them. They had not made up any time from Edinburgh. He did not examine the track after the accident.

Signalman David Johnstone at Morpeth said that the train had passed him at 1.42 a.m. at the usual speed. He had then heard a sound as if something had broken under his cabin, and replacing his signals went down with a hand lamp to investigate, when he met the guard, who asked if he had blocked the road. He had then examined the permanent way and found a broken fishplate at the joint near the water tank close to his box.

Assistant Locomotive Inspector John Huwell said that he had examined the engine, No 901, after the accident. He could find nothing wrong with the locomotive which could have caused the accident. At first, he had thought that broken engine springs found after the derailment could have something to do with it but having found the under the engine after the accident it was clear that they were the result and not the cause of it. Driver Shipley was an experienced driver who regularly drove the same engine.

The inspecting officer, Captain Henry Tyler, suffered from the disadvantage that by the time that he reached the scene the wreckage had been cleared away and the track re-laid, so he had to rely on examining what had been saved of the fishplate and so on involved and the statements of those of the company's officers who had examined the site when it was fresh, as it were. He said that it was clear from the suddenness of the derailment that the engine had been the first part of the train to leave the track, and that its broken springs had resulted from its falling over onto its side. It was clear, he stated, that the leading wheels of the engine had mounted the off rail on the outside of the curve at the joint opposite the water tank. For this to have happened, the joint must have been in

[88] Ibid, 27th March 1877

a defective condition and the damage to the track, including the loss of the fishplates, must have occurred before the train arrived. The facing points were too far away from the point of derailment to have had anything to do with it.

It would be better, he added, if the sharp curve on one of the principal main lines in the country could be removed (not that excessive speed over the curve had anything to do with the accident). Meanwhile to reduce the risk of damage to the permanent way being caused by passing trains he recommended that permanent way be strengthened, and the gauge eased on the curve, and he had no doubt that the North Eastern Railway Company would take such a course.[89]

The inquest jury returned a verdict that the deceased had been killed 'by an engine running off the line, and that the engine so ran off in consequence of a fish-plate having got detached from the rails, but how or in what manner the fish-plate became detached the evidence is insufficient to show. And the jury recommend that the speed be reduced to a low rate while passing round the curve at Morpeth.'[90]

Bad weather in the form of a torrential rainstorm led to a bad accident on the Pickering Valley Railroad short line in Chester County, Pennsylvania, United States on Thursday 4th October 1877. A train under the charge of Conductor Charles Guldin, and consisting of an engine and tender, two passenger coaches and a milk car, with Engineer Frank Kenney and Fireman George Griffith, left Phoenixville at 6.20 p.m. It was carrying about 130 passengers, including many members of the Pennypacker family returning from a reunion at Schwenksville, on the centenary of the Battle of Germantown, fought during the American War of Independence.

The train had gone about three miles and was near Kimberton, when it ran into a washout caused by the excessive downpour. A newspaper reported: 'The engine and cars dashed down a ditch nearly 100 feet deep, which was caused by a very heavy wash-out during the storm. In an instant the work of destruction was done. The engine struck the wash-out and plunged headlong to the bottom, instantly killing the engineer and fireman. A passenger coach came next, shot across the opening, broke the foremost end, and fell upon the engine. The cars then piled in one upon the other, and in the terrible darkness of the night the scene that followed was more fearful than pen can picture. Physicians were summoned from miles around, and the work of searching for bodies was continued all day. The dead were taken into Phoenixville. Deputy Coroner Hunt has impanelled a jury. The last report is that fifteen bodies have been recovered.

Despite having just stated that 15 bodies had been recovered, the report then listed only nine dead. Apart from Engineer Kenney and Fireman Griffith, they were passengers Isaac Tustin and his son Jonas, of Chester Springs; William Hallenin and his unnamed wife of East Pikeland; Squire Nathan Pennypacker, of Uwehland; Henry Smith and an unidentified beautiful young woman who was terribly mangled, dressed in plum-coloured silk. Brakeman Michael Corbett and passenger Mrs Albert Pennypacker died not long afterwards from their injuries.[91]

---

89 Board of Trade report

90 The Times, London, England, 12th April 1877

91 Bloomington Pantagraph, Bloomington, Illinois, United States, 12th October 1877

Another report added that: 'During the day the rain had fallen with persistent force. The streams rose rapidly, and the ground became thoroughly saturated. At the place of the accident is an embankment, which on the side next to the hill is probably fifteen feet deep, and on the other side possibly forty feet. The natural formation of the ground caused the water on the upper side to gather at that point in a large pool, and on the evening in question this water, of greater volume than usual, worked its way through the embankment, and washed out a gap some sixteen feet deep and sixty feet in length. The time of the train at that point is 6.05, and the night being dark and stormy the engineer rushed his train with the freight of humanity with fearful certainty into the gulf of death open to receive him. Conductor Golden, Brakeman Major and Baggageman Ganewell were in the baggage car at the time and escaped with little or no injury, the car remaining on the track. Brakeman Corbett was in the front car and went down in the wreck.

'Mr Golden went to Evans' farm, gave the alarm, and was brought to town in a waggon, summoning medical and other assistances. About half-past nine, two cars were run up by the "pusher," in which were Drs Hawley, Porter, Yerkes, Williams and Kulor, with the necessary appliances for the occasion. The storm still raged with unabated fury, and the run to the scene of the wreck was most carefully made. The scene which presented itself is happily not often witnessed. Spanning the chasm were the two passenger cars, the top part of the first being telescoped into the second car, and the lower portion having fallen to the bottom., while the first part of the second car rested on the western bank of the opening, its hind end resting on or at least even with the track.'[92]

This might appear to have been a pure accident, but the coroner's jury decided that despite the exceptional rainfall the railway company were still at fault for not having provided sufficient drainage and for running the train in an improper order, as the engine was running tender first and then the gentleman's car, occupied on this journey by both sexes, then the combination ladies' and baggage car with the milk car at the rear, in contravention of the company regulations.

Following the jury's verdict, the Pennsylvania Supreme Court awarded damages of $3,500 against the railway on the grounds that the accident was the railway company's fault.[93]

Almost exactly a year to the day after the Kimberton accident, on 8th October 1878, the United States had a worse one at Wollaston, Massachusetts, on the Old Colony Railroad. On the day of the accident a major sporting event was being held at Silver Lake, the national single scull championship race, between M.F. Davis, of Portland, Maine, and local Patrick Reagan, of Boston. The Old Colony Railroad ran a special train for the large number of Bostonians who wanted to see the event. Davis won the race and hence the championship together with a handsome purse of $2,000. The sports fans were in no great hurry to return to Boston after the end of the race and the returning special was thus considerably delayed in setting off on the trip back to the city.

In the opposite direction a local freight train in charge of Conductor Charles Hartwell had left Boston at about 6.30 p.m. stopping to shunt at various points along the line including at the Wollaston Iron Foundry. The line was double track, and the siding to the

[92] Ibid.

[93] Wikipedia – Pickering Valley Railroad

iron foundry was reached via a facing connection from the inbound track, the one on which the returning excursion train was running. For the local freight train to reach the siding it therefore had to be set back from the outbound to the eastbound track and thence from the inbound track to the siding. The mains crossover was about 100 feet north of the junction with the siding.

The excursion train, had it been running to time, would have arrived back in Boston before the local freight had even started out, and although the crew of the local freight could not have seen it pass they assumed that it had done so and proceeded to occupy the inbound track without sending out a flag or doing anything to protect their train against any inbound ones, in direct violation of the rules.[94]

While the shunting was in progress, Engineer Hurlburt, of the local freight, suddenly saw the headlight of the returning excursion train. Realising the danger, he opened his regulator and just managed to get his train back over the mains crossover on to the outbound track. It was a very close-run thing and Engineer Charles Westgate of the excursion train had initially thought that the accident was due to his train colliding with the rear waggon of the freight train.

At this point the incident was nothing more than a near-miss, a warning perhaps to the crew of the freight train to be a lot more careful in the future. The excursion train had probably damaged the points of the mains crossover by running through them when they were wrongly set but no great harm was done and apparently the foundry siding was long enough for the excursion to have pulled up safely after it was diverted into it. Instead, in a belated effort to rectify the situation, Conductor Hartwell caused a disaster. He rushed to the lever working the points leading into the siding and attempted to reset the road for the main line. He was a few seconds too late. The leading engine of the excursion was on the points just as they were being moved and was derailed. Although it was only travelling at about 20 miles an hour, despite being over an hour late, the train was only fitted with old-style Westinghouse brakes, which worked more quickly on the vehicles close to the front of the train than on those at its rear, so that the rear cars forced themselves onto the ones in front. The two engines overturned to the left, the leading baggage car was thrown across the track, the next two cars continued forward, the fourth overturned and the next three piled up onto it. Most of the 21 passengers who lost their lives were in the fourth to seventh cars. The fatalities included Patrick Reagan, the losing rower – it really was not his day. About 65 passengers were injured. The only railwayman hurt was Fireman Dennison on the leading engine who was badly scalded.[95]

'News of the accident soon spread throughout the city {of Boston}, and thousands of people began flocking to the station of the Old Colony road, and by 1 o'clock the crush was fearful, and the police, with difficulty, made a clearing around the station. The first train, conveying the dead and wounded, arrived about 11 o'clock, and for two hours the streets were filled with ambulances, undertakers' waggons, and express waggons; the latter containing beds hastily improvised of straw and bed clothing brought from homes in the neighbourhood of the station. The United States Hotel proprietors sent all the spare mattresses at their disposal, while medical aid poured in from every quarter. The scenes

[94] A History of Railroad Accidents, Safety Precautions and Operating Practices by Robert B. Shaw, pgs. 109-11

[95] Fort Wayne Daily Sentinel, Fort Wayne, Indiana, United States, 9th October 1878

in and around the depot were harrowing in the extreme. Relatives of friends on the ill-fated train were clamouring within the lines established by the police, and in many cases where they were assured that death or injury had befallen those whom they sought, gave vent to their agony in the wildest grief.'[96]

Conductor Hartwell was subsequently charged with manslaughter.

The accident might have been prevented had the excursion been run as two shorter trains, as according to Charles Adams, author of *Notes on Railroad Accidents*, published the following year, 12 cars was the limit for the simultaneous application of the original Westinghouse brakes.

The automatic block system, with continuous track circuiting, might also have prevented the accident. A track circuit had been installed in England by William Sykes as early as 1864 and a failsafe track circuit was invented in the United States by William Robinson in 1872. Although, 1878 was therefore still an early date for its use it was in fact installed on the line, but as a matter of economy certain little used sidings, such as that to Wollaston Iron Foundry, were bypassed, and thus left unprotected. Had the siding been included the mere operation of the points leading into the siding would have caused the protecting signal to revert to danger. Track circuits were not used widely in Great Britain until the 20th century; in other countries progress was probably equally slow, for example in the Netherlands the first track circuit was only installed in 1898. When they were in place, they could be used to detect not only obstructions such as waggons left on the line but also broken rails.[97] [98]

News of a major disaster in India on 31st October 1879 reached the *Times* newspaper in London via a telegram to the secretary of the Madras Railway Company, who stated that an accident had occurred to the mail train near Arconum (now Arrakonam) in Tamil Nadu. Nineteen had died, including three Europeans, and 45 were injured, six of them Europeans. Seven of those injured were in a serious condition. The cause of the accident was the collapse of an embankment following heavy rain.[99]

A further report stated that the Europeans injured were Sergeants Price, Roberts and Terry of the Sappers, the wives of Price and Roberts and a passenger (presumably meaning a civilian passenger) named Smith. Those killed were Mrs Terry and children of Terry and Price.[100]

---

[96] Ibid.

[97] A History of Railroad Accidents, Safety Precautions and Operating Practices by Robert B. Shaw, pgs. 109-11

[98] Spoorwegongevallen in Nederland 1839-1993 by R.T.Jongerius, pg 112

[99] The Times, London, England, 5th November 1879

[100] Ibid, 8th November 1879

CHAPTER 2

DERAILMENTS - BROKEN TYRES, BROKEN AXLES AND OTHER VEHICLE DEFECTS

This chapter will deal with train derailments caused not by problems with the track but by defects in the trains themselves. In the early days there were frequent cases of trains suffering from broken tyres, wheels and axles. Such accidents were to a large degree unavoidable at this period and diminished only with improvements in technology.

What was probably the earliest documented passenger fatality on a railway was due to what might loosely be described as a defective vehicle. It occurred on the Kilmarnock and Troon Railway in Scotland on 24th August 1811. The railway (which was more correctly a plateway, using track consisting of L-shaped iron plates and flangeless wheels) did not officially open until 6th July 1812, but it was already carrying passengers in horse-drawn waggons over part of its length a year earlier.[101]

A local newspaper reported that, 'A melancholy accident happened six miles from Kilmarnock on Saturday last, the waggon was returning with a number of people from Troon by the railway when harnessing gave way and owing to the darkness of the night, the waggon was hurled down precipice seven or eight feet perpendicular. One man was unfortunately killed and some so severely hurt that their lives are despaired of.'[102]

Shortly after the opening of the Liverpool and Manchester Railway, a Mr Henry Hunter, the landlord of a pub at Eccles, England, had got into the habit of riding on the engines or waggons of trains carrying red earth from Eccles to Chat Moss to help strengthen the embankment at the latter spot. On 21st October 1830, the day of the accident, he had positioned himself on steps at the rear of an engine tender of a train returning with empty waggons to Eccles, 'notwithstanding the remonstrances of the engineer.' The engine pulling the train was none other than the famous *Rocket* of the Rainhill trials. The train was running tender first, and proceeding at about nine miles an

[101] Wikipedia – Kilmarnock and Troon Railway

[102] The Glasgow Herald, Glasgow, Scotland, 30th August 1811

hour, when the axle of the tender broke, and the body of the tender, with its water barrel, fell on the landlord and he was killed. The engine driver suffered a severe wound to his forehead and several bruises.[103]

Another derailment resulting from a broken axle caused the first passenger fatalities in any train accident in the United States. It happened on 8th November 1833 to the second section of a train on the Camden and Amboy Railroad, which had been delayed due to an engine problem and was apparently running fast to avoid getting too far behind the first section, leading to suggestions that excessive speed had led to the accident. This was denied by the company officials who stated that no train was permitted to cover the 35 miles between South Amboy and Bordentown in two and a quarter hours, at an average speed allowing fifteen minutes for stops of 17½ miles an hour, and with no train in any case allowed to travel at more than 20 miles an hour. The company had even gone to the length of having agents riding on their trains armed with stop watches, whose sole duty as to ensure that they did not exceed the prescribed speeds. The casualties in the accident occurred not in the car which left the road but in the one behind, which was upset and run into by the following cars. There were two brakemen with the train, but one was riding on the car which was upset and was thrown off it and the other, at the rear of the train, had gone to the baggage car to extinguish a fire which had coincidentally broken out in a bale of cotton.

There were 24 passengers in the crushed coach. Those in the rear compartment were unhurt but those in the front were less fortunate and a Mr J.C. Stedman, a jeweller, of Raleigh, North Carolina, was so severely hurt that he died within a few hours. Several other passengers were also badly hurt, and a Mr Lex or Rex of Lebanon, Pennsylvania, or New Lebanon, Ohio, died later as well. A Rev. Mr West, of Washington, D.C., suffered a broken leg. Uninjured passengers included ex-US president John Quincy Adams, who was in the car with the broken axle, and future railroad magnate and founder of the New York Central Railroad Cornelius Vanderbilt. The derailment occurred two miles north of Hightstown, New Jersey.[104] [105]

Another broken axle derailment in the United States, three years after Hightstown, probably resulted in three passenger fatalities. It occurred on Sunday 2nd October 1836, nine miles south of Lancaster, Pennsylvania, on what the newspaper report described as the Philadelphia and Columbia Railway. A train of passenger and burden (freight) cars was travelling at about 15 to 18 miles an hour when the axle of one of the passenger cars broke about midway between the wheels and one of the broken ends of the axle was forced up against the floor of the flimsy wooden car, which collapsed. A Mrs Gibson and her two-year-old child fell through the opening and she was killed outright and the child so badly injured that it died about two hours later. A third passenger, an unnamed man, was very seriously hurt, the train wheels running over his thighs, and the newspaper reporter noted that 'he is probably dead at this time.' Having witnessed the death of his

103 The Times, London, England, 2nd November 1830

104 Republican Compiler, Gettysburg, Pennsylvania, United States, 19th November 1833

105 A History of Railroad Accidents Safety Precautions and Operating Practices by Robert B. Shaw, pgs. 16-18

wife and child, Mr Gibson was stated to have become a raving maniac. After the accident the engine was uncoupled and proceeded to Lancaster for assistance, but meanwhile some of the passengers had placed several of the injured in one of the undamaged cars and began to push that towards Lancaster. The returning engine and the car met on a sharp curve in a minor collision.[106 107]

What was probably the worst railway accident in the world at the time, and for some years afterwards, occurred between Bellevue and Meudon, France on Sunday 8th May 1842. On that day, public celebrations were being held in the Gardens of Versailles in honour of King Phillipe. As the celebrations drew to a close, in the evening a train departed from Versailles station at 5.30 p.m. for Paris Montparnasse. The very heavy train of 16 to 18 carriages was supplied with three engines, two at the front and one at the rear, and was carrying between 640 and 770 passengers. As was customary at the time all the carriage doors were locked.

The train was proceeding at 25 miles an hour when it reached a falling gradient near Meudon. Here the speed of the train began to increase, and the leading engine, the four-wheeled *Matthew Murray*, began to "wobble" until it suffered a broken axle. The leading engine then left the track, the train engine, still travelling at speed, overrode it and was badly damaged and the leading three carriages piled up in turn on the two engines and were smashed. The broken wooden carriages were then set alight by burning coals from the fireboxes of the two engines. Some of the passengers in these carriages were able to escape through the windows but many were unable to do so and were burnt to death. The passengers in the fourth and following carriages escaped the fire, but many were injured in the collision.

A passenger who had been in one of the rear carriages described the scene. 'I felt a violent shock. Two others followed, then stopped. First there was a dead silence, then the most dreadful cries. Still I thought the danger passed, as nothing moved; but it was only after I had, with much trouble, succeeded in forcing my way out of the window, that I found the dreadful position we were in. The engines, the coal carriage, and the four or five first waggons were an immense heap of ruins, under which more than 100 passengers lay buried. Some succeeded in disengaging themselves from the ruins and ran about in agony. But soon the spectacle ... [was] more terrific; this heap became a furnace. The burning coal, excited by the violence of the wind, had spread itself to the mass of overturned carriages, in which the passengers were burning.'

When the Prefect of Police heard of the accident, he sent doctors and guards on horseback to the scene. Wounds were dressed, amputations performed 'in the field' and the casualties taken to homes in the neighbourhood. Nothing, however, could be done for those trapped in the burning carriages. There was no water available to extinguish the fire and it continued to burn until after 11 p.m. Most of the victims were unrecognisable and their jewellery and money was found fused together by the heat.

At least one victim who survived was terribly mutilated. A M. Brioche, a merchant from Nantes, lost both his legs and both his arms in the accident as well as dreadful lacerations to his mouth caused by the hook of the pole used to drag him out of the

---

[106] Star and Republican Banner Gettysburg, Pennsylvania, United States, 10th October 1836

[107] A History of Railroad Accidents Safety Precautions and Operating Practices by Robert B. Shaw, pg18

wreckage. The friends and relatives of other victims had the sad task of identifying their belongings amongst the items recovered from the ashes. 'Already one gentleman has found among these relics a chain and medallion which belonged to his young wife, to whom he had very recently been married. One of the chief clerks in the office of the Minister of the Finances has ascertained the loss of a young person to whom he was affianced, by seeing among these remnants the case of a watch he had presented to her. A mother has also found the fragment of a box, and a ring belonging to her only daughter, who went on Sunday to Versailles with her cousin, but neither of whom has returned or been heard of.'

There was at least one case of a dramatic rescue; a 69-year-old man wrote that he, his daughter, his three granddaughters, and their maid were involved in the conflagration but were saved when a gentleman wearing a white hat appeared. He 'broke the panels, saved first the three children … then their mother, who had fainted, and myself … He then left us to relieve our unfortunate maid whom he brought on his back, she being unable to walk from having her legs burnt.' They tried to discover his name, to which he jokingly replied, 'I am called Arthur Trois Etoiles; I have nothing to fear from fire.' He was last seen limping away, and the saved man said, 'I owe to him the lives of myself and my family.'

The victims included Admiral Jules Dumont d'Urville, a well-known explorer of Australia, New Zealand and Antarctica, and his wife and family. Estimates of the total number of dead vary from 46 to 200, but the most generally accepted figure seems to be 57.

The catastrophe led to the abandoning of the practice of locking carriage doors, leading presumably to an increase in the number of passengers falling from the open doors of moving trains but at least enabling their escape in the case of fire. It will be noted that in Great Britain, at least in the 21st century, the wheel has turned full circle and central door locking, with emergency releases, has been reintroduced. Other measures taken by the Chemin de fer de Paris à Versailles rive gauche Company after the accident included withdrawing from service their six four-wheeled locomotives and the introduction of a new steam brake whistle with a distinctive sound.

On 22nd November 1842 the trial of six company officials for manslaughter began but their defence counsel argued that the disaster was 'a misfortune but not a crime' and eventually the Paris Criminal Court decided that they could not be incriminated for either the excessive speed of the train or the state of the material and they were acquitted.

A commission appointed to investigate the accident recommended that axles should be tested and replaced after they had travelled a certain distance. The accident also led to the beginning of research into the problem of metal fatigue.[108] [109] [110] [111]

A serious derailment occurred in Belgium on 17th May 1844. A fast train, not timed to stop at Vieux-Dieux station, became derailed due to a luggage waggon, the second

---

[108] The Times, London, England, 11th, 12th, 13th, 17th and 24th May and 9th December 1842

[109] Wikipedia – Versailles Rail Accident

[110] Wikipedia - Liste des accidents ferroviaires en France au XIX siècle

[111] Webpage – Versailles Railway Accident of 1842 – Geri Walton

vehicle in the train, suffering a broken axle, derailing it and the following three vehicles which were thrown against a waggon of sand standing on an adjoining line and smashed to pieces. The initial report on the accident, from *The Times* correspondent, ended by saying that the cause of the derailment had not been ascertained, and that while some attributed it to a broken axle, others ascribed it to the condition of the track following some recent alterations. A second report, quoting a Belgian newspaper, seemed to confirm that a broken axle was the cause.

The first report stated that a woman passenger had been killed on the spot, another woman and a baker from Hoboken had died in hospital during the night, and the lives of eight or ten other persons were in danger. Some were taken to their own homes and others to various Antwerp hotels. A Bruges surgeon, who was among the passengers, rendered medical assistance, although hurt himself, with a severe cut on his head which was bleeding profusely, and he was aided by doctors from neighbouring villages, while the Governor of Antwerp, the Procurer du Roi (public prosecutor) and principal railway officials were soon on the scene 'and made every necessary arrangement to alleviate the state of the sufferers.'[112]

A letter published in *The Times* from a correspondent who was on the ill-fated train described the accident in detail, 'A lamentable accident on the Antwerp and Brussels Railroad took place on Friday afternoon last, at about a quarter before 6 o'clock, within three miles of the terminus at Antwerp. I was one of the passengers who, with a friend of mine, escaped without personal injury. I may state, for your information, that where the accident took place a high road from Antwerp runs across the line, and both my friend and myself have no doubt that, by the constant traffic of heavily-laden carriages across the line, the rails must be liable to become displaced or sufficiently shook from their correct and accurate line as to occasion the late and similar accidents. We were in the third carriage from the engine, and all of a sudden, we experienced a heavy bumping sensation. In exclaiming there must be something wrong, our conviction was confirmed by a crushing noise which soon reached our carriage; we fortunately jumped out in time to escape injury. A more distressing and heartrending scene could not be imagined. The carriages in arrear of ours were all crushed one upon another, and thrown completely across the railroad, the last being upside down. On looking into the cause and origin of the accident, I found that the engine and tender had not got off the line, and I believe one carriage followed, but the second was partially off, and the connecting bolt from the force was broken right in half. A luggage carriage was wrenched out, turning it off the line to the right swinging the remainder round to the left. Two persons were killed on the spot, two died within an hour afterwards, and, on leaving Antwerp on Sunday morning, four more deaths had taken place from amongst the wounded, which in all amounted to about 40, and half of these were seriously hurt- broken limbs etc. I have no hesitation in saying, the comparative slight structure of the carriages, as compared with those on our Great Western line, contributed to the extent of the calamity, as also the injudicious structure of the carriages, the wheels being placed underneath, with a projection of nearly two feet of carriages over on each side, rendering them more liable to tilt over. In concluding this short narrative, I cannot refrain from expressing my gratitude for the system adopted, in leaving the carriage doors unlocked, whereby many were enabled to make their escape, and amongst the number myself and friend. Should you think it worthwhile to this, it may

[112] The Times, London, England, 21st and 22nd May 1844

perhaps, induce the Prussian authorities to adopt the same plan; but, at present, all passengers are rigorously locked in to abide the full result of all accidents.'[113]

That it was not just axles which could fail with serious consequences was shown by an accident in the United States on 24th July 1846. The 6.30 a.m. New York and Erie Railroad train from Middleton was in charge of Conductor James Lytle and Engineer Joseph Meginnes and consisted of four passenger cars and three milk cars. It had been strengthened that morning to accommodate a party of excursionists to the Elysian Fields at Hoboken and had about 200 passengers, with further excursionists joining at New Hampton, Goshen, Chester, Oxford and Monroe. About a mile east of Monroe, New York, a bridge twenty feet long and twelve feet high carried the line over an arm of Seaman's mill pond, and it was here that the train was derailed. The engine and two leading milk cars kept the rails, but the third milk car was derailed, and the following passenger cars were derailed and piled up behind it, except the rear one which kept the track. Considerable telescoping between two of the passenger cars added to the destruction. The accident reportedly occurred at 8 a.m. though 7 a.m. seems more likely, as Monroe is only about 12 miles from Middletown.[114]

Conductor Lytle despatched the engine to Piermont, from where news could be sent by fast steamship to the company headquarters in New York, and a man on horseback to get assistance from Middletown, with instructions to get fresh horses at Chester and Goshen. Meanwhile, the rear of the train was transformed into a temporary hospital, though later the dead and injured were taken on a trolley to Stickney's Hotel in Monroe. Dr Boyd of Monroe was soon on the scene and later doctors from Chester. Company president Loder arrived at the scene with surgeons from New York at about the same time as a relief train from Middletown.

Two passengers named Ira Crane and George Stephens were killed outright and Charles Monnell died the next day. A number of other passengers were seriously injured, and it seems likely that three more died later.

Examination of the line showed that the accident had been caused by a broken wheel on the third milk car, the wheel being of the Winans type, with cast spokes.

The accident caused something of a sensation in the country, being the first of its kind as well as the worst accident so far to a New York and Erie train. The company came in for much criticism for ignoring previous problems with defective equipment, though locally they were exonerated from all blame at a public meeting held in Middletown. The company immediately abandoned the Winans wheel in favour of solid wheels and other companies followed suit. In addition, four wheeled vehicles such as the milk van were replaced by eight wheeled ones and a system of testing wheels during journeys by tapping them with a hammer evolved. Work also began on the filling in and strengthening of trestles. The accident cost the company over $100,000.[115]

The Great Western Railway suffered an unlucky accident at Southall, England, on 25th

---

[113] Ibid, 22nd May 1844

[114] Webpage – Catskill Archive: First Fatal Accident to a Passenger Train on the Erie

[115] Ibid.

January 1847. The up express from Exeter, due in London Paddington at 4.30 p.m., had just passed the station when the tyre of the right driving wheel of its engine, *The Queen*, broke. As chance would have it, this happened just as the express was passing a down passenger train, the 4 p.m. departure from Paddington, and a large fragment of the tyre, at least 7 feet long and weighing about 240 lb, was thrown into the air and landed on the leading second-class carriage of the down train, killing two men named Henry Bishop and Henry Holt. The two were cattle dealers from Cowley and Colnbrook respectively, and were returning from the Middlesex Sessions, where Holt had just won an appeal against a conviction for animal cruelty. Several other passengers were injured, and a fragment derailed the engine of the down train. The express stopped down the line and then continued to Paddington at reduced speed.

At the inquest into the accident, Driver James Almond testified that *The Queen* had also thrown a tyre seven or eight months previously. Great Western locomotive engineer Daniel Gooch also gave evidence at the inquest, the jury of which returned a verdict of "Accidental death" with a recommendation that 'tires should be fastened on the wheels in the manner recommended by the engineers who had been examined.'[116]

An accident at Brookline, Massachusetts, United States was also one which might have ended much less seriously than it did. In this case, on 6th November 1847, a Boston and Worcester Railroad train from Springfield to Boston became derailed because a brake rod dropped loose on one of its cars. The car remained more or less in line but sufficiently outside of the gauge to foul a bridge abutment and was then crushed by the following vehicles, killing six passengers. Boston and Worcester Railroad cars already had their brake rods secured by safety chains to guard against just such an event but the vehicle in question belonged to the Western Railroad, which did employ such measures.[117]

Poor workmanship was the cause of an accident on the Caledonian Railway main line north of Rockcliffe station, England, late in the evening of 10th February 1849. A down mail train from London Euston to the north left Carlisle at 9.35 p.m. that day. Just after it passed Rockcliffe station its engine and tender became separated from the remainder of the train and the whole train, apart from the engine but including the tender, left the rails. The tender, leading luggage van and a following first-class carriage for Glasgow stayed more or less upright on the line but the next seven vehicles, comprising a second-class carriage for Glasgow, travelling Post Office van, three horse boxes, first-class carriage for Edinburgh and composite or second-class carriage for Perth (*The Times* report and the Board of Trade report disagree) were all thrown down a 15 ft embankment. The second-class carriage was turned upside down and crushed by the heavy Post Office van and five passengers in this carriage were killed and a sixth seriously injured. The victims were 'all evidently, from their style of dress, labouring men.' A few other passengers were less badly hurt, and an unstated number of horses were also killed, only three being saved.[118 119]

116 The Times, London, England, 26th, 27th and 28th January and 2nd February 1847

117 A History of Railroad Accidents Safety Precautions and Operating Practices by Robert B. Shaw, pg.318

118 Board of Trade Report

119 The Times, London, England, 13th, 15th, 19th and 23rd February 1849

Mr Councillor Drummond, a Glasgow police commissioner who was a passenger on the train, described it to a newspaper reporter, 'The mail train left Carlisle at 25 minutes to 10 o'clock. When about four miles on our way, I felt, by the violent shaking and bumping of our carriage, that it was off the road. The engine almost immediately stopped. My friend and I got out by the window of our carriage and found that we had been disconnected from the carriage in our rear, and some of those along with the post-office had upset over the bank, which is about 12 feet high. Those who were able immediately proceeded to see what damage was done and found that the carriage which was next to ours had been broken to pieces and completely upset, and the post-office carriage was resting on the top of the far side of it. We soon saw that many of the passengers who had occupied it were dreadfully hurt, and several were under it. Steps were immediately taken to get them removed, when it was found that four men were dead, and several others in a fearful state; one of them died in Carlisle Infirmary soon after being taken there. One man, who had lost his foot, had to get his leg amputated below the knee. Poor fellow! He stood it with great firmness. Many others were more or less hurt and were removed to the houses of friends. There was only one lady in the train, going to Perth, who fortunately escaped unhurt, though she was in the carriage next to the post-office, and went over the bank. A messenger was sent to Carlisle for assistance and medical aid, and nearly all the passengers had returned before 12 o'clock. Those who were able used every exertion to render assistance and relief to the sufferers. The mail carriage having been deranged, and the manager hurt, was taken back to Carlisle, and came in at 9.16 a.m. on Sunday. A special train was despatched at 3 a.m. with information to Edinburgh and Glasgow, by which two of the passengers also went, and Captain Coddington left Edinburgh early on Sunday. The rails soon got replaced, and the mail train, which left Carlisle at 9.16 a.m. on Sunday, came along the line and arrived here {Glasgow} at the usual hour…'[120]

By examining marks found on the track at the site of the derailment and having conducted some experiments at Carlisle, Captain J.L.A. Simmons, the inspecting officer for the Railway Commissioners, was able to determine that it had probably been caused by a loose wheel on the tender of the engine, the result of poor workmanship on the part of the man who had fitted the wheel to the axle in the Company's factory at Greenock. The unnamed man had been discharged prior to the accident for a similar failure in the fitting of another wheel. Captain Simmons referred to a witness describing similar cases of faulty axles on the Paris and Rouen Railway, and recommended, in order to prevent a recurrence of the accident, the construction of a slight shoulder on the axle to prevent a wheel moving to a dangerous extent, and the examination of existing wheels and axles to detect any such faults.[121]

At the conclusion of the inquest on the victims of the accident, the jury returned a verdict in accordance with Captain Simmons' findings, adding that 'It does not appear that blame attaches to any of the officers of the Caledonian Company, unless to the party who had the charge of the construction and fixing of the wheels and axles of the tender.'[122]

---

120 Ibid, 15th February 1849

121 Board of Trade Report

122 The Times, London, England, 23rd February 1849

A different sort of engine defect caused an accident south of Hampton, England, on the London and North Western Railway, on 3rd August 1852. As the 9.15 a.m. up passenger train from Birmingham was passing through Berkswell Cutting, a mile or so from Hampton, its driver, William Jenkinson, suddenly felt a violent oscillation, and immediately shut off steam and applied the brake, and the guard in the front van, finding that a shower of stones and sand was being thrown up against it, did likewise. The engine and tender were soon brought to a stand but the leading van and a following second-class carriage had meanwhile become detached from the tender and thrown across to the opposite running line. By an unlucky chance, the 9.15 a.m. passenger train from Leamington to Birmingham was close at hand at that moment and struck the two vehicles. Two passengers, a Mr John Beddington, a Birmingham optician, and a youth named William Floyd, who had been on his was to Oxford University, were killed and a third passenger, Mr Eggard Brooks, a Birmingham gun-maker, was seriously injured. No other passenger was hurt.[123]

Seeing what had happened behind him, Driver Jenkinson took his engine on to Coventry for assistance, and from there information of the accident was telegraphed to Birmingham, so that assistance soon arrived from both places. Once the single injured passenger had been removed and provision made for the forwarding of the other travellers, attention turned to the cause of the accident and an examination of the train and the track soon made it obvious that the derailment had resulted from the ash-pan having become detached from the engine, No. 153, and having fallen onto the track had fouled and derailed the following van. All that therefore remained for the inspecting officer, Captain R.M. Laffan, to determine was exactly how the ash-pan had come loose.

The initial theory was that the ash-pan had struck some obstruction on the track but the inspecting officer ruled this possibility out because a careful examination of the line afterwards had failed to discover any such obstruction, the ash-pan itself showed no signs of having received such a blow, and the fastenings by which the ash-pan had been secured to the engine were very loose, and one of them was defective and unsound. Captain Laffan stated that the conclusion was irresistible that the accident was not due to chance but to the defective fitting of it in the workshop together with a failure to detect the problem afterwards by examination of the engine.

In conclusion, the inspecting officer stated that 'a better, a safer, and a more convenient mode of fastening the ash-pan might easily be adopted, and I would strongly recommend that the fastenings at present adopted in these six engines built for high speed be changed without delay.'[124]

In contrast to the following century, when a good number of famous people were to be killed in either road accidents or air crashes, very few famous names seem to have lost their lives in train accidents in the nineteenth century. There were, however, some close escapes, such as on 6th January 1853, when General Franklin Pierce, the United States president-elect was involved in a derailment on the Boston and Maine Railroad near Andover Massachusetts. The train involved was the 12.15 p.m. express from Boston, timed to run the 26 miles to Lawrence in 40 minutes. Not far from Andover, one of its

---

123 Ibid, 3rd, 5th, 10th, 12th and 26th August 1852

124 Board of Trade Report

passenger cars left the track and fell down a 20 ft embankment to be smashed to pieces. It contained 60 or 70 passengers but only one, who happened to be the 12-year-old President-elect's son, was killed outright, with two more passengers succumbing later. Ironically, the train had just made a special stop at Andover to pick up the President-elect's family, which had been to Concord to attend the funeral of an uncle of Mrs Pierce and had then stopped off to visit relatives at Andover. They had only been on the train about five minutes. Despite the accident and his bereavement, President Pierce served four years in office from March 1853 until March 1857.

The derailment was attributed either to a broken wheel flange or a broken axletree.[125]

On Monday 22nd August 1853 a York and North Midland Railway passenger train left Scarborough at 11.45 a.m. for Hull. All went well until the train was within a mile and a half of Cottingham, England, when the passengers riding in a third-class open carriage next to the guard's van were alarmed by its wheels making a strange noise, immediately after which they discovered it was off the rails. Some of the passengers wanted to jump but were prevented by a Mr Whitton, a stonemason, who shouted to the guard, George Clayburn. The latter applied the brake in his van and signalled to his driver, William Wood, who immediately reversed his engine and applied his brake, so that the train was pulled up fairly quickly. By that time, unfortunately, the train had passed over a culvert and the derailed carriage broke the culvert's timbers and then smashed up against its stone abutment, pulling a second carriage after it and the two ending up in a ditch. A woman passenger in the open carriage named Margaret Ake was killed instantly and 20 or 30 other passengers were injured, one of whom, another woman, died some time later. Miss Ake had been returning to Hull from a visit to her father at Beeford (a village about 8 miles east of the nearest station at Driffield).[126]

The derailment was blamed on a broken tyre on the third-class carriage which had a flaw due to a faulty weld.

The only criticism made by Captain George Wynne, the inspecting officer who investigated the accident, was of the absence of a second guard and brake van at the rear of the seven carriage train. He had no comment to make on the conveyance of passengers in open carriages, something still perfectly acceptable, it would seem, in 1853.[127]

The cause of a derailment on the Chicago and Rock Island Railroad near Minoka, Illinois, United States on Wednesday 1st November 1854 was not entirely clear at first, as after the accident the body of a dead horse was found at the scene, but the engineer and fireman of the passenger train involved were confident that it was due to the breakage of the forward axle of the locomotive truck, as both saw the wheel fly off. The line was fenced but there was a level crossing close by where the horse could have got onto the line. The appearance of the horse corroborated the statement of the engine crew as its body was not mangled, as it would have been had it been struck full on by the train, possibly it was hit by one if the derailed vehicles.

---

[125] A History of Railroad Accidents Safety Precautions and Operating Practices by Robert B. Shaw, pgs. 42-3

[126] The Times, London, England, 24th and 26th August 1853

[127] Board of Trade report

The passenger train involved consisted of an engine and tender, baggage car and seven passenger cars in charge of Conductor van Buskirk. It had left Chicago at 11 p.m. and was travelling at about twenty miles an hour. Following the breakage, the engineer attempted to reverse his engine and whistled for brakes, but the truck immediately settled down on the track and the engine was derailed. The derailment broke the coupling between the engine and tender and the engine went off the line on one side and the tender and baggage car on the other. The leading passenger car went past the locomotive and no one was hurt but the next car landed on top of the locomotive, the steam dome broke through the floor of the car, the steam valve was broken or opened, and scalding steam enveloped the fifty passengers in the car. Engineer William G. Brown escaped with a few bruises but his brother, the fireman, suffered a broken leg (another account stated that the engineer was badly scalded).

Conductor Buskirk made his way to Joliet after the derailment and an engine sent from there drew back the rear three cars, in which the injured victims were placed. Mr N.D. Elwood of Joliet, the company secretary, turned over a new stone building there for the use of the injured and Dr McArthur of Joliet was 'unremitting in his exertions.'

The locomotive involved was described as a new one of the best class, constructed by Messrs Breese Kneeland and Co of Jersey City, and its axle was half an inch larger than was normal. It had been carefully inspected and tested without the discovery of any imperfection.[128]

The number of dead was initially only two but a few days later had risen to eight and reportedly eventually reached 16 with the deaths of more and more of the scalded passengers.

The derailment of a train near Gran, in Hungary, on the night of 17th May 1856 was apparently due to the breakage of some part of the engine, which as it hung down lifted one of the rails from its place. The engine, tender and leading baggage van kept on the line but two or three of the following carriages of the train fell down a 35 ft embankment. Two passengers were killed outright and ten very seriously hurt, one of whom died as he was being taken to hospital.[129]

An engine problem was also the cause of a derailment near Nantyderry, Wales, on the Newport Abergavenny and Hereford Railway on 12th November 1856. The 12.35 p.m. down express passenger train from Hereford to Newport was less than a quarter mile north of Nantyderry station when its engine suddenly left the track on an S curve and went off to the left. The coupling behind the tender snapped and two carriages were thrown off to the right across the up line. Approaching on the up line was the 12.45 p.m. up goods and mineral train from Pontypool Road to Hereford, The Nantyderry stationmaster, who had witnessed the accident to the passenger train, threw the Nantyderry down distant signal to danger and signalled with his hands for the down goods to stop, upon which its driver, Stephen Lake, shut off steam, braked, reversed his engine and whistled for the guard's brake before jumping off, but a collision with the

[128] Weekly Wisconsin, Milwaukee, Wisconsin, United States, 8th November 1854, from The Chicago Press

[129] The Times, London, England, c25th May 1856

derailed carriages could not be avoided. According to the Board of Trade report, the collision completely destroyed the two carriages and two passengers named Edmund Hands and Mark Hicks, one of them a traveller for a Wolverhampton boot and shoe manufacturer, were killed and two other passengers and two railwaymen seriously injured.

The derailment of the express engine was found to be due to the fact that it had lost its left-hand leading spring. Its driver, Nathan or Nathaniel Sargeant, noticed the loss of the spring when the train arrived at Abergavenny and mentioned the fact to his guard, asking the guard if he had seen it, but the latter replied that he had not. The spring was later picked up from the side of the line eight miles north of Abergavenny. Notwithstanding the loss of the spring, Driver Sargeant decided that it was safe to continue the journey, but the train left the rails on the first really sharp curve that it encountered after Abergavenny. Driver Sargeant gave as his reasons for continuing that he had often travelled before with broken springs, that having mentioned the problem to the guard it became the guard's responsibility, and that he had had no screw jack with which to pack the engine. The running shed foreman at Hereford, George King, admitted that he knew that the engine had gone out without a screw jack. Driver Sargeant also alleged that the track was defective at the point of derailment, a claim which Lt-Col Yolland, the inspecting officer, dismissed.[130]

Driver Sargeant was a young man of only 22 years of age, who was usually employed as a fireman and could neither read nor write.

A prolonged inquest, lasting six days, was held on the deaths of the two passengers, at the conclusion of which the jury returned verdicts of manslaughter against both Driver Nathaniel Sargeant and Running Shed. Foreman George King, added recommendations that the locomotive superintendent should personally examine all candidates for the position of driver and that all drivers should be able to read and write.[131]

Lt-Col Yolland stated that he could not entirely agree with the propriety of the jury's verdict and that he considered that the full responsibility for the accident lay with the management of the Company, who had employed an insufficient number of experienced drivers to cater for any unexpected event and hence had had to use a young man of 22, who could neither read nor write, and had shown by his actions in leaving Hereford without the proper tools on his engine and by continuing from Abergavenny without the leading spring that he was entirely ignorant of his proper position and of his duty as an engine driver. As Sargeant had not been proved to be a responsible and efficient driver he could not be held responsible for the accident. As for Running Shed Foreman King his conduct in permitting the engine without the proper tools was exceedingly reprehensible and he appeared to be quite unfit for the post he held.[132]

On 12th March 1857 the derailment of an engine due to a broken axle was the initial cause of a major catastrophe near Hamilton, Canada, but as the principal feature of the disaster was the collapse of a bridge it is discussed in the chapter on bridge accidents.

---

[130] Board of Trade report

[131] The Times, London, England, 15th, 17th, 20th,21st and 28th November and 1st December 1857

[132] Board of Trade report

During the summer of 1858 the South Eastern Railway ran a daily excursion train from London to Ramsgate and Margate, leaving London Bridge station at 3.30 p.m. On 30th June 1858 the train was within a short distance of Chilham station, England, between Ashford and Canterbury, when one of the engine crank axles fractured, and one of the eccentrics then struck and cracked the bottom of the engine boiler, enveloping the engine in a cloud of steam. Alerted by the oscillation of the engine and the peculiar motion of the right-hand driving wheel, Driver George Wilkinson shut off steam, while his fireman applied the tender brake. The engine came to a stand in about 400 yards with neither the driver nor the fireman being aware of what had happened behind them. They soon discovered that their train had been wrecked. The damaged engine had displaced the rails throwing the whole of the train off the line. The fourth carriage, a first-class one, collided with a wood and earth buffer at the end of a siding, was thrown into a field and smashed to pieces. The obstruction thus caused broke the couplings between the carriages and those in the middle of the train were likewise thrown into the field and badly damaged, while the vans at the front and rear of the train and the carriages adjoining them escaped comparatively lightly. About 20 passengers were injured, three of whom subsequently died. Guard White was also quite badly hurt, being thrown against a quickset hedge which badly cut his face.

The accident was witnessed by Platelayer William Ballard from the gatehouse at French-mill Crossing and by the Chilham stationmaster, Henry Wood, so help was soon at hand. A servant of Mr Hamilton, the Chilham surgeon, was also present and set off at once on horseback to fetch his master. The injured were initially treated on the platform of Chilham station, placed on cushions taken from the carriages. The most badly hurt were then forwarded to Canterbury Hospital.

Captain Tyler, the investigating officer, stated that the initial cause of the accident was clearly due to the failure of the axle, from ordinary wear and tear, for which no one could be blamed. What he was more concerned about was the way that the coaches had been so badly thrown about due to the damage to the track caused by the engine. There was, he said, a strong feeling that excessive speed was one of its principal causes (and the jury at the inquest had returned a verdict of "Accidental death" with a recommendation that the speed of trains be limited to 25 miles an hour between the whistle board for French-mill Crossing and Chilham station). The inspecting officer did not think that the speed of the train had been too high, or that the long but gentle curve required any great restriction of speed. He was less happy, however, with the strength of the permanent way, which consisted of 70 lb double-headed rails resting in cast iron chairs with wooden keys, the chairs being secured to the sleepers by wooden trenails and with wider joint-chairs at the rail joints. He recommended that wrought-iron spikes be substituted for the wooden trenails, that fishplates be used at the rail joints, that the outer rail be raised, and extra sleepers be inserted on the curve, adding that then neither somewhat excessive speed or the fracture of a crank axle should then cause such a bad derailment.[133] [134]

The 7.30 a.m. up passenger train from Cambridge to London had almost reached its

133 Ibid.

134 The Times, London, England, 2nd, 7th and 13th July 1858

destination on Monday 20th February 1860 when it was derailed as it passed through Tottenham station, England, at about 9.20 a.m. By chance, the accident was witnessed by both Mr Boys, the Tottenham stationmaster, and by Mr Sinclair, the Eastern Counties Railway engineer and locomotive superintendent. The engine mounted the platform and overturned, and the train piled up behind it and was wrecked, a first-class carriage, the second vehicle in the train, being especially badly damaged. The accident was a bad one; Fireman Barber was found dead under his engine and Driver Rowell and five passengers died of their injuries, while about a dozen other passengers were hurt, though the guard and his companion in the leading van escaped almost uninjured.

The derailment was caused by the fracture of the tyre of the left leading wheel of the engine of the train, No. 103. Captain Henry Tyler, the inspecting officer, found that the failure of the tyre had resulted from an undetected flaw in it. He was critical of the fact that the Eastern Counties had no record of from whom it had been purchased or who had carried out the faulty weld which led to its failure and he referred to an accident at Tilbury Junction on 17th August 1859 from a broken axle, which he had also inquired into, and where again the Company could give no information as to the axle's history.

Captain Tyler went on to refer to the apparently stronger tyres used on the South Devon Railway, which were steeled during manufacture and bolted onto the rims, to the improved method of securing tyres to wheels patented by Mr Beattie, locomotive superintendent of the London and South Western Railway, and especially to the introduction by Mr Sinclair of the Eastern Counties of tyres made of Prussian steel, rolled into a circular form at once, so as to eliminate any risk of faulty welding.

Finally, Captain Tyler touched on a subject which was much commented upon by Board of Trade inspectors over the years – the brake power of the train. The train involved in the derailment had consisted of ten vehicles and had been travelling at 30 or 40 miles an hour at the moment the tyre shattered. There had only been one brake van, immediately behind the tender, and therefore no brake power at the rear of the train. The guard had chanced to see what happened at once and begun to brake straight away and the engine brake had presumably been applied, but the proportion of brake power, inadequate in any case, had proved quite insufficient in the emergency, and a number of people had lost their lives because of the violence with which the coaches were thrown against the engine. Given the distance over which the train could have been braked, judging from where the pieces of broken tyre were picked up, much more could have been done with adequate brake power.[135] [136]

On 4th January 1861 the 12.40 p.m. down train from Shrewsbury to Hereford, on the then still independent Shrewsbury and Hereford Railway, was between Dinmore and Moreton stations, England, when the whole of the train behind the engine and tender left the rails and after travelling a short distance went down an embankment into a lineside ditch containing water five or six feet deep. A panic ensued among the passengers in the leading compartment of the second-class carriage and two women in it drowned before they could be rescued.

The treatment of the subject of the unlucky passengers was remarkably different in

[135] Board of Trade Report

[136] The Times, London, England, 21st and 22nd February 1860

The Times and the Board of Trade reports. *The Times* reporter described in detail the gallant action of Marines sergeant Wilcox, 'a tall powerful fellow,' and William Watkins of the Flintshire Militia, who had both behaved most gallantly, each saving several people, including a lady, a nurse, a child and an old man, but who had unfortunately not been able to save a Miss Sophia Lowe, of Chester, en route to Bristol, or a Mary Jones, aged 15, of Breinton, Herefordshire, who had been on a Christmas visit to Leominster. The reporter also mentioned the rescue by Driver Jonathan Cheeseman and his fireman of Guards Brown and Child from the van in which they must otherwise have surely drowned, by smashing the van's windows with hammers to get them out. In contrast, Colonel W. Yolland, the Board of Trade inspecting officer, commented that, 'it is quite manifest that had greater presence of mind been exercised, and the parties in this compartment had mutually assisted each other, no life need have been lost.'[137]

Examination of the vehicles after the accident showed that the right leading wheel of the second-class carriage was broken to pieces, and the trailing axle of the same carriage was also broken. Further examination disclosed that the initial event was the breakage of the tyre of this wheel into four pieces, the breaking of the axle being a secondary event caused by the derailment. The tyre had originated from Messrs Jackson of Manchester and had been fitted onto the wheel the previous May. It was made of best Staffordshire iron, and had no appearance of any flaw. The very cold weather at the time of the accident might have contributed to the failure.

Colonel Yolland criticised the continual use by companies of tyres fixed by the old method of bolts or rivets, when so many better methods of fixing tyres had been developed so that they did not fly off even if they fractured. He mentioned the improved methods patented by Messrs Beattie, of the London and South Western, Gibson of the Great Western and Craven of the London Brighton and South Coast Railways, and the use of a dovetail to avoid making holes, any of which if they had been employed the previous May would have prevented the accident. Like Captain Tyler in his comments on the Tottenham accident, Colonel Yolland mentioned the subject of brake power, the fact that the rear brake van was not the last vehicle of the train, and that there had been no means of communication between the guard and the driver or the passengers and the guard, concluding that, 'It is unfortunate for the public that the law does not make someone criminally liable for such neglect.'[138]

A broken axle on a car belonging to the Milwaukee and Mississippi Railroad caused the derailment of a Chicago and North Western Railroad train carrying Union soldiers – a portion of Colonel Barstow's Third Wisconsin Cavalry – at a point between three and four miles out of the railroad's Chicago depot and a mile beyond the Milwaukee and Mississippi crossing on Wednesday 26th March 1862. The defective vehicle was the fifth in the train, behind four freight cars. When it derailed, due to the breakage of its front axle, it overturned, as did the following car and two more cars left the road but remained upright. Ten soldiers of A and G companies died, and fourteen others were injured. Some of the men who lost their lives drowned in a lineside ditch. Immediately after the accident, the engine was detached and proceeded into Chicago for assistance,

[137] Ibid, 7th and 9th February 1861
[138] Board of Trade Report

accompanied by Superintendent Dunlap and General Ticket Agent Robinson who had been riding on the train. Surgeons were procured, and a special train was sent to the scene. Meanwhile the uninjured soldiers were busily at work rescuing their colleagues. A newspaper reporter wrote, 'To add to the horror of the scene, it was pitch dark, and the water on both sides of the track was nearly waist deep. As the dead bodies were brought out, they were laid upon the snow, which but lent its shroud-like pallor to the ghastly and sickening scene. Here was a soldier, a tall, robust, handsome looking specimen of manhood, stiff and stark in the embrace of death. A quiet smile upon his features, and not a bruise or mark to be seen upon his person. By his side lay one frightfully disfigured. His head and face were smashed into a pulpy form utterly beyond recognition; one of his feet was crushed to a jelly, splintered bones, tangled flesh and torn sinews protruding from his ensanguined uniform. The other foot was so nearly severed at the ankle that, but a mere strip of skin held it. One of his legs was half cut through just below the knee. It was a sight so appalling as not even the battle field may rarely furnish its parallel. And this poor fellow, Walter Snell, of Co. G, only an hour or two before had parted from his tearful wife who accompanied him to Howard Station. She had given him up to die, if need be, upon the battle field of his country, little dreaming that a violent and frightful death was so near. And yet her sorrow will be but the type of keen anguish that this calamity will carry to other homes, which will miss the love-light of husbands, fathers, brothers and sons. God help the widowed and fatherless.' As noted earlier, reporters of the time did not spare their readers all the grisly details. One wonders if Mrs Snell saw the gory description of her husband's death in the paper.

At the subsequent inquest at Chicago, Conductor John Hull, who had been in charge of the train, stated that when the derailment had caused the freight cars to separate from the passenger cars due to couplings breaking this had made the bell ring and the engineer to stop the train. The train had been travelling at thirty miles an hour for part of the journey but had commenced to brake for the stop at the Milwaukee and Mississippi crossing when the accident happened. He had gone back after the accident and seen marks on the ties where the wheels had run off. He had also examined the broken axle but did not know why it had failed as he could see no obvious flaw in it. The train had left Janesville, Wisconsin, 91 miles from Chicago, at 4.45 p.m., the accident occurring at about 10 p.m.

Engineer Nelson Barns, driving the *Ontonagon* engine, told the inquest that his had been the first of two troop trains. The first he had known of the derailment had been feeling a shock and hearing the ringing of the bell. He had stopped his train and going back had found some of the cars off the road and smashed to pieces, whereupon he had taken his engine with the freight cars into Chicago for assistance. Like the conductor, he had gone back three-quarters of a mile and had found where the wheels of one of the vehicles had cut through the ties; the final derailment had, he thought, occurred at a cattle guard which a displaced wheel had struck. He attributed the derailment to a broken journal and according to him the axle was worn so smooth that it was impossible to say whether or not it had had a flaw.

Brakeman R.A. Nugent said that he had also seen the marks on the ties when he went back to protect his train.

The coroner's jury subsequently went to the scene of the accident by a special train. They found that the track had been re-laid but the wrecked cars were still strewn about on either side of the line. Back in the coroner's office they returned a verdict that the deaths had resulted "by the breaking of the axle of one of the cars, while running at the usual speed - 23 to 26 miles an hour."[139]

Very brief reports appeared in several American newspapers reproducing a story in the *Memphis Daily Appeal (*Grenada, Mississippi) newspaper of an accident on 2nd November 1862 to a train carrying a large body of soldiers on the East Tennessee and Virginia Railroad near Cleveland, Tennessee, United States. Seventeen of the Confederate soldiers were killed in the accident and about fifty injured. The train was derailed when one of its cars suffered a broken axle.[140]

On Friday 21st July 1865 an eastbound Central Ohio Railroad train which left Columbus at 4.28 a.m. was derailed near Columbia Center station, Ohio, United States, between Pataskala and Summit. The train was tightly packed with discharged Union soldiers of the 17th Ohio regiment, returning to their homes in Fairfield county. Because of the crowding of the train, with many soldiers unable to find seats, the platforms of all the cars were filled with soldiers, making it difficult or impossible for the brakemen to reach the brake wheels when the train left the road. Conductor A.H. Condit had repeatedly warned the soldiers not to crowd the platforms but in vain, as they had nowhere else to go. The derailment resulted from the breaking of one of the leading wheels of the engine. As soon as it happened, as the train was accelerating down a falling gradient, the engineer had whistled for brakes, but the brakemen were unable to comply. And the sudden stoppage of the engine threw the baggage car, Adams express car and two passenger cars off of the track and into one another. Six soldiers and the chaplain of the regiment, named McConn, died in the accident together with one civilian. Twenty-two others were more or less badly injured. One of the injured died a few hours after the accident and it was feared that others would not survive.[141]

A broken wheel caused another bad accident on the Pennsylvania Railroad, four miles west of Lancaster, Pennsylvania, United States, on the afternoon of Saturday 14th October 1865. One of the wheels of the leading bogie of the third passenger car broke, six or eight feet of the floor of the car were torn out, the passengers in it were thrown out and then run over by the rear wheels of the car. The following four cars were detached from the train, derailed and somewhat damaged but no one in them was much hurt. Nine passengers died in the accident, including the mother of three young children, who said that their father had died in the South, and they were on their way to visit a relative in Philadelphia, whose name they could not give. They were being looked after in Philadelphia pending someone claiming them. The car was a relatively new one, and the accident was regarded as having been purely accidental.[142]

---

[139] Weekly Gazette and Free Press, Janesville, Wisconsin, United States, 4th April 1862

[140] Indianapolis Daily Journal, Indianapolis, Indiana, United States, 15th November 1862

[141] Ibid, 25th July 1865

[142] Philadelphia Age, Philadelphia, Pennsylvania, United States, 16th and 18th October 1865

The Paris-Orleans line (Chemin de fer de Paris à Orléans) in France suffered a serious derailment on 4th September 1866. Train No. 20, which had left La Rochelle at 5.55 p.m. for Paris, was running at about 40 kilometres an hour when it left the rails near Saint-Maixent station at about 8.40 p.m. The axle of the second-class carriage next to the postal van broke and four carriages went down a three-metre-high embankment, only the engine, tender and leading waggon remained on the track. The first and last of these suffered little damage but the middle two were wrecked and six passengers were killed and 15 injured. Locals from Saint-Maixent and soldiers from a local garrison rushed to the aid of the victims and two doctors named Rondeaiu and Pelé were either on the train or lived close by and were able to treat the injured.[143] [144] [145]

A major disaster occurred at Angola, New York, United States, on 18th December 1867, which was initially caused by a defective vehicle, but which was much more serious because the train fell from a bridge. It is therefore dealt with in the bridge accidents chapter.

A mixed train from Limoges to Poitiers on 27th May 1870, due at the latter place at 9.47 a.m., was derailed on a 25-metre-high embankment overlooking the River Clain at Mauroc, near Saint-Benoit. The axle of an empty coal waggon, sixth from the engine, snapped, and a first-class, second-class and three third-class carriages that were following were pitched down the slope. Jean Brodu, a soldier of the 86th of line, on a fortnight's furlough, was killed outright, M. Boutillier du Retail, a former mayor of Château-Larcher, died in a few minutes and M. Brouillard, curé of Notre-Dame at Montmorrilon, died from his injuries some time later. Several other passengers were seriously injured while some had lucky escapes, thus, 'A sporting dog, covered with blood, was seen emerging from a carriage in search of its master, whom it came upon only slightly hurt,' and 'A family party, composed of father, mother, and an infant in the arms of its nurse, had a miraculous escape.' The first to come to the rescue of the victims were young men belonging to the Poitiers Seminary for Priests, who happened to be walking in the neighbourhood.[146] [147]

In June 1870 English railways experienced two fatal accidents within a fortnight due to defective vehicles. The first was a relatively minor affair, the second a major disaster. On 7th June an excursion train was run by the Lancashire and Yorkshire Railway to the growing seaside report of Blackpool, leaving Cleckheaton at 8 a.m. A carriage examiner tapped the wheels of the carriages with a hammer at Sowerby Bridge and found nothing wrong, but between Pleasington and Hoghton stations the rear guard noticed a piece of wood fly up from under one of the carriages and the front guard and the engine crew felt a jerk. The train was pulled up in about 600 yards, by which time the seven rear carriages had become detached and the leading one of these, in which most or all of the casualties

143 Webpage – Il y a 147 ans à Piozay le train express dẽraillait

144 Wikipedia - Liste des accidents ferroviaires en France au XIX siècle

145 The Times, London, England, 8th September 1866

146 Ibid, 29th May and 1st June 1870

147 Wikipedia - Liste des accidents ferroviaires en France au XIX siècle

occurred, had plunged down an embankment about 10 ft high. About 30 passengers were injured and although *The Times* could report the next day that it was not aware of any fatalities two of the injured subsequently died.[148] The accident was found to have been caused by a broken tyre on the third-class carriage which preceded the one which had gone down the embankment. The tyre which broke was a new one, of Bessemer steel, manufactured by Messrs Cammell of Sheffield in November 1869. The Board of Trade inspecting officer, F.H. Rich, had little comment to make on the accident other than to point out that the method of fastening the tyres to the wheels by rivets, as employed in this case, was not the best.[149]

The derailment between Newark and Claypole on the Great Northern Railway on 21st June 1870 was a much more serious affair than the Pleasington-Hoghton accident. Again, an excursion train was involved, in this case a return working of one from Leeds to London. The return train left London Kings Cross on time at 9.20 p.m. and consisted of an engine, a tender, 23 carriages and brake vans and was carrying 370 passengers. The accident occurred without any warning as the train was approaching the three arch Clay Lane overbridge, three-quarters of a mile south of Newark station, at 1.15 a.m. The guards of the train, and doubtless the passengers, suddenly felt a series of violent shocks and were much thrown about. The train, having come to a stand, they alighted and found the engine and tender partially overturned against the slope of the cutting, the leading brake van partially destroyed and a second and a third-class carriage completely broken up, while five of the remaining carriages were much damaged and all had their off sides grazed.

Approaching Newark on the up line had been the midnight up goods train from Doncaster to London, which was not booked to stop at Newark. As it passed under Clay Lane bridge its driver, Edward Brewer, felt a sudden jerk, and realised that his engine had parted from the 29 waggons and brake van of his train. He gradually reduced speed, wary lest the waggons, which he supposed were following, should run into his engine, and saw the headlights of the down excursion train coming towards him. He held out his arm as a signal to the driver of that train but doubted if it could be seen. Finding that his train was not following, he brought his train to a stand at Barnby-road Crossing, 760 yards beyond the overbridge. He then ran back on foot, with his fireman following cautiously with the engine, and finding the leading waggon about 300 yards back, as he thought, coupled it to the tender, thinking that his train was now complete. Just then Guard Thomas Walton, in charge of the York portion of the excursion train, ran up and told him that a part of his train had broken loose and fouled the down line, causing a very serious accident.[150]

Guard Frederick Beasley of the goods train felt a sudden stoppage as his train ran under Clay Lane overbridge and was thrown against the handle of his brake. Recovering himself, he got out of his van and found that one of the waggons of his train had fouled the down line. He then saw the down excursion train coming and heard its engine whistle for the Midland Railway level crossing. He just had time to show its driver a red light when its engine collided with the waggon lying across the down line, forcing it against

148 The Times, London, England, 8th June 1870

149 Board of Trade report

150 Ibid.

half a dozen other derailed waggons of his train and a pier of the overbridge.

Fourteen passengers and Driver Thomas Hurst of the excursion train were killed more or less outright. Fireman Joseph Watson of the excursion train died a couple of hours after the accident and two more passengers had died by the end of the next day. Forty other passengers were injured, together with Guard Hulford, in charge of the Leeds portion of the train.

The derailment of the goods train was due to a broken axle on Manchester Sheffield and Lincolnshire Railway waggon No. 3238, loaded at Crowle with rather less than six tons of potatoes for London. It had been examined during the course of its journey at Doncaster and Retford, but neither examination could have detected the flaw in the axle which caused the accident. The axle dated from 1852 and was smaller than those of modern (1870) design. The only comments which Henry Tyler, the inspecting officer, could make were that greater care should be taken in using waggons with such older axles and that they should be replaced as soon as possible.[151]

The coroner's jury at the inquest into the victims of the accident returned a verdict of "Accidental death" while expressing the opinion that the up goods train had been driven too fast from Retford to Newark and recommending that there should be a limit to the amount of time that axles were allowed to be in use and some means devised for their periodical testing, 'the jury being of the opinion that the present system is defective.'

A considerable fuss was created by allegations published in the *Leeds Mercury* newspaper and subsequently reproduced in other papers that looting of the dead and injured had taken place after the accident, allegations which were later shown to be completely without foundation.[152]

On Saturday 6th August 1870 an up train on the Chesapeake and Ohio Railroad was derailed at Jerry's Run, near White Sulphur Springs, West Virginia, United States. Climbing a very steep grade and having just negotiated a trestle work and a curve, the train was only moving at six miles an hour when the rear truck of the penultimate car jumped the track. Discovering that the car was derailed, the conductor pulled the alarm cord and at the low speed that it was travelling the train was stopped very quickly, but despite this the derailed car tipped over the high embankment at that point and falling about 100 ft was smashed to pieces. Twelve passengers were killed outright and about twenty more injured. The cause of the accident was stated to have been a broken wheel.[153]

Back in England, the Great Northern Railway was unlucky in having another bad accident on its main line, this time due to a broken tyre, before the end of the year. This second accident happened on Boxing Day, 1870, at Marshmoor Crossing, just south of Hatfield, to the 4.25 p.m. down fast passenger train from London Kings Cross. Driver John Fuller, in charge of the engine of this train, told the subsequent Board of Trade

---

[151] Ibid.

[152] The Times, London, England, 22nd, 23rd, 24th, 25th, 27th, 28th, 29th June and 2nd, 6th and 8th July 1870

[153] The Philadelphia Inquirer, Philadelphia, Pennsylvania, United States, 8th August 1870

inquiry and the inquest that there had been nothing wrong until he had felt a slight oscillation just before Marshmoor Crossing, nothing that much, but then he had felt that his engine was gaining speed and looking back found that he had lost his train. He had stopped at Red Hall signal cabin, three-quarters of a mile beyond the crossing, and begun to cautiously set back, preceded by his fireman, but they shortly met their guard, Thomas Richards, who told them that there had been an accident, when he had gone forward again to Hatfield for assistance.

Guard Richards, in the brake van at the rear of the train, there being another brake van, but no guard, at the front of the train, spoke of noticing the train slowing up as they neared Marshmoor Crossing and then hearing a crackling or breaking of carriages and they came to a standstill. He had not felt much of a shock and had kept his feet. Getting down, he had met a signalman named Town, who lived at the crossing, and telling him to go and protect the down line he, the guard, had proceeded down the line towards Hatfield, and managed to stop an up coal train.

Signalman Henry Town, who worked at Bell Bar cabin, half a mile on the London side of the crossing, had been looking out of the door of his cottage and had seen the train derail and its carriages strike a stack of sleepers and then bound across and strike the cottage of a platelayer named Hawkins. He had run to the line, had seen some of the dead and injured passengers and then met Guard Richards, who had handed him some fog signals, and he had gone back to Bell Bar box to warn the signalman there, thence to a neighbouring farm and to Welham Green village for assistance, and thence back to Marshmoor to assist at the scene. His mate, the man on duty at Bell Bar, was presumably Signalman Oswin, whose wife and sister, Mrs Kershaw, were two of those killed in the accident while they were walking along the line.[154]

The leading brake van, the third-class carriage behind it and the second-class carriage behind that were the vehicles which sustained the most damage in the accident. The leading brake van and third-class carriage were fortunately both empty, there being only a small number of passengers on the train. The six passengers who lost their lives were all in the second-class carriage, No. 385. They were James Pullen, a footman to Mrs Lewis, of Sandy; Thomas Bransom, a cattle dealer, of Glatton, near Stilton; Mr Blake, a travelling hawker for the British and Foreign Bible Society; R.P. Reynolds, the 12 year old son of a permanent way inspector, travelling with Mr Blake to Wheathampstead; Mr G.A. Potter, a tobacconist, of Clerkenwell, London; and 18 year old Harry Rawlings, a pastrycook's apprentice of Worthing, bound for Peterborough.[155]

The accident resulted from the fracture of the tyre of the left leading wheel of the front brake van, the pieces of which were found after the accident by Signalman Town. The tyre was a weldless steel one, supplied by Messrs Cammell of Sheffield in February 1870 and put on the wheel at Doncaster Works in September 1870. It was secured to the wheel by Beattie's patent fastening, supposedly a safeguard against a broken tyre leaving the wheel. Lt-Col C.S. Hutchinson found it hard to ascribe a cause for the failure of the tyre but attributed it to a combination of the extreme rigidity of the permanent way due to a severe frost, the tyre probably having been fixed somewhat too tightly and the metal used having been somewhat too hard. The use of Beattie's patent tyres had been discontinued

---

[154] Board of Trade Report

[155] The Times, London, England, 28th and 29th December 1870

by the Great Northern Company, except for van wheels, following a derailment at Little Bytham in 1863, and Mansell's patent adopted. The Great Northern locomotive superintendent stated that tyres of Mansell's patent had not proved suitable for van wheels, an opinion at variance with the experiences of the London and North Western and South Eastern companies, which used Mansell's tyres indiscriminately on all types of rolling stock.

The inspecting officer urged the Great Northern and other companies to search for a method of *really* preventing broken tyres from leaving wheels, a matter of all the more necessity in view of the almost universal adoption of steel tyres, which were more liable to fracture in severe weather and cause accidents to passenger trains, as four had done that winter.[156]

At 10 a.m. on the morning of 3rd January 1871 the axle of a baggage car of a Mississippi and Tennessee Railroad passenger train broke as the train was crossing an iron bridge at Nonconnah Bottom, five miles south of Memphis, Tennessee, United States. A second-class passenger car was dashed down an embankment and immediately caught fire, its stove having been upset. Conductor Peter Kirby was standing by the stove as the car went down the embankment and the stove overturned onto him, burning him badly, but he managed to struggle free and climbed out of a window, after which he worked heroically to save the negros trapped in the car, but the fire spread rapidly and nearly all the injured were burnt to death. Another passenger car smashed into the bridge and was badly damaged but few of its passengers were severely hurt. The engine and tender were not involved in the accident and the passengers in the rear cars escaped with a shaking. A number of surgeons and others went from Memphis to the scene and everything was done to help the sufferers.

The exact number of dead was unclear, but Sallie Ames, Mary Docherty, a youth named Davis, of Horn Lake, Mississippi and four unnamed coloured persons were fatally burned, and three coloured persons were slightly burned. C.C. Thompson, of Choulahoula, Mississippi; Felix Deney and Sam River, of Horn Lake, Mississippi; W.B. Wallin, Charles Ecole and omnibus agent William White were rescued by the conductor and escaped the fire but were seriously injured.[157]

A month after the Tennessee accident another derailment due to a broken axle was the precipitating event of a complex American disaster which also involved a collision, a fire, the collapse of an overloaded drawbridge and the heroism of the engineer of a passenger train.

The disaster at New Hamburg, or Wappinger's Creek, New York, on the New York Central and Hudson River Railroad, began with the derailment on 6th February 1871 of a car in the middle of a southbound train consisting mostly of waggons carrying crude oil. The derailment was discovered by the train crew and the train was brought to a halt in about a quarter mile, stopping on a small wooden drawbridge over Wappinger's Creek, immediately south of New Hamburg station. By the time that the train had been stopped

[156] Board of Trade Report

[157] New York Herald, New York, United States, 4th January 1871

other cars had left the track, one of them a flatcar loaded with two large vats of crude oil, which stopped foul of the opposite running line. What was, so far, only a minor accident became a disastrous one because the northbound *Pacific Express* was close at hand. There was no time for the crew of the freight train to properly warn Doc Simmons, the engineer of the express, and although a red light was waved at him, he was unable to pull up his train in the 500 yards which was all that was available. The collision with the flatcar smashed it to pieces and the oil was instantly set alight, wrapping the locomotive, cars and bridge in a sheet of flame. To make matters worse, the shock of the collision then proved too much for the wooden drawbridge, which collapsed under the strain, precipitating the engine into the ice-covered creek, and the baggage and express cars piled up on top of it, the lot crashing through the ice and half disappearing in the icy water beneath. The Chicago sleeping car behind the baggage car was a sheet of flame and, although desperate efforts were made to rescue its passengers, it was thought that not one got out alive. In ten minutes the supports of the remaining part of the drawbridge burnt through its superstructure and fell onto the half-submerged cars in the water, taking with it the Chicago sleeping car and its occupants. The two sleeping cars in the rear of the Chicago car and several freight cars were also destroyed by fire, but the passengers in these cars had time to escape. The rear four passenger cars were uncoupled by crew and passengers and pushed back out of harm's way. Eighteen passengers probably lost their lives, together with Engineer Simmons, Conductor Peter Varsburgh of the Chicago sleeping car and a brakeman supposedly named Bhrine {sic}. There was some doubt as to the number of dead, with Conductor Cossum stating that from the number of sleeping cars tickets unmatched with survivors there were only 13 passengers missing, while passengers estimated that there had been 25 or 30 people in the Chicago car who must have died.[158]

The locomotive brake having been applied and the whistle sounded for brakes when they saw the red light, which they naturally supposed indicated that the drawbridge was open, Engineer Simmons and Fireman Tallon had had ample time to jump, and the fireman did so, but the engineer chose to 'go down with his ship' as it were and stuck to his engine. Describing the accident in his book '*Notes on Railroad Accidents*' in 1879 Charles Adams' wrote that, "it was a good way to die. He went out of the world and out of the sight of men with his hand on the lever, making no reply to the suggestion that he should leave his post, but 'looking ahead and watching his business.'" To the present author, it seems rather a pointless sacrifice, since it seems doubtful if by his remaining a few moments longer on his locomotive he was able to do anything useful to minimise the effects of the disaster.[159]

The Times correspondent in Berlin reported on an accident on 21st June 1871 which had occurred on the Leipsic – Bitterfeld line between Rackwitz and Zachastau stations to a train carrying Pomeranian troops who were returning from France following the conclusion of the Franco-Prussian war the previous month. Four French carriages were thrown from the track in consequence of one of them being defective. 'According to information of the catastrophe received here at present two mon-commissioned officers

[158] New Albany Daily Ledger, New Albany, New York, United States, 7th February 1871

[159] Railroad Accidents by Charles Francis Adams

and 21 privates were killed, and one officer and 40 men injured, some seriously.'[160] The German version of Wikipedia mentions the accident very briefly but implies that it was a collision between a train and an engine rather than a derailment.[161]

The French Chemin de fer de Paris à Lyon et à la Méditerranée (PLM) line was in trouble on 16th September 1871 with a bad derailment near Pont-sur-Yonne station. Eleven passengers died and 32 more were injured in a derailment caused by the breakage of the axle of a coal waggon which had been marshalled in the train between two passenger carriages, contrary to regulations. In a subsequent court case and appeal, the chief operations manager of the company was first found guilty and then acquitted on appeal. The State controller was dismissed. In another development, the French Government forbade the further use of four-wheeled passenger carriages.[162 163]

The Grand Trunk Railway of Canada suffered a major disaster in the early hours of 22nd June 1872. An eastbound express from Toronto to Montreal was running about 20 minutes late, five minutes having been regained on the run to Belleville only to be lost there loading more wood fuel. The train consisted of an engine and tender, a combined baggage and express car, a smoking car, a second-class car, two-first class cars and a sleeping car at the rear. Riding in the first-class cars were mainly Anglican clergymen returning from a synod, the second-class passengers were mostly lumbermen heading for Quebec, together with some telegraph builders and three families.

The train left Belleville at 12.20 a.m. and was running down an incline at 30 miles an hour, skirting the northern shore of Lake Ontario near the village of Shannonville, when the leading truck of the engine derailed. Engineer John Hibbert whistled for brakes and tried to reverse his engine, but the engine immediately fell over onto its side. The leading baggage and express car were shot past the engine, but the smoking and second-class cars telescoped one another and, what was worse, tore off the safety valve and dome of the engine, so that a mass of scalding steam entered those cars. There were only a few first-class passengers and lumbermen in the smoking car, but the second-class car was much fuller. Twenty-five people were said to have been literally boiled alive and another 45 were terribly scalded in the half minute that it took for the contents of the boiler to escape. An American army surgeon present at the scene remarked that he had never seen anything like it in 25 years working in a military hospital.[164]

The two first-class cars were derailed but remained upright, the sleeping car was not even derailed. Conductor Nielson and a businessman named Robert Roddy led the rescue effort, using crowbars provided on trains after previous disasters, and brakeman John Rundle ran back with a red handlamp to protect the line. Doctors were telegraphed for from Shannonville and ten arrived from Bellville in a special train, while the sleeping car was used as a temporary first-aid station. Only two men were killed outright, one of them Engineer Hibbert, but 65 more people had been severely scalded, and, as in the

[160] The Times, London, England, 23rd June 1871

[161] Wikipedia – Liste von Eisenbahnunfällen in Deutschland

[162] The Times, London, England, 20th and 21st September 1871

[163] Wikipedia - Liste des accidents ferroviaires en France au XIX siècle

[164] Salt Lake Daily Tribune, Salt Lake City, Utah, United States, 4th July 1872

aforementioned Minoka disaster, the death toll rose greatly in the following days, so that by 2nd July the number of dead had risen to 34.

A very quick and perfunctory inquest on the victims of the accident was heard and concluded within a couple of days. There were two witnesses who spoke of problems with the track, but the jury quickly brought in a verdict that the derailment was an accident, due to a broken flange on the right-hand front wheel of the engine. This was not the end of the story, though, for there were protests that the inquest had been a 'farce' and the government agreed to hold a second inquest, while local Conservative and Liberal newspapers sniped at one another over the affair.

The second inquest was more comprehensive, with discussion over the speed of the train and whether the broken wheel flange found after the accident was the cause or the result of the derailment. In the end, although there was some evidence that the track was not as good as it might perhaps have been, the verdict of the second jury was the same as that of the first – that the derailment was the result of a broken wheel flange.[165]

A major disaster at Prospect, Pennsylvania, United States on Christmas Eve 1872, began with a broken wheel but involved a train falling from a bridge. It is dealt with under bridge accidents.

The Shrewsbury and Hereford line, which we have met already in January 1861, suffered another accident on 8th May 1873. By now the independent company had become the joint property of the Great Western and London and North Western Railways. On the day named, the 9.20 a.m. passenger train from Hereford to Shrewsbury left the track just after passing through Condover station, England. The train was not booked to call at Condover, and on the driver whistling for the Condover distant signal it was cleared and the train ran through the station at about 30 miles an hour. As it was doing so, its driver, hearing a rumbling noise, shut off steam and reversed the engine, while the fireman applied the tender brake. The engine came to a stand 345 yards beyond the station and it was only then that its driver realised that it had separated from the train. The carriage immediately behind the tender was still on the line but the next three composite carriages had fallen down an embankment into a field five or six feet below the track. One of these three carriages was turned upside down and crushed and four passengers in it were killed. The remaining vehicles of the train separated from the three composite carriages and went down the embankment but were less badly damaged. Several other passengers were more or less injured. The four passengers killed were George Yorke and Francis Turner, both going to a situation at High Bank Farm, Gorton, Manchester; Thomas Gibbs, returning to his job as a farm labourer at Horsforth, near Leeds, after a home visit to Bromyard (could he not get employment as a farm labourer nearer than Leeds?) and a Miss Hughes, of Shrewsbury.[166]

The accident was traced to the breakage of the leading axle of the Great Western Railway engine (No 183), which, on examination, contained a flaw in the shape of a half moon. The axle had been made at Stafford Road Works, Wolverhampton, in August

[165] Wreck! By Hugh A. Halliday pgs. 39-45

[166] The Times, London, England, 9th, 12th, 17th and 20th May 1873

1866. It had failed progressively in four stages, the first sometime previously, the second and third more recently and the fourth at the time of the derailment.

Other than making some suggestions as to slight alterations in the design of axles to hopefully make them stronger, Lt-Col F. H. Rich, the inspecting officer, had little to say regarding the prevention of such failures.[167]

On 21st June 1873 the Midland Railway excursion train left Leeds at 7 a.m. for Buxton, the trip being the annual treat for the officers and men of the Midland goods department in Leeds. It was carrying 274 Company servants and 17 members of the public. It travelled as far as Stretton station without incident. It was severely checked by signals there, and its driver remembered applying steam when the signals at Stretton South cabin were cleared for him. He could also recall shutting off steam for the Shirland Siding cabin, and reapplying it when that signal was cleared, but could not remember if he had passed that signal when he heard a sudden "crack" and his engine began jumping about wildly. He had sounded the whistle and touched the reversing lever when his engine went down the embankment, and the next thing that he remembered was being carried in a chair to a nearby cottage.

The fireman of the train also heard a crack and got two or three turns on the brake wheel, but he did not think that the engine went more than 30 yards before it plunged down the embankment. He was trapped between the engine and the tender with coals and boards thrown on top of him.

The head guard was riding in a brake van at the rear of the train. He saw ballast flying up about three-quarters of a mile beyond Stretton South cabin and managed to get his brake fully on before the train stopped. He got out and found a saloon carriage lying across the down line and other carriages derailed and, giving his flags and detonators to a goods guard travelling on the train, sent him forward towards Wingfield to protect the down line.

The under guard, riding in a van behind the tender, saw dust flying up but had no time to do anything before his van was derailed and he was thrown out of it onto the slope of the embankment, while his van was broken up and fell over a bridge parapet to the River Amber and road below. He sustained some broken ribs and a broken right arm.

Four carriages were derailed in the accident and fell down an embankment. Two passengers, a Mrs Ritchie and a Mr George Coverdale, both of Leeds, were killed, and 12 others were injured.[168]

The derailment was due to the fracture of the tyre of the nearside leading wheel of the engine, No. 661. The tyre had been fitted to the engine in 1868 by Messrs Kitson and Company of Leeds, but according to the specifications of the Midland Company, it was secured to the wrought iron wheel by eight bolts. Captain H.W. Tyler was highly critical of the Midland Company for still using bolts to tyres, pointing out that occasional failures of tyres were unavoidable but the danger arose when they flew suddenly and violently off of the wheel, 'When a tyre fails and flies from the leading wheel of an engine, the guiding power of the engine and of the train is gone; and an accident such as is above

---

167 Board of Trade Report

168 The Times, London, England, 27th June 1873

recorded is only too likely to result, the only marvel being that so few people comparatively are killed and injured, and so many escape. The risk being so great, and the remedy so simple, involving no extra cost worth a moment's consideration, it does appear surprising that there should have been so much difficulty in inducing railway companies to adopt the remedy. Having procured the best tyres they can manufacture or purchase, they too often proceed to bore great holes in them, and so to weaken them, in order to fasten them to the rims of the wheels, when there are superior methods of attaching them to the wheels without making such holes in them, and by means of which they may wear them safely down to the lowest desirable dimensions, without frear of accidents in the event of fracture, and with less risk of fracture occurring. The Midland Company were so free of accidents for many years that they did not see the necessity or improved tyre-fastenings; but at length, just as had happened at other times with other companies, a period came upon them in which they were very unfortunate in this respect; and it is only to be regretted that some time must necessarily elapse before the whole of their stock can be so fitted.'

Captain Tyler also sent out a circular letter to all the major railway companies enquiring as to what mode they adopted to fix tyres to the wheels of carriages and engines, and in conclusion he recommended the use of a double-clip system, with continuous clips with square shoulders on each side of the tyre, the strength of the clips being made proportional to the strains likely to be brought upon them.[169]

An unusual derailment at Trinity Junction, near Hartlepool, England, on the North Eastern Railway, on Tuesday 2nd September 1873, was complicated by the collapse of a footbridge. Trinity Junction was the point where the lines from West Hartlepool, and from Ferryhill and Hendon, came together to run to East Hartlepool station. As the passenger train which had left Hendon at 11.15 a.m. was approaching Trinity Junction the engine, tender and three leading carriages left the rails and the engine collided with the supports of a wooden footbridge, bringing the bridge down upon the tender and leading carriage. Two passengers in the carriage, and a boy who had been standing on the footbridge, were fatally injured and a second boy on the footbridge was less badly hurt. The newspaper report on the event noted, 'Had the accident happened at the dinner hour, when the bridge is usually crowded, the loss of life would have been appalling.'[170]

Examination of the engine after the accident showed that the tyre of the nearside right leading wheel was quite loose. The tyre had been put on in October 1871 in place of a steel tyre that had broken. There being no steel tyres in the store at Gateshead at the time that would fit the wheel, a wrought-iron tyre, which was too big but was the best fit, was used, packed with a ring of sheet iron ⅛ thick and the tyre fastened with six bolts. The tyre later being reported loose again, a second sheet of sheet iron packing $\frac{1}{16}$ thick was put in, and the tyre re-secured with six rivets. The engine had left the rails because the packing had dropped out, the rivets had given way and the tyre had become entirely loose, so that there was nothing to guide the near leading wheel of the engine round the curve at the junction. The Board of Trade inspecting officer, Colonel F.H. Rich,

[169] Board of Trade Report
[170] The Times, London, England, 3rd September 1873

commented, 'The manner in which the tyre was put on was very wrong. I should not have thought it possible that a large wealthy company like the North Eastern would have used an engine thus patched up with their passenger trains. An engine with a tyre put on in such a manner is not fit to run on any passenger railway either with goods or passenger trains.' The locomotive foreman, who packed the tyre the second time, pleaded that he had not been instructed not to do it in that manner and the wheel had been sent to him already packed from the chief workshops. Mr Fletcher, the chief locomotive engineer, said that the packing was quite contrary to his instructions.[171]

On Christmas Eve, 1874, a derailment due to broken tyre near Shipton-on-Cherwell, England, caused the worst accident in the history of the Great Western Railway. Thirty passengers were killed outright, four more died later, and 65 passengers and four railwaymen were injured, some of them very seriously.

The train involved in the disaster was the first portion of the 10 a.m. narrow gauge express from London Paddington to Birkenhead. (That is standard gauge, as opposed to the Great Western's 7 ft 0¼ in gauge.) The train left London with two brake vans and ten carriages, one brake van being at the head of the train and the other in the eighth position, with four carriages behind it. The train was two or three minutes late leaving Paddington and lost more time on the journey to Reading and at the latter station, where a carriage and a luggage van were added to the rear of the train. The express left Reading 25 minutes late and lost another five minutes to Oxford. A third-class carriage, which had been intended to form part of the 12.40 p.m. train to Wolverhampton, was added to the head of the train at Oxford and the engine which had worked the train from London was taken off and two engines were added as the weather was bad, two more carriages were likely to be put on at Leamington and it was not thought that one engine would be able to get the heavy train up Hatton Bank (2½ miles averaging about 1 in 105 between Warwick and Hatton). Oxford was left at 12.14 or 12.15 p.m., 34 or 35 minutes late. The train was fitted with Harrison's communication cord, extending from the rear carriage to the head guard's brake van, now ninth in the train, and from that van to a bell or gong on the tender of the train engine, but the passengers could not communicate with the under guard in his van second in the train. The cord was tested at Oxford to the extent of the boy who fitted it pulling the cord about 14 inches from the gong.

The train was between Woodstock Road and Kirklington (Kirklington was renamed Bletchington in 1890 when Woodstock Road became, confusingly, Kidlington) stations when Driver Henry Richardson on the train engine saw the communication cord shaking and snow and dirt being thrown up from the right side of the train. At the same time, James Hill, his fireman, saw that the leading third-class carriage was off the rails and Fireman William Money of the pilot engine saw a man with his head and arms out of the window of the same carriage, the man waving his hat and shouting "Wow, Wow!" Driver Richardson immediately shut off steam, whistled for the guard's brakes, reversed his engine and applied steam again while Fireman Hill applied the tender brake, and Driver William Butler of the pilot engine likewise reversed his engine and Fireman Money applied his tender brake. Neither driver had looked back to see what vehicles were off the

[171] Board of Trade report

line and the reversals took place before the train crossed the bridge over the River Cherwell.

Shortly after the two engines were reversed the drawbar between them and the leading third-class carriage broke and they ran ahead. The carriage was thrown off the line, run over and completely destroyed by the following vehicles, the leading brake van, with Underguard William Hill, ran down an embankment and was followed by a composite carriage and a third-class carriage, all coming to a stand in reverse order on the south side of the Oxford Canal, all badly damaged. The fifth and sixth vehicles of the train struck the parapets and wooden beams of the underbridge over the Oxford Canal and went down the embankment on the north side of it, the fifth being entirely destroyed and the sixth badly damaged. The remaining vehicles of the train were also derailed with the exception of the rear third class carriage and the luggage van preceding it, which remained on the rails.

Immediately after the accident the head guard, John Price, and a couple of passengers made their way back to Woodstock Road station in time to stop the following "cheap" train; an all too necessary precaution as the line was still worked on the time interval system. Meanwhile the pilot engine had been detached and ran forward to Kirklington to stop any up trains. After some unavoidable delay, the accident having brought the telegraph wires down, assistance was obtained from Oxford, as well as from a Mr Pearson and his employees, of Hampton Gay mill, near the scene of the disaster.[172] The first doctor on the scene was Mr Mallam, an Oxford surgeon, who had been on a visit to a Shipton-on-Cherwell churchwarden with a broken leg when a servant announced that he was sure that there had been an accident on the railway, as he could hear the screams of the victims.[173]

The names of those who had lost their lives in the accident were listed in *The Times* account of the first inquest into it (as was often the case there was more than one inquest, in this case three). They included cousins John Richards and John Hughes, a surgeon and a barrister respectively, travelling from London to Ruabon to spend Christmas; Mary White, an Oxford lodging-house keeper, also going to Ruabon to see her husband, a butler; Samuel Busby, travelling from Witney to Birmingham to spend Christmas with his daughter-in-law; Elizabeth Holmes, mistress of a national school at Ryde, Isle of Wight, going to visit her parents at Lichfield, Staffordshire; journeyman carpenter Humphrey Williams, bound from London to relatives at Dollgelly; and Sarah Yates, travelling with two young children from Fairford to friends at Wolverhampton. One of her children had survived, the other had not at that time been traced.[174]

Examination of the line after the accident led to the discovery of pieces of a broken tyre which proved to have come from the right leading wheel of third-class carriage No. 845, the one which had been put on the front of the train at Oxford. The carriage had been constructed in 1855 for the Newport Abergavenny and Hereford Railway by Messrs Wright and Son of Saltley Works, Birmingham. The tyre itself was manufactured by the Lowmoor Iron Company in 1865 and put on the wheel in January 1868. It was secured to the rim of the wheel by four countersunk rivets. It had broken into two pieces. The iron

---

[172] Ibid.

[173] Wheels to Disaster! by Peter R Lewis and Alistair Nisbet pg. 109

[174] The Times, London, England, 25th, 26th, 28th, 29th, 30th, and 31st December 1874, 1st, 2nd, 4th, 5th, 6th, 7th, 8th, 9th, 15th, and 29th January 2nd, 3rd, 4th, 5th, 8th, 9th, 11th and 15th February and 17th March 1875

from which it was manufactured was said to be of the usual good quality Lowmoor iron.

The Board of Trade report into the accident was comprehensive, the inspecting officer, Colonel W. Yolland, being assisted by a barrister, W.W. Ravenhill. In his conclusions to the report, Colonel Yolland first went over the events leading up to the accident, and pointed out that if the passengers in the leading carriage, Lewis, Garner, Proctor or Rudd, had been in possession of an efficient and instantaneous means of communication with the head guard and the drivers, the engine and guard's brakes might have been applied together, but he was not prepared to say, given the small and insufficient amount of brake power, and the position of the brake vans, there would not have been a serious accident. Although the initial cause of the accident was the failure of the tyre, its disastrous consequences were the grievous result of the fatal mistake, made with the best of intentions, by the two drivers and firemen, in reversing their engines and applying the brakes, without knowing what was actually wrong. At the point of the derailment, the train was running around an easy left-hand curve, and the pressure on the buffers of carriage No. 845, which had lost a tyre from the right-hand leading wheel, would be greatest on the left-hand side, and the immediate check on the speed, due to the brake application and reversal of the engines, would tend to push the left wheel farther away from the left rail. Had the train been going around a right-hand curve no further mishap would have ensued. Again, if the two drivers had just shut off steam the train would probably have just eventually come to a stand. Likewise, had the train been fitted with continuous brakes throughout its whole length, applicable by the driver by turning a handle to open a valve, as in Westinghouse's air and Smith's vacuum brakes; or by pulling a cord, as in Clark's brake, the train could have been brought to rest without casualty.

Colonel Yolland stated that in the absence of rules and instructions for their guidance, he did not blame the railwaymen involved for their actions in stopping the train as quickly as possible, even though that was the cause of the accident being so serious, but the absence of such instructions was a serious defect, and he urged all railway companies to adopt the instructions already found in the regulations of some companies: "But if any of the intermediate parts of the train be off the rails, the allowing of the carriages to stop themselves has in some cases kept up a disabled carriage when it is probable that if the front brakes had been applied the carriages behind would have forced themselves over the disabled one."

The train had not been properly made up when it left Oxford, with six vehicles in the rear of the rear brake van, and the brake power had been insufficient.

The want of an efficient means of communication between the passengers and the servants of the company, as required by the Regulation of Railways Act, 1868, section 22, certainly contributed to the magnitude of the accident and possibly permitted it to occur.

The fastening of the tyre, which had broken by rivets, was a dangerous and insecure method of fastening which the company had determined to do away with 20 years previously, and it was a grave mistake for the tyre to be fixed in this way in 1868 and the pair of wheels placed under the carriage in December 1873. Captain Tyler's comments on the methods of fixing tyres made in his report on the June 1873 Wingfield derailment had been quite correct.

As far as recommendations derived from the accident were concerned, the inspecting officer noted that it appeared that there were still many tyres in use secured only by rivets or bolts, although railway companies had switched to more secure means of fastening at least 20 years previously. He considered it highly desirable, through Parliamentary intervention, for companies to be required to send in returns showing the number of tyres in use secured by each method, so that the magnitude of the risk to the travelling public, using unsafe and defective methods of fastening, could be determined.[175]

Before the Shipton disaster, in June 1874, a Royal Commission on Railway Operation had been established in Great Britain in response to public concern over the increasing number of railway accidents, and as a result comparative trials of brake systems were held at Newark on the Midland Railway in July 1875. The Saunders automatic vacuum and Westinghouse automatic air brake systems showed themselves to be the best, and the Westinghouse brakes acted more quickly than the Saunders ones. The Westinghouse brakes were unfortunately very complicated and hence expensive, as well as requiring more maintenance. Following the report of the Royal Commission in 1876, the London Brighton and South Coast Railway quickly adopted the Westinghouse brake, as did subsequently the Great Eastern, North British, North Eastern and Midland. Other companies such as the Great Western and the London and South Western adopted the automatic vacuum brake, which the Midland later switched to. Companies, such as the London and North Western, Manchester and Lincolnshire and London Chatham and Dover, under the management of personalities such as Richard Moon and James Staats Forbes, continued to use inferior types of brake because they were cheaper.[176]

The rural Rothbury branch, in the county of Northumberland, England, on the North British Railway, was the location of a serious accident on 3rd July 1875. The noon passenger train from Rothbury to Morpeth started on time and arrived at Longwitton station at about 12.33. There it became a mixed train; eight empty stone waggons being attached to the front of the train. The train left Longwitton at 12.39, 12 minutes late, for the 3½ mile stretch to Scots Gap. The line was on steep falling gradients for much of the way to Scots Gap, varying from 1 in 60 to 1 in 127 but mostly at 1 in 75. As the train neared a bridge over the Delph Burn, about 1¾ miles from Longwitton, and descending a gradient of 1 in 60, the second waggon from the engine left the rails, followed by the six waggons behind it and five of the six carriages of the train, and fell down a 30 ft embankment, the third-class brake carriage at the rear of the train being the only one to remain on the line. The driver and fireman were unaware that they had lost the greater part of their train until their attention was attracted by the shouts and waves of three sisters named Winship who had been standing in a field close to the line and saw the accident happen.

The engine, tender and waggon came to a stand 500 yards on the Scots Gap side of the Delph Burn, and the rear brake third 140 yards south of the burn. The front brake composite was turned upside down and the guard in it, Matthew Lilley, was crushed to death, the two following third-class carriages were completely wrecked and the two composite carriages behind them badly damaged. The front end and footboards of the

---

175 Board of Trade Report

176 Railwaymen Politics and Money by Adrian Vaughan, pgs. 208-11

rear brake composite were also damaged as were all the waggons to a greater or lesser extent. Two of the passengers, William Fenwick, a mason of Shafto Moor, Deanham, and George Shaw, a patternmaker, of Lyker, were killed outright and George Thompson, a contractor from Stanhope, suffered a fractured skull and died during the afternoon. Many other passengers were injured, some seriously. Guard Henry Swimburne, 'an old servant of the company' in the rear brake composite escaped with a shaking and a bump on the head. After the accident, assistance arrived from nearby Rothley village and the engine of the train proceeded to Scots Gap and returned with some carriages in which the injured were forwarded to Morpeth, where they were attended by all the doctors in the town.[177]

Colonel F.H. Rich investigated the accident for the Board of Trade and determined that the derailment had been caused by the draw-bar at the rear of the leading waggon of the train having been pulled out, when it had acted as a lever to lift the second waggon off of the rails. The third waggon had then been pulled off of the line in turn, and every succeeding vehicle except the last one. The falling of the draw-bar was in turn due to the failure of a cotter. The use of single cotter fastenings was common among the different railway companies, but they were not secure fastenings and should be supplemented using screw nut fastenings, as he was happy to report the North British Company were in the process of doing.

Colonel Rich was remarked on the dangers of running mixed trains and stated that when it was unavoidable the goods waggons should be placed behind the passenger carriages. The guard had examined the waggons at Longwitton but could not examine the drawbars and cotters without going underneath them, which he was not supposed to do, and in any case would probably not have found anything wrong.

The inspecting officer concluded his report by thanking 'the Misses Winship' for alerting the engine crew, assisting the injured after the accident and giving him a most accurate description of what they had seen.[178]

---

[177] The Times, London, England, 5th, 6th and 7th July 1875

[178] Board of Trade Report

CHAPTER 3

OTHER DERAILMENTS

Previous chapters have dealt with train derailments caused by defects with the track or by defects of locomotives or rolling stock and following chapters will cover those resulting from deliberate human action involving vandalism or sabotage. Derailments can, however, also result from a variety of other causes such as excessive speed, accidental obstructions on the line (including animals) and traction shocks. This chapter will examine some of these accidents.

The first derailment of any significance which resulted neither from a defect in the train or track nor from deliberate action appears to have been one which happened between Auchterhouse and Pitpointe, in Scotland, on 15th June 1834, on the Dundee and Newtyle Railway. At about 2.15 p.m. the engine *Trotter*, with Driver William Grey and Fireman James Pirrie departed southbound from the top of the Hatton incline with a mixed train, the engine running tender first. It duly stopped at Auchterhouse station, and here a local miller named John Anderson got onto the tender, it was supposed with the intention of making his journey without paying a fare. The driver and fireman stated that the miller had not spoken to them, but it seemed likely that they were used to letting acquaintances such as the miller travel free. It proved to be the last journey that the latter would ever make, for between Auchterhouse and Pitpointe the tender and then the engine left the track after the tender had struck some obstruction on the line. Driver Grey had shut off steam as soon as he felt the tender running roughly but that was about all he could do, not having an effective brake that he could apply. Feeling that it would be too dangerous to jump off, both driver and fireman stayed on the engine. After running some yards on the ballast, the engine overran the tender and overturned it. The driver and fireman escaped more or less unhurt, but John Anderson was critically hurt by the locomotive passing over both of his legs and one of his arms and he died some hours later from his injuries.[179]

---

[179] The Dundee and Newtyle Railway by Niall Ferguson pg. 37

A serious derailment near Frankfort, Kentucky, United States on 16th March 1836 was blamed by one newspaper report on the train's excessive speed, its engineer having refused to slow down despite the entreaties of his passengers, until the train was thrown from the track. Another newspaper report was less certain of the cause of the derailment, stating that it had not yet been accounted for. Whether the speed of the train was really excessive, or whether it merely seemed so to passengers who, in a pre-railway age, had never travelled so fast before in their lives, it is, of course, impossible to say.

The 6 a.m. departure from Lexington became derailed two miles from Frankfort while crossing one of the highest embankments on the line, twenty or thirty feet high, and the engine fell to the bottom of the embankment on one side and the carriages were precipitated down it on the opposite side and smashed to pieces. Of the fifteen or twenty passengers on the train only one escaped unhurt. A Mr Braxton Wilson of Madison Indiana and a Mr Tinder of Woodford County, Kentucky, were killed and a number of the passengers seriously injured, as was Engineer Kinkead. A child of a Mr Tutt suffered a broken leg and died later.[180]

A bad accident in England on Friday 7th August 1840 became the very first to be subject to an official inquiry on behalf of the Board of Trade. It occurred at Howden, on the Hull and Selby Railway. At 7 a.m. on the morning of the day of the accident, a passenger train left Leeds for Selby and Hull via the Leeds and Selby Railway. On arrival at Selby it was strengthened by the addition of carriages destined for Hull, belonging to the Hull and Selby and York and North Midland companies, and also by a waggon loaded with a casting and other iron work, which had arrived from Leeds the previous day and was bound for Hull. On leaving Selby, the train consisted of an engine and tender, the waggon loaded with the casting, a first-class Leeds and Selby carriage, second and third-class Hull and Selby carriages, third and first-class York and North Midland carriages, a second-class Hull and Selby carriage, two third-class Leeds and Selby carriages, a Leeds and Selby luggage waggon containing a horse and two second-class and a first-class Leeds and Selby carriages. The casting had been 12½ feet long and the waggon upon which it was loaded was only slightly over 10½ feet in length, so that it must have overhung the truck at one or both ends, but the persons who loaded it could not recall exactly how it had been placed.[181]

The train had proceeded normally until within three-quarters of a mile of Howden station, where the casting fell off and falling across the track derailed the leading first-class carriage, which in turn dragged the four following carriages off the line. The leading carriage was smashed to pieces and that behind it partially broken up, but both had been empty. The carriages behind up to and including the sixth sustained varying amounts of damage. The sixth, the Hull and Selby second-class carriage, was not that badly damaged but had contained four and possibly all five of the passengers who lost their lives in the accident. Those who died were Mr Nicholas Veltman, a Hull merchant, Mr Thomas Stead jun., a York coach-maker and Elizabeth Lee, a 35-year-old single woman, all of whom were killed outright, plus a Mr Thomas Cragg, of Beverley and a Mr Aldersmith, a York

[180] Madison Republican and Banner, Madison, Indiana, 24th March 1836
[181] Board of Trade report

guilder, both of whom died later the same day. Several other passengers were injured.[182]

The investigation into the accident was conducted by Lt-Col Frederic Smith of the Royal Engineers. Guard William Greaves told the inspecting officer that he had not checked the waggon before the train left Selby and was therefore not certain whether or not the casting had been lashed to the truck, but that he believed that it had been, because after the accident he had seen at the scene fragments of rope such as were commonly used by the railway company, and these looked as if they had been chafed by friction. Mr Gutteridge, Leeds and Selby Railway superintendent, who saw the train at Selby, believed that the casting had been lashed to the truck, Mr Robert Grubb, the goods department foreman, could not be sure. Johnathan Dunn, a Hull and Selby policeman, stated that he had picked up at Howden some pieces of rope lashing which appeared to him to have been cut by chafing.

The Leeds and Selby waggon loader at Selby, J. Daws remembered the waggon coming from Leeds the previous day and was positive that the casting had been lashed to the waggon but admitted that he had not examined whether it had been safely secured.

Lt-Col Smith stated that it was the practice of the Hull and Selby company, in the interest of passenger safety in mixed trains, to place goods waggons between the engines and passenger carriages, and in first-class trains to interpose empty waggons or carriages. In conclusion, he stated that had the casting been properly secured by chains, ropes or wood framing the accident would not have happened, and there had clearly been great and unpardonable neglect on the part of those responsible at Selby to check that the casting was properly lashed and secured. The waggon concerned had only a small ledge around its sides and ends and the safety of the passengers therefore depended entirely upon the security of its lashings. He did not consider in any case that lashings alone should be entirely depended upon, but that the load should also be enclosed by a substantial framework. He was also doubtful of the wisdom of placing goods waggons next to the engines in mixed trains, as at speed such waggons tended to jump the rails and pull the following carriages off with them.

The inspecting officer found the driver of the train, Richard Hesketh, and the guard, William Greaves, to be free from blame. The fault, he said, lay with those who had attached the truck to the train. It should be the duty of the goods department foreman to see that all articles were firmly lashed or fixed before they were placed on a passenger train, and it was desirable that before the departure of a mixed train the foreman should deliver to the chief clerk of the station a written report that he, *personally,* had checked that all such loads on waggons were properly secure and safe.

Finally, Lt-Col Smith noted that he had examined the Hull and Selby passenger carriages and had no great objections to them now, except that they were only equipped with four-wheels.[183]

Because the passengers had died at different times and places two separate inquests were held. The jury of one found a verdict of 'Accidental death' with a deodand of £500 upon the engine and carriages (Deodands, objects or instruments forfeited because they had caused someone's death, or financial penalties imposed in place thereof, were finally

[182] The Times, London, England, 10th and 12th August 1840

[183] Board of Trade report

abolished by Parliament in 1846).[184]

After the termination of the first inquest, the coroner, Mr Pearson, immediately proceeded to Howden for the other inquest on Messrs Cragg and Aldersmith. Here the jury again returned a verdict of 'Accidental death,' this time imposing a deodand of 50 guineas. They added that, 'in giving this verdict, the jury that sum by no means sufficient to express their indignation, and they consider it their duty to observe that great and inexcusable negligence has been manifested in the packing of this casting; and the jury think it consistent, in connection with this awful event, to state their deep regret that the directors of the Leeds and Selby railway and the Hull and Selby railway should, by their conduct, sanction the violation of the Sabbath.' Given that the accident had occurred on a Friday it is hard to see what the rights or wrongs of Sunday travel had to do with it.[185] [186]

Less than a fortnight after the Howden accident, on 19th August 1840, a fatal derailment occurred on the line of the Eastern Counties Railway near Brentwood, England. Situated 17½ miles from London, Brentwood was at this time the furthest point to which the line had been opened. At 7 a.m. on the 19th August a passenger train departed there for London, consisting of an engine and tender, four first-class carriages, two second-class carriages, two third-class carriages and two flat waggons loaded with the *Norwich Times* and *Colchester Defiance* stagecoaches. The train was in charge of Guard John Collier with Driver George Foster and Fireman William Austin and was carrying 35 passengers. From Brentwood the line descended for slightly over two miles at 1 in 103. Driver Foster, having put on steam as the train departed, was soon travelling at a fast pace and by the time it had travelled about two miles it was reportedly running at over 60 miles an hour. At that point, a location known as Fulwells-gate, the train began to sway badly, and its passengers soon realised that an accident was about to happen. The next moment the engine left the rails and went down an embankment about ten feet high on the eastern side of the line and overturned, followed by the vehicles of the train apart from the two waggons carrying the stagecoaches. The first-class carriages were smashed to pieces and the second and third-class carriages badly damaged. One passenger, John Eastman, a labourer for the company, was killed outright and about ten passengers and the driver and fireman were seriously injured. A second passenger, George Emsworth, a Fleet-street, London, optician, suffered concussion and was severely scalded; he died of his injuries, as subsequently did both Driver Foster and Fireman Austin.[187]

Lt-Col Thomson investigated the accident on behalf of the Board of Trade. He was sure that the accident was due to excessive speed, though he was not sure of the exact mechanism of the derailment. Evidence before the coroner, Mr Flowers, at the inquest on two of the victims mentioned a broken chair, but it had not been found. Driver Foster, in a dying deposition, had spoken of the possibility of a wheel becoming loose on an axle, but an examination of engine No. 6 had failed to reveal any such defect. The embankment on which the derailment had occurred was in good condition.

Lt-Col Thomson recommended the establishment of a school for the education of

---

[184] Wikipedia – Deodand

[185] The Times 14th and 15th August 1840

[186] Observer, London, 17th August 1840

[187] Board of Trade

engine men to teach them the names and uses of the engine parts and how to start, reverse and stop them, with an examination by a competent person and the issuing of a certificate before a driver was permitted to take charge of a passenger train on any railway. In addition, maximum authorised speeds should be laid down for every railway, all passenger carriages should be fitted with buffer-springs, all engines should have six wheels, all railway lines should be provided with milestones or iron castings so that any passenger so disposed could time their train and heavier chairs should be used on the Eastern Counties Railway.[188]

At the conclusion of the inquest into the deaths of the four people killed in the accident, the jury returned a verdict of 'Accidental death,' with a deodand of £500.[189] [190]

Christmas Eve 1841 saw a serious accident on the Great Western Railway in Sonning Cutting, just east of Reading, England. *The Times* newspaper wrote of a 'Frightful Accident on the Great Western Railway.' What was referred to as a goods train, though in reality it was a mixed train, left London Paddington at the ungodly hour of 4.30 a.m. It consisted of the engine *Hecla*, two third-class carriages, a station truck, and 17 goods waggons, with First Guard George Hassam, Second Guard George Ayres, Driver Thomas Reynolds and his unnamed fireman. There were 38 passengers on the train, looking forward to the Christmas holidays, 'chiefly of the poorer class.' The train started ten minutes late. During the first part of its journey, in the five miles between Twyford and Reading, it had to run through a cutting at Sonning almost two miles long through clay and gravel with a maximum depth of 60 ft.

Driver Reynolds told the inquest on the victims of the accident that he had reached Twyford at about 6.40 a.m. He had seen nothing of any obstruction until his engine reached the overbridge carrying the road from Worley to Sonning, when his engine ran into some loose earth which had slipped from the side of the cutting onto the rails. His train had been travelling at about 18 miles an hour at the time. The impact with the mass of earth threw his engine from the rails and the carriages were thrown against each other and his engine. Half a mile back the line, at a place called Land's End, a policeman had shown him a white light to indicate 'All-clear.' It was still completely dark at the time of the accident.

Eight passengers were killed outright in the accident and 17 injured, another passenger died later. Lt-Col Frederic Smith investigated the accident on behalf of the Board of Trade. He found that the permanent way was in excellent condition on both sides of the slip and the slopes of the cutting were in as good a condition as could be expected given the recent wet weather. The depth of the cutting at the point where the slip occurred was 58 ft and its slope one in two. The inspecting officer did not consider that any engineer would think it necessary to give a cutting a greater slope than one in two, and he did not, therefore, consider that there had been any error in its construction. He did, however, question whether the large mass of spoil dumped on the land adjoining the cutting had not contributed to the slip. The slip must have occurred between the passing of the previous down goods train at 1 a.m. and about 6.45 a.m. Small slips had happened

[188] Ibid.

[189] The Times, London, England, 21st,22nd,24th,25th,28th and 31st August and 4th and 5th September 1840

[190] Observer, London, London, 23th August 1840

previously, but none had covered the rails and there had seemed no necessity to have someone watch the track at night.

The accident would have proved less serious, the inspecting officer continued, had the carriages been provided with spring buffers, and less serious again had the station truck and at least some of the waggons also been so provided. Quite apart from the question of their buffers, the third-class carriages were not of such a construction as the travelling public had a right to expect. Their seats were 18 in high and their sides only 2 ft high, so that a person standing up was in danger of being thrown out, as must those sitting near the sides of the carriage, while the exposure to cutting winter winds must be very injurious to the traveller going from London to Bristol, a journey of ten or twelve hours, mostly in the night-time. He recommended that the Great Western and other companies with low-sided carriages increase their height to 4 ft 6 in. He repeated a suggestion he had made in a previous accident report that all locomotives should carry a reflector white light on their buffer beams. These might be sufficient to enable the drivers of trains moving at moderate speeds to have time to apply the brakes and whistle for the guard's brakes on seeing an obstruction. One wonders how effective this would have been given the low intensity of 1840s lamps. Finally, Lt-Col Smith did not consider that passengers should be carried by heavy goods trains.[191]

The jury at the inquest into the deaths of eight of the passengers returned a verdict of 'Accidental death' with a deodand of £1,000 on the engine, tender and carriages (effectively a fine on the Great Western Company). They also recommended that the cutting should be watched more carefully for possible slips and that the passenger carriages should be marshalled further from the engine. The inquest on the ninth passenger, Richard Wooley, who had died on 29th December, returned a similar verdict, with a deodand of £100.[192]

Asked in 1825 by a Parliamentary committee in London, England, examining the proposed bill for the construction of the Liverpool and Manchester Railway, of the possible danger attending a train colliding with a cow which had got onto the railway the famous engineer George Stephenson had replied that 'It would be very awkward indeed – for the cow!'[193] This was usually true, but the train did not always get off so lightly either. Animals, of course, are not capable of understanding warnings not to trespass on the railway, and large animals, such as cows, are in fact quite massive enough to derail a train. In countries such as Great Britain the railway lines are fenced off, thus to a large extent preventing animal incursions, but in countries such as the United States, where railways were built cheaply, unfenced lines were the norm. The first fatal accident involving a derailment due to a collision with an animal appears to have occurred in 1845. It happened near Herbertsville, New Jersey, on the Camden and Amboy Railroad, on Tuesday, 30th September 1845. A passenger train was approaching a bridge over the South river about seven miles from South Amboy at its usual speed when a cow, probably alarmed by the noise, sprang upon the track in front of it. The engine and tender passed over the animal and remained on the line, but the two forward deck cars were

[191] Board of Trade report

[192] The Times, London, England, 25th, 27th, 28th,29th, and 30th December 1841 and 1st January 1842

[193] Life of George Stephenson (Railway Engineer) by Samuel Smiles, pg. 232

thrown off by the impact and dragged the length of the bridge with their axles on the rails. At the far end of the bridge, the cars, which were broken up, were thrown down the embankment. The jarring motion also caused two of the crates containing the passenger's baggage into the river. A Mr Briant of Philadelphia, described as 'an old pensioner,' who had been standing on the platform, was thrown onto the track and killed and Messrs Johnson and Talbot and Dr Cross, of Lexington, Kentucky, were seriously injured. 'We understand that all the persons seriously injured were standing upon the platforms, contrary to the express injunctions of the company, a notice of which appears in a conspicuous place upon every car.'[194 195]

On 8th July 1846 the Nord railway company of France (Chemins de fer du Nord) suffered a major disaster at Fampoux, on the main line from Paris to Lille, then almost new, having only been open a few months. At 7 a.m. on 8th July a passenger train left Paris Nord station, consisting of engine No. 44, with Driver Petiet and Fireman Dutoit, engine No. 48, with Driver Hovell and Fireman Bolu, and 28 vehicles. It was carrying 220 passengers, including General Nicholas Oudinot, a famous French soldier and politician.

At about 3 p.m., 22 minutes after the train had left Arras and as it was crossing a viaduct over the Scarpe river, its passengers felt a sudden shock as it became derailed. The train became divided into several portions. The two engines and leading baggage car ran a further 200 metres down the track before coming to a standstill, the second locomotive and baggage car slightly derailed. The second baggage car, a flat waggon loaded with the diligence of a M. Guérin, a third-class carriage and a second-class carriage fell a height of seven metres down an embankment into a peat bog which was flooded with water three or four metres deep. In the fall, the order of the vehicles was reversed. The majority of the fatalities in the accident occurred in this part of the train, two guards in the baggage car and twelve passengers, three of them from the stagecoach, losing their lives in these vehicles. A third portion of the train, consisting of two first class, a second class, a third class and another vehicle fell into the pond to the right of the second portion of the train but was not greatly damaged. The fourth portion, beginning with the carriage containing General Oudinot, fell onto the second portion. The general escaped injury but his aide-de-camp was badly hurt. All thirteen vehicles went down the embankment. The remainder of the train also left the track. Fourteen people were killed outright and three more died later of their injuries.[196]

The Nord company had only taken over responsibility for the line on 1st April and were therefore happy, it would seem, to blame the accident on the condition of the track. A company representative told the *Messager du Nord*, 'We have been told of a very great danger that exists on the road from Lille to Roubaix. We have been told that in a certain part of this road, built on a loamy soil, at a little distance from Roubaix, the carriages, when passing, tip so much to one side, that the movement frightens all the passengers.' To investigate the cause of the accident the Department of Public Works despatched a bridges and roads inspector, Pierre-François Frissard to the scene. He submitted a report

194 Brooklyn Daily Eagle and Kings County Democrat, Brooklyn, New York, 3rd October 1845

195 Logansport Logan Chief, Logansport, Indiana, United States, 11th October 1845

196 Wikipedia – Accident ferroviaire de Fampoux

on 13th July which must have been something of a disappointment, as he declared himself unable to account for the accident, or even to make a reasonable guess at its cause. He found a broken rail, but apparently ruled this out as the cause of the accident, stating that there was nothing wrong with the condition of the road at the point of the derailment. He had not found a broken wheel or axletree and while a broken coupling chain might have caused the derailment it was impossible to say so. The speed of the train did not appear to have been excessive, having taken 22 minutes to cover the eight kilometres from Amiens to Fampoux, though he conceded that the train might have been going faster as it was on the falling gradient approaching the viaduct.[197] He put forward a theory that the compression occurring between the parts of a long train as the front was slowing down climbing the incline beyond the viaduct while the rear portion was going fast downhill might have caused the derailment.

Despite the negative findings of M. Frissard's investigation, the two drivers and two firemen were charged with having caused the derailment but after a trial between 11th and 15th November at Lille Criminal Court all were acquitted due to lack of evidence. This was not the end of the matter for following a ministerial appeal there was a re-trial on 21th December at the Royal Court in Douai. Driver Petiot acknowledged having attained a speed of between 39 and 50 kilometres an hour, in violation of an order which limited the maximum speed of double-headed trains to 24 kilometres an hour. He and his fireman, Dutoit, were therefore found guilty this time, presumably of involuntary manslaughter or some similar offence. Both were sentenced to fifteen day's imprisonment and ordered to contribute to the cost of the trial. Driver Petiot was in addition fined 3,000 francs. Driver Hovell and Fireman Bolu, who had been on the train engine, were again acquitted.[198]

It would seem possible that the accident had resulted from a combination of excessive speed and track defects. An English engineer, James Baverstock, had written to *The Times* newspaper in London regarding the accident. He had travelled over the line, just before it happened, and 'I said it was not safe in my opinion, in the apparent state of the rails, either to run a heavy train at medium speed, or an express train at high velocity, and added that before long an accident appeared to me inevitable.'[199] *The Times* also noted that, 'The *Univers* states, that in consequence of the recent accident on the Northern Railroad, parapets sufficiently high to prevent the waggons from falling over the banks, and strong enough to resist a violent shock, are to be established on the bridges and other dangerous passages along all the railroads of France.'[200]

As the Fampoux disaster demonstrated, the cause of derailments was often difficult to ascertain, and they were sometimes due to a combination of factors. Such was the case regarding an accident in England, again by chance on the Hull and Selby Railway, on Sunday 21st February 1847. At 6 p.m. on the evening of that day the London mail train left Hull, consisting of a luggage van, three fish waggons, eight passenger carriages and another fish waggon, drawn by two engines, the *Kingston* or *Kingstown* as the pilot

---

197 The Times, London, England, 16th July 1846

198 Wikipedia – Accident ferroviaire de Fampoux

199 The Times, London, England, 15th July 1846

200 Ibid, 16th July 1846

engine and No. 63, named *York*, as the train engine. The latter engine was one of Stephenson's long-boiler type. The train left Hull between two and four minutes late and was allowed an hour and forty-five minutes to cover the 42 miles to Castleford, with eight intermediate stops to pick up or set down passengers as necessary. It only got as far as a point about one and a half miles short of the first station at Hessle when the second engine jumped the rails and in turn threw off seven of the carriages, and 'smashing them to atoms.' One of the passengers, Mr James Brown, of Hessle, a tin-plate worker, was killed outright and a second, Mr James Waring, of Dewsbury, was fatally injured, dying at the Coburg Hotel, Hessle, the following evening. Eight other badly injured passengers were lying at the time of the report at various Hessle inns, while the slightly injured and uninjured passengers were taken back to Hull.[201]

An inquest upon the two men was unable to determine the cause of the accident and the coroner's jury returned a verdict of 'Accidental death,' adding that they did not believe that blame attached to any particular party. Captain J. Coddington then undertook a detailed investigation on behalf of the Board of Trade. He concluded that the derailment had begun with the second engine, which had not been closely coupled to the pilot engine, acquiring considerable lateral oscillation, such as engines of that type, because of their peculiar construction, were especially prone to do. The permanent way, which was seven years old, had not been designed to bear engines of the size and weight of No. 63, and was insufficient to resist the locomotive's oscillation. Engine No. 63 had bent a rail and although the loosely coupled leading vehicles managed to get past the bent rail and the carriage further back the train did not, and coming to a sudden stand on the ballast, the following carriages had piled up against it, apart from the last two.[202]

The derailment, Captain Coddington deduced, had arisen from as combination of causes, and it was difficult to say how much the railway company were to blame in the circumstances. Almost every railway used two engines on one train, and it was not considered unsafe. In this case, had the leading engine not restrained the second one the accident would probably have been much more serious. There again had the leading waggons been of proper construction and tightly coupled they would have been the first to go off and borne the brunt of the collision. The track was certainly too light to bear the weight of heavy engines travelling at high velocities, and although he did not trace the accident to the presence of common goods waggons in the train, he thought they should not have been there, and the looseness of the fish waggons might have contributed to the increased oscillation of engine No 63. If fish waggons were to be sent on passenger trains they should, like carriage-trucks, horse-boxes, luggage vans and brake vans, be provided with the same springs and fastenings as passenger carriages.[203]

The construction of a railway in the South American colony of British Guiana had been authorised in July 1846. Almost a year before its first section had been officially opened it suffered an accident which cost the lives of two passengers. On Monday 24th January 1848 a cow wandered onto the track at High Street Crossing, Georgetown, one of a number of unguarded level crossings on the line. A train had departed from

[201] The Daily News, London, England, 27th February 1847

[202] Board of Trade report

[203] Ibid.

Georgetown that day, going along the line as far as it had been opened and carrying a party of gentlemen on a tour of the route. The newspaper report did not specify who the gentlemen were, presumably they were officials connected with the line, or possibly local dignitaries. The train was on its way back and had almost reached the terminus when its engine collided with the aforementioned cow. The carriages or trucks of the train were derailed, and a Mr Alexander Wishart was either thrown or jumped from the train and a Mr Joseph M'Claverty unwisely decided to jump. Both were fatally injured, Mr Wishart being run over by the tender of the engine, which broke both his thighs, and Mr M'Claverty suffering internal injuries.

The writer of the letter which described the accident added that, 'This is a sad commencement to this promising enterprise, which can scarcely fail to depreciate the price of shares at home; and very properly too, for if the company do not take more care, the passenger traffic will be very small, people will prefer taking to their feet or to their horses and gigs, rather than run the risk of meeting incautious cows on every mile of the line.'[204]

Two collisions with animals on the line, both leading to serious derailments, occurred in just over a month in the summer of 1854 in Canada West, on the Great Western Railway. On 2nd June a train collided with a cow near Lobo and on 6th July one struck a horse near Thorold. A number of accidents occurred upon the Great Western Railway within a short length of time, among them the disastrous collision at Baptiste Creek at the end of October 1854, and it was subsequently decided to appoint a commission of inquiry to investigate all of them.

The commission found that the Lobo accident had occurred at about 2.30 p.m. An eastbound express train struck the cow. The engine and tender and two baggage cars passed over its carcass without leaving the rails but the following vehicles, a freight car, a second-class and a first-class carriage were thrown from the track. The last two carriages were thrown partly down the 30 ft embankment upon which the accident happened, and their occupants escaped serious injury, but the freight car fell to the bottom of the slope and was broken up. It had been loaded with Norwegian emigrants and their baggage and five of its inmates were killed, as was a sixth passenger who had jumped from the platform of the car only to be crushed under a wheel.

The jury at the inquest upon the victims of the accident returned a verdict that their deaths had occurred because the railway company had failed to erect fences along the line and that emigrants should not be placed in the same cars as their heavy luggage. The commissioners considered that the jury had failed to consider the most important feature of the accident, the conduct of the engine driver. In charge of the engine of the train had been Driver Thomas Horton, accompanied by Fireman Thomas Bostwick. The evidence made it clear that the cow had got onto the line some time before the train had arrived, and the line being straight and it being daylight it must have been visible a long way off. It should therefore have been possible for Driver Horton to pull up his train or at least greatly reduce its speed before reaching the animal, but instead he had accelerated with a view to throwing the cow off the line, 'a most dangerous thing to do at a dangerous place.' He had also failed to sound his whistle 'because he found from experience that it

[204] Nonconformist, London, England, 12th April 1848

tended to irritate the cattle,' and thus his conductor had had no indication to apply the brakes. The commissioners noted that in collisions with large animals the engine and tender, because of their weight, were rarely thrown off the line, rather it was the lighter carriages behind which suffered. The passengers might well therefore be grievously injured while the author of the mischief, in most cases, escaped unhurt. That the rule book did not make any provision for such a situation they did not feel excused the driver and they suggested that some legal action should be taken against him in view of his recklessness.[205]

The Thorold accident occurred at about midnight on 6th July 1854 to a westbound express from Niagara Falls, which struck two or three horses at a point about a mile east of a bridge over the Welland Canal at Thorold. As at Lobo, the engine, tender and leading baggage car remained on the track. The following second-class passenger car was partially derailed and then ended up at right-angles across the line, and the following second-class car dashed through it, killing seven inmates, all Norwegian emigrants, and injuring 14 more.

The commissioners noted various points in relation to the accident:

Firstly, although the line was fenced at the actual point of the accident, there was no fencing a little further on, and cattle guards at the spot where the horses were believed to have got onto the line were unfinished.

Secondly, the tracks were unballasted, the sleepers or ties being laid on the surface of the roadbed.

Thirdly, although it had been night-time there was a testimony that it had been a clear moonlit night, the line was straight, and the driver admitted that he had seen the horses from some way off.

Fourthly, from the fact that the train had been unable to pull up it could be inferred that it had been travelling at an excessive speed at a location where, according to both the law and the company rules, it should have been going slowly, via approaching a bridge over a canal and a station.

Fifthly, it appeared that the tools which Conductor Matthews had been supplied with to employ after the accident had occurred, and those on the engine and tender, had been very inadequate.

The commissioners came to the conclusion that, as at Lobo, the driver, named Collinson, had either attempted to run past the horses or else to run them down, rather than stopping or at least reducing speed. It was, however, fair to add that this was only an inference, the accident having taken place at night when there were no witnesses or bystanders. This time the jury at the inquest had returned a verdict of 'Accidental death' and exonerated the railway company from blame, merely adding that they hoped that the fencing would soon be completed throughout the length of the line.

The commissioners suggested that the legislature should follow the lead of the American state of New York and compel the railway companies to carry axes, saws,

[205] Reports of the commissioners appointed to inquire into a series of accidents and detentions on the Great Western Railway, Canada West, by commission bearing date Nov 3, 1854

hammers, screw-jacks and crowbars at convenient points on every train for use in the case of accident.[206]

What was reportedly a very bad accident on the Mississippi and Missouri Railroad near Wilton, Iowa, United States received only a brief mention in a couple of American newspapers. Even the date of the accident is not very clear, though it probably occurred on Wednesday 30th April 1856. An evening express train from Davenport to Iowa City ran off the track just beyond the junction with the Muscatine branch, and ten or twelve persons were reportedly killed, and many others hurt. The entire train, consisting of engine, baggage car and four passenger cars, were derailed while the train was travelling at full speed.[207] [208]

On 6th May 1856 a major disaster occurred on the Panama Railway, which had been opened across the isthmus from Aspinwall (later Colón) on the Atlantic coast to Panama City on the Pacific coast the previous year. At this date Panama was a province of the republic of New Granada. With the opening of the first transcontinental railroad across the United States still some years in the future, this rather roundabout route was, at this date, the easiest way of getting from one side of the United States to the other.

The steamer *George Law* had left New York on 21st April and arrived at Aspinwall on 5th May with about 1,150 passengers. On 6th May these left there by train for Panama City, thence to take a steamer up the west coast of the United States to San Francisco. Three trains in succession left Aspinwall that morning, the first, at 7 a.m., having seven or eight cars loaded with mail, express matter and the passenger's luggage. This train went off the track at Obispo Bridge, 31 miles from Aspinwall, and despite several hours effort by its crew it proved impossible to re-rail it. The second train, carrying the cabin passengers from the steamer, left Aspinwall at 8 a.m. and consisted of a baggage car and nine passenger cars, each carrying about 84 passengers. (Cabin passengers were, as the name implies, provided with cabins and travelled in better conditions than the steerage passengers, who travelled in the lower decks of the ship in cramped and unhygienic conditions.) The engine of this train was not working very well and after about thirty miles it was decided that it could pull the train no further, and the train was put into a siding at Matachin, one and a half miles short of Obispo Bridge. The third and final train carrying the steerage passengers from the *George Law* left Aspinwall at 8.30 a.m. and again consisted of a baggage car and nine passenger cars. On arriving at Matachin, where the first passenger train had stopped, it was decided to combine the two trains, which then proceeded to Obispo Bridge as a single train hauled by both engines with two baggage cars and eighteen passenger cars. Arriving at the latter place it was found impossible to proceed past the derailed mail train. There was no accommodation at Obispo Bridge and no food could be obtained there for the train passengers, so at about 2 p.m. the combined train set off back to Aspinwall again with just one engine, the other remaining behind to assist with the re-railing operation. It was on this return journey that inconvenience was

---

206 Ibid.

207 Evansville Daily Journal, Evansville, Indiana, United States, 12th May 1856

208 Auburn Weekly American, Auburn, New York, United States, 14th May 1856

transformed into disaster.[209] [210] The passengers were soon commenting among themselves that the train was going dangerously fast and sure enough at a point about nine miles from Aspinwall, between Gatun's Bridge and Monkey Hill Grave Yard station, the train left the track on a curve, the first vehicle to become derailed being that behind the leading baggage car. This carriage and the following seven were thrown from the track and badly damaged, it was these carriages where the deaths and injuries occurred. The derailment happened just over a culvert, where there was impenetrable jungle on either side of the line and the ground was low, wet and marshy, with deep ditches on either side of the track filled by water after recent rain into which some of the wreckage fell. The rear vehicles of the train remained on the line, as did the engine, which was therefore able to proceed to Aspinwall for assistance.

A relief train returned from Aspinwall at 4 p.m., very fortunately with a supply of water, as many of the injured passengers were suffering greatly from thirst and had been reportedly forced to resort to quenching their thirst by drinking the bloodstained muddy water alongside of the track. Strangely no doctors accompanied the relief train, the only medical assistance at the scene being provided by a Dr R.F. Paris of Sacramento and his brother, who by chance had been passengers on the train. The train returned to Aspinwall with 53 of the injured passengers. A second relief train arrived from Aspinwall at 1 a.m., the next morning and returned with more of the injured. The passenger who provided this account of the accident, who had assisted in removing the bodies of some of the dead from the wreckage in the darkness, also returned by this train. Some of the injured and the bodies of the dead were still left at the scene. Those passengers who were uninjured remained at the scene, wet and without food, until they were finally forwarded to Panama City on the afternoon of the 7th. There were reports of the looting of dead and badly injured passengers and a baby whose parents had apparently both died was given to a group of women, none of whom wanted to take care of it and it was passed from one to another before being eventually found on the seashore at Aspinwall, and taken care of by the wife of a French merchant. The treatment of the victims back at Aspinwall was also the subject of much criticism. Some of the hotels refused to receive the injured, the railway company would not permit the names of the known dead and injured passengers to be transmitted to New York and none of the uninjured passengers were permitted to return to New York on the *George Law*, though many wished to do so.[211] [212]

No less than 43 persons reportedly died in the accident and about 50 were seriously injured and another 50 slightly hurt. The carriages of the 8.30 a.m. train ended up in front of those of the 8 a.m. train for the return journey, many of the victims were steerage passengers, including a party of six German emigrants from New York. Sixteen brakemen and other railwaymen from Jamaica and the Isthmus, presumably riding as passengers, were also victims. The bodies of the dead were buried on the morning of the 7th in pits at the site of the accident, without shrouds, coffins or a funeral service.

The accident was blamed on the running of an over heavy and unmanageable train,

[209] New York Daily Times, New York, New York, United States, 17th May 1856

[210] Brisbane Courier Mail, Brisbane, Queensland, 26th August 1856

[211] New York Daily Times, New York, New York, United States, 17th May 1856

[212] Brisbane Courier Mail, Brisbane, Queensland, 26th August 1856

which the driver was forced to run at some speed on the level stretches to surmount ascending gradients, though in general the line descended from a summit 37 miles from Aspinwall towards the Atlantic terminus. The condition of the track was also not good, 'The loose nature of the soil, the decaying effect it readily has upon the wooden material used in the construction of the road, and its constant absorption of moisture, endanger every train unless run at a very moderate speed and the danger is increased by the fact that where the ties begin to soften with rot, destructive insects lend their aid in weakening and consuming the durability of the wood. Combined, these everlasting agents have destroyed the adhesiveness of the tie spikes, consequently the rails are neither permanent nor capable of any considerable jar or strain; hence, when the long train had attained a velocity of twenty-eight miles an hour - which was the case at the time of the accident – the ties refused to hold, the rails spread apart, the forward trucks of the second car sank into the slight embankment, and the dire event of the day was the result.' The commentator added that the Panama Railway lacked proper care in keeping its line in repair and caution in the management of trains which were always too full of people. 'There was a palpable fault, an unpardonable wrong, in allowing the unfortunate train on the 6th to be run at the speed it was. No excuse can be offered at least calculated to palliate the wrong committed or remove the opinion that the horrible slaughter was accompanied by a recklessness demanding prompt and unanimous public condemnation.'[213]

A train carrying a battalion of Austrian troops, numbering about 1,200 men, who were on their way to relieve the garrison at Mayence (Mainz) ran off the line at Laufach in Bavaria on Friday 22nd August 1856. Six soldiers were killed and a number (different reports said 14 and about 30) of officers and men were injured. According to a German writer, there was a large movement of Austrian troops from north to south at this time; consisting in all of 112,228 men, 30 artillery pieces, 1,381 waggons and 11,501 horses, for reasons which are not entirely clear, but the men were probably being transported from Schleswig-Holstein, where the German Confederation had been defeated by Denmark in 1852, to meet a likely Italian War of Independence following the Congress of Paris which had ended the Crimean War.[214] [215] The cause of the derailment was not stated.

On 24th September 1857 the 9.30 a.m. up express train from Manchester to London Kings Cross, train No 13, became derailed near Tuxford station, England, on the Great Northern Railway main line. The train consisted of an engine and tender, second-class carriage, composite carriage, first-class carriage, second-class carriage {?} and a brake van, in charge of Guard William Dyson with Driver John Shaw and Fireman Samuel Setchell. As the train was crossing a bridge over a turnpike road 2½ south of Tuxford station the engine became separated from the train and ran forward and the vehicles behind also separated into two portions. The leading two vehicles made a complete somersault and landed right side up in a cottager's garden, relatively undamaged, while the other three collided with the bridge abutment, fell into the turnpike road and were

---

[213] Ibid.

[214] Eisenbahn-Katastrophen in Deutschland by Hans Joachim Ritzau, pg. 18

[215] The Observer, London, England, 31st August 1856

badly broken up. Four passengers lost their lives and several others and Guard Dyson were seriously injured. It was not thought at the time that the guard would recover. Those who died were the Hon. W.M. Windsor Clive, brother of the Hon' Robert Clive, M.P., who was also on the train; Miss Latitia Paget, of Garston, Liverpool; Mrs Heaton, of Lancaster and Mrs Pitman, wife of a Great Northern Company officer.[216]

What makes this accident unusual is that despite a thorough investigation, Lt-Col George Wynne, the inspecting officer, was unable to determine how it had occurred. He was unable, despite a minute examination, to detect any problem with the rolling stock – the badly bent axle of a brake van was clearly the result of the derailment – nor could he find anything wrong with the permanent way. Driver Shaw stated that he had been running at 50 or 55 miles an hour, but the line was perfectly straight and there was no reason not to believe that it was not perfectly safe for high speed. The most probable conjecture appeared to be that as the line was not provided with fishplates between the sections of rail that a key at the joint having been out or partially out the rail had sprung sufficiently to allow the wheel of the brake van to pass through the opening in the joint. Lt-Col Wynne expressed surprise that the local ganger, Robert Brooks (who witnessed the accident), was able to throw so little light on the matter. There was no doubt that the line in general was in very good order.[217]

The inquest into the deaths of the four who lost their lives ended with a verdict of 'Accidental death' but the jury clearly believed that the main cause of the accident had been the high speed of the train, and appended to their verdict the remarks that 'The jury, in giving their fullest consideration to the details of this melancholy accident, cannot separate without stating that it appears that there is not sufficient caution given to the engine-drivers in working express trains, that they are entrusted with the exercise of a very large and dangerous amount of discretion, and that there should be a *maximum* speed which the drivers of them should not, under any circumstances, exceed. And we further desire the coroner to communicate the verdict to the Board of Trade and request them to bring the subject under the consideration of the Government.'[218]

The inspecting officer disagreed with the coroner's jury and gave it as his opinion that whatever the cause of the derailment it would have occurred the same if the speed had been that of an ordinary train. He pointed out that of the 13 derailments of express and mail trains during 1855 and 1856 only four could *possibly* be ascribed to excessive speed, to one out of 5,500 trains run. He was, however, of the opinion that the average speed which express trains were timed to run at was too high, because it was likely to involve excess speeds around curves, through stations, junctions and so on, also that the high speeds of express trains probably contributed to the frequency of their suffering broken tires.

Lt-Col Wynne recommended that the Great Northern Company provide all its track with fishplates. Finally, he criticised the lack of a van or empty carriage between the tender and the other carriages, and the fact that although there was an apparatus allowing communication between the guard and the driver it had not been in working order at the time of the accident.

---

[216] The Times, London, England, 25th September 1857

[217] Board of Trade report

[218] Leicester Journal and Midland Counties General Advertiser, Leicester, England, 2nd October 1857

Five people reportedly died when a train on the Pittsburgh Fort Wayne and Chicago Railroad collided with a horse and was derailed near Wooster, Ohio, United States on 15th June 1859.

At about 11 a.m. on 1st November 1859 Chicago and North Western Railroad excursion train consisting of an engine and tender, baggage car and twelve passenger cars left Watertown, Wisconsin, United States, for Chicago, in charge of Conductor Hobart, with Engineer George McNamra. The engine was named *Perry H. Smith*, after the President of the company. Five of the cars of the train were from Oshkosh, four from Fond du Lac and three from Watertown and the train was carrying 600 passengers, two-thirds of them women. The train had covered eight miles from Watertown, and was crossing a bridge over Johnson's Creek, near Bellview station, when Engineer McNamra saw two cattle by the side of the line. The train was travelling at about 18 miles an hour. Suddenly, one of the oxen jumped up onto the track right in front of the locomotive. The engine was fitted with a cowcatcher which threw the animal twenty feet down the track, but it remained in the middle of it and when the cowcatcher struck the animal a second time it rolled beneath the engine and lifted it off the line. The collision happened at the end of a short bridge over a stream, and the line was surrounded by low swampy ground with ditches on either side containing about a foot of water. The derailed engine broke a rail on the far side of the bridge and was thrown into a ditch. Engineer Mack remained on it and was unhurt, his fireman, James McCabe suffered a spinal injury and four broken ribs. The tender followed the engine into the ditch and the baggage car and leading passenger car telescoped one another. Further back the train, the small fourth passenger car was telescoped for half its length by the larger adjoining car. All five passenger cars were wrecked.[219]

Twelve or thirteen passengers died in the accident and a number of others were badly injured.[220] Everyone agreed that the tragedy had been entirely unavoidable and that no one was to blame.

A branch line from Kings Lynn to Hunstanton, England, had opened in 1862 and by the following year was already attracting considerable excursion traffic. On Monday 27th July 1863 the 7.45 p.m. departure from Hunstanton to Lynn had consisted of two locomotives and no less than 34 carriages, causing problems at intermediate stations with the train being too long for the station platforms. On the following Monday, 3rd August, it was therefore determined to run two shorter trains over the branch in place of the 7.45 p.m. departure, as had been done with the 9.15 a.m. morning departure from Lynn, which had been supplemented by a 9.30 a.m. working. The first evening train duly left Hunstanton at about 7.45 p.m. with the second following at about 8 p.m. (the departure times of the two trains varied according to the evidence of different witnesses).[221] The driver of the first train, James Beaumont, glimpsed an animal, referred to as a heifer in the Board of Trade report but a bullock at the inquest, grazing by the side of the line, its

[219] Janesville Morning Gazette, Janesville, Wisconsin, United States, 3rd November 1859

[220] Milwaukee Daily Sentinel, Milwaukee, Wisconsin, United States, 2nd and 5th November 1859

[221] Board of Trade report

body partially in the dike and partially in the field, at a point between Wooton station and Lynn. He did not think the matter of any importance and said nothing about it to anyone when he arrived at Lynn.

The second train consisted of an engine and tender, fifteen carriages and a brake van. According to the Wooton stationmaster it passed through his station without stopping at 8.35, eleven minutes behind the previous train, and was travelling at 40 miles an hour (the driver and guard said 25 to 28 miles an hour). Two miles from Lynn, its driver saw the animal on the line 40 yards in front of him and shut off steam, shouted to his fireman to apply the tender brake and whistled for the guard's brakes. The engine and leading three carriages passed over the animal but remained on the track but the fourth, sixth, eighth and ninth carriages were thrown from the track. Feeling the jerk as the carriages left the line, the driver told his fireman to release the tender brake in the hope of keeping the couplings tight and preventing the carriages running into one another but one of the couplings broke and the carriages shortly came to a stand. The fourth, first-class carriage (No. 78) struck a piece of timber at a road crossing and was turned onto its side. The sixth, third-class carriage, No. 279, ended up on its side with its front end badly damaged and the eighth, third-class carriage, No. 55, was smashed to pieces. Five passengers were killed outright, a sixth died shortly afterwards and the Board of Trade report states that a seventh succumbed later; thirty passengers were more or less seriously injured. Five of the dead had been riding in the front compartment of carriage No. 279. Those who lost their lives at or close to the time of the accident were a Mr John Laird, a Lynn builder; Mrs Clarke, of Brandon; Miss Clarke, aged 73, of Lynn; Miss Palmer, of Walpole, near Wisbech; Mrs Brown of Lynn and John Dennis, a Lynn shipowner, who died a little later. The final death was probably that of Mrs Laird, the wife of John Laird, whom *The Times* of 24th August reported to be still in a critical state. Two and probably three inquests were held upon the victims of the accident, they having expired at different times and places. Verdicts of 'Accidental death' were returned in the two reported in the newspaper. The jury in the first, held on five of the victims, also blamed the railway company for the inadequacy of the fences, for not putting the animal off the line and for 'the disgraceful state of the carriages,' and also found fault with the inspecting officer for passing the line for opening the previous year. The jury on the death of Mr Dennis contented itself with criticising the fences and carriages and found that Captain Tyler had been perfectly justified in approving the opening of the line.[222]

The Board of Trade inspecting officer, Captain H.W. Tyler, attributed the serious consequences of the accident to the fact that carriage No. 78 had struck the crossing timber, thus coming to a sudden check and causing the following carriages to run forward upon it and upon each other. But other factors were the want of stronger carriages and a lack of brake power, so that the carriages had travelled over 200 yards from the point where the animal was first seen by the driver. A greater amount of brake power would have enabled the guard to do more to stop the train and prevented the carriages from running into one another, but one brake van with one guard was insufficient for a train of 15 carriages. A van at each end of the train with brakes working continuously on two or three carriages would have stopped the train quickly.

The animal which caused the accident had escaped from a field owned by a Mr Edward Bloomfield, which was separated from the line by a post and rail fence and

---

[222] The Times, London, England, 5th, 6th, 11th, 13th,14th, and 24th August 1863

smaller and larger ditches, the latter intended to act as a wet ditch, which it had done at the time of the inspecting officers survey of the line before its opening the previous September. Unfortunately, during the dry summer, the ditches had dried up and cattle had periodically strayed onto the railway. The Great Eastern Company, who worked the line, were aware of the problem, and had for some months past been adding a second stronger fence inside of the original one to make the line more secure. Meanwhile, they had requested Mr Bloomfield not to keep any cattle in the field adjoining the railway. This Mr Bloomfield had been unable to do and so a platelayer had been appointed to watch the spot and prevent any cattle getting onto the line. On the day of the accident this platelayer had been present until about 5.30 when his ganger, seeing no cattle and assuming that they had been moved from that field, had shifted the platelayer to another point about a mile away.[223]

Serious derailments caused by trains striking animals on the line continued with one in the United States on 3rd March 1865. The American Civil War was coming to an end as a train carrying Union soldiers of the 33rd Illinois regiment from Terre Bonne parish struck a horse and was derailed four miles from New Orleans, Louisiana, on the New Orleans Opelousas and Great Western Railroad (Shaw states in his account that the animal was a mule, but the newspaper reports all speak of a horse).[224] The accident probably occurred on or near a level crossing. Ten soldiers died in the accident and 39 were injured, 'several of whom will probably die.' [225]

An unexplained derailment occurred on the Great Indian Peninsula Railway near Titwalla, India, on the morning of Sunday 30th July 1865. A passenger train from Sheagaum to Bombay on the North Eastern line, carrying between 200 and 300 passengers, was involved. The train was about 2½ miles from Titwalla when one of its carriages left the track and dragged the remainder of the train with it. This first carriage was completely smashed up and all the deaths and most of the injuries happened to passengers who had been riding in it. The engine remained on the line and ran to Titwalla, from where news of the accident was telegraphed to Bombay, and then proceeded to Callian for medical aid. The district traffic manager, Mr Hart, at once proceeded from Callian to the scene of the accident with doctors and the train returned with the injured to the hospital at Tanna, which was reached at 4 p.m. Eight passengers riding in the wrecked third-class carriage were killed and seventeen injured, only two seriously. The only first-class passenger on the train, a Mr J. Wainwright, had a very lucky escape as he was in the carriage next to the smashed third-class one and was thrown into it. Unhurt, he had climbed out through a window and assisted those injured.

The train had been running on a straight stretch of line at its ordinary speed and it was difficult to assign a cause for the derailment. An investigation into the accident was begun the next day but merely came to the obvious conclusion that 'the deceased persons had died through a railway accident.' Coincidentally, a major disaster had occurred at a point only a short distance away just over a year earlier, in this case due to heavy rain

---

[223] Board of Trade report

[224] A History of Railroad Accidents, Safety Precautions and Operating Practices by Robert B Shaw, pg.283

[225] Boston Post, Boston, Massachusetts, 11th March 1865

damaging the track foundations.[226]

The Austrian Empire and the Kingdom of Prussia engaged in a brief war in the summer of 1866. Prussia invaded Bohemia, then a part of the Austrian Empire, and won a decisive victory at the battle of Königgrätz on 3rd July. The railway between Lundenburg and Prague in Bohemia was subsequently taken over by the Prussians and used by them as an artery of communication to carry the materials and men of three Prussian armies. Most of the normal rolling stock used on the line, which consisted of four-wheeled carriages able to cope with the line's many sharp curves, had been withdrawn by the retreating Austrians, and the Prussians were therefore forced to bring in mostly six-wheeled carriages from Saxony and even Prussia itself, vehicles designed for lines which were straighter and had easier curvature. Thus, it was that a six-wheeled carriage of a train which was probably running faster than was appropriate for the location ran off the track on the part of the line which twisted and turned as it ran up the valley from Brünn to Prerau. The carriage was overturned and formed a barricade against which and one other twelve following carriages were piled up. Five men died and eight were seriously hurt. Many horses also fell victim to the accident, including seven belonging to the King. This unfortunate accident affords, *The Times* correspondent in Prague wrote, 'a moral, inasmuch that it shows that not only must the permanent way be entire and safe, but that the rolling stock used must be suited to the particular line if railways are required to afford in time of war not only powerful, but also rapid means of transport.' [227]

Various causes were suggested for a derailment between Koslow and Riäsan, in Russia, on approximately 20th February 1867. Ten people died and 30 were injured when the entire train overturned. Excessive speed and the breaking of one of the axles of the locomotive were suggested as the causes of the disaster.[228]

On 28th June 1869 a Bombay Baroda and Central India Railway passenger train from Ahmedabad to Bombay, consisting of an engine and tender and about sixteen carriages had covered about eight miles from Baroda and was within four miles of Etola, India, travelling at 20 or 25 miles an hour, when it ran into a buffalo. Four third-class carriages were thrown off the line, two of them falling down an embankment 14 or 15 feet high and landing upside down, the third overturning on the line and the other remaining upright. The leading second-class carriage was also drawn down the embankment and landed on one of the overturned third-class carriages. The next carriage, a first-class car, was also drawn over but held in place by another second-class carriage behind it to form a bridge from the top of the embankment. The application of the brakes by the guards prevented the rest of the train following over the embankment and the engine and tender and leading brake van remained upon the rails. The engine driver, two guards, a rolling stock inspector and two European passengers then attempted to assist the injured passengers, while the majority of the Indian passengers 'sat quietly down or merely looked on, requiring repeated calling to lend a hand even to carry off their wounded or dead

---

226 Allen's Indian Mails, London, England, 6th September 1865

227 The Times, London, England, 13th August 1866

228 Ibid, 25th February 1867

countrymen.' On the other hand, 'Railway porters and platelayers called from along the line, and coolies, rendered great assistance. The engine went forward to Etola, from where Baroda and other stations could be telegraphed, but the writer complained that they had had to wait three hours before a train arrived from Baroda, though it was only eight miles away and doctors could have been sent on trollies. By the time that help arrived all the injured and all but three of the dead bodies had been extricated; the remaining bodies could not be got out with the screw jacks which were the only equipment available.[229]

Eleven passengers lost their lives in the accident and seven were injured, one seriously. A report by Colonel Strachey on railway accidents in India in 1868 had recently been published in which it was commented that, 'The fencing on these lines requires attention. It is known to have been deliberately neglected on the Bombay Baroda and Central India Railway.'[230]

Towards the very end of the Franco-Prussian war, at 9.30 p.m. on New Year's Day 1871, a French troop train was derailed at Saint-Sornin in the Allier department of central France. The ten soldiers and one sailor who died in the accident to train No. 8 on the Montluçon line came from different regiments and from all over France. They had been wounded in battle or prisoners of war and were returning to their regiments or going home on leave. Ten of them were more or less killed outright and the eleventh probably died later in hospital. The national press, preoccupied with the war and the siege of Paris, failed to mention the accident but a commemorative stone was later erected in the town.[231] [232]

A train, supposedly of the Austrian State Railway, ran off the line not far from Pesth, Hungary, on about 4th May 1873. Six carriages were smashed, 21 persons killed and another 40 injured. 'The Mayor, accompanied by surgeons, proceeded at once to the scene of the disaster.' [233] The cause of the derailment, if known, was not stated in the brief press accounts of the accident.

A mail train ran off the track near Jubbulpore, India, on the Great Indian Peninsula Railway, on 15th August 1875. One report spoke of five deaths, a second, from *The Times* correspondent in Calcutta, stated that six 'natives' had lost their lives together with the European driver and a European platelayer, while two or three 'natives' had been injured. Different explanations were put forward for the cause of the derailment, it being variously attributed to the malicious placing of an obstruction on the track and to excessive heat buckling the rails.[234]

A Barcelona – Zaragoza mail train was involved in a major disaster in Spain in the

---

[229] Allen's Indian Mail, London, England, 4th August 1869

[230] The Times, London, London, 28th July 1869

[231] Webpage – T Comme train – le train Montluçon – Moulins – le blog d'une génélogiste

[232] Wikipedia – Liste des accidents ferroviaires en France au XIX siècle

[233] The Times, London, England, 8th May 1873

[234] The Times, London. England, c13th and 15th September 1875

summer of 1876. On 24th June the train consisted of an engine and tender, two first-class, three-second-class and three third-class carriages, a mail van, three luggage vans and a guard's van. The journey went normally until the train was negotiating a 700-metre radius curve on a 17-metre-high embankment on the approach to Tarrega station. The train was reportedly travelling at full speed at this point when a third-class carriage, placed sixth from the engine, left the rails and dragged the following five vehicles with it, all six falling down the embankment and being badly smashed up. The mail van and a luggage van (brake van?) at the rear of the train remained on the rails. Fourteen passengers were killed outright, 19 seriously injured and another 35 less badly hurt but well enough to continue their journeys. Three of the most badly hurt victims later died, raising the number of deceased to 17.

The forward part of the train ran to Tarrega to obtain help and a large number of the local population turned out to assist the uninjured passengers in the rescue work. Telegrams were sent in all directions from Tarrega and a rescue train soon arrived from Lerida with civil guards, soldiers, doctors and medical attendants. A single doctor who had either been on the train or lived at Tarrega, was already doing his best to care for the injured.

The cause of the derailment was unclear. After the accident there was considerable animosity towards the train crew, who, the writer who described the accident insisted, had been perfectly innocent of the catastrophe, a view apparently shared by local newspapers. It was suggested that the carriage which had initially become derailed had been defective in some way, but it was too badly broken up in the accident to yield any evidence of a problem. The bridges, rails and embankments of the line were also in poor condition, no serious maintenance having been done on them for upwards of three years because of threats and interference from the Carlists, despite repeated calls from Government engineers for the line to be put in better order. The fact that the train was travelling at full speed around a curve also suggests to the present writer the possibility of excessive speed having been the cause of the disaster.

The line was soon repaired, and traffic resumed. 'Subscriptions have been opened, as many of the sufferers belong to poor families.' The writer who provided the description of the disaster given above stated that it had occurred on 25th June but the correspondent to *The Times* in London implied that it actually took place on 24th June.[235] [236]

Christmas 1876 saw a heavy snowfall in Denmark and in order to keep the line open for traffic on 26th December (Boxing Day) a snow-clearing train was sent out from Århus, consisting of a snowplough, two goods locomotives and some coaches carrying labourers for manual snow-clearing. According to a railwayman who saw the train at Hansted, near the point where the accident occurred, there was half an engine's gap between the two machines at that point, suggesting that unbeknown to the engine crews the coupling between them had snapped during the clearing operations. The leading engine then ran into a heavy snow drift and stopped abruptly, whereupon the second locomotive collided violently with it. Nine people died in the accident and another 26

[235] Bendigo Advertiser, Bendigo, Victoria, 18th September 1876

[236] The Times, London, England, 26th and 27th June 1876

were seriously injured.[237]

An accident occurred on the Cape Midland line of the Cape Government Railways in Cape Colony, southern Africa, on 8th May 1878. A train was ascending a steep incline between Klein Poort and Barkley Junction when a coupling reportedly broke and one or more vehicles were precipitated into a valley 70 feet deep. Seven people were killed and ten seriously injured. One of those killed was a Gilbert Gladstone Stewart, whose name as given by *The Times* without any explanation, so he presumably was a person of some importance at the time, though his name seems to have been forgotten.[238]

Seven people died when a train ran off the line on the Minho Railway in Portugal on about 12th October 1878. It has not been possible to ascertain any further details of the accident.[239]

An express passenger train on the Rostow – Vladikavkas Railway in Russia was derailed on about 19th December 1878 and fell from a height of 15 metres in the Caucasus. Eighteen passengers, 'including General Hientze and several functioning officers serving in the Caucasus' were killed along with the driver, fireman and guard. Thirty-eight other persons were injured.[240]

Another serious derailment occurred in Spain on 9th May 1879. A Madrid – Cádiz mail train ran off the line near the latter place, killing five people and injuring eight more. The van containing the mails for Cuba (at this date still a Spanish colony) was smashed. The letters for Cuba were saved but the departure of the Cuba mail steamer was delayed for some hours. The cause of the derailment was unknown. There were further troubles for the Spanish railways a couple of days later with the failure of a tunnel at Despenoperos.[241]

A regular afternoon passenger train on the narrow-gauge Pacific Coast Railroad picked up three extra cars at the Big Tree switch, California, United States, on Monday 24th May 1880, the cars being loaded with 1,235 holidaymakers returning from an Independent Rifle picnic there to Santa Cruz, a distance of only five miles. Engine No 3, with Engineer Elliott, was pulling the train. The extra cars were open vehicles with high railings at their sides and ends and were provided with seats.

All went well on the trip until the train passed through No. 7 tunnel, near Santa Cruz powder works. As it passed through the tunnel, on a falling gradient, and reached a curve in the line, the leading car left the track to the left-hand side of the line and the next car to the right-hand side. The third car was not derailed. The passengers in the first two cars

[237] Webpage: Liste de største togulykker i Danmark

[238] The Times, London, c3rd and c12th June 1878

[239] The Times, London, England, 14th October 1878

[240] The Times, London, England, 23rd December 1878

[241] The Times, London, England, c12th May 1879

were piled into a heap and thrown out of the cars onto the track, and the wheels of the cars passed over them. Other passengers jumped from the cars in their panic and fell beneath the wheels. The passengers in the third car were thrown off their feet but unhurt and at once went to help their less fortunate fellow travellers, while calls for assistance were sent to Santa Cruz and to Felton.[242]

By the following day the number of dead had reached sixteen and a number of others were injured. The president of the railroad company said that Engineer Elliott had told him that had had the engine reversed, had whistled down brakes and had been only going at a speed of twenty miles an hour when the accident happened. The engineer was unable to explain the derailment but suggested the possibility of an obstruction on the line. Other parties asserted that the train had been going at an unsafe speed for a curve of such a sharp radius. Some of the railroad employees suggested that passengers climbing up on the racks on one side of the car had caused the wheels on the opposite side to lift off the track. It was also pointed out that there had only been one brakeman on the train and that it was not equipped with air brakes, so it could not be easily handled.

Features of the aftermath of the accident, which it shared with a number of others, were that there had been attempts at looting the victims, which had been 'promptly checked,' and that some of the indignant survivors had talked of hanging or drowning the engineer and fireman on the assumption that had been responsible for the disaster.[243]

Early in the afternoon of Wednesday 11th August 1880 it was a Midland Railway express train from Leeds to Lancaster and Morecambe which was involved in a bad accident at Wennington Junction, England. Here the Furness and Midland Joint Railway line to Carnforth diverged from the main line to Lancaster by a facing junction and it was at the junction points that the derailment took place. The results of the accident were serious; the leading brake van of the train was badly broken up and the following composite and third-class carriages were smashed to pieces, while following carriages were less badly damaged. Seven passengers were killed outright, an eighth died later of his injuries and 23 others were more or less severely hurt.

Relief Signalman Edward Booth in Wennington Junction cabin told the Board of Trade inquiry into the accident that the normal position of the facing points was towards Lancaster and that he had last shifted them after the passage of a down goods train to Carnforth at 12.59, about an hour before the accident. The points had then been working normally, had there been any obstruction in them he would have been aware of it from his lever not working properly. He had received the 'Be ready' signal for the express involved in the accident from Low Bentham box at 2.01 p.m. and acknowledged it, the 'Train on line' signal had followed at 2.06 and the train had passed him at 2.11. The first that he had known of the accident was seeing a carriage towards the rear of the train off the road and running over the sleepers. A moment later the carriages collided with an overbridge and came to a stand. He could not see the front of the train because it was hidden around a curve.[244]

The 12.15 p.m. express from Leeds was in the charge of Guard John Morris, with

---

242 Terre Haute Weekly Gazette, Terre Haute, Indiana, United States, 27th May 1880

243 Boston Post, Boston, Massachusetts, United States, 25th May 1880

244 Board of Trade report

Driver John Whiteoak and Fireman John Fletcher. Driver Whiteoak said that his train had consisted of an engine and tender, seven carriages and two brake vans, one immediately behind the tender and the other at the rear of the train. There was only one guard. His train had stopped at Shipley, Keighley, Skipton, Hellifield, Giggleswick, Clapham and High Bentham and was running 17 minutes late when it left the latter station, but he was not trying to make up time. He estimated that he had been travelling at 30 miles an hour at Wennington. He was aware that the platelayers were doing something to the track there but did not know that it had been recently relaid and had received no notice about it. Steam had been shut off before he reached the station, and all went right until his engine was a few yards beyond the crossing (junction points) when he felt it jump a bit and realised that it had left the rails. He had reversed his engine and applied the Westinghouse brake on the tender. The engine, tender and part of the leading brake van had passed through an overbridge before coming to a stand, the remainder of the train stopped behind the bridge. After the accident, he had gone back and had found a mark on the rails were the derailment had occurred and had gauged the track and found it slightly narrow to gauge. He had no personal interest in arriving punctually or regaining time and had never been told to negotiate the curve at a lower speed or else there would be an accident. His train was late because it had left Leeds 16 minutes late. Driver Whiteoak added that he would not reduce speed merely because he knew that the platelayers were working on the track, but only if they gave him a green caution flag.

Foreman Platelayer John Bee told the inquiry that he had lifted the crossing and bedded it up by putting some ballast under the sleepers on the day before the accident. On the day of the accident, nothing had been done to the crossing; he and his men had been working lifting the track 30 yards on the Lancaster side of it. He saw the engine of the express jump the track at the crossing. There had been no need to flag the trains as the line was in good order. He had gauged the crossing before the accident, and it had been all right. He had never remarked that trains came around the curve too fast. He had never known an engine leave the line at that point before.[245]

Considerably more evidence was given at the Board of Trade inquiry and the derailment was clearly not one of those for which there was a simple, obvious cause, but having considered the evidence and the results of his examinations Colonel W. Yolland concluded that the action of Foreman Platelayer Bee the day before the accident in taking out the "slacks" and lifting and packing the crossing, had contributed, with the absence of cant or lowness of the outer rail of the crossing, the tightness of the gauge, and the want of an efficient check rail, to a very heavy engine, travelling at a speed of at least 30 miles an hour, mounting and striking the fixed point and fracturing the box crossing, thereby causing the derailment.

Contributing to the severity of the accident was the fact that the only brake power on the train was a Westinghouse brake on the tender and a hand brake in the rear van. Had the train been fitted with a good continuous brake under the control of the driver, the train could doubtless have been stopped much more quickly and the accident would have been much less serious. Colonel Yolland complained that it was all very well for the Midland Company to plead that they were now busily fitting continuous brakes to all their passenger trains, but the Board of Trade had been telling railway companies that passenger trains needed a larger amount of brake power for 20 years, and with the

[245] Ibid.

exception of a very few companies the efforts of the Board of Trade, which they had no legal powers to enforce, had been resisted.

The Midland Railway had issued instructions to drivers to restrict their speeds through certain junctions, but Wennington was not one of them. 'I trust therefore that the Midland Railway Company will lose no time in having this particular junction put into proper order, and with a cant or super-elevation of the outer rail of each line, and further direct that special instructions may be given to the engine-drivers as to a modified rate of speed for running through this junction, as the average rate of 37 miles an hour, which this Express train must run in order to keep time, is far too great.' [246]

The coroner's jury at the inquest into the victims of the accident returned a verdict of 'Accidental death,' adding a series of remarks clearly based upon the conclusions of the Board of Trade inspecting officer; namely that the speed of the train had been too high, its brake power had been inadequate, that drivers should receive instructions as to the speed which they should pass through all junctions and crossings such as that at Wennington, that there should be a more complete inspection of platelayer's work by an authorised person, that crossings and guard rails, especially those on curves such as at Wennington, should receive more careful attention and that continuous brakes should be fitted to all passenger trains as soon as possible.[247] [248]

[246] Ibid.

[247] The Times, London, England, 12th,13th,18th and 21st August 1880

[248] Lloyd's Weekly Newspaper, London, England, 22nd August 1880

CHAPTER 4

REAR COLLISIONS

The risks associated with one train running into the back of another were less in the formative years of the railways when trains were few and train speeds were low, though on the other hand the brakes available were primitive, the wooden rolling stock did not stand up very well to a heavy blow from behind and signalling was either primitive or non-existent. As the nineteenth century progressed, train speeds and traffic intensity grew faster than advances in brake power, rolling stock construction and signalling and as a result serious rear collision became common.

A rear collision of a sort occurred near Parr, England, on 15th September 1830, the opening day of the Liverpool and Manchester Railway, at an earlier hour than the famous accident to Mr Huskisson, but in this case the first train, hauled by the engine *Phoenix*, had become derailed before it was subsequently run into, without any very serious consequences.[249]

What might be termed the first 'pure' rear collision seems to have taken place at an unstated location on the Stockton and Darlington Railway in England on 4th August 1831. At this date, traffic on the line was only partly worked by locomotives, horses hauling many of the trains. Numerous trains of horse drawn waggons tended to encumber the working of the line and as a solution the measure was adopted of having the horse-drawn trains run in groups, at first of five and then of four. On the day in question, the driver of one of these horses left his steed and climbed into the dandy-cart of the set in front of him, where he promptly fell asleep. No longer urged forward by its driver, the horse first slowed up and then stopped. There was no light at the rear of the now stationary waggons and at about 1 a.m. the *Globe* engine came along and ran into them, throwing them off the line and disrupting traffic for two hours. The use of horses was abandoned over the

---

[249] Wikipedia – Opening of the Liverpool and Manchester Railway

next couple of years.[250]

The next accident occurred at Newton, England, on the Liverpool and Manchester line on 22nd February 1832. On that day the 7 a.m. passenger train from Liverpool caught up a preceding goods train and had to crawl along behind it until it could be shunted out of the way into a siding. The passenger train then proceeded as far as Newton station, where, as usual, it stopped for several minutes. Before it had set off again, the goods train, whose driver was not running as cautiously as he might have done as he expected the passenger train to have gone, ran into the back of it, considerably damaging the rear mail carriage, which was fortunately unoccupied. The morning was very foggy, a factor which one might have thought would have made the goods train driver all the more careful.[251]

The Liverpool and Manchester Railway suffered another rear collision in Olive Mount cutting, near Liverpool, in June, and then on 23rd November 1832 there was a more serious affair at Rainhill, described by *The Times* down in London as a 'Shocking railway accident.' As in the Newton accident fog, that bane of railway working, was a factor. The 7.15 a.m. second-class train from Manchester, having been delayed on its journey 'owing to different causes' arrived at Rainhill at about 9 a.m. Here it was delayed further taking on passengers' luggage and it was still there at about 9.10, when the 8 a.m. second-class train from Manchester was heard approaching. The Rainhill gateman ran back down the line to try and stop it, others cried out loudly for the driver to stop, and the 7.15 a.m. train was got into motion. The fog prevented the driver of the 8 o'clock train from seeing as far as he might otherwise have done, and as at Newton he expected the 7.45 train to be further on its journey than it actually was. When he realised the danger, he shut off steam and applied the brakes, but the rails were greasy, and the distance was too short to prevent a violent collision. The hindmost carriage of the front train was broken to pieces, the next, closed, carriage, little damaged but the remaining three coaches were more severely affected. The collision threw the engine of the second train against the station house, the front of which was completely carried away. One passenger, a recruit, en route to join his regiment in Ireland, suffered a severe head injury and died shortly after the accident, another recruit sustained a broken arm and other passengers suffered cuts, bruises and contusions.[252]

No further rear collisions seem to have occurred until 9th September 1837, when there was one at Kenyon Junction, England, again on the Liverpool and Manchester Railway. The 6 p.m. train from Manchester that Saturday evening had been stopped at Kenyon Junction for less than a minute when a Grand Junction Railway train from Birmingham was seen approaching. It was assumed that the driver of the Grand Junction train would pull up, but he continued at an estimated 30 miles an hour and collided with the Manchester train, smashing to pieces the rearmost of the Bolton carriages, which had just been attached to the Manchester train. One-woman passenger, a widow named Alice Hunt, who was travelling from Bolton to Wigan, lost her life and others were seriously

250 Tomlinson's North Eastern Railway by William Tomlinson, pgs. 383-5

251 The Times, London, England, 29th February 1832 –reproduced from the Liverpool Albion

252 Ibid, 26th November 1832

injured. The child of another woman was thrown from her arms and through a window by the violence of the collision but was caught by a bystander and little hurt. As at Newton and Rainhill the driver of the Birmingham train was not expecting to find the Manchester train there and would not have done so had not the Manchester train been more than an hour late, but as *The Times* pointed out that was really no excuse, as it was his duty to be on the lookout. The only signals normally displayed at most stations at this time were flags normally hoisted to indicate that there were passengers waiting to join the train and policeman Walburton stated at the inquest that it was not customary to hoist a flag for the Birmingham train, as it did not stop there. However, from 1833 it had been decreed that a red flag should also be hoisted in case of an accident or another train on the line.[253] At any rate none was shown, and all the policeman could do was wave his hat, shout to the passengers in the stationary train to jump out to save their lives and put down ashes onto the rails. The Birmingham train was still travelling at about 15 miles an hour at the moment of impact. At the close of the inquest on Mrs Hunt the coroner's jury found a verdict of manslaughter against the driver and fireman of the Grand Junction train.[254]

The rear collision on the Eastern Counties Railway at Old Ford, England (Coburn Road from 1879) on 14th September 1840 was the first accident of its kind to be investigated by an inspector appointed by the Board of Trade, but it was only mentioned briefly as a sort of footnote to the report on an August derailment at Brentwood. Again, there was one fatality and the inquest upon the unfortunate victim was fully detailed in *The Times* newspaper.

On the day of the accident a regular Eastern Counties passenger train left the Shoreditch terminus one minute late at 2.01 p.m for Romford and was followed in five or six minutes by a special train consisting of an engine, No. 5, tender and a single first-class coach belonging to the Northern & Eastern Railway, carrying a party of engineers (Messrs Stephenson, Frederick and John Braithwaite, Bidder and others) to Stratford to examine the junction there. The regular train stopped at Devonshire-street and Mile-end stations and then at Old Ford (confusingly also referred to as Bow in the paper) and had not left the latter point when it was run into by the following special, 'literally dashing one of the third-class open vehicles to atoms.' Four of the passengers in this coach were injured; two of them were able to walk home after remaining several hours at the station and the two more seriously injured, James Mason, a mechanic, and an elderly man named John Perry were taken to the London Hospital in Mile-end road. Two days later, John Perry, identified as a 57-year-old silk-dresser of Spitalfields, died in the same hospital from his injuries.

At the subsequent inquest into the accident George Miles, a friend of John Perry who was another of those hurt, told of Mr Perry seeing the special coming up as they were stopped at Old Ford, holding out his hand and crying out "They'll be upon us" immediately before the crash. Miles himself was briefly knocked out in the accident and the next thing he remembered of his friend was seeing him in a cart with his face strapped up about to be conveyed to hospital and appearing to be much injured. After George Miles had completed his evidence the coroner remarked that he did not see any necessity

---

253 Operations on the Liverpool & Manchester Railway by Thomas Donaghy, pg. 131

254 The Times, London, England, 12th, 13th and 15th September 1837

for such a rapid succession of trains, to which Mr Hall, the Eastern Counties Railway managing director, made the rather meaningless reply that 'the second was not a train' meaning presumably that it was not a regular working, as if that made the slightest difference when it collided with the train in front of it. It also transpired that Old Ford station had opened for the first time on the day of the accident. The inquest was then adjourned, and when it resumed experiments were carried out at Shoreditch as to how quickly engines could be stopped in emergency, in the course of which Mr Frederick Braithwaite said that he had started the test engine three minutes after the regular train, and he would have had no objection to starting only one minute after it! He thought that the accident had resulted from the driver of the special train, William Moore, not keeping a good lookout, but nevertheless spoke highly of the driver, and said that he would have no objection to riding with him again. Driver Moore himself gave evidence that he had not been aware that the passenger train would stop at Old Ford.

The inquest jury returned a verdict of "Accidental death" at the same time adding that they did not consider Driver Moore to be wholly free from blame. 'The Coroner accordingly cautioned Moore in a very impressive manner.' [255]

Lt-Col R. Thomson, the inspecting officer for the Board of Trade, was of the opinion that the collision was due to Driver Moore's disobedience of orders and culpable negligence in not keeping a good lookout. He added that the directors had dismissed Moore, and that he should be rendered incapable of working as a driver for any other railway company. In gross contrast to Mr Braithwaite's idea of running trains only a minute apart, the inspecting officer also suggested that there should be a minimum period of half an hour between the starting of trains from a station – hardly a very practical proposal as the traffic intensity increased.[256]

On 17th October 1840 it was the turn of the London and South Western Railway to have a rear collision at Nine Elms, England, which at that time was the terminus of the main line from Southampton (the line was extended to London Waterloo in July 1848).[257] The 1.30 p.m. departure from Southampton was due to arrive at Nine Elms at 5.16 p.m. but did not arrive until 5.49 p.m. due to heavy traffic. Before it had quite come to a stand it was run into by the 3 p.m. fast train from Southampton, due at Nine Elms at 6 p.m. According to *The Times* several carriages of the first train were smashed in the collision and a maidservant, named Elizabeth Andrews was killed. The Board of Trade report stated that the carriage in which the deceased was sat was not broken in the collision and implied that instead she was thrown out of an open carriage. She and her fellow servant Sarah Conelly had been given a day's holiday by their employer, a woman living in Edgeware Road, and they were returning from a visit to Hampton Court Palace.[258]

Even at this early date there was some rudimentary signalling at Nine Elms, the station being protected by a fixed signal capable of showing a white light, denoting that the station was clear, a green light for caution or a red light for danger. In charge of the signal and nearby switches was a pointsman named John Turner, while a notice bell and

[255] Ibid, 16th, 21st and 26th September and 1st October 1840

[256] Board of Trade Report

[257] Wikipedia, London Waterloo station

[258] The Times, London, England, 17th October 1840

further switches nearer the station were in charge of a second pointsman named William Greenaway. Had Turner placed the fixed signal at danger immediately upon the arrival of the first train the accident should have been prevented. Unfortunately, it transpired that Greenaway had in fact been absent all day and Turner had been doing his work as well as his own, so that on the arrival of the first train Turner had gone to the points which Greenaway should have been controlling to work them and before he had time to return and place the signal at danger the second train had arrived. The reason for Greenaway's absence was not stated.

Lt-Col Frederic Smith recommended that machinery should be provided to put the signal to caution as soon as the first carriage of a train had passed it, together with a trustworthy servant to subsequently put the signal to danger, also that the station clerks should preserve and interval of 15 minutes between trains, that all engine drivers should be provided with watches and timetables, and that all trains should approach Nine Elms 'with the greatest caution.' [259]

The third and last bad rear collision in England in 1840 occurred at South Milford Junction, otherwise known as Taylor's Junction. This was the point where the York and North Midland Railway line from Leeds to York via Castleford had a connecting spur to the Leeds and Selby line, and at this time all passenger trains between Selby (and Hull) and Leeds ran via this spur and the York and North Midland line via Castleford. Most of the passenger trains from Leeds, York and Hull had portions for the other two places, so that this was an important junction station with much interchange of traffic. At 6.40 p.m. on the evening of Wednesday 11th November 1840 a passenger train left Leeds for Milford Junction, consisting of three carriages each for York and for Selby and Hull. It stopped at Woodlesford, Castleford and Burton Salmon and arrived at Milford Junction at about 6.50 p.m. The York and Selby and Hull carriages had just been separated when the latter were run into by a 'luggage' (goods) train, killing a Mr Henry Patteson and a Mrs Sarah Cousins and seriously injuring several other passengers.

The staff at Milford Junction were unaware that a goods train was following the passenger train. In contrast to the situation at Nine Elms, there were no fixed signals at Milford Junction, though the porters were provided with revolving hand lamps capable of showing white, green or red. The station was staffed by three porters, who were there to assist the guards with the making up of the trains and with operating the points. The senior porter, John Baynes, told the Board of Trade inquiry into the accident that he and the other porters had been busy with the passenger train when they had heard the sound of the luggage train approaching. He had shouted to the passengers to get out and had run towards the advancing train with a hand lamp.

Driver John Watkins senior was in charge of the engine of the goods train (his son was driving the passenger train). He said that coming upon the straight on the approach to Taylor's Junction at Fryston-bridge he had spotted a white light on the up line, which he had assumed was the headlight of an engine on that line. He denied seeing any tail lamp at the rear of the passenger train, though there was evidence that it had been there when the train had left Leeds, and moreover that it had still been burning after the collision. He knew that the passenger train was ahead of him, and that it had three intermediate stops to

[259] Board of Trade report

make, but assumed that because it was travelling considerably faster it would have left the junction before he got there. Even so, he had told his fireman, Thomas Smith, to keep a good lookout, and approaching the junction they had both been doing so but had neither of them seen any danger until they suddenly perceived a red light 200 yards from the junction (this was no doubt Head Porter Baynes' hand lamp) when he had shut off steam and the fireman had applied the brake. They could get no assistance in stopping the train from Brakesman Joseph Scott because he was seated on top of the goods on the fourth waggon, out of reach of the brake handle. This was because the train had previously arrived from York without a tail lamp, and there being no tail lamps at Leeds, the brakesman had been told to ride on the fourth waggon exhibiting his hand lamp as a substitute tail lamp.

Lt-Col Frederic Smith investigated the collision and while he found Driver John Watkins senior to be principally to blame for want of care, he found much else to criticise and made a number of recommendations. He found the goods foreman at Leeds much to blame for permitting the train to leave without a tail lamp, and for allowing, if not ordering, the brakesman to sit on top the waggon. He criticised the lack of any signalling at South Milford Junction and the fact that the speeds of trains and the departure times of goods trains were not fixed. He was not happy that passenger and goods trains were despatched from different Leeds stations by different superintendents who did not seem to be in close communication with one another. He recommended that the passing times of trains should be noted in the guard's journals and at stations, that trains should be kept at least ten minutes apart at stations, and red flags or lamps used to enforce those intervals. No guard should ride other than in a brake carriage or waggon, all passenger carriages should be fitted with spring buffers, and a waggon provided with buffer-springs should be provided at the end of every train to act as a kind of fender. He suggested the addition of a second tail lamp placed at roof level. In addition, there appeared to be a lack of police, the ballasting of parts of the line was deficient and mileposts needed to be erected! Inspecting officer's reports were nothing if not thorough.[260]

The jury at the inquest into the death of Henry Patteson returned a verdict of "Accidental death" with a deodand of £500 on the engine *Zetland* of the goods train. Sarah Cousins must have died elsewhere and was the subject of a separate inquest where the jury almost came to blows over whether the deodand should be £16 or £1,060 and they were eventually dismissed.[261]

A goods train from Manchester left Warrington at 11.50 p.m. on 10th February 1841 bound for Birmingham via the Grand Junction Railway and hauled by the *Wildfire* engine. Subsequently, two goods trains from Liverpool reached Warrington at 12.25 a.m. and 12.34 a.m. on the 11th February. They were combined into one train and left Warrington, also for Birmingham, and drawn by the *Briareus* and *Charon* engines, at 1.30 a.m., thus an hour and 40 minutes after the preceding train. The second double-headed train was however running a good deal faster than the first one, so that it was only 13 minutes behind it at Whitmore, England, 34 miles to the south.

Since the onset of severe winter weather in December, the practice had been adopted

[260] Ibid.

[261] The Times, London, England, 16th, 18th and 23rd November 1840

of giving the drivers and firemen a cup of coffee at Whitmore station. During the day the porters took the coffee to the engine crews but at night the drivers and firemen had to themselves go to the station office for their drinks. Thus it was that Driver Hurst was absent from his engine for a short period, in contravention of the regulations which stated that he was never to leave his engine unattended on a running line. It was during this brief of absence that the combined Liverpool train ran into the back of the Manchester one. The results of the accident would probably not have been as serious as they were had it not been that the Liverpool train was propelling two waggons loaded with pigs and one goods waggon, and it was these that felt the full force of the collision. Riding with the pigs were two drovers, one of whom, James Nester, of Tulsk, County Roscommon, Ireland, was fatally injured and died soon after the accident. The other drover was also hurt and was taken to Stafford Infirmary.

Lt-Col Frederic Smith blamed the accident on the carelessness of Driver Ireland on the *Briareus*, the leading engine, who, on the falling gradient of 1 in 390 and knowing that the rails were likely to be icy, should have shut off steam much sooner than he did. That the three waggons in front of his engine might have obscured Ireland's view of the tail lamp was no excuse given that he had to stop at the station anyway for water and had the collision not have occurred must have overshot it. The inspecting officer also blamed, to a lesser degree, Driver Hurst for leaving his engine and that station clerk for letting him do so under *any pretence whatever* (italics in the Board of Trade report).

The coroner's inquest returned a verdict of "Accidental death" with a deodand of 1 shilling on the *Briareus*.[262 263]

The first rear collision to involve fatalities outside of England occurred near Wellesley, Massachusetts, United States, on 9th January 1843. A steamboat train on the Boston and Worcester Railroad connecting via the Norwich and Worcester Railroad with Norwich line steamers to New York overtook a regular train in thick fog, crushing its rear coach. The events which led to the collision began with the derailment of a freight train at Newton. This in turn delayed the 3 p.m. accommodation train from Boston to Worcester, which when it could finally proceed beyond Newton was not far in front of the steamboat train, which had left Boston at 4 p.m. A message was left at Newton that the accommodation train was only a short distance in front, but it was either not delivered or was misunderstood. An early report on the accident stated that no one had been killed, though a child and a coloured person had been badly hurt. The small child apparently subsequently died of the scalds that it had sustained, and five other passengers were hurt.[264 265]

Three days after the Wellesley accident, a passenger was killed on the Midland Railway in an accident at Barnsley, England (Cudworth from 1856). A mixed train which had left Leeds at 5 a.m. stopped at Barnsley just before 7 a.m. on Thursday 12th January 1843 and had been standing there about twelve minutes when it was run into by a goods

---

[262] Board of Trade Report

[263] The Times, London, England, 11th February 1841

[264] The Globe, Washington, D.C., United States, 17th January 1843 - reproduced from the Boston Courier

[265] A History of Railroad Accidents, Safety Precautions and Operating Practices by Robert B. Shaw pg. 75

train from Normanton. The morning was very foggy, and Porter Joseph Towndrow ran back towards the goods train waving a red hand lamp. Driver Edward Jenkins reversed his engine and told his fireman to apply the brake as soon as he saw the porter's signal but on the slippery rails, he was unable to stop, and three or four carriages of the passenger train were 'smashed to atoms.' A commercial traveller for a Glasgow firm named Robert Harvey was killed. Jenkins had only been employed as a driver for three weeks. The coroner's jury at the inquest found him guilty of manslaughter and he was committed to York Castle for trial. He does not seem to have been any more guilty than the drivers involved in earlier collisions; clearly, he was unlucky with his jury.[266] The Board of Trade seems to have contented itself with detailing the evidence given at the coroner's inquest with no separate investigation by an inspecting officer.

Glasgow Queen Street station, Scotland, is approached down a very steep incline of 1 in 42, and in the early days all incoming trains were stopped at Cowlairs station, rather more than a mile outside of the terminus, their engines were detached, and special brake waggons were placed at the heads of the trains. During this time the tickets were collected, and the train engine was moved to the rear of its train to push it to the start of the incline. On 1st August 1850 the Scottish Central Railway advertised a special train from Perth at 6 a.m. for Glasgow for the Highland Agricultural Society Show, conveying to passengers the very reduced fares. The Edinburgh and Glasgow Railway company, over whose tracks the Scottish Central train would reach Glasgow, was advised of this train, but not of the decision taken on 31st July, because of the number of intending passengers, to allow them to travel also by the 6.45 a.m. ordinary train from Perth. The 6 a.m. special from Perth arrived in fact at Greenhill Junction, where the Scottish Central trains came onto the Edinburgh and Glasgow, in three sections, the first with 27 carriages arriving at 8.40 from Perth and going forward to Glasgow, reaching there at 9.25. Twelve carriages then arrived at Greenhill Junction from Stirling at 9.15 and seven carriages from Larbert at 9.21, and these carriages were attached to the Edinburgh and Glasgow ordinary train which had left Edinburgh at 8 a.m. The massive train of 39 carriages then proceeded towards Glasgow and reached Cowlairs at 10.22 a.m.

The 6.45 a.m. ordinary train from Perth left there at 6.53 and arrived at Greenhill Junction at 9.35 a.m. It had become completely full by the time that it reached Greenloaning, midway between Perth and Greenhill Junction, and many passengers had resorted to climbing onto the roof. Despite the exhortations of Mr Harrison, the Scottish Central locomotive superintendent, they had stubbornly refused to get down and wait for the next train, so eventually three cattle waggons and three sheep trucks had been attached to the end of the train to accommodate them. The sheep trucks were low-sided goods waggons without spring-buffers or spring draw-bars. Into these 'carriages' the roof riders climbed, and the train set off at 9.44. Another train arrived at Greenhill at 10.05 from the Scottish Central line and was taken forward by Driver James Brown at 10.10. Stationmaster Ash at Greenhill had not expected to have to need so many engines and Brown's engine had to be ordered from Cowlairs at the last minute. There was no time to turn it when it arrived at Greenhill without causing unacceptable delay, and hence it returned to Glasgow running tender first, 26 minutes behind the ordinary Perth train, operated by a driver named Cardwell.

---

[266] Board of Trade report

The ordinary train had to stop at all stations except Bishopbriggs. Driver Cardwell could see the Edinburgh train ahead of him at Croy, got a red signal which was then altered to green at Campsie Junction and was cautioned by the porter at Bishopbriggs. He proceeded cautiously to Cowlairs and there found the Edinburgh train standing at the ticket platform. There were conflicting accounts as to exactly where Cardwell had stopped his train at Cowlairs relative to the stop signal protecting that station, but at any rate he passed the signal and Driver James Pirrie of the Edinburgh train asked him to set his train back so as to clear a mains crossover and thus permit Pirrie's engine to get to the rear of his train. This Cardwell did, but he then became anxious about protecting the rear of his train, which now stood well outside of the signal, Cardwell first asked a pointsman to go back, but the latter replied that he had other things to do. He then told his fireman to get the guard to go back, and as the fireman was going back to tell the guard to do so Brown's train appeared around the corner. Cardwell immediately applied steam and had just got his train moving when the collision occurred about eight minutes after he had first arrived at Cowlairs.[267]

Brown's train had run non-stop from Greenhill to Cowlairs. It was nine minutes behind Cardwell's train at Bishopbriggs, where Signal Porter John Fulton said that he had given Brown a caution hand signal, which Brown denied having seen. He was travelling at 25 miles an hour beyond Bishopbriggs and as soon as he came around the corner and saw Cardwell's train he had whistled, reversed his engine and applied the tender brake, while the brakes of two carriages were applied by the guards, but the distance of 400 yards was too short to stop in. The sheep truck at the very end of Cardwell's train was the one in which all the fatalities occurred. When the passengers riding in these open waggons realised their peril many were able to jump out of them but two men, two women and a child in the last truck were caught and killed; three of the five came from Dunblane. Nine other passengers, mostly from the last truck, were seriously injured.

Lieut. Douglas Galton investigated the collision on behalf of the Board of Trade. He found that Driver Cardwell has disobeyed the Edinburgh and Glasgow regulations in the first place in passing the stop signal at Cowlairs and had then been guilty of gross negligence in setting back without first having sent someone back to protect his train. Guard Gallackly, in charge of Cardwell's train, was, the inspecting officer continued, likewise guilty of most culpable negligence in not having even made an effort to protect the train, excusing himself by saying that he had no time and that Cardwell should have told him to go back. John Fulton, the Bishopbriggs porter, was found fault with for not having shown a green signal to both trains instead of giving Drivers Cardwell and Brown a hand signal which neither admitted to having observed. It was, the inspecting officer said, to be regretted that Brown's engine was running tender first, as this had worsened the effects of the collision. Likewise, the deaths and injuries sustained by the passengers would have been much reduced had they not been carried in sheep trucks lacking spring buffers and 'peculiarly ill-adapted for the conveyance of passengers.' The staff at Greenhill Junction should not have attached the sheep trucks to the train but made the passengers wait for a following service. Finally, there was damage to two carriages in the middle of Cardwell's train because their buffers were of different heights and thus had overridden one another.[268]

---

[267] Ibid.

[268] Ibid.

A most unusual and very likely unique accident occurred on the Midland Railway's branch from Ashchurch to Tewkesbury, England, on 14th December 1850. Since the previous month, the goods trains on the branch had been operated by a locomotive, but the passenger trains had reverted to being operated by horsepower, one horse pulling a single passenger carriage or omnibus. The main line passenger train due at Ashchurch at 4.04 p.m. on 14th November 1850 was half an hour late, and the connecting omnibus left there for Tewkesbury at about 4.40. As it departed, a storm was brewing and, as the horse and omnibus emerged from a cutting onto an embankment one and a quarter mile from Ashchurch, the wind and rain became so severe as to blow the horse off the track and it fell down. As it was December it had also become dark and the omnibus had no tail lamp, it not being usual to put one on at Tewkesbury for that train and none being kept at Ashchurch.

Meanwhile the driver of the goods engine had seen the passenger omnibus leave, had performed some shunting operations, and on instructions from the goods guard the engine and tender had set off for Tewkesbury 15 minutes after the passenger train, proceeding at about 10 miles an hour. The omnibus was overtaken shortly after the horse had fallen, the engine driver only seeing it at the last moment, when he reversed his engine and applied the brakes. The omnibus was thrown down the embankment by the collision and broken to pieces, and a woman in an end compartment and a man who had got out to assist in helping the horse get up were both killed. The driver of the omnibus, who was not a railwayman but an employee of the innkeeper who supplied the horses, had seen the engine approaching as he was dealing with the fallen horse and had run back 50 yards to try and stop it, to no avail.

Lieutenant Douglas Galton noted that had the omnibus driver gone back as soon as his horse had fallen or had the omnibus been supplied with a tail lamp the accident would probably have been averted. He thought the accident was partly due to the horse omnibus traffic not being worked with the same strictness as the locomotive traffic, partly perhaps because it was undertaken by a contractor, and was also critical of the fact that the branch had been worked for five weeks without any instructions having been issued, while it was desirable that tail lamps should be kept at both Ashchurch and Tewkesbury stations.[269]

On the evening of Monday 19th May 1851, the 9.05 p.m. Midland Railway train from Derby to Leeds, England, departed 15 minutes late and having made additional stops at Duffield and Wingfield stations was about 24 minutes late at the north end of Clay Cross tunnel. Just after it had passed Clay Cross station, north of the tunnel, its driver, John Sheldon, heard a loud crash, and stopping his engine found that the connecting rod working the pump had broken and damaged the pump. The driver worked to remove the broken parts as the guard, John Thompson, alighted from his brake van, went up to the engine and stood there for about for a minute or two watching the driver. Driver Sheldon then seems to have told the guard that they were likely to be stopped several minutes and the guard had best go back and protect his train, whereupon the guard went back in a leisurely fashion. When the guard had got back 30 or 40 yards he heard, he said, someone shout "All right," at which point he turned to re-join his train. Just as he was doing so, he

---

[269] Ibid.

heard the sound of another train coming, ran back to his train, mounted the footboard of the last carriage and waving a lamp shouted for his driver to move on. Driver Sheldon had in fact just finished removing the damaged parts and his fireman, seeing the approaching train, put on steam and had just got the train in motion when a 16-waggon goods train ran into it, destroying the last two carriages. Two passengers were killed outright, a Mr John Meynell, a Chesterfield magistrate and Mr John Blake, of the Sheffield manufacturing firm of Blake and Parkin; while almost all of the other passengers more or less severely injured, including a Mrs Tennant, of Charlestown, United States, who suffered a broken thigh.[270] [271]

The goods train was due to leave Derby at 9.10, its engine in charge of driver of Driver Samuel Stretton. It had actually left at about 9.20, just behind the delayed passenger train, but had been held 10 minutes at Derby North Junction, just outside of the station, and stopped another five minutes at Belper, where its guard told his driver to go steadily to Ambergate. At Ambergate the porter told the driver to look out at Wingfield, where the passenger train was going to make an extra stop, but the signals at Wingfield and Stretton stations had been at "All right." At the telegraph station at the south end of Clay Cross tunnel the train as given a green "Caution" signal and Driver Stretton slackened his speed through the tunnel. The train passed the telegraph station at the north end of Clay Cross tunnel, where he also received a caution signal, at 10.16, ten minutes after the passenger train and 15 minutes before the goods train was due at Clay Cross station. Seeing the white "All right" light at the station Driver Stretton then put on steam again until, as he was passing the station, he saw the three red tail lamps at the end of the passenger train and a fourth light being waved by Guard Thompson.

Concluding his report into the accident, Lieut. Douglas Galton, who was again the inspecting officer for the Board of Trade, outlined various factors which he considered caused the accident, beginning with the unpunctuality of the passenger train, often due it seemed to delays at Rugby, at the junction with the London and North Western line, which were beyond the Midland company's control, and accentuated on the day of the accident by the train having to make two extra stops. In addition, the goods train was running before time; the drivers of goods trains, but not apparently passenger trains, considered that they had to run faster to make up time if they were late, and the goods train driver had neglected to adhere to the times set down in the working timetable by passing Clay Cross fifteen minutes early. Finally, the guard of the passenger train had been very dilatory in going back to protect it; had he even gone back 200 yards he would have rounded a corner and been able to give Driver Stretton considerably more warning, and it seemed likely he would have had time to reach the platform at Clay Cross station and put the signal there to "Danger." (Clay Cross station closed at 7 p.m. and the station signal was left at "All clear" during the night-time in the manner of a signal box which was switched out.) [272]

After an inquest on Messrs Meynell and Blake, lasting several days, the coroner's jury reached their verdict. They condemned the practice of the goods train being started five minutes after the passenger train without sufficient measures to keep them apart, the great negligence of not having a brake van at the rear of the passenger train, the fact that Clay

[270] The Times, London, England, 21st, 23rd, 24th and 27th May 1851

[271] Board of Trade Report

[272] Ibid.

Cross station was unstaffed at night and the stopping of passenger trains at places not set down in the timetable. Nothing was said in their verdict about the neglect of guard Thompson to protect his train, instead they found a verdict of manslaughter against Driver Stretton for reckless driving.[273]

The return working of the last cheap excursion train of the season from Berlin to Madgeburg was in trouble near Burg, Prussia, on the evening of 28th September 1851. A spring of a carriage broke and while the train was stopped for examination it was run into by a following train 'in spite of cries and signals.' No one was killed but an unknown number of passengers were injured, some it was said severely, and 'the greatest confusion prevailed, as none of the carriages had lights in them.' The injured passengers were conveyed to Burg station, where a doctor who chanced to be present could immediately treat the casualties. The train due in Berlin at 11 p.m. did not finally arrive until 5 a.m. the following morning.[274]

On the morning of 14th February 1852, a goods train from Hof ran into the back of the 6 a.m. passenger train from Hof at Mehltheuer, Saxony, on the Sächsisch-Bayrischen-Staatseisenbahn, thick fog preventing the driver of the goods train from seeing the passenger train in time. Most of the carriages of the passenger train were damaged and 12 persons were injured. Even the leading postal van, the ninth vehicle of the passenger train, was pushed up into the air by the impact.[275]

On 15th September 1852 a New York and Erie Railroad accommodation train passed a freight train at Andover and the freight train followed it towards Hornellsville, New York, United States. Before the accommodation train got to Hornellsville its conductor, afraid that they would meet a late running up train from New York on the single line, stopped his train at Almond station. He neglected, however, to send a flagman back to stop the following freight train, which un-warned ran into the accommodation train. Five people were killed, four seriously if not fatally injured and about twenty slightly hurt.[276]

On 7th October 1852 the New Hampshire, United States, state fair was being held at Meredith Bridge (Laconia) and for the convenience of residents of the coastal part of the state trains were run from Dover to Alton Bay, from where visitors could take the steamboat Dover across Lake Winnepesaukee to Meredith Centre, travelling thence by the Boston Concord and Montreal Railroad the short distance to Meredith Bridge. Two trains were needed to carry all the passengers off the steamer and duly set off for Meredith Bridge, the second train leaving ten minutes after the first. The first train had only covered a mile, but had disappeared around a corner, when it came to a stand, probably because of a problem with the engine, on a long trestle bridge over a body of water near The Weirs. No precaution seems to have been taken to warn the following

[273] The Times, London, England, 31st May 1851

[274] Ibid, c1st October 1851

[275] Eisenbahnunfälle in Sachsen by Claus Schlegel, pg. 42

[276] LeRoy Gazette, LeRoy, New York, United States, 22nd September 1852

train, which was shortly heard approaching. The engineer hastily applied steam to get his train moving and its brakes not having been released a coupling snapped, leaving the rear coaches behind to be run into by the second train, which crushed them, killing five or six passengers and injuring a dozen more. 'The wounded persons were all taken down to Lake Village, where everything was done for them that could be. A large number of physicians at the Fair rendered their professional aid and were occupied until late at night in amputating limbs and dressing the wounds.'[277] Many of the passengers reportedly jumped into the water, which was 15 feet deep. Governor Martin of New Hampshire was in the first train but was not hurt.[278]

The failure to protect a train stopped in mid-section at Mount Union, Pennsylvania, United States, on 4th March 1853 was due to the person charged with undertaking the task falling asleep. An eastbound mixed train on the Pennsylvania Railroad, consisting mainly of freight cars but with two emigrant cars attached to its rear, came to a stand near Newton Hamilton, according to the newspaper report, because it had run out of fuel. That its engine should run out of coal, or more likely wood, seems rather unlikely, possibly it ran out of steam on an incline. At any rate Conductor William Mercer took a lantern and went back, but at a bridge below Mount Union he sat down in a cabin and dozed off, so that his train was left without protection. The engineer of a following eastbound passenger train, running at 30 miles an hour, thus received no warning of the obstruction ahead. The rear emigrant car was broken up and the second car lifted up and lodged on top of the engine of the passenger train. Two passengers, said to have been drovers, were killed outright, and three men, a woman and a child were fatally scalded. The engineer of the passenger train was slightly hurt. The accident obviously occurred in the early hours of the morning, for the passenger train was due in Harrisburg, not many miles ahead, at 4 a.m., but it seems strange even so that the conductor should be so tired as to fall asleep; possibly he was working excessive hours. The passenger train did not get away from the scene of the accident, after the wreckage had been cleared, until 6 p.m., by which time it was therefore over 14 hours late.[279]

On 9th March 1853 the Midland Railway night mail train from Birmingham to Bristol, England, started 20 minutes late at 2.30 a.m. It left Gloucester at 4.58 a.m., 28 minutes late, and Charfield, its last stop before Bristol and 17 miles therefrom, at 5.47, 26 minutes late. All went well from Charfield until the driver saw fire flying from his engine and brought the train to a stand just short of Mangotsfield station, six miles out of Bristol. Getting off his engine, the driver, Thomas Trueman, found a problem with a loose pin abutting the cotter of the connecting rod, and told the head guard, Abraham Perkins, who had come up to the engine, that the problem would take a minute or two to fix, and he had better send a signal back. The under guard, William Maycock, said at the Board of Trade inquiry that he had set off back as soon as he was instructed by the head guard to do so. The driver soon fixed the problem and told the head guard that they were ready to

[277] The Democratic Banner Davenport, Iowa, 22nd October 1852 – reproduced from the New Hampshire Patriot

[278] Weekly Wisconsin, Milwaukee, Wisconsin, United States, 20th October 1852

[279] Republican Compiler, Gettysburg, Pennsylvania, United States, 14th March 1853 – reproduced from Harrisburg Item of 5th March

go, but Guard Perkins said that they must wait for the under guard to return. With that the driver and guard heard a noise and looking round saw the chimney of an engine, which immediately ran into the train. The morning was very foggy. The driver of the engine, James Pollock, saw Guard Maycock's lamp, 100 yards from the rear of the train, and the tail lamp of the train, at the same moment as he came under a bridge. The train had come to a stand at 6.12 and the collision occurred at 6.14 or 6.15, smashing the rear carriage of the train and forcing the mail van, the only other vehicle of the train, against the tender. Two passengers in the mail train, William Antill and John Jones, were killed.[280 281]

The light engine which ran into the mail train had left Bristol the previous evening assisting the heavy 8 p.m. goods train to Gloucester. The goods train had become derailed at Berkeley Road, whereupon the engine had continued to Gloucester for assistance, returned to Berkeley Road and was then on its way back to Bristol to work the 7.15 a.m. passenger train from there. The engine left Berkeley Road at 5.50, 15 minutes behind the passenger train, and Wickwar 14 minutes behind it. Signalman Daniel Young at Westerleigh Junction, rather less than three miles short of Mangotsfield, stated that the mail train had passed him at 5.57, after which he had exhibited a danger signal for five minutes and a caution signal for another five minutes, the engine passing him under clear signals at 6.10. Lieut. Galton did not, however, believe signalman Young's times, which would have meant that the mail train had covered the nine miles from its stop at Charfield in 10 minutes at an average speed of 54 miles an hour. It came to light that signalman Young had tried to cover up the fact that by agreement with his mate they had made an unauthorised alteration to their shifts, whereby Young had worked from 6 p.m. the previous evening until noon on the day of the accident, and Galton considered that as Young had lied about this he had probably also failed to properly regulate the running of the trains, and had not cautioned or stopped the engine as he should have done, because it was running too close behind the mail train. Evidence given by some platelayers and by a man walking to work along the line both indicated that the mail train and the light engine had been seven minutes apart, presumably at a point near Westerleigh, but none of them had a watch and the inspecting officer did not give their statements too much weight.

As for the light engine, it did not appear to have gained much on the mail train in the seven miles from Berkeley Road to Wickwar, over which distance the line mostly climbed at about 1 in 264, but it had only taken 24 minutes (from 5.50 to 6.14) to cover the 16 miles from Berkeley Road to Mangotsfield, at an average speed of 40 miles an hour, so must have gained considerably on the mail train on the level or falling gradients beyond Wickwar, and been travelling at a speed which was not justifiable given the foggy conditions and the fact that its driver knew that the mail train was in front of him. The accident was due therefore to the signalman at Westerleigh Junction failing to caution the light engine, the driver of the engine running too fast and the under guard of the mail train not going back quickly enough to protect his train. The collision also would have happened, Lieut. Galton pointed out, had the mail train been on time. It had left Normanton 20 minutes late, awaiting mail from the North, and been further delayed at Tamworth waiting for the mail bags (doubtless from the London and North Western Railway's Trent Valley line) but had gained 13 minutes with a short stop at Birmingham. The Post Office was therefore largely to blame for the mail train's unpunctuality.

---

[280] Board of Trade Report

[281] The Times, London, England, 21st March 1853

Lieut. Galton recommended that the directors of the Midland Railway should follow the practice adopted by many other companies of having their signalmen record the passing times of trains and the hours at which they came on and off duty.[282]

The coroner's jury placed all the blame for the accident on Guards Perkins and Maycock with a verdict of manslaughter against them. They also recommended that the caution signal should be exhibited for ten minutes instead of five.[283]

A rear collision near Straffan, Ireland, on 5th October 1853 was easily the most serious accident of its type up to that date and would remain the worst train accident in Ireland till the Armagh catastrophe in June 1889. A Great Southern and Western Railway express from Killarney and Cork had left Cork at 12 noon for Dublin, where it was due at 6 p.m. By the time that it left Sallins station, 18 miles from Dublin, it was 16 minutes late, and having run another four and a quarter mile it had to be brought to a stand owing to the breaking of one of the engine pistons. A police sergeant, Henry O'Connor, who was acting as a pay clerk first went back 300 yards with a red flag. The engine then whistled to recall him, and he stuck the red flag in the ground and returned to the train. The engine driver, having found that he could not proceed Guard Berry, then went back several hundred yards with a hand lamp, which he waved at the following train. At the time that the train stopped it was light enough for one of the passengers to read in his carriage, but in the twenty minutes that the train was stopped dusk had fallen, and there was also a misty rain falling. The passenger train should have been carrying both a tail lamp and two side lamps at its rear, the latter being carried, according to the regulations, on all trains likely to be running after dark. A tail lamp was affixed at Kildare but no side lamps and by the time that the train had stopped near Straffan the tail lamp had either gone out or fallen off, so that it was showing no warning lights to the rear.

Behind the passenger train, a pick-up goods and cattle train had arrived at Sallins station at 5.56 and was allowed to leave there at 6.06p.m., being then fifty-one minutes late. The goods train proceeded at about 30 miles an hour, and the line in the rear of the stopped passenger train being dead straight for several miles, the passengers could see the goods train approaching when it was still some distance off. They were not, however, alarmed, but rather pleased that another train was coming to push them on to their destination. They could not tell how fast the goods train was running and some apparently thought that the guard had boarded it as it had slowly passed him. No one seems to have been concerned that the only thing to stop the heavy goods train running into them was the red hand lamp of the guard, hard to see in the misty twilight. There were in fact three powerful red lights on the engine, but no one thought of using them to stop the approaching train. Driver James Gass of the goods train failed to see the guard's hand lamp early enough and crashed into the back of the express, killing 14 passengers outright and seriously injuring several others, two of whom died later, though three ladies in the first class coupé at the very end of the train survived the accident. The effects of the collision were accentuated by the fact that the buffers of the goods engine were at a lower height than those of the express train's carriages.[284]

---

282 Board of Trade Report

283 The Times, London, England, 21st March 1853

284 Board of Trade Report

The coroner's jury found Driver Gass and his fireman, John O'Hara, guilty of the manslaughter of Thomas William Jolly, one of the passengers.[285] Lieut. H.W. Tyler, who conducted the Board of Trade investigation, stated that he could not at all concur with this verdict. To begin with, the fireman O'Hara was a mere servant of the driver, and although the regulations required him to keep a good lookout, he had other duties to perform and could not be considered responsible for the conduct of the train. Driver Gass was certainly responsible for keeping an unceasing lookout but could not be blamed to the same extent when inadequate signals were displayed, nor could the speed he was travelling at, 30 miles an hour, be considered excessive when the preceding faster train, was twenty minutes ahead and his own train was fifty-one minutes late. From experiments which he {Lieut. Tyler} had conducted, it did not seem likely that the goods train could have been stopped, with greasy rails, even if the guard had gone back 830 yards, the greatest distance that the guard's lamp could have been seen. The deficiency in the brake power of the goods train relative to the speed at which it was expected to travel was also highlighted.[286]

Turning to the question of tail lamps, Lieut. Tyler said that tail and side lamps had been sent to Kildare, but without a definite order to the Kildare stationmaster to affix the side lamps, and so he had not done so. An improved mode of trimming the lamps was also required, as there had been several cases of their going out. As to the company's regulations, he considered the practice of having the rules and regulations for different types of servants issued on separate pieces of paper to be most objectionable; all the rules and regulations should be kept together in one book. The regulations for policemen and switchmen regarding the protection of trains did not specify the use of fog signals, while those for guards did, but specified that upon being recalled he should pick up the fog signals that he had put down, thus leaving his train unprotected. In view of the imprecise and uncertain regulations the inspecting officer said, he was not inclined to blame Guard Berry as much as he otherwise would have done. Lieut. Tyler said that he must also comment upon the stupidity of Police Sergeant O'Connor, whose placing of his red flag in the ground when he was recalled in the gathering darkness was a useless exercise and who, on his return, did not think to check once in 20 minutes whether there was a tail lamp at the rear of the train, despite the fact that he constantly travelled on them

Pointing out that no tail lamps at the rear of a train, however bright, could be seen around a corner, the inspecting officer stated that it seemed to be the absolute duty of a guard to go back 600 or 700 yards to protect his train, and recommended that every guard carry at all times a pouch containing six detonators, and in place of accidental stoppage place three of these on the rails, a short distance apart. The guard should also be instructed to check that the tail lamps were showing before going back and should leave the detonators in place when recalled to his train, or alternatively a red light constructed to burn for 10 or 15 minutes.[287] [288]

The year 1854 had almost passed when there was a collision on the Eastern Counties

---

[285] The Times, London, England, 19th October 1853

[286] Board of Trade report

[287] Ibid.

[288] The Times, London, England, 7th, 8th. 10th, 11th,12th,13th, 14th, 17th, 18th and 19th October 1853

Railway's Ely-Norwich main line at Two Mile Bottom, England, late in the night of 30th December. The up-mail train had left Norwich at 9.45 p.m., three minutes late. It was four minutes behind time at Attleborough, where its guard saw a special cattle train which had been shunted out of its way, and left Thetford station at 11.07 p.m., now six minutes late. Two and a half miles beyond Thetford, at Six Mile Bottom, Driver John Barton signalled to his guard that there was a problem, Guard Colman applied his brake and the train was brought to a stand. When he got to the engine, Guard Colman was told by Driver Barton that they were likely to be stopped a quarter of an hour and the guard had better go back and protect his train. The gib cotter rod of the small end of one of the engine's connecting rods had broken and, having sent his guard back, Driver Barton got down underneath his engine to fix the problem.

Guard Colman, it would seem, did not hurry back, stopping first at his van to collect some detonators and also to put on his great coat as protection against the cold. He had managed to get back 450 yards and put down two fog signals, had gone on, and was about to lay some more when he heard the special cattle train which they had passed at Attleborough coming. Instead of putting down more detonators, he then ran forward towards the oncoming train waving his hand lamp, but Driver Drury of the cattle train only saw his signal at the last moment. The distance covered by Guard Colman after placing his detonators was disputed, the guard saying that he had got back another 120 yards and the driver that he saw the guard and exploded the detonators almost simultaneously. The cattle train consisted of 17 waggons of cattle and five of grain. It was travelling at 30 miles an hour down a 1 in 277 gradient, and the distance was too short for it to be able to pull up. A carriage truck, presumably the last vehicle of the mail train, was smashed and a composite carriage and a first-class carriage were slightly damaged. Driver Barton, who was still beneath his engine, was killed outright and one passenger, William Meager, an assistant undertaker of Long-acre, was fatally injured. Two other passengers and Goods Guard Cornell were injured. The assistant undertaker was returning to London with the hearse after having taken a body from there to Norwich for internment.

Major George Wynne investigated the collision for the Board of Trade. and stated that the night porter at Thetford had said that the mail train left there at 11.07 and the cattle train had passed at 11.28 while Guard Colman stated that the mail train had stopped at Two Mill Bottom at 11.15. Allowing four minutes for the cattle train to cover the 2½ miles to Two Mile Bottom, it would have reached there at 11.32, so even according to Guard Colman's *own* testimony he had had 17 minutes to go back and protect his train and yet he had only got back 570 yards with no time to put down further fog signals, implying that should a train stop out of course in the section even 17 minutes was not sufficient to protect it. The inspecting officer concluded that it could not be doubted that the guard must have loitered very much in carrying out his duty. The fog signals at 450 yards were quite inadequate to stop a cattle train travelling fast on a falling gradient and the light of the hand lamp was too feeble to be of much use as a signal, especially on a bright moonlit night. The regulations of the company, which only directed the guard to go back 600 yards, were also defective; most companies stated a minimum distance of 800 yards. The driver of the cattle train was also travelling too fast to enable him to pull up short of an obstruction visible ahead, or alternatively was not keeping a good lookout, as the tail lamps of the mail train were burning brightly.

Major Wynne considered that the basic cause of this and most other collisions was, however, what he termed "impatience" on the part of railwaymen generally, and that the solution was the spacing of trains by means of the electric telegraph, in place of the time interval system. The concept of separating trains by a space interval using the telegraph had been first suggested by W.F. Cooke back in 1842 but the idea was not adopted very quickly. The only other observation that he had to make was that the brake carriage had been in the middle of the mail train instead of where it ought to be, at its rear.[289] [290]

A major disaster occurred in France at 4 a.m. on Sunday 21st October 1855 on the Chemin de fer de Paris à Lyon line between Thomery and Moret. A train of 64 waggons was bringing cattle and sheep to the Monday market at Sceaux, and the last vehicle of the train, instead of being empty or loaded with goods, was a carriage carrying 26 drovers. The train travelled slowly but was running to time at Montereau. Beyond there, the line climbed to Fontainebleau and the train made heavy work of the ascent and lost much time. *The Times* correspondent wrote that, 'The guard, stationed at the end of the heavy train, endeavoured by every means in his power to avert the impending danger, but neither his cries nor his signals were understood.' What endeavours the guard made were unfortunately not specified – was his train going sufficiently slowly for him to jump off and lay fog signals? Was he equipped with flares? Did he just wave a red lamp at the following train? At any rate the driver of the express train which had left Lyons at 7 p.m. the previous evening did not perceive the train in front until he was within 30 metres of it. Several, if not most, of the drovers were asleep when the collision occurred. Three of those who were badly injured but not killed outright were trapped under the tender of the colliding engine and it was some time before they could be released. Another survived because he had crept, with his dog, under the seat in order to sleep more comfortably. In all, 16 of the drovers lost their lives. Two of the Post Office clerks on the express suffered slight bruises, the remainder of the express passengers escaped unhurt. Those injured were taken to hospital at Moret and the dead buried the next day at Veneux.[291]

The first train accident in The Netherlands to cause the death of a passenger occurred near Schiedam on the line of the Hollandsche Ijzeren Spoorweg Maatschappij (HIJSM) on the evening of Sunday 10th August 1856, when the last train of the day to Rotterdam ran into the rear of the previous one. Three passengers lost their lives and five were injured. The two trains were booked to leave Den Haag at 9.58 p.m. and 10.30 p.m. The earlier train was a through one from Amsterdam, the second a local service from Den Haag only. The first train was in fact a through service only inasmuch as it ran the whole length of the line, as both trains called at every station. Due to the heavy summer Sunday traffic, the first train did not leave Den Haag until 10.11. By the time this train reached Schiedam after further delays it was 11 p.m. The second train was lightly loaded, kept its time and reached Schiedam only four minutes later. The Schiedam passengers alighted, and it left there at 11.07, having been kept there until the previous train had gone seven minutes by Stationmaster Berckenhof. He assumed that this interval would be sufficient as trains usually only took eight minutes to reach Rotterdam.

---

289 Board of Trade Report

290 The Times, London, England, 2nd, 3rd and 16th January 1855

291 Ibid, 23rd and 24th October 1855

Shortly after leaving Schiedam, at a distance of 500 or 600 metres from the station, Driver Van Dijk, in charge of the engine *Haarlem* pulling the first train, felt a violent shock, the cause of which he did not know, and brought his train to a stand. Driver Van Dijk then spent five or six minutes examining his engine together with Conductor De Wit, who waved a red lamp in the direction of Schiedam station as a warning to the following train. Meanwhile Conductor Amiabel showed another red light from the lookout portion of the baggage van. While the driver was still busy examining his engine he glimpsed the following train starting away from Schiedam station and at once jumped up onto his engine, sounded the whistle and set his train in motion. The conductors were, however, concerned about some of the passengers who had got out when the train stopped. One of the conductors had just stepped onto the footboard of the last carriage when he saw the following train close at hand, and he had just time enough to jump off again and blow his whistle to warn its driver.[292]

Driver Hagendijk, on the engine *Hercules,* pulling the second train, saw the obstacle in front of him at the last moment and reversed his engine but the collision was unavoidable, and the two rear third class carriages of the first train were forced into one another and smashed. It was in these that the deaths and injuries occurred. Some of the passengers in the front carriages of the first train remained unaware of what had happened behind them.

The inquiry into the accident soon discovered that the initial shock felt by Driver Van Dijk of the Amsterdam train had been due to his engine having collided with a carriage which had been stabled all day at Schiedam station as a reserve coach in case it were needed to strengthen a train. The force of the collision had thrown this carriage completely off the line. The darkness of the night had presumably prevented Driver Van Dijk seeing the coach before he collided with it. How the carriage came out onto the running line was unclear, but Stationmaster Berckenhof was blamed for it having ended up there. In order to understand why the Schiedam station staff had failed to see the first train stopped down the line, and Driver Hagendijk had only seen it too late, the events of the accident were restaged three days later. These tests showed that a red light shown from the baggage van was not visible from the station, nor could the whistle of the *Haarlem* engine be heard there, while the first train was only visible from the engine of the second at a distance of 72 metres, and the minimum braking distance to stop short of it was 163 metres. A red hand lamp could be seen from the rear of the train for 190 metres, but apparently was not being displayed at the moment the train left the station. No suggestion seems to have been made that one of the conductors should have gone back down the line to protect his train.

The tail lamps of trains at this date were only meant to indicate that the train was complete and not to serve as a warning to the driver of a following train. They usually showed a white light, a red light indicated that an extra train was to follow. The HIJSM had introduced the electric telegraph in 1845 but had not yet begun to use it to control train movements, still relying on the time interval system.

Stationmaster Berckenhof was dismissed and Chief Conductor Amiebel reduced in rank.[293]

---

[292] Spoorwegongevallen in Nederland by R.T. Jongerius, pgs. 33-7

[293] Ibid.

What seems to have been a rear collision occurred near what a newspaper report referred to as 'Long Mountain' tunnel, presumably Gallitzin tunnel, near Altoona, Pennsylvania, United States, on the Pennsylvania Railroad, when a freight train ran into the rear of an emigrant train. Reports of the accident are scanty and even the date when it happened is unclear, but it was probably Thursday 26th February 1857. One newspaper stated that two passengers had been killed instantly and two had their legs broken, other papers that there had been four or six fatalities. One report added that 'We learn that Wm A. Stokes Esq, Attorney for the Company, has been on the ground endeavouring to compromise with those who have lost their wives and children. We fear that some negligence has been the result {sic} of the accident.' Any compassion felt by Mr Stokes towards the victims of the accident was probably tempered by his wish to avoid the railroad company having to pay out too much in damages.[294] [295]

The first major disaster on a railway in England occurred on the late evening of 28th June 1857, ironically on the South Eastern Railway, a company which had already discarded the primitive system of time interval working in favour of an absolute block system, worked by electric telegraph instruments. The accident happened at Lewisham Old station on the North Kent line when the 9.15 p.m. passenger train from Strood, standing between the home and distant signals, was run into by the 9.30 p.m. passenger train, also from Strood, 12 passengers losing their lives.[296] The line at Lewisham was worked with Tyer's instruments and, in common with most British railways at this date, the fixed signals stood normally at "Clear," except at junctions such as Lewisham Junction, between Lewisham Old station and New Cross. Lewisham Old station was protected by a semaphore signal at the station and a distant disc signal, 302 yards in the rear.

The 9.15 train left Blackheath station at 10.32 or 10.33. Its driver, Thomas Hill, found the Lewisham Old distant signal at danger but nevertheless passed it to stop just short of the station signal, which was also at danger. According to the regulations he should have stopped at the distant signal, but he told the subsequent Board of Trade inquiry that trains frequently passed this signal at danger and it was never cleared for a train to draw inside it, though the distant signals at some of the other stations were. The train stopped at Old Lewisham station at 10.35 or 10.36. It was not booked to call there, but the stationmaster, John Nelson, asked the night-watchman, David Wild, who acted as the night signalman, to go and see if there was any room in the train for the many passengers waiting on the platform, while he, the stationmaster, called over a porter, Samuel Saxby, to attend to the signals during Wild's absence. As the Board of Trade inspector, Lt-Col Yolland pointed out the stationmaster might just as well have asked Porter Saxby to see if the train could take on passengers. The second guard told Wild there was no room, whereupon the stationmaster, anxious to get passengers away, enquired himself, getting the same answer. Driver Hill asked Wild how long the previous train had been gone and, being told seven minutes, moaned about the Lewisham Junction signalman having given

[294] The Adams Sentinel, Gettysburg, Pennsylvania, United States, 16th March 1857

[295] The Ohio Repository, Canton, Ohio, United States, 25th March 1857 – reproduced from The Hollidaysburgh Whig

[296] The Times, London, England, 29th and 30th June 1857

priority to a Beckenham (Mid-Kent) train. Driver Hill then asked the stationmaster twice if he could draw up at the platform but the stationmaster, having been told there was no room in the train, said no as he did not want passengers trying to force their way onto the already full train.[297]

With the 9.15 train stopped where it was, its rear should have been 138 yards inside the distant signal and, with a curve in the line, its tail lamp was visible for 276 yards. Head Guard William Wyley told the Board of Trade inquiry that the tail lamp had been burning brightly. In one statement he said that he had gone back 150 yards, in another statement he mentioned a greater distance, and that he had waved his hand lamp and whistled towards the advancing 9.30 train. Some of the passengers, however, said that he had remained close to the rear of his train.

The 9.30 train arrived at Blackheath station at 10.52, and two minutes later, with the station work complete, the head guard, George Abbott, asked the stationmaster, Edward Chapman, if they were clear to go. The stationmaster referred to Porter Signalman John Griffiths and asked him if it was "All clear up" to which Griffiths had replied "All clear up." Guard Abbott shouted "All right" to Under Guard John Bartram and the train set off. The guards agreed that they had applied their brakes when they saw the red light of the Old Lewisham distant signal before their driver whistled for the brakes.

Driver Thomas Perry, on the engine of the 9.30 train, said in a statement that he had not seen the Old Lewisham distant signal until he reached the Love Lane overbridge, when he had seen a pale white light, he had shut off steam between that point and a second overbridge. Before his engine reached the second overbridge, he had seen that the signal was more red than white, had whistled for the guard's brakes and told his fireman, Edward Whiffen, to "Hold on." He did not see the tail lamp on the 9.15 train until he was within two carriage lengths of it. He had reversed his engine before passing the distant signal.[298]

Fireman Whiffen more or less agreed with his driver's evidence, including his assertation that he had first seen a white light and only when closer to it did it become clear that it was in fact a red one. The fireman stated that he had not seen Guard Wyley, nor heard his whistle.

Driver Hill of the 9.15 train estimated the speed of the 9.30 train at 10 or 12 miles an hour. The collision forced the rear brake van of the front train onto the adjoining open third-class carriage, containing 30 or 40 passengers, destroying the upper framework of the carriage. All the fatalities and serious injuries to passengers occurred to those in this carriage, several of them dying from suffocation.

Night Watchman (Signalman) Wild stated that he had signalled to Blackheath that it was "All clear up" for the 9.15 train, which was answered by one stroke of the bell, and a minute or so later Blackheath had given him two strokes for the 9.15, he had replied by one stroke on the gong and turned his needle to "Stop all up." (A bell was used to signal forward up trains, the reply being given on the gong and vice versa for down trains) He had then been away about three minutes attending to the 9.15 train, during which time Saxby had minded the instruments, Saxby saying that no signals were given or received while he was there. After Wild had returned, Blackheath had given two stokes on the bell,

---

[297] Board of Trade report

[298] Ibid.

which Wild had not answered, nor had he altered the position of his instrument, which was still showing "Stop all up." Wild's statement was strongly supported by the testimony of a Mr Thompson, a barrister (and a railway enthusiast).

Porter Signalman Griffiths at Blackheath had been arrested on a charge of manslaughter along with Driver Perry and Fireman Whiffen and did not give a statement. Porter Jupp had been on the up platform collecting tickets near the signal box and had not heard any bells ring about the time at which Griffiths would have had the 9.15 train signalled as having cleared Lewisham Old, but Griffith's book recorded such a bell signal as having been received at 10.48 with the 9.30 train signalled as departing to Old Lewisham (the two strokes Wild said he had heard) at 9.53. Lt-Col Yolland said that it was impossible to be certain, but the weight of evidence was in favour of the mistake having been made at Blackheath. He added that since writing most of his report he had become aware of bells being rung and the positions of needles being altered independent of the signalmen during severe thunderstorms, but there had been no such storm on the night of the accident.

Turning to the delay to the 9.15 train at Old Lewisham, the inspecting officer stated that the Lewisham Junction signalman was not permitted to clear trains back to Lewisham Old or Lewisham Junction New stations until they had been cleared back from New Cross. The up train preceding the 9.15 was the 9.56 p.m. train from Woolwich, which had left Lewisham Old for Lewisham Junction at 10.28. It was signalled back from New Cross at about 10.43 and the Lewisham Junction signalman, had he been aware that the 9.15 train was stopped at Lewisham Old, could have accepted it then. He did not know it was there waiting, however, and the practice adopted at some places, of a signalman who had a train waiting giving one stroke to the box in advance, was not used at Lewisham Old. Thus, the Lewisham Junction signalman did not clear back to Lewisham Old at 10.34 but instead accepted a Beckenham train from Lewisham Junction New on the Mid-Kent line at 10.47. Colonel Yolland said he did not blame the Lewisham Old signalman but thought it extraordinary that Stationmaster Nelson had not taken steps to ascertain the cause of the detention of the 9.15 train.

Concluding his report, Colonel Yolland recommended improvements to the system of telegraph working, that the positions of the outdoor and telegraph signals should correspond, that drivers should always stop at distant signals, that the brake power of the trains should be improved and that the use of open third-class carriages should be discontinued. The report ended with a considerable volume of correspondence between the Board of Trade and the South Eastern Railway management, who disputed most of the inspecting officers' conclusions and recommendations.[299]

At the conclusion of the inquest upon the victims of the disaster Driver Perry, Fireman Whiffen and Porter Signalman Griffiths were all committed to stand trial for manslaughter.[300] At the Central Criminal Court on 16th September they were all found "Not Guilty." [301]

The United States Hudson River Railroad suffered two rear collisions on Thursday

299 Ibid.

300 The Times, London, England, 1st,2nd,3rd, 4th and 7th July 1857

301 The Morning Post, London, England, 17th September 1857

19th January 1860. The first and less serious one in the morning occurred in or near Yorkville tunnel, New York, was due to the fact that thick smoke prevented the driver of a New Haven train from seeing the signals made by the flagman of a Hudson River train stopped in front. One passenger, a Mr Palmer, of White Plains, was seriously hurt and others were slightly hurt.

The same afternoon as the Yorkville accident, the rear car of the Albany Express was demolished by a Sing Sing train near Tarrytown. The train had come to a stand due to a problem with its engine and its conductor was criticised by a newspaper for assisting the engineer rather than looking out for the safety of his passengers. This was not really fair comment as the conductor had sent his flagman back to protect the train. The flagman, on the other hand, seems to have been particularly irresponsible, for despite remonstrances, presumably from passengers, he failed to go back more than 600 feet, to the start of a sharp curve, stopping there to amuse himself by skimming stones over the surface of a frozen river. The engineer of the Sing Sing train thus received no warning of the train in front in time, and a violent collision ensued, with a tragic outcome. Only one person was killed outright; she, Mrs Thomas Field, had been married only two hours before. He husband was stated to have become frantic with grief, and it was for a time feared that he would lose his mind. A Mr William Cory was stated to have been dangerously wounded and a couple of other passengers were badly hurt.[302]

The short-lived Racine and Mississippi Railroad suffered a bad accident on 14th September 1860 at Delavan, Wisconsin, United States. A passenger train which had arrived at Delavan at 7.30 p.m. from Elkhorn was run into there by a following freight train. The passenger train was a regular service, but was carrying more passengers than usual that day, being filled with visitors to the Walworth county fair, which had been held at Elkhorn that day, returning to villages along the line as far as Beloit. A passenger on the train, Mr Eddy, told a newspaper reporter that the train had run quite fast after leaving Elkhorn, and he (Mr Eddy) had been told by the conductor that there was a freight train behind which they were keeping out the way of. The train had stopped at Delavan about 20 rods (330 feet) from the depot to take water and had not been stopped more than a minute when the freight train had come around a sharp curve and into them. The passenger train consisted of a locomotive, baggage car, two passenger cars and a wood car at the rear. The platform of the wood car had been crowded with men and boys, but they had had time to jump clear. The impact of the collision drove the wood car right through the rear passenger car and partly into the car in front. Seventy-five or eighty people were in the rear passenger car, five of whom were killed and twenty or thirty more injured, of whom it was supposed eight or ten would die. Who was to blame for the collision was not known at the time of the report; it was alleged that the passenger train conductor had left word at Elkhorn that the freight should be held back a while.[303]

Boxing Day 1860 saw wintry weather in France and an accident on the Chemins de fer du Nord. A passenger train from Erquelines was stopped for three quarters of an hour

---

[302] The Berkshire County Eagle, Pittsfield, Massachusetts, United States, 26th January 1860

[303] Janesville Daily Gazette, Janesville, Wisconsin, United States,15th September 1860

at Fresnoy-le-Grand waiting for the line ahead to be cleared of snow. While it was still stopped at the station it was run into by a goods train, the driver of which had failed to see a signal 'which was imperfect.' Three third-class carriages were smashed to pieces. At the time of the report one passenger had been killed on the spot, a second had died later at the scene and a third, the wife of a small proprietor at Paris, had died on arriving at St Quentin hospital, having suffered broken arms and legs and a fractured skull. Her husband was in the same hospital with broken legs and had yet to be told of his wife's death. Several other injured passengers were also in the hospital including two soldiers.[304]

The 25th August 1861 saw the Clayton tunnel disaster, a rear collision on the London Brighton and South Coast Railway in the long tunnel under the South Downs which cost 23 lives. This accident is fully described in the chapter on tunnel accidents and so will only be dealt with briefly here. The line through the tunnel was already worked on the block system, but the remainder of the line was not. Three passenger trains were despatched from Brighton by the stationmaster close behind one another and a misunderstanding between the signalmen at the south and north end of the tunnels meant that the third train was permitted to enter the tunnel before the second had left it and indeed was actually setting back to the south end. Contributing to the accident were the failure of an automatic signal at Clayton Tunnel South box to work properly and the excessive hours which the signalmen were expected to work.[305]

The French Chemin de fer de Paris à Lyon et à la Méditerranée Company had no less than three accidents in two days on 23rd and 24th August 1863, though the first, on 23rd at Beaucaire, France was admittedly due to vandalism by children. On 24th there followed a collision at Anse and this was followed by a more serious rear collision at La Fouillouse, which, like the Beaucaire 'accident', cost four lives. The La Fouillouse accident occurred at about 6 p.m. when a late running express bound for Paris ran into an omnibus train stopped in the station.[306] A sequel to the accident was the conviction on 7th November by the Saint Etienne Correctional Court of no less than nine railwaymen judged in some way to be responsible for the accident; two stationmasters, a guard, a driver, a fireman, and four crossing keepers, who were sentenced to between 15 months' and 15 days' imprisonment and fines of between 1,000 and 50 francs.[307]

Another serious rear collision happened in France, early in 1864, again on the Chemins de fer du Nord. The engine of train No. 24, an omnibus service from Calais, due at Paris Nord at 4:13 p.m. on 5th January suffered a broken connecting rod between Gonesse and Pierrefitte and its driver was able to get to Pierrefitte only with great difficulty. Just as it reached there, train No. 26, a semi-fast service from Brussels, due to pass Pierrefitte 35 minutes after the Calais train, appeared in sight, and was clearly not going to pull up in time. A station inspector, seeing the danger, told the station staff to open the doors of the omnibus train and get out as many passengers as possible, and in

304 The Times, London, England, c2nd January 1861

305 Board of Trade report

306 The Times, London, England c25th August and 5th October 1863

307 Wikipedia – Liste des accidents ferroviaires en France au XIX siècle

doing so doubtless saved a number of lives. Even so, five passengers in the omnibus train were killed and 19 more or less seriously injured. The passengers on train No. 26 experienced a violent shock but none were more than bruised. The engine of the colliding train mounted the rear carriages of the omnibus service and set them alight, while the shock of the collision detached the engine of the front train, which ran more than a kilometre, almost to St Denis, before it was stopped. The injured passengers 'received all requisite attention from the company's surgeons in the neighbourhood, who came in all haste to the scene of the accident.' When the injured arrived in Paris 'all the pupils of the Hospital Lariboissiè came to the railway terminus to assist in removing them to the hospital.' [308]

On 29th and 30th January M. Auclois, the Pierrefitte stationmaster, and Dombrowski, Neufsel and Houssay, the driver, fireman and guard of the Brussels train, stood trial at the Tribunal of Correctional Police for having caused the accident. The evidence showed that the station staff had done all that they could to stop the Brussels train and the stationmaster was acquitted, but the other three were found guilty and Driver Dombrowski was sentenced to three years' imprisonment and a fine of 300 francs. Fireman Neufsel and Guard Houssay were each sentenced to six months' imprisonment and an identical fine. The three were likewise condemned, conjointly with the company, to pay 500f a year for life to the widow Dubois, and 300f yearly to each of her two children, as a compensation for the death of their husband and father; also a sum of 3,000f to the passenger named Robinet, as well as all costs.' [309]

A collision at Egham, England, on the London and South Western Railway, on 7th June 1864, took place at a time of heavier than usual traffic on the first day of Ascot Races, though the traffic to Ascot from London was stated not to be anything like as heavy as on Cup day. Egham station was situated about eight miles on the London side of Ascot and was protected by a station signal and a distant signal, the latter being 513 yards on the Ascot side of the former. There was also a semaphore signal at Rusham Road crossing, about 1,210 yards from Egham station.

Special regulations for the working of the line had been issued by Mr Scott, the traffic manager: no train was to exceed 16 vehicles, and each must have two brake vehicles and two guards. Special care and caution were to be exercised when approaching all junctions. On Tuesday (the day of the accident) and Thursday platelayers were to be stationed along the line between London Waterloo, Ascot and Reading and must show a red "Stop" signal for five minutes after the passing of each train, and a "Caution" signal for another five minutes, then an "All right" signal. The guards were responsible for ensuring that every train was equipped with spare couplings, tail lamps and so on. Train drivers were to run with more than usual caution, and to pull up promptly when given a "Stop" signal.

Lt-Col Yolland, who conducted the Board of Trade inquiry into the accident, pointed out that the race traffic was operated under great disadvantage, as the racegoers would set off at the last possible moment they could to get to the course for the start of the racing and then at the conclusion of the day's events would crowd the station wishing to return

[308] The Times, London, England, 8th and 9th January 1864

[309] Ibid, 3rd February 1864

to London as quickly as they could, so that in order to prevent accidents at the starting stations train must be despatched as fast as possible. The platelayers employed to keep an interval of five minutes between trains were not furnished with watches or timepieces (and presumably could not afford to buy their own) and the drivers were instructed not to actually stop on getting a "Stop" signal from a platelayer but to continue very cautiously until they actually saw an obstruction.[310]

Fourteen special trains returned from Ascot on the day of the accident, the first at 4.45 p.m., at intervals of not less than five minutes, and Mr Scott, the traffic manager, or Mr Godson, the superintendent of the line, handed the guard of each a paper on which were pencilled the names of the stations at which each train was to stop, and the drivers were generally told the stations at which the preceding train was to call. As the trains had all left Waterloo with their engines chimney first and there was no turntable at Ascot, they had necessarily to all run tender on the return journey.

A special train left Ascot for Waterloo at 7.09 or 7.10 p.m. and its guard was given a piece of paper denoting stops at Egham, Twickenham, Barnes, Clapham Junction and Vauxhall. It reached Egham at 7.26 or 7.27 and was held there slightly longer than intended while some card sharpers who had been the subject of a complaint were moved to a different carriage. It appeared certain that the station and distant signals had been placed and maintained at danger after the arrival of the train. The delay was not a long one and the train was just on the move again, at 7.29 according to the Egham porter, when it was run into by the following one, travelling at speeds estimated at between seven and 20 miles an hour. The rear brake van of the first train was pushed into the second-class carriage in front, partly or wholly destroying its rear and front compartments while sparing the centre compartment. The rear compartment of the second-class carriage next in front was also stove in. Five passengers were killed outright, one was fatally injured, and another less badly hurt; all the victims were supposed to have been from these two carriages.

The colliding train had left Ascot at 7.15 or 7.16 with a first stop at Staines. Its driver asserted that he had not been told that the preceding train was to stop at Egham. The driver, fireman and guard of the 7.15 train agreed that it was not checked at all at either Sunningdale or Virginia Water, as it should have been because it was less than 10 minutes behind the train in front, but that they received a green "Caution" signal from the platelayer at Stroud Lane crossing, rather more than a mile and a half from Egham. The driver, Thomas Lee, and fireman, John Trainer, of the colliding train alleged that the Rusham Road Crossing signal had been clear for them and they had only shut off steam, reversed the engine and applied the brakes when they saw the Egham distant at danger. Head Guard John Redshaw and Under Guard Demsey of the train both stated, however, that the Rusham Road Crossing signal had been at "Danger" when they had first seen it and they had each applied their brakes. The crossing keeper at Rusham Road, Frederick Oliver, said that his signal had been at "Danger" when the second train came in sight, but he had then lowered it to "Caution" because the previous train had been gone five minutes. The bulk of the evidence was that the driver of the second train only sounded his whistle when his engine was at or near the Egham distant signal. At that point it was clear that driver, fireman and both guards had done all in their power to stop the train but given the available distance its speed was too high and the brake power insufficient. Some

[310] Board of Trade report

witnesses had stated that the second train was travelling very fast, even 50 miles an hour, but the inspecting officer did not believe that its speed had been excessive, nor did he believe the guard's statements that they had applied the brakes before passing Rusham Road Crossing. Had the train been provided with continuous brakes, Lt-Col Yolland continued, like the North London Railway, over which the South Western trains worked daily, the accident might have been averted, but he complained that it was almost useless for him to urge the adoption of such measures, as his recommendations were almost certain to be entirely disregarded. A similar if less serious collision had occurred at Virginia Water on Ascot Cup day 1859, and in his report on this accident the inspecting officer had commented on the inadequate brake power. The inspecting officer had also been told then that the electric telegraph was being laid down. This had now been done but it was not apparently used to regulate trains.

Another point mentioned by Lt-Col Yolland was that on the London and South Western the guards had long had a means of communicating with the drivers, termed the communicator, by which they could request the driver to apply his brakes. But the system could not work if the engine was running tender first. He urged the company to install a turntable and extra siding accommodation at Ascot.

In conclusion, Lt-Col Yolland stated that the accident was clearly the fault of driver Lee in not having reduced his speed as soon as he passed Rusham Road but that he was working under great disadvantages; having inadequate brake power, not having been told that the previous train was stopping at Egham, not having the advantage of communication with the guard because he was running tender first and working under the objectionable time interval system, by which false signals were constantly given. Experience had amply shown that the promise of safety given by the five-minute interval between trains under the time interval system had not been fulfilled and absolute block working using the electric telegraph should be adopted. It had come out in the evidence that the down Royal train had been permitted to follow another train from Egham in only four minutes, and on its return journey the up Royal train had been only three minutes in advance of another race train at Egham, and the latter was not checked by signals. If a burst tube or other problem had delayed the first train in either of these cases it was impossible to say that a collision would not have occurred.[311] [312]

On 24th June the coroner's jury found Driver Lee and Fireman Trainer guilty of manslaughter but, doubtless following the inspecting officer's lead, added that they had been working under great disadvantage, and recommended they bring into use of a turntable at Ascot and the adoption of electric telegraph working, and criticised the trains being despatched at only five minute intervals and the fact that Driver Lee was not informed that the previous train was calling at Egham.[313] As might be expected when their case came to trial at the Guildford Assizes on 3rd August Driver Lee and Fireman Trainer were acquitted.[314]

Neglect of the crew of a freight train to shunt out of the way of or to protect their train

[311] Ibid.

[312] The Times, London, England, 7th, 8th, 9th, 15th, 18th and 22nd June 1864

[313] Ibid, 25th June 1864

[314] The Express, London, England, 4th August 1864

in the rear against a following eastbound Pennsylvania Railroad passenger train, which was running at its regular time, was given as the cause of a collision at Thomson's or Thompsontown station, Pennsylvania, United States, at 4.30 a.m. on the morning of Wednesday 21st September 1864. The engine of the passenger train was described as demolished and its front baggage car was driven into the following passenger car. Fire then broke out and three passenger cars were burnt. A passenger on the train, Mr L. Convers of Ohio, stated that there had been about 60 passengers in the front car and but seven or eight were known to have escaped because the doors were locked, and that more lives could have been saved had they not been. His statement that there were 60 passengers in the car and less than ten escaped would imply a death toll of about 50 but another report stated that only six had died, among them Conductor Jim Mullison and Brakesman P. Limerick.[315 316]

An accident on the Great Indian Peninsula Railway at Tanna, India, 20 miles from Bombay (now Mumbai) on the night of 5th November 1864 resulted in the death of three 'natives.' A goods train ran into the passenger train, which was stopping at the station, killing the passengers and injuring several others. 'The driver of the goods train has been committed for trial on a charge amounting to culpable homicide.'[317]

At 8.5 a.m. on the morning of 23rd December 1864 the regular mixed train left Nässjö for Malmö, Sweden, with a marker board on the rear vehicle to indicate that an extra train was to follow it. The first booked stop was at Sandsjö, 17 kilometres south of Nässjö. The extra train followed 20 minutes after the regular service and consisted mostly of empty vehicles. It was driven at high speed, in excess of that permitted, and its driver was unable to pull up at the semaphore home signal protecting Sandsjö station. On the approach to the station was a level crossing, the gates of which, in accordance with the normal practice, had been closed across the line. The extra train smashed through the crossing gates, then ploughed into the rear vehicle of the mixed train. Of the 16 passengers in the rear carriage seven were killed and the remainder injured. Two other people on the train were also hurt. The victims were mostly workers returning from the construction of the Katrineholm - Norrköping line.[318]

A minor mishap near Bristol, Pennsylvania, United States, led to a serious accident in the early hours of 7th March 1865. Some Adams Express cars became derailed just outside the station at about 2 a.m. and as a result the midnight Camden and Amboy Railroad express from Philadelphia to New York, the "Owl Line," was brought to a halt behind them until they could be re-railed. The "Owl Line" was already running an hour late, having had to wait at Philadelphia for the southern passengers, and consisted of eleven well-filled cars. Whether or not it was properly protected in the rear was not clear, but the passengers blamed their conductor for not sending a flagman back despite the fact that the 6 p.m. train from Washington D.C. was nearly due. The engineer of the

---

315 Janesville Daily Gazette, Janesville, Wisconsin, United States, 22nd September 1864

316 Burlington Weekly Hawkeye, Burlington, Iowa, United States, 24th September 1864

317 The Times, London, England, c6th December 1864

318 Wikipedia - Sandsjöolyckan

Washington train thus failed to realise that there was an obstruction on the line and, coming around a curve, ploughed into the rear car of the Philadelphia train, which was filled with Union soldiers, the colliding engine being encased by the remains of the passenger car. Those passengers trapped in the wreckage of the car could only be extricated by cutting it to pieces. According to one report, one passenger was taken out dead and two others were so badly injured that they died in a few minutes, while one of the two firemen of the Washington train also lost his life. Another account stated that there were five killed; three soldiers, the fireman and a civilian, and 25 to 30 severely wounded, some of them fatally. Two other cars were badly damaged as was the engine of the colliding train. The wrecked vehicles caught fire, but the blaze was extinguished by Bristol firemen. The force of the collision threw the cars of this train backwards, 'leaving some, among them the postal car, on the Bristol bridge, where they were detained seven hours.' The soldiers were mostly returning prisoners of war who had left Washington D.C. at 4 p.m. the previous afternoon. 'The ladies of Bristol turned out with refreshments for the sick and wounded- of the wounded three were so crushed that their recovery is hopeless.'[319] [320]

A collision in Connecticut, United States, on 15th August 1865 was described by the *Titusville Morning Herald* as 'The Housatonic R.R. Murder' and 'the late railroad slaughter' while the Providence Press suggested that 'it is time somebody was hung' for causing such an accident.

The accident happened on the Housatonic Railroad, which at this date only ran two passenger trains daily in each direction. As far as the northbound trains were concerned, these were a freight train in the early morning and a passenger train which left Bridgeport at 10.45 a.m. On 15th August the company had just received delivery of a new locomotive, the *Fairfield*, which they were keen to try out. Despite having all day to do so, Master Mechanic Andrew Winstow and Engineer Edmund Lyman set out on their test run only 15 minutes behind the passenger train. The crew of the passenger train were apparently unaware that it was planned to make this test run, although Rule No12 of the company stipulated that, *A red flag by day and a red lantern by night must always be exhibited upon an engine when an engine other than that of a regular train is to follow*, which might be taken to imply that they should have been told. Asking why this had not been done, a newspaper reporter was told that it was 'not customary' when an engine was going out on a trial run for 'a short distance.'

The passenger train had only covered four miles when it caught up with the morning freight train, which was stopped due to a problem with its engine near Trumbull. Now, Rule No 5 of the company directed that the crew must '*Send out flags in both directions when train is halted*' and the conductor of the freight train had duly flagged the passenger train, but the crew of that train clearly imagined that there would be nothing further along the line all day and that this rule was therefore superfluous. Worse still, the decision was taken to couple the passenger train to the disable freight and set back the combined train to Bridgeport. Thus, it was that engine and backing train met on a sharp curve 'near the Pequennock mills,' 'where a view of not 10 rods {55 yards} could be commanded.' The

[319] Burlington Weekly Hawkeye, Burlington, Iowa, 11th March 1865

[320] The Adams Sentinel, Gettysburg, Pennsylvania, United States, 14th March 1865

violent collision demolished several of the cars and caused the boiler of the Fairfield to explode as it reached the penultimate car, scalding some of the passengers and 'making awful havoc.' Twelve passengers in the rear car of the train were killed and about twenty severely injured. The only house nearby was a small one belonging to a German. Whether he was at home and gave assistance was not stated but at least some of the injured were taken to the shade of the nearby woods to await the arrival of a relief train, which was despatched about noon with Drs Hubbard, Nash and Burrett. The *Springfield Republican* reported that 'The President of the Road, Mr Charles Hunt, of Falls Road, Ct, was on board the train, and did not escape entirely unharmed. It is to be hoped that Mr Hunt, the responsible manager of the road, will learn something from this accident and give future travellers along the road the benefit of the lesson.' [321] [322] [323] [324]

Whitby, Canada West (soon to become Ontario), Canada, was the location of a bad accident in the early hours of Thursday 23rd May 1867. A Grand Trunk Railway fast freight train, known as the *Merchant's Dispatch*, had reached the station there at 2.50 a.m. Although it was basically a freight train, two second-class passenger cars were attached to its rear immediately in front of the caboose. The passenger cars contained 220 German emigrants who had landed in Canada East (Quebec) a day or so earlier and were on their way to Milwaukee and the Western United States. A special freight train was following the Merchant's Dispatch, ten minutes behind it. Morning was just breaking. The engineer of the second train was keeping a lookout but the station's distant signal had not been replaced to danger by the switchman, Wallace Fraser, after the arrival of the first train and on the steep falling gradient approaching the station, the freight train engineer was unable to pull up in time when he discovered that there was a train stopped in the station. The engine of the second train smashed into the caboose of the first and drove it forward into the passenger car in front of it, killing three women, a man and a child and injuring three other passengers. The victims had all been asleep at the time. On learning of the accident, the passenger superintendent, Mr James Stephenson, left Toronto for the scene at 3.15 a.m. The man, one of the women and the child who lost their lives were all members of the same family and a boy of about 10 was left as an orphan. He was taken care of by the Toronto police during the day and later continued west with the other emigrants, except those who were too badly hurt to immediately continue their journeys. The special freight train had been following at the regulation interval and the accident was blamed entirely on the negligence of the pointsman in failing to exhibit the semaphore signal at the east end of the station.[325]

A coroner's jury quickly decided that Switchman Fraser was responsible for the accident and he was committed to stand trial for manslaughter. [326]

A broad gauge express of the Cincinnati Hamilton and Dayton Railroad, due in

---

321 Titusville Morning Herald, Titusville, Pennsylvania, United States, 29th August 1865

322 The Adams Sentinel, Gettysburg, Pennsylvania, United States, 22nd August 1865

323 The Daily Index, Petersburg, Virginia, United States, 17th August 1865

324 A History of Railroad Accidents Safety Precautions and Operating Practices by Robert B. Shaw pgs. 76-7

325 Janesville Gazette, Janesville, Wisconsin, United States,30th May 1867

326 The Daily Intelligencer, Atlanta, Georgia, United States, 8th June 1867

Cincinnati at 6 a.m. was delayed on 21st November 1867 at Lockland, Ohio, United States, waiting for a northbound freight train to clear the single line. Before its flagman could get back, it was run into by a following freight train. (The otherwise 4ft 10inch gauge railroad had laid a third rail in 1865 to give a 6ft gauge line between Dayton and Cincinnati to accommodate Atlantic and Great Western cars and thus connect with the Ohio and Mississippi Railroad for St Louis.[327]) The rear vehicles of the express were from a Toledo train and had been attached at Dayton, the very last vehicle being a sleeping car from Toledo. Fire broke out following the collision and four women and a man who had been in the sleeping car were burnt to death. The victims were described as three sisters named Morgan travelling from Detroit to Louisville, a very large woman aged about thirty from Detroit and a man named Jackson from Boston. The fire arose from the breaking of the oil lamp on the front of the colliding engine, the spilt oil causing the fire to spread very quickly. One of the Morgan sisters reached the door of the sleeping car and broke a window but could not get out and while efforts were being made to help her a sudden burst of flames struck her, and she fell down dead.[328]

Almost a year passed after the Lockland accident before the next bad rear collision, which happened in England. The 5 p.m. Great Western Railway up mail train left New Milford on time on 5th November 1868 but lost four minutes waiting for the single line train staff at St Clears and another four minutes taking on a carriage at Carmarthen. It was still eight minutes late at Newport, but only five minutes down at Lydney, where the policeman, William Morris, shouted something which the guard, David Betty, did not catch, about Bullo Pill. Guard Betty had just finished sorting his bills and tying up his letters when, as the train was passing under an overbridge, he saw three bright red lights in front of him. He threw himself to the floor of the van and only felt a slight shock.

In charge of the engine (*Rob Roy*) of the mail train was Driver James Turner. He had had no indication of danger until his engine passed under Awre bridge, a mile or so west of Bullo Pill, when, from his position on the engine, he saw a single red light and shouted "Hold on" to his fireman. At about the same moment his fireman, Henry Tyrrell, saw first one and then three red lights and shouted "Wo" to his driver. Driver Turner did not think that he had had time to do anything, but his fireman remembered him shutting off steam. Neither driver nor fireman were hurt in the collision which followed.

The train that the mail train had run into was a special cattle train from Carmarthen to Gloucester which had left Carmarthen at 1.37 p.m. and which consisted of 20 loaded cattle trucks and a third-class brake carriage pulled by the engine *Tantalus*. The brake carriage carried the guard and eight drovers. Under the instructions of the night foreman and Guard William Probert, but against the better judgement of Driver Robert Seys, six waggons of general goods were added to the train at Newport, the maximum load for the engine being from 25 to 30 waggons according to their weight and the weather. As all the waggons were loaded, 26 was close to the maximum. The train had to stop at various stations, and to be divided to get up Chepstow bank. It left Chepstow, according to the driver, at 9.35, intending to run the 16½ miles thence to Bullo Pill, where it would shunt out of the way of the mail and take water. The mail train was due at Chepstow at 9.51.

---

[327] The Cincinnati Hamilton & Dayton R.R. by Charles E. Fisher

[328] The Ohio Democrat, New Philadelphia, Ohio, United States, 29th November 1867

The eight miles to Lydney were done in 25 minutes, the train passing there about 10 p.m. Driver Seys said that he had whistled at Lydney to ask the signalman (policeman) if he might proceed to Bullo Pill or must shunt there for the mail and had been shown a white light, so he had kept going. From Lydney, the line climbs for about a mile at 1 in 264 and 1 in 198 and his engine began to slip in mounting this incline. He had picked up a bit of speed past Gatcombe. but his engine lost more speed on the subsequent two miles at 1 in 316 on the approach to Bullo Pill, while even on the preceding level stretch white frost on the rails had caused slipping. His train was going at eight or ten miles an hour when the mail train ran into them.[329]

The effects of the collision were little felt in the mail train but were much more serious in the cattle train. The *Rob Roy* ploughed through the third-class carriage, mounted two of the cattle trucks and came to rest partly on top of a third. Guard Probert and the drovers were thrown out of the carriage to the left of the line and the guard and drovers David Thomas and Evan Morgan were killed, while drover John George was fatally injured and died later in Gloucester Infirmary. The other five drovers were more or less seriously injured, and 40 cattle were killed.

The exact time of the collision could not be established. The Lydney policeman stated and had recorded in his book that the cattle train had passed him at 9.58 and the mail train at 10.20. He had received a printed notice stating that the cattle train would leave Carmarthen at 1.30 p.m. and had been told by the day policeman that the 11 a.m. train from New Milford, which had passed at 6.20, had carried a signal that a special train was to follow, but he did not know when to expect it. The regulations required that a passenger train be cautioned if after dark a goods, cattle or mineral train had passed less than 20 minutes previously, but here the interval was 22 minutes. Another regulation demanded that no slower train be started from a station unless it was clear that it had time to reach the next point at which it could be shunted, but this was a through train.

The Board of Trade inspecting officer, Captain H. W. Tyler, pointed out that the drovers would have been in a safer position had they not been travelling in the last vehicle of the train, an arrangement adopted for the convenience of being able to accommodate the drovers and guard in the same carriage. The principal lesson of the accident, though, was the inadequacy of the time interval system for preventing collisions. An interval of space was the only true remedy. 'On certain railways in this country, trains are continually following one another in quick succession at half-mile intervals from year to year, under what is called the "block system" of working by telegraph, with comparative safety, while on others continual risk is incurred for want of an extension of the telegraph system, although with more perfect arrangements a sufficient interval of space might without difficulty be secured.'

The inspecting officer pointed out that in this instance an interval even of 22 minutes had not proved enough. A powerful engine was unequal to the weight of a too heavy train and slippery rails. It had not been going fast enough to keep ahead of the mail train, but just too fast to enable the guard to alight and protect it. Its three bright tail lamps had been practically useless because the collision had occurred on a curve. The passengers in the mail train had only escaped because the destructive force of their train had been expended in destroying the rear of the cattle train and its unfortunate inmates.

---

[329] Board of Trade report

Captain Tyler recommended the introduction of block telegraph working between Awre Junction and Bullo Pill, by day and night, with no train being allowed to leave one place in either direction until the previous train had been reported clear of the other, and noted that the same system might with great advantage be extended, more especially over this and other lines where the view was obstructed, and where trains ran at high speeds without making frequent stops.

Because drover John George had died later, two separate inquests were held, both returning verdicts of "Accidental death".[330] [331]

A major disaster occurred in Bohemia, then a part of the Austro-Hungarian empire, at 5:50 a.m. on the morning of 10th November 1868. An overnight passenger train from Zbirov (now Karizek) was carrying 250 soldiers of the 26th Hungarian Regiment, who had completed their drill with breechloading rifles and were going on furlough. The train was not excessively heavy, but the weather was severe, with a violent snowstorm causing snow to pile up a metre or more deep in some of the cuttings as well as making the rails very slippery. The train was an hour and a half late when, not having been preceded by a snow plough to clear the line, it got stuck in a snowdrift near Horowitz. Its tail lamps were burning, but were hard to see in the drifting snow, which had also brought down the telegraph wires, so no warning that it was stopped could be sent back. Its guard ran back to the second guard-house and both he and the guardsman made signals to a following goods train, which was then seen approaching, to stop. Their signals were observed, and the brakes applied but with slippery rails, only hand brakes and a falling gradient of 1 in 100 it was impossible for the train to pull up in time. The line was worked on the time interval system, with a required interval of only five minutes, but presumably in view of the conditions the goods had been held for 15 minutes at Pilau, the previous station. The driver and fireman of the freight train jumped from their engine and survived. The collision wrecked the three hindmost carriages of the passenger train, which were filled with soldiers. Four days after the accident the wreckage had still to be cleared away and the number of casualties remained uncertain. Nine soldiers were known to have been killed outright, four had since died and twenty were believed to be still buried in the wreckage, which would indicate 33 dead. Seventy others were more or less badly injured. Another source states that 22 soldiers died at the scene and seven more later in hospital, with 52 more seriously injured. The fourth carriage from the rear of the train, a first-class one containing the officers and some civilians, was badly damaged but no one in it was hurt. Only one railwayman, a brakesman, was injured.

The Times correspondent added that 'But one asks why did not the passengers get out of the carriages when the train stuck fast, and the guard was aware that another train was coming! If there was time for him to run back to the second guard-house surely there ought to have been time for every man to get out of the train before the collision occurred. Somebody, or several bodies, must evidently have lost their heads.' The correspondent seems rather unrealistic here. Would the soldiers be very keen to leave the relative warmth and comfort of the coach to stand outside in the bitter cold and the teeth of a blizzard in order to avoid being involved in an accident which they did not expect to

---

[330] Ibid.

[331] The Times, London, England, 7th, 10th, 14th and 21st November 1868

occur anyway? [332]

Subsequently both train drivers, the Pilau signalman, the guard of the passenger train and a crossing keeper were all charged with having caused the collision. The signalman and the guard were acquitted, the driver of the freight train was sentenced to a year's imprisonment with every fourth day fasting, and the driver of the passenger train and the crossing keeper each to six months imprisonment with every fourth day fasting. The crossing keeper had given stop signals 200 metres from the rear of the stopped train, one wonders if he was regarded more as a signalman than merely a crossing keeper and hence the court felt that he should have done more to protect the stalled train.[333]

A Pennsylvania and New York Railroad way train (that is one stopping at intermediate stations) halted for water at Athens, Pennsylvania, United States, on the morning of 20th September 1869. Conductor S.C. Mean appears to save sent back a flagman to protect his train but there was a thick fog and, the flagman's signals not having been seen by the engineer of a following express train, it collided with the rear of the way train. Two of the passengers on the express, Nelson Ackley and William Mabee, were killed, a Mrs Zelinda Spalding and George Beers, the fireman of the express, were seriously injured, and other passengers slightly hurt. One of the brief newspaper reports describes Mabee as an engineer and although he is listed as a passenger it seems possible that he was actually the engineer of the express.[334]

A coroner's jury at Athens censured conductor Means and John Rohm, the dispatcher at Towanda and it was reported that Means had been arrested, though if the coroner's jury did not actually bring in a verdict of manslaughter it is hard to see on what grounds.[335]

Fog was also a key factor in a bad accident at Long Eaton Junction, England, on the Midland Railway, on 9th October 1869. The accident occurred late on a Saturday night, and involved a mail train and an excursion train.

Long Eaton Junction was 6½ miles west of Nottingham and the point where the lines to Trent (Derby and London) and North Erewash Junction (Chesterfield and the North) diverged. A signal cabin was provided at the junction, which worked a distant signal 634 yards off towards Nottingham. The cabin was provided with telegraph bells by which trains could be signalled to and from the three adjoining cabins, Mansfield Junction (towards Nottingham), Trent North Junction and North Erewash Junction, but with neither block instruments nor speaking telegraph instruments.

When a fog came on, certain specified platelayers were required to report themselves to the signal cabins to work under the direction of the signalman. The platelayers were to provide themselves with flags or lamps and fog signals and station themselves at the distant signals, placing two fog signals a yard apart in the rear of the distant signal. Should the signal be cleared for an approaching train, the platelayer (fogsignalman) was to remove the fog signals if possible and at any rate show the driver a green flag or lamp.

332 Ibid, c19th November 1868

333 Eisenbahnunfälle in Europa by Erich Preuβ pg. 20

334 The Evening Gazette, Port Jervis, New York, United States, 21st September 1869

335 Ibid, 23rd September 1869

Should the signal remain at danger he was to exhibit a red flag or lamp. The fog signals were to be replaced on the rails as soon as the train had passed. Only two fogsignalman were provided at Long Eaton Junction to man the distant signal from Trent North Junction and Mansfield Junction. Should the signalman want the North Erewash distant signal fogged, he had to call in the Mansfield Junction fogsignalman by moving the arm of the distant signal up and down, that fogsignalman thus having a double duty.

Saturday 9th October 1869 was the day of the Nottingham Goose Fair, a big event in the area, and no less than eleven special excursion trains had been put on, running to Nottingham from Lincoln, the Erewash Valley, Leicester, Burton on Trent and Derby. Six of these trains were timed to return westwards from Nottingham at 9, 9.40, 9.50, 10.10, 10.20 and 10.45 together with an additional untimetabled train to Leicester. There were also three regular passenger trains at 9.50, 10.20 and 11.30 and three goods trains at 9.45, 9.55 and 10.[336]

Signalman Joseph Barker came on duty at Long Eaton Junction at 5 p.m. A thick fog came on between 6 and 7 and two platelayers, Horobin and Marshall, arrived at the cabin just after 6 p.m. Horobin was sent out to the down distant signal from Trent and the signalman kept Marshall near the cabin until 9, when, there being only one more train due from North Erewash Junction, he sent Marshall out to the Mansfield Junction distant with 18 detonators (fog signals) and his lamp. The Mansfield Junction and North Erewash Junction distant signals were 950 yards apart, and trains had passed the former at 6.56 and 7.29 and the latter at 6.54, 7.10 and 7.22 (according to Barker's train register) so it was impossible for Horobin to have been at the distant signals for all of them.

The first return excursion train reached Long Eaton Junction for the Erewash Valley at 9.23, followed by two for Leicester at 10.1 and 10.17, a regular passenger train for Leicester at 10.23 and a goods train for Trent at 10.37. Then came an excursion train for Burton at 10.50, a regular passenger train for Trent at 11.6, an excursion for the Erewash Valley at 11.35 and a goods train for Derby at 11.50, the last time being from the signalman's train register, the guard of the train said 11.40. So far all had gone smoothly except that the departure times from Nottingham were becoming later and later relative to those set down.

The last-named goods train had left Lincoln at 7 p.m., and was due to leave Nottingham at 9.45, but delayed by the excursion trains did not pass Mansfield Junction until about 11.25. According to its guard it took 19 minutes to reach Long Eaton Junction, where the distant signal was off, and was three-quarters of the way through the junction when its van was bumped by the engine of a following train, which the guard heard whistle for the junction signal and caught sight of just in time to jump from his van. The bump uncoupled the goods train at three points and its driver, realising that there was something wrong, had pulled up at Trent station, while the passenger train came to a stand in 30 yards.[337]

The passenger train, for Burton, had 25 coaches, including three brake vans. It was timed to depart from Nottingham at 10.20 and actually left according to the Nottingham stationmaster at 11.20 and to its driver and guard at 11.25. According to the Mansfield Junction signalman, it passed there at 11.30 and he gave it a green "Caution" lamp as the

336 Board of Trade report

337 Ibid.

goods train had been gone between five and ten minutes. The semaphore signal at the intermediate station at Beeston, 2½ miles from Mansfield Junction and almost three miles from Long Eaton Junction, was at danger but the porter there gave the driver another green hand signal and said, at the Board of Trade inquiry, that the goods train had been gone seven minutes. The driver of the 11.25 train found the Long Eaton distant signal off for him (the signalman not having replaced it as he was unsure if the preceding goods train had passed it) and continued, when he saw the tail lamps of the goods train. It was very foggy, with visibility down to 10 to 12 yards. Reversing his engine, the driver was able to avoid anything more than a slight collision which broke a piece off one of his buffers. After three or four minutes he continued to Trent, propelling the detached waggons of the goods train before him. He estimated the time of this slight collision as 11.40, which if correct, Colonel F.H. Rich, the inspecting officer, remarked showed that he must have travelled too fast from Mansfield Junction on such a foggy night with a heavy train in front of him.

When the goods train stopped, its rear guard went back with his hand lamp and detonators and put fog signals down at 400 yards, 600 yards and 1,200 yards (the last further out than the regulations required) and then returned to his van, where Fogsignalman Horobin came up him and told him he could return to his train. Asked twice if the line ahead was clear, Horobin first said yes and then that he was not sure and would check how the distant signal stood. The guard remained where he was and shortly a special train for Leicester (presumably the 10.45 p.m.) came up. The guard gave its driver a red lamp and having also passed over the guard's fog signals its driver pulled up very sharply so that the guard could get up on the engine and tell the driver, "There is something in front, go up cautiously" to which the driver had replied, "Go back as quick as you can, and give four more fog signals, as the mail is close behind us." The goods guard said that he had then gone back as fast as he could and had reached the back of the Leicester train when he heard the mail train whistle and he had time to put down one fog signal and show his red light before the mail train passed him and collided with the rear van of the Leicester train. The guard put the speed of the mail train at 17 or 18 miles an hour. After the collision he went further back to protect the mail train.

The Leicester train consisted of 35 carriages, including four brake vans, and had left Nottingham at 11.35. The signals at Mansfield Junction had been clear for it, and it had run into the fog near Beeston. The Beeston signal had been at "Danger" but was lowered to "Caution" upon the driver whistling. The Beeston stationmaster said that it had passed there at 11.57, implying an average speed of only nine miles an hour from Nottingham. The Attenborough station signal (lit but not operated at night) was white and shortly afterwards two fog signals had been run over, speed was reduced, and its driver saw a man with a red lamp (the aforementioned guard of the previous train) who jumped on the engine and told the driver to go cautiously. The driver told the man with the lamp to go back and stop the mail and proceeded towards the junction, where he was beckoned on by a man showing a white light (Platelayer Horobin). Shortly afterwards there was a collision, which on the engine felt like little more than a push, but which smashed the rear composite brake carriage of the train and part of the carriage in front of it, killing seven passengers and seriously injuring several others.[338]

The mail train was booked to leave Nottingham at 11.30. There was a discussion

---

[338] Ibid.

between the Nottingham stationmaster and one of the station inspectors as to whether to let the Leicester train or the mail go first, but as the platform was crowded with Leicester passengers and the mail train was booked to stop at both Beeston and Trent stations, and the Leicester train at neither, it was decided to let the Leicester train go first. The mail train, of eleven vehicles, two of them brake vans, accordingly, did not depart until 11.50, fifteen minutes after the Leicester train, and was checked briefly at Mansfield Junction, but then reached Beeston already at 12.01, only four minutes after the Leicester train, having run twice as fast as the latter from Nottingham. Some passengers alighted at Beeston, and the train left there at 12.05, eight minutes after the Leicester train, but the Beeston stationmaster and porter told the mail train driver that the Leicester train had only been gone five minutes, to induce him to proceed more cautiously in view of the fog. An inspector put the highest speed of the mail train after leaving Beeston at 25 miles an hour, the driver and others at 16 to 18 miles an hour. The driver had had barely time to reverse, and the fireman to get three turns on the brake, between the time that the van of the Leicester train was seen, and the collision occurred.

Colonel Rich calculated that the collision, 480 yards on the Mansfield Junction (Nottingham) side of the Long Eaton Junction distant signal and 2¼ miles from Beeston, had occurred at 12.12. The average speed of the mail train from Beeston must therefore have been about 20 miles an hour, and that of the Leicester train, before it was checked by the fog signals of the goods guard, must have been 12 miles an hour. Had the Leicester train not therefore been checked it would have been well with the protection of the Long Eaton distant signal before the mail train overtook it. The immediate cause of the accident, the inspecting officer concluded, was the failure of Fogsignalman Horobin to protect the goods train. Had he done so the first minor collision would have been prevented and the goods guard would not have needed to check the Leicester train. Horobin pleaded that he had had to go to the cabin for a fresh supply of detonators but Colonel Rich thought it more likely that because of the bad practice of expecting one fogsignalman to look after two distant signals, which was at times physically impossible, the lax habit had grown up of the fogsignalman leaving his post at other times, and Horobin had only hurried back to his post when the first collision took place. Even if his statement were true, the company were at fault as he should have been kept supplied with fog signals without having to leave his post for 10 minutes to fetch them.

Another cause of the collision, Colonel Rich continued, was the lack of caution of the driver of the mail train, even after he found that he had almost caught up the Leicester train at Beeston. A third cause was the despatching of the heavy Leicester train in front of the mail. If the driver of such a heavy train ran cautiously, as the Leicester driver had done, for fear of losing control of his train, there was a chance that he might be overtaken by a faster train. It had been a grave mistake also, on a night with so many extra excursion trains, to run any goods trains between Nottingham and Trent.

The special timetable appeared to have been drawn up carelessly, with a special train for Leicester given the same departure time from Nottingham, 9.50, as the regular train to that pace, and both only ten minutes after a previous train. Again, a special train for Derby and Burton was booked to depart at the same hour, 10.20, as the regular train and both ten minutes after the previous train, while goods trains were interpolated five minutes before and after the 9.40 and 9.50 trains.

Having gone through what he considered to be the immediate causes of the collision

Colonel Rich, like Captain Tyler in his report on the Bullo Pill accident eleven months earlier, concluded that the prime cause of the accident was the 'completely illusory protection' offered by the time interval system. He had been informed that in the previous June the directors of the Midland Railway Company had decided upon the introduction of the block system from Bradford to London and Chesterfield to Bristol, including Trent to Nottingham. It was, he said, only to be regretted that this decision had not been come to earlier.[339] [340]

Appended to the Board of Trade report was a letter from James Allport, the Midland Railway general manager, in which he disputed most of Colonel Rich's findings, stating for example that the Long Eaton Junction distant signal from North Erewash Junction was never manned by a fogsignalman as it was very close to North Erewash Junction cabin. The fogsignalman were experienced men, well conversant with their duties, and did not need extra supervision. The 18 fog signals taken by Horobin initially should have been sufficient for many trains. The speed of the mail train had not been excessive. It had not been a mistake to despatch the Leicester train in front of the mail train, and so on.

The inquest on the seven passengers who lost their lives seems to have been a somewhat arbitrary affair which ended in a verdict of "Accidental death." It was felt by many, including one of the jurors who wrote to *The Times*, that the causes of the accident should have gone into much more thoroughly by the inquest but that the jurors had been overruled by the coroner.[341]

New Year's Day 1870 saw an accident on the Caledonian Railway in Scotland. New Year's Eve, or Hogmanay, is of course a very important celebration in Scotland, and it is perhaps not surprising therefore that Stationmaster McCurragh left his post at the wayside station of Forgandenny, south of Perth, at 7.30 p.m. on 31st December, doubtless to go and take part in the festivities. Left in charge of the station was Porter Baynes, with a surfaceman being instructed to help him if necessary. The Hogmanay traffic was very heavy, and the trains were running very late. At about 8 p.m. the first portion of a train from Edinburgh and Glasgow, which trains were normally combined at Larbert, stopped at Forgandenny, almost an hour and a half late. Porter Baynes assisted in handing the alighting passengers and was informed by the guard that another train was following. Baynes said that this was the only way that he had known of the second train. Caledonian Railway Rule No. 48 stipulated that a second green tail lamp was to be affixed to the rear of a train to denote that an extra or special train was following but the only green tail lamp at Larbert had been sent north two days previously and had not come back, so the Larbert stationmaster had instead affixed a second red tail lamp. The second train reached Forgandenny not long after the first had gone. There was then a gap in the down trains until a third arrived at 10.15 p.m., an hour and twenty minutes late. Baynes attended to the rear of the train, while the surfaceman and the stationmaster's son attended to the front of it. Baynes took two parcels from the guard, then, as the train was pulling out, asked the guard if that was it for the night as regards to trains stopping there. Receiving, as he stated, no reply, and seeing no green tail lamp, he assumed that it was, locked the

339 Ibid.

340 The Times, London, England, 11th October 1869

341 Ibid, 12th, 15th and 20th October 1869

station up, put out the signal light and the platform lamps and went off bed at his house near the station. The surfaceman and the stationmaster's son had not noticed, or did not bother to tell him, that the train had had no front van or front guard, from which he might have inferred that another train was to follow.

The last train that Baynes had dealt with was in fact just the Edinburgh portion of a train for the North. It had left Edinburgh 20 minutes late and the heavy traffic had meant that it was 50 minutes late at Larbert, arriving there at 8.05. Having learned by telegraph that the Glasgow portion of the train had not yet reached Greenhill, the Larbert stationmaster had determined to send the Edinburgh portion on separately, as he had the previous two trains, and still lacking a green tail lamp had again affixed an extra red one, positioned such that it might appear, as it did to porter Baynes, simply as one of the sidelights. The guard of the train, Peter Matthews, was unaware that his train did not have an extra green tail lamp and was not told by the stationmaster to give notice at each station that there was another train following. He said that he had done so anyway at five of the eleven stations that another train was following but had not had time or had forgotten at the others, including Forgandenny. His version of the conversation with Baynes was that Baynes had remarked, "You're here the night yourself, Peter," to which he had replied, "Yes" and had understood from that that Baynes was aware that there was another train to follow. Baynes, however, stuck to his original story. The train left Larbert at 8.10, took on three more carriages at Stirling, lost more time at each stop and was 80 minutes late by Forgandenny.[342]

The Glasgow portion of this train, due to leave Queen Street station at 6.20 p.m., was not despatched till 7 p.m. and then only as far as Cowlairs, where it was strengthened by extra carriages brought up from Queen Street at 7.56, and finally left Cowlairs at 8.15, about an hour and three-quarters left, and consisting of two engines and tenders, brake van, 32 carriages and another brake van. Larbert was reached at 9.28, 2 hours 13 minutes late, and problems with a broken coupling chain and the detaching of eight carriages meant that Stirling was not left until 10.18, 2 hours and 33 minutes late. Problems with another coupling chain at Greenloaning meant that a carriage had to be detached and the train left there 2 hours 47 minutes late. The Greenloaning stationmaster had not noticed the extra tail lamp but had been told by a porter that there was no front van, had enquired of the guard and, finding that there was another train following, had kept his station open. More couplings broke at Forteviot, the last station before Forgandenny, another carriage had to be detached and the train left there at 11.54, 3 hours 7 minutes late. The Forteviot stationmaster was also absent without leave but the porter left in charge had observed the extra tail lamp. The Forteviot porter had put his station and distant signals to "Danger" and then "Caution" and was just about to put them to "All right" with the elapse of the second five minutes after passing of the Glasgow train when the down mail passed.

The Glasgow portion reached Forgandenny at midnight. The station was in darkness and deserted. Only two passengers alighted, and after a minute the train was ready to restart when the rear guard signalled that there was a problem, the train stopped again, and it was found that yet another coupling had given way. The rear guard had just attached the spare coupling chain when he heard the whistle of the mail train and just had sufficient time to regain the platform, while the head guard got back a short distance waving a red light.

[342] Board of Trade report

The train which caused the collision was the down mail, consisting of portions from London Euston, Edinburgh and Glasgow and due to leave Stirling at 10.17. Crowds of passengers and a lack of carriages meant that its Glasgow portion left 52 minutes late, and further time lost on the road meant that the combined train left Stirling 63 minutes late, at 11.20. It was not booked to stop between Stirling and Perth ticket platform except to detach the engine which assisted it from Perth to Kinbuck and was allowed 63 minutes for the 33 miles from Stirling to Perth. Contradicting the statement of the Forteviot porter, the driver of the mail train stated that the station and distant signals there had both shown poor lights, but they had both been white. He estimated his speed at 40 miles an hour on the falling gradients to Forgandenny, and stated, "when the best part of ¾ of a mile from Forgandenny, I caught sight of the tail lights of a train, these did not at first come distinctly in sight, but seemed obscured by smoke or vapour between them and me in the cutting. On first distinguishing them, I gave the break whistles repeatedly, had the tender break applied, shut off steam, reversed the engine and put on back steam. This had all been done, and the driving wheels had skidded, before the distant signal (380 yards from the van) was reached. The speed was, I think, reduced to about 14 miles an hour when I struck the van." The driver attributed his failure to stop to the slippery rails and the lightness of his tender at the end of the journey. The guards stated that they had also applied their brakes though this seemed inconsistent with the speed of the train at the moment of the collision.[343]

Little damage was caused to the mail train but most of the 24 carriages of the Glasgow train were more or less damaged and a man in the twelfth carriage from the engine and another who was in the act of jumping out were killed. Both men were carpenters from Dundee. Several other passengers were hurt.

LT-Col Hutchinson inquired into the accident. He found that the immediate cause of the collision was the absence of signals to protect the Glasgow train while it was stopped at Forgandenny. This was due to the lack of a green tail lamp on the Edinburgh train thanks to the inadequate number of such tail lamps at Larbert, the negligence of Guard Matthews in not informing all stations that another train was to follow, together with the failure of the Larbert stationmaster to remind him to do so, and the unauthorised absence of the Forgandenny stationmaster, who would have been more likely than an inexperienced porter to either see the extra tail lamp or notice the absence of a front van.

The inspecting officer did not blame the head guard of the Glasgow train for not protecting his train at once when he found Forgandenny station unstaffed but thought he should have done so more quickly on finding that his train was going to be delayed.

The impolicy of running such huge trains, composed partly of inferior and worn rolling stock, Hutchinson found fault with, also the fact the brake power of the Glasgow train was very inadequate. That a "Caution" signal was exhibited at Forteviot for the mail train the inspecting officer doubted; in any case the signals were very dim, and the porter should have exhibited a hand signal. The hours worked by the Forgandenny porter (from 4.30 a.m. till 9 or 10 at night) were too long and either there should be more staff, or a younger stationmaster and a more experienced porter employed there, and the hours divided between them. Finally, the Forteviot distant signal should be re-sited twice as far

[343] Ibid.

from the station.[344] [345]

A mixed train from Constance entered Murg station, Baden, on the Grand Duchy of Baden State Railway, at 4 p.m. on about 2nd June 1870 (exact date unclear). It was followed by an engine and tender from Waldshut, bound for Basel. The usual signals had been displayed behind the passenger train, but the driver of the light engine had been drinking and failing to observe the signals entered the station, ran into the rear of the mixed train, and 'a fearful scene of suffering and death was produced.' Three carriages filled with passengers were reportedly completely crushed, the rearmost being forced into that in front. When the smoke and dust had cleared it was found that nearly 60 passengers were more or less buried in the wreckage. One young girl was taken out dead, and 12 others were dangerously injured. A man with both his thighs broken underwent amputation but died a few hours later and a Dr Ruff, of Waldshut, was also fatally injured and died later. The 30 or so injured were taken to hotels where they were attended by surgeons summoned from nearby towns. The local townspeople were so enraged by the negligence of the drunken engine driver that he was almost lynched but was ultimately taken away to prison to await trial.[346]

A serious collision occurred at 4 a.m. on the morning of 20th September 1870 at Plessis, near Tours, France. A passenger train was stopped for the collection of tickets when it was run into by a special troop train, killing eleven passengers and seriously injuring about 20. The fatalities included a M. Duval, initially identified as the editor of the newspaper *Journal des dèbats* but later stated to be only a well-known writer in the paper.[347]

The French railways had been severely damaged in many places during the Franco-Prussian War of 1870-71 and traffic on many lines was only gradually restored during 1871. On 9th (some sources state the 8th or the 10th) March 1871 it was, however, a train of wounded Prussian soldiers returning home who were the victims of an accident at Puteaux, France, on the Chemins de fer de l'Ouest. A late-running suburban train stopped in Puteaux station at 7 p.m. Behind it a hospital train from Le Mans to Pantin and thence on to Germany, consisting of 32 waggons each occupied by 20 to 25 men, and totalling 430 passengers in all, driven by an Ouest Company crew, managed to pull up safely, but this second train was then run into by a following (Prussian) freight train travelling at 32 kph, whose driver had overrun signals. Nineteen vehicles of the hospital train were broken to pieces and the engine and five or six waggons of the colliding freight train were also demolished. Ten German soldiers were killed instantly, and a large number seriously injured. An Ouest employee, presumably a conductor, also lost his life. On 12th March *The Times* correspondent in Paris reported that the number of dead had risen to 20 with 60 injured. The Ouest railway company stressed that the accident was entirely the responsibility of the Prussian authorities, rather than themselves, as under the terms of the

344 Ibid.

345 The Times, London, England, 3rd January 1870

346 Ibid, 6th June 1870

347 Ibid, 21st and 22nd September 1870

armistice of 28th January the Prussians were responsible for regulating the trains belonging to them.[348] [349] [350]

An Erie Railroad westbound freight train on 20th May 1871 included an emigrant car placed next to the caboose at the rear of the train. At Griswold station, New York, United States, a coupling broke, leaving the passenger car and the caboose behind. The station signal was placed at "Danger", the front of the train was set back, it was recoupled and restarted, and the signal was cleared again. The coupling then broke a second time and, before warning could be given, an extra freight train had run into the back of the caboose. Five children in the emigrant car were killed and about 20 adults were injured. 'The killed and wounded were brought to this city {Buffalo} and the wounded cared for at the Sisters of Charity's hospital.' [351] [352]

Brief reports in two English newspapers spoke of four deaths in a collision at Ablon station, France, on the 'Orleans railway' (Chemins de fer de Paris à Orléans) a few kilometres from Paris, on Sunday 4th June 1871.[353] [354]A longer report in *The Times*, spoke, however, of 24 women being killed on the spot and about 50 other persons injured, some very seriously. *The Times* article explained that due to the destruction of a viaduct at Dravel in the Franco-Prussian War, which had then just ended, trains from Montargis, which had formerly run to the station on the Boulevard Mazas in Paris, were being diverted at Javisy onto the Orleans line to run into the station on the Boulevard de l'Hôpital. On the day of the accident, a passenger train from Corbeil, running at full speed, had run into the rear of another passenger train from Vendome, which was 50 minutes behind time, crushing its rear four coaches. All the deaths and serious injuries were in the Vendome train, the occupants of the Corbeil train escaping with a severe shaking and some contusions. A signalman was also injured when his box was struck by a broken wheel.[355]

The next rear collision to be considered, at Revere, Massachusetts, United States, on the evening of 26th August 1871, is as well remembered in the United States as say the Tay Bridge disaster in Great Britain, for reasons that are not entirely clear. Its notoriety probably stems from its heavy death toll, the fact that it was the first bad accident to occur anywhere near Boston, at the time regarded as the cultural centre of the nation, and the fact that its victims included a prominent pastor, the Reverend Ezra Gannett, of the Arlington Street church in Boston. The accident also highlighted a general incompetence of higher management, rather than simply the recklessness or negligence of an individual

---

[348] Webpage - Chemins de fer et dȇplacements autour de Paris pendant la crise de 1870-1871

[349] Webpage – Racines en Seine – Puteaux

[350] The Times, London, England, 11th, 13th and 14th March 1871

[351] Bangor Daily Whig and Courier, Bangor, Maine, United States, 13th May 1871

[352] The Coshocton Age, Coshocton, Ohio, United States, 19th May 1871

[353] The Daily News, London, England, 7th June 1871

[354] Lloyds Weekly Newspaper, London, England, 11th June 1871

[355] The Times, London, England, c7th June 1871

railwayman.[356]

The accident occurred on what was then the Eastern Railroad. That August evening, the railroad was having to deal with an unusually large amount of traffic, the normal extra holiday traffic being supplemented by that resulting from a military muster and two large camp meetings (Protestant religious meetings with day-long sermons, conversion experiences, religious hysteria and so on). The number of trains entering and leaving the Causeway Street station had thus swollen from a normal 152 to 192. The main line from Boston to Salem was double track but the branches were single and, although the telegraph was installed, Eastern Railroad superintendent Jeremiah Prescott refused to permit its use for train despatching, a refusal which was to be an important factor in causing the accident, as it prevented alterations to train schedules to be made in the event of breakdowns or delays.

Four trains were scheduled to leave Boston early that Saturday evening, with trains for the Saugus branch – basically a single line loop leaving the main line at Everett and re-joining it at Lynn – departing at 6.30 and 7, the main line Beverly accommodation train to Lynn at 7.15 and an express to Portland at 8. Excessive traffic and confusion resulted in the first Saugus train not leaving until 7 p.m., then the main line accommodation train followed it at 7.40 in front of the second branch train, which left at 7.53, the express following just after 8. So far, there had just been delays but a further problem arose when the first Saugus branch train reached Everett, the junction with the single line branch, for the 6 p.m. southbound train from Lynn, which was scheduled to cross the northbound train from Boston, was itself an hour and a half late starting due to an engine problem. A simple use of the telegraph would have enabled the crew of the northbound train to learn of this delay and a crossing could have been arranged at a convenient point along the branch but use of telegraph for such purposes was not permitted. When the situation had occurred previously, the signalman at Everett, lacking a siding where a waiting train could be put clear of the main line, had simply let the train proceed a short distance down the branch or onto the other running line, under flag protection, strictly speaking probably an infraction of the rules but a common-sense solution to the problem. Unfortunately, on 29th August a different man was on duty who either lacked common-sense or else was a stickler for the rules, and thus the Saugus branch train was kept waiting on the main line, with following trains stopped behind it. At about 8.10 the southbound branch train finally appeared and the 7 p.m. northbound branch train left the main line for the branch.

The conductor of the accommodation train seems not to have realised that the second Saugus branch train was still behind him and thought that the express was immediately behind his train and therefore aware of its position, while the engineer of the express failed to deduce that both branch trains and the accommodation train between them must have been held up at Everett by a delayed southbound branch train. Thus, when he saw the second branch train immediately in front of him leave the main line, he assumed that he had a clear road and that the preceding accommodation train was long gone. The Boston stationmaster had warned him to look out for trains ahead, but he had assumed that the warning applied only to the second branch train.[357]

---

[356] A History of Railroad Accidents, Safety Precautions and Operating Practices by Robert B. Shaw, pgs. 77-81

[357] Ibid.

The accommodation train after leaving Everett stopped first at Chelsea station and then at Revere. Revere was the junction for the old main line to East Boston, by 1871 relegated to a freight branch, and a signal was erected to show whether the points were open for the branch or the main line, a red light showing that they were set for the branch and no light that they were set for the main line. The engineer of the express, William Brown, said that he had been peering through the mist trying to see this signal, rather than looking lower down. When he had finally looked down to the track, he had seen the tail lamps of the accommodation train, which "seemed to spring right up in his face." The express was then travelling at 30 miles an hour, and the accommodation train was 800 feet away. Engineer Brown whistled for brakes, but with hand brakes only available and slippery rails, the train was still travelling at about 10 miles an hour at the moment of impact, sufficient for the engine to bury itself two-thirds of the distance into the rear car of the accommodation. Damage to the boiler of the engine then resulted in the passengers in the crowded car being immediately engulfed by searing steam and boiling water, killing many who might otherwise have survived with injuries. The accommodation train had just been setting off, but all its cars were damaged. The coupling broke in front of the penultimate car, and the platforms of the baggage and smoking cars were locked together by the collision, but not so badly as to prevent the passengers in the smoking car from getting out, which was fortunate as engine coals ignited oil which had spilt on the upholstery from broken kerosene lamps, and the flames spread rapidly to the remainder of the forward part of the train. All the 29 fatalities, however, and the majority of the injuries occurred amongst the passengers in the rearmost car. Most of the dead apparently died from scalding. The engineer and fireman of the express jumped from their engine and escaped.[358]

The Revere accident provoked an especially strong reaction from an outraged public. There was an "indignation" meeting at nearby Swampscott and noted anti-slavery campaigner Wendell Phillips went as far as to call the accident "deliberate murder" which seems to be putting matters a little strongly, to say the least. Contemporary author and Commissioner of Railroads, Charles Adams, did not consider that Wendell Phillips had contributed anything to railway safety by his outburst. Adams asked Superintendent Prescott if the use of the telegraph might not have prevented the accident. Prescott replied that he did not think so. Some years previously, apparently, because of non-use of the telegraph, an Eastern Railroad freight train had waited all night at Salem for an opposing extra passenger train, and the passenger train all night at Ipswich for the freight! A conductor also said that he had complained to both Superintendent Prescott and President Browne of the difficulty of stopping using hand brakes but had merely been told to 'do the best he could.' New England newspapers took up the campaign against the Eastern Railroad and pamphlets were issued attacking the company. Company President Browne was forced to resign. Prescott remained another year, but under the charge of a new general manager, Charles Hatch, who introduced the use of the electric telegraph for despatching.[359] The Railroad Commissioners severely criticised the company for laxity in discipline, rolling stock deficiencies and attempting to do too much in the excursion business. The company had to pay out about half a million dollars in damages, was for

[358] The Titusville Morning Herald, Titusville, Pennsylvania, United States, 28th, 29th, 30th, and 31st August and 1st and 2nd September 1871

[359] The Story of the American Railroads: From the Iron Horse to the Diesel Locomotive by Stewart H Holbrook pgs. 279 and 281

years afterwards on the verge of bankruptcy and was eventually absorbed by the Boston and Maine Railroad.

Train accidents seem to have frequently occurred to passengers who were either excursionists or, especially in North America, immigrants, and it was the latter who were the victims of a collision one mile east of Muir, Michigan, United States, on the evening of 28th August 1873. A westbound night express of the Detroit and Milwaukee Railroad consisted of a baggage car, two coaches, a sleeping car and two emigrant cars. The train was disabled at the point of the accident after its locomotive threw off a driving wheel, and a following through freight train ran into it, killing five women and a child and slightly injuring ten other passengers. 'Medical assistance was at once procured from Muir.' All the victims were Icelandic immigrants,[360] [361]who must have been among the first from their country, as immigration from Iceland did not begin until 1873 with the setting up of the first immigration agency.[362]

A rear collision at Kildwick, England, on the Midland Railway, exactly two years after the Muir accident, on 28th August 1875, again emphasised the apparent propensity of trains carrying excursionists and immigrants to be involved in bad accidents. This time it was an excursion train. The 8.50 p.m. return excursion train from Morecambe to Bradford was run into by the 10.15 p.m. fast passenger train from Ingleton to Leeds. Two carriages and a brake van at the rear of the excursion train were knocked off the rails and the two carriages were nearly destroyed. Seven passengers in them were fatally injured and 39 other passengers and the head guard in the brake van were hurt.

The line where the accident happened was still worked by the time interval system, supplemented by telegraph bells between the various signal boxes, by which the trains were described from cabin to cabin. The excursion train was started from Skipton station at 11.12 p.m. and was not booked to stop at Kildwick. The Ingleton fast (mail) train was despatched five minutes after at 11.17 p.m.; this train was booked to call at Kildwick. The starting signal at Skipton South cabin was lowered to "Caution" for both trains; it was never lowered beyond "Caution." Signalman Henry Cockshott at Robinson's Stone Siding cabin, just south of Skipton station, testified at the subsequent Board of Trade inquiry that he had also given the mail train a green "Caution" signal, as the excursion train had only been gone five minutes. That Cockshott had showed the mail train a caution signal was confirmed by Signalman John Smith at Skipton South. Cockshott was sure that five minutes had elapsed since the passing of the excursion train before the mail train passed him. Three red lights (a tail lamp and two side lights) were burning at the end of the excursion train.[363]

The next cabin to Robinson's Stone Siding was at Cononley station. Signalman Jonathan Baldwin was on duty there on the night of the accident. He said that the Morecambe excursion train had been signalled to him by Cockshott, he had belled it forward to Kildwick cabin and it had passed him at 11.19 p.m. As it did so, he noticed

[360] Daily Republican, Decatur, Illinois, United States, 30th August 1873

[361] Morning Oregonian, Portland, Oregon, United States, 30th August 1873

[362] Wikipedia – Icelandic Americans

[363] Board of Trade report

that although the two side lights were showing the centre tail lamp was out, and he therefore sent the seven beats "Stop and Examine Train" bell signal to Kildwick cabin. He then received two beats, the code for a fast passenger train, from Robinson's Stone Siding and lowered his distant signal to "Clear" and his home signal to "Caution," knowing that it was the mail train which was due to stop at Kildwick anyway. He did not think it necessary to stop the mail train, as the excursion train had been gone four minutes and he could see the Kildwick distant signal at "Danger", but he showed the mail train driver a green light from his cabin. The mail train was running fast, and he remarked as much to Porter Edward Newbold, who had come into the cabin while he waited for some excursion trains from Liverpool.

Signalman Richard Staveley was working Kildwick cabin. He remembered receiving two beats from Cononley for the excursion train at 11.18, immediately followed by the "Stop and Examine Train" signal, upon which he had replaced his home and distant signals and called to Porter George Quincey to get the driver of the excursion train to draw into the station and to find out what was amiss. He saw the excursion approaching, gave it a green signal with his hand lamp, heard Cononley give him three beats for a stopping train and shortly saw the mail train appear.

Driver Harrison Palfreeman was in charge of the engine of the Ingleton mail train. He stated that after getting the usual green "Caution" signal at Skipton South cabin, he had seen nothing but white lights at Robinson's Stone Siding and Cononley cabins. He had only seen the Kildwick distant signal at "Danger" when about 100 yards from it, had shut off steam, whistled for the guard's brakes and his fireman had applied the tender brake. Seeing the tail lamp of the train in front and feeling that his train was not pulling up very quickly, he had then reversed his engine, told his fireman to put on the brake more tightly if possible, whistled again for the guard's brakes and applied his water brake. He admitted that according to his (June 1871) rule book he should have stopped at the distant signal. He did not know that the excursion train was in front of him.

Palfreeman's fireman, William Dobson, stated that he had not seen the signals at Robinson's Stone Siding but the Cononley signals had shown white lights. The night had not been foggy, but it was wet, and the rails were in a bad condition for stopping.

Guard William Wiggins of the mail train said that he had been sorting letters between Skipton and Cononley, but he had seen a white signal light at the latter place. He did not think that they had been going too fast passing the Kildwick distant signal.

The inspecting officer, Captain H.W. Tyler, pointed out that there was a conflict of evidence as to whether the signals at Robinson's Stone Sidings and Cononley had been green or white, but that in any case a green light, if exhibited, would only have told Driver Palfreeman that the excursion train was not more than five minutes ahead of him. It was clear, though, that the Cononley signalman should have shown the mail train a red light after sending a signal to stop the previous train at Kildwick, rather than clearing his signals at less than the minimum five minutes. Equally, Driver Palfreeman should have approached Kildwick station more cautiously after finding the distant signal at "Danger." Both men were therefore to blame for the accident, but it was right to point out that the causes of the accident were more or less incidental to the time interval system of working together with trusting to the distant signal to protect against an obstruction between it and the home signal when the drivers of fast trains were permitted to pass such distant signals at such a speed as they in their discretion considered safe. Under such circumstances a little extra speed or a slight want of caution on the part of a driver was likely to lead to a serious collision. Had the block system been in operation and properly worked, the mail train would have been stopped at Cononley and no accident would have happened.

In conclusion, the inspecting officer remarked that it was satisfactory to learn that the Midland Railway Company were planning to introduce the block system on the line, and such a system was already in operation on 627 miles of their system, with plans to adopt it over an additional 176 miles.[364] [365]

Because six of the passengers died at Kildwick but a seventh, a William Taylor, died later at Bradford there were two inquests into the accident. At the first inquest at Kildwick, the coroner, in his summing up, more or less suggested to the jury that they should find driver Palfreeman guilty of manslaughter, and they duly did so. The coroner at the second inquest, at Bradford, was more impartial and the jury there returned a verdict of "Accidental death."

At Palfreeman's trial at Leeds on 25th March 1876 the judge, Mr Justice Mellor, told the jury that he did not think that there was a particle of evidence that Palfreeman was guilty of manslaughter, and he was duly acquitted.[366]

Fog was a factor in a collision in the suburbs of Paris, France, at 7 p.m. on the evening of 3rd February 1880, when an omnibus train bound for Argenteuil was run into by a Paris Ouest-Paris Nord circular service train at Clichy – Levallois station. The accident left 13 dead and 69 injured, those killed including the driver of the colliding train and the conductor of the omnibus service. *The Times* correspondent in Paris reported that 'The train to which the accident happened was a suburban train containing several financiers, merchants, clerks and businessmen living at a short distance from Paris, who were returning to their homes. The names of several of the victims have been made public, but it is difficult to verify the exact number of dead, as several of the wounded were carried to private houses and hospitals in the neighbourhood' and 'M Puyserrat, head of a firm of brokers on the outer Bourse, and M. Dusargues de Colombies, the director of the Bulletin des Halles, are among the persons killed in the accident.' [367]

On 10th July 1880 the Paris court decided that the dead driver and conductor were mainly to blame for the accident but sentenced two signalmen whom it considered also to be partly responsible each to four months' imprisonment and a fine of 50 francs.[368]

August 1880 saw another major disaster in the United States, at Mays Landing, New Jersey, on the West Jersey and Atlantic Railroad. An excursion was arranged for Wednesday 11th August 1880 to convey 1,500 parishioners of St Anne's Roman Catholic Church in Pennsylvania from that city to the seaside at Atlantic City, New Jersey, two special trains being run. The outward trip went smoothly but disaster struck on the return journey. The first section left Atlantic City at 6 p.m. and the second section was intended

364 Ibid.

365 The Times, London, England, 30th August and 3rd and 20th September 1875

366 Webpages – The Kildwick Rail Crash of 1875-Parts 1, 2 and 3

367 The Times, London, England, 5th February 1880

368 Wikipedia – Liste des accidents ferroviaires en France au XIX siẽcle

to follow at an interval of ten minutes, though Superintendent Sewall stated that the second section had been allowed to depart only three minutes after the first and Assistant Superintendent Hackett that it had left five minutes after the first but had been instructed to run slowly until the interval had increased to ten minutes. The second section was in any case allowed to come within five minutes of the first at locations where they were timetabled to cross opposing trains, as was the case at Mays Landing, where they were scheduled to cross the express train from Philadelphia at 6.40 p.m.

The passengers on the platform of the rear car of the first section saw the second section coming up behind them and realised that it was not going to pull up in time. Some ran back into the interior of the car and others jumped off, though the rear car was standing on the Great Egg Harbor bridge. The locomotive ploughed halfway through the car and one of its cylinders was damaged, so that scalding water and steam instantly filled the car in a scene reminiscent of that at Revere, nine years earlier. Some of the women passengers had gone down to the seaside in light summer outfits which provided scant protection from the searing steam. A violent thunderstorm was raging at the time of the accident with heavy rain, so all the car windows were closed, trapping the steam inside. Twenty-six passengers died in the accident, all from the rear car.

The president of St Ann's literary institute, T.B. Judge, gave a newspaper reporter a description of the scene after the accident. He said, 'I was on the last section and was in the rear car when the crash occurred. The shock was a strong one and several persons were thrown from their seats. There was the appearance of a panic. I did all in my power to quiet it. I partly succeeded, and then the people began to leave the car to see what was the matter. I went with the rest and when I reached the head of the section, I saw a sight which I cannot adequately describe. The injured passengers were being drawn from the car and the water on each side of the track was filled with persons who jumped from the train and were up to their waists in the river. A boat soon put out and picked them up. On the embankment of the railroad, which is seven feet high, were sufferers from the crushed car, groaning in agony, and the dead body of Sweeney. I think the railroad company guilty of gross negligence in running sections so close together.'

By this date the train was provided with air brakes, though train crews were still expected to apply the hand brakes if necessary. A newspaper stated that 'It is the general opinion that the engineer of the colliding locomotive did all he could to check his engine, and when it struck the engine, had been reversed and the air brake put on.' This was apparently not the view of those at the scene immediately after the accident for engineer Edward Aitken was almost lynched at the scene and some railwaymen were assaulted back at Camden. Engineer Aitken himself stated that the accident had been due to the air brake failing to work and he had not had time to apply the hand brakes. Assistant Train Master Mills believed the brakes had definitely been tampered with, whether by accident or design he could not say. He thought that someone must have interfered with it by fooling with it in one of the closets of the cars. However, no proof that the brakes had really been tampered with was found.[369] [370]

The inquest into the accident focused upon the death of one victim, James Sweeney,

[369] The Evening Gazette, Port Jervis, New York, United States, 14th August 1880

[370] A History of Railroad Accidents, Safety Precautions and Operating Practices by Robert B. Shaw, pgs. 86-87

presumably the Sweeney mentioned by T.B. Judge, and possibly one of the organisers of the excursion, but he was not further identified. Engineer Aitken testified that they had been ten minutes behind the first section at Pleasantville (about 10 miles short of Mays Landing) and five minutes behind at Mays Landing. He had been running at about 25 miles an hour down the hill and approaching Mays Landing had called for 'down brakes' and applied sand but had not been able to pull up in a mile and a quarter due to the rails being wet. He does not seem to have referred to any failure of the brakes at the inquest. Conductor Hoagland testified that he had tried the brakes after the collision and had found them all turned on the same way. The jury returned a verdict of "Accidental death" but the newspaper reported that Engineer Aitken and Conductor Hoagland were still being held on bail to await the action of the grand jury.[371]

The Mays Landing disaster was followed two months later by a worse one at 28th Street Crossing, Pittsburgh, Pennsylvania, United State, on 9th October 1880. In this case, there was no doubt that the engineer of the colliding train was at fault. Traffic on the Pennsylvania Railroad was especially heavy on the evening of the day because of a large Democratic Party demonstration and the closing of an exposition and the passenger trains were extremely crowded, with passengers standing in the aisles and crowding the platforms of the cars, including that of the rear car of the first section of the eastbound Wall's accommodation train, which was run in two sections to deal with the traffic. This train left Pittsburgh Union station at 11.31 but got no further than 28th Street Crossing, where it was stopped at an automatic signal because the block section ahead was occupied. It was followed by the second section of the train which left the Union station a few minutes later. The automatic signals were standing at "Danger" behind the first section, but the night was slightly foggy and Engineer Julian Hughey of the second section overran them. The last line of defence was the tail lamps on the rear car of the first portion, but these were obscured by the crowd of passengers clustered there. Finally, Flagman William Penrod was busy collecting tickets and Conductor John Routh, probably confident in the protection offered by the automatic signalling, did not order him back to protect the train as he was meant to do after the train had been standing five minutes.

The colliding train was only travelling at about 12 to 15 miles an hour but its engine buried itself in the rear car of the first section as far as the cab windows and yet again, as at Revere and Mays Landing, damage to the boiler of the locomotive meant that the passengers in the rear car, if they had not keen killed directly by the impact, were likely to be scalded to death by searing hot water and steam. By 1 p.m. on 11th October there had been 27 deaths and the final toll was 32. Engineer Hughey and Fireman Rambo jumped from their engine before the collision without reversing their engine or making any other endeavour to prevent the collision. They escaped unhurt. Some of the passengers stood on the platform of the rear car who saw the second train coming were also able to jump off and escape.[372] [373] [374]

[371] The Evening Gazette, Port Jervis, New York, United States, 17th August 1880

[372] Oshkosh Northwestern, Oshkosh, Wisconsin, United States, 11th October 1880

[373] Daily Gazette and Bulletin, Williamsport, Pennsylvania, United States, 12th October 1880

[374] A History of Railroad Accidents, Safety Precautions and Operating Practices by Robert B. Shaw, pgs. 236-237

A coroner's jury found Engineer Hughey, Conductor Routh and Flagman Penrod all guilty of criminal negligence.[375] It has not been possible to trace any report of a trial.

A crowded excursion train bound for the North Carolina state fair was run into from behind by a freight train at Youngsvile, North Carolina, United States, on the Raleigh and Gaston Railroad, on 20th October 1880. Three men were killed and 14 injured, three of them seriously.[376] 'The conductor and engineer of the freight train which telescoped the passenger have been arrested.' [377]

[375] The Marion Daily Star, Marion, Ohio, United States, 20th October 1880
[376] Davenport Daily Gazette, Davenport, Iowa, United States, 21st October 1880
[377] The Helena Independent, Helena, Montana, United States, 22nd October 1880

CHAPTER 5

SIDELONG HEAD-ON DOUBLE AND FOULING COLLISIONS

Trains can come into collision with one another in a variety of ways. Accidents involving one train running into the back of another, trains meeting one another head-on on a single track and trains colliding on rail/rail level crossings are detailed in other chapters. This chapter will deal with other types of collision such as those involving trains running into one another at junctions, meeting head-on on multiple tracks and colliding with vehicles left in sidings which are foul of the running lines.

An early example of a fouling collision occurred on the Great Western Railway on the morning of Saturday 10th November 1838. The 10 a.m. train from London Paddington consisted of the engine *Neptune*, four carriages, two stagecoaches loaded onto flat waggons and a private carriage. It had almost reached the station at Maidenhead, England, at this date the end of the line, when the mishap happened. A driver named Gregory was at Maidenhead engine shed with his locomotive, the *Venus,* waiting to work the 12-noon train up to town. Due to carelessness, he contrived to move his engine to a position not more than three inches from the offside rail of the running line, leaving insufficient clearance for the down train to pass in safety, and the two engines collided 'with great force.' The footboards of the passenger train were torn away, and the *Venus* lost its buffers and suffered considerable damage to its tender. No one was hurt, though the guards of the train, who were just leaving their seats to assist the passengers in alighting, had a narrow escape from being crushed. The newspaper report on the accident added that 'A gentleman passenger loudly complained that the accident had been solely occasioned by neglect, and two of the directors, who happened to be present reproved Gregory for having placed the engine in that dangerous situation.'[378] Whether Driver Gregory escaped with a telling off, or whether he subsequently lost his job, was not stated.

---

[378] The Times, London, 13th November 1838

A further accident on the Great Western Railway, again at the western extremity of the line, was more serious. By Sunday 25th October 1840 the railway had been extended as far as Faringdon Road, England, between Didcot and Swindon. What was described as the down night luggage train, which had left London Paddington the previous Saturday evening, was actually a mixed train which included a passenger carriage, containing at the end of its journey four hardy passengers. It was in the charge of Guard James Marlow and was hauled by the *Fire King* engine, with Driver John Ross and Fireman James G James. The trip from London had gone normally until the train had passed Steventon and was nearing the terminus at 5 a.m., when Guard Marlow realised that the driver had neither shut off steam nor reduced speed at the usual point. The guard shouted to his driver to pull up, as did people waiting at the station, among whom were the famous Isambard Kingdom Brunel, the chief engineer of the company. There was no response from the driver and his engine ploughed into the doors of an engine shed, which was situated at the end of the line, and collided with a waggon loaded with coke within. The *Fire King* ended up overturned in a field beyond the engine shed and its tender and all the vehicles of the train but the last were smashed up. Driver Ross and Guard Marlow were both killed instantly, and the four passengers were all injured to a greater or lesser degree. Fireman James somehow escaped unharmed. The passengers were William Welling, a slepseller(?) and Thomas Kitchen, an earthenware dealer, both of Cheltenham; labourer Robert Edwards, of nearby Fernisham and a drover, Richard Bolhead, of Barnstaple, Devon, who had been to Brentford fair. The labourer suffered a broken thigh and the drover a broken collar bone, the other two were less badly hurt.

Fireman James was the principal witness at the inquest into the deaths of the driver and guard but could offer little explanation of the accident. He testified that Driver Ross was a very steady and sober man and had been acting normally up to the moment of the accident. He (the fireman) had apparently only realised that they were close to the station at the last moment when he had shouted to his driver and jumped to the tender brake. The inquest jury returned verdicts of "Accidental Death." Given the fireman's evidence, and the hour that the collision occurred, it would seem clear that Driver Ross had dozed off at the crucial moment, perhaps suffering a microsleep. According to Fireman James, his driver had been instructing him as to the management of the engine only a minute or so earlier.[379] No question seems to have been raised at the inquest into whether the driver had been working excessive hours.

On 29th July 1844 a passenger train bound for Cette, France, on the Chemin de fer de Montpellier à Cette was accidentally diverted into a siding at 6.15 a.m. just before Villeneuve-lès-Maguelone station. The train collided with some waggons standing in the siding, was derailed and its carriages were thrown down an embankment. Three people died, and four others were injured.[380]

Measures taken to assist a failed train led to a serious collision, also in France, on 1st

[379] Ibid, 27th October 1840

[380] Wikipedia – Liste des accidents ferroviaires en France au XIX siècle

March 1846, on the Chemin de fer de Saint Étienne à Lyon. The engine of the noon passenger train from St Ètienne to Lyons broke down at Vernaison. Messages for assistance were set in both directions, forward to Lyons and back to Givors. An engine detached from another train and sent from Givors arrived first and was coupled to the rear of the failed train, which by then was an hour and a half late. It promptly set off once more and was soon running fast to try and regain at least a small amount of time. As the train was approaching Pierre Bénite tunnel, the crew of its leading engine saw, about 80 metres away, an assisting locomotive which had been despatched from Perrache (Lyons) emerge from the tunnel and coming towards them on the same line. M. Laurent, referred to in *The Times* as the chief driver of the Lyons engine, and his colleagues on the locomotive, had assumed that the passenger train would be waiting for them at the siding at Irigny, and were likewise hurrying there in order to avoid further delay. A violent head-on collision thus ensued in which both locomotives were reportedly destroyed, and the leading carriages of the passenger train were all more or less wrecked, the fifth carriage suffering the most. The number of passengers in the 14-coach train was variously stated as about 200 and about 500 and the number of deaths as between nine and 13, with about 30 more injured. The two engine crews jumped at the last moment and saved their lives but were injured. Reference was made in a newspaper report on the accident to the recent Versailles catastrophe, and it was stated that it was very fortunate that in this case fire had not added to the disaster, especially as six of the carriages had been telescoped into one another.[381] Nevertheless, the accident led to debate both in the Academy of Sciences and in the French Parliament on the dangers of railway travel and the necessity for more regulation of railway working.[382]

Faringdon Road station, England, the location of the 1840 accident described above, coincidentally suffered another collision of a different nature on Monday 21st September 1846. By this date, the Great Western Railway main line had been opened through to Bristol, and Faringdon Road had become just another wayside station. The 4.45 p.m. down passenger train from London Paddington on this day consisted of the *Elk* engine, seven carriages and two luggage vans in the charge of Guard Simon Cosgrave, with Driver Robert Patterson. There had been a fair held at Reading and as a result the train was very full when it left that station. All went well until it reached Faringdon Road, at which point it was running seven minutes late, but as it was passing under a bridge, close to the station, Guard Cosgrave felt a violent concussion and was thrown to the floor. When recovered himself, he found that the luggage van adjoining the tender had been forced into the following second-class carriage. Two passengers, Paul Sargeant, 'a respectable farmer,' and James Bonner, a mason, were fatally hurt in the accident. The lengthy inquest into the deaths of the two men was unable to work out how the accident had happened, but its jury returned a verdict of 'Accidental Death.' [383]A brief official report on the accident concluded that the engine and tender had become detached from the vehicles of the train and the tender brake having been applied the two portions of the train had then come together, shattering the carriage.[384]

381 The Times, London, England, 7th and 9th March 1846

382 Wikipedia – Liste des accidents ferroviaires en France au XIX siècle

383 The Times, London, England, 24th and 25th September 1846

384 Board of Trade report

The 1846 French accident at Villeneuve-lès-Maguelone was followed by a very similar one the next year at Wolverton, England, on the London and North Western Railway main line. On Saturday 5th June 1847 the 8.45 p.m. down mail train left London Euston at about 8.55 for Liverpool, a long train of 19 assorted carriages. Rather more than a quarter mile short of Wolverton station were two signal posts and a set of facing points leading into a down siding, the signals and points being in charge of a policeman named Barnard Fossey. As the mail train approached the station, the signal stood at clear for the main line. Driver William Manners, in charge of the engine of the mail train, Stephenson 'long-boiler' locomotive No. 153, had no idea that anything was wrong until his engine gave a jerk as it was suddenly diverted into the siding. Policeman Fossey, having cleared the signal for the main line, had walked down to the points and set them for the siding. Driver Manners immediately reversed his engine and applied the brakes as he felt the train going into the siding, and the guards applied their brakes, but with a heavy train and slippery rails the train could not be quickly pulled up and it was still travelling at about 10 miles an hour when it struck some coal waggons in the siding, a full 540 yards from the junction points. The engine, tender and leading vehicles of the train were little damaged in the collision but the lightly loaded second-class carriage situated fifth in the train was lifted on end and the sixth carriage, also second-class, was forced under it. Seven passengers in the sixth carriage, all men, lost their lives, and four other passengers were injured. Driver Manners and his fireman, Edward Bradbury, escaped unhurt.[385]

The coroner's jury at the inquest found Policeman Fossey guilty of manslaughter. He was unable to explain his mistake, but it transpired that just the previous Monday, the 31st May, he had done the same thing, diverting an excursion train into the siding. In this case, the train had only been travelling very slowly and no accident had resulted. For this misdemeanour he had been fined 1s and ordered to be severely reprimanded, but the reprimand had not been administered at the time of the accident. The coroner's jury considered that Fossey should have been immediately dismissed after the first incident and were also critical of the amount of brake power available on the train.[386] The interlocking of points and signals, so that the signal could not be cleared for the man line with the points set for the siding, was of course still some years in the future.

Captain J.L.A. Simmons, who investigated the accident for Board of Trade, pointed out that the consequences of the collision would have been much less serious had the heights of the buffers of the fifth and sixth carriages not been different, so that one had overridden the other, and suggested that the dimensions of the buffers of railway companies which exchanged traffic should be standardised. The accident would also not have occurred without the presence of facing points, but he conceded that on busy lines of railway these were to some extent unavoidable.[387]

On 10th May 1848 it was the Great Western Railway which was in trouble again, and the accident happened only a few miles down the line from Faringdon Road, at

[385] The Times, London, England, 7th, 8th and 12th June 1847

[386] Ibid, 16th June 1846

[387] Board of Trade report

Shrivenham, England. *The Times* newspaper in London reported that on the evening of that day there had been concern in the capital at the non-arrival at Paddington station of the up noon express from Exeter. It should have come in at 4.30 p.m., but up until 6 p.m. there was no news of it. Then, shortly after that hour, a message forwarded by the electric telegraph from Slough spoke of a disastrous collision at Shrivenham, and the worst fears of those awaiting its arrival were confirmed.

The up express had left Exeter 13 minutes late due to delays on the South Devon Railway, and consisted from there of the engine *Sultan* and tender, and five carriages. The attaching of a sixth carriage at Bristol Temple Meads cost another four minutes and the slipping of the engine in Box tunnel cost more time, so that it was 21 minutes late leaving Swindon, from where it was made up of a luggage van, three second-class and three first-class carriages. It was timed to run at between 45 and 50 miles an hour, a higher average speed than that of any other train in the country.

At the time of the accident, Shrivenham station had a siding on the up side passing through a goods shed and equipped with a waggon turntable serving two short spurs situated at right angles to the main siding. The station was protected by standard disc and crossbar signals, and the regulations stipulated that these were to be maintained at 'Danger' until a train was due, when, if the line was clear, they should be altered to 'All right.' The up express was due at 3.03 p.m. and Policeman Pargetter duly placed the up-station signal at 'All right' for it just before then and crossed to the down line to attend to a down train. After the down train had departed at 3.17 p.m. the two station porters, James Waybury and William Willoughby, proceeded to shift a horse-box and a cattle waggon along the siding clear of the waggon turntable to move another waggon which was in one of the short sidings via the turntable to a position where it could be conveniently unloaded. In doing so, because of the short distance between the turntable and the up main line, the cattle truck and horse-box were necessarily placed foul of the latter. It had apparently not occurred to the two porters that the up express had not yet passed, and they had failed to notice that the signal was standing off for it, while the policeman had failed to notice that the porters had fouled the running line.

Misled by the station signal standing at 'All right' and by a hand signal from Policeman Pargetter, Driver Roscoe of the express only whistled for his guard's brakes when he saw the obstruction, but by then his train was only 150 yards away and the train was travelling at 55 to 60 miles an hour, so very little could be done to avert disaster. The *Sultan* struck the cattle waggon and horse-box and forced them against the station platform, and the remains of the smashed trucks and the masonry of the platform were thrown against the carriages of the express. The engine was little damaged in the collision and the damage to the carriages was confined to the parts above the level of the platform. The second-class carriage immediately behind the luggage van suffered the most, its near-side, seats and partitions being completely crushed. Four passengers were killed outright in the accident and two more died later, Captain Blair, R.N., lingering until Thursday 18th; about a dozen other passengers were injured.

The coroner's jury at the inquest into the accident found a verdict of 'Manslaughter' against the two porters, who were committed to Reading Gaol to await trial, but do not seem to have thought that Policeman Pargetter was as much to blame, though *The Times* reported that he had been suspended by the railway company.[388] Captain J.L.A. Simmons pointed out that Pargetter had not only cleared the station signal at 3 p.m. but had furthermore given an 'All right' hand signal to Driver Roscoe. The policeman had alleged at the inquest that the two waggons had only been placed foul of the running line a moment before the collision but from other evidence this was clearly untrue. It was his especial duty to see that the line was kept clear for the late-running express, and Captain Simmons considered that his negligence had significantly contributed to the accident. The porters had no printed instructions to guide them and had been verbally instructed by the Shrivenham station clerk, who himself did not clearly understand the working of the signals. Written instructions should be provided, the inspecting officer stated, and the porters should be obliged to obtain permission from the signalman before fouling the main line. The turntable was to be re-sited and a stop block should be provided to protect the running lines. The up and down station signals were good and efficient but should be placed closer together, so as to lighten the load of the signalman, who also had to operate the level crossing gates, while the accident emphasised the importance of trains, especially expresses, running with regularity.[389]

For our next accident we travel for the first time in this chapter to the United States, to Salem, Massachusetts. In early November 1848 a presidential election was being held in the United States and the accident on Friday 3rd November involved trains carrying supporters of the two rival parties, the Democrats and the Whigs. Just outside of Salem on the main line of the Eastern Railroad to Lynn and Boston was the junction with a short branch to Marblehead. The trains involved were a six-coach special carrying Whig party supporters home to Salem from a torchlight procession and speech by Daniel Webster at Lynn, and a two-coach special filled with Democratic Party adherents returning to Marblehead from Salem, where they had gone to hear General Caleb Cushing, the Democratic candidate for governor.[390] When the outgoing Marblehead train reached the junction with the branch, its engineer found that the facing points were wrongly set and it was unexpectedly continuing along the main line towards Lynn instead of down the branch. Before the engineer could pull his train up, it came into a violent head-on collision with the special from Lynn. Both trains were travelling at some speed and the two engines were reportedly almost destroyed by the shock, while the heavier Lynn train overwhelmed the lighter Marblehead train, the two cars of which were 'thrown into a heap and broken to pieces.' Five passengers in the Marblehead train were killed outright and six badly hurt, one of whom died later. The engineer, fireman and conductor of the Marblehead train were injured, but not seriously. The collision occurred just after midnight. Following the accident, the Whig Party abandoned a proposed torchlight procession to be held in Salem on the Saturday.[391] The Whigs went on to win the presidential election, Whig General Zachary Taylor defeating Democratic Party Senator Lewis Cass to become the 12th leader of the country.[392]

388 The Times, London, England, 11th, 12th, 13th, 15th and 20th May 1848

389 Board of Trade report

390 A History of Railroad Accidents, Safety Precautions and Operating Practices by Robert B Shaw, pg. 65

391 Boston Post, Boston, Massachusetts, United States, 4th November 1848

392 Wikipedia – United States presidential election 1848

A rash excursion by a party of young people led to a tragedy on the evening of 6th May 1852 near Pierrpont Manor, New York, United States on the Rome and Watertown Railroad. Assuming that the last train of the day had gone, nine of them, two married couples and five young women, decided that it would be fun to take a ride along the line from Pierrpont Manor to Sandy Creek on a hand car, disregarding the warning of the stationmaster not to do so as a special train or engine might be running. They had only covered half a mile and were rounding a curve when they were met by an engine and tender backing up from Sandy Creek. There was little warning but one of the men managed to seize his wife and the other his wife and one of the young ladies and they jumped from the car. The other four young ladies, perhaps paralysed by fear, remained on the car; three of them, Miss Wright, of Oswego, Miss Williamson and Miss Mary Allen were thrown under the tender and crushed to death and the fourth, Miss Emily Andrus of Pierrpont Manor, received a probably fatal head injury. After the accident, Engineer D. Smith regretted that he had postponed placing a light on the tender of his engine before he reached Pierrpont Manor, but clearly the blame for the tragedy lay with the young people themselves.[393]

Burnley, England, was the scene of the next accident to be considered. On the morning of Monday 12th July 1852 the Lancashire and Yorkshire Railway company ran a two monster excursion trains which between them carried many Sunday school children, teachers and friends to York and Goole. The York excursion, which had 45 carriages and 1,000 passengers, left first, at 6 a.m., with two engines and three guards. The Goole excursion, chartered by the Wesleyan Methodists, consisted of 35 carriages carrying 800 passengers and again had two engines but only two guards. The carriages for the two trains had arrived at Burnley the night before and the departure of the morning trains was under the control of the superintendent of Manchester station, assisted by an inspector and five porters. All these railwaymen left on the two excursions and for the return of the trains only one porter, John Parker, and the night watchman, William Grant, were assigned to the station.

The approach to Burnley from the Goole direction fell very steeply for the last four miles to the station at an average gradient of about 1 in 70. The line was single until a point about 400 yards from the station, where it became double track at what were known as the 'Top Points.' The station platform itself was on a short siding at the far end of the up line, with a solid masonry wall forming the wing of a bridge at the end of the siding. Beyond the station platform, the main line became single track again and continued to join a branch of the East Lancashire Railway. As the station platform could only accommodate six carriages, the passengers had boarded the two excursion trains in the morning while they were drawn up on the main line. On the return of the York excursion at about 10.30 p.m., Nightwatchman Grant was stationed at the 'Top Points' and, in accordance with Parker's orders, directed the drivers to uncouple their engines there and take them down the straight line, allowing the carriages to descend by gravity down the loop line leading to the station platform. The guards of the train, however, who were theoretically bound to obey Parker's orders, wisely used their discretion and instead the whole train was allowed to run 140 yards farther down the straight line, and its passengers alighted safely and went home. When the Goole train returned an hour later its engines were detached at the 'Top Points' and brought forward on the straight line as

[393] Brooklyn Daily Eagle, New York, United States, 12th May 1852

per Parker's instructions, and Grant having checked with the guards that they had sufficient brake power to control the carriages on the falling gradient, they were allowed to run into the up (loop) line, with the idea that they would regain the main line at the East Lancashire points at the end of the double track and continue into the East Lancashire siding. The points at the junction with the station siding were situated about 140 yards from the buffers at the ends of the platform. They were weighted to stand open for the platform line unless someone was at them to hold them open for the East Lancashire siding. Porter Parker, having assigned Nightwatchman Grant to the 'Top Points,' and being himself occupied in giving hand signals to the train, had enlisted a non-railwayman friend named James Crabtree, a calico printer, to attend to the East Lancashire points for the two trains. Crabtree in turn, before the Goole train arrived, left the points in charge of another non-railwayman named Thomas Bridge, a blacksmith, who was assisting in collecting the tickets. Bridge held the points as directed until, before the carriages of the Goole train reached them, the engines came down the straight line and a driver shouted for him to turn the engines into the shed. Keen to oblige, he let go the handle of the East Lancashire points to do so, the weighted points reset themselves for the platform line, and before he had got back the carriages were past the points. The guards of the train applied their brakes hard when they realised what was happening but could not stop the carriages in time. The collision with the buffer stops only occurred at walking pace but the fifth and sixth carriages reared up on end; the buffers of these carriages were nine inches lower than those of the ordinary carriages.[394] Four excursionists lost their lives in the accident and a few others were injured, some seriously. Those who lost their lives were Daniel Greenwood, aged 34, who left a wife and two children; and three children, Ellen Riley, aged 10, Elizabeth Taylor, aged 10 and Margaret Hartley, aged 12.

Porter Parker was arrested after the accident but the jury at the inquest on the four victims eventually returned a verdict of 'Accidental Death' while criticising the fact that the Goole excursion train had only two guards, the decision to detach the engines, the lack of staff at the station to deal with the returning excursions and the selection of carriages with unequal buffer heights, while, 'in conclusion, the jury beg to submit to the directors of this railway, that in their opinion the siding into which the ordinary arrival-trains run, is fraught with danger, in consequence of being situated at the bottom of an incline, and terminating against a strong dead buffer.' [395]

Captain George Wynne, who investigated the accident for the Board of Trade, ascribed it to several causes. Firstly, the entire want of definite instructions for bringing a train of unusual dimensions down a steep incline and the absence of all arrangements for the safe discharge of 800 passengers who, because of insufficient accommodation at the station, had to be let out onto the side of the line on a dark night.

Secondly, the arrangements being left to an individual who was incompetent to determine the effect of gravity on a train consisting of six or seven times the number of coaches of a normal train.

Thirdly, the utter lack of discipline among the railway staff at Burnley such that they made unauthorised arrangements with irresponsible non-railwaymen, ignorant of railway

[394] Board of Trade report

[395] The Times, London, England, 15th,17th,23rd and 26th July 1852

working, and of the responsibility attached thereto, who by neglecting their duties had caused the accident.

Captain Wynne did not attribute the accident to insufficient brake power, for had the engines not been detached the train could have been brought down safely. Nor did he think John Parker culpable, as he considered that the porter had been anxious to carry out what he thought was the best arrangement. Like the coroner's jury, the inspecting officer was critical of the fact that the Goole train had only two guards. He considered that the arrangement whereby James Crabtree was roped in as a porter and pointsman on alternate Sundays (with the incentive of the tips he received from passengers for carrying luggage) and Porter Parker acted as guard on those days was a result of the fact that Ashton, the only regular guard on the line, was otherwise forced to work every Sunday. The practice, the inspecting officer stated, whereby the engines of ordinary trains were detached as Townley station, three-quarters of a mile from Burnley, had been discontinued after the accident and should not be resumed. Finally, the composing trains of carriages with buffers of different heights was most objectionable and should be discontinued.[396]

On 18th November 1854 a freight train and a wood train collided near Dowagiac, Michigan, United States, on the Michigan Central Railroad. The type of collision involved is not clear from the brief newspaper reports, but the accident was blamed on the engineer of the wood train disregarding signals. He and his fireman jumped off their locomotive and escaped injury; the engine itself was reportedly destroyed in the collision. Adjoining the engine was a box car containing several Irishmen, who were huddled around a stove for warmth. A wood car was driven into this box car by the impact, killing six of the men outright and seriously injuring several others, one of whom had both legs broken. 'One of the persons killed was named Quigley, of New York, and had on his person a certificate of deposit for $700 in one of the Chicago banks. A brother was also seriously hurt.' No one on the freight train was injured. The accident occurred during a severe blizzard which probably impeded the vision of the engineer of the wood train. There was also a foot of snow on the ground.[397] [398]

France suffered another bad accident on 9th September 1855, this time within a kilometre of the Paris Montparnasse terminus on what was now the Chemins de fer de l'Ouest. At about 7.15 p.m. on that day a passenger train from Versailles-Rive-Gauche was approaching the terminus when a wrongly placed set of facing points diverted it into the goods yard at Vaugirard, where it collided with a goods train which was just waiting to depart. One carriage of the passenger train was crushed to pieces and a second greatly damaged. Nine passengers were killed outright and another 17 seriously injured. One of those killed was an infant, only eighteen days old, who died in its mother's arms. The mother escaped with a broken limb and the father, sat opposite, with slight bruises. The Procureur Imperial (public prosecutor) visited the scene of the accident and ordered a full judicial inquiry. A newspaper reporter noted that, 'The left bank of the Versailles Railway has been singularly unfortunate, many dreadful accidents having occurred upon

[396] Board of Trade report

[397] Aurora Weekly Standard, Aurora, Indiana, United States, 23rd November 1854

[398] Goshen Democrat, Goshen, Indiana, United States, 29th November 1854

it within the space of a few years.' [399] [400] [401]

Only three days after the Paris accident, there was trouble across the channel at Reading, England, on the South Eastern Railway's branch from Redhill. A new passenger station had been opened at Reading a few days earlier, closer to the Great Western one, and connected to the old double track by 600 yards of single line. At the junction of the old and new lines a double semaphore signal was provided, one signal controlling trains from the new station and the other access to the main line from the goods and engine sheds, turntable, sidings and old station. The signals and points were worked by porters for the passenger trains, but for the sake of economy the drivers and firemen of goods trains were left to work the signals and points themselves, while shunting operations were carried out without reference to the fixed signals. The signals were also left unlit after 7 p.m.

Like Burney station a couple of years earlier, Reading South Eastern station was understaffed and on the day of the accident, 12th September 1855, an excursion train was run in two parts from Reading to Brighton and two of the staff, the ticket collector and telegraph clerk, and a porter, were taken away to act as guards on the excursion train. To take their place, the stationmaster resorted to using his nephew, a boy of not quite 15, who was expected to work unpaid, plus one extra man to act as a porter. The running of the excursion trains also upset the engine workings, the engine which normally worked the 6.30 p.m. down passenger train from Guildford to Reading being attached instead to one of the excursion trains. To obtain an alternative engine to work this train onward to Reading the locomotive foreman at Redhill telegraphed to Driver Joseph Crossley at Reading 'Come to Guildford to meet 6.30 down train.' The stationmaster's nephew, who was acting as telegraph clerk, took the message at 6.15 p.m., while only he and a porter were at the station, the stationmaster being away having his tea. Neither the telegraph clerk nor the porter was inclined to leave their post to hunt for the driver and the stationmaster advised his nephew to send a cleaner, but the telegraph clerk being unable to find one, the porter eventually undertook the task. The porter did not know where Crossley lived but after enquiries at the goods shed and then at the *Cannon* and *Spotted Dog* public houses was eventually directed to Crossley's house and gave the driver the message. With all this delay, by the time that Crossley reached the engine shed it was already 7.15 p.m., and his engine, which was to work the 7.52 p.m. train from Guildford (the 6.30 p.m. train from London), was still at Reading, 25 miles away. He was thus in a hurry to get going and as his fireman was not available, he asked the cleaner, Jesse Ferguson, to take the regular fireman's place, which Ferguson did. Before they set off, the lamps at the chimney end of the engine and the gauge lamp were lit but Driver Crossley told Cleaner Ferguson that they could light the other lamps on the road to Wokingham, rather than wasting more time doing so before they set off. As the engine was running tender first it therefore set off with no lights at its front end.

In getting from the engine shed to the up main line to Guildford, the locomotive had to be crossed over from the line that it was on via one or the other of two crossovers, but this fact seems to have escaped both Driver Crossley and Cleaner Ferguson, probably

---

[399] The Times, London, England,11th September 1855

[400] The Express, London, England, 11th September 1855

[401] Wikipedia – Liste des accidents ferroviaires en France au XIX siècle

because the driver was preoccupied with tending the fire and filling his oilcan and the cleaner with trimming his lamps. There was of course no signalman at the junction to direct them onto the proper line and they therefore proceeded towards Guildford up the down line, well-aware that the 4.40 down passenger train from London was due but blissfully unaware that there was any danger. The failure of either man, and especially Driver Crossley, to realise that they were running on the wrong line was possibly further explained by the fact that up until a fortnight previously only the down line had been in use in both directions, the up line having been in the hands of the contractors for some months.[402]

The light engine and the passenger train collided on a curve a mile and a half from Reading at a combined speed of perhaps 40 miles an hour. The two engines 'were literally dashed to pieces' and a luggage van and a second-class carriage were entirely demolished, though a first-class carriage was little damaged. Three passengers and Driver Crossley were killed outright, a fourth passenger was fatally injured, and several other passengers, Cleaner Ferguson and Driver William Graham and Fireman Henry Harrison of the passenger train were less badly hurt.

At the inquest into the victims of the accident, the jury found Driver Crossley guilty of manslaughter and that he 'did kill and slay himself.' They recommended the appointment of a regular switchman at the junction.[403]

Captain H.W. Tyler, of the Board of Trade, who investigated the accident very thoroughly, considered that Driver Crossley was by no means the only person to blame for it. According to the company regulations, Driver Crossley should not have left Reading without first communicating with the stationmaster, but Captain Tyler thought that the telegraph message given to Crossley directing him to leave at once and the driver's knowledge that the stationmaster was at tea absolved him in this respect. That Driver Crossley had passed the junction signals at danger was hardly surprising given that they were regularly passed at danger during shunting operations and, furthermore, they were unlit. Likewise, the driver would not have proceeded up the down line had there been a signalman at the junction. As with regards to other matters, the inspecting officer found that the station staff were unjustifiably overworked, even on ordinary days, while on a day such as that of the accident when extra trains were run, instead of the staff being increased men were taken away to work the specials. In addition, Captain Tyler pointed out the deficiency of the arrangements for calling up Driver Crossley, the need for a locomotive foreman at Reading, the delay in the transmission of the telegraph message from the Redhill locomotive foreman to Reading, and the fact that the telegraph instrument and the clock at Reading were at the old and new stations respectively, 300 yards apart, while the train book was at the old and the telegraph message books at the new. The telegraph boy had therefore to constantly run between the two stations and the times of the train signals were often necessarily entered by guesswork.

Continuing a very lengthy report, the inspecting officer pointed out that the state of affairs which had arisen at Reading derived from problems with the management of the company, and a conflict between the general manager and the superintendent. By the time that the report was written, the general manager had been dismissed and the

---

[402] Board of Trade report

[403] The Times, London, England, 14th and 18th September 1855

situation at Reading had been improved with the placing of a permanent signalman at the junction and the addition of two porters at the station.[404]

On 4th May 1859 a major disaster occurred at San Michele, close to Porto Vescovo station near Verona, now part of Italy but at this time in the Kingdom of Lombardy-Venetia and part of the Austro-Hungarian empire. A troop train coming from Venice collided, probably head-on, with a stores and ammunition train; the leading waggon of the latter, loaded with ammunition, was shattered, and an explosion resulted. Strangely it was the carriages in the middle of the troop train which were broken to pieces and 23 soldiers in them were killed and 121 injured. The officers, riding in a carriage immediately behind the engine, were unhurt. The train was carrying troops of the 17th Regiment of the Line, 'which recruits at Laybach, in Carniola,' and a 12-pound battery.[405] 'The inhabitants of San Michele behaved in a most exemplary manner, receiving the unfortunate sufferers into their houses, and tendering them every assistance in their power.'[406]

Russia suffered a collision on its Riga – Dünaburg Railway, at the time still in the course of construction, on Friday 28th September 1860. Although it was not yet complete, it would seem possible that the line was already carrying passengers, for three adults and two children lost their lives in the accident. Five more people were hurt, among them an Englishman, Mr T.C. Watson, C.E. who had to undergo a below-knee amputation of his right leg due to the injuries that he had received. The brief newspaper report added that, 'We are happy to be able to inform his numerous circle of friends that he is progressing very favourably.'[407]

Atherstone, England, on the Trent Valley main line of the London and North Western Railway, was the location of the first major accident involving the shunting of a slower train out of the way of a faster one. At about 1.30 a.m. on 16th November 1860 a special up cattle train from Holyhead to Kings Lynn was permitted to leave Tamworth in front of the up Scotch limited mail from Scotland to London, due there at 1.56 a.m. The Tamworth Night Foreman, William Moorcroft, stated that the cattle train had left there at 1.30 but John Matts, the driver of the cattle train, did not believe they had started until 1.35. Foreman Moorcroft stated that he had intended the cattle train to shunt out of the way of the up postal train at Nuneaton, 13 miles beyond Tamworth. Special notices had been issued by the District Superintendent giving the times of the cattle train at different points: Stafford 12.30 a.m., Tamworth 1.28 a.m., Atherstone 1.47 a.m., Nuneaton 2 a.m. and Rugby 2.30 a.m. The points where the cattle train was to shunt for faster trains were not stipulated in the notices but were left to the discretion of the local stationmasters or their subordinates.

The cattle train slipped for about a mile and a half after leaving Tamworth so that it took twelve minutes to cover the first three miles from there and it was 2.05 a.m. by

---

404 Board of Trade report

405 Beacon and Christian Times, London, England, 11th May 1859

406 The Magnet, London, England, 23rd May 1859

407 The Times, London, England, 18th October 1860

Driver Matts' watch by the time that he reached Atherstone station, still 5¼ miles short of Nuneaton, and he decided that he must shunt there. Driver Matts stated that he had shouted to the Atherstone policeman, Meakin, and the pointsman at the south end of the station, named Jenkins, that he needed to shunt for the mail train. Pointsman Jenkins had followed his train to change the points and, having done so, had signalled for him to set back into the siding. He was just in the process of doing so when he saw the lights of the mail train coming and a violent collision shortly occurred, the effects of which were disastrous. Nine drovers, who had been riding in a brake van at the rear of the cattle train and Fireman James Cherry of the mail train, lost their lives.[408]

The up line at Atherstone was protected by a semaphore stop signal placed at the north (Tamworth) end of the station platform close to a public road level crossing (with the former Watling Street) and by an auxiliary or distant signal placed 433 yards in the rear of the main signal. The two signals were worked by levers placed close together and under the control of the policeman. The main signal was gas lit with a revolving lamp casing to show white ('All Right') or red ('Danger') while the auxiliary signal was oil lit with spectacle glasses to show the same two indications. 266 yards north of the auxiliary signal was Merivale's Crossing, and 626 yards of Merivale's was Whittington Crossing. Each level crossing was worked by a crossing keeper and provided with semaphore signals for daytime use and lamps at night, while a quarter mile north of Whittington Crossing was its own up distant signal. The gates of Merivale's and Whittington Crossings were left locked at night while the crossing keepers were off duty, but the lamps of the crossing signals were left burning. The lamp of the up main signal could be seen, even in unfavourable conditions, from 60 yards north of Whittington Crossing, and that of the auxiliary signal from 200 yards further back.

Policeman Meakin stated that he had replaced the auxiliary signal to 'Danger' as soon as the cattle train had passed it and, likewise, the main signal as soon as the cattle train had passed that. He had not been able to make out what Driver Matts had shouted to him, but he then heard the noise of the approaching mail train and shouted to the pointsman that the mail train was coming, whereupon Jenkins had told him that the cattle train was going to shunt there. He could see the auxiliary signal lamp burning at that point, but the mail train ran past it with steam on and he stepped into the four-foot of the up line to wave a red hand lamp at it. Steam was still on when the train was 20 yards from him and he shouted to its driver that 'he would be into them,' but the driver appeared to take no notice. The train passed him travelling at its ordinary speed of 40 or 45 miles an hour and the collision ensued. After the collision, Pointsman Jenkins assured him that the down line was being protected and he (Meakin) went back to protect the up line. When he got to the auxiliary signal it was still burning, showing a red light towards Tamworth. He put down four or five fog signals between his post and Merivale Crossing and thereabouts met the following Manchester mail train, showed it a red light and it was stopped. He then continued to the two level crossings where he roused the crossing keepers and the signals at the crossings were placed at 'Danger.' He added that the light of the auxiliary signal had gone out two or three times during the previous 12 months in frosty weather, when he had sent out a man with a lamp to a point in the rear of the signal, it being no part of his duty to relight the signal lamp. Pointsman Jenkins confirmed Meakin's evidence.

---

408 Board of Trade report

In charge of the engine of the mail train was Driver Edward Barber, described by Mr McConnell, the London and North Western Railway Southern Division Locomotive Superintendent, as a first-rate driver. Driver Barber said that his train had left Stafford right on time and had stopped as usual three minutes at Tamworth for water. Nothing had been said there about the cattle train being in front of them, though they were generally given such information, and they had left Tamworth a minute before time at 1.58. Approaching Atherstone there had been no light showing at the auxiliary signal, while that at Merivale's Crossing and the stop signal at the station were both showing white lights. When he was about 40 yards from the station, the main signal was suddenly turned to red and he saw a red handsignal being displayed, whereupon his fireman, the deceased James Cherry, had applied the brakes and he had reversed his engine, but had not whistled for the guard's brakes. He then saw the three red lights at the end of the cattle train, but it was too late to stop, and the collision took place. His engine crushed the rear guard's van, the van in which the drovers were sleeping and a cattle waggon and then fell over on its left side down the embankment, burying his fireman. Head Guard Charles Garlick and Underguard James Porter of the mail train likewise stated that they had not seen any light showing at the auxiliary signal, but they could not swear that there was none. Driver Tinkler and Fireman Watson also stated that the light of the auxiliary signal had been out, and the two guards of the train said that they had not seen it.

In view of the conflicting testimonies of the different witnesses, the Board of Trade inspecting officer, Colonel W. Yolland, had some difficulty in deciding who was to blame for the disaster. He pointed out that the evidence of the auxiliary signal having been lit was confined to the statement of Policeman Meakin, while the fact that the main signal was at 'Danger' was supported by the testimonies of the policeman, the pointsman, Jenkins, Head Guard Garlick of the mail train and of an independent witness, a nurseryman named Thomas Robinson, who lived near the railway and being ill had been up at the time of the accident, and who saw the light of the main signal change to red five minutes before the collision. Colonel Yolland stated that it would have been very hard to decide where the truth lay had not the crew of the Manchester mail train not testified that the auxiliary signal light had still been out when they had returned to Atherstone at about 4 a.m., after having been sent to Tamworth for assistance. A great number of people, having no interest in screening Policeman Meakin, had testified that the signal had then been showing a good light, among them the landlord of the White Hart Inn at Atherstone, the Merivale's and Whittington gatekeepers, an Atherstone porter, Incline Brakesman Tebbatt of Tamworth, Mr Cosgrove, the Tamworth stationmaster and railway police inspector Cole. The inspecting officer decided that he must therefore reject the statements of the Manchester enginemen altogether and regard that of Driver Barber with the utmost suspicion. After careful consideration, Colonel Yolland decided, he had formed the opinion that both the main and auxiliary signals at Atherstone had been lighted and had been placed at 'Danger' promptly behind the cattle train. Even were it admitted that the auxiliary signal was out, it was clear that the main signal was burning, and the accident must therefore have been due to the neglect of Driver Barber, whom he had no doubt had taken for granted that he had a clear line to Rugby, and probably did not react until he saw the three rear lights of the cattle train.[409]

Turning to other causes of the accident, the inspecting officer considered that Foreman

---

[409] Ibid.

Moorcroft had shown great neglect in not informing Driver Barber of the preceding cattle train. The foreman pleaded that he had given no orders for the mail train to start, but that it had been started by the guard. It was also very desirable that the directors of the company should extend the use of the electric telegraph, at that time already in use between London and Rugby, as a prelude to working the line on the absolute block system, thus keeping an interval of space between trains.

Colonel Yolland went on to mention the insufficiency of brake power on the mail train and the lack of any means of communication between guards and engine drivers. In addition, drivers should be instructed to treat the absence of any signal where one should be exhibited as a 'Danger' signal and the up distant (auxiliary) signal should be re-sited further out, preferably at Merivale's Crossing. Finally, he noted that the Atherstone stationmaster had not been at the station on the night of the accident and wrote 'I have no doubt that the directors of the company will specially inquire into the circumstances connected with his absence.' [410]

The jury at the inquest on the victims of the disaster found Foreman Moorcroft at fault for not telling Driver Barber of the cattle train and Driver Barber at fault for failing to observe the main signal at Atherstone but returned a verdict of 'Accidental Death.' The jury recommended that the company should not permit its officers to despatch fast trains so close behind slow ones, that the drover's vans should be placed in front of the cattle trucks rather than behind them, that additional signals should be erected at Atherstone and that there should be telegraph communication between Atherstone, Tamworth and Nuneaton both day and night.[411] One cannot but help think that Driver Barber was lucky not to have had a manslaughter verdict returned against him.

Rather less than a year after the Atherstone collision, on Monday 2nd September 1861, another collision with a shunting train occurred at Kentish Town, England, on the Hampstead Junction Railway in north London, resulted in multiple fatalities. Coming only a few days after the Clayton Tunnel disaster, the accident gave rise to what *The Times* referred to as a 'painful excitement in the metropolis.' The collision involved an up-excursion train, returning from Kew to Bow, and a train of empty ballast waggons in the process of crossing from the up line to the down line. The excursion train engine and six of its carriages were thrown off of the line by the impact and three of the carriages were precipitated over the wing-wall of an underbridge and fell into the road below. Fifteen passengers and one railwayman were killed or fatally injured, twenty more people were badly hurt and no less than 301 were shaken or bruised.

The Hampstead Junction Railway was five miles long and connected at its western end with the North and South Western Junction Railway and at its eastern end with the North London Railway. It was leased by the London and North Western Railway, who supplied all the stationmasters, signalmen and so on but the North London Railway supplied the train crews. At Kentish Town station there were stop and distant signals, all worked from the up platform. The up distant signal was 473 yards from the platform, placed on the same post as the Hampstead Heath down distant signal. At the east (Camden Town) end of the station extensive coal sidings had been under construction for

[410] Ibid.

[411] The Times, London, England, 19th and 26 November 1860

some months, and trains of ballast for them were being brought from Bushey, coming onto the Hampstead Junction line at Willesden Junction. Traffic over the Hampstead Junction Railway was worked by the electric telegraph between Bow and Willesden Junction, with separate instruments for the up and down lines, and a speaking instrument. Unless the line had been reported clear by telegraph from the station in advance, no train was permitted to leave a station without being cautioned. On the North and South Western Junction Railway, however, there was no telegraph working, though trains were not permitted to leave Kew without the sanction of the stationmaster.

The station staff at Kentish Town consisted of the stationmaster and two men who acted as both signalmen and porters. One of the men was supposed to come on duty at 7.30 a.m. to work until 7 p.m., the other starting at 9 a.m. and working until 10 p.m. or until the last train had gone. The men were to work alternate early and late turns. In practice, the men had altered the arrangement for their own convenience, so that they might have alternate short and long days, one man working from 7.30 a.m. until 10 p.m. and the other from 9 a.m. until 7 p.m. or until the passing of the 6.35 p.m. up regular train from Kew, due at 6.57.

On the day of the accident, a young man named Henry Raynor was working the long shift and at 7 p.m. he relieved his colleague, the more experienced George Fessey, who went home. Before he left, Fessey had put the up and down signals to danger to cover trains which had just departed in either direction. A ballast train of 19 waggons and a brake van had meanwhile arrived at Kentish Town from Bushey at 5.30 p.m. and had been put into the sidings for unloading. The ballast train driver, Charles Perkins, took his engine to Camden Town for water and arrived back at the sidings at 7.9 p.m. He then attached the brake van to the rear of the ballast train and after the 6.35 regular train had departed, he had whistled for permission from the signalman to come out onto the up line. He asserted that he had received a wave from Signalman Raynor to come out and all the station signals had been at danger. He had brought the waggons and van out onto the up line, had run around his train via the two mains crossovers, and was just about to draw his train over onto the down line, prior to returning to Bushey, when his engine was run into head-on by that of the up-excursion train, the running of which none of the crew of the ballast train was aware.[412]

There were in fact five excursion trains put on that day by the North London Railway, mainly for the benefit of company employees, their friends and relatives. The return trains were intended to leave Kew between 8 and 9.30 p.m., but some of the cards issued to the excursionists stated wrongly that they must return by 7 p.m. A large number, therefore, appeared at Kew station before that time. Some left on the regular 6.35 p.m. train but after that the station was becoming crowded with excursionists, and the next regular train was not due to leave until 7.35 p.m. There were empty carriages waiting in the sidings, so, exercising his discretion, Stationmaster James Morse elected to send the first return excursion train off at 7 p.m., an hour early and 25 minutes after the regular train, leaving ample time for the latter train to reach Willesden Junction, beyond which it should have been protected by the telegraph. The regular train passed Hampstead Heath at 7.02 p.m., Kentish Town at about 7.04 and reached Camden Road at 7.06. The excursion train stopped for three minutes at Edgware Road, from where it was telegraphed forward to Finchley Road and thence to Hampstead Heath. The Hampstead

[412] Board of Trade report

Heath signalman having received the 'Line Clear' signal from Kentish Town for the 6.35 regular train the excursion train was allowed to pass there without being cautioned at between 7.12 and 7.15. Signalman Fessey stated that he had not cleared back the 6.35 train before he left, that he had cautioned Raynor to have all his signals at danger before letting out the ballast train, and that it was not normal practice there to give the blocking back signal before obstructing the line with a shunting move unless the weather was unfavourable.

Driver George Scott on the engine of the excursion train stated that the Kentish Town up distant and home signals had both been at 'All Right' for him. He had whistled as he approached the station and had then seen the engine of the ballast train, but due to the curvature of the line could not determine whether it was on the up or the down line. No danger signal of any kind was shown to him but being uncertain as to what the ballast train was doing there, he had sounded the whistle and told his fireman, George Bolton (who was one of those killed) to apply the brakes. He had only realised that the ballast engine was on his line when it was about 20 yards away and only heard it whistle at the last moment. He was only running at 15 or 16 miles an hour and thought that if the ballast train had been set back down the incline a collision could have been averted. Driver Scott suffered a broken arm in the accident.

Guard Joseph Brydges of the excursion train had riding with him in his van two North London Railway signalmen, Charles Cox and George Woodley. The guard said that he saw no red light at Kentish Town, only a white one (on the engine of the ballast train) but one of the signalmen had exclaimed, "Good God, Joe, there is something with that engine across our line!" whereupon he had put on the brake, but too late to be of much use. The guard and two signalmen were all badly injured, Signalman Cox fatally so.[413]

At the inquest, Signalman Raynor said that he was unable to remember many key points connected with the accident, but in a statement made to his superiors on the night of the accident he admitted having let the engine of the ballast train run round its train and seeing that it was then ready to go. Just at that moment, Hampstead Heath had telegraphed the excursion train 'On Line' and he had gone outside and cleared his up signals. Seeing, then, that the ballast train was still on the up line he had thrown his signals back to danger, and he alleged that they had only been standing clear for a moment or so. He had also telegraphed the excursion train forward to Kentish Town Junction. He alleged that it had been travelling very fast. The collision had occurred between 7.15 or 7.18 p.m.

Colonel W. Yolland stated that, considering the facts, he had no difficulty in concluding that the direct cause of the accident was the neglect of Signalman Raynor in clearing his signals for the excursion train while the empty ballast train was being shunted from the sidings to the down line. He did not think that it had been clearly proved that Raynor gave the crew of the ballast train permission to leave the sidings, but that as the 6.35 train had passed, the station signals had been at danger and as they had had no knowledge of the early running of the excursion train they might have left the siding without permission, and Raynor had not shown them a red flag to stop them.

The inspecting officer made a number of recommendations.

---

[413] Ibid.

Firstly, the telegraph system should be extended to Kew, and the North London system or obtaining 'Line Clear' for each train should be adopted in place of the less safe London and North Western system of merely cautioning following trains if the line was blocked.

Secondly, a more experienced signalman than Raynor, 19 years of age and paid 14s or 15s a week, should be employed.

Thirdly, sidings such as that near Kentish Town station should be provided with their own signals and signalman.

Fourthly, that no excursion or special or excursion train should be allowed to run at other than the booked time, unless the previous regular train was carrying a special tailboard or lamp, as per the procedure on the London and North Western Railway.

Finally, that excursion trains should be provided with another guard with a set of continuous brakes at their rear.[414]

At the inquest, the jury were highly critical of the company not employing more experienced signalmen, considered that it was highly improper for stationmasters to start special trains either before or after their booked times, and that shunting operations should be protected by the telegraph as well as by outdoor signals. They then, somewhat reluctantly, returned a verdict of manslaughter against Signalman Henry Raynor, adding a recommendation for mercy on the grounds of his youth, but the coroner informed them that such a recommendation was a matter for another place.[415]

Another collision occurred on the outskirts of Paris, France, at 10 p.m. on the evening of Thursday 26th September 1861. This time the accident happened on the Chemin de fer du Nord, at the junction of the line from Soissons with the main line to the north, just outside of Paris Nord station. A suburban train which had left Paris at 9.50 for Dammartin was involved in a sidelong collision on the junction with an express coming from Lille, and five of the passengers on the suburban train were killed, three seriously injured and four slightly injured. The boiler of one of the engines reportedly exploded following the collision and its driver was thrown into the air but landed on the track not much hurt. The Amiens train was little damaged. After the accident, some of the injured passengers were brought to Paris and others were taken to the Lariboissière Hospital. A judicial inquiry was opened into the accident.[416]

Scanty reports in the British press spoke of a collision between two trains 'on the railway to Sienna' in Italy on or about 20th April 1862. One of the trains, containing infantry soldiers, was damaged and there were 'about 100 dead and wounded.'[417]

No further major collisions of the type dealt with in this chapter happened until early in 1865, when what was then the British colony of Ceylon experienced what was aptly

[414] Ibid.

[415] The Times, London, England, 4th,5th,6th,7th,9th,10th,11th and 12th September 1861

[416] Ibid, 28th and 30th September 1861

[417] Ibid, 22nd April 1862

described as a 'fearful railway accident,' on the first railway to be opened in the colony, the line being at that time still only partly completed. On the afternoon of Saturday 7th January 1865, a clerk named Streeramaloo Naido set off from Ragame (contemporary spelling – now Ragama) on a trolley, pushed along the line by coolies, to the house of George Ingham, one of the engineers of the new railway, who resided in a small dwelling close to the line at Gonegittote, between seven and eight miles from the Colombo Marendahn terminus. Engineer Ingham had written to the clerk that morning asking him to come to Gonegittote, bringing sufficient money with him to pay the coolie's wages. Meanwhile, at about 5.30 p.m., a train had set off from Colombo for Ragame carrying Ingham and three other engineers named Manning, Harding and Hodges, an overseer named De Neys and 128 coolies, who were mostly Cingalese, riding in sixteen trucks which were being propelled by the engine. The clerk had left the trolley on the line 300 or 400 yards on the other side of Ingham's house, awaiting the train from Colombo, which he expected to stop there, but instead it kept going and collided with the trolley. The accident occurred at a point where the line was on a curve, which meant that the trolley could only be seen from about 20 yards away, at a point where the railway ran on an embankment 15 feet high with swampy fields on either side. Oozy mud in the swamp was reportedly fifteen feet deep and covered by a layer of water. The train was travelling at 20 miles an hour and the leading truck was almost smashed to pieces. The momentum of the train doubled the second truck up against the remains of the first, the third was thrown upon it and so on down the train, until all but three of the sixteen trucks had been piled up in a confused mass and turned upside down into the swamp, burying the coolies underneath them. Most had no chance to escape, though Engineer Harding, who had been riding in the fourth truck, managed to save himself by jumping back from truck to truck until he was eventually thrown down the embankment into the mud but out of reach of the overturned waggons. The number of people who lost their lives was variously put at between 40 and 80, though the lower figure seemed to be more likely. Engineers Manning, Harding and Hodges and overseer De Neys escaped with their lives but Engineer Ingham was among the fatalities. He was a Scotchman and left a wife and four children. That more lives were not lost was due to the promptitude of the driver of the train, who as soon as he saw the danger reversed his engine, so that it and the three rearmost trucks remained on the line, the coupling breaking between the fourth and third trucks from the rear. Some of the other coolies survived by either jumping or being thrown from the trucks. A number of those who thus escaped reportedly at once left the scene instead of remaining to try and help their less fortunate colleagues.

After the disaster, it was reported that 'Hundreds of women, the wives, sisters, and mothers of the dead and wounded coolies, had surrounded the spot, and were bewailing their misfortune in a piteous manner.' Various senior officials went to the scene of the accident and the assistance of the army was promised to help in the removal of the bodies of the dead and the clearing of the line. An inquest was opened on the spot by M.D. De Saram and later adjourned to Colombo. The verdict of its jury was 'That the deceased persons came to their death by the upsetting of the train, caused by its upsetting against a trolly, used without authority and left by the clerk Streeramaloo Naido, on the railway line. And they beg to recommend that the use of trollies be discontinued, except under more stringent rules; and that telegraphic communication be opened between stations where the train has begun to run.' A warrant on a charge of manslaughter was issued against the clerk, who had absconded after the accident.[418] [419]

Several newspapers stated that an initially relatively minor collision on the Ohio and Mississippi Railroad at Loogootee, Indiana, United States, had ended up as a major disaster due to the negligence of the conductors of the two trains involved. At about 1 a.m. on Thursday 22nd June 1865, the story went that an eastbound freight train collided head-on with a westbound troop train, killing the engineers and firemen of both trains. The American Civil War had ended the month before and the troop train was carrying soldiers of Illinois and Missouri regiments who were returning home. Matters were then reportedly made much worse after the accident by the conductors of the two trains becoming involved in a very heated argument as to who had been to blame. While they were so preoccupied, they had completely forgotten about protecting their trains and as a result a westbound freight train ran into the rear of the stalled troop train. Fifteen soldiers were said to have died in the second collision with 150 hurt, many of whom were expected to succumb to their injuries.[420] However, officers of the railway company subsequently stated that no such accident had taken place, though there had been one at Carlisle, Indiana, on the same day, in which the engineer, fireman and two soldiers were killed, and eight or ten soldiers injured.[421] Whether or not the Loogootee disaster really happened is of course, at this length of time, almost impossible to say.

A mistake by a pointsman led to a collision between an excursion train from Halberstadt and a goods train from Dortmund at Buckau, Prussia, either late on 16th or early on 17th July 1865, Thirteen people were at first stated to have died in the accident but the number of dead was later corrected to five or six, comprising Driver Thiele, Assistant Driver Bothe and three or four passengers. A large number of other people were injured. The pointsman had reportedly left a set of facing points wrongly set so that the passenger train, running at speed, was diverted into the path of the goods train. He was arrested.[422] [423]

Two trains collided near Fort Deposit, Alabama, United States, on the Mobile and Alabama Railroad, on about 18th July 1865. Two locomotives were damaged, three cars smashed, five passengers died, and several passengers were injured. [424] [425]

A collision at Cawnpore (now Kanpur), India, on the East Indian Railway, at about 3 a.m. on 2nd August 1865 was the result of negligence on the part of several railwaymen. The down mail train was still travelling at considerable speed when it arrived at the goods

[418] Overland Ceylon Times, 13th January 1865

[419] Singapore Free Press, 2nd February 1865

[420] Indianapolis Daily State Sentinel, Indianapolis, Indiana, United States, 24th June 1865

[421] Sandusky Daily Commercial Register, Sandusky, Ohio, United States,26th June 1865

[422] The Times, London, England, 18th July 1865

[423] The Standard, London, England, 21st July 1865

[424] Philadelphia Age, Philadelphia, Pennsylvania, 31 July 1865

[425] Galveston Weekly News, Galveston, Texas, United States, 26th July 1865

yard, three or four hundred yards from the passenger station. Here it was diverted into a siding and collided head-on with the engine of a goods train which was being unloaded prior to departure. The two locomotives ended up locked together and three carriages of the mail train and three waggons of the goods train ended up piled up on one another. Six passengers were killed on the spot and a seventh died shortly afterwards. The latter victim was an honorary local magistrate named Kinder Sing and two of the others were his servants. One of those killed was decapitated. A man who had been unloading oil casks from the goods train was crushed and not expected to live. There were only two English passengers on the train, one of whom, Captain Sir William Hamilton, of the Royal Artillery, received severe bruises about his head and body.

A newspaper correspondent complained that the jemadar (a junior official, presumably the person in charge of the station at the time), who ought himself have satisfied himself that the two pointsmen were on the alert, had thought it sufficient to tell a chuprassie (subordinate official) to see to the matter, and the chuprassie in turn had contented himself with a shout from near the passenger station, not bothering to obtain a reply. The Cawnpore magistrate, on the other hand, who immediately inquired into the accident, completely exonerated the jamadar and chuprassie and placed all the blame on the two pointsmen, who were committed on a charge of culpable homicide. They had set the points ready for the goods train to depart and gone back to sleep, presumably forgetting about the mail train.[426]

Frustratingly brief details appeared in the London Times newspaper of an accident on the night of 21st September 1866 on the railway between Odessa and Balta at Kolontaievska station, Russia, about 30 miles from Odessa. A passenger train and a goods train were in collision. Fourteen passengers lost their lives and 51 were injured.[427]

Right at the end of 1866, on Friday 28th December, another collision due to incorrectly set points occurred between Zwickau and Werdau in Saxony. At about 9 a.m. a passenger train from Zwickau to Hof was diverted onto the wrong line and collided head-on with another passenger train running from Werdau towards Chemnitz. Both engines and 23 carriages were wrecked, the leading carriage of one train ended up mounted on top of an engine and its passengers were scalded by escaping steam from the damaged locomotive. Four were killed outright, four fatally injured and another 27 were hurt, some seriously. On 9th and 10th May 1867 an assistant pointsman and the driver of the train from Zwickau stood trial accused of negligence but they were acquitted, as it was unclear, for one thing, who had wrongly set the points. The court case brought to light several deficiencies at Zwickau, above all the lack of staff and the long working hours (cf for example the Reading collision of 1855). The train driver was exonerated because the accident had occurred during a raging blizzard when visibility had been practically nil.[428]

A signalman's error was also responsible for a serious collision on the London and

---

[426] Allen's Indian Mail, London, 27th September 1865

[427] The Times, London, England, c12th October 1866

[428] Eisenbahnunfälle in Sachsen by Claus Schlegel, pgs. 48-9

North Western Railway at Walton Junction, England, on Saturday 29th June 1867. This spot was just over a mile south of Warrington station and the point where the Birkenhead Lancashire and Cheshire Junction Railway diverged from the London and North Western main line, with a facing junction for up (southbound) trains. There was also a further junction, 170 yards south of the above-named, where the Arpley branch of the Warrington and Stockport Railway joined the Birkenhead line by a trailing connection for up trains. Both junctions were under the control of a signal box placed in the fork of the junction between the North Western and Birkenhead lines. The box worked two double-armed semaphore signals situated close by, with associated distant signals, and indicators were provided to show the signalman whether the facing points had properly responded to his levers.

On the morning of the accident, an up-coal train for London was despatched from Warrington at 11.25 a.m. in front of an up-passenger train which should have proceeded it, but which was running late, with instructions to shunt out of the way of the passenger train at Walton Junction. Arriving there, the coal train was sent forward onto the Birkenhead line, where it came to a stand clear of the junction. The passenger train, from Liverpool for Birmingham and London, left Warrington at 11.35, ten minutes after the coal train. Its driver found the Walton Junction up distant signal at danger, whereupon he whistled for the main line and the signals were immediately cleared. He approached the junction at 15 or 20 miles an hour and only realised at the last moment that the points were set to divert his train onto the Birkenhead line on which the coal train was standing. There was only time to close his regulator and hang on as his train crashed into the brake van at the rear of the coal train. Both he and his fireman were knocked unconscious and badly hurt. The collision threw the brake van off the track and more or less wrecked five coal waggons but the brakesman somehow escaped injury. The engine and tender of the passenger train were derailed, the leading brake van was partly overturned, a composite carriage was badly damaged at one end and two third-class carriages were smashed to pieces. Of the 300 passengers on the train, five were killed outright or died within an hour, three more died later and 70 were more or less injured, as well as the driver, fireman and second guard.

A shunter at Walton Junction stated that he had noticed that the points had been left set for the Birkenhead line just as the passenger train was approaching and had shouted to the signalman, John Rowson, "Oh, my God, Rowson, them points!" Signalman Rowson told the Board of Trade inspecting officer, Colonel W. Yolland, that after the coal train had gone onto the Birkenhead line, he had reset the road for the main line before clearing his signals for the passenger train, securing the points lever handle with a stick as usual, and implied that a piece of coal must have fallen between the points blades and held them open. Colonel Yolland did not believe him, as apart from the shunter's statement had an obstruction got into the points it would not have been possible to move the handle of the points lever sufficiently to insert the stick as the signalman had alleged that he had done. It was very clear, the inspecting officer stated, that Signalman Rowson had forgotten to reset the road after admitting the coal train to the Birkenhead line. The coroner's juries at two inquests on the victims thought the same and the signalman was committed to stand trial for manslaughter at the next Chester Assizes.[429]

Considering the question of facing points, Colonel Yolland noted that accidents could

---

429 Board of Trade report

occur at them in three different ways. Trains could become derailed on them because they were not properly closed, signalmen could move them beneath a passing train or, as here, trains could be diverted onto the wrong line at them. As with regards to Walton Junction, Colonel Yolland noted that there had been a previous accident there in 1862, which had been inquired into by Captain Tyler. The latter had made a number of recommendations for the improvement of the signalling at the junction included the lighting of the distant signals with gas, the provision of extra sidings and most crucially the interlocking of facing points and signals so that the signals could not be cleared for one line when the points were set for another, as had happened in the present accident. Nothing, however, had been done to carry out these recommendations in the five years between the two collisions. Had they been carried out the accident which had claimed eight lives could not have occurred.[430]

Signalman Rowson was acquitted of manslaughter at the Chester Assizes on 7th August.[431]

Like the driver involved in the Saxony accident nearly three years earlier, the engineer blamed for the collision at Mast Hope, Pennsylvania, United States on 14th July 1869 was charged with having caused it, but was acquitted at his trial, though for different reasons. The circumstances of the Mast Hope collision were simple, and the collision was clearly the fault of Engineer James Griffin of first extra freight train No. 39, which had received orders from the despatcher to take the siding at Mast Hope to allow two westbound passenger trains, No. 7 and 3, to overtake it.

Conductor Judson Brown told the subsequent inquest on the victims of the accident that his freight train had accordingly taken the siding at 10.32 p.m. Train No. 7 had passed safely at 10.50 or 10.51., two or three minutes late, and shortly afterwards he had heard the whistle of train No. 3, which was 25 or 27 minutes late. At that moment his own train began to move, and he assumed that it was just pulling up to the end of the siding.

Fireman Milton Dean stated that as Engineer Griffin had started away, he had picked up the fire rake and then he had seen No. 3 only four or five car lengths away and had shouted to his engineer, "Griffin, where are you going?" The engineer had not answered. and the fireman had dropped his rake and jumped for his life.[432]

Train No. 3 was the 6.30 p.m. departure from New York. The Erie Railroad express consisted of four sleeping cars, a ladies' car, smoking, postal, baggage and express cars, and was well filled with passengers.[433] Its engineer, Charles Coffee and fireman, Perry Hoyt, had little or no warning of what was about to happen before their engine struck that of the freight train just forward of its tender. The impact threw the express engine to the left down the embankment, the tender was thrown to the right and the postal and express cars piled up against the engine and set on fire by burning coals from its firebox. The baggage car was smashed into the smoking car, driving seats and passengers to the rear end of the latter, where the passengers were trapped by the broken seats and unable to escape the flames which soon spread to these cars. Eight passengers lost their lives in the

---

430 Ibid.

431 The Express, London, England, 8th August 1867

432 Port Jervis Evening Gazette, Port Jervis, New York, United States, 17th July 1869

433 Wellsville Free Press, Wellsville, New York, United States, 21st July 1869

collision; the Rev B.B. Halleck, an elderly Universalist minister; a David Baer; all five members of a family of German emigrants, husband, wife and three children; and an unidentified person, supposedly another German emigrant. Only a handful of passengers were injured. The two engineers and two firemen escaped with cuts and bruises. The Rev. Halleck, who was trapped by his legs, remained conscious as the flames crept up on him and calmly directed his rescuers, but the wreckage held him tightly and they were unable to release him in time.

The accident fell short of being a major disaster, but it attracted great attention at the time, partly because it occurred exactly fifteen months and only two miles from the scene of the Carr's Rock disaster, and partly because it coincided with a 'war' of epic proportions between the different railroad tycoons, Drew, Fisk, Gould and Vanderbilt, for control of the Erie company.[434]

Engineer Griffin was clearly the sole person responsible and he candidly admitted that he had dozed off while waiting for the meets and being awoken by the passing of train No. 7 had set his engine in motion while still half asleep and not really thinking what he was doing. Not surprisingly, the inquest jury found that he alone was responsible for the collision and guilty of manslaughter. He was subsequently tried not for manslaughter but under a state law which made it a misdemeanour for a railway employee to disregard the regulations of the company such as to endanger the lives of the public. His defence lawyer more or less abandoned any hope of getting him acquitted, and Judge George Barrett directed the jury to find him guilty, whereupon against all the evidence they promptly acquitted him! Needless to say, the judge was not at all pleased and addressing the jury told them that he was astonished at their verdict, that it was against the law, against justice and an outrage against humanity, adding that 'I trust that the spirits of the dead, dying, bleeding and burnt victims of Mast Hope will rebuke you as long as you live.' He had, however, to accept the verdict and decided that a prosecution proposal to try the engineer on a charge of manslaughter instead was futile.[435]

The Halle – Nordhausen line of the Madgeburg – Halberstadt Railway of Prussia was constructed as cheaply as possible, and as a result suffered numerous accidents. The most serious was a collision between a troop train and a stationary goods train at Wallhausen, near Sangerhausen, on 9th December 1869. Ten died in the accident and 47 were injured.[436]

France and Prussia had been at war from mid-July 1870 and, by 4th October 1870, France was close to defeat. At the beginning of September, it had lost the Battle of Sedan and its capital, Paris, had been under siege since the middle of that month. To add to its misfortunes, shortly after midnight on 4th October a French troop train was involved in a bad accident at Critot station on the Montérolier – Buchy (Amiens – Rouen) line. Having suffered a defeat at the hands of the Prussians at Motteville, a company of the 20th Foot and Field Battalion was retreating by train towards the River Loire when the train was

434 A History of Railroad Accidents, Safety Precautions and Operating Practices by Robert B Shaw, pgs. 133-4
435 Ibid, pgs. 414-5
436 Eisenbahn-Katastrophen in Deutschland Bd 1 by Hans Joachim Ritzau, pg. 103

accidentally turned into a siding and ran either into the buffer stops or off the end of the siding, if no buffers were provided. Fourteen or fifteen soldiers were killed and over 100 injured, a number supposedly fatally. Happening in wartime and involving a troop train, news of the accident was apparently censored in France, though brief accounts of it appeared in the English press.[437 438]

At about 4.51 or 4.52 p.m. on the afternoon of 26th November 1870 a train of empty goods waggons destined for Rugby and the Midland Railway left the London and North Western Railway sidings at Willesden behind the 4.35 p.m. down passenger train to St Albans. It was running in the place of a Wolverton goods train which was booked to shunt at Harrow station at 5.05 p.m. out of the way of the 5 p.m. down express from London Euston to Liverpool and Manchester, due to pass Harrow at 5.17. The train was in charge of Brakesman John Hubbard, with Driver William Croshaw, and consisted of an engine and tender, 44 empty private owner waggons and a brake van. Between Willesden and Harrow were Brent and Wembley telegraph stations, both of which were only provided with one signal for the down line. Driver Croshaw found the signal at Brent at green. Beyond Brent, it became increasingly foggy, but the driver was able to see the Wembley signal at red from 30 or 40 yards. He stopped his train as soon as he could, but Brakesman Hubbard shortly shouted for him to go on, having received the all clear from the Wembley signalman, Charles Robinson, and he set off again. At Harrow the down distant signal was white or 'All clear' but the main signal was at red and the signalman, Thomas Walker, was waving a red hand lamp from his cabin and calling to him "to shunt." He gradually put on steam to draw his train forward beyond the siding points but then realised that it had become divided. Brakesman Hubbard, meanwhile, had heard someone shouting from the station that his train had come apart and first ran back to warn the signalman and then forward towards the engine. He had recoupled the train, which had divided between the seventh and eighth waggons, and it was just moving forward again to get clear of the siding points when it was run into from behind. The effects of the collision were disastrous; the pilot engine of the colliding train was overturned to the left, the train engine was derailed and ploughed up the ballast and four or five of the leading vehicles piled themselves up against the two locomotives. Six passengers were killed outright, and another died of his injuries; 41 more were injured, some seriously. The driver of the pilot engine was also killed, and his fireman and two guards badly hurt.

Foreman Thomas Smerdon had been taking charge of the station while the stationmaster, John Dawson, was having his tea. Seeing that it was becoming foggy, he had just asked a porter to call the platelayers to act as fogsignalmen when he heard the coupling break and called to Signalman Walker that the train had broken loose. The fog became much worse as the empty train arrived.

Signalman Walker stated that he had replaced his signals behind the empty waggon train and had not given "Line Clear" back to Wembley as the train had not yet gone into the siding. On being told that the goods train was divided, he had sent the "Line Blocked" signal to Wembley in accordance with Rule 58 and had then disconnected the wire to

[437] Wikipedia – Liste des accidents ferroviaires en France au XIX siècle

[438] The Daily News, London, England, 5th October 1870

Wembley so that the needle dropped to the vertical position (Rule 60). After several minutes, Signalman Walker said that he had managed to get hold of Porter Thomas White, who had been occupied calling the platelayers for fogsignalmen, and had told him to run up the line with fog signals.[439]

Porter White said that he had been given some detonators by the Harrow stationmaster and had run up the line. He had time to put two down fairly close together, 330 yards from the tail of the empty waggon train, and then ran forward and tried to put down another, but the express was upon him and he had no time to properly clip it to the rails. He had then waved a red hand lamp at the engine of the express and shouted to the driver of the leading engine as it passed him

Signalman Robinson at Wembley confirmed that he had received the six beats "Line Blocked" signal from Harrow at 5.13 or 5.14 and the needle of his instrument to Harrow had then become vertical, which he had understood as a repetition of the obstruction signal. His semaphore signal was still standing at "Danger" after the normal three-minute interval when according to the rules it could be lowered to "Caution," because it had become foggy. He had been standing outside his cabin waving a red lamp at the express as it approached but its drivers took no notice of either his hand lamp or the fixed signal. He had then belatedly attempted to put down a fog signal, but the express was by then too close. He did not inform the Harrow signalman that the express had run past him.

The Liverpool and Manchester express consisted of two engines and tenders and 19 assorted vehicles, including four brake vans and carriages for Leamington, Birmingham, Shrewsbury, Holyhead, Birkenhead, Liverpool, Manchester and Rochdale. Only two guards, George Gemmell and George Latham, accompanied the train, seated in the brake vans placed fourth and last in position. The pilot engine was in charge of Driver William Shelvey with Fireman John Claridge and the train engine was in charge of Driver Robert Stone with Fireman William Rutter. Driver Shelvey was no longer able to explain his actions of course but his fireman said that he had not seen the Wembley signal, where the fog was very thick, but Driver Shelvey had said nothing about it. He had then been busy breaking up coal when his driver had said, "Look out for the signal, mate," and he had looked out and seen the tail lights of a train in front as they ran over a fog signal. His driver had then reversed his engine and the collision had occurred.[440]

Driver Stone of the train engine stated that according to the rules the driver of the pilot engine was responsible for observing signals, though the driver of the train engine was to assist him as much as possible. He said that he had seen the signal at Brent at "All Clear" but because of the fog and the steam from the pilot engine had missed seeing the Wembley signal, though he saw a white hand lamp in the cabin. He had also missed seeing the Harrow distant signal. A fog signal had exploded just before reaching an overbridge approaching Harrow station, when he had shut off steam, reversed his engine and put steam back on while his fireman had flown to the brake. He had not been concerned about missing the Harrow distant, trusting that Driver Shelvey had seen it. Fireman Rutter confirmed his driver's evidence. The guards of the train were sorting parcels and letters and had not been looking out for signals.

Lieutenant-Colonel C.S. Hutchinson, the Board of Trade inspecting officer, found that

---

[439] Board of Trade report

[440] Ibid.

the primary causes of the accident were a want of caution on the part of Driver Shelvey and a want of judgement on the part of Signalman Robinson at Wembley. He considered that possibly Driver Shelvey had been misled by the white hand lamp in Wembley cabin or by the white back light of a signal. The want of judgement of the signalman was no doubt due to his only having five weeks service; he should not have been employed in such an important position.

As to secondary causes, the inspecting officer considered that one was the breaking of the waggon coupling. He recommended that a test should be introduced to test the strength of couplings, and they should be stamped with the date that the test was made. Another cause was the time taken to recouple the goods train and the delay in sending anyone back to protect it, for which he blamed both Foreman Smerdon and Signalman Walker. Again, the brake power of the express had been very insufficient. Furthermore, every telegraph station should have distant as well as stop signals, backlights in signals close to cabins were unnecessary and care should be taken not to show white lights from cabins which might be mistaken for the indications of fixed signals.

Well ahead of his time, Lt-Col Hutchinson thought that a means could be devised for providing an audible warning at distant signals, perhaps by using two wires. The accident also demonstrated, he said, the fallacy of the time interval system, and the inadequacy of any permissive block system which did not secure an interval of space between following trains. Much had been said about separating passenger and goods trains, as was done by the third up line between Bletchley and London, but a complete separation would not happen for a long time and an absolute block system was a far more useful measure.[441]

The inquest jury placed the blame for the accident on the late Driver Shelvey, while censuring Signalman Robinson. They recommended the separation of passenger and goods trains in the neighbourhood of the metropolis, the introduction of the absolute block system and the reduction of signalmen's hours.[442]

Brockley Whins, England, on the North Eastern Railway, was the location of a serious collision on 6th December 1870. Brockley Whins station was originally on the Brandling Junction Railway, and at the end of 1870 had what Colonel W. Yolland, the Board of Trade inspecting officer, described as an objectionable layout. The line was double track, but the station had only a single side platform to the south of the up line and immediately west of the junction of the lines from South Shields and Sunderland. Down trains accessed the single platform by means of a facing crossover worked by a lever situated between the crossover and the pointsman's box. On the morning of the 6th December a down mineral (coal) train consisting of an engine and tender and 12 waggons left Gateshead at 10.15 a.m. for Tyne Dock. As it approached Brockley Whims station at about 10.40 a.m. its driver, Hodgson Turnbull, found the signals cleared for him to run through the station on the down line. Arriving at the crossover, however, he found the facing points set to divert his train over to the up line. He whistled, reversed his engine and shouted to the pointsman who was stood outside his box but realising that a head-on collision with an up train was unavoidable he and his fireman, William Summers, then

441 Ibid.

442 The Times, London, England, 28th, 29th and 30th November and 1st, 2nd, 3rd,7th, 8th, 14th and 16th December 1870

jumped from their engine.[443]

The up train with which Driver Turnbull's train collided was the 10.30 a.m. up express from Sunderland to Newcastle, booked to run non-stop between Sunderland and Gateshead. It consisted of tank engine No. 150, a brake van, a first-class, two second-class and two composite carriages in that order, in charge of Guard Herbert Tapley, with Driver William Garnett. Like Driver Turnbull, Driver Garnett had seen that the crossover points were standing wrong as he had neared them and had reversed his engine and whistled to attract the attention of the pointsman, but the points were not altered, and the two trains collided about the middle of the crossover while traveling at a combined speed of about 18 miles an hour. The two engines, No. 150 and 350, and the vehicles of the express were badly damaged and four passengers and Guard Tapley lost their lives and 57 passengers and Driver Garnett and his fireman were injured, many very seriously. The passengers killed were Frederick Younge, of London, described as a comedian and manager of the "Caste" company; Henry Richardson, a Sunderland paper manufacturer; William Ogden, a retired Sunderland solicitor and Robert Turnbull of Sunderland, an office clerk.

The inquest into the victims of the accident found that it had been caused by the negligence of Robert Hedley, a platelayer employed as a temporary pointsman, in clearing his signals for the two trains having omitted to reset the facing crossover. The jury recommended that proper signalmen be employed and commented on the impropriety of Goods Shunter John Adamson or any other person being in the pointsman's cabin, potentially distracting him from his duties. The coroner, Mr J.M. Favell, cordially agreed with the jury that no other man should be allowed in the pointsmen's cabins and said that he hoped that the company would devise some means of preventing the practice.[444]

Colonel Yolland was not impressed with the jury's verdict. He wrote that, 'This verdict and the subsequent remarks supply further confirmation, if any were needed, of the fact that coroner's inquests, as generally conducted, are singularly ill-calculated to ascertain the real causes of railway accidents; but they are supposed to be sometimes serviceable, as in this instance, to the railway companies, in concealing the mismanagement of the railway company from the public.' The inspector pointed out that Goods Shunter Adamson had in fact a duty to be in the pointsman's cabin as three goods trains were shortly due at Brockley Whins and it was his duty to advise the pointsman as to how they should be disposed of. He noted that Pointsman Hedley had subsequently stood trial for manslaughter at the Durham Assizes and had, in his opinion, been very properly acquitted.

The inspecting officer declared that it appeared to him that the company's management was wholly to blame for the accident. It was almost 15 years since he had first pointed out the dangers of pointsmen heedlessly placing facing points in the wrong position and more than 14 years since a means had been devised of preventing such accidents. For upwards of ten years, inspecting officers had insisted on the interlocking of points and signals on new lines before they would sanction their opening, in the face of strenuous opposition from important railway companies. They had no power to interfere

---

443 Board of Trade report

444 The Times, London, England, 8th and 9th December 1870

with the arrangements on lines already open. Had the facing crossover at Brockley Whins been interlocked with the signals so that the signals could not be cleared with the points wrongly set the pointsman could not have caused the collision. His fellow inspecting officers, Captain Tyler, in 1863, and Colonel Rich, in 1865, had pointed out the desirability of interlocking points and signals in their reports on accidents, but it was not until a bad accident at Thirsk in May 1869, due to a mistake by a signalman similar to that of Pointsman Hedley, that the North Eastern company had finally actively begun to pursue the interlocking of points and signals.[445]

On 28th January 1871 the French, soundly beaten by the Prussians, signed an armistice at Versailles which suspended the hostilities of the Franco-Prussian war on all fronts. The French Army of the East, however, was excluded from the armistice and its commander, General Clinchant, decided to request internment in Switzerland as the alternative to capitulation to the Prussians. This the Swiss agreed to and the French troops remained in Switzerland until 13th March, when, the peace treaty between France and Prussia having been ratified, they were permitted to return to France, which most did from Geneva by train.[446] One of the trains carrying the returning internees was involved in a major disaster at Colombier, Switzerland, on Wednesday 22nd March 1871; possibly one of three accidents to occur to trains carrying the returning internees, with less serious ones at Kirchdorf and Morges. In the Colombier disaster, a train of 17 carriages, carrying about a thousand soldiers, collided while running at full speed with a coal train standing in a siding. Twenty-four men died and 61 were injured, three carriages being completely crushed. The driver and fireman spent ten days in Pourtalès hospital. The funeral of the victims saw their bodies carried on three hearses, escorted by Neuchâtel troops and attended by Government delegates, with military music and Chopin's funeral march.[447]

On 4th September 1871 the Lille correspondent to *The Times* in London reported that a frightful accident had happened at 10.30 p.m. the evening before, Sunday 3rd September 1871, at Seclin, France, not far from Lille on the Chemin de fer du Nord. A Paris - Lille express, train No. 29, had collided with a local train which was being shunted out of its way but was still foul of the main. The local train, from Douai, was 47 minutes late. Three carriages of the local train were crushed, and many passengers were scalded by steam and hot water escaping from the wrecked locomotive of the express. The number of dead was initially stated in *The Times* as ten, but a second report spoke of only four dead and 71 more or less injured. Another report spoke of nine dead.[448] [449] [450] It was not clear whether there was a signalling error or if the driver of the Lille express had been overrunning signals, though the reports suggested the former.

A mistake by a signalman in a busy box not yet fitted with up-to-date equipment lead to

445 Board of Trade report

446 Webpage – The arrival of Bourbaki's army at Les Verrières by François Bugnion, Article No 311 of the International Review of the Red Cross, published 30th April 1996

447 Webpage – L'emouvant passage des Bourbakis en 1871

448 The Times, London, England, 5th and 7th September 1871

449 The Magnet, London, England, 11th September 1871

450 Wikipedia – Liste des accidents ferroviaires en France au XIX siècle

an accident at Agecroft Sidings, England, on the Lancashire and Yorkshire Railway on 3rd August 1872. The sidings, on the down side of the line, were protected by home and distant signals but there was no block working yet in operation. A coal train arrived at sidings on the morning of that day from Clifton Hall, consisting of an engine and tender, just four waggons and a brake van. The train was backed into the sidings to attach more coal waggons and was then ready to proceed but Signalman Harrison was unable to let the train go for 55 minutes due to the pressure of traffic. Doubtless anxious to get the train away, the signalman finally told Driver Saxon of the coal train to be ready to leave after the Fleetwood to Manchester passenger train, which had just been telegraphed to him from Clifton Junction. As soon as this train had passed, he replaced his signals both ways and immediately showed Driver Saxon a green flag to leave the sidings. Driver Saxon did so, and his engine had just reached the down main line when he saw the engine of the 11.30 a.m. express from Manchester approaching. The driver of the Manchester express, and the signalman saw the danger at about the same time, but the collision was unavoidable. The impact threw both engines off the line and the leading two carriages of the express and some of the goods waggons were broken up. The Board of Trade report spoke of only four fatalities, but *The Times* spoke of the more or less immediate deaths of five people; William Hilton, of Preston; Thomas Holmes and Emma Crossley of Manchester and two unidentified persons, and added that a sixth person, James Cummings, a Hulme shoemaker, had died later of his injuries. About ten other passengers, the fireman of the coal train and the driver and fireman of the express were also injured.

The collision had occurred because Signalman Harrison had been too hasty to signal the goods train to depart after replacing his down signals so that the driver of the express had already passed the distant signal at "Clear," while his view of the home signal was obscured by an overbridge. Ironically, had the signalman waited for the express to pass he would have had almost a full quarter hour free of trains to let the coal train out.

The Board of Trade inspecting officer, Lieutenant-Colonel F.H. Rich had little to say with regard to recommendations but noted that the Lancashire and Yorkshire Railway company had been engaged for some time in providing the Manchester – Bolton line with new signals, locking equipment and telegraph instruments, which should prevent many accidents.[451]

At the inquest on the victims of the accident, the jury severely censured Signalman Harrison for his lack of care but stopped short of a manslaughter verdict, most of them believing that there was not satisfactory evidence of criminal negligence. They also recommended the laying of additional lines of rails.[452]

It was the mistake of a stationmaster, rather than a signalman, which led to a disaster at Kirtlebridge, Scotland, on the Caledonian Railway main line, on 2nd October 1872. At this date Kirtlebridge station was, like many other places in Great Britain at the time, in a sort of transition state between the early days of the railway when there were no signal boxes or signals and a few years later when everything was fully signalled. There was a signal cabin at the north end of the station, at the junction with the Solway Junction line, where the signals and points were interlocked, but sidings and a mains crossover at the

451 Board of Trade report

452 The Times, London, England, 5th, 9th and 12th August 1872

south end of the station were worked by levers on the spot, independently of the signal box and not interlocked with the running signals. There was a single needle telegraph instrument in the station buildings but none in the cabin for the main line, only one for the Solway line. The main line was not yet worked on the block system.

On the morning of the accident, the 9 p.m. down express from London Euston, due to leave Carlisle at 6 a.m., was very late because the line had been blocked by a minor derailment at Grayrigg, on the London and North Western Railway. It finally started from Carlisle at 7.50 a.m. A down goods train which would normally have left Carlisle well behind the express was thus instead in front of it, and having called at Floriston, Gretna and Kirkpatrick had arrived at Kirtlebridge at 7.55 a.m. There were waggons to leave and pick up at Kirtlebridge, plus it had to be shunted onto the up line out of the way of the down express. The driver of the goods train had been told by the Gretna signalman that the express had not left Carlisle at 7.27 a.m., and thus began shunting operations immediately on arrival at Kirtlebridge. Two waggons were first left in a siding on the down side and the train was then crossed over onto the up side and began a series of shunting operations there. The shunting was done under the instructions of the goods guard. No permission was asked from the signalman to let the goods train occupy the up line, but the guard stated that he had checked that the up signals were at "Danger" and they had trusted to the station officials to stop them if they were doing wrong. The signalman, Robert Ramsay, had meanwhile replaced his signals to "Danger" when he saw the goods train arrive to protect it. When the goods train had finished shunting on the down side and had crossed to the up side, he had lowered his down signals to "All Right" for the 6 a.m. passenger express from Carlisle, which he knew had not passed, though he did not know where it was. He considered it a distinct understanding that the goods train would not re-cross to the up line unless his signals were at "Danger." He had no means of preventing the danger that might occur from the goods train commencing to cross at the moment he lowered his signals. His down signals had been standing off for 10 minutes before the down express arrived.

The Kirtlebridge stationmaster, Alexander Currie, had been on the up platform when the goods train arrived. The brakesman of that train informed him that the 6 a.m. passenger train was late, and he had realised for the first time that it had not passed, having heard a train pass while in his office or his house which he had assumed was the 6 a.m., but was actually the preceding down limited mail, which was also running late. He had then given instructions to the goods guard as to what work needed to be done and went to his office to enquire of his telegraph clerk as to the running of the passenger train, but the clerk was not there. Coming outside again, he had seen the goods train setting back along the up line and assuming that there was more work to be down on the down side had operated the points of the mains crossover to put the train across. He had not noticed that the down home signal was standing off for the express and could not see the distant signal. He said that he had received no verbal, written or printed instructions as to how shunting was to be carried out, but that the general practice had been to indicate to the signalman by a flag or lamp that the line was clear for a passenger train to pass and that signals were to be kept at "Danger" until whistled off by the driver of an approaching train. On this occasion he had had no communication with the signalman and had been under the impression that the signalman would not lower his signals until advised that the line was clear. When he worked the crossover, he had forgotten for the moment that the 6 a.m. train had not passed.[453]

The down express consisted of two engines and tenders, a brake van, composite carriage, third-class carriage, composite carriage, third-class carriage, composite carriage, two empty carriage trucks, an empty horse box, a family saloon carriage, a composite carriage, a third-class carriage, a composite carriage, a brake van, a composite carriage, a third-class carriage, a composite carriage and a brake van; eighteen vehicles in all. The driver of the leading (pilot) engine of the express, John Gordon, was one of those killed in the accident, but his fireman, Wilfrid Lawson, stated that they had been approaching Kirtlebridge station travelling at about 40 miles an hour, when he had seen a porter apparently making stop signals. Looking ahead he had then seen some waggons moving through the crossover onto their line and had told his mate, or had intended to, he could not recall which, to "Pull up." He had dropped from the tender onto the ballast as the collision had occurred and had escaped with relatively minor injuries. The driver of the train engine saw nothing, his fireman saw the waggons at the last moment and went for the brake. The leading engine of the express overturned onto its side and the tender mounted the down platform. The second engine stopped nearly in its proper position with its tender reared up against its near side, and the front vehicles of the train were all damaged, especially the second composite carriage and the third-class carriage following it. Ten passengers were killed outright or died from the injuries in the hours that followed. About 15 passengers were injured, and one of them, Evan Kennedy, a shepherd, succumbed to his injuries in Dumfries Infirmary on 22nd October, increasing the number of passenger fatalities in the accident to 11.[454]

Captain H.W. Tyler conducted the Board of Trade investigation into the accident. He noted that the signalman had disobeyed the regulation which directed that the signals at junctions should not be cleared until whistled off by the driver of an approaching train, but that on account of a curve and obstructions he might not be able to see an approaching train, nor would he be able to hear the whistle if the wind was blowing in the wrong direction, and it was important not to further delay an important train which was already very late. The whole onus for the collision, therefore, rested upon the stationmaster, who was reportedly a very hard-working and careful man. The stationmaster was, however, working under the disadvantage of good means of communication with the signalman, and the crossover points, 330 yards from the junction cabin, were worked completely independent of it and not interlocked with the signals.

In summary, Captain Tyler stated that there were four points to be considered in connection with the collision: -

1. The forgetfulness of the stationmaster as to the 6 a.m. train not having passed.
2. The lack of a signal cabin at the south end of the station from which the siding connections and mains crossover could be worked, and interlocked with the signals
3. The want not only of a block system but of notice by telegraph of the overdue passenger train.
4. The unpunctuality of the passenger train.

The inspecting officer stressed that in the working of a railway such as the Caledonian main line, with fast non-stop passenger trains running day and night through stations

[453] Board of Trade report

[454] The Times London, 3rd,4th,5th,7th,8th,10th and 23rd October 1872

where shunting was undertaken, it was essential to have the two systems which Board of Trade inspecting officers had been recommending for many years, namely the absolute block system and the interlocking of points and signals. Under an efficient block system trains might follow one another upon the same line of rails within a few minutes, almost without risk. Under an efficient interlocking system, they could pass through complicated junctions and cross the directions of one another at complicated stations at intervals of a few seconds with a surprising degree of safety. Neither system need cause obstructions or delay in working except through a lack of cabins or of siding accommodation. Regarding the Kirtlebridge accident, it would have been prevented by a system of interlocking alone and had the crossover been under the control of a signalman the stationmaster would have been relieved of the duty, while the efficient working of the block system depended on the signalmen being able to operate the points of sidings, junctions, crossovers and safety points as well as the signals. During 1870, of 122 accidents, 60 had been due to defective point and signal arrangements and a lack of interlocking and 43 to want of telegraphic communication or want of a good block system. The figures for 1871 were respectively 53 and 32 cases out of 159 accidents.[455]

As the accident happened in Scotland, there was no inquest but an inquiry by the Procurator-Fiscal, conducted in private. Stationmaster Currie was subsequently committed for trial on a double charge of manslaughter and violation of duty.[456] In the event both Stationmaster Currie and Signalman Ramsay stood trial on the two charges at the High Court of Justiciary in Edinburgh on Monday 13th January 1873. They were both acquitted, a verdict which was greeted with loud applause.[457]

We remain in Scotland for the next disaster, which occurred on the line of the Caledonian Railway's great rival, the North British Railway Company, at Bo'ness Junction, on 27th January 1874. Like the Kirtlebridge accident it was a collision between an express passenger train and a shunting goods train. Bo'ness Junction was situated on the Edinburgh and Glasgow main line, rather more than a quarter mile west of Manuel station, and was the point when a spur climbed steeply to join the Slammanan and Borrowstowness (commonly called Bo'ness) Railway. The branch joined sidings on the north (up) side of the main line, the sidings having trailing connections with both main lines. There were further sidings on the south (down) side of the main line and there was a trailing crossover between the main lines. Manual station was provided with a down distant signal, 623 yards east of the lever working it, which was outside the station signal box at the east end of the station platform. Bo'ness Junction box worked another down distant signal, 615 yards away, and situated between the station platform and the station distant signal. Finally, there was a repeating signal worked from the junction box by the same lever and wire and situated 30 yards west of the station platform. Neither station nor junction box had home signals. The Bo'ness Junction down distant signal could supposedly be seen for 1,620 yards and its repeating signal for 1,857 yards but this was only under very favourable conditions. There was no interlocking of points and signals, the work of introducing it having been ordered in September 1872 but the contractor being very busy with jobs elsewhere.

455 Board of Trade report

456 Lloyd's Weekly Newspaper, London, England, 6th October 1872

457 Ibid.

On the day of the accident, a mineral train consisting of a pilot engine and 12 coal waggons arrived at the up sidings from Causeway End on the Slammanan and Bo'ness Railway at about 7.05 a.m. It waited in the up sidings until the 6.30 a.m. down express from Edinburgh to Glasgow had passed, at 7.10 a.m., and was then crossed over to the down line to take nine waggons from the down sidings to the up side and deposit the 12 Causeway End waggons in their place. Bo'ness Junction Signalman John Gordon and Second Guard John Anderson were operating the points while First Guard William Neilson coupled the waggons. The nine waggons had been removed from the down sidings and pushed across to the up sidings and the engine was in the act of pushing the 12 Causeway End waggons along the down line into the down siding when they were run into by the 6.35 a.m. Edinburgh to Perth express.[458]

The Perth express consisted of a tender and engine, a horse box containing two horses, a third-class Caledonian carriage, second-class Caledonian carriage, first-class Caledonian carriage, Caledonian van, East Coast Joint composite carriage and East Coast van, in that order, and was in charge of Guard Barbour, with Driver Robert Allan and Fireman Sutherland. It was still travelling at almost full speed when it struck the mineral waggons with terrific force at 7.22 a.m. The rearmost mineral waggon was shattered to pieces and much more significantly so was the third-class carriage, in which a number of passengers died. In all, 14 passengers were killed outright, two more succumbed to their injuries and 28 were hurt. Driver Allan was buried by coal which fell from the tender, pressed up against the tender and scalded to death by steam from a broken gauge glass, while Fireman Sutherland was seriously injured.[459]

The Perth express was due to pass Bo'ness Junction at 7.07 a.m. and the collision had therefore occurred fifteen minutes *after* it was due. Rules 56, 213 and 229 prohibited trainmen from occupying the main line without the permission of the signalman and directed that goods and mineral trains must be shunted clear of the main line fifteen minutes *before* an express or fast train was due. Traffic Inspector William Cumming told the Board of Trade inquiry that he had been surprised when Signalman Gordon had again brought the goods train out onto the down main line after it had 'kicked' the nine waggons across to the up side and had told him "to be quick, as the north express would be down upon them." Gordon had replied that "he was protected by his signals." When he had seen the Perth train coming at full speed Cumming said that he had shouted to Gordon, "Here is the express; see what we have done! Show him the red!" (the red hand lamp) and he (the inspector) had run towards the express waving his handkerchief and shouting to the driver to pull up. The inspector pleaded that with the density of traffic on the Edinburgh and Glasgow it was impossible to obey Rule 229 and clear the main line fifteen minutes before a fast train was due. This was disputed by Mr Mason, the North British general manager, but Colonel W. Yolland, the investigating inspecting officer, found that the 6.30 a.m. Glasgow express had been delayed eight times at Bo'ness Junction by red signals since the 11th June 1873 and the 6.35 a.m. Perth express six times since 21st October 1873, but on different days. Allowing 15 minutes for shunting a clear interval of half an hour would be required to obey Rule 229, and from the timetable there were only fifteen such intervals in the whole 24 hours, and none at that time of the day, so it was exceeding probable that the traffic inspector was correct, and the general

---

[458] Board of Trade report

[459] The Times, London, 28th January 1874

manager mistaken.[460]

As to the position of the Bo'ness Junction signals, the weight of the evidence was that they had been at "Danger" when the 6.35 express passed them. Inspector Cumming said that he had seen Signalman Gordon work the lever to place them to "Danger" (Signalman Gordon, along with First Guard Neilson and Driver James Brown of the mineral train had been recommended by their legal adviser not to give evidence at the Board of Trade inquiry as they had been committed on a charge by the Procurator Fiscal and they did not do so). Two ploughmen named Alexander Hook and James Borthwick, who had been working in a field close to the line, also gave very decided evidence as to the down repeating signal having been at "Danger" and of remarking between themselves that the Perth train was running past signals. Manual Telegraph Clerk James Anderson, Signal Repairer Robert Paterson, who was in Manual station box and Foreman Surfaceman Alexander Beaton all stated that they had seen the repeating signal at red after the crash.

Guard Barbour of the express said that he had seen the Manual station signal at "Clear" and, satisfied with this, had returned to sorting his parcels when he heard two sharp whistles. They were travelling at ordinary speed, about 40 miles an hour, and there was no perceptible attempt to stop the train. Hearing the whistles, he had just begun to apply his brake when he was thrown down by the impact. Afterwards he ran back and saw the signals at red. Conductor Titmus of the express had been busy sorting parcels and could offer no evidence as to the position of the signals.

Fireman Sutherland stated that he had seen the Manual station distant signal was white and that he had then begun to break up some lumps of coal as they were passing the signal and then to use the bar pricker on the fire. He had seen neither of the Bo'ness Junction signals. He had just put some small coal on the fire when Driver Allan had called out "Brake, brake, that is the red signal, but it is too late!" His driver had reversed the engine and whistled for brakes and he (Sutherland) had put down his shovel and applied the tender brake, then seeing the waggons close at hand he had jumped. The fireman said he was badly hurt in the accident and could say nothing as to the position of the signals afterwards.

Colonel Yolland stated that a possible explanation for the accident, and the one which most agreed with the evidence, was that Driver Allan, having seen the Manual station distant at "Clear," had thereafter failed to maintain a good lookout, until he had seen one of the side lights of the backing goods train. In that case, the absence of a home signal which would also have been red was most unfortunate. The alternative explanation, suggested by altered entries in the Bo'ness Junction train register book, was that Signalman Gordon had mistaken the Glasgow express for the Perth one and failed to replace his signals until Inspector Anderson had shouted that the express was coming. The inspecting officer said that he could not pretend to say which was the true explanation.

The primary explanation for the disaster, the inspecting officer stated, was the direct disobedience of the Company's regulations, coupled with the inadequacy of the signalling. Gordon had had only eight months' service and had only worked as a signalman since 1st December. What was more surprising was that a traffic inspector with 22 years' service had aided and abetted the young signalman to do what was expressly forbidden and between

[460] Board of Trade report

two express trains due to pass only seven minutes apart from one another.

It was admitted, Colonel Yolland continued, that the accident would not have occurred had the line been worked on the absolute block system, and had the points and signals been interlocked the uncertainty as to the position of the signals would have been removed. The block system was finally introduced 10 days after the accident.

Driver Allan would probably have survived, the inspecting officer remarked, had the engine not been running tender first. Mr Wheatley, the North British locomotive superintendent, said he could not understand why the engine had not been turned. It was true that the turntable at Larbert was not large enough, but engines usually ran to Greenhill to turn. A new, larger, turntable had been ordered for Larbert. Another North British driver told Yolland that the Caledonian line to Greenhill was so congested that it sometimes took engines two hours to run there and back, although it was only three miles. It had also been stated that the Caledonian Company had declined to allow North British engines to turn at Greenhill.

Finally, Colonel Yolland stated that he did not accept the excuses of the guards and fireman for failing to observe signals when passing a dangerous point in the line that they had been occupied sorting parcels or breaking up coal.[461]

Extreme weather conditions and a design of signals which proved unable to cope with them were factors in a disaster at Abbots Ripton, England, on the Great Northern Railway main line on 21st January 1876. The weather on the evening of that day was very severe with a blizzard, 40 miles in breadth, spreading across England in a north-easterly direction, accompanied by gale force winds. The storm had already caused problems on the Great Western and London and North Western Railways before reaching the Great Northern system at about 5 p.m. and lasting until midnight. The snow fell in large flakes and adhered to whatever it landed upon.

At 5.53 p.m. an up coal train, consisting of an engine and tender, 33 loaded waggons and a brake van, left New England Yard, just north of Peterborough. It was in charge of Guard William Hunt, with Driver Joseph Bray and Fireman Edward Faulkner. The train joined the main line at Crescent Junction, south of Peterborough, at 6.5 p.m., 21 minutes in front of the 10.30 a.m. up Scotch express from Edinburgh to London Kings Cross, and by the time it reached Holme station it was only 16 minutes in advance of the express. Signalman John Osborne at Holme had intended to shunt the coal train for the express and, as he thought, had kept his signals at "Danger" but the train did not stop. He duly gave the 'Train on Line' signal to Conington, the next box but sent for the stationmaster, Mr Gregory, to tell him that the coal train had run past his signals, and then wrote in his train register book "Ran past signals." The stationmaster told him that the starting signal was at "Danger" but the down-home signal was not working properly. The Conington signalman cleared back the coal train at 6.25 and the Scotch express passed his box at 6.37, the signals having been cleared for it. Signalman Osborne said at the Board of Trade inquiry into the accident that he had not been worried that there would be an accident, only that the express would be delayed.

Stationmaster Gregory said at the inquiry that after Signalman Osbourne had told him

[461] Ibid

of the goods train running through his signals, he had examined them and found the up signals to be at "Danger" but the down home partly off. He saw the platelayers knocking large quantities of snow off of the signals. He had never seen such an accumulation of snow on the signals before. As Conington box had cleared back the coal train he pleaded that he did not consider any further precautions were necessary or that the Scotch express should be stopped, even though the coal train running past the signals was irregular.

Driver Bray of the coal train said that after leaving New England the signals had all been showing white "All Clear" lights up to and including the Abbots Ripton distant. He had been expected to have to shunt at Abbots Ripton so was looking towards the signal cabin, where he saw the signalman waving a hand lamp and had shouted, 'What is it, Bobby?' 'Siding' the signalman had replied and then, as he was setting back, 'Come on back; look sharp; you are stopping the express.' The signalman had no sooner spoken when the Scotch express had run into them, striking the fourth, fifth or sixth waggon from the engine. It took him five or six minutes to recover himself, when he was joined by Mr Usher, a relief clerk, who suggested that they go on to Huntingdon for assistance. After getting permission from the signalman to do so they had set off, accompanied by Guard Hunt and a passenger, and picking up Fireman Faulkner, who had gone up the line with fog signals. As they were restarting after picking up the fireman, he saw a down express coming and opened his whistle to warn it, while Guard Hunt showed it a red light.

The down express was train No. 206, the 5.30 p.m. from London Kings Cross to Leeds. Its driver, William Wilson, had seen a white light at the Abbots Ripton down distant signal. He was then alarmed by running over two fog signals (these had been laid by a foreman platelayer, John Jolley, who had run up the line after the initial collision in front of the fireman), shut off steam and told his fireman, James Falkinder, to apply the tender brake. He was travelling at 40 or 50 miles an hour. An instant later he heard sharp whistles from an engine on the up line and saw a red lamp being waved. He reversed his engine, reapplied steam, and his train collided with the tender of the engine of the Scotch express and then with some of the carriages. Thirteen passengers died in this double collision and 53 were more or less injured. The driver, fireman and front guard of the Scotch express and the driver, fireman and front guard of the Leeds express were also injured.

Driver Enoch Catley of the Scotch express had left Peterborough at 6.25 p.m. without any particular instructions or warning. Like Driver Bray in front of him, he had seen nothing but white lights after starting and the first thing that had happened out of course was when he had pitched into the coal train. It was snowing very heavily approaching Abbots Ripton. The Abbots Ripton was a bright white light, not an imperfect signal, visible at 700 yards. He had seen no red light at the rear of the coal train before striking it at 40 or 45 miles an hour.

Between Holme and Abbots Ripton stations there were two boxes at Conington and Wood Walton, which were simple block posts with no sidings or even a mains crossover, so they could not shunt trains. Signalman Timothy Jakes at Conington had the coal train pass him at 6.25 and it had cleared Wood Walton at 6.32. He had had no idea that it had run past the signals at Holme. The Scotch express had passed him at 6.38 and cleared Wood Walton at 6.41. The first he had known of a problem with his signals was when he had heard a slow passenger train approaching at some speed at 7.10 p.m. despite his signals supposedly being at danger because he had not got line clear from Wood Walton.

He had stopped the train with a red hand lamp and the driver asked why he had been stopped when the signals were clear. He told the driver that his levers were over in the frame but found that the driver was right. He had then blocked back to Holme with five beats on the telegraph bell and had fastened half of a railway chair to each home signal to keep then at 'Danger.'

Signalman Charles Rice at Wood Walton had had the coal train go by at 6.31 according to his train book; it was not cleared back from Abbots Ripton. He had received 'Be Ready' for the Scotch express from Conington at 6.35 and 'Train on Line' at 6.37. It passed him at 6.40, although Abbots Ripton had not cleared back the coal train and as far as he was aware his signals were at 'Danger'. He could not see that his signals were off because the windows of his cabin were covered by snow and he did not show a red hand lamp to the Scotch express because he did not hear it coming over the noise of the gale. He managed to stop the following Manchester express with a hand lamp.

Signalman Charles Johnson at Abbots Ripton had received the 'Be Ready' for the coal train from Wood Walton at 6.25, 'Train on Line' at 6.32 and it had reached him at 6.41 according to his train register. He told the Board of Trade inquiry that he had signalled its driver to shunt and that as the engine was passing his cabin he had shouted to the driver, 'Shove them back, the Scotchman is standing at Wood Walton.' He had turned to his instruments when he heard the points lever shaking and assumed that the train was off the road. He then saw the engine of the coal train shoot forward and it occurred to him that the Scotch express had run into it. He had received no signals for the express other than the 'Be Ready' at 6.39. After the collision he had replaced his down signals to 'Danger.' Like the other signalmen it had never occurred to him that his signals were not working properly. The Leeds express had been telegraphed to him as leaving Tempsford at 6.35. The first collision had occurred at 6.44.

Unfortunately, after the first collision Signalman Johnson seems to have lost his head, or, as he said, 'become confused.' He wasted about eight minutes vainly trying to telegraph Huntingdon to inform them of the accident before doing what he should have done at done at once and sending the five beats 'Line Blocked' bell signal to Stukeley cabin, 2½ miles up the line. (Like Conington and Wood Walton Stukeley was a simple block post and was not on the telegraph circuit). Signalman William Trowell at Stukeley finally received the 'Line Blocked' signal at 6.52, just as the Leeds express had passed his box, and too late for him to do anything to stop it.

Having detailed the circumstances of the accident in considerable detail in his report Board of Trade inspecting officer Captain H.W. Tyler went on to discuss its causes. The most obvious cause, he wrote, was the failure to act of the fixed signals. The jury at the inquest into the deaths of the 13 passengers had returned a verdict of 'Accidental death' while attributing the accident to 'the block system having proved inefficient in this case of emergency.' This verdict, Captain Tyler stated, while natural enough to those not thoroughly versed in the subject, arose from a confusion of ideas. The block system, the fixed signals and the means of telegraphic communication were three distinct things which could be employed irrespective of and independently of one another. Fixed signals had been in use for years without any block system, which their failure had nothing to do with. Likewise, the telegraph could be used to give notice of the running of trains or of obstructions without the use of the block system. In the interest of railway safety, it would be wise not to immediately accuse the block system of inefficiency should a collision occur but to carefully scrutinize the defects which led to it. In the present disaster the defects which led to it were all too evident. The fixed signals, through which the signalmen communicated with the drivers whether or no the line was worked on the block system, had failed to work properly during a severe snowstorm, showing white lights instead of red because snow and ice loading their arms and the signal wires had prevented the arms and spectacle glasses from being worked by the levers in the signal cabins. The engine drivers were thus deceived by false clear indications. In the words of the inspecting officer, 'The most subtle ingenuity could hardly devise means more misleading, or more certain of success, for luring engine-drivers, with their trains loaded with precious human freight, forward to inevitable destruction.' [462]

The company servants in charge of the trains, Captain Tyler went on, could not be blamed for the disaster, misled as they were by false signals. The Holme stationmaster, on the other hand, might have prevented the initial collision if, after being informed that the coal train had overrun the signals and having found that his signals were affected by frost and snow, he had 'displayed greater activity and a moderate amount of foresight and caution.' He knew that the coal train could not be shunted before Abbots Ripton, at the end of an incline of 1 in 200, he could see what the weather was and might have inferred that other signals were affected. Having no telegraph communication with Conington or Wood Walton stations he should have stopped and cautioned the Scotch express. Signalman Johnson at Abbots Ripton was clearly not to blame for the first collision, but evidently then lost his presence of mind to some extent. He ought, undoubtedly, to have given the obstruction signal to Stukeley at once. It was clear, however, from the evidence of several witnesses, that Johnson had tried to telegraph Huntingdon South cabin a minute after the collision, using the urgent S.P. (special) code but Signalman Richard Maddison at Huntingdon had not replied until 7.05, 20 minutes later. Maddison's excuse for not replying at once was that the message had not been preceded by the code time and was thus 'unbusinesslike.' Then again Signalman Rice at Wood Walton should have laid fog signals on the rails when he was advised of the approach of the Scotch express by Conington when the coal train had not been cleared back from Abbots Ripton. Failing that he could at least have shown a red lamp from his cabin, as he did afterwards to the Manchester express. Also, the platelayers might have averted the disaster had they attended the signals to ensure that they were working properly, as per their printed instructions, though the inspecting officer conceded that the men had just finished their day's work and had been just going home for tea when the storm came on, preparing, if necessary, for a long night stood at the signals in the snow.

A number of witnesses at the inquiry had related previous instances of the fixed signals not working properly in such weather. The signals themselves were worked by wire from levers or wheels in or near the signal cabins, which when pulled lifted balance weights and drew the semaphore arm down to a vertical position in the post, while the red spectacle glass was from the front of the lamp. When the lever was replaced, the signal arm was returned to the horizontal by the balance weight and the red spectacle glass moved back to the front of the signal lamp. The design was chosen for simplicity and efficiency but ice and snow overloading the signal arms, coating wires already contracted by cold and interfering with the pulleys through which the signal wires ran could all too easily cause them to fail. Various suggestions had been made as to deal with the problem

---

[462] Ibid

such as the use of rodding instead of wires, the covering of the wire run throughout its length and the use of double wires. Signals might also be kept at danger except when cleared for a train instead of being left normally at clear.

The Abbots Ripton signalman had not, in the conditions, accepted the Scotch express from Wood Walton because the coal train was still shunting, but in clear weather he was permitted by the regulations to do so. Under such a system of block working the interval of space was reduced to the short distance between the shunted train and the home signal post, in this instance 68 yards. The safety of the block system then became a delusion.

The regulations did not require platelayers to be stationed at signals which were not at stations or junctions during fog or snowstorms and there were thus none at Conington or Wood Walton. The signalmen at these cabins were required to place two detonators on the rails *as far as practicable* in such conditions. It would be useful, Captain Tyler suggested, if signalmen were enjoined to exhibit flags or handlamps from their cabins to confirm the aspects of the fixed signals. The Holme signalman might thereby have stopped the coal train, and the Wood Walton signalman the Scotch express.

Having questioned the advisability of attempting to run trains normally in such abnormal conditions, the inspecting officer turned to the question of brakes, pointing out that had the Leeds express been equipped with continuous brakes under the control of its driver it could easily have been pulled up in time and the second collision averted.[463] [464]

The Berne correspondent to *The Times* newspaper in London telegraphed on 7th July 1876 that ten persons had died, and many had been injured in a collision that day near Palézieux, Switzerland. The collision, presumably head-on, involved trains from Berne and Lausanne.[465] Other reports suggested that the number killed was only four.[466]

On the evening of Tuesday 31st October 1876, the express train from Ostend to Basle (known as the Indian mail) collided with a goods train near Sarreguemines, Belgium. Four persons were killed and nine injured.[467]

An accident occurred on the network of the Chemins de fer de l'Ouest company at Gagny, France, at 8.20 p.m. on Monday 5th March 1877. Some waggons of a goods train which were being shunted were accidentally directed onto the main line into the path of an express train from Paris. The locomotive of the express struck the waggons and was derailed, and the leading carriages of the train piled up against it. Four people died and 16 were injured. The goods train had been loaded with animals for the Paris market and 'a flock of sheep in the goods train were cut in pieces.' Eight months later the Seine correctional tribunal found the signalman and another railwayman responsible for the

[463] Ibid

[464] The Times, London, England, 24th, 25th, 26th, 27th, 28th, and 29th January and 1st, 3rd and 4th February 1876.

[465] The Times, London, England, 8th July 1876

[466] Wikipedia – Liste des accidents ferroviaires en Suisse

[467] The Times, London, England, 3rd November 1876

accident and sentenced them to six months and 15 days imprisonment respectively.[468] [469] [470]

On 27th May 1877 the Times correspondent in Bucharest reported an accident on the railway between Bacau and Valea Seca, in Romania, on Friday 25th May. Romania had been a province of the Ottoman Empire, but Russia had declared war on Turkey on 24th April 1877 and Russian troops were allowed to pass through Romanian territory in accordance with a treaty signed between Russia and Romania in Bucharest on 16th April, Russia, in turn, respecting the integrity of Romania. Romania declared itself independent of the Ottoman Empire on 22nd May.[471] The accident involved a hospital train, carrying Russian soldiers and their horses. The train became divided, the driver stopped the front portion and the rear portion ran into it. Six waggons were crushed, killing four men and five horses.[472]

This was not by any means the only accident to occur on the Romanian railways in a short time, and a little later M. Gilloux, General Director of the Romanian Railways, was summoned by telegraph to the Russian headquarters in Plojesti to explain why so many accidents were happening, and it was stated that the Russian military had decided to rake over the whole running of the Romanian railways.[473]

The accidents must, nevertheless, have continued, for on 21st August it was reported that, 'On account of numerous railway accidents on the Romanian lines, an agreement has been made by which the railway will pay 4,000f for every soldier hereafter killed, and 12,000f for every officer.' [474]

The author has a note of a sidelong collision at Lübben, Germany, on 3rd February 1878, in which five lives were lost, but cannot now trace any reference to the accident or any further information on it.

An accident at Sittingbourne, England in the summer of 1878 was very reminiscent of the Kirtlebridge disaster in Scotland, almost six years earlier. On Saturday 31st August 1878 many families were returning to London from the seaside and to deal with the heavy traffic the London Chatham and Dover Railway Company arranged for a relief train to be put on to supplement the 11.30 a.m. train from Ramsgate. The relief train started first, leaving Margate at 11.40 a.m., the 11.30 a.m. train from Ramsgate following shortly behind. The Margate portion of the train ran through Sittingbourne safely and continued its journey to London, but the Ramsgate portion was not so lucky.

Sittingbourne station was the junction for the branch to Queenborough and Sheerness and there were sidings on either side of the main lines. The main line was protected by home and distant signals in each direction, worked from levers placed close to the signal

[468] Ibid, 6th March 1877
[469] The American Register, London, England,10th March 1877
[470] Wikipedia – Liste des accidents ferroviaires en France au XIX siècle
[471] Wikipedia – Romanian War of Independence
[472] The Times, London, England,28th May 1877
[473] Ibid, 1st June 1877
[474] Ibid, 22nd August 1877

cabin which was situated at the east end of the up platform. None of the points was connected to or worked from the signal cabin and they were not interlocked with the signals. At the east end of the station there were three sets of points on the down line. Two of these were a set about 220 yards from the signal cabin led to the up sidings and a set 275 yards from the cabin led across the down line to the down sidings.

At about 12.18 p.m. on 31st August the 6.50 a.m. goods train from Stewart's Lane to Faversham arrived at the station and was then occupied in extensive shunting operations on both down and up sides of the line. About eight or nine minutes before the Ramsgate portion of the "cheap up fast" was due, Driver Albert Love of the goods train drew his train clear of the up line and whistled three times to indicate to Signalman Edward Gambell that the up line was clear. He then continued with the shunting operations on the down side and the signalman cleared his up line signals for the Ramsgate train. During the shunting operations, the train came to a stand on the down line with Head Guard Jacob Moden riding on the fifth waggon and Under Guard Charles Clark riding on the twelfth and last waggon. Guard Moden then uncoupled behind the fifth waggon and signalled his driver to set back, intending to run back and set the points so as to let the seven loose waggons run into the up sidings, where the Yard Foreman, William Burden was waiting to distribute them into the different sidings. Before Guard Moden could get back to the points he saw Under Guard Clark working the set of points 55 yards closer to his train which turned the waggons through the mains crossover onto the up line. At the same instant he saw the up passenger train coming along at full speed and only 200 yards away.[475]

The 11.30 Ramsgate of the fast train consisted of a third-class carriage, two first-class, a second-class, a luggage van and a luggage brake van (the Ludgate Hill portion of the train), followed by a third-class, first-class, two third-class, a first-class and a composite carriage and a brake van (the Holborn portion). In charge of the train was Head Guard Thomas Chapman, with Second Guard John Mogford, Driver Edward Elliott and Fireman William Dayland. Driver Elliott told the Board of Trade inquiry into the accident that he had found the Sittingbourne distant and home signals off for him and his first warning of danger had been seeing Guard Moden standing with both his arms raised. He had shut off steam and whistled for the guard's brakes and had then seen that the home signal had been thrown back to "Danger" by the signalman, whereupon he had reversed his engine and had found that some waggons were being pushed back from the down to the up line into the path of his train. His engine then struck the waggons, he did not see quite which as he was sheltering in his cab from the impact. His engine was slightly damaged, but the tender was badly damaged as were several of the leading carriages of the train.

Five passengers lost their lives in the accident and about 40 were more or less hurt. Those who died were Mr George Corbett, of Edgeware-road, London, apparently a prison warder; Mrs Brown, of Oxford; Mrs Over, of Brixton; Miss Robinson, servant to Mr Horton, of Rugeley, Staffordshire and a boy named Frederick Knivett, of Fitzroy-square, London.

Signalman Gambell stated that the passenger train had still been outside of his distant signal when he had thrown it back to "Danger" but the Board of Trade inspecting officer, Colonel Rich, was satisfied from the evidence of other witnesses that the passenger train

[475] Board of Trade report

had in fact been close to the up home before the signals were reversed. The passenger train had been travelling at 35 to 40 miles an hour, and even had it been provided with continuous brakes it was not likely that the driver, taken by surprise after seeing clear signals, could have pulled up in time, though the effects of the collision would have been lessened. The accident had been caused by the neglect of the head guard and the mistake made by the under guard. Head Guard Moden should not have signalled his driver to set back until he had set the points for the sidings and Under Guard Clark should have been more careful that he was operating the right lever, he knew that the waggons were destined for the up sidings. He was inexperienced, with only 13 weeks railway service.

The accident could not have happened, the inspecting officer continued, had the points been interlocked with the signals and worked from a properly placed cabin. Sittingbourne station had opened for passenger traffic before the interlocking of points and signals was introduced, and the Board of Trade had no powers to enforce the resignalling of existing as opposed to new lines. He was informed that the London Chatham and Dover Company were now taking the interlocking of Sittingbourne station in hand.[476]

The coroner's jury at the inquest in the victims found a verdict of manslaughter against guards Moden and Clark.[477] They were tried at Maidstone in January 1879 when it took the jury 'only a few moments' to find them 'Not Guilty.' [478]

In 1878 Pontypridd station, South Wales, on the Taff Vale Railway, was the junction for company's branch to Treherbert (the Rhondda branch), which diverged at the north end of the station platforms. There was also a double line curve connecting the main line and the branch, called the north curve, which left the main line north of the station at North Curve Junction and met the branch at Rhondda Cutting Junction, with signal boxes at each point. The curve, together with the main line and Treherbert branch, thus formed a triangle by which trains could be turned without the use of a turntable, and the procedure had been adopted of passenger trains from the Llantrisant and Cowbridge branch, south of Pontypridd, running after unloading their passengers to the North Curve box, then setting back on the wrong road to Rhondda Cutting box, from whence they could return to Pontypridd station on the down branch line. As the empty train was being pushed back along the north curve in the wrong direction there were no fixed signals to warn its driver if anything was wrong.

On Saturday 19th October 1878 the 2.15 p.m. up passenger train from Cowbridge, consisting of an engine and tender and five carriages, in charge of Guard Peter Griffiths, with Driver John Morgan and Fireman Morgan Aubrey, having dropped its passengers, proceeded from Pontypridd station to the North Curve box, arriving there at 4.15 p.m. The train having cleared the junction points at the North Curve, Signalman William Roberts set them for the train to come back, then came down from his box to hand a package to Driver Morgan. Afterwards, he returned to his box, but before re-entering it he shouted 'All Right' to Guard Griffiths, who in turn gave a hand signal to his driver, and the train began to back around the curve at a speed of four or five miles an hour. Signalman Roberts told the Board of Trade inquiry into the accident that as he was returning to his box, he thought that he had heard the two rings 'All Clear' bell signal

---

[476] Ibid.

[477] The Times, London, England, 2nd,3rd,4th,5th,14th, 16th, 19th, 24th and 29th September 1878

[478] The Magnet, London, England, 20th January 1879

from Rhondda Cutting box and so had given the guard permission to proceed. When he got back in the box, however, he had found the Rhondda Cutting block instrument showing 'Junction Blocked' and had then realised that he had made a serious mistake, but at that point there was nothing that he could do to stop the Cowbridge train.[479]

The 3.35 p.m. down passenger train from Treherbert consisted that day of an engine and tender, a brake van, six third-class carriages, a brake carriage, a first-class carriage, a second-class carriage, two third-class carriages and a brake van, with Driver David Thomas and Fireman Shadrach Okey. It was due in Pontypridd at 4.16 p.m. and was running over the Rhondda Cutting junction points at about 11 miles an hour when Fireman Okey shouted to his driver, "Look out, Davy, the Cowbridge train is into us!" The Cowbridge train struck the leading brake van of the Treherbert train, grazed the first two third-class carriages and severely damaged the third and fourth third-class carriages.

The exact number of fatalities in this accident is unclear. *The Times* newspaper reports seem to imply that there were 13 people killed, while the Board of Trade report lists 12 dead and an unidentified passenger, a publican from Vaynor, who was not expected to live. Both the newspaper and the Board of Trade report list the victim's names, but these do not entirely correspond in the two sources. In addition to the dead, over 100 passengers were injured.

The investigating Board of Trade inspector, Colonel W Yolland, commended Signalman Roberts for having candidly admitted his mistake, and said that had he not done so it would have been very doubtful if the collision had resulted from a mistake made by Roberts or had been the fault of the Rhondda Cutting signalman, Stephen Ellis. This was because Signalman Ellis had written prematurely in his train register book that he had accepted the Cowbridge empty stock 'to save time' and had also made erasures in the book to excuse himself. This, Colonel Yolland stated, was 'a very irregular and improper proceeding,' which he believed was frequently done, but was most objectionable, and every endeavour should be made to stop the practice.

Signalman Roberts, the inspecting officer continued, had been 35 years with the company, and a signalman for 30 years, and had never had a serious mishap before. He could not have caused the accident by a mistake in block working had an improper and dangerous mode of working not been introduced by the company in 1875. The enormous advantage gained by the travelling public by the introduction of interlocked points and signals was thus altogether lost. He recommended that the mode of working should be abandoned without any unnecessary delay and a mains crossover should be installed at the North Curve Junction to enable empty trains to be crossed over to the down line to obviate the wrong direction working.[480]

Signalman Roberts was found guilty of manslaughter by the inquest jury on the accident victims.[481] At his trial at Cardiff in February 1879, the jury were unable to agree and he was bound over to appear at the next assizes, but no further report of a trial has been found and it would appear likely that at this point the case was dropped, it being

---

[479] Board of Trade report

[480] Ibid.

[481] The Times, London, England, 22nd,23rd, 25th and 29th October 1878

felt, perhaps, that he had suffered enough already.[482]

Some days after it had happened, on 10th January 1879, news reached London of a major disaster in Russia on the Warsaw Railway 50 versts (about 33 miles or 55 kilometres) from St Petersburg at Suide station. A passenger train from St Petersburg was entering the station, situated eight versts from Gatchina, when it collided with a wood train which had been shunted off a siding. Both locomotives were smashed to pieces and the leading third-class carriage of the passenger train, containing 40 people, was completely destroyed, as were the two following carriages, which were half empty. The remaining carriages of the passenger train were all thrown off the line and the passengers in them more or less hurt. Twenty people were killed outright and 32 injured, 21 of them dangerously so. Several of the victims trapped in the broken timbers of the wreckage were severely frost-bitten as it was intensely cold, the temperature reportedly being 15 deg. Réaumur (this must have been -15 Réaumur or -19 Centigrade, 15 Réaumur being a comfortable 19 Centigrade). Only one of the guards riding in the front of the train survived, the others being found dead along with the drivers and firemen among the debris of the engines. The report on the accident added, 'It is regarded as a very fortunate circumstance that the Warsaw mail train was a few minutes arriving at Suide, otherwise a second collision would have taken place, and the loss of life would have been, in consequence, proportionately greater.' [483]The accident was blamed on 'the obstinate conduct of the stationmaster in causing the wood train to be shunted in sight of the advancing passenger train.' [484]

Early on the morning of Friday 10th December 1879 a switch engine was making up a train at the east end of the yard at Air Line Junction, Jackson, Michigan, United States, on the Michigan Central Railroad. The westbound Pacific Express, consisting of an engine, baggage and express cars, four passenger coaches, mainly occupied by emigrants, and seven Wagner sleeping cars was due at 12.35 a.m. and the main line had to be kept clear for it. Engineer Robert Jones was asked by Yardmaster Evander Colwell (or Caldwell) to fetch some cars from the north side of the yard and take them to the south side, which would mean crossing the main line. A freight train had just arrived, running in the time of the express, and Engineer Jones was unwilling, he said, to foul the main line before the express had passed, the crew of the freight having told the south side switchman, named Meyers, that the Pacific Express was twenty minutes behind time. Switch engines were supposed, according to the rules, to be clear of the main line 15 minutes before the expected arrival time of a train. When he told Yardmaster Colwell that he wished to wait for the express the latter went to Operator Henderson and asked him how late the express was. The express had been reported as having passed Dexter, 25 miles east of Jackson, at 12.25 a.m. and was 45 minutes late at that point, but for some reason Operator Henderson told the yardmaster at 1 a.m. that the express was 45 minutes late 'now' rather than making it clear that it had been so at Dexter 35 minutes earlier. The yardmaster in turn told Engineer Jones that the express was running 45 minutes late, and thus reassured the

[482] The Magnet, London, England, 17th February 1879

[483] Ibid, 20th January 1879

[484] Lloyds Weekly Newspaper, London, England, 26th January 1879

engineer began the shunt move, although if the express had still been 45 minutes late by the time the yardmaster reached him from the operator's office he would only have had about eighteen minutes to make the shunt, or three minutes to complete the shunt and be clear of the main line 15 minutes before the express was due at 1.20 as per the regulations. As matters turned out, the time available proved too short, and the express collided head-on with the switch engine as the latter was moving slowly forward with nine freight cars. Both engines were demolished, the baggage and express cars were telescoped; the first passenger coach more or less escaped but the second and third coaches behind them were wrecked. None of the Wagner sleeping cars was at all damaged, and many of the sleeping car passengers did not even wake up until they were rudely disturbed for sheets and bedding to care for the injured, when they were not best pleased until they learnt of the extent of the disaster. Express engineer Milton Gilbert and his fireman, C.B. Smith were both killed together with at least 14 passengers, and a large number of passengers were injured.[485] [486] [487]

Although the express was thus three-quarters of an hour late at Dexter, its engineer, Gilbert, had run his train very hard to make up time. According to its sleeping car conductor, "The train ran like the devil – all of 45 miles an hour – to make up its time" (the scheduled time for the express between Detroit and Jackson involved a speed of only 27.6 miles an hour) and it so arrived at Jackson only two or three minutes late, still running at not less than 35 miles an hour. Engineer Gilbert was therefore also responsible for the accident as the regulations forbade him from passing any switch at a station where he had not stop at more than fifteen miles an hour.[488]

In late October an inquest jury returned a verdict of criminal negligence against Yardmaster Colwell and censured Switchman Joseph Sawyer (presumably the same person as the switchman named as Meyers in an earlier newspaper account) and Engineer Robert Jones. They also urged the propriety of providing additional danger signals and safeguards at Jackson.[489]

Carelessness on the part of a signalman cost the lives of five passengers and four railwaymen and resulted in more or less serious injuries to 48 passengers and two railwaymen in a collision at Brickfield Sidings, near Burscough Junction, England, on 15th January 1880.

Burscough Junction was on the Lancashire and Yorkshire Railway Liverpool – Preston line, and was the terminus of some local passenger trains from Liverpool, which then returned whence they had come. The station had two main platforms and there was a short bay platform on the down (to Liverpool) side at the Liverpool end. There were sidings at each end of the station and two mains crossovers allowed the engines of terminating trains to run around their coaches before starting again for Liverpool.

On the evening of 15th January, the 5.10 p.m. up train from Liverpool, which terminated at Burscough and returned from there as the 6.05 departure to Liverpool., was

485 Fort Wayne Weekly Sentinel, Fort Wayne, Indiana, United States, 15th October 1879

486 Boston Weekly Globe, Boston, Massachusetts, United States, 14th October 1879

487 A History of Railroad Accidents, Safety Precautions and Operating Practices by Robert B. Shaw, pgs. 72-3

488 Boston Weekly Globe, Boston, Massachusetts, United States, 14th October 1879

489 Fort Wayne Weekly Sentinel, Fort Wayne, Indiana, United States, 22nd October 1879

somewhat late, having been stopped by signals at Aintree Goods Yard cabin because a goods train had had to be shunted onto the up line at Maghull out of the way of the 3.30 p.m. express from Skipton to Liverpool. There was also a carriage from Southport at Burscough waiting to be attached to the following 5.30 p.m. passenger train from Liverpool to Preston, and to avoid delay the decision was made to start the 6.05 train from the wrong (up) platform, passing it over to the right (down) line via No. 8 crossover at the Liverpool end of the station. Stationmaster Roger Sayer stated that about one train in six was thus started from the wrong side of the station. The Burscough signalmen also changed over at 6 p.m. and Signalman Anthony Melia was slightly late in taking over because he had arrived from Ormskirk, where he lived, on the delayed 5.10 train. What exactly happened when he arrived in the cabin at about 6.05 to take over from his mate, Signalman, John Spencer, depended on which of the signalman's accounts was the most reliable. The Board of Trade inspecting officer who inquired into the accident, Major-General C.S. Hutchinson thought it probable that Signalman Melia, in his anxiety to clear his signals for the 5.30 p.m. Liverpool fast train, for which the 'Be Ready' signal had already been received from the Brickfield Sidings cabin, had restored lever No. 8, working the mains crossover, before the departing 6.05 train had reached it. The 6.05 train had thus continued along the up line in the wrong direction until it collided head-on with the 5.30 train near Brickfield cabin, almost half a mile from the station.[490]

Driver Thomas Looney and Fireman Robert Sanderson of the 6.05 train were both killed in the accident, as was their guard, Edward Sutton. It seemed likely that Driver Looney had belatedly realised that he was running on the wrong line, as his engine had been reversed, though it was strange that neither driver not fireman had realised that they had not passed through the crossover. John Bullfield, the driver of the 5.30 fast train, had seen the headlights of the 6.05 train from some distance away but naturally it did not occur to him that the train was on his own line until they were close together; he and his guard, Moses Smith, were seriously injured, while his fireman, Robert Clarkson, was killed. Burscough Foreman Porter Thomas Whalley had noticed that the 6.05 train was going out on the wrong line and said that he had shouted as loudly as he could and had run after it waving a red light to try and stop it, but to no effect.[491]

Major-General Hutchinson stated that although the immediate cause of the accident was a signalman's mistake other causes had contributed to it. The absence of a bay platform at Burscough capable of accommodating the 6.05 train was the cause of its being started from the wrong side of the station, while had there been proper siding space at Maghull the down goods train would not have needed to shunt to the up line and so delay the 5.10 train. Until Burscough station could be enlarged interlocked signals should be provided to control trains starting from the wrong side of the station and a man should be deputed to accompany them as far as the crossover. The accident also demonstrated the extra risk incurred in running tank engines, such as the 6.05 train was hauled by, coal box first and the necessity for tool boxes on engines to be properly secured.[492]

Signalman Melia was committed for manslaughter following the conclusion of the

---

[490] Board of Trade report

[491] The Times, London, England, 16th, 17th, 19th, 20th, 24th and 29th January 1880

[492] Board of Trade report

coroner's inquest on 28th January but was acquitted at his subsequent trial.[493]

A rather unlucky accident occurred on the morning of 17th February 1880 on the Candahar Railway, near Mittri, India, twelve miles from Sibi. A heavy gale blew some goods waggons along the line and they were run into by a train from Jacobabad, carrying the Poona Horse cavalry regiment and some civilians. Seven passengers were killed and four injured, but the Poona Horse escaped.[494]

The Magdeburg – Halberstadt Railway suffered a serious collision involving two passenger trains on Saturday 20th March 1880 at Halle, Germany. Several carriages were reportedly shattered to pieces and about seven passengers killed and many injured.[495] A contemporary German account of the accident seems to have concerned itself mainly with the damages that the railway company would have to pay as a result of the accident and remarked that the railway company were fortunate that the dead and injured victims came from the lower ranks – fourth-class passengers. A watchmaker who lost his wife received a single payment of 2,500 marks, and children whose mother had died received 300 marks a year until maturity. A passenger who suffered a crushed hand was awarded 900 marks a year and those with slight injuries received 400 marks. The Berlin correspondent to *The Times* newspaper in London stated the day after the accident that, 'The pointsman by whose carelessness the deplorable accident was caused lost very little time in hanging himself, preferring death to the five or six years' imprisonment which the German law inflicts on such offenders.' It was proposed to pay his widow a sum of 6 marks per week, though this apparently was not confirmed.[496] [497]

Locomotive Junction, Nine Elms, England, on the very busy London and South Western Railway main line, just outside of its London Waterloo terminus, was the location of a bad accident on Saturday 11th September 1880. The 10 p.m. down passenger train from London Waterloo to Hampton Court consisted of a tank engine, third-class, second-class, and first-class carriages, a brake van, three first class, a second-class and a third-class carriage. It was running at nearly 30 miles an hour when it collided with a light engine at about 10.8 p.m. The locomotive had been standing on the main line for about six minutes waiting to be let into the locomotive yard via a set of trailing points. The impact drove the engine back about 50 yards and it was badly damaged, but not derailed. The engine and all the vehicles of the passenger train left the rails and the leading carriage was 'telescoped and broken all to pieces' while the ends of the following second-class carriage were badly damaged. Five passengers, Fireman George Arnott of the passenger train and Driver Henry Watson of the light engine were either killed outright or died of their injuries and 42 passengers were injured, as were Driver Samuel Taylor and Guard John Smith of the passenger train and

---

493 The Times, London, England, 29th January 1880

494 Ibid, c21st February 1880

495 Brief: The Week's News, London, 26th March 1880

496 The Times, London, England, 22nd March 1880

497 Eisenbahn-Katastrophen in Deutschland Bd 1 by Hans Joachim Ritzau, pg. 104

Fireman Thomas Woodbridge of the light engine.[498]

Locomotive Junction box was very busy and was staffed by two signalmen plus a boy from 7 a.m. until 11 p.m. Like the Burscough accident earlier in the year, the collision occurred just as the signalmen were changing over. Signalman Henry Almond told the Board of Trade inquiry that he had been on nights that week and had come on duty a few minutes before 10. His colleague, George Davis, who was the senior man, came on at 10 or a minute or so after. He (Almond) had received the two beats warning signal for a train on the down main from Wandsworth Road box at 9.59 p.m. and the on line signal, three pause two beats, at 10 p.m. for a locomotive to go the yard. He had put over the switch which raised the semaphore arm of the block instrument at Wandsworth Road and cleared the home signal, No. 31. The engine had passed his box and come to a stand beyond the connection into the yard at about 10.02 p.m. and he had replaced the home signal. Signalman Davis had arrived in the box at about the same time. He was then dealing with an up train and the light engine had passed out of his mind. He had not seen its tail lamp or heard it whistle. At 10.05 he received the warning signal from Wandsworth Road box for another train on the down main and said to Signalman Davis, who having just come on knew nothing of the engine, "Here's the warning for the Hampton Court train." Davis told him to accept the train, and either he or Davis had cleared the signals for it, having had it accepted by Queens Road box. It had passed at 10.08 with steam on and he had immediately heard the sound of the collision. It was a very dark night and raining hard and he had seen nothing of the engine standing near his box. Engines waiting there did not usually whistle.

The inspecting officer, Major F.A. Marindin, stated that the forgetfulness of Signalman Almond was clearly the principal cause of the accident, and the signalman had not attempted to throw the blame onto anyone else. Others were, however, not entirely free from responsibility. Although the rule that drivers standing on the main line awaiting admission to the locomotive yard should whistle was probably not always attended to, it showed great carelessness on the part of the unfortunate Driver Watson to allow his engine to stand for over five minutes on a busy main line without trying to attract the attention of the signalman, even if he was not aware that the Hampton Court train was due. The booking boy, Albert Kempston, had also contributed to the accident by entering in the train register book that the engine had gone into the locomotive yard, having taken it for granted that it had done so. Although, he had not directly misled Signalman Almond as the latter had not consulted the book before accepting the Hampton Court train, yet had he been aware that the engine had not entered the yard, he could have warned the signalman. Moreover, neither of the signalmen had taken sufficient care to check the line was clear before lowering the signals. There was no doubt that the tail lamp of the light engine had been burning and visible from the box if carefully looked for. Major Marindin stated that, 'it is a far too common practice for signalmen to trust almost entirely to the block telegraph instruments, and the interlocking of their signal levers, which are after all only aids to safe working, and not to make certain by actual observation that the line within their view is clear and unobstructed before they take off their signals.'

The inspecting officer urged the railway company to adopt a system of electrical interlocking such as that used on the London Chatham and Dover and Metropolitan

---

[498] The Times, London, England,13th, 14th, 15th, 16th, 21st, 23rd, and 24th September 1880

District Railways which would make it impossible for the signalman, after having lowered his home signal for an engine to go to the yard, to lower his signal again until the points had been set and the signal lowered for the engine to go into the yard and subsequently restored. The position of the signal box might also be improved, and it be arranged for light engines to reach the yard via an independent line, while meanwhile firemen should be ordered to wait by the box until their engines had got in clear. The number of fatalities would probably have been less had there been a brake van or dummy carriage next to the engine.[499]

The inquest jury found a verdict of 'Accidental death' by a majority of 12 to 2.[500]

499 Board of Trade report

500 The Times, London, 28th September 1880

CHAPTER 6

SINGLE LINE COLLISIONS

The operation of trains in opposite directions over a single line of rails brought the obvious danger of head-on collisions between them. When the first railways opened in Great Britain, they used horse-power and the risk of a serious head-on collision did not exist, but with the advent of faster and more powerful steam locomotives the situation changed. The early main line railways in Great Britain, such as the Liverpool and Manchester, London and Birmingham and so on were all provided anyway with double track and the single-track branch lines which were opened mostly only saw a handful of trains a day. When railways spread to less densely populated parts of the United Kingdom and to other parts of the world where a double-track railway could not be economically justified, the problem of single line collisions greatly increased. Adding to the problem in the earliest days was the fact that the fastest means of communication between two points was the train itself. Crossings between trains on a single line could be set down in a timetable but should one train be delayed for some reason there was no means of communicating the fact to the other and allowing it to advance further along the single line. There was also the problem of different locations having their own local times, which led the railways to adopt a standard railway time at an early date.

To deal with the problems caused by one or both trains being late, various regulations were devised, such as stating that a train which was not likely to arrive at a crossing place within a given time after it was scheduled to do so lost its right of way and must wait; while an opposing train waiting at the crossing place gained the right of way after the same specified interval. Thus, if we consider the stretch of single line between two stations, let us call them Tiverton and Cadeleigh, should a train from Tiverton due to cross an opposing train at Cadeleigh at 4.30 p.m. be delayed so that it was not likely to reach Cadeleigh 15 minutes after its allotted time, that is by 4.45, it lost the right of way and must be shunted clear of the main line at the first possible point, while the train from Cadeleigh to Tiverton must wait at Cadeleigh until 4.45, after which it gained the right to proceed. Should both trains be delayed, they must run at great caution. If the trains subsequently met one another between stations the one which had the shortest distance to go must set back, taking into consideration factors such as gradients, weights of trains and suchlike. At times, it would seem, the decision as to which train must give way was also decided by a fight between the train crews. Before the introduction of the electric telegraph such regulations were the best, and indeed the only, method of working single lines, but they could lead to significant delays, and there was a great temptation for the crews of trains which were delayed to hurry to reach the next crossing place rather than wasting time waiting at the previous one for an opposing train.

The first recorded single line collision occurred three miles above Burlington, New Jersey, United States, on the Camden and Amboy Railroad, at about noon on 24th February 1836. A train from Amboy, carrying passengers from New York, collided with a freight train in a fog which limited visibility to a few hundred yards. Either the freight train was going more slowly, or its engine crew were more alert, for it had almost stopped, whereas the passenger train was still going at some speed. The locomotives were 'broken into many pieces' and the baggage cars at the front of the passenger train were damaged, but the two engineers and the two firemen jumped and saved themselves and the passengers escaped with bruises and shock. After the accident, some of the passengers walked to Burlington, some were brought there on sleighs, some stayed at a nearby farmhouse and others remained at the scene. The newspaper reports blamed the accident entirely on the thickness of the fog.[501]

The newspaper report on the next collision was more definite in blaming it on the negligence of one of the train crews. It occurred on the Boston and Providence Railroad at Roxbury, Massachusetts, United States, on Wednesday 22nd June 1836. A passenger train of eleven cars coming from Providence, Rhode Island, and carrying a large number of sailors and marines going to join their ship, the *U.S.S. Boston*, at Boston, collided with an engine and single freight car bound for Dedham. The crew of the Providence train had looked at their watches and calculated that they had time to reach Boston before the freight train for Dedham left. The staff at Roxbury were also blamed by the newspaper, it being stated that they must have heard the train from Providence approaching but had failed to stop the Dedham train. The trains had been running at full speed when they met on a curve. The collision was described as dreadful. The two front cars of the Providence train were shattered to pieces and the engine much damaged. Nine of the sailors, together with an officer and a midshipman, were seriously injured, some suffering broken legs. The engineer of the passenger train jumped and escaped, but, his fireman, in attempting to do so, suffered two broken legs. The injured passengers were carried on litters into neighbouring houses for treatment and then taken into Boston in carriages. The 1 p.m. departure from Boston for Providence was delayed by the accident until 2.45 p.m.[502]

What might be regarded as the cavalier attitude of some of the early railroad staff was illustrated by an accident on the Baltimore and Ohio Railroad near Halltown, West Virginia, United States, on Friday 5th August 1837. A train filled with passengers 'on a party of pleasure' was to be run from Harpers Ferry to Halltown. Its engineer was aware that a train with 100 passengers was due in the opposite direction from Winchester but considered that he could reach Halltown, only four miles distant, in the time available. He managed to cover three of the four miles before the trains met on a curve. One of the

---

501 National Intelligencer, Washington, D.C., United States, 29th February 1836

502 Boston Morning Post, Boston, Massachusetts, United States, 30th June 1836

engineers was able to reduce the speed of his train before jumping off in what later became the time-honoured fashion. No one lost their lives and no limbs were broken but a good number of the passengers, including several ladies, were fairly badly hurt. One gentleman, a stranger whose name the reporter could not obtain, was unable to leave the place and was taken to a nearby house. The other passengers eventually arrived back in Harpers Ferry with bandaged heads and eyes and tied up limbs.[503]

The immunity from fatalities in single-line collisions could not last and ended less than a week after the Halltown collision. At 8 a.m. on Friday 11th August 1837 a regular Portsmouth and Roanoke Railroad train left Portsmouth, Virginia, United States, with nearly 200 passengers, most of them returning home after having been on a steamboat excursion trip. The train had called at Suffolk station and proceeding from there had covered another mile and a half when the accident happened. As it was leaving a 35 ft high embankment and was approaching a sharp curve it met a lumber train with 15 cars heavily laden with staves, running fast down an incline. The engineer of the passenger train was able to bring it to a halt, but the lumber train could not pull up. The collision pushed the passenger engine against the first car of its train, which was pushed in the second, and the second was forced into the third, but most of the casualties were in the second car, which was overridden by the one in front. Many of the young and active passengers had time to jump to safety but three girls were killed, and a number of other passengers were seriously injured. Those who lost their lives were Miss Elizabeth McClenny, of Nansemond county, Virginia; Miss Margaret Roberts, of Isle of Wight county, Virginia and Miss Jemima Ely, also of Isle of Wight county. None of the passengers in the first car was hurt.[504]

In another five minutes, the newspaper report pointed out, the lumber train would have reached Suffolk and no accident would have happened. The residence of a Mr Richard Goodwin lay within 100 yards of the accident site, and the dead and injured were taken there, the injured to beds in the house, which took on the appearance of a hospital, and the bodies of the three girls to a shed in Mr Goodwin's turpentine factory. Two of the three girls were shortly to be married. The fiancé of one of them was sat next to her in the car and refused to jump and save himself as he could not save her. Remaining in his seat he was very badly injured. The fiancé of the other girl was not on the train but arrived at the scene some hours later upon hearing of the accident and 'by the expression of his grief too well told the wounds of his heart' as the newspaper reporter expressed it.

A further tragedy marred the day. An engine and coach carrying the directors of the company and a number of doctors proceeded from Portsmouth upon hearing of the collision to render assistance. Later in the day, this relief train had to return to Suffolk for wood and water for the engine, and in doing so, on a dark evening with heavy rain, it ran down a Mr James Woodward and a Mr Richard Oliver, two local citizens who were walking along the line. Mr Woodward was killed and Mr Oliver probably fatally injured.

A single inquest was held into both accidents. The collision, its jury found, had been the result of the wilful mismanagement and gross negligence of Conductor (referred to at the inquest as the captain) Gordeus Etheridge and Engineer Williams of the lumber train

---

[503] Maryland Republican, Annapolis, Maryland, United States, 15th August 1837

[504] Maryland Gazette, Annapolis, Maryland, United States, 17th August 1837

in 'running down a rapid descent on a curve of said road with great velocity, at a time when they might reasonably have expected to meet the regular train.'[505]

There was a gap of almost three years before two passenger trains again disputed possession of a stretch of single line, this time on the Boston and Worcester Railroad, two miles on the Worcester side of Westboro,' Massachusetts, United States. The ascending train which left Boston at (according to the newspaper) 4 a.m., being the fourth departure that morning, well filled with passengers, collided with the 7 a.m. descending train from Worcester. The engineers and firemen and some of the passengers had time to jump before the collision. No one was killed, but several passengers were injured, the worst hurt being a Mr Samuel Brookes of Brighton and the son of M Ostinelli, a musician. The accident was later blamed on the crew of the descending train having continued at full speed after failing to meet the ascending train at Grafton, as per the timetable.[506] [507]

A disastrous single line collision occurred in Russia on about 25th August 1840 (conflicting dates given in different newspaper reports) on the St Petersburg – Czarskojeselo Railway. At 11 p.m. a passenger train of 13 carriages driven by a Mr R. Maxwell set off from Czarskojeselo (also spelt Tsarskoye Selo) for St Petersburg. Mr Maxwell had been told in the morning that he must stop at a certain point to cross a train coming from St Petersburg but stated after the accident that he had forgotten about this message. He thus failed to stop his train at the designated point and instead continued towards the Russian capital until his train collided with the outgoing one. Both train drivers attempted to stop their trains, but a violent collision ensued, and the carriages of Mr Maxwell's train were crushed. According to the first report, six passengers (four men and two women) on the train from Czarskojeselo were killed and 21 injured, six seriously.[508] A further report from St Petersburg on 2nd September stated that the number of lives lost in the accident was about 100, which if true would make it by far the most serious railway accident to have happened anywhere in the world at this time.[509]

The first recorded single line collision in Great Britain occurred in Scotland, between Clocksbriggs and Forfar on the Arbroath and Forfar Railway, on 12th February 1841. The 8.30 a.m. up passenger train from Arbroath collided with a down ballast train. Drivers William Barrie and John Adamson were seriously injured and Trainman George Porter, labourer Robert M'Guire and Mr D. Campbell, the governor of Dundee Prison, were also hurt. At the time of the report, the accident was still being investigated by the company directors and the Sheriff of the county of Forfar.[510]

On 5th October 1841 the Western Railroad eastbound passenger train due to arrive at

---

505 Ibid.

506 Boston Morning Post, Boston, Massachusetts, United States, 18th June 1840

507 A History of Railroad Accidents, Safety Precautions and Operating Practices by Robert B Shaw, pg.27

508 The Times, London, England, 8th September 1840

509 Ibid, 25th September 1840

510 Board of Trade report

Pittsfield, Massachusetts, United States at 10.05 a.m. was scheduled to cross the 12 noon westbound train from Springfield at Westfield at 12.25. Fearing that the eastbound train might well be late and not wishing to delay the westbound one, Trainmaster Barnes at Springfield wrote out an order that morning instructing the eastbound train, if running late, to wait one hour over time for the 12-noon westbound one. To enable the order to be delivered to Conductor Warren of the eastbound train, it was handed to Conductor Lee of the early morning westbound train out of Springfield, which was due to cross Warren's train at Pittsfield, where the morning westbound train was also due at 10.05. Unfortunately, Lee's train did not arrive at Pittsfield until 10.25 and Warren's was not there until 10.35. There was therefore some urgency in getting both trains away, to avoid further delay, while Conductor Lee, busy dealing with passengers and other duties, did not at once pass the order to Conductor Warren. The station staff were then hassling Lee to start and unable to immediately spot Conductor Warren, Lee handed the order to a man handling the luggage, whom he supposed belonged to Warren's train, telling him to be sure that Warren received the order. Lee's train then set off westwards and Warren's eastwards. Lee's train continued to the then terminus of the railroad at State Line, and later in the day returned to Pittsfield, where the baggage-handler gave him back the order, telling him that he had forgotten, or had been unable, to give the order to Conductor Warren.

Conductor Warren's train thus continued eastwards and was running fast to try and make up time, but waited an extra ten minutes at Chester Village, as per the regulations, while the westbound train, in charge of Conductor Moore, had waited twenty minutes over time at Westfield. The trains collided head-on four miles west of Westfield. There were two fatalities in the collision, one of which was Conductor Warren, victim of an accident for which he was not to blame. The other fatality was a passenger, a son of Mr Bloodgood, of Albany. Two other sons of the same gentleman were among a large number of injured passengers, some of whom were seriously hurt. The person clearly at fault was Conductor Lee, who should have ensured that he gave such an important order directly to Conductor Warren, even if it meant a little more delay to his own train, rather than entrusting it to an unknown third party.[511] [512] [513]

A misunderstanding by an engine driver was the cause of a collision near Llandaff, South Wales, on the Taff Vale Railway, on 31st March 1842. The indirect cause of the accident was the disruption to traffic caused by a tree being blown down across the line during a severe gale. Because trains were running out of course, Mr Bage, the company official at Cardiff, thought it best to remind the driver of a down empty coal train that he must wait at Taff's Well, where there was a long crossing place, for an up passenger train. The Taff's Well policeman duly turned the coal train into the siding, but the engine driver, who had misunderstood what Mr Bage had said, thought that he had been instructed not to wait, as he normally would have done, at Taff's Well, and told the policeman there that he had orders to continue, whereupon the policeman let his train back out onto the main line. Fortunately, the drivers of the two trains saw one another in time to reverse their engines and the four carriages of the passenger train were only carrying a few passengers. The collision occurred at about 1.10 p.m. Two passengers and

---

[511] Weekly Argus and Rough Hewer, Albany, New York, United States, 9th October 1841

[512] Boston Morning Post, Boston, Massachusetts, United States, 7th, 8th and 14th October 1841

[513] A History of Railroad Accidents, Safety Precautions and Operating Practices by Robert B Shaw, pgs. 27-8

the guard of the passenger train were slightly hurt.[514]

On Tuesday 8th October 1844 a head-on collision occurred near South Shields, England (referred to simply as Shields in the Board of Trade report) on the Brandling Junction section of the Newcastle and Darlington Railway. This was in effect a single-line collision, for although the line was double track it was worked for some strange reason as a pair of single lines, with one line reserved for passenger trains in both directions and the other for coal traffic. At 5 a.m. that day, a special train for butchers left Shields for Newcastle, the train consisting of one second-class carriage drawn by the *Nelson* engine. The train would normally have been worked by the *Leopard* engine, and two miles from Shields the passenger train coming up collided head-on with the latter, which was belatedly going to Shields to work it. The collision occurred in a cutting on a curve, with limited visibility, but the drivers had time to reverse their engines and with their firemen jump off their engines. Unfortunately, Driver Matthew Burdis, on the *Nelson* engine, did not have the presence of mind to shut off his steam, and after the collision the engine began to propel its carriage back towards South Shields, attaining an estimated speed of 30 miles an hour. Many of the 29 butchers on the train jumped out at this point and all of those who did so escaped more or less unhurt. Ten were still on the train, however, when it collided with some empty carriages at South Shields. One of them, John Brown, was killed outright, and a second, Joseph Burrell, was fatally injured. The other eight were all seriously injured.[515]

The accident arose in the first instance from the sorry state of the Brandling Junction Railway, which had always been in a perilous financial position, and whose engines were in a constant need of repair. A month prior to the accident, the company had been sold to the Newcastle and Darlington Junction Railway. The day before the accident, Monday 7th, the passenger engines all being laid up for repair, the passenger trains were worked by the *Nelson*, a coal engine, which finished the day at South Shields. On Monday evening the *Leopard*, a passenger engine, having been repaired at Gateshead, Mr Bell, the foreman of the engine department, ordered it to proceed to South Shields on the Tuesday morning and bring up the butcher's train. Mr Bell also sent a verbal message to Shields for the driver of the *Leopard*.

Two trains were timetabled to leave South Shields at 5 a.m. on what were effectively the two parallel single lines, the butcher's train and a train of empty coal waggons. The message from Mr Bell to the driver of the *Nelson*, which passed through three persons, was that he was to bring up the 5 a.m. coal train, but it was delivered to him stating that he was to bring up the butcher's train, which as he had been working the passenger trains the previous day and was unaware that the *Leopard* had been got into working order, he was hardly surprised by.

Captain J Coddington, the inspecting officer who investigated the accident, pointed out that another contributing factor which had led to the accident was the peculiar way in which what should have been a double line was worked, as far as the three miles nearest South Shields were concerned, as two single lines. Due to the absence of turntables, points and sidings the present owners of the line could not immediately convert it to an

---

514 Board of Trade report

515 Ibid.

ordinary double line. The accident would still not have happened had the *Leopard* engine arrived at South Shields before 5 a.m., as it should have done, but the driver who had taken it out for a trial run the night before had failed to restock it with coke and water afterwards, so that it had to be 'fed' in the morning and was late starting out for South Shields.

Concluding his report, Captain Coddington stated that the new company were doing their best to order new engines, install sidings, crossings and turntables and generally improve the line.[516 517]

With alert engine drivers, slow moving trains, a straight bit of track and probably an element of luck, a head-on collision could sometimes be avoided altogether. In 1844 there was a steamship service between Brighton and Shoreham, England and Dieppe, France. At an unspecified date during this year bad weather prevented the steamship from calling at Brighton Pier, so the London and Brighton Railway ran a special train to convey the passengers from Shoreham, where it could berth, to Brighton. Between Hove and Southwick this train met the regular passenger train from Brighton to Shoreham coming in the opposite direction along the single track. Both engines were carrying red headlamps, their respective drivers saw one another while they were still a good distance apart, they shut off steam, reversed their engines and whistled for the guard's brakes. They were thus just able to pull up short of one another and disaster was averted.[518] There was presumably a thorough inquiry by the railway into what had gone wrong.

A single line collision of a different sort occurred at Ringley, England, on the Manchester and Bolton Railway, on 5th July 1845. In contrast to the situation at Shields, the railway between Manchester and Bolton was a normal piece of double track, but on the night of the accident part of the down line was being re-laid, and trains were being temporarily worked in both directions over the up line. Instructions were given to Driver William Williams, in charge of a spare engine running from Manchester to Bolton, to proceed as normal along the down line to the point where the relaying was in process and to there wait for an up luggage train to pass before crossing to the up line and continuing on that line to Bolton. Instead, he set off from Manchester on the up line and collided with the luggage train at Ringley. Driver Williams jumped from his engine before the collision and escaped with slight injuries. His fireman, John Ryder, who was on the tender at the time, also jumped, but misjudging his leap fell under the wheels and was killed. Driver Williams excused himself by stating that he had expected that the luggage train would have been shunted out of his way onto the down line, though quite why he had this idea is not clear. An inquest on the fireman, held the same day, returned a verdict of 'Accidental death,' and the railway company informed the Board of Trade that Driver Williams had been suspended, to which the Board of Trade replied that they considered this punishment to be quite inadequate for the gravity of his offence.[519 520]

---

516 Board of Trade report

517 The Times, London, England,10th,11th,12th and 22nd October 1844

518 The Railway Magazine, 1955, pg. 668

519 Board of Trade report

520 The Times, London, England, 9th July 1845

The problems with the lack of a means of communication between stations in the pre-telegraph days were well demonstrated by a newspaper report of an incident in July 1846 on the Eastern Counties Railway. The Governors of Christ's Hospital had chartered a special train to take them from the London terminus at Shoreditch to Hertford. The train set off at 9.30 a.m. and travelled as far as the junction with the branch to Hertford at Broxbourne, but here the driver refused to go any further, fearful of meeting the 10.45 a.m. up train from Hertford on the single line. Meanwhile the clerk at Hertford was holding the 10.45 a.m. departure awaiting the arrival of the down special, as per his orders. Eventually, a man was dispatched on horseback from Hertford to Broxbourne to find the cause of the delay and the down special belatedly reached Hertford at 1.30 p.m. Finally, on their return journey, the Governor's train was delayed an hour and a half at Broxbourne due to a burst engine boiler tube. By the end of the day they were doubtless less than impressed with the Eastern Counties Railway.[521]

On 20th September 1849 there was trouble on what was at that date the Lancashire and Yorkshire Railway Company's single track Lytham branch, which ran for five miles from the Preston and Wyre Railway at Kirkham, England to the little seaport. Ordinary traffic on the branch was worked by one engine shuttling between the junction and the terminus, supplemented in the summer months by local trains between Lytham, Kirkham and Fleetwood, worked by a second engine. Between working the passenger trains, the branch engine worked goods and coal trains over the branch and occasionally, if there was a great demand for coal in Lytham a third engine would be sent from Preston to work an extra train. Such extra trains always immediately followed a regular passenger train from Kirkham or Lytham, and the passenger train on arrival at the other end would inform the staff there that a special train was following directly behind it, thus ensuring safety on the single line.

The accident coincided with the holding of a big agricultural meeting at Lytham, with a show of implements and cattle, an event held only at three-year intervals. To cope with the extra traffic the Lytham station staff were supplemented by a locomotive foreman and a goods superintendent from Fleetwood. The last up passenger trains for Fleetwood and Kirkham Junction normally departed one behind the other at 6.50 p.m., the branch engine returning to Lytham and the Fleetwood engine remaining at Fleetwood. On this day it was planned to run two extra passenger trains from Lytham at 9 p.m., and thus at 6.30 p.m. the locomotive foreman instructed the Fleetwood driver to return to Lytham from Fleetwood, bringing with him some carriages to make up the special trains, and telling him that the line would be kept clear from 8.30 until 9. The foreman then upset the arrangements he had made by delaying the departure of the 6.50 trains until 7.40 p.m., and in addition letting the Kirkham train depart in front of the Fleetwood one, thus delaying the Fleetwood train further at Kirkham Junction while the first train was disposed of. Thus, held up, the Fleetwood driver could not get from Lytham to Fleetwood and back to Lytham before 9 p.m., but the foreman gave him no fresh instructions.

Even so, the Fleetwood driver, when he returned from there to Kirkham at 8.40 p.m. would probably have waited at the junction for the first of the two 9 p.m. special trains

---

[521] The Times, London, England, 7th July 1846, reproduced from the Herts Mercury

from Lytham had he not assumed, seeing no carriages left at Lytham when he had left there at 7.40, and having been given a notice directing him to "bring back empty carriages to help make up the special trains," that no special trains could leave Lytham until he arrived there with the carriages to make them up. To be sure though, he checked with the goods superintendent at Lytham, who, making the same assumption as the driver, told him to "return with as many carriages as he could get together, as they had not got a single carriage to go on with, *and to make all haste back*." Unfortunately, neither the Fleetwood driver nor the goods superintendent was aware that there were in fact ten empty carriages stabled on the Dock line at Lytham, about a mile from the station. The locomotive foreman, who knew nothing of the goods superintendent's instructions to the Fleetwood driver, was aware of the carriages, and determined to use them to form the first of the specials.[522]

Thus, the Fleetwood driver arrived at that place at 8.35, quickly made up a train of 10 carriages and hastened to Kirkham Junction, arriving there at 9.20 p.m. There he asked the pointsman if the latter had any new instructions for him. No, the pointsman replied, and reminded the Fleetwood driver that a special was due to leave Lytham at 9, whereupon the driver answered that he knew that, but it could not start until he got to Lytham with the rolling stock for it, and he had been told to bring them there as quickly as possible. Meanwhile, the locomotive foreman at Lytham had made up the 9 p.m. special with the ten carriages from the Dock line, together with three carriages brought from Kirkham by the branch engine. Delayed by the large number of passengers, the up special left Lytham at 9.20 p.m. at about the same time as the Fleetwood driver started from Kirkham Junction.

The two trains met head-on near Moss Side station in what must have been only a slight collision, as neither train was very much damaged and there were only as few minor injuries.

Captain R.M. Laffan investigated the accident on behalf of the Board of Trade. He attributed it firstly to the failure of the locomotive foreman to give the Fleetwood driver fresh instructions when the delay in starting his train made it impossible for him to adhere to the original ones.

Secondly, he blamed the mistaken zeal of the goods superintendent in venturing to direct the passenger trains without consulting the locomotive foreman or telling him what he had done. A classic case, one might think, of the old adage of too many cooks spoiling the broth.

Captain Laffan recommended the adoption of the electric telegraph for working single lines.[523] Sir William Fothergill Cooke had written back in 1842 in a work entitled "Telegraphic Railways" that as well as increasing the safety of double line railways by permitting an interval of space between following trains the electric telegraph could be used to make the working of single lines both safe and flexible, adding that he saw no reason to make "country lanes as wide as metropolitan thoroughfares." The British railway companies had been slow to introduce working by the electric telegraph but in 1858 the Board of Trade had announced that it would refuse to sanction the opening of any new single lines unless they were worked by 'one engine in steam', which in practice

---

522 Board of Trade report

523 Ibid.

was taken to include working with the aid of the telegraph. The use of the electric telegraph also gradually superseded working by the timetable outside of Great Britain; in the United States the use of the telegraph was reportedly pioneered by Charles Minot, Superintendent of the Erie Railroad, in 1851.[524]

A report appeared in the British press in mid-October 1851 of a 'frightful accident' in Russia on about 20th September 1851 at an unstated location on the St Petersburg – Moscow railway. In what was presumably a single line collision, a long train carrying soldiers who had participated in the Coronation Fête in Moscow and were returning to St Petersburg collided with a train coming from the latter place. Many persons were stated to have been killed but the exact number had not been published. Many others were injured. The passengers killed on the train from Moscow included some civilians who had been granted permission to travel on the troop train.[525]

Another accident of the type involving a head-on collision during temporary single line working occurred at Port Meadow, near Oxford, England, on 3rd January 1853, with serious consequences. The London and North Western Railway branch from Bletchley to Oxford was double track, but part of Wolvercote tunnel, three miles from Oxford, had collapsed, and to expedite the removal of a large mass of earth from one end of the tunnel and the conveyance of building materials to its other end, the down line between Islip and a point just over a quarter mile from Oxford was given over to the contractor engaged in the tunnel repairs, the up line only being used for general traffic. Through the tunnel itself, however, but one line was in use for both the ordinary railway and the contractor's traffic, a policeman being stationed at the points at each end of the tunnel. The electric telegraph was installed along the line.

On the evening of 3rd January, the 5.30 p.m. up passenger train was waiting to depart from Oxford but had to await the arrival of a down coal train from Bletchley, due at 5.20. The coal train did not normally run right into the passenger station but was shunted into a siding at the ticket platform 300 yards out. It was difficult to see from the station platform when it had arrived, and Stationmaster Alfred Blott had been instructed that if there was any doubt as to whether or not it had done so he should get Foreman Porter William Hayes to walk out to the swing bridge, 200 yards from the station, to see. At about 5.28 p.m. Stationmaster Blott advised the guard of the passenger train, Joseph Kinch, that the coal train had let Islip and that he must not start the passenger train till it arrived. The stationmaster then returned to his office. The guard passed on the information to his driver, John Tany. About the same time, Foreman Porter Hayes went along the train to close the carriage doors and said the same thing to the driver and guard, then set off towards the swing bridge to check if the coal train had come in.

At about 5.25, the same time that the coal train was expected, a train arrived, but this was not the coal train but a contractor's engine, running tender first, pushing a single waggon, which it placed in a siding and then went into the engine shed road. Foreman Porter Hayes, not having seen the coal train, was then walking back to the station platform when he saw the passenger train setting off. He shouted to Driver Tany that the

---

524 The Railway Magazine 1959 pgs. 745-51

525 Morning Chronicle, London, 11th October 1851

coal train was not in, but Tany's engine was blowing steam through its cylinders and the driver did not hear him. Guard Kinch said that he heard the foreman porter shout something and, although he could not catch what it exactly was, it came to him that it must mean that the coal train was not there, and he had applied his brake and waved a red light to his driver, but in vain, and his train kept going. Rather less than a mile out of the station it collided head-on with the coal train, which consisted of two engines, 44 waggons and a brake van.

The passenger engine was thrown onto its side into the ditch separating the London and North Western and Great Western lines and the leading engine of the coal train was thrown alongside it, the two being locked together. The first-class carriage and guard's van of the passenger train were little damaged but the second and third-class carriages were shattered to pieces. The collision killed Driver Tany and his colleague, Driver Samuel Wilcox of the coal train, Firemen Robert Bugden, Thomas Landon and Robert Laws and passengers Julia Norman, a nursery maid; and an unidentified labourer. A third passenger died later.[526]

Exactly why Driver Tany had been under the impression that the coal train had arrived was a mystery. It was naturally supposed that he had mistaken the contractor's engine for that of the coal train, but the former carried a white headlight, while the coal train should have carried a green headlight. As the contractor's engine had been travelling tender first its headlight would not have been visible anyway and Captain George Wynne, who undertook the Board of Trade investigation, concluded that Driver Tany must have relied upon hearing the sound of the contractor's engine. If the coal train had been carrying a green headlight, it was also strange that the driver of the passenger train, in the words of the Board of Trade report, had 'rushed madly on to destruction.' Captain Wynne was not convinced by the evidence of the stationmaster, policemen and porters that the passenger train had started off at an abnormally high speed, which was not in agreement with that of a very intelligent passenger, and he thought that the station staff were trying to throw all the blame for the accident on the recklessness of the dead driver. Moreover, statements by Guard Kinch and other staff that the goods train had been showing a green headlight were contradicted by two independent witnesses, a Great Western Railway watchman named Brooke and a poulterer named Stringer, both of whom stated that there had been no headlight at all on the coal train engine.

Captain Wynne said that it would probably never be cleared up who had actually started the train, and perhaps it was not that important. The working of a temporarily single line was obviously even more dangerous than that of an established line, and if the company were going to rely on using just the electric telegraph, without the ordinary precaution of a pilot engine, it was essential that all trains should be despatched by the stationmaster *personally*, or in his unavoidable absence to an appointed assistant. The company had drawn up written instructions for the working of the single line, but, the inspecting officer considered, they had been loosely and carelessly drawn up, and placed no responsibility on any single individual. A printed circular from the general manager made it the duty of the guard to start his train from roadside stations but Captain Wynne did not consider that it applied to terminal stations, though the traffic superintendent considered that it did, and that the stationmaster did not need to personally despatch the train but could depute a subordinate to do so. When the passenger train had departed

[526] Board of Trade report

neither the stationmaster nor the foreman porter had been on the platform and it had been left to the discretion of the train crew when to start.

Captain Wynne was not impressed and stated that it was his opinion that there was no security for the public if the superior officers could at any time relieve themselves of responsibility by stating that their subordinates had disobeyed orders, and then, when asked to produce such orders, had pleaded that they had been verbal or understood. He therefore placed the entire responsibility for the accident and its consequences upon the company.[527]

The coroner's jury, at the inquest upon the seven people killed outright in the collision placed the blame on Guard Kinch, upon whom they considered the duty of starting the train had devolved, while blaming Driver Tany for starting off too fast. They recommended that a pilot engine should be employed, that a policeman should be stationed between the swing bridge and Wolvercote tunnel, that London and North Western and Great Western trains should carry distinctive marks to prevent a train on one line being confused with a train on the other and that a means of communication between guards and drivers should be devised.[528]

A misunderstanding of telegraph messages almost led to a second single line collision further up the line between Bicester and Winslow only two days after the Port Meadow one.

A serious collision occurred on the Camden and Amboy Railroad near Old Bridge, New Jersey, United States, on 9th August 1853. A passenger train which had left Camden, New Jersey with a 2 p.m. connection from Philadelphia, Pennsylvania, was in collision with one which had left Amboy with a 2 p.m. connection from New York at a point eight miles from Amboy at 5.30 p.m. The six car Camden train was only travelling at five or six miles an hour but the five-car train from Amboy, which was late and was trying to make up time, was going considerably faster. The collision occurred on a sharp curve and the Camden train had just emerged from a wood. The train crews of both trains just had time to jump off before the impact. The locomotives of each train, Nos. 40 and 41, both nearly new, were locked together and the leading and second cars of each train telescoped into one another. Several passengers on the train from Camden were injured but they were luckier than those on the opposing train, as four passengers on the train from Amboy were killed outright. About twenty passengers were injured. Those who lost their lives were the nurse and a child of Mr D.H. Dustin, of New York, an unknown Irish woman and an unknown German man, apparently on his way home from California.[529]

The regulations of the railroad company were that if trains arrived at meeting point and the opposing train was not there, they were able to proceed at much reduced speed, looking out for the other train. The trains were due to meet at Old Bridge. The Camden train, whose crew's watches had been correct, had arrived at Old Bridge a trifle after the appointed time and finding the Amboy train was not there had proceeded cautiously. The watches of the engineer and conductor of the Amboy train were two and a half minutes slow and its engineer, supposing that he had another minute to reach the siding, had been

---

527 Ibid.

528 The Times, London, England, 5th,6th,7th,8th,12th,13th and 18th January 1853

529 New York Daily Times, New York, New York, United States, 10th August 1853

running at at least twice the regulation speed.[530]

By a majority verdict, the inquest jury found that the accident was due to the gross negligence of Engineer John Anderson of the train from Amboy in running his train at excessive speed (said to have been 40 miles an hour) and that Conductor Edward Muschamp of the same train was highly censurable for failing to check his watch, and that of the engineer, with the standard clock of the Camden and Amboy Company in New York. The crew of the train from Camden and the Camden and Amboy Railroad Company were completely exonerated from blame. Six of the 21 members of the jury considered that the directors of the company were also to blame for the accident.[531]

An accident in England two days after that at Old Bridge took place on a single-track branch of nearly five miles in length which was worked in a very casual manner. The Byers Green branch of the West Hartlepool Harbour and Mineral Railway was a mineral branch, but a passenger train was run on Saturdays to enable the miners and their families to get to the market. There were no regulations to prevent more than one engine or train being on the single line at the same time and there were usually two or three and sometimes as many as seven. Speeds over the branch were limited to seven or eight miles an hour and there were five sidings, situated at three-quarter mile intervals, where trains could cross one another. Empty trains going up the branch were meant to give precedence to loaded trains coming down, and light engines to empty waggon trains. There was a signalman at the junction of the branch with the main line, but he had no instructions for working the branch and there were no signalmen at the sidings on the branch itself.

At the time of the accident, some large works were in progress at a point on the branch three miles from the main line junction for a new Whitworth colliery and the railway company had provided a carriage which was stabled at the junction and used to convey workmen up the branch to the works in the morning and back down in the evening, attached to one of the empty waggon or loaded trains. The workmen were not charged for riding in this carriage, it was not attached to any specified train and no guard was provided to accompany it. On the morning of Thursday 11th August 1853 an engine left Ferryhill station at 5.25 a.m. for the branch and arriving at the branch junction its driver allowed the workmen to attach the carriage to the front of the engine. Empty waggons were taken on behind the engine and the train proceeded up the branch, propelling the carriage before it. It was a foggy morning and after the train had proceeded some distance and was near Tudhoe Iron Works, it met a light engine coming down which had just taken empty waggons up the branch and was now bound for the workshops at Stockton. The drivers could not see one another very quickly because of the fog and the collision was violent enough to smash the workman's carriage, which was carrying nine men and several boys. All were hurt and three were killed outright.[532]

Captain Douglas Galton, the inspecting officer, noted that the Saturday market train did not appear to be worked with any greater care. He remarked that the fatal consequences of the collision would have been avoided had the engine not been propelling the carriage, but that a higher degree of blame must rest upon the very lax

[530] Ibid, 16th August 1853

[531] Richmond Morning Mail, Richmond, Virginia, United States, 16th August 1853

[532] The Observer, London, England,15th August 1853

system upon which the branch was worked and that the directors of the company bore a very heavy responsibility for the accident by allowing the men to travel on the branch without having issued any regulations which required the most ordinary precautions to be taken for their safety. He was informed that the traffic over the branch had increased considerably and he recommended that the branch be doubled, and some safe method of working introduced.[533]

Back in the United States, yet another single line collision, the third such accident in either England or the United States in three days, occurred on Friday 12th August 1853 near Valley Falls, Rhode Island, on the Providence and Worcester Railroad. Half a mile south of Valley Falls station the single-track Providence and Worcester line joined the Boston and Providence line and thence southward to Providence there was double-track. The trains involved in the accident were the regular up train and a down excursion train, which was also a regular working during the summer months, carrying passengers from Whitinsville, Uxbridge and other points along the line to Providence. The excursion trains were run in connection with excursion boats plying Narragansett Bay and were very popular with mill workers and other local residents seeking a respite from the summer heat. The excursion train involved in the accident was packed with 475 passengers in eight heavily loaded cars.

According to the timetable, the two trains were scheduled to meet at the end of the double track at Boston Switch at 7.32 a.m. and the excursion train had the right of way, but were it more than five minutes late it lost its right of way and both trains had to proceed cautiously. The up regular train waited the allotted time at Boston Switch and then proceeded slowly towards Valley Falls, entered a curve and collided with the down excursion train only a short distance from the beginning of the single line. The excursion train was running fast, reportedly at 40 miles an hour, in order to reach the double track, and the two collided violently. The two engines were totally demolished, and the cars of the excursion train were badly damaged; that next to the engine, which contained about sixty passengers, all from Whitinsville, was shattered, and its occupants mostly killed or injured. The second car suffered less but was driven back twenty feet into the third car, like a spy-glass according to a contemporary report, causing heavy casualties. The remaining cars were less affected, and the two coaches of the regular train, a first-class and a second-class, were not greatly damaged. None of the passengers in the regular train was killed or severely injured. Thirteen persons were killed and about 40 injured. The dead were all passengers, mostly factory operatives, except for John Parkins, of Uxbridge, the fireman of the excursion train, who was scalded to death. His driver, Ephraim Gates, had time to jump and survived. The first assistance came from those on board of the 7.30 train from Boston, which arrived at Boston Switch on the Boston and Providence line five minutes after the collision. After the accident, the injured were taken to Pawtucket, a little south of Boston Switch, and the bodies of those killed and those thought to be dying were taken to Valley Falls.[534]

A Boston newspaper headlined the accident as 'Twelve or Fourteen Persons Murdered by a Railroad Collision;' a somewhat libellous statement but presumably one which reflected the outrage of its readers who had already been shocked by news of the Old Bridge accident only three days earlier. The paper attributed the accident to the

[533] Board of Trade report

[534] New York Daily Times, New York, New York, United States,13th and 15th August 1853

carelessness of the conductor of the excursion train, who had been anxious not to delay his train and so miss the steamboat connection at Providence.[535]

An inquest on the accident was opened the same day in the Town Hall at Valley Falls before Coroner George Dana and a jury. The first witness, Owen Smith, stated that the excursion train had left Valley Falls at 7.34 a.m. by his watch. He had heard the conductor, Frederick Putnam, ask Engineer Gates, if he could reach Boston Switch in four minutes, and tell him that that was all the time that he had. He had not heard the engineer's reply, but the train had started at once. Hearing the sound of the collision he had run to the site of the collision, reaching there at 7.37. As soon as he got there, Conductor Putnam had checked his watch with that of the baggagemaster of the regular train; the conductor's watch had read 7.35, and thus was two minutes slow.

Depot Master Johnathan Chase at Valley Falls stated that the excursion train was due there at 7.29 but did not arrive until 7.36. He had remarked to Transportation Master Stevens of Providence as the train was about to start that it was too late. Stevens told him later that he had not heard him.

At the resumed inquest on Saturday 13th, Brakeman David Pott testified that he had seen the conductor and Mr Stevens compare watches at the previous stations but not at Valley Falls. When he had seen the regular train, he had applied the brake and then jumped off.

Engineer Gates testified that his train had left Whitinsville at 6.25 a.m. According to his watch it had arrived at Valley Falls at 7.32. There, Conductor Putnam had told him that they had four minutes to reach Boston Switch and he had replied that that was enough. The regular train was due at 7.32 and according to the regulations must wait five minutes beyond that time for them, giving them five minutes to get to the switch. He had run his train up to Boston Switch at the usual speed. On seeing the regular train, he had reversed his engine and whistled for brakes but had not been able to prevent the collision. Subsequent to the accident he had given his watch to the ticket master at Province to check its accuracy and it had been right, not varying with the Company clock by more than five seconds. He stressed that although he had been under the orders and direction of the conductor, he would not have started his train from Valley Falls unless he had felt that he had had sufficient time to reach Boston Switch.[536]

The inquest concluded on Wednesday 17th. In a lengthy verdict, the jury found that the accident had immediately resulted from the culpable carelessness, inexperience and want of judgement of Conductor Putnam, in leaving Valley Falls with only four minutes to reach Boston Switch and in not providing himself with a suitable and correct watch, the timepiece having lost two minutes since it was compared with the Company clock at Providence the night before. The jury blamed the company management for appointing such a young and inexperienced man as conductor of such a difficult train to work, for not checking that every conductor had a suitable and correct watch and for not being aware that the Uxbridge train had been almost invariably late since it had been in charge of Conductor Putnam and applying a remedy. With, one would have thought, much less reason the jury blamed the conductor of the regular train for permitting Engineer Freeman and the brakeman to be away from their posts so that the regular train was a

---

[535] The Boston Post, Boston, Massachusetts, United States,13th August 1853

[536] New York Daily Times, New York, New York, United States,15th August 1853

minute late in starting and thus met the excursion train on a curve rather than on a straight stretch of line. Finally, the jury concluded that the whole management of the railway was bad and that there should be a signal within view of Valley Falls station to warn Uxbridge (down) trains of the approach of a Providence (up) train.

Conductor Putnam was arrested for manslaughter after the inquest. He was said to be deeply affected. He was apparently only temporarily acting as a conductor, on a brakeman's wages, and not having been provided with a watch by the company had had to borrow one from a milkman. On a previous occasion, he had apparently reached Boston Switch with an excursion train only one minute ahead of an up train. Conductor Putnam, Engineer Gates and Master of Transportation Stevens were dismissed.[537]

It has not been possible to trace any report of Conductor Putnam's trial, if indeed one took place.

On Sunday 18th September 1853 the Chemin de fer de Paris à Orléans mail train which had left Paris at 7.15 p.m. collided head-on with a double-headed goods train between Poitiers and Ligugé at Saint Benoit. The mail train was travelling at full speed when the trains met on a sharp curve. The mail train engine and the two engines of the goods train were smashed to pieces and two drivers and two firemen were killed, though M. de Sezenay, an inspector riding on the locomotive of the mail train, was thrown off the engine by the shock and escaped almost unhurt. The conductor of the mail train, riding in the luggage van immediately behind the engine, sensing that something was wrong, had opened the sliding door of his van to look out as the collision occurred and was decapitated. About twenty passengers were injured, one of whom also subsequently died. The fact that there were six goods waggons and luggage vans at the front of the mail train probably saved the lives of a number of passengers. The blame for the collision was initially placed on the driver of the goods train, which was running several hours late, who had tried to reach Poitiers before the mail train left there and paid for his mistake with his life.[538] [539] [540] It would seem that others were also judged to be at fault though, for on 28th November the Poitiers Criminal Court sentenced four officials to terms of imprisonment ranging from two years to six months and fines ranging from 1500 to 300 francs.[541]

Exceptionally severe winter weather and mistakes made by senior officers led to a serious head-on collision between Thetford and Harling Road stations, England, on the Eastern Counties Railway, on 5th January 1854. However, it will be seen that strictly speaking this was not a collision during temporary single line working inasmuch as the collision did not actually occur on what was meant to be the single line in use, but even so this chapter seems the most appropriate place in which to deal with the accident.

A heavy snowfall over the night of 3rd and 4th January had blocked the lines of the Eastern Counties so that traffic was stopped on some sections and various telegraph wires

---

[537] New York Daily Times, New York, New York, United States, 15th and 18th August 1853

[538] Morning Post, London, England, 23rd September 1853

[539] Bell New Weekly Messenger, London, England, 25th September 1853

[540] The Magnet, London, England, 26th September 1853

[541] Wikipedia – Liste des accidents ferroviaires en France au XIX siècle

were brought down, stopping communications. Mr Peter Ashcroft, the superintendent of works, was snowed up at Yarmouth and the line was badly blocked at Thetford. With no communication with Yarmouth, the job of clearing the lines devolved to the local permanent way inspector, named Howard, who was at Wymondham. Howard began by telegraphing for two engines to be sent from Norwich, at the same time being informed that an up goods train was snowed up at Thetford. Having cleared the Dereham branch from Wymondham, Inspector Howard proceeded on an engine to Harling Road, where he told Stationmaster Charles King to let nothing past until the latter heard from him, and on towards Thetford, but found both lines blocked four miles further on. He enlisted the help of some platelayers and by the end of the 4th the down line had been cleared, but the wires were down at Thetford, so he sent a guard on an engine to Harling Road with a verbal message to telegraph Norwich from there and inform them that only the down line was open between Harling Road and Thetford.

Howard then had a few hours well-earned sleep at Brandon, the next station on the London side of Thetford, but at 6 a.m. on 5th January he was informed that the up mail train from Norwich was stuck in the snow on the *up* line between Harling Road and Thetford. Stationmaster King at Harling Road explained later that he had been told by a guard that *both* lines were clear. Howard set about freeing the mail train and while he was there first a parliamentary train and then an express arrived on the up line behind it. These two further trains did not get stuck in the snow and Inspector Howard accompanied them as they returned in the wrong direction over the up line to Harling Road. From there Howard despatched the express to Thetford over the down line, accompanied by Guard David Briggs who was ordered not to allow any train to leave Thetford until he (Howard) reached there. Inspector Howard shortly followed to Thetford with the parliamentary train. As down trains were waiting at Thetford Inspector Howard and Guard Briggs then returned to Harling Road on an engine, and the 11.45 a.m. up train not having arrived there either Inspector Howard left Guard Briggs at Harling Road to pilot the 11.45 train to Thetford over the down line, and himself went back to Thetford on an engine to clear the up line. At 1 p.m. Guard Briggs arrived at the point where the up line was being cleared on an engine and told Howard that the 11.45 had still not arrived at Harling Road and that he (Briggs) was going to Thetford to see if any down trains were waiting there. Inspector Howard went with Guard Briggs to Thetford and they found a number of passengers there, upon which Howard sent them to Harling Road with Briggs on the engine, impressing upon Briggs not to let any trains up over the down line until he (Howard) returned. Stationmaster King at Harling Road had also been instructed again that traffic was only to be worked over the down line, and that all trains must be piloted by Guard Briggs.[542]

No sooner had Briggs left Thetford for Harling Road than two down trains arrived at Thetford. Finding that the telegraph was still not working to Harling Road, Howard, determined to work the two down trains over the up line in the hope of being able to force a way through the snow which had not been cleared, depending for safety on the instructions, had left with the Harling Road stationmaster. Meanwhile Mr Ashcroft, the works superintendent, had got the line cleared at Yarmouth with a large force of men and had come with his men from Norwich to Harling Road to help clear the blockage there. Both Stationmaster King and Guard Briggs protested at Mr Ashcroft continuing towards

542 Board of Trade report

Thetford on the up line as it being contrary to the very specific instructions that they had received from Inspector Howard, but Mr Ashcroft, perhaps presuming that the order did not apply to work trains, elected to continue, accompanied by Permanent Way Inspector John Latham and the body of men.[543]

The two trains met on a long curve about midway between the two stations, each driver assuming, until the other train was close at hand, that it was on the other line, though it might have occurred to those on the up train that as the pilotman, Guard Briggs, was still at Harling Road there should not have been a train on the down line. The collision was a violent one and the up train suffered most. The bodies of Fireman Hipperson, Platelayers Baldwin and Underson and a carpenter named Smith were found in the wreckage and a passenger, the Rev Mr Joseph Bell, curate of Bunwell, Norfolk, died of his injuries on Sunday 9th. Two other passengers, Mr Phillips, a Birmingham silversmith, who broke both legs, and Mr Ellison, a Norwich ivory turner, were seriously injured, one of them probably died subsequently. Inspector Howard, who jumped from the engine of the down train, and Mr Ashcroft escaped injury, but Inspector Latham was seriously hurt. Twenty labourers were also more or less hurt.[544]

The inquest juries returned verdicts of manslaughter against Inspector Latham and Mr Ashcroft.[545] The Board of Trade inspecting officer, Captain George Wynne, declined to voice an opinion as to whether Inspector Howard, who had taken a train of passengers in the wrong direction over a line which was virtually closed, without the normal precautions; or Mr Ashcroft, who although travelling in the right direction, had forced the stationmaster to allow them to proceed after the stationmaster had been very specifically ordered by Inspector Howard not to let anything proceed from his end, was most to blame. He observed that it was contrary to usual practice to work trains over the line which was supposedly out of use during temporary single line working without giving notice to the other end of the section, especially in the wrong direction. On the other hand, disobedience of precise orders, such as those given to Stationmaster King and Guard Briggs by Inspector Howard could in no way be defended. There was, however, no written record of Inspector Howard's order.

Captain Wynne was of the opinion that the arrangements for single line working made by Inspector Howard had been defective in a number of ways. The orders that he had issued had all been verbal instead of written and that intended to be telegraphed to Norwich had never been sent. Furthermore, a pilot engine should have been provided to allow Guard Briggs to travel from one end of the section to the other as was required. The inspecting officer found it strange that a local inspector with many years' service should have been content to rely on verbal orders and that the Harling Road and Thetford stationmasters had been so willing to acquiesce in the matter.[546]

Disaster struck an excursion train returning from an Independence Day celebration at Rider's Grove to Baltimore, Maryland, United States on 4th July 1854. The fourteen car Baltimore and Susquehanna Railroad excursion train had reportedly been carrying two

---

543 Board of Trade report

544 London Mail, London, England, 24th January 1854

545 The Magnet, London, England, 16th January 1854

546 Board of Trade report

thousand passengers when it collided head-on with a regular passenger train from Baltimore to York, Pennsylvania, at Riderwood, near Relay, at about 5.30 p.m.

Three excursion trains had been run from Baltimore to Rider's Grove. One returned to the city already at 2 p.m., the second departed from Rider's Grove at 5 p.m. and the third, the one involved in the accident, at 5.15. The regular accommodation train, consisting of four or five cars, had left Calvert station, Baltimore, at 4.25 p.m., under Conductor William Scott, with orders to cross excursion trains from Rider's Grove at Relay, and that if the trains were not there at 5.10 to wait until 5.20 and then allow ten minutes 'grace time' before proceeding, when the excursion trains would keep out of its way. The accommodation train went onto the Green Springs branch to cross a delayed express from York and waited for the second excursion train, consisting of nineteen cars filled with excursionists. It then proceeded but had only travelled three-quarters of a mile when it met the third excursion train. The cars of the latter train were being propelled by the engine and in the collision the locomotive of the accommodation train 'crushed and shivered' half a dozen of them to atoms. After the collision, 'Axes and crowbars were brought into requisition, and those alive and unhurt made superhuman efforts…the cries for water from the sufferers was continual, and several persons were engaged constantly in supplying them.'[547] The last of the injured could not be extricated until 7.30 p.m. Though most of the deaths were on the excursion train, one or two of those on the York train also lost their lives, including Baggagemaster Benjamin Merryman, killed as he was screwing down the brake. Eight hundred or a thousand uninjured passengers walked the nine miles into the city, suffering much from the heat and dust but thankful to have escaped death or injury. By 6th July 28 people had died; the final number of dead was 34 and many others were injured.[548]

As was usual at this time, a coroner's jury was quickly convened and found that the accident was due to the carelessness of Conductor Scott and the gross negligence of the company in failing to give their rules in more explicit language.[549] Conductor Scott was dismissed by the company and died at the relatively young age of 49, still haunted by memories of the disaster.[550]

The Riderwood or Riders Grove disaster was to be followed by one across the border in Canada less than four months later. Like many lines in the United States and some elsewhere the Great Western Railway of Canada was built on the cheap, while its construction was also a rushed job. The result was that when it opened the ballasting of its tracks was insufficient, meaning very rough riding trains and necessitating very slow running to avoid derailments. In an effort to improve the situation, the company employed private contractors, though it supplied the locomotives, rolling stock and train crews. In order to maximise their profits, the contractors needed to operate their trains as much as possible, something which was likely to interfere with the ordinary scheduled trains, especially in the absence of telegraphic communication along the line. One source of ballast used by the contractors was a ballast pit at Stony Point, about 20 miles west of

---

547 Peru Miami County Sentinel, Peru, Indiana, United States, 13th July 1854

548 Cooper's Clarksburg Register, Clarksburg, West Virginia, United States, 12th July 1854

549 A History of Railroad Accidents, Safety Precautions and Operating Practices by Robert B Shaw, pgs. 65-7

550 The Story of the Northern Central Railway by Robert L. Gunnarrson, pgs. 36-7

Chatham. Most of the ballast trains run from there were under the charge of Conductor T.D. Twitchell and by late October 1854 he was sending them out well before dawn in an endeavour to complete the ballasting work before winter set in.

A Great Western passenger train left Suspension Bridge (Niagara) at 2 p.m. on October 26th, in the charge of Conductor G.F. Nutter with Engineer Thomas Smith, consisting of an engine and tender, an express and mail car, a baggage car, two second-class and four first-class coaches. It was due to pass Baptiste Creek, about five miles on the Chatham side of the ballast pit and the point where the accident occurred, at 10 p.m. arriving at Detroit, Michigan, United States at 11.20 p.m., but it was much delayed. One and a half hours were lost at St George, where a gravel train was derailed, then more time was lost following a slow freight train to Princeton. It was 1 a.m. on 27th October before the train left London, and three or four miles beyond there an engine cylinder burst, and another engine had to be sent from London to draw the train back to there before it could continue with the fresh engine. With further slow running the train was no less than seven hours late when it approached Baptiste Creek at 5.10 a.m.

At 3.30 a.m. on 27th October the night watchman at the ballast pit, Patrick Pine (or Pyne) had woken up Conductor Twitchell and the engineer of the gravel train, named Kettlewell, then prepared the engine of the ballast train while they ate breakfast. There was contradictory evidence regarding enquiries made of the night watchman as to what trains had passed, it being one of Pine's duties to note the passing of regular trains during the night and so inform the ballast train crews. Conductor Twitchell alleged that he had asked Pine about westbound trains and Pine had replied that one had passed about 11 o'clock, but the night watchman denied that the conductor had asked him at all. Pine confirmed that Engineer Kettlewell had asked him and said that he had told the engineer that a train had passed going eastwards between midnight and 1 a.m., though the engineer said that he had enquired as to *westbound* trains. The next westbound train after the one involved in the accident was not due until 8 a.m., the next eastbound not until 12.30 p.m.

Thus, reassured that there were no trains due for some hours, the crew of the ballast train set off eastwards from Stony Point at 4.30 a.m., with the engine pushing fifteen cars loaded with wet gravel. Conductor Twitchell, holding a red lantern, and a negro 'helper' rode on the front of the leading car. It was still dark and there was a thick fog. Suddenly, the westbound 'express' (it was only travelling at 20 miles an hour) loomed out of the fog. The backing gravel train was moving at 10 or 12 miles an hour. There was little chance for either train to reduce speed and the collision was a violent one. The two second-class carriages, filled with German immigrants, suffered the most, though the front of the leading first-class carriage was also smashed, and the passengers in that half of it were killed or injured. Engineer Smith and his fireman survived and Conductor Kettlewell jumped and also escaped but the negro boy was killed, as was a brakeman on the ballast train. All the others killed were passengers. Forty-seven people died at the scene, one of the injured died on the way to hospital and nine more later, so there were at least 56 dead, though as many of the passengers were immigrants, difficulties of language, combined with official indifference and the desire of the survivors to get away from the scene may have meant that some deaths went unrecorded.

At least one newspaper covering the accident a few days later placed the blame for the accident on Night Watchman Pine, whom it as alleged had fallen asleep and later, when asked if the express had passed, had stated that it had on the assumption that it must have done while he was sleeping.[551] The subsequent investigation failed to produce any evidence that this had happened, and the focus then shifted to Conductor Twitchell. It emerged that on 12th October the conductor had ordered the gravel train out when a superior train was due and Engineer Kettlewell had remonstrated with him and later complained to his superintendent. The superintendent had reminded the engineer that the conductor had charge of his train but had written to Twitchell advising him that he would be held to blame should any accident occur as a result of any further transgression. Furthermore, after the accident Twitchell had hurried to the house of a friend and asked him to hide him until he learnt how things had gone and had exclaimed, "Great God, I have killed forty or fifty poor souls who have never done me any harm, and what in God's name shall I do?"

The jury at the first inquest on the victims of the disaster could not agree on whether to blame just Twitchell or both him and the Great Western management. They were dismissed on 3rd November and the next day a second jury found both Conductor Twitchell and Engineer Kettlewell guilty of manslaughter for having taken their train out in a thick fog and not having ensured themselves that the express train had passed. The company was censured for its failure to keep a watchman at switches and crossings.[552] [553] No report has been traceable of a trial of the conductor and engineer so possibly the manslaughter charges were later quietly dropped.

The collision, along with several other less serious accidents and incidents on the Great Western Railway, was the subject of a special investigation by a board of commissioners. They found that the person most directly to blame was Conductor Twitchell, that the action of Engineer Kettlewell in complaining to the superintendent after the near-miss of 12th October was most commendable, and that Twitchell should have been dismissed at that time; that it was not fair to expect Night Watchman Pine to keep a note of what trains had passed during the night and instead a proper switchman should have been provided at the ballast pit, that the company should have undertaken the ballasting operations themselves rather than leaving them to a contractor, and that the regulations should have prohibited the running of ballast trains during the hours or darkness or during a fog.[554]

Bad weather and the absence of a specific regulation prohibiting more than one engine being on a stretch of single line at one time caused an accident on the short Manchester Sheffield and Lincolnshire Railway branch between New Holland and Barton-on-Humber, England, on 14th February 1855. Snow was falling heavily and as the 3.50 p.m. from Barton, due at New Holland at 4.02pm, had not arrived by 4.10, when the London express was due to depart, Stationmaster William Lewis became concerned, and suspecting that the train had become stuck in a snow drift on the branch, went to the engine shed and discussed with Mr William Gutzmer, the fitters foreman, and mentioned the idea of going down the branch with an engine to find what had happened to the missing train. Before anything could be done, however, the train arrived, and its driver,

[551] Morgantown Monongalia Mirror, Morgantown West Virginia, United States, 11th November 1854

[552] Wreck! By Hugh A. Halliday, pgs. 17-22

[553] A History of Railroad Accidents, Safety Precautions and Operating Practices by Robert B Shaw, pgs. 67-9

[554] Reports of the Commissioners Appointed to Inquire into a Series of Accidents and Detentions on the Great Western Railway, Canada West, by Commission Bearing Date Nov 3, 1854

George Woolley, told Stationmaster Lewis that he had been held up by snow on the line, especially at the level crossings. The stationmaster then asked Mr Gutzmer for an assisting engine to attach to the branch train but was told there was none to spare. Guard Robert Garbutt and Driver Wooley of the Barton train also separately asked Mr Gutzmer for engine No. 56 in place of the less-powerful branch engine, No. 10, but their requests were refused.[555]

The 3.50 Barton train eventually arrived at New Holland at 4.25 and set off again for Barton again at 4.45 p.m. instead of the booked 4.20. There was no delay on the trip down the branch and then, at 5.30 p.m. the train set off from Barton station once more for New Holland. It appeared that Mr Gutzmer had not wanted to part with engine No. 56 because he believed that he might need it to assist a train from Lincoln which arrived at 5.05 p.m. When that train appeared, and he found that it did not need the assistance that he had expected, he belatedly decided to use engine No. 56 to assist engine No 10 on the up-branch train from Barton. Meanwhile, he had let the driver and fireman of engine No. 56 go for their tea and so set off down the branch, unbeknown to the stationmaster, with himself driving the engine and with Henry Rogers, the stoker of a stationary engine, as fireman. The up-passenger train and the light engine collided at 5.35 p.m., about a mile out of Barton. The branch was dead straight, with good visibility in clear weather and daylight, but it was snowing hard and dusk and the drivers only saw one another at the last moment. Two passengers in a third-class carriage second from the engine were fatally injured and three passengers and two railwaymen were more or less injured. The passengers who died were Mr Thomas Judd, a young clerk with the *Hull Advertiser* newspaper, and Mr James Young, a Hull butcher, who later succumbed to his injuries.

At the inquest on the victims of the accident, Mr Gutzmer stated that he had understood that the driver of the up passenger train was going to wait at Barton for him, that he had entered the branch at 5.12 and should have reached Barton before the passenger train departed and had only travelled at four miles an hour all the way; his fireman, whose statement the Board of Trade inspector considered closer to the truth, said they had left at 5.18 and initially had been going at eight miles an hour.

Captain Henry Tyler, the inspecting officer, was critical of the fact that the railway company had no regulations specifically pertaining to the working of trains over a single line and considered that the company should issue a special regulation prohibiting the use of two engines on the branch at the same time. Otherwise, he considered that Mr Gutzmer had simply made an error of judgement, though he had no doubt taken the engine down the branch with the best of intentions. An inquest jury were less charitable and found Mr Gutzmer guilty of manslaughter.[556] [557] [558]

Mr Gutzmer was found 'Not Guilty' at his trial at the Lincoln Assizes on 10th March, a verdict with which the judge, Mr Baron Alderson, concurred, remarking that if he had been in Mr Gutzmer's place he should have understood that Woolley was expecting him.[559]

---

555 Board of Trade report

556 Ibid.

557 The Times, London, England, 19th February 1855

558 The Patriot, London, England, 26th February 1855

559 The Express, London, England,16th March 1855

The worst railroad accident in the United States, up to that time, occurred at Camp Hill, Pennsylvania, on the North Pennsylvania Railroad, on Thursday 17th July 1856. On that day, two excursion trains were to be run from the Market Street depot in Philadelphia the 14 miles to Fort Washington; conveying pupils, parents and priests of St Michaels Roman Catholic Church on a Sunday School trip, with a picnic at the destination. The first excursion was due to depart at 5 a.m. in the charge of Conductor Alfred Hoppel, with Engineer Henry Harris in charge of the locomotive, named *Shackamaxon*. It naturally took some time to load 600 or so passengers, many of them children, into the thirteen cars and the train left almost half an hour late. Conductor Hoppel of the excursion train had calculated that his train could reach Camp Hill in time to cross the regular train if they ran fast but Engineer Harris was only able to make up a few minutes of time.

Meanwhile, the regular southbound train, the 6 a.m. departure from Gwynedd, just two cars hauled by the *Aramingo* engine, in the charge of Conductor William Vanstavoren with Engineer William Lee, had left Fort Washington one minute early at 6.14 a.m. It reached Camp hill four minutes later. When it stopped there, several of its passengers left the train for breakfast at the adjoining hotel, assuming that the train would wait there for the excursion, but Conductor Vanstavoren immediately called them back and ordered Engineer Lee to proceed. The engineer, aware of the excursion train, was reluctant to do so, and said to the conductor, "Very well," adding "You had better not." The conductor replied, "All right, but sound your whistle pretty freely." [560] The train then set off, slowly, and whistling continuously, Engineer Lee feeling that he had no choice but to obey the order of his conductor. His train had only covered 300 yards when it collided with the excursion train.

All the casualties in the collision occurred in the ill-fated excursion train. The three leading cars of the excursion were crushed into the *Shackamaxon* engine, which locked together with the *Aramingo* and climbed up into the air. The wrecked cars instantly caught fire and were completely gutted together with the next two in the train, while the boiler of the *Aramingo* exploded, with a noise that could be heard five miles away. A bucket brigade taking water from a nearby stream could do little to quench the blaze, which was finally extinguished by the Congress Engine and Hose Company from Chestnut Hill. Engineer Harris was among the dead, but Engineer Lee and his fireman jumped and escaped and none of the twenty passengers on the regular train were more than shaken. The heroine of the disaster was a Mary Ambler, a Quaker widow, mother of nine children and the operator of her husband's fulling mill since his demise. She worked for hours on end in the sweltering heat to tend the injured without stopping for food or drink while neighbours carried injured passengers to her house on stretchers fashioned from carriage windows, she having offered her house for use as a makeshift hospital. After her death in 1868, a suburb and a station were named after her. The total number killed in the accident is uncertain, some sources state 59, others 60 or 67, with over 100 injured.

Terribly distressed as he watched the removal of the bodies of the victims from the wreckage, Conductor Vanstavoren eventually hired a carriage to carry him home and

[560] New York Daily Times, New York, New York, United States, 22nd July 1856

feeling that he was to blame for the catastrophe he swallowed a large dose of arsenic and was dead within a few hours. Ironically, the coroner's jury acquitted him of blame, Coroner Daniel Jacobs writing that 'We deem it to the memory of William Vanstavoren, the conductor of the down train, since deceased, to state that he was running his train at the time of the collision in accordance with the regulations of the road.' The timetable issued for the excursion train provided for it to reach Fort Washington at 6 a.m., and that if it was running more than 15 minutes late then the regular train had the right of way and the excursion must give way to it. Engineer Lee estimated that the collision had occurred, near Camp Hill, at 6.23 a.m. The excursion train should according to its schedule have reached Camp Hill at about 5.57 a.m. so at the time of the collision it was clearly more than 15 minutes late and should have waited at the previous siding at Edge Hill, or at the very least have gone forward very cautiously, as the regular train had done.[561 562]

Coroner Jacob's summing up continued, 'We find that William L. Lee, the engineer of the down train, is free from all blame, and did all in his power to avert the catastrophe after the first warning of the impending disaster.' As for Conductor Hoppel, the coroner wrote, 'We find that the immediate cause of the collision was the gross carelessness of Alfred F. Hoppel, the conductor of the excursion train, in running is train contrary to his instructions after it had gone so far behind time as to render it his duty to keep the track clear for the regular down train.'

Conductor Hoppel was later charged with voluntary manslaughter but was acquitted in February 1857 on the grounds that the railroad had no clear-cut regulations for controlling the running of excursion trains.[563]

The North Pennsylvania Company provided compensation for the survivors of the disaster.

By October 1857, the principal stations of the South Wales Railway were provided with the electric telegraph, using double-needle instruments. The more important stations at Bridgend and Port Talbot, Wales, were thus so equipped but not the less important roadside station between them at Pyle. Rather less than two miles east of Pyle was a goods station at Stormy, where there was a junction with the horsepower worked Llynvi Mineral Railway, and to protect trains on the main line from shunting movements on this line from August 1857 there had been established separate circuits with single-needle telegraph instruments between Bridgend and Stormy and between Stormy and Port Talbot.

At about 10 a.m. on 14th October 1857 the driver of a down Irish cattle train arrived at Port Talbot with his engine and only part of the train and reported that a waggon had left the road about three miles back up the line. Stationmaster Henry Turner at Port Talbot immediately told one of his porters to telegraph to Stormy to stop all down trains until further notice and without waiting for a reply at once proceeded up the down line with men and tools to clear the obstruction. When he reached the scene of the mishap, he learnt that a down train had been stopped at Pyle. The clearing up operation was completed about noon, whereupon Stationmaster Turner returned to Port Talbot, and

561 A History of Railroad Accidents, Safety Precautions and Operating Practices by Robert B Shaw, pgs. 69-71

562 Wikipedia - Great Train Wreck of 1856

563 Great Train Wrecks of Eastern Pennsylvania by Charles J. Adams III and David J. Seibold, pgs. 15-27

finding three up trains waiting there without going into his office at once despatched the first one, a passenger train, and the second train was also despatched behind it after the usual time interval. Very soon afterwards news was received that the first up train had collided on the up line with the down train which had previously been stopped at Pyle.[564]

The collision was a serious one. The two locomotives remained on the rails, but the tenders of each crushed the front ends of the coaches behind them. A little girl of four named Sarah Ann Harman, returning home to her parents at Bridgend in the company of her aunt, was killed outright, and half a dozen or so of the other passengers were more or less injured, as were Driver Robert Palmer of the up train, the 7.05 a.m. from Milford Haven, and Fireman J.S. Lawson of the down train. Two more of the passengers, George Ashman, a Cardiff pork butcher, and Elizabeth Israel, of Pyle, later died from their injuries.[565]

Stationmaster Charles White at Stormy had received the message on the single-needle instrument from Port Talbot "Stop all down trains" but the 6.30 a.m. passenger train from Cheltenham to Milford Haven had already just passed so he telegraphed back "The train has passed." Leaving his office, he had then met the guard of the down passenger train who had told him that it had been stopped at Pyle by someone who had gone back to protect the rear of the partially derailed cattle train. The stationmaster had then gone into his office and tried to telegraph to Port Talbot to ask them how long they expected the obstruction to last but could not get them to understand the first word of his message, "How." He had then tried "Can down train go on up line?" The answer had come back, "Send down train on up line," to which he had replied, "Good, I will do so." He had then proceeded to Pyle and had accompanied the down passenger train as it was backed along the down line to Stormy. Returning to his office again, Stationmaster White had written out an order authorising Driver Thomas Gunnell of the 6.30 train to proceed to Port Talbot down the up line and had attempted to telegraph to Port Talbot "Train has left on up line," but could not make them understand. Notwithstanding this, the stationmaster had then set off down the up line with the train, desiring Policeman John Days to persist in attempting to send the message to Port Talbot. Policeman Days had only been able get Port Talbot to understand "Down passengers" when they had stopped his message to send "Up passengers now left." This message was received at Stormy four minutes after the down train had left there.[566]

Porter Henry Barney at Port Talbot could read the single needle instrument but Clerk Edgar Evans, who had only just started there, could not, though he was familiar with the double line instrument. Both men strongly denied having had any difficulty in understanding any message from Stormy or of receiving any message from Stormy that a down train had left there on the up line, as was stated by Stationmaster White at both the Board of Trade inquiry and the inquests on the victims of the accident. Unfortunately for them, an, as it may be supposed, independent witness at Port Talbot, a Miss Catherine Llewellyn, of Baglan Lodge, a passenger waiting for an up train, had heard them reading off the message, "C,a,n,p,a,s,s,e,n,g,e,r,s,c,o,m,e,o,n." and then saying "Can passengers train come on?" followed by a discussion as to what was to be done and the expressions "What can he mean?" and "Here is a fine mess." Upset by what she had heard she had

---

564 Board of Trade report

565 The Times, London, England, 15th October 1857

566 Board of Trade report

remarked to the clerk, "I think Mr Turner ought to be here." Clerk Evans admitted that Miss Llewellyn had said this but denied the remainder of her evidence.[567]

The different inquests at Neath and Swansea found verdicts of manslaughter against Porter Barney, Clerk Evans and Stationmaster White while the Swansea jury also found Stationmaster Turner highly blameworthy for having started the up trains from Port Talbot before he had checked whether any further telegraph messages had been received from Stormy.[568] Perhaps because the manslaughter trials of the three railwaymen had not yet been held Captain George Wynne of the Board of Trade declined to give an opinion as to whom he believed to be at fault but merely commented that 'the transmission and reception of messages involving matters of life and death were entrusted to men who from their position could hardly be expected to appreciate the responsibility involved in their acts, and who were, moreover, imperfectly educated to the use of the instrument intrusted to them.'

The inspecting officer noted that the Company had issued excellent rules for working the telegraph, directing that the message should first be written down and the number of words counted, and that the same procedure should be adopted at the receiving station. Had this been done, there should have been no misunderstanding, but the company had not enforced the rules. He was satisfied that the telegraph was a most valuable aid to safety when systematically and carefully used but it became an instrument of danger when used by persons imperfectly educated in it and anxious to return to their normal duties.

In conclusion, Captain Wynne noted that the casualties had occurred in the carriage next to the tenders of each train, the buffer carriages having slipped under the tenders, and that had the recommendation made many years earlier of placing an empty carriage next to the tender, the severity of the accident would have been much reduced.[569]

The grand jury threw out the manslaughter charges against Porter Barney and Clerk Evans, but Stationmaster White stood trial at Swansea on 27th February 1858 and, furthermore, the jury found him guilty, with a strong recommendation to mercy. Baron Bramwell sentenced him to a month's imprisonment.[570] In the author's view, at any rate, he was unfairly treated, being no more at fault than the Port Talbot porter and clerk, or perhaps Stationmaster Turner.

A major disaster in the Kingdom of Sardinia on 25th July 1859 was an indirect result of the Austro-Sardinian War (or Second Italian War of Independence) which had been fought by the Kingdom of Sardinia, aided by France, against Austria and which had ended a fortnight earlier on 11th July.[571]

The accident happened two miles from Turin in the San Paolo area on the Susa line (Victor Emmanuel Railway). The trains were travelling at full speed when they collided. The southbound (or eastbound) train, from Susa, was loaded with Piedmontese artillery which the government had sent to the Alps at the beginning of the war, at the end of

---

567 The Times, London, England,17th,19th,20th,21st,23rd and 24th October 1857

568 The Express, London, London, 12th November 1857

569 Board of Trade report

570 The Standard, London, England, 2nd March 1858

571 Wikipedia – Second Italian War of Independence

April, to forestall a possible Austrian attempt to cut off communication with France. It also contained some passengers from France and was drawn by a single engine. The northbound (or westbound) train was loaded with artillery which had been captured from the Austrians at the Battle of Solferino, also some waggons loaded with horses and officer's baggage, and was double-headed. The scene of the accident was described as presenting a shapeless mass of smashed locomotives, waggons reduced to fragments, crushed and mutilated horses and human bodies dispersed in a horrible manner. At the time of the report, the number of fatalities was put at 11 and of injured at 30, but it was feared that more dead and injured lay buried in the wreckage.

Upon news of the collision reaching Turin, several surgeons hurried to the scene and the chief surgeon of the French army, which was encamped outside of the Susa Gate, was also soon at the site with everything needed for the treatment of the injured, while 'the inhabitants of the neighbouring cottages vied with each other in the zeal with which they went to the aid of those unfortunate persons.' A French officer was expecting his wife and two children by the train from Susa and upon learning of the disaster had hastened to the scene 'in a state of dreadful anxiety.' He found his wife alive, but his children had been 'dashed to pieces.' It was stated that many of the passengers had been injured from jumping from the trains when they saw a collision inevitable. The dead included six artillerymen. The two engines were so crushed and welded together that another locomotive had to be brought from Turin to pull them apart.[572] How the collision had happened was not stated.

On 27th August 1860 a collision occurred near Craven Arms, England, on the Shrewsbury and Hereford Railway. Rather more than 50 miles in length, the Shrewsbury and Hereford line was at this time single track throughout, with crossing places. The points where train were to pass one another were set out in the timetable, and alterations to the crossings could only be undertaken by application by the stationmasters by telegraph to a designated responsible officer. Since 1st August the responsible officer had been Mr Jeffreys, the locomotive superintendent at Shrewsbury.

The circumstances which led to the accident began with the failure of No. 2 down goods, due to leave Shrewsbury at 5.15 am, near Woofferton, 32 miles out, due to an engine problem. A report of its failure was telegraphed to Shrewsbury from Ludlow at 9 a.m. and given to Mr Finlay, the traffic manager, the locomotive superintendent having left at 8.15 on an excursion train to Hereford. Traffic Manager Finlay forwarded the message to the locomotive superintendent at Craven Arms at 9.30. The locomotive superintendent then had to make complicated arrangements for rearranging the traffic; sending telegrams from Craven Arms to Leominster, Woofferton, Hereford, Church Stretton, Ludlow and Shrewsbury. Superintendent Jeffreys then proceeded southwards from Craven Arms, towards Ludlow, and shouted to the Onibury stationmaster, Frederick Kerr, as he passed that No. 4 up goods was discontinued, that No. 8 down passenger train might pass Onibury, and its driver and guard would receive instructions at Ludlow.

Arriving at Ludlow Superintendent Jeffreys telegraphed Stationmaster Patchett at Shrewsbury: *"No. 4 up goods discontinued, No. 8 down passenger can proceed from Salop* {a commonly used abbreviation for Shrewsbury} *on arrival of Nos 1 and 2 up*

[572] Milwaukee Daily Sentinel, Milwaukee, Wisconsin, United States, 23rd August 1859

trains; instruct guard and driver to pass Onibury, reply to Leominster." According to the timetable, No. 4 up goods to Shrewsbury was due to cross No. 8 down passenger to Hereford at Onibury, 23¾ miles from Shrewsbury, at 1.36 p.m.; the passenger train having previously crossed No. 3 up goods at Church Stretton, 12¾ miles from Shrewsbury, at 1.10 p.m. The telegram from Superintendent Jeffreys was intended to indicate to the crew of No. 8 passenger train, the 12.40 p.m. departure from Shrewsbury, that No. 4 up goods train would not run that day, and therefore they did not need to wait to cross it at Onibury but could pass there. Unfortunately, Stationmaster Patchett at Shrewsbury, in copying out the message to give to Guard Robinson of the passenger train, inserted an "at" between the words "pass" and "Onibury" so that the message now read: *"No. 4 up goods discontinued, No. 8 down passenger can proceed from Salop on arrival of Nos 1 and 2 up trains; instruct guard and driver to pass at Onibury."* A locomotive foreman, an inspector and Guard Robinson, who all read or heard read the message, presumed that it meant that the passenger train must cross some down train, and that as No. 4 goods was cancelled, it must be No. 3 which they were to cross at Onibury, rather than at Church Stretton as per the timetable. By chance, both Traffic Manager Finlay and the Stationmaster Patchett travelled by the passenger train, Guard Robinson told Traffic Manager Finlay that they were to cross No. 3 up goods at Onibury, and the latter assumed that those must be the correct instructions from the locomotive superintendent.[573]

Stationmaster Morgan at Church Stretton, where the passenger train should have crossed up goods No. 3, was much surprised at Guard Robinson having an order to cross No. 3 at Onibury, as he had heard nothing about it, and he told the Board of Trade inquiry that he had been not at all inclined to let the train proceed, to which Captain Henry Tyler retorted that he had had no right to do so without an order from the locomotive superintendent. Some discussion apparently took place as to whether the passenger train should start from Craven Arms, but matters were decided upon Traffic Manager Finlay telling the driver, Peter Lawson, "Go on Peter, it is all right," or words to that effect, and after a stop of only a couple of minutes the train set off again. Driver Lawson understood from his guard and the traffic manager that he was to proceed. He had asked for and was handed the note from Stationmaster Patchett but being practically illiterate had handed it to his brother, the fireman, to read, but the latter, who had been busy affixing a red flag to the front of the engine to denote that they were a special, did not read the message either. The train thus set off from Church Stretton at 1.47 p.m. without anyone on it suspecting that anything was wrong, though as Captain Tyler pointed out by using the telegraph, they could have discovered that not only had No. 3 goods passed Onibury, but that it had left Craven Arms, three miles closer to them, two minutes before they had started from Church Stretton. Ironically, No. 3 goods could have stopped at Onibury, where Stationmaster Frederick Kerr had a waggon to be taken on, but Driver Peter Cooper, knowing that he would have a full load at Craven Arms, had whistled the Onibury signal off, and the stationmaster, aware that there was no urgency for the truck to go forward, had let the goods go on without it.[574]

The goods train was a heavy one, and was going uphill, so it was not going very fast when it met the passenger train, but the passenger train was going down the incline of 1

573 Board of Trade report

574 Ibid.

in 106 at 35 or 40 miles an hour when the trains collided 18 miles from Shrewsbury, between Marsh Brook station and Craven Arms. The passenger train consisted of an engine and tender, two second-class carriages, a first-class carriage, a guard's van, a composite carriage and a carriage truck. The goods train consisted of an engine and tender, 26 loaded waggons and a brake van. The trains were about 600 yards apart when they came in sight of one another and the speed of the passenger train was reduced to about 15 miles an hour before they met, while the goods train was almost at a stand. In the collision, twelve passengers were more or less severely hurt as were Driver Lawson, who jumped from his engine, Guard Robinson and a porter on the passenger train. One of the passengers, Mr Jobson, a well-known Shrewsbury corn merchant, later died from his injuries.

Analysing the causes of the accident, Captain Tyler noted that in the first instance the order from Superintendent Jeffreys had not been as explicit as it might have been. The accidental insertion of the word "at" by Stationmaster Patchett had rendered it ambiguous and obscure and the presence of the traffic manager on the train had overridden the authority of the Church Stretton stationmaster. The illiteracy of Driver Lawson was another factor. A more basic cause for the accident, the inspecting officer concluded, was the want of more stringent regulations for working the single line. On many lines in the country trains were not permitted to leave one station for another, even regular ones, without permission having been obtained by telegraph from the station in advance. Had this system been in use on the Shrewsbury and Hereford the goods train would not have been allowed to leave Onibury until the stationmaster there had telegraphed Church Stretton, and the Church Stretton stationmaster would then not have let the passenger train depart. Other lines used train staffs, differing in colour and shape for different sections, which each train must carry through the section.

Finally, Captain Tyler stated that given the magnitude and nature of its traffic, it was time that the Shrewsbury and Hereford line was doubled.[575] The line had in fact been constructed with future doubling in mind, and this was done in 1862, except for Dinmore tunnel, which must wait until 1893.[576]

The inquest at Craven Arms on the death of Mr Jobson was lengthy but eventually, on Saturday 22nd September, returned a verdict of "Accidental Death" with recommendations that more stringent rules be laid down for crossing trains and that the line should be doubled.[577]

The American Civil War was in full swing at the time of the next accident to be considered. It occurred in the Confederacy near Ponchatoula, Louisiana, on the New Orleans Jackson and Great Northern Railroad on 27th February 1862. An up train of thirteen cars was filled with soldiers of the Seventh Mississippi regiment, who had been stationed at Pass St Louis for the previous six months and were being transported to Jackson. At about 7 a.m., on a straight stretch of track but with visibility restricted by fog, it was running at full speed when it collided with a down lumber train. The effects of the collision were very serious; it was reported that every man in the second car of the troop

[575] Ibid.

[576] Wikipedia – Shrewsbury and Hereford Railway

[577] The Morning Chronicle, London, England, 5th, 10th and 24th September 1860

train (the leading vehicle was presumably a baggage car) was killed or injured, many losing limbs, and one man having both legs cut clean off. This report put the number killed outright at 22, with 48 badly injured. All the members of one company, except eight, were killed or injured and fully 100 men were rendered unfit for service.[578] Another report stated that the Dahlglen Rifles, from Pike county, had suffered the most, with not more than 25 of their company escaping, and that the Quitman Rifles had also suffered very severely. Contradicting the other report, this account stated that it had been the first and third cars of the troop train which had suffered the most, the second car being little damaged. This account put the number of dead as over 30, while a third report gave the number as 34, and added that the engineer of the troop train had reversed his locomotive and done his best to minimise the effects of the accident, but that the lumber train had come on at full speed. The report added, 'It is positively stated that it was a premeditated act on the part of the engineer of the lumber train. It is said that just before the trains struck, he put on all steam and jumped off and ran for the woods. The conductor, as soon as he could do so, also jumped off, and followed him, with pistol in hand to shoot him, but at last accounts he had not been caught, but we hear that he is being followed up with dogs.'[579] Apart from it sounding unlikely anyway, a problem with this story of the lumber train engineer's alleged actions is that if his train was really travelling at high speed how did he manage to just jump off his engine, keep his feet and be able to run away? It would seem more likely that his train was actually travelling much more slowly and that he had jumped off and fled fearing the wrath of the soldiers, who believed, rightly or wrongly, that the accident was his fault.

Three young women lost their lives in a collision near New Glasgow on the Albion Mines Railway, in the province of Nova Scotia, on Monday 28th July 1862. One of them, a Miss Jane Smith, was a niece of stagecoach tycoon Hiram Hyde. Being an industrial branch used principally for coal trains operation of the six-mile line from Stellarton to Dunbar's Point was basic. It had a single track with two intermediate sidings where trains could pass one another. There were no timetables and train conductors were expected to make their own meets with opposing trains. One the day of the accident, all the crews of the two trains except one were inexperienced, but nevertheless their conductors had been given no special instructions by the superintendent as to meeting other trains. Three passenger trains were run daily over the line at 8.30 a.m., 12 noon and 3.30 p.m., the trains running to correspond with the departures of the company's steamer *Pluto,* which operated a ferry service between Dunbar's Point and Pictou. The passenger accommodation consisted of one small stagecoach type car which was propelled by the engine of a coal train. The coach had a nominal capacity of six but on the day of the accident had been carrying fifteen passengers - eight men and seven women. The accident 'caused a tremendous hubbub' and the jury at the inquest on the deaths of the three women returned a verdict which criticized and severely censured the officials of the General Mining Association (the English company which owned the coal mines and railway).[580]

578 Arkansas State Gazette, Little Rock, Arkansas, United States, 1st March 1862
579 Southern Federal Union, Milledgeville, Georgia, United States, 18th March 1862
580 Webpage – History of Nova Scotia with special attention to transport and communications: Chapter 10

Scotland was the scene of a major disaster involving a collision during temporary single line working on 13th October 1862. The up main line (towards Edinburgh) of the Edinburgh and Glasgow Railway's main line had been closed to traffic since 15th September from Pardovan Siding, 2½ miles east of Linlithgow, to Priest Inch, ¾ of a mile west of Winchburgh, for relaying. All the traffic between Edinburgh and Glasgow and between Edinburgh and the north thus had to run over the down main line for 2 miles and 600 yards. A pointsman was stationed at the double to single line connections at Priest Inch and Pardovan and a pilot engine was provided to follow accompany every train over the single line, following a quarter mile behind it. Each pointsman was provided with a hut and a distant signal and speed through the connections was limited to four miles an hour.

On the evening of 13th October, the 6 p.m. passenger train from Edinburgh to the north consisted of a tank engine, two third-class and two first-class carriages, a guard's van and a horse-box, with forty third-class and four first-class passengers. It had to wait three minutes at Priest Hole for the pilot engine to return from the other end, and then proceeded, its driver, Robert Morgan, having no reason to doubt that the line was not clear. He had covered just over a mile when he saw a train coming towards him. He shut off steam, reversed his engine and whistled for the guard's brakes. The next moment, his train collided with the 5 p.m. passenger train from Glasgow to Edinburgh, and he was one of those killed. His train almost invariably had precedence over the Glasgow train through the single line section. The engines did not mount one another in the collision but jammed together, the slightly smaller tank engine of the Edinburgh train running in between the framework of that of the Glasgow train. The leading third-class carriage of the Edinburgh train was smashed to pieces and the tender, leading van and next two third-class carriages of the Glasgow train were jammed together, and the carriages destroyed. In addition to Driver Morgan, his fireman, A. Cochrane, Fireman John Cunningham of the Glasgow train and fifteen passengers lost their lives in the immediate aftermath of the accident and 33 other passengers were more or less injured. Driver Archibald Neill of the Glasgow train, who had jumped and was the only survivor of the four-engine crew, was also seriously hurt.[581] There was at least one more late fatality, Mr James Shaw, a passenger, dying of his injuries on 30th July 1863.[582]

The 5 p.m. Glasgow train consisted of an engine and tender, two third-class, two first-class, and a third-class carriage, a third-class brake composite and another third-class carriage, with about 112 passengers. It left Linlithgow seven minutes late at 6.24 p.m. and Driver Neill found the Pardovan distant signal at danger and reduced speed as he saw first a down train and then an engine pass him. He could not recall if the engine was attached to any trucks. The distant signal was then cleared to a green light, a white handsignal was shown from the pointsman's hut and he took his train onto the single line, increasing his speed once he had passed through the points. His train was travelling at 15 or 20 miles an hour when he saw the lights of the Scottish Central Railway train coming towards him and he found time to reverse his engine before jumping. Like his two guards he had assumed that the light engine which he had seen had been the pilot engine, No. 9.

Engine No. 9 had followed the 5.15 p.m. train from Edinburgh westward over the single line, picked up a truck there and pushed it to Linlithgow, carrying platelayers, then

581 The Times, London, England,15th October 1862

582 The Express, London, England, 3rd August 1863

returned to hurry back over the single line to meet the 6 p.m. train from Edinburgh, still pushing the same truck as there was not time to detach it. Instead, they found an empty cattle train and a ballast train and piloted them to the west end, stopping twenty or thirty yards short of the pointsman's hut, then returned once more to Priest Inch for the 6 p.m. Edinburgh train. This train they had duly followed in turn, being far enough behind it to be able to pull up when the collision occurred. The crew of the pilot engine stated that they had stopped close enough to the Pardovan pointsman's hut to see his handlamp through the steam and smoke.

The Pardovan pointsman was named George Newton. He had worked previously as a platelayer, then as a farm labourer, then again as a platelayer. He should have taken up duty as a pointsman on Sunday 12th but had begun a day earlier to accommodate the previous pointsman, John Robinson, who had been anxious to get away to look for a new job. Having worked Saturday night and Sunday night he had come on duty at 6.10 p.m. on the Monday, only a few minutes before the accident. The pilot engine and truck, returning to Priest Inch, must have passed him shortly afterwards. Next, he held the points for the train of empty cattle trucks, which was followed by the ballast train, consisting of an engine pulling a waggon and a carriage, and the pilot engine. It was here that the fateful mistake occurred. The ballast train stopped at about the same place as the pilot engine did, and also dropped off some platelayers before continuing, as the pilot engine was in the custom of doing, and had done earlier that evening, to Linlithgow. Newton did not see the pilot engine stopped in the darkness behind the ballast train and failed to see it returning to Priest Hole, nor did it occur to him that the pilot engine should not be pulling a carriage. The Board of Trade inspector, Captain Henry Tyler, made much in his report of the fact that pilot engine, No. 9, was a four-wheeled coupled goods engine with a dome over its boiler, very different in appearance to a careful observer to the ballast engine, No. 57, a "single" passenger engine with 6 ft 8 in driving wheels and no dome, but on a dark night the difference would not be so obvious, and Newton doubtless saw what he expected to see. Thinking that he was simply doing his duty he then let the Glasgow train proceed and only realised his mistake when some of the platelayers from the ballast train told him of his error. Shortly afterwards the hind guard of the Glasgow train returned with a red lamp to stop the pilot engine he expected to be following and to report the collision.[583]

Captain Tyler remarked that Pointsman Newton would not have made his mistake had the usual small pilot engine not been replaced by a larger one; had the pilot engine not been habitually employed on other jobs such as shunting and conveying platelayers to and from Linlithgow; or if the driver of the pilot engine had followed the ballast train through the connection before returning eastwards. Nor had the pilot engine carried special lights to distinguish it at night, 'and the want of better arrangements, stricter discipline, and more precise rules for the guidance of the engine-driver and signalman, contributed in an important degree to produce the accident.'

Single lines of railway could be worked, the inspecting officer continued, if they were short branches, by one engine in steam, otherwise the alternatives were the use of a train staff and the employment of a pilot guard of pilotman. The use of a pilotman was the alternative most suitable in the present instance, but the pilot engine and even the telegraph should be used only as a matter of convenience to allow the pilotman to be at

[583] Board of Trade report

the correct end of the section to accompany the next train over it. To ensure safety, the pilotman should travel with each train through the section, except when he had two or more trains to follow one another from one end, in which case he should start the first and subsequent trains and ride on the last one. In the accident under consideration, a pilot engine was used but without a pilotman, the engine followed the trains instead of preceding them, and as he had previously stated a lax system was employed, with vague instructions. The pilot engine did not always follow its train through the connections, but instead stopped short of them and then returned to the other end. In addition, according to the Priest Inch pointsman trains repeatedly passed through the connection there at 20 or 30 miles an hour.

Captain Tyler recommended that a pilotman should be provided with a distinguishing dress and lamp for day and night-time, likewise the pilot engine should be equipped with peculiar flags and extra lamps, so they were undistinguishable under all circumstances. Main signals as well as distant signals should be provided at each end of the single line and the pointsman's huts provided with clocks.[584 585]

Pointsman Newton, Driver Davidson in charge of the pilot engine, Superintendent of the Line J.B. Thompson, Permanent Way Inspector Alexander Rennie and Assistant Permanent Way Inspector Alexander Forrest were all committed for trial on a charge of culpable homicide at the end of October.[586] By mid-January the prosecutions had been abandoned except that of Superintendent Thompson and of Mr Latham, the general manager, whose name had not been mentioned earlier.[587] The two officials were acquitted at the Edinburgh High Court of Justiciary on 16th February 1863 after an eight-hour trial.[588]

Less than eight months after the Ponchatoula disaster, another occurred on the lines of the Confederate states near Duck Hill station, Mississippi, United States, at about 2.30 a.m. on 19th October 1862. The accident occurred this time on the Mississippi Central Railroad and involved two passenger trains. A down (southbound) train in charge of Conductor Peter Kirby and hauled by an engine named *James Brown* collided with an up (northbound) train of twelve cars in charge of Conductor Shelton, hauled by an engine named *A.M. West*. The southbound train had stopped for some time at Duck Hill and then set off slowly. It had only covered half a mile when it was struck by the northbound train, which was trying to reach Duck Hill before the southbound train left. The engineer of the southbound train saw the other approaching, reversed his engine, whistled for brakes and jumped. The engineer and conductor (fireman?) of the northbound train also jumped. The locomotive of the southbound train was wrecked, and its tender forced into the leading vehicle but only one man on this train was killed and a few persons slightly hurt. The northbound train suffered much more, especially as two flat cars, loaded with soldiers who had been unable to find room in the crowded coaches, were marshalled immediately behind the tender. These cars were demolished, and many soldiers killed. The following

---

[584] Board of Trade report

[585] The Times, London, England, 16th,18th and 20th October 1862

[586] The Lady's Newspaper, London, England,1st November 1862

[587] The Leeds Intelligencer, Leeds, England, 24th January 1863

[588] The Express, London, England, 17th February 1863

baggage car and front passenger car piled up onto the wreckage. Another five passenger coaches, sleeping car and two platform cars were less badly damaged.

Thirty-three or thirty-four passengers (all soldiers) died in the accident, together with the unnamed black firemen of the trains. Two white men and three black men travelling with the soldiers were unidentified. Forty or fifty others were injured, the worst injured were taken to nearby Canton. The northbound train had been under the military command of Colonel Horace Miller, of Vicksburg, of the 20th Mississippi Infantry Regiment. The dead soldiers came from a variety of Alabama, Arkansas, Mississippi, Missouri and Texas regiments. Local doctors and Duck Hill residents rushed to lend help after the accident and assistance was also offered by the Mississippi and Tennessee and Memphis and Ohio Railroad Companies.[589] The collision due to the crew of the northbound train 'running on short time' rather than side-tracking at a previous point for the southbound train.

Considerably less serious than the Duck Hill accident, but bad enough all the same, was a collision on 28th October 1862 at a location variously described as near Wake Forest and between Huntsville and Forestville, North Carolina, United States. This was stated to have been the first accident of any magnitude to have occurred on the Raleigh and Gaston Railroad. The trains involved were a northbound gravel train and a southbound accommodation train. The gravel train should have left Raleigh behind a freight train, but its departure was delayed and instead it left just behind a northbound mail train. In the collision, the tender of the accommodation train was forced back through the leading passenger coach, causing the deaths and injuries. Those who lost their lives were Lieut. H.J.B. Clark, of the 2nd State troops; Mr Nesbitt, of Concord and Mr J.J. Berryhill, of Charlotte, all of whom were killed outright, and Mr Thomas Rowland, of Stanly Court, who succumbed to his injuries two hours after the accident. Fifteen other passengers were injured.

The fireman of the accommodation train was the first to see the gravel train and shouted a warning. Engineer Holleman reversed his engine, whistled for brakes and jumped, as presumably did his fireman. Conversations with Engineer Holleman and Conductor Williams of the accommodation train showed that they were not to blame, and it transpired that Engineer Sidney Hinton and his unnamed conductor had been at fault, as they were running on the time of the accommodation train. Engineer Hinton reportedly manifested much emotion at the suffering that he had caused and then made his escape, having been seriously threatened by the passengers.[590]

A digression here to outline the system of working single lines in the United States and Canada which gradually evolved following the introduction of the electric telegraph. By the mid 1860s the use of the electric telegraph was widespread, and the use of a train dispatcher had grown up; as mentioned earlier, what was effectively the first train order was sent by Charles Minot of the Erie Railroad in 1851 and the latter company was the first to introduce a proper train order system as early as 1854.[591] Control of a considerable

---

[589] Webpage – Tragedy at the Duck Hill Station, Collision of the James Brown and the A.M. West by Norman L. Ezell

[590] Weekly Standard, Raleigh, North Carolina, United States, 5th November 1862

[591] Trains June 1992 pgs. 20-3

length of line was placed in the hands of a dispatcher who was in telegraphic communication with the operators at all the different stations within his area and who was charged with working the traffic as expeditiously as possible by issuing orders to the different operators to hold trains or advance them against opposing trains, or where there was double track perhaps to permit them to use the opposite running line to overtake a slower train. In the absence of orders issued by the dispatcher, the order of trains over a given stretch of single track was decided by their superiority, trains of a superior class having an absolute right over trains of a lower class. In the case of trains of the same class, superiority would be decided arbitrarily, by, for example direction, southbound trains, say, being superior to northbound trains. A train arriving at a crossing place where it was scheduled to meet a superior train must not proceed until the superior train had arrived, except on the basis of an order issued by the dispatcher giving it the right to the road. Any train not in the timetable (distinguished by two white flags or lights at the front),was designated an "extra" and could only proceed on orders. Some railroad companies designated all freight trains as extras. Trains could also, in times of heavy traffic, run in more than one section (each except the last carrying green flags or lights to show that another section was following). The crew of an opposing inferior train had to ensure that all the sections of an opposing train had passed before proceeding, again unless they had an order from the dispatcher giving them the right to the line. At stations manned by operators, a train order signal was provided advising the crews of trains if there were any orders for them to pick up. Orders were of two types, the 31 order, which the train crew had to sign for so that the dispatcher could be sure that they had received it, and the 19 order, which did not need to be signed for and could be collected on the move using a hoop.[592] [593] The system was unfortunately liable to a variety of errors; dispatchers could inadvertently give two trains orders to use the same stretch of track at the same time (a so-called 'lap order'), operators could write down orders incorrectly or illegibly or fail to deliver them, train crews could misread or forget orders or fail to wait for all the sections of an opposing train, and so on. The dispatcher system was also used in some other countries, for example in Great Britain on the Shrewsbury and Hereford Railway, as noted above, on the Salisbury and Yeovil Railway and on the Somerset and Dorset (see Foxcote, 1876, below) but never as extensively in Britain as in North America. In the United States, the rules for working train by the dispatcher system were made the same for all companies by the introduction of the Standard Code in July 1889.[594] [595]

Confederate troops were involved in another United States disaster on 15th July 1864, but this time as prisoners of war of the North. The passenger train involved in the accident near Shohola, Pennsylvania was carrying 833 prisoners, many of whom had been captured at the recent battle of Cold Harbor. The prisoners were being transported from Point Lookout, Maryland to a new prisoner of war camp at Camp Rathburn, Elmira, travelling first by steamer to New Jersey and thence by train over the Erie Railroad. Accompanying the prisoners were 128 Union guards of the Veteran Reserve Corps. The train consisted of 17 passenger and freight cars hauled by engine No. 171. Departure

---

[592] Ibid.

[593] Webpage - Train Order Accidents

[594] The Railway Magazine 1959, pgs. 749-51

[595] Webpage - Train Order Accidents

from Jersey City was scheduled for 4.30 a.m. but was delayed until 6 a.m. because the guards had to search for three missing prisoners who had hidden themselves on the ship. There was then further delay waiting for a drawbridge to be closed. By the time that the train had reached Port Jervis it was four hours late.

Beyond Port Jervis, the single-track line hugged the Delaware River and at Lackawaxen there was a junction with the Hawley branch. The operator on duty at Lackawaxen was Douglas "Duff" Kent. The previous westbound train, No. 23, had carried flags to indicate that an extra train, the prisoner of war train, was following it. At about 2.30 p.m. a coal train of 50 loaded trucks hauled by engine 237 with Engineer Samuel Hoitt and Fireman Philo Pretis reached Lackawaxen from the Hawley branch and stopped clear of the junction points. Conductor John Martin got down from his caboose and going to Operator Kent asked him if it was clear for the coal train to proceed to Port Jervis, whereupon the operator, who had either failed to notice the flags warning of the following extra or had forgotten about them, assured him that it was. The conductor returned to his train and Brakeman G.M. Boyden operated the switch to let the train out onto the main line.[596]

The prisoner of war train had passed through Shohola station at 2.45 p.m. and the two trains collided shortly afterwards at King and Fuller Cut where there was a curve giving a visibility of only 50 or so yards. In the violent collision, the wood in the tender of engine No. 171 was thrown forward into its cab, killing Fireman Tuttle outright and pinning Engineer William Ingram against the split boiler plate so that he was slowly scalded to death, he warning rescuers not to try and help him as there was a danger of the boiler exploding. Fireman Pretis and Brakeman Boyden were likewise crushed to death by wood thrown forward from the tender of their engine but Engineer Hoitt just had time to jump and survived. The leading car of the prisoner's train was crushed to a length of less than six feet and 36 of its 37 occupants were killed, the sole survivor having fallen through the platforms between the cars to the ground. Three cars in all were destroyed and seven or eight more so badly damaged as to be totally useless. Two Union guards were placed on the platforms adjoining the doors at each end of each car and some of the prisoners were stood there also talking to them. All but one or two of these men were killed.[597]

As soon as possible after the collision, the survivors set to work to extricate the dying and injured and word was sent to Shohola, from where assistance was telegraphed for from Port Jervis, and Superintendent Hugh Riddle shortly left for the scene with a relief train carrying three surgeons. A second relief train was apparently also sent from Port Jervis. The doctors worked through the night and 100 injured men were taken to Shohola and quartered at the station or the Shohola Glen Hotel. Five prisoners also took the chance to escape in the chaos after the accident. An initial count after the last of the victims had been removed from the wreckage gave a count of 16 Union and 44 rebel (Confederate) dead. Associate Judge T.J. Ridgeway of Pike county soon reached the scene and, after a consultation with Superintendent Riddle and Union officers, held an inquest on the spot. A trench 76 ft long, 8ft wide and 6 ft deep was dug and the dead were buried by the lineside, the Confederates, including four more who had died overnight, four to a coffin in coffins constructed from the wreckage, but the Union soldiers in

---

[596] Webpage – The Great Shohola Train Wreck

[597] Daily Journal, Indianapolis, Indiana, United States, 20th July 1864

individual pine coffins brought to the site. The bodies of prisoners who died at Shohola were also taken to this trench to be buried, but rebels John and Michael Johnson, who had been taken to the house of a Mr Hickock and had died overnight there, were buried in the cemetery of Shohola church. One of the Confederate injured, a North Carolina infantryman named Albert G. Smith, wrote to his wife later from Elmira, "I got heart {sic} coming up hear {sic} by the cars running together but I am not confined. We are faring very well and are treated very kind {sic}, more so than I thought we would be." All the injured prisoners had been forwarded to Elmira within a week. Estimates of the final number of persons killed varied from 65 to 74, the latter figure comprising 51 prisoners, 19 Union soldiers and four railroad staff.[598]

The inquest found that the accident had been caused by the negligence of Operator Kent, who disappeared on the morning after the tragedy and was not heard of again. There were reports that he had been intoxicated before the accident, and that he had not taken the crash very seriously, going to a dance in the evening after it had happened. Then, when his responsibility for it had become clear and he had feared the consequences, he had fled.[599] An alternative suggestion, that the accident was caused by the crew of the coal train attempting to reach Port Jervis in front of the prisoner of war train, has also been put forward, but seems much less credible.[600]

A collision on the Macon and Western Railroad between Barnesville and Milner, Georgia, United States on 1st September 1864 was another which was very much a direct result of the American Civil War, now in its latter stages. On the day of the accident, the Confederate forces under General Hood were in the process of evacuating the city of Atlanta in front of the advancing army of William T. Sherman, who was to occupy it the following day. According to Robert B. Shaw, the troops commandeered the locomotive *Dispatch* and some cars which they loaded with wounded soldiers and despatched southwards over the Macon and Western towards Macon.[601] It would appear however from the reminiscences of its engineer, Bill Mitchell, that it was not a special train but the regular passenger train which the troops had charge of, consisting of eight passenger cars and five box cars. Nor, as implied by Shaw, was it according to Michell the fault of the crew of the passenger train, which was running late, or of a failure to advise of its running, but rather the failure of the opposing train to wait for it at Barnesville, their designated crossing place. The northbound working was a commissary train hurrying supplies northwards to Hood's army, hauled by an engine named *Governor*, in charge of Engineer Lum Carr.[602] Engineer Mitchell had of course a definite interest in blaming the crew of the other train, so his account might not be entirely accurate.

When he saw the opposing train, Engineer Mitchell reversed his engine and jumped, escaping with a broken arm. He wrote in 1882, 'The passenger cars were smashed into kindling wood; the engines were totally demolished; the commissary train was a mass of ruins; one car loaded with bulk peas was stood on its end, and six soldiers, who were

---

598 A History of Railroad Accidents, Safety Precautions and Operating Practices by Robert B Shaw, pgs. 71-72

599 Webpage – The Shohola Train Wreck of 1864: A Telegrapher's Negligence

600 Webpage – The Great Shohola Train Wreck: Was the Coal Train in a Hurry?

601 A History of Railroad Accidents, Safety Precautions and Operating Practices by Robert B Shaw, pg72

602 Webpage – Engineer's Account of Civil War Train Crash Here – Barnesville.com – reproduced from The Atlanta Journal, Atlanta, Georgia, 12th November 1882

riding in there, were smothered in the peas. Colonel Hulbert was riding on the engine with me, and just before the engines struck a negro kicked him off, and thus saved his life. Out of the wreck we took thirty-two dead bodies and twenty-two died in a few days from the effects of injuries received.[603] Other accounts give the number of dead as 31 soldiers and one lady – a nurse named Saffen, from Memphis [604] – but Mitchell's account would make the final death toll 54. An Augusta newspaper reported that after the accident everything possible was done to help the victims not only by the local residents but also by eighteen Yankee (Union) prisoners who had been on the down train and had been unhurt. The paper recommended that these men's actions should be remembered and that they should be given the first priority when it came to an exchange of prisoners.'[605]

The entire railroad was destroyed by Sherman's forces three months later in his March to the Sea but according to Shaw this did not prevent the management of the Macon and Western from criticising General Hood in its annual report, stating that "had the orders given from this office {superintendent's office} the collision would not have occurred."[606] The present author believes that this was a reference to the running of the supplies train.

The Lafayette and Indianapolis Railroad passenger train due at Lafayette, Indiana, United States, at 4 p.m. on 31st October 1864 was running twenty minutes late when it left Culver's station, eight miles from Lafayette and its last stop, but its conductor, Edward Wright, informed a newspaper reporter that it had orders to proceed to Lafayette. It had only travelled another mile and was rounding a curve when it came into head-on collision with a cattle train. The engineers of both trains reversed their engines and the speed of the cattle train was considerably reduced but both were heavy trains, the nine-car passenger train being crowded, with all its cars, aisles and platforms filled; and the cattle train having eight cars of beasts. The most serious effect of the violent collision was the telescoping for two-thirds of their length of the baggage and express car and the leading passenger car, which was filled with soldiers. Charles Clisbee Esq, a Cassopolis, Michigan attorney who was a passenger on the train, informed a reporter that 'the cries and groans of the poor fellows were most pitiable. Many begged that an end might be put to their sufferings.' Within an hour and a half all the victims had been extricated from the car and it was then found that there were twenty-eight dead and twenty-five more injured. All the fatalities were in the leading passenger car of the train.

Engineer George Lamb of the cattle train jumped from his locomotive before the collision and ran away. 'Had he remained, he would most probably have suffered violence at the hands of the excited soldiers, who were in no mood to accept explanations,' the newspaper wrote, adding that it was not prepared to say who had been at fault, but the matter would be thoroughly investigated by the local coroner, Dr Ingersoll.

---

603 Ibid.

604 Southern Recorder, Milledgeville, Georgia, 6th September 1864

605 Webpage – Engineer's Account of Civil War Train Crash Here – Barnesville.com – reproduced from The Augusta Constitutionalist, Augusta, Georgia, 7th September 1864

606 A History of Railroad Accidents, Safety Precautions and Operating Practices by Robert B Shaw, pg72

From a vote taken on the train shortly before the accident it was carrying 508 passengers (the newspaper did not explain the purpose of the vote). Most of those who lost their lives were soldiers, but some were civilians, including a Squire Ross, who lived near Frankfort and had been returning from Indianapolis, where he had been 'filling the quota.' Again, the newspaper does not explain what this meant. Several of the passengers, including Squire Ross, had been carrying large sums of money, one having $30,000 upon his person. The soldiers were from several Western states and were mostly returning home after serving their time, looking forward to seeing their families in a few hours. 'One poor fellow said it was hard to meet the enemy in battle twenty times and escape unhurt yet die by such an accident almost in sight of home. He died before the train reached town.' [607]

Following the Culver's disaster, the United States was free of significant single line collisions for some months, but then suffered two serious accidents within a few days in August 1865. The first of these was near Titusville, Pennsylvania, on the Oil Creek Railroad, a little before noon on Thursday 24th August 1865. The express train from Corry due at Titusville at 11.45 a.m. was running to time when it collided with an extra freight train a mile from Titusville depot. The engine crews of both trains jumped and saved themselves. The locomotive of the express was knocked to pieces, and its tender reportedly 'dovetailed into the baggage car, and the latter into express and mail car.' Six passengers were killed outright, and a number of others badly hurt.

An inquest was held the same day, and its jury found that the collision was caused, 'by the culpable negligence of Joseph Fink, engineer, and Patrick Carey, conductor, as by evidence shown, said Funk {sic} and Carey having control of the aforesaid extra freight train, and having neglected to properly flag the line, when the said express passenger train was then and there due, was approaching. And we, the aforesaid jury, therefore, request the coroner to issue his warrant for the apprehension of the said Joseph Funk and Patrick Carey.' The engineer of the freight train had disappeared after the accident; the newspaper suggested that had he been found he would have been summarily dealt with.[608]

The New York Tribune newspaper, noting that the Titusville accident, like the majority of other recent ones, had occurred on a sharp curve, suggested that as a safety precaution all trains should be ordered to come to a dead stand before negotiating such curves, as trains on the New Haven and other railroads did already before crossing certain drawbridges.[609]

On Monday 28th August 1865 what the Brooklyn Daily Eagle the next day sarcastically called 'quite a respectable accident' occurred on the Long Island Railroad. Four people had been killed outright, one or two could hardly be expected to survive and half a were 'mangled and maimed for life.' According to the initial report on the accident, the express from Greenport due to arrive at Jamaica, New York state, at 9.30 a.m. was

---

[607] Logansport Journal, Logansport, Indiana, 12th November 1864 – reproduced from Lafayette Journal, Lafayette, Indiana, 1st November 1864

[608] Philadelphia Age, Philadelphia, Pennsylvania, United States, 29th August 1865, reproduced from Titusville Herald, 25th August 1865

[609] Philadelphia Age, Philadelphia, Pennsylvania, United States, 29th August 1865

there due to cross the accommodation train which had left Hunter's Point at 8.30 a.m. but if the up (Hunter's Point) train was late it should wait for the express at Winfield station. On the day of the accident the accommodation train was fifteen minutes late in leaving Hunter's Point and losing more time en route should have stopped at Winfield, but its conductor, Richard Rice, instead chose to push on to Jamaica even though the train was 35 minutes late leaving Winfield, on the assumption that the express could also be behind time, as it usually was.[610]

By 30th August a fifth passenger, a German named George Laule, had died of his injuries. The inquest on the accident, before Coroner Hendrickson, lasted several days. Engineer James White of the express train told the inquest that his train had been travelling at 15 or 20 miles an hour when it met the accommodation or mail train on a curve; he had reversed his engine and whistled when he saw the other train. He attributed the accident to one of the trains being off its time. Fireman John Cronan said he knew nothing of the accident other than seeing the mail train coming.

Engineer Daniel Chase stated that he had left Winfield at 9.20 and would have taken 17 minutes to reach Jamaica. In his opinion, the accident was due to the express leaving Jamaica before its time, had it waited four minutes the accident would not have happened. His fireman, John Palmer, could add no additional information.

Mr Robert Charlick, President of the Long Island, agreed that the express should have waited four minutes at Jamaica, stated that all trains were required to exercise great caution in going around curves, and that the express being on time should have waited four minutes more, but stressed that the conductor of the mail train had no right to use this extra time.

A passenger on one of the trains told the newspaper that the mail train was timed to leave Jamaica at 9.02 a.m. and cross the express at Queens, three miles east of Jamaica, at 9.17. According to him the express had arrived at Jamaica at 9.22, not 9.30, allowed passengers for the New York Central train to alight, refilled its water tank and departed. The passenger mentioned an incident three weeks earlier when the afternoon mail train had not been at Hicksville, its booked crossing place, for the express, and the express had gone on, whistling long and hard approaching Farmingdale and going around a curve at high speed. It was also suggested that had the last train from Hunter's Point on Saturday, which formed the Monday morning express, had more carriages no passengers would have had to stand on the platforms of the express and there would have been no fatalities. The Saturday train had had standing passengers to Deer Park, 36 miles out.[611]

On the last day of the inquest, Acting Superintendent Robert White stated that the mail train was usually late three times a week, and the express twice a week. It was the duty of the conductor to ensure that there were sufficient seats for all the passengers, taking on extra cars at Hunter's Point if necessary, and of the conductor of the mail train to wait at Winfield if his train was late.

Charles Carpenter, 'a practical engineer,' stated that the Long Island road used old fashioned wooden brakes, which were never now used on well-regulated roads. Patent brakes might have stopped the trains within 200 feet, and there should be telegraphic communication with each station to ensure the safe working of traffic. There was also

[610] Brooklyn Daily Eagle, Brooklyn, New York, United States, 29th August 1865

[611] Ibid,30th August 1865

evidence that the trains had insufficient brakemen, with only one for six cars, that the general management of the railroad was very poor and that there was a great scarcity of locomotives and cars.[612]

The inquest jury found both Conductor Rice of the accommodation/mail train and Engineer Chase responsible for the accident and they were arrested under a coroner's warrant.[613]

What could have been a various serious accident occurred in England on the Bristol and South Wales Union Railway (which was worked by the Great Western Railway) on 3rd November 1866. As it was, the company had a very lucky escape. Direct rail communication between Bristol and South Wales via the Severn Tunnel was still some years in the future but a single line branch ran from Bristol to New Passage station, from where a ferry connected with Portskewett Pier on the opposite side of the Severn estuary, providing a direct route between Bristol and South Wales which avoided the roundabout journey via Swindon and Gloucester. The 12-mile branch was worked by staff and ticket combined with the electric telegraph, with an intermediate staff station at Patchway. This improvement on simply relying on the electric telegraph alone entailed every train carrying either a staff engraved with the names of the stations between which it applied or, if more than one train was to proceed in one direction before one in the opposite direction, the driver of the first train, having seen the staff, would be given a ticket marked with the names of the stations, the staff being retained for use by the next or last train in that direction. No train was to pass through a section without carrying either the staff or a ticket. The tickets were kept in a locked box, the key to which was on the staff, so that a ticket could not be issued unless the staff was present to unlock the box. The system was combined with the electric telegraph to maintain an interval of space between following trains.

At 11.10 a.m. on 3rd November a down ballast train arrived at New Passage but its driver failed to give up the train staff to the stationmaster and the latter failed to ask for it. The stationmaster then failed in turn to give the staff to the driver of the 11.50 a.m. up train from New Passage to Bristol and the driver of this train did not think to ask for it. Having arrived at Patchway station at 12.10 p.m. without the staff, the driver of the passenger train could not give it to the Patchway stationmaster but gave the official the Patchway – New Passage ticket that he had been given on a previous down journey, which should have been given up at New Passage! The train was then allowed to proceed to Bristol with the Patchway – Bristol staff. Arriving at Bristol the driver of the 11.50 up train told the driver of the next (12.45 p.m.) down train that he had not brought the New Passage – Patchway staff to Patchway but remarked that it would probably be sorted by the time that the 12.45 train got to Patchway.[614]

Accounts of what was said at Patchway when the 12.45 train reached there at 1.05 p.m. varied somewhat between the driver and the Patchway stationmaster. The driver told the Board of Trade inquiry that the stationmaster had assured him that although the Patchway – New Passage staff was not there he (the stationmaster) had made it all right

---

612 Brooklyn Daily Eagle, Brooklyn, New York, United States, 2nd September 1865

613 Madison Courier, Madison, Indiana, United States, 6th September 1865

614 Board of Trade report

for him to go on. The stationmaster, who did not attend the Board of Trade inquiry, having been dismissed, told the internal railway inquiry that the driver had said that he supposed that it was all right to go on, but did not deny that he had let the train proceed. The Patchway stationmaster also stated that after the 11.50 up train had arrived without the staff he had tried to communicate with New Passage via the telegraph, but the New Passage stationmaster had been away at dinner and no one else there could work the telegraph.

The New Passage stationmaster realised, when he returned from dinner at 12.55, that he had forgotten to give the staff to the 11.50 up train. Knowing that the staff would be needed at Patchway for the 12.45 down train and finding that the driver of the ballast train had it in his possession he directed the latter to take the staff as fast as he could to Patchway, but to watch out for the level crossing gates at the intermediate station at Pilning, which were likely to be closed across the line. He then sent the 'Line blocked' signal to Pilning on the line instrument and attempted, without success, to contact Patchway on the general instrument.

The Pilning stationmaster heard the ballast engine whistling at his distant signal and cleared it to let the engine up to the station, keeping his home signal at danger. He also waved his arms for the driver to stop, but the engine kept going.

The passenger train had just left the 1246 yard Patchway tunnel and was running down a steep incline of 1 in 90 at 35 or 40 miles an hour when its driver saw some steam which he and his fireman first assumed came from the nearby Cattybrook brickworks, but they then spotted the light engine coming towards them. The passenger train driver was able to reduce his speed to 15 miles an hour, while as it was going uphill the light engine managed to almost stop and the collision was thus not very serious. Locomotive Superintendent Mack, riding as a passenger, was fairly badly hurt; eleven other passengers were slightly hurt; the passenger train driver suffered a broken arm and rib and the ballast driver, and the two firemen were shaken and bruised. Had the collision occurred in the tunnel it would, of course, have been much worse.

As might be supposed the inspecting officer, Colonel W. Yolland, was not at all happy about this gross example of sloppy working and wholesale rule breaking by several different railwaymen. He considered that the New Passage stationmaster had greatly neglected his duty by failing to collect the ticket and staff from two different trains, by permitting the 11.50 up train to depart without the staff and by sending the ballast engine to Patchway without first having communicated with the latter station. Even more reprehensible was the conduct of the Patchway stationmaster in letting the 12.45 down passenger train go knowing that the staff was at the other end. Both stationmasters had been dismissed by the company. Again, the driver of the 11.50 up train was grossly careless in leaving New Passage without the staff and the driver of the 12.45 train was even more culpable in knowingly leaving Patchway without the staff. This driver alleged that he was unaware that he risked dismissal for leaving a station without a staff. The two drivers were fined but kept their jobs. The ballast train driver was also at fault, Colonel Yolland added, and the Pilning stationmaster had made an error of judgement in clearing his distant signal prematurely.

The accident would not have happened, the inspecting officer decided, had not the regular branch drivers been sent to work elsewhere, and new drivers been put on the line without there having been given any instruction as to how it was worked. The driver of the 12.45 passenger train had only worked the branch once before, two years earlier.[615]

The branch was doubled as far as Narroways Hill Junction in 1874 and to Pilning in 1886 (with a new alignment beyond there) upon the opening of the Severn Tunnel.[616]

The luck which had attended the collision between Patchway and Pilning was absent in France on Sunday 23rd December 1866 on the Paris Lyon Méditerranée line (Chemins de fer de Paris à Lyon et à la Méditerranée) when a passenger train from Belfort to Dijon collided with a goods train between Franois and Besançon. M Fricot, the assistant stationmaster at Franois, who had been advised of a special goods train which was approaching his station but through forgetfulness or inattention permitted a Belfort - Dijon passenger train to proceed before the goods train had passed him. The passenger train had covered about a kilometre when it collided with the goods train near Dannemarie-Velesmes station at about 6 p.m. The collision left the line blocked 'by an indescribable mass of wreck of locomotives and carriages', as the newspaper report described it, though the track itself was not damaged. Thirteen persons were killed outright and about 20 injured, one of whom later died. Those killed included the conductor of the train, M. Kuntz and M. Godefroy, assistant stationmaster at Besançon, a wine merchant, a potter, a joiner and five Baden soldiers. The injured included the two drivers, named Pidancet and Kleine, Fireman Feriot, Conductors Gener and Taillard and ten German soldiers.[617] What were ten German soldiers doing on a French train?

Assistant Stationmaster Fricot was arrested after the accident and at Besançon Criminal Court on 5th January 1867 was sentenced to five years' imprisonment and fined 2,000 francs as well, which seems an unusually harsh penalty.[618]

There seem to have been no major single line collisions in 1867 or 1868 but the United States suffered another one towards the end of 1869, this time on its west coast. A few minutes after 9 a.m. on Sunday 14th November 1869 a six-car eastbound Western Pacific Railroad train, the 8.30 a.m. departure from Oakland, consisting of an engine and tender, an express and mail car, a baggage car, a smoking car, two passenger coaches and a sleeping car, collided with a San Francisco and Alameda Railroad train from Hayward consisting of an engine and tender, a freight car, a box car and two passenger cars. Each train was travelling at about twenty miles an hour when they collided a mile and a half south of Brooklyn Park station. The surrounding countryside was flat and the line straight, but a fog prevented the engineers of the two trains from seeing one another. The collision had disastrous results, 17 persons losing their lives and another 20 being injured. The dead railroad staff were Conductor Robert Owen and Fireman Charles Martin of the Alameda train, Engineer Edward Anderson and Fireman George Thompson of the Western Pacific train and Brakemen Carroll, Hallon, B. Milliken and P.A. Owen. The passengers who lost their lives included Alex Baldwin, District Judge of Nevada and a Mr Boult, principal of the young ladies' seminary at Oakland. Those injured included Judge William Campbell, US district attorney for Nevada and two Chinamen who were

615 Ibid

616 Wikipedia – Bristol and South Wales Union Railway

617 The Times, London, England, 24th and 29th December 1866

618 Wikipedia – Liste des accidents ferroviaires en France au XIX siècle

badly hurt. Those killed and injured were mostly on the Western Pacific train, the first passenger car of which was completely telescoped by the smoking car. The occupants of the sleeping car escaped unhurt. The two locomotives were wrecked, the box car of the Alameda train was thrown up into the air and its two passenger cars, which were almost empty, telescoped one another for three-quarters of their lengths. It was fortunate that neither train was that well-filled.[619] [620] [621]

The reactions of the passengers after the collision varied, as the *Virginia City Enterprise* told its readers, 'When the passengers who had escaped injury emerged from the cars their ears were at once saluted with the moans of the dying and the piteous appeals of the wounded. Some, when they attempted to render assistance, fainted at the sight of the torn and mangled, and were for a time unable to extend any aid. Others seemed bewildered or affrighted, and were therefore inefficient, while others were so entirely different to all save themselves that they took their carpet sacks and walked away as rapidly as possible. There were others of the passengers, and some who were quite severely wounded, who were perfectly cool and acted with advice and deliberation. Several procured horses and started to Oakland, Alameda and San Leandro for surgical aid. All the ranchmen of the neighbourhood rallied promptly to the succour of the sufferers, bringing axes and implements of various kinds. They and the passengers sided in cutting away the timbers of the wreck wherever necessary to remove dead bodies or relieve those who were living. While this work was going on many heart-rending incidents occurred. One man, who was horribly mutilated, begged a bystander to shoot him to put him out of his misery. A dead man and a wounded one were wedged in between the timbers in such a way that the wounded one could not be removed for some time; he begged a bystander to cut off the legs of the dead man in order that he might be extricated. The operation was performed with an axe. It was nearly 11 o'clock before all the bodies were removed from the wreck. Justice Clement, of San Francisco, empanelled a jury with a view of holding an inquest, but it was afterwards decided to remove the dead bodies to Oakland and postpone the inquest until today. The most of those who were wounded severely or dangerously were taken to the Alameda Asylum.'

Various causes were given for the accident. It was alleged that the Western Pacific train had been running early and should have waited for the Alameda train or alternatively that it was the fault of the person named Smith at the junction switch where the Alameda line diverged, variously described as the stationmaster and the switch tender. It was stated that Alameda train being behind time he had remarked that he was dammed if he would wait for it and had gone into his house for his breakfast. While he was inside, he had heard a train pass which he had assumed was the Alameda train, but which was in fact a gravel train. The Western Pacific train had then come up and its conductor asking him if it was OK to proceed, he had replied, "All right, go ahead." The switch tender was a new man who had only been appointed three or four days earlier. The track was quickly repaired after the accident and 'the hotel express and the Atlantic express came in on time.' [622]

---

[619] Wisconsin State Journal, Madison, Wisconsin, United States, 16th November 1869

[620] The Deseret News, Salt Lake City, Utah, United States, 24th November 1869

[621] Buffalo Express, Buffalo, New York, United States, 26th November 1869, reproduced from Virginia Enterprise, Virginia, Nevada, 16th November 1869

[622] Daily State Sentinel, Indianapolis, Indiana, United States,16th November 1869

A disastrous collision occurred on 12th May 1870 about half a mile east of Eureka station, Missouri, United States between an eastbound Missouri Pacific Railroad express passenger and a westbound freight train. The engines of the two trains were driven into one another by the force of the collision and the baggage car and two leading passenger cars of the express and several box cars of the freight train were wrecked. Nineteen passengers were killed outright and 30 injured, half of them seriously and eight dangerously.[623]

The passenger train had been an hour and ten minutes late at Jefferson City, Missouri but had regained a considerable amount of this time. According to its conductor, George Dubois, when it left Eureka it had an order authorising it to run 30 minutes late on its schedule. It was scheduled to leave there at 4.40 a.m. and actually departed at 5.15, having crossed there the regular westbound freight, and leaving 35 minutes after its scheduled time. The train with which it collided was an extra from St Louis to Gray's Summit. Conductor William Oder of the extra freight stated that his train had started from St Louis at 2.30 a.m. and had waited for the express at St Paul's, five and a half miles from Eureka. His head brakeman, H.D. Sullivan, had remarked to him there that the express was approaching, but this train turned out instead to be a through stock train, which informed them that the express had orders to run 30 minutes late. Conductor Oder then ordered his train to leave St Paul at 4.45 a.m., allowing it 25 minutes to cover the five and a half miles to Eureka and arrive there at the revised departure time of the express. Conductor Oder blamed Engineer Jackson of the express and added that his train had nearly been in a collision with one driven by Jackson at Eureka a fortnight earlier.[624]

The general view, however, was that the crew of the freight train were at fault, a view reinforced by the fact that Conductor Oder and Engineer Tracy had left the scene of the accident immediately afterwards, whereas Conductor Dubois and the crew of the express had remained to help with the rescue work. Conductor Oder and his engineer blamed the accident on a difference of watches, others believed that he had either misunderstood the order or deliberately run on short time in an effort to reach Eureka before the express left there.[625] He was arrested following the accident and subsequently charged with fourth degree manslaughter.[626] No report of his trial has been found.

At about 6.15 p.m on Sunday 26th February 1871 a Chemins de fer de l'Ouest passenger train from Rennes collided head-on with a train from Nantes carrying soldiers and their horses on the Rennes – Redon line between Beslé and Fougeray-Langon, France. Nine of the soldiers died in the accident.[627]

A Toledo Wabash and Western Railroad way freight train of 33 cars laden with grain collided with a construction train loaded with sand at 5.30 p.m. on Tuesday 25th July

[623] Janesville Gazette, Janesville, Wisconsin, United States, 13th May 1870

[624] North Vernon Plain Dealer, North Vernon, Indiana, United States, 26th May 1870

[625] Philadelphia Inquirer, Philadelphia, Pennsylvania, United States, 16th May 1870

[626] The Evening News, Indianapolis, United States, 1st July 1870

[627] Wikipedia – Liste des accidents ferroviaires en France au XIX siècle

1871 about a mile and a half from Edwardsville, Illinois, United States. Both trains were running at full speed and the accident occurred on a curve with woods on either side of the line further obscuring the view. Both engineers reversed their engines, whistled and jumped, along with their firemen, but 25 labourers riding on a flat car in the construction train remained ignorant of the danger until it was too late. The two locomotives jammed into one another and went over the embankment, completely wrecked apart from their boilers. Freed of the restraint offered by the locomotives, the trucks of the construction train shot forward into those of the freight train. Five of the labourers were killed outright and 16 or 17 injured, some of whom died later.

News of the accident quickly spread, and a large number of citizens were soon on the scene to help the injured and extricate the dead. Doctors Armstrong, Pogue and Weir went to the scene by a special train and worked there with their sleeves rolled up until midnight. A little girl who lived near the scene could see both trains from her vantage point and ran towards one of them waving her handkerchief to warn the engineer of the danger. Unfortunately, it was a common practice for people along the line to wave to trains and her warning signals were disregarded. The conductor of the way freight reported to the depot after the accident, then disappeared and had not been seen since, implying that it was that train which was at fault.[628] [629]

The Philadelphia and Erie Railroad suffered an accident on Saturday 26th (also reported as Sunday 27th) August 1871 near Westport, Pennsylvania, United States. A mail and passenger and a freight train collided, killing Conductor G.W. Hyman and Fireman Thomas Gannon of the mail train, Conductor J.W. Ward and Engineer William Killinger of the freight train and passengers B.C. Bryan of Lock Haven and R. Winslow. Conductor Hyman and Engineer P McConnell of the passenger had been issued orders to wait at Westport for the freight train but had not done so. The order was found in the pocket of the badly injured engineer after the collision. A Port Jervis newspaper was not at all happy with an editorial in another newspaper, *World*, which had mixed up the Philadelphia and Erie with the Erie Railroad and had directed its "righteous wrath" at the latter company, which the Port Jervis paper wrote did not have a line within miles of the accident site.[630]

The westbound mail train had left Renovo forty minutes late with orders to cross the first section of the eastbound freight train at Westport, then continue five miles to Clinton and cross two more sections of the freight train there. All the trains were more or less late and were running "on orders." Instead of waiting at Westport, the mail train had stopped there only briefly for passengers to leave and join it, then had continued its journey at speed. Both engineers, both firemen and the conductor of the freight train had all jumped, which had not prevented three of them being killed. Both locomotives were completely wrecked, the express car of the mail train was overturned and broken to pieces and the baggage car telescoped the leading passenger car to within two feet of its rear. The conductor and passengers were killed in this car and about ten passengers and Baggagemaster Frank Boon were injured. Fireman Alonzo Lyon of the freight train, one

[628] The Alton Telegraph, Alton, Illinois, United States, 28th July 1871

[629] Wisconsin State Journal, Madison, Wisconsin, United States,26th July 1871

[630] The Evening Gazette, Port Jervis, New York, United States, 29 August 1871

of those who had jumped, was also injured. The rear car of the mail train was undamaged, and its passengers escaped with a few bruises. The leading three cars of the freight train were loaded with lumber which piled up on the tender. Eight of the freight cars were wrecked.[631] [632]

The Rockford Rock Island and St Louis Railroad southbound passenger train No. 4 bound for St Louis and due at Alton Junction at 6.20 a.m. and St Louis at 7.10 a.m. on Wednesday 7th February 1872 collided with a northbound freight train, No. 7, three-quarters of a mile below the station at Upper Alton, Illinois, United States. The results were horrific. The baggage car of the passenger train telescoped the leading passenger car and fire almost immediately broke out in the wreckage due to an overturned stove at the front of the car. Four of the passengers were trapped in their seats and, unable to escape the flames, were slowly burnt to death. The *Alton Telegraph* newspaper wrote, 'When the collision occurred, Mr Rains, his wife and child, were sitting together on one seat, the child between its parents. The force of the collision was such that the seats were jammed together, and these unfortunates were caught in the wreck. Mr Rains finally succeeded in extricating himself, although terribly burned. Mr Franklin Grover tried his utmost to rescue Mrs Rains, and almost tore her clothes off in his efforts, but all in vain. Seeing that his exertions to save her would not be successful, he next tried to save the child, and succeeded in doing so, though at great personal risk.' To add to the difficulties of those attempting to rescue the passengers trapped in the wreckage no axes or other tools were available to the rescuers to cut away the sides of the coach and some of the victims had to be pulled out with broken legs and broken arms. About a dozen passengers were injured, some of them very seriously with severe burns and broken limbs. The baggage car and leading coach were totally consumed by the fire but the second coach and sleeping car escaped damage. The front freight cars were badly jammed together but considered repairable, the fire not having extended to them. The *Wm Green*, the locomotive of the passenger train, was badly damaged; the *Astoria*, hauling the freight train, suffered damage only to its headlight and cowcatcher. The passenger train had been travelling at 15 or 20 miles an hour at the time of the accident, the freight train rather faster according to one source, another gave its speed as not less than 12 miles an hour.[633] [634]

Conductor J. Fitzgerald of the passenger train told a newspaper reporter that his train had been on time and the newspaper reported that, 'Every one seems united in laying of this terrible calamity on Mr Baker, the conductor of the freight train, as his train was behind time, and furthermore a passenger has always the right of way over every other train, and as the passenger train was on time, there seems to be no possible shadow of an excuse for him to shield himself, or to justify his case.'

Conductor Frederick Baker stated later than on the arrival of his train at Alton Junction, he had compared time with his engineer, and they decided that they had 15

---

[631] Philadelphia Enquirer, Philadelphia, Pennsylvania, 28th August 1871

[632] Ibid, 30th August 1871, reproduced from The Williamsport Bulletin, Williamsport, Pennsylvania, United States, 28th August 1871

[633] Alton Weekly Telegraph, Alton, Illinois, United States, 9th February 1871

[634] Delphi Journal, Delphi, Indiana, United States, 21st February 1872 reproduced from St Louis Republican, St Louis, Missouri, United States, 8th February 1872

minutes to reach the next switch and had started off.[635]

The freight train had left East St Louis at 5 a.m. and was due at Upper Alton at 6.06 a.m. but was several minutes late. Notwithstanding this, Conductor Baker had ordered his engineer, Patrick Halpine, to proceed the 1¾ miles to Upper Alton rather than waiting for the passenger train at Alton Junction, which the engineer did, though he had expressed misgivings about being able to reach Upper Alton before the passenger train was due to leave there. It would seem also that Engineer Halpine, and probably the conductor, were working grossly excessive hours, the engineer having only reached East St Louis at midnight on Tuesday 6th after having been on the road for eighteen hours. Engineer Halpine had also only been over the line once before.

At the inquest on the victims of the collision, Conductor Fitzgerald stated that his train had consisted of the engine *Wm Green*, a baggage car, passenger coach and two sleeping cars, and had left Upper Alton at 6.11, two minutes late. This part of his statement was confirmed by Engineer Cessford and Fireman Sage of the passenger train. The conductor testified that he had been standing in the forward end of the sleeping car when the trains had collided at 6.13. There had been thirty-five passengers on the train, about fifteen of them in the passenger coach. He knew of seven passengers who had escaped from this coach. This would seem to suggest that eight passengers had died but only four bodies or parts of bodies were subsequently it would appear that either there were less passengers in the coach to begin with than the conductor thought, or others got out whom he had not noticed in the confusion. The conductor remembered that there had been two distinctive shocks, the first had caused the baggage car to telescope ten feet into the passenger coach, the second had telescoped it fifteen feet further. He had seen the fire break out four minutes after the collision from the baggage car stove, which had been pushed fifteen feet into the coach.

Conductor Fitzgerald stated that he knew Conductor Baker, who had a reputation of being a competent man; he did not know Engineer Halpine. The freight train was due at Wood's station (on the far side of Upper Alton) at 6.1 a.m. The rule of the road was that freight trains must wait indefinitely for passenger trains unless orders were issued to the contrary. He had received no orders from the train despatcher at Rock Island. The first warning that he had had of danger was a whistle for "Down Brakes" and the collision had occurred a few seconds later in a cut 40 ft deep, where there was a visibility of 200 yards. He had found his engine reversed.

The inquest jury returned a verdict that the deaths of Joseph Tweisel, of Le Claire, Iowa, Elizabeth Rains of Salem, Iowa, and two unknown persons had been caused by the criminal negligence of Conductor Baker and Engineer Halpine.[636] The conductor and engineer disappeared soon after the accident and the railroad company offered a reward of $500 for their apprehension.[637]

At 6.19 p.m. on Thursday 5th September 1872 a Baltimore and Ohio Railroad northbound express train collided head-on with the second section of an excursion train returning from the Ohio State Fair at Mansfield about half a mile north of Independence, Ohio, United States. Engineer Pollock of the express, his fireman and Engineer Douglass

[635] Alton Weekly Telegraph, Alton, Illinois, United States, 9th February 1871

[636] Alton Telegraph, Alton, Illinois, United States, 16th February 1872

[637] The Republican, Henry, Illinois, United States, 22nd February 1872

of the excursion train all jumped and were slightly injured. The fireman of the excursion train, unaware of the danger until too late, remained on his locomotive but escaped with a few bruises. Neither of the train conductors was hurt either but a baggage car and eight coaches were wrecked, several being completely telescoped, and it was too much to hope that the passengers would be equally fortunate. Four of them were killed outright and twenty-seven injured, three of whom soon died and one or two more of whom it was thought would not survive. The only passenger on the express hurt was a Mrs Eliza Francis, of Bellevue, who upon hearing the engine's alarm whistle had rushed from her seat out on to the platform and jumped off. The excursion train was very crowded, with lots of passengers riding on its platforms, this being the location of all those killed and most of those injured. The leading coach was telescoped into the baggage car but four men riding in the latter had a miraculous escape. After the accident physicians were telegraphed for from all the neighbouring towns and the citizens of Independence threw open their doors to the injured passengers. The last dead and injured were removed between 2 and 3 the next morning and the wreckage was cleared by 8 a.m. Following the accident, the company cancelled all excursion trains, preventing large numbers from travelling to Mansfield.[638] The accident was blamed on a misapprehension of a telegraph order by Conductor Fitzpatrick of the up train.[639]

At 3.43 a.m. on the morning of Thursday 2nd January 1873 the up Macon and Western Railroad passenger train which had left Macon, Georgia, United States, at 10 p.m. the night before collided head-on with a down train from Atlanta at the 74 milepost, between Bear Creek (Later renamed Hampton) and Lovejoy's stations and three miles from the former. The down train had left Atlanta ten minutes late and was still ten minutes late at Lovejoy's but as its conductor, Robert A. Harkie testified later when it left there it still had a minute of its schedule time (that is, a minute short of the time that it was due at Bear Creek) plus the 15 minutes which rule 6 of the company obliged the up train to wait at Bear Creek beyond the scheduled arrival time of the down train. The train therefore had 16 minutes to cover the five miles to Bear Creek, a run for which the time allowed was 13 minutes.

In charge of the train from Macon was Conductor Charles M. White, who at the time of the accident was engrossed in a conversation with a railroad official, a Mr Kendrick, in one of the cars.

Unfortunately, he was so preoccupied with the conversation that he failed to notice that his train had passed Bear Creek station. His engineer, Alexander Johnson, Fireman Louis Tisero and Wood Passer Charles Horn having all fallen asleep. The first intimation Conductor White had of the accident was feeling the shock of the collision.

Engineer F.M. Micklejohn on the down train saw the up train at the last moment and had time to whistle for "Down Brakes," shut off steam and jump. The engines, *Emerson Foote*, of the up train and *Victory*, of the down train were demolished and the tender of the up train was forced half-way through the following coloured car, which was filled with negroes who were emigrating westwards, and seven of them were killed outright. The dead were the wife and two of the six children of Peter Lewis, of Americus; brothers Ben, Joe

---

[638] Sandusky Daily Register, Sandusky, Ohio, United States, 7th February 1872

[639] The Democrat, Defiance, Ohio, United States,14th September 1872

and Pleas Buckner, of Griffin, Spalding county and the wife of Richard McPherson. Many other passengers were injured and there were unconfirmed rumours that two more had died of their injuries. The baggage car of the down train was wrecked; its contents included a fine piano of the Strakosch concert troupe. None of the troupe was hurt.

At the subsequent committal proceedings of Engineer Johnson, Conductor Harkie testified that after the collision he had spoken to Engineer Johnson and asked him what in the name of God did he mean by passing Bear Creek? The engineer replied that he did not think that he had, at least not by more than 100 yards. Conductor Harkie pointed out the 74-mile post, whereupon Engineer Johnson admitted that he must have fallen asleep.

Macon and Western Railroad President A.J. White stated that Engineer Johnson had also confessed to him that he had fallen asleep. The up train was due to arrive at 3.33 a.m. and depart at 3.35 if the down train was on time. The down train was due to arrive there at 3.32.

Conductor White was also arrested and committed for trial.[640] [641]

The Chicago and Alton Railroad suffered a major disaster on 16th August 1873 when two trains collided between Willow Springs and Lemont, Illinois, United States. Passenger train No. 4 consisted of engine No. 141, an express and baggage car, a smoking car, two passenger coaches and the sleeping car, *Louisiana*. It had left Chicago at 9 p.m. and was eight minutes late when it arrived at Willow Springs, 18 miles out, but, according to the rules of the road, had priority over an opposing coal train which should have waited for it at Lemont, but at about the time that the passenger train left Willow Springs the coal train set off from Lemont.

A Chicago newspaper reporter described what happened, 'At 10.20 p.m., when three miles north of Lemont and 22 miles south of this city, at a curve in the road, they suddenly approached a coal train of 35 cars running at the rate of 20 miles an hour an hour. The passenger train was running at the rate of 30 miles. Neither engineer saw the other train until it was too late to reverse, a fact partly accounted for by a curve in the road and partly by a dense fog. The two came together with fearful force, there being a curve, however, each engine left the track and passed each other, that attached to the coal train striking a baggage car a few feet from the end and shoving it off the track, and breaking the coupling between it and the smoking car, which the engine struck square in the end, with such force as to throw the forward end into the air, so that the engine ran under it, tearing the floor completely out, and hurling the 50 or 60 unfortunates who were in the car in a struggling mass to the lower end, whence there was no chance of escape, and now ensued a scene, the horrors of which cannot be adequately described. The smoke stack and dome of {the} engine were knocked off by the collision, and the broken timbers of the smoking car penetrated the boiler, letting loose and dense volume of hot steam, which poured into the car, blinding and scalding the helpless inmates, who with heart-rending shrieks struggled vainly to extricate themselves.' [642]

---

[640] Union and Recorder, Milledgeville, Georgia, 8th January 1873, reproduced from Telegraph and Messenger, Macon, Georgia, 4th January 1873

[641] Telegraph and Messenger, Macon, Georgia, United States,11th January 1873

[642] Wisconsin State Journal, Madison, Wisconsin, United States,18th August 1873

The same newspaper report added, 'The following is a statement from Dr Pierce, of Lemont: The freight train was in charge of a conductor who had been discharged from the company's service some time since, but who had been reinstated some two or three weeks ago. The freight was fully fifteen minutes on the passenger's time, and the agent begged the conductor to lay over at Lemont until the passenger was in. He refused and said he should go on. The agent remonstrated with him and told him it would be a risky business as he was sure to meet the other train. He pulled out of the depot, and a very short time afterwards the shock of the collision was plainly discernible at Lemont, and the report was heard by a number of people. The agent procured an engine which was on a side track, and went, accompanied by the doctor, to the scene of the disaster. This was at a place called Sag Bridge, at the end of a long curve, at the end of which the express had suddenly come upon the freight in the fog. When they reached the spot the scene was horrible, it was very dark, and a fire had been built on the track out of some of the ruins and some rail fencing. Many of the wounded had been placed around this fire, and the groans were awful. The engine of the passenger train was swung off the track and was badly smashed, and the engine of the freight train was literally smashed to fragments and was partly jammed into the baggage car and the fore part of the smoking car. The whole debris was indistinguishable mass of wheels, broken iron, and general ruin. As far as I could judge, and we paid particular attention, we found two persons killed on the ground and thirty-nine wounded. Five men died on the train coming to Chicago. The smoking car was broken in two pieces. Those only were injured who were in the smoking car and the first coach behind. They were principally injured with the scalding steam, and I have no doubt that more will die. O'Neil, the engineer of the passenger train, jumped from the engine in such a state of excitement that it overcame his reason. The last thing he was heard to say before he became perfectly crazy was," I didn't see him," two or three times over. He began to mutter incoherently and relapsed into a state of insanity, in which condition he remains at Dr Pierce's house in Lemont.'[643]

Somewhat remarkably, a little after the accident a reporter for the *Chicago Tribune* newspaper managed to interview Joshua Puffenberger, the engineer of the coal train. Asked who was to blame for the collision the engineer replied that he supposed that the reporter already knew that it was his (Engineer Puffenberger's) conductor, Edward Bean, who had completely forgotten about the passenger train. Asked if he, the engineer, should not have also known of the passenger train the engineer replied that his and the conductor's watches had been different, his being fifty minutes slow. The watches had been compared at Joliet, and there showed the same time, but he supposed that in winding his at Lockport the hands had got caught, so that afterwards it was nearly an hour behind time. The passenger train was due at Willow Springs at 9.50 p.m. and according to his watch they had left Lemont at 9 p.m., so that he had had no objection to setting off on the order of the conductor, supposing that they had fifty minutes to cover the five miles.

Continuing, Engineer Puffenberger said that they had been delayed at Joliet and had put four cars off at Lemont, where the conductor had plenty of time to decide if it was safe for them to proceed. He did not suppose that the Lemont agent had realised that they were not going to wait there for the passenger train until it was too late. It was a dark foggy night and they met the passenger train on a sharp curve; he did not see it until it

---

[643] Ibid.

was about five car lengths apart, when he had jumped. He could not recall if he had shouted a warning to his fireman (who was killed).

Engineer Puffenberger said that he had only been on the Chicago and Alton a couple of weeks but had been an engineer on other roads previously. He did not know Conductor Bean very well but considered him nervous and excitable.[644]

It was subsequently reported that the engineer had been arrested while the conductor was missing, and a search was being made for him. At the coronial inquiry, Conductor Henry E. Russell of the passenger train testified that it had left Chicago perhaps two minutes late and had been detained 18 minutes at Willow Springs taking a sleeping car (the *Missouri*) from train No. 1. They had left Willow Springs at about 10.08.p.m. He had been collecting tickets in the sleeping car when he had been thrown to the floor. Getting up he had gone to the rear of his train and send a brakeman back to protect it, gone forward to the engine and spoken to Engineer O'Neil, who was 'rather flighty' (suffering from shock perhaps) but told him that they had collided with train No. 23, gone back and sent Mr Lindsay, a fellow conductor, back to Willow Springs to rouse the operator there and get assistance, then made his way towards Lemont and met an engine and car bringing a physician to the scene, and having left the doctor there returned to Lemont on the engine to telegraph details of the accident to Bloomington. Clearly, he had earnt his wages that night. Ending his testimony, the conductor explained his that train had had exclusive right of road from Chicago to Bloomington and it had been unnecessary to make inquiries at Willow Springs. The coal train had had no right over his train at all.[645] The conductor's statement clearly illustrates the method of working single lines in the United States outlined earlier. The passenger train, being a superior class of train to the coal train, had priority over it. The crew of the coal train would have had a working timetable detailing the times at which the passenger train was due at different points and it was their responsibility to keep out of the way of it and not to leave any station where the passenger train was due until it had arrived there, unless they had received definite orders from the dispatcher to do so.

Conductor Bean was subsequently located and both he and Engineer Puffenberger were found guilty of criminal negligence at the conclusion of the inquest on the victims of the disaster. The engineer was found 'Not Guilty' of manslaughter at his trial at Chicago at the end of October, a verdict which was applauded by some members of the court, leading to their severe censure by Judge Williams.[646] Conductor Bean opted for a separate trial. He was due to appear in court at Joliet in January 1874, but his trial was indefinitely postponed[647] and in August 1874 the indictment against him was quashed.[648]

The final number of dead in the accident was no less than 19.[649]

The circumstances of a collision on the Cornwall Railway in England might have been considered to have had a humorous element had they not resulted in the death of an

---

[644] Fort Wayne Weekly Sentinel, Fort Wayne, Indiana, United States, 27th August 1873

[645] Daily Leader, Bloomington, Illinois, United States, 20th August 1873

[646] Wisconsin State Journal, Madison, Wisconsin, United States, 27th October 1873

[647] Alton Weekly Telegraph, Alton, Illinois, United States, 23rd January 1874

[648] The Monmouth Review, Monmouth, Illinois, United States, 14th August 1874

[649] The Daily Freeman, Kingston, New York, United States, 29th October 1873

engine driver and the injury of several other railwaymen. The accident in question happened early on the morning of 2nd December 1873. At this date, the main line of the Cornwall Railway from Plymouth to Truro had still only a single track apart from a short stretch between Plymouth and Devonport. The line was equipped with the electric telegraph, by means of which the crossing places of trains set out in the timetable could be changed as necessary, but it was not a mistake in the exchange of telegraph messages which led to the accident.

On duty at Menheniot station on the night of 1st to 2nd December was Porter Signalman Pratt. The late running 2 a.m. down goods train from Plymouth arrived at his station at 4.45 a.m. and had to stop there as an up goods train, the 3.15 p.m. (the previous day) from Truro, had not yet arrived. St Germans, the next station up the line towards Plymouth, then advised Pratt on the telegraph instrument that another down goods train, the 3.30 a.m. from Plymouth, was at St Germans, and Pratt, unsure of how long the up goods train would take to reach him, decided to permit the 3.30 a.m. train to come down from St Germans and cross the up goods train at Menheniot. It will be noted that the method of working single lines on the Cornwall Railway was completely different to that commonly employed in the United States. Instead of the altered crossings of trains being decided by a central despatcher at say Plymouth, or Truro, who issued orders to the different operators, the decisions were made locally by the men on the spot.

Shortly after Pratt had given ‘Line Clear’ to St Germans for the 3.30 a.m. down goods train, the 3.15 p.m. up goods train arrived from Liskeard. The porter signalman cleared the up signal near the west end of the up platform and showed a red hand lamp to the drivers of its two engines. He also unblocked the needle of his block instrument to Liskeard for the up train and asked for and obtained ‘Line Clear’ from Liskeard for the 2 a.m. down goods.

According to Pratt’s evidence at the Board of Trade inquiry, Richard Wills, the head guard of the 2 a.m. train, was standing at the door of his brake van, and Pratt shouted to him, “All right Dickey for the down train.” The guard shouted back, “All right, Jimmy,” and shouted to his driver, “All right, George.” Pratt then had to walk down to the Liskeard end of the crossing loop to operate the points, the goods train following him down. Having set the points, Pratt was walking back up the line when he saw the up goods train setting off as well and Head Guard Richard Scantlebury walking along the up platform. “Wherever are you going?” he had shouted to the guard. “Right away, isn’t it?” Scantlebury had replied.” Oh, bless the man,” Pratt had cried, “the 3.30 a.m. down goods is off St Germans 10 minutes.” “Oh my God,” the guard had replied and the two of them had run after the train shouting for it to stop but it had kept going. Pratt then returned to his office and telegraphed to St Germans, “How long down goods off St Germans?” “Nearly 10 minutes,” St Germans replied. Pratt telegraphed, “3.15 p.m. up goods off Menheniot without orders.”[650]

The St Germans signalman entirely confirmed Pratt’s statement as to his telegraph messages, though he said that, at first, he did not believe them. Going out onto the platform he could hear the down train running, then heard a brake whistle and the crash of the two trains. He then found the telegraph line to Menheniot was dead and he telegraphed to Plymouth for assistance, as Pratt had likewise done to Truro. Pratt said that

---

650 Board of Trade report

when he had shouted to Guard Wills, he had not seen that Scantlebury was the guard of the up train; he knew Scantlebury's Christian name was also Richard.

The 3.30 a.m. down goods train consisted of a tank engine and 19 vehicles, including two or more brake vans. The 3.15 p.m. up goods train had two engines, 22 waggons and two brake vans. The trains met in a small cutting on a curve at one of several summits between the stations at about 5 a.m. There was very little visibility and the crews of the two trains could have had very little warning of danger, though according to the St Germans porter signalman someone had time to sound a brake whistle. The leading engine of the up train was destroyed and the other two engines and 23 waggons more or less badly damaged. Driver Richard Hocken was fatally injured and died the next day; the other two drivers, all three fireman and a guard were more or less hurt.

The statements of Guards Wills and Scantlebury as to what happened at Menheniot were somewhat at variance with that of Porter Signalman Pratt. Guard Wills stated that when the up goods had arrived, he had spoken to Guard Scantlebury and asked him why they so late, to which Scantlebury had replied that they had had a lot of work to do and a heavy train. The porter signalman had then appeared and shouted to him (Guard Wills) "All right, Dick" and "Right away, Dick". As he was going to his driver to tell him to start, he had heard Guard Scantlebury ask, twice, "Is it all right for me?" and Pratt (perhaps mistaking whom he was talking to in the darkness) had replied, "All right, Dick."

Guard Scantlebury had been dismissed by the company and did not appear at the Board of Trade inquiry but at the inquest on Driver Hocken he confirmed the conversation with Guard Wills as to why his train was so late and of Pratt shouting "Right away, Dick," of his asking Pratt if it was all right and Pratt confirming that it was. Guard Scantlebury said he had walked up towards his engine, waved a white lamp top his driver and had been making his way back along the platform when the porter signalman had come towards him showing a red light and had asked, "You are not off, are you?" He had said, "Why, what is the matter?" and Pratt had answered, "Line is not clear for you." "You don't mean to say the 3.30 is off St Germans?" "Yes, about ten minutes," Pratt had replied, and he had exclaimed, "My God!"

Colonel W. Yolland investigated the collision. He noted that Guard Scantlebury had made a great mistake in not immediately jumping into his van and screwing his hand brake down hard before waving a red lamp at his drivers. The up train had run through the points at the St Germans end of the loop, but this apparently did not alert its engine crews. Scantlebury said that he was called "Dick" at almost every station. It had not occurred to him to ask where the 3.30 a.m. down goods was, nor the 4.50 a.m. down mail, which was not yet due at St Germans, and which he thought might be in front of the 3.30 goods. The inspecting officer was of the opinion that as the up goods was running over four hours late it was all the more important that the head guard should have carefully and specifically made enquiries as to the whereabouts of down trains before authorising his train to proceed. Since the accident, the Superintendent of the line, Mr C.E. Compton, had issued instructions aiming to prevent a repetition by instructing guards to enquire at every station as to what trains they were to cross, station officials to plainly and distinctly tell the guards the stations at which they were to cross other trains and so on. The directors of the Company had also ordered the erection of starting signals at every station.

Colonel Yolland concluded his report by recommending the addition of staff and ticket working to the use of the electric telegraph (as described in relation to the 1866 Patchway accident) as a safer means of working single lines.[651]

Ironically, only six months after the Cornwall Railway accident, in the report on which Colonel Yolland had recommended the addition of the staff and ticket working and providing greater safety, a collision occurred in Scotland on a line using just that system, a demonstration that although the system might be safer, it was by no means perfect.

This accident occurred on the Glasgow and South Western Railway near Dalbeattie, on a line over which the Caledonian Railway Company had running powers, and the 11.30 a.m. Caledonian train from Dumfries to Stranraer was one of those involved in the collision, the other being a Glasgow and South Western ballast train returning from a ballast pit 1½ miles from Dalbeattie. The trains collided three-quarters of a mile on the Stranraer sider of Dalbeattie station. The fireman of the goods train was killed outright, the driver of the ballast train was fatally injured, and two platelayers were seriously and six slightly injured. Both engines were more or less completely destroyed, and several waggons more or less knocked to pieces.

The ballast train had set off for the ballast pit at about 11 a.m. with the Dalbeattie – Castle Douglas staff, its crew intending to be back at the station in time for the Caledonian goods to depart for Castle Douglas at 12.17. The ballast train consisted of an engine and tender, four empty waggons and a van with a driver, fireman, permanent way inspector, foreman platelayer, ballast guard and six platelayers. By about noon the waggons were almost filled, and the train started back for Dalbeattie at 15 or 20 miles an hour. Permanent Inspector McKean saw the goods train coming towards them when it was about 200 yards distant, shouted, hesitated a moment and then jumped, followed by the fireman. The driver said "Aye" but did not jump.[652]

The Caledonian Railway train consisted of an engine and tender, nine loaded waggons and a brake van. The first single line section it had to traverse began at Lochanhead. Caledonian Driver John Robb received a ticket there and was not sure whether he had seen the staff or not, he told the Board of Trade inspector that he did not consider it his duty to ask to see the staff if it was not shown to him. He likewise received a ticket at Kirkgunzeon and did not see the staff there either. The Dalbeattie signals were lowered for his train when he whistled for them and a lad came out of the station office and handed him a ticket. He asked the lad if it was all right and the lad replied "Yes, go on." He did not enquire about the staff; he usually left Dalbeattie with a ticket with this train. He stated that more often than not he had passed Dalbeattie with a ticket without seeing the staff; this would be when the ticket was handed to him by a porter or a clerk; the stationmaster usually showing him the staff. He repeated that he did not consider it his place to ask to see the staff if it was not shown to him. He had seen the ballast train 150 or 200 yards away, reversed his engine and clung to the reversing lever.

The guard of the train stated that he had seen the train staff at both Lochanhead and Kirkgunzeon but admitted that he had not seen it at Dalbeattie.

651 Ibid

652 Ibid

Dalbeattie Stationmaster James Carmichael stated that he had given the staff to the foreman platelayer at about 11 and then at 11.40 had gone into town on company business without leaving anyone in particular in charge of the station, nor had he specifically told anyone that the ballast train had gone out with the staff. He heard of the collision as he was returning to the station. He stoutly denied having ever given out a ticket without showing the staff or allowing any of his subordinates to do so, though he acknowledged that he was aware that the ticket box was sometimes opened with a ruler to enable returns of the number of tickets issued to be made out.

Dalbeattie Booking Clerk William Douglas stated that he had heard the goods train approaching and had considered it his duty to give it a ticket. He found the ticket book lying on the desk in the booking office and without noticing whether or not the staff was attached to the ticket box filled in the ticket with his name and the time, 12.11, took it out and gave it to the driver without them saying anything to one another. He had not known that the ballast train was on the line. He had always shown the staff previously but had been in a hurry. He had frequently seen Porter McKee (who had absented himself the day before the accident) give out a ticket without displaying the staff and it was constant practice to open the ticket box with a ruler or a poker when necessary, when the staff was not there.

Porter Thomas Hutton said that he always found the ticket book lying outside of the box, but he had never issued a ticket without showing the staff.

The 14-year-old parcels clerk acknowledged opening the ticket box with a ruler when he had to in order to make out a return of tickets issued.

Lt-Col C.S. Hutchinson stated that the very serious collision was primarily the fault of Clerk Douglas and Driver Robb but was only the natural result of a laxity of practice in the issuing of tickets and the showing of the train staff for which the stationmaster was mainly responsible. The latter was also much to blame for leaving the station without anyone being in charge and for not advising anyone that the ballast train had gone out with the staff. The Caledonian guard was also to blame to a lesser degree.[653] It was later rumoured by Glasgow and South Western men that after the collision the Caledonian crew had tried to take to train staff from the unconscious driver of the ballast train in an effort to show that they were not to blame.[654]

'Last night one of the most appalling accidents which ever happened in the Eastern Counties occurred on the Great Eastern Railway between Norwich and Brundall.' Thus, did *The Times* in London begin its account of a single line collision which happened on Thursday 10th September 1874 when the 5 p.m. down express from London Liverpool Street collided with the up mail train from Reedham Junction, the latter composed of portions which had left Lowestoft at 8.20 p.m. and Yarmouth at 8.40 p.m. The accident resulted in the death of 22 passengers and four railwaymen and more or less serious injuries to over 70 passengers and two railwaymen.[655]

The line from Norwich Thorpe Junction, a quarter mile outside of Norwich Thorpe

---

653 Ibid

654 The Railway Magazine 1941 pg. 19

655 The Times, London, England, 11th September 1874

station, over the 5¼ miles to Brundall, the station before Reedham Junction, was still single track, though work was in progress to double part of it, with plans to eventually double the remainder. It was worked by telegraph and crossing order.

On the evening of 10th September 1874, the down London express was running late, as was often the case. It was due to leave Norwich Thorpe at 9.10 p.m. but at 9.17 had not arrived and Night Inspector Alfred Cooper was becoming anxious that it might delay the up mail train if the latter had to wait for the express to reach Brundall. A telegraph message should have been received from Wymondham if the express was more than 15 minutes late, but they had not had one. The night inspector went to Stationmaster Sproule's office to get his orders for the night and asked him, "What about having the mail up, Sir?" The stationmaster asked, "What time is the mail due at Brundall?" "9.25 p.m." he was told. The stationmaster answered, "We will not have the mail up, certainly not." The night inspector then remarked, "You know, Sir, there is an order allowing us to detain the down train as late as 9.35." to which Sproule replied, "All right, we will soon get her off," and Cooper then left the office. The stationmaster stated that the night inspector had not come to his office until about 9.17 but Cooper testified that he had got there some minutes earlier. More importantly there was apparently a misunderstanding between Sproule and Cooper as to whether the down express or the up mail should be given priority, Cooper testifying that he had understood the stationmaster's last words had been, "You had better do so," or words to that effect. Under the impression that the stationmaster had agreed to his proposal to have the mail train up from Brundall to Norwich first, Cooper had, according to his evidence at the Board of Trade inquiry, then gone to the telegraph office, knocked on the closed wicket, which Telegraph Clerk John Robson had opened, and had said to Robson, "The express is not in, we had better have the mail up," or words to that effect. He had waited outside for a minute or so at which point he had seen the down express belatedly running into the station. Returning to the telegraph office he stated that he had told Robson not to order the mail up as the express was just coming in and Robson had replied "All right, Captain." [656]

Robson's account of what had happened varied considerably from Cooper's. He testified that Cooper had knocked on the wicket and said to him, "Tell Brundall to send the up mail train on before the down train leaves." Robson said that he wrote down the message and immediately sent it to Brundall. Crucially, he had not waited for the night inspector to sign the message as Cooper had promised that he would return shortly and do so. He had never sent an unsigned message before. There was a bell which he could have rung to call Cooper back to sign the message, but he did not use it. He sometimes rang the bell to call inspectors back and then they complained. He had waited from 9.25 to 9.32 for the night inspector to return and sign the message. Robson denied that the night inspector had said anything to him about not ordering the mail up. He admitted that he had telegraphed the message to Brundall to send the mail up ending 'Signed A. Cooper,' which of course it was not. He had sent the message to Brundall at 9.24.

Night Inspector Cooper's next action is extremely hard to understand. Day Inspector William Parker should have left duty at 9.10 p.m. but stopped to collect the tickets from the down express. Knowing that the train was late, he asked Cooper if the latter had ordered the mail up. Depending, presumably, upon the fact that he had not signed the message that he had given Robson a couple of minutes earlier about doing just that,

---

656 Board of Trade report

Cooper had replied, "No, certainly not. Let us get the train away as soon as possible." Inspector Parker had handed the driver of the express, Thomas Clarke, a starting order and the express had steamed out of the station at about 9.31. According to Cooper, he then returned to the telegraph office to advise Reedham that the express was over 15 minutes late and when he got there, before he could say anything, Robson had told him, "I have ordered the mail up," and he had exclaimed, "Good God, Robson, how dare you order the mail up when I distinctly told you not to do so." Robson had replied. "You did nothing of the kind," and he had retorted, "How dare you assert such a lie," and "A lie stands for nothing."

Robson's account of the matter was that at about 9.30 Police Inspector Trew had come to the wicket to enquire about a message for Walsingham and opening the wicket he (Robson) had noticed that the express was not standing in the station and had said to Trew, "Where is the down train?" or "Has the down train gone?" Trew had replied that it had gone. He asked Trew if the mail had left the yard and at that moment Inspector Cooper had rushed up, looking very excited, and told him to telegraph Brundall and stop the up mail train. Robson had run to the telegraph instrument and sent "Stop mail" to Brundall. It had then been 9.32. Brundall had immediately telegraphed back, "Mail left." Robson said that he had then asked the inspector to sign the message sent to Brundall to order the mail up, but Cooper had refused to do so, saying, "No, I never gave you the message." To this Robson had responded, "Yes you did, or why, if you did not give me the message, do you now come back to cancel it?[657]

Stationmaster William Platford at Brundall produced the message at the inquiry that he had received from Norwich regarding the mail-

"*Send the up mail on to Norwich before the 9.10 p.m. down passenger train leaves Norwich. Signed A. Cooper. Time received 9.26.*"

He said that had the message not been signed he would not have let the up mail proceed. He had telegraphed back-

"*I will send the up mail train on to Norwich before the 9.10 p.m. down passenger train leaves Norwich. Signed W. Platford. Time received 9.26*"

The trains collided a mile and three-quarters from Norwich on a dark and wet night. Engines No. 218 of the express and No. 54 of the mail train were very badly damaged and Driver Clarke and Fireman Frederick Sewell on the express and Driver John Prior and Fireman James Light of the mail train were all killed outright. It appeared that both drivers had had time to shut off steam and both firemen had applied the tender brakes. Thirteen vehicles in one or other train were destroyed and four more damaged. Slightly mitigating the effects of the disaster were the fact that the front of the express had just cleared a wooden bridge over the River Yare and that there had been three empty carriages and a horse box at the front of the express and a fish waggon and a brake van at the front of the mail train. The Lowestoft portion of the mail train was crowded, as there had been a flower show there, but fortunately the less crowded Yarmouth portion was marshalled in front of it. Heavy rain probably also prevented the wreckage from catching fire.

A Dr Eade, who had been a passenger on the up mail train, told a newspaper reporter

---

[657] Ibid.

that he had been riding in a first-class carriage. As they were coming down the incline from Postwick, he had felt a tremendously heavy blow. The thought had occurred to him that perhaps the bridge over the River Yare had given way, that the engine had struck the far bank and had fallen into the water. The occupants of his carriage were thrown violently from one side of it to another and the carriage then seemed to split into two. The doctor had managed to get out of the carriage through the opening onto the marshy ground adjoining the line. A passenger who had joined the train at Brundall said that the mail train had not stopped more than two minutes at Brundall and that he had remarked to a fellow passenger that the down express would not be departing punctually. The mail train had been running at a fair speed, and shortly after passing the Thorpe Lunatic Asylum a heavy blow had thrown all the passengers forward, there had been a powerful recoil and then the carriage had completely overturned. When the passengers in his coach had recovered themselves somewhat, they saw that their carriage was a lot closer to their engine than it had been earlier. The shock had put all the lamps in the carriage out and they were fearful of being scalded, as the engine was blowing off steam at a terrific rate, but they were eventually rescued.[658]

The down express had consisted of engine No. 218 and tender, a Stockton and Darlington Railway horse box, a second-class carriage, open third-class carriage, first-class carriage, third-class carriage, two composite carriages, two third-carriages and a brake van with Guard George Read (the Yarmouth portion); and a second-class carriage, a first-class carriage, a third-class carriage and a brake van with Guard Black (the Lowestoft portion). The rear carriages kept to the rails. They came to a stand on the Yare bridge, which lacked railings, and it was doubtless thanks to Guard Black, who had edged his way along the outside of the carriages with his lamp calming the terrified passengers, that no one had jumped out and fallen into the river.[659]

The up mail train consisted of engine No. 54, a fish truck, a brake van, two composite carriages, a third-class carriage, a mail van and a composite carriage (the Yarmouth portion); and a brake van, a third-class carriage, a first-class carriage, two second-class carriages and a third-class carriage (the Lowestoft portion).[660]

Captain Henry Tyler investigated the accident. He noted that this was the most serious single-line collision that had ever occurred in the country. According to the method of operation, no train was allowed to depart from either Norwich or Brundall without its driver receiving written instructions. At Norwich the duty of starting trains devolved to the inspectors, at Brundall it was the duty of the stationmaster alone to both despatch trains and operate the telegraph instruments but at Norwich there was a telegraph clerk who sent and received messages. It was the duty of the inspector to sign every telegraph message received and sent.

As with regards to the accident, in the first instance there had clearly been a misunderstanding between the inspector and the stationmaster, while the immediate blame for the collision rested between Inspector Cooper and Telegraph Clerk Robson. On the one hand, it was difficult to believe that the inspector, almost immediately after he had left the telegraph office wicket, would have informed Inspector Parker that he had

---

[658] The Times, London, England,12th September 1874

[659] Webpage – Broadland Memories Blog – The Thorpe Railway Disaster 1874

[660] The Times, London, England, 15th September 1874

not ordered the mail up unless he was satisfied that his order to Robson to do so had been delayed because it was unsigned or cancelled. Equally, it was difficult to understand why, if Robson had heard the inspector say, as Cooper maintained, that he had changed his mind about having the mail up, he had not cancelled the order to Brundall or had afterwards sent it regardless. Obviously there had been a misunderstanding between the two and both were to blame. Cooper admitted having originally told Robson to order the mail up and, having done so, he should have been very careful to ensure that the message had not been sent, or that it had been cancelled if already sent. In his defence it could be argued that he had not signed the message, which ought therefore not to have been transmitted. Robson admitted that he had sent the message to have the mail up to Brundall as signed by the inspector, though it had not been, and having failed to summon the inspector to sign it as he could have done. He had also had no less than four of his mates in the office at the time who might have distracted him.[661]

The inspecting officer then embarked on a lengthy review of the different means of single line working employed; by telegraph alone by local stationmasters, by train staff and ticket, by train staff and ticket combined with the block telegraph, and, in North America, by the use of a central train despatcher who issued orders to the local stations over a large area. None of the systems had shown itself capable of being worked entirely free of accidents, though the train staff and ticket combined with the block telegraph appeared to be the safest. Whatever system was used, after a period of time it tended to lead to a laxity on the part of employees, which could only be avoided by the maintenance of rigid discipline by constant, efficient and irregular supervision. The important lesson to be learnt from the disaster was that whatever system was used 'it should be fenced about with safeguards precisely expressed and carefully observed.'[662]

The last passenger to die of injuries sustained in the collision was Jane Faulkner, a domestic servant, who survived until 28th October 1874 before succumbing in Norwich Hospital.

The inquests into the deaths of the victims of the accident predictably found both Inspector Cooper and Telegraph Clerk Robson guilty of manslaughter. At their trial at Norwich Assizes in April 1875 Robson was acquitted but Cooper was found guilty, with a recommendation to mercy, and sentenced to eight month's imprisonment.[663] [664]

A few months after the Norwich Thorpe disaster, a letter in *The Times* described a similar, though much less serious, accident which had occurred in Italy. W.S.P. and his wife were passengers in an express which left San Remo, on the line from Nice to Spezzia, at 6.30 p.m. on 19th December 1874. All went well until they had just passed Porto Maurizio station, when their train came into head-on collision with a goods train from Oneglia on the single line, killing the drivers and firemen of both trains and heaping the engines and several carriages in 'inextricable confusion.' The writer stated that the

661 Board of Trade report

662 Ibid.

663 The Times, London, England, 14th,15th,16th,21st, 22nd,23rd, 28th and 29th September 1st,2nd,3rd,5th,10th,12th,14th,16th,20th,21st,24th and 31st October; 19th and 28th November and 4th December 1874 and 12th January and 24th and 25th February 1875

664 West End News and London Advertiser, London, England,10th April 1875

accident would have been much worse had the express not been going slowly, or had it happened a few yards further on where the line ran on a high embankment overlooking the sea. As at Norwich a mistake in the working of the telegraph had caused the accident. The telegraph clerk at Oneglia had enquired of Porto Maurizio if the express had arrived. The Porto Maurizio clerk had replied that it had not and that he would hold it for the goods train. At that moment the express did arrive at Porto Maurizio and before the telegraph clerk had had time to tell the stationmaster that he had accepted the goods train the stationmaster had let the express proceed.[665]

What was probably a typical American 'cornfield meet,' as single line collisions were commonly termed in the United States, occurred on the Burlington and Missouri Railroad, two and a half miles east of Tyrone, Iowa, United States, early on Sunday 4th April 1875. Westbound express No. 1 and eastbound express No. 2 usually crossed one another at Tyrone but sometimes, if the westbound train was late, they crossed at Albia, the next station eastwards. On the night in question, it would appear that the despatcher sent an order to Operator Harrington at Tyrone to hold train No. 2, which was itself a little late, at Tyrone for the westbound train, No. 1. Harrington had his train order signal displayed to stop train No. 2, it pulled up and Conductor H. L. Miller went into the operator's office. The operator handed the conductor several orders applying to other trains. Miller read them, then asked, "Is that all?" Harrington replied that it was. Conductor Miller signalled to his engineer, A.J. Rugh, that they were clear to depart, and the train started away. The tail lamps of the train had only just disappeared from view when Operator Harrington spied the order that he should have delivered to Conductor Miller for his train to wait there for train No. 1 lying on the table.

There was a freight train waiting on a side track for train No. 1 and, as quickly as possible, its engine was uncoupled and set off at speed in vain pursuit of Conductor Miller's train, which by then was at least a mile ahead. The freight engine had not got very far before the two expresses collided and the engineer of the freight train reversed his steed and returned to Tyrone.[666]

The two engines were wrecked, and the baggage car of train No. 2 was badly damaged and partly telescoped by the leading passenger car. Conductor Miller, who had been riding in the baggage car, was killed as were Express Messenger W.A. Gardner on train No. 2 and Fireman C.M. Pickle on train No. 1. Engineer Rugh of train No. 2, Engineer Mark Strassler of train No. 1 and Mail Agent N.J Miller were injured, Strassler and Miller seriously. Strassler was also pinned under his engine and in danger of drowning for some time before he could be extricated. No passengers were hurt and none of the mail or of a considerable quantity of gold and silver being carried on the train was lost.

Having done all they could at the accident site, District Superintendent Potter and Colonel Hepburn went to Tyrone and confronted Harrington, who at first attempted to justify his actions, but was told curtly that there was no time for excuses. The superintendent did not reveal the extent of the accident to Harrington but just told him, "You have got us into trouble, now you must help get us out of it." Harrington then sent off a large number of messages written by Colonel Hepburn. 'Cool, collected and steady,

[665] The Times, London, England, 24th December 1874

[666] Chariton Leader, Chariton, Iowa, United States, 10th April 1875

he sent off these accounts of his own forgetfulness and its fatal consequences. Never a mistake in his sending or a tremor in his writing; surely a man with such perfect control of himself at a time when one would imagine in a perfect agony of excitement, surely cannot be a habitually careless of forgetful man.'[667] Another newspaper was much more critical of the operator, writing that he, 'is a reckless sort of a young man, and no more fit for such a responsible position than he would be to guard the gates of Paradise.'[668]

Harrington sent in his resignation a few hours after the accident.

In 1875, the 4th July was a Sunday so the Independence Day holiday in the United States was Monday 5th July and the Southern Railroad of Long Island put on a number of extra trains to convey passengers to Rockaway Beach, New York, with a new timetable in force. The 12.30 departure from Eighth Street, Brooklyn, left considerably after time and its six cars were crammed with passengers. At the junction with the single-track branch to Rockaway Beach at Valley Stream, the Brooklyn passengers changed into another train, consisting of the engine *Norwood* and tender and seven cars in charge of Conductor Jesse Hibbard, a veteran employee of the company. This train was also badly overcrowded, with a number of passengers standing on the platforms of the cars. It left Valley Stream 15 minutes late and was going down a grade between Lawrence and Far Rockaway stations when it met an empty train in the charge of Conductor Augustus Holdridge, which was on its way to Valley Stream. This was an extra working put on to deal with the heavy holiday traffic. Despite the combined efforts of the two engineers, the collision was a severe one and Conductor Hibbard, Fireman William Flynn, a brakeman and eight passengers were killed outright, half a dozen passengers possibly fatally injured and about twenty less badly hurt.[669]

According to the regulations in force, up until noon trains from Rockaway had the right of way for ten minutes beyond the scheduled departure time, with an additional allowance of five minutes to allow for a difference in watches, though this extra five minutes was not to be used by conductors to advance to the next crossing place. After noon the same rule applied to trains from Valley Stream. It was thus the duty of the train from Valley Stream, not having arrived there until 1.39 p.m., within one minute of the time that it should have crossed the train from Rockaway at Far Rockaway, to have waited at Valley Stream, although it could have gone as far as the siding at Woodsburg without danger. Mr H.C. Pappenhausen, the general manager of the company, stated that while at Far Rockaway he had received a telegram from Valley Stream that the train had left there 15 minutes late, Conductor Holdridge, knowing that after waiting fifteen minutes he had the right of way, had shown no recklessness in starting his train at 1.55 p.m., especially after receiving the telegram that the Valley Stream train was 15 minutes late. The empty stock train consisted of the engine *Montauk* and five cars. It had covered less than 1,000 feet from Far Rockaway when the Valley Stream train appeared around a corner a short distance away, running at speed. The collision, at 1.57 p.m., smashed the *Norwood* engine to pieces and the smoking car of the Valley Stream driven was driven beneath the *Norwood's* tender and several of its cars telescoped. The *Montauk* engine

---

[667] Ibid.

[668] The Patriot, Chariton, Iowa, United States, 7th April 1875

[669] The Philadelphia Inquirer, Philadelphia, Pennsylvania, United States,7th July 1875

also suffered badly, but less so than the *Norwood.* Mr Pappenhausen said it was very evident that Conductor Holdridge had not been going at excessive speed and had followed the rules implicitly, if anything waiting a fraction of a minute longer than he needed to.[670]

It was noted that a similar collision had occurred on the line on 4th July three years earlier, near Long Meadow, but that the management of the road had changed since then.

Early on the morning of 15th November 1875 two Swedish expresses, train No. 1 southbound from Stockholm to Malmö and its opposite number, train No. 2 northbound from Malmö to Stockholm, collided head on at Lagerlunda, this being the name of the ancestral estate of friherre (Baron) Lagerfelt, which adjoined the single line between Malmslätt and Bankeberg stations. The trains involved were two of the most prestigious on the Swedish system, carrying only first and second-class passengers and equipped with gas lighting, lavatories and luxurious seating. Travelling on train No. 2 this day were Baron Anders Koskull, the Greve and Grevinna Wachmeister, Edouard Anspach, former Belgian ambassador to Sweden, and Edward Joseph, a prominent London art dealer. Despite their status, the expresses were not equipped with continuous brakes, which had not yet been introduced on the Swedish railways.

The two expresses normally crossed one another at Linköping, where train No. 1 was booked to arrive at 12.27 a.m. and train No. 2 at 12.28 a.m. but on the morning of 15th November train No. 1 was on time but its northbound twin, train No. 2, was considerably delayed, partly by heavy snow on the line and partly by a problem with its locomotive, the *Svea.* Trainee Stationmaster Michal was notified by telegram from Nässjö at 10 22 p.m. that it had left there 52 minutes late, and another telegram from Tranås at 11.35 p.m. stated that it was still 51 minutes late there. Somewhat surprisingly perhaps, especially given the late hour, but probably due to the status of trains and the fact that he was only a trainee, Michel did not simply take it upon himself to change the crossing places of the two trains, but instead sent a messenger with the telegram to the home of the traffic superintendent, Adolf Sjöstedt, who received it at about midnight. The traffic superintendent reached the station at 12.20 a.m. at about the time that train No. 2 was approaching Mjölby Junction, its last stop before Linköping.[671]

At 12.30 a.m. a telegram was sent not to Mjölby Junction, but to Bankeberg, where extra worker Jakob Jakobsson was doing his first unsupervised night shift. Jakobsson had been on duty since 5 a.m., was reportedly not especially bright and had little training in the use of the telegraph, so it was not surprising that it was another ten minutes before Stationmaster Björkelund got to the telegraph to receive the message, from Michal that, 'There was talk of a train meeting, but it is probably too late now.' The meaning of this telegram is not really clear to the author as it was by no means too late to arrange a crossing at Bankeberg, though it was too late to advise Driver Andersson of train No. 2 of the crossing at Mjölby Junction, which station the train was just leaving. Presumably that was what Michal meant. At any rate, at Linköping, Traffic Superintendent Sjöstedt was advising Driver Sundqvist of train No. 1 that he would be crossing train No. 2 at Bankeberg, and at 12.56 a.m. a telegram was sent to Bankeberg '*Detain train No. 2 until*

---

[670] Ibid, 7th and 8th July 1875

[671] Webpage – The Lagerlunda collision and the introduction of colour vision testing

train No. 1 has arrived, Sjöstedt' and Stationmaster Uno Björkelund telegraphed back at 12.58, *'Director of traffic, Lp Train No. 2 will be held until train No 1 Arrives. Ubd'* Train No. 1 left Linköping at 1.3 a.m.

Roadside stations such as Bankeberg were only provided with a single semaphore signal by the station house with arms for each direction; there were no separate signals for starting trains (cf the 1873 Menheniot accident). The station semaphore was cleared from red to green for a train to enter the station and the departure of the train was authorised by the stationmaster. There was an additional measure of protection in the form of a banvakt or track guard, one of which was provided every few kilometres and whose duty it was to monitor their stretch of track and signal to trains that the line was clear. The banvakter belonged to a separate division independent of the train or station staff and had their own supervisors. At Bankeberg, 19-year-old banvakt Oskar Johansson was stationed at the level crossing at the north (Linköping) end of the station and should have been there when a train arrived. With expresses due shortly from each direction, Stationmaster Björkelund sent Jakobsson to the east end points to set the road for the loop for train No. 1. Earlier, the banvakt had borrowed Jakobsson's lantern to examine the track. Jakobsson now needed it back and so Johansson left his post to set off for his lodgings at the west end of the station for his own lantern.

According to the regulations, Stationmaster Björkelund should have changed his semaphore from red to a green caution signal and then stopped the express with a hand signal but he stated afterwards that he had kept the semaphore at red longer to be sure of stopping the train. Train No. 2 approached at a considerable speed and the stationmaster waved his green lantern at it. Driver Andersson whistled and Conductor Laurentius Martin interpreted the whistle as a signal for him to apply the brakes, which he did, assisted by Baggagemaster Anders Andersson. Seeing that the train was not stopping very quickly Stationmaster Björkelund changed his lamp to red, then after the driver had whistled again and appeared to be pulling up, he changed it back to green, knowing that the express should not leave his station without his explicit permission. The train finally stopped with its front end over the level crossing and beyond the end of the crossing loop, having run through the eastern loop points.[672]

Conductor Martin then asked Jakobsson if the level crossing had been unattended by the banvakt, which Jakobsson confirmed, but he did not think of telling the conductor that they were to cross train No. 1 there. The banvakt, meanwhile, had lit his lantern and joined the stationmaster on the platform. Then, shortly after it had come to a stand the express started off again. Bahnvakt Johansson began to run after it and the stationmaster shouted for Jakobsson to do the same but the latter did not react until the banvakt arrived at the points. By that time the tail lamps of the train were fast disappearing.

It was not clear why Driver Andersson had restarted but a problem with the small locomotives of the time in Sweden was that they were narrower than the carriages they were pulling, meaning that the drivers had to lean out some way to see a signal from the rear, or else rely upon a tip from a railwayman in the leading vehicle of the train. Baggagemaster Henrik Pamp, who was with Oiler Larsson in the leading luggage van of train No. 2, said that Larsson had suggested after the first set of whistles that they had been given because the station staff had overslept. When the whistles were repeated

---

[672] Ibid.

Larsson had looked out of the window and said that he had seen a man running towards their train waving an 'All clear' signal and had then shouted three times to Driver Andersson "It's the all clear," and Andersson had set off. Conductor Martin and Baggagemaster Andersson stated that they had looked behind for a kilometre after the brakes had been released but had seen only green lights, not the red light waved by Johansson.

The trains met four kilometres from Bankeberg. Stationmaster Björkelund heard the sound of the collision and wrote a telegram, '*T.D. Lp When train No. 2 arrived stop was shown with red lantern, but train kept going and there must have been a terrible collision as train No. 1 had already left Lp. Björkelund.* Unfortunately, the collision had brought the telegraph wires down and the message could only be sent to Mjölby and by a roundabout route via Katrineholm, reaching Linköping at 2.52. Stationmaster J.A. Petersen despatched a rescue train from Mjölby at 2.40 a.m. and another was eventually sent from Linköping by Traffic Superintendent Sjöstedt at 4 a.m. after Driver Sundqvist had arrived back there to confirm the accident. Five passengers lost their lives in the accident, along with Engineer Andersson and his fireman, named Söderberg; Oiler Larrson and Fireman Wallner of train No. 1. Driver Sundqvist survived because he had jumped from his engine. Several second-class passengers were injured and Baron Koskull suffered a broken rib.

Following the accident, Traffic Superintendent Sjöstedt, Conductor Martin, Station Worker Jakobsson and Stationmaster Björkelund all found themselves in court. Banvakt Johansson was not charged with any offence, probably because he was more useful as a witness. Sjöstedt, Martin and Jakobsson were all acquitted and Stationmaster Björkelund received all the blame in what appears to have been a very unfair judgement, which was criticised at the time. He was sentenced to six month's imprisonment, lost his job and was ordered to pay 125,683 kronor in compensation to the railway company, 4,000 kronor to the widow and children of Driver Andersson and 125 kronor yearly to each of the underage orphans of Assistant Traffic Director Franz Anrep (killed in the accident along with his wife). His salary as a stationmaster had been 1,200 kronor a month. He was truly made a scapegoat.[673]

There were suggestions that the accident had been caused in part by the colour-blindness of some of the railwaymen involved, a theory first proposed almost a year after the collision in a letter to the State Railway Board from noted ophthalmologist Frithiof Holmgren, who speculated that either Driver Andersson or Oiler Larsson was colour-blind. Tests of railwaymen for colour-blindness were subsequently introduced on Swedish railways, followed by other European and North American railways and then worldwide. The role played by colour-blindness in the accident has been disputed; possibly Oiler Larsson, Conductor Martin and Baggagemaster Andersson were all colour-blind, though this seems unlikely.

The 1874 disaster between Norwich Thorpe and Brundall was followed rather less than two years later by another one in England, this time on the recently opened extension of the Somerset and Dorset Joint Railway to Bath. Late on the evening of August Bank Holiday Monday, 7th August 1876, a down special passenger train from

[673] Ibid.

Bath to Radstock collided head-on with an up relief passenger train from Bournemouth to Bath just on the Bath side of Foxcote signal box. Twelve passengers and a head guard lost their lives and 28 passengers, and six railwaymen were more or less badly injured.

The Somerset and Dorset lines were at this date almost entirely single track. Train movements were controlled and alterations in the crossings of trains were arranged in the manner of a North American railway by the employment of a crossing agent (equivalent to a despatcher) at Glastonbury, who was advised of train movements by the different stations along the line and arranged crossings accordingly. All the stations which had crossing places were in telegraphic communication with the crossing agent at Glastonbury but Foxcote signal box, which was simply a block post on the single line between Wellow and Radstock stations, was not.

As might be expected, traffic on the Bank Holiday was considerably heavier than on a normal day, and a number of extra trains had been put on. In addition, the superintendent of the line, Abraham Difford, having travelled down on a train in the morning on the 8.11 a.m. train from Glastonbury to Poole and found it undesirably overloaded, determined to put on another train in the reverse direction to relieve the 6.10 p.m. up train from Bournemouth, and having ensured that he could find an engine and rolling stock sent a message to Caleb Percy, the Glastonbury crossing agent, that an up relief train would precede the 6.10 ordinary train from Wimborne (the point where the Somerset and Dorset line began at a junction with the London and South Western line to Bournemouth). It then became Percy's task to arrange crossings between the up-relief train and opposing down trains and the various stations were in turn supposed to report the passing of the up-relief train to him so that he could monitor its progress. Thus, Evercreech Junction station reported that the relief train had left there at 10 p.m., but that was the last that Percy heard of it. It reached Radstock station at about 11.4 p.m. and consisted of an engine and tender, ten carriages, two horse boxes and two brake vans.

In the opposite direction, another down special (No. 7) had been arranged to leave Bath at 9.15 p.m., run to Radstock, thence return from there to Bath, en route picking up 500 or 600 waiting passengers from Midford station, the first one out of Bath. Percy had not arranged a crossing between the up Wimborne relief train and this down Bath special because the latter should have got to Radstock and returned to Bath in front of the relief train. Matters did not go according to plan because the departure of the down special was much delayed, firstly because insufficient rolling stock could be found to make it up and then because it had to wait for another up train (No. 18, ordinary) to reach Bath over the single line. It finally started at 10.45 p.m., an hour and a half late, and reached Wellow at 11.08 p.m. It consisted of a tank engine and 14 carriages, two of them brake composites.

Stationmaster James Sleep had returned to Wellow on the special from Bath. Also, on duty there were Telegraph Clerk Arthur Hilliard and Signalman William Gillard. At Radstock were Stationmaster John Jarrett, Telegraph Clerk William John and Signalman Isaac Horsey, while Signalman Alfred Dando was alone at Foxcote.

Captain Henry Tyler, the Board of Trade inspecting officer, pointed out that according to the spirit, though not the letter of the signalling regulations, the up relief train could not leave Radstock without an order from Crossing Agent Percy at Glastonbury and without being put on the block to Foxcote. Similarly the down special from Bath could not leave Wellow without a crossing order from Percy and being put on the block to Foxcote, while the Foxcote signalman could not accept trains from Radstock and Wellow at the same time; in fact, as there was no loop at Foxcote, there would be no point in his doing so. As with regards to the immediate cause of the collision, it must have been due either to a mistake by Stationmaster Sleep and Telegraph Clerk Hilliard at Wellow or Signalman Dando at Foxcote.

Stationmaster Sleep at Wellow told the inquiry when that the previous up train, a special from Burnham, had arrived at Wellow, Hilliard had announced its arrival to Foxcote, had obtained permission from Foxcote for the Bath special to proceed and the latter train had left Wellow at 11.10 or 11.11. Telegraph Clerk Hilliard alleged that at 11.16 he had seen the needle of the block to Foxcote standing vertical. About a minute later Glastonbury had telegraphed, 'Where is No. 7 down special?' and he had replied 'Down over T.' He had then taken on the up relief train from Foxcote, put his block instrument to 'Train on line,' entered 11.18 in the book, and advised Glastonbury, 'Up special left T,' whereupon the stationmaster had remarked, 'Surely that train cannot have got to Radstock and the up one left in that time?' or, according to Sleep, 'I should think that train had not had time to leave Radstock.'

Signalman Dando at Foxcote stated that the up relief train from Wimborne had been stopped at his box because the previous up Burnham special had still been in the section to Wellow. After it had been reported as having arrived at Wellow there had been a delay of two or three minutes before he could get the attention of Wellow, then Wellow had accepted the up relief train. It had started off again at a time he booked as 11.06 (subsequently shown to be incorrect) and the collision had shortly occurred. He had had no notice of the down Bath special.[674]

Reviewing the conflicting testimonies, Captain Tyler noted that that of Sleep and Hilliard presented many points of difficulty. If the down Bath train had left Wellow as stated at 11.10 or 11.11 it should have got considerable further than Foxcote, less than three miles from Wellow, by the time of the collision, fixed at 11.21. Furthermore, it appeared that the Burnham special had not been admitted to Wellow until 11.14 so the Bath train could not have left until then. There was also the evidence of Guard George Upward and Brakesman William Cooper of No. 16 up goods train, which was waiting at Wellow to follow the Burnham special to Midford. At 11.19 or 11.20 they had seen that the up relief train was on the block from Foxcote and had remarked that if the down Bath special had got to Radstock then she must have been very quick for the Burnham special had not yet reached Midford. The deceased guard of the Bath train, Thomas Willis, had booked his train as leaving Wellow at 11.13. There were also alterations and erasures in the Wellow record book and the times of the alleged telegraph messages to Glastonbury were disputed by Glastonbury.

The evidence of Signalman Dando, the inspecting officer stated was wrong and contradictory as regards the times of the various occurrences. But Dando had never varied in his statements at the time, his written memoranda and his subsequent evidence in stating that he had known nothing of the down Bath special, and it had never been offered to him by Wellow.

Captain Tyler concluded that having regard to all the evidence, actions and circumstances of the case, there could be no reasonable doubt that the blame lay with the stationmaster and telegraph clerk at Wellow, who had let the down Bath special go

[674] Board of Trade report

without putting it on the block to Foxcote, and in then accepting the up relief train before the Bath train could possibly have reached Radstock.

Discussing why the crossing system had failed, Captain Tyler noted that it had been the duty of the crossing agent at Glastonbury to arrange their meets, and of the stationmasters to advise Glastonbury of the train's running and not to alter the crossing places without issuing the drivers with proper crossing passes. These duties had been wholly neglected at Wellow and Radstock. Crossing Agent Percy had been advised that the up special had left Evercreech at 10 and that the down special had left Bath ticket platform at 10.48. He had neither asked for or received any further information until 10.59, when his clerk attempted to contact Wellow. Wellow had finally replied 'Down over T, up on from T' at what must have been 11.21 (T was telegraph code for Foxcote). Percy said that he considered it the duty of Radstock and Wellow to advise him, rather than his clerk having to inquire of them. Herbert John, the telegraph clerk at Radstock, stated that when he saw Signalman Horsey open the level crossing gates for the up special, he had telegraphed Glastonbury. Intending to ask, 'Must we keep up special here, as she has left Midsomer Norton?' he had only got as far as 'Must we' when his message was cut short by 'O.A.', which was 'an exceptional mode of refusing to receive a message.' Percy and his clerk, John Locke, denied stopping the message from Radstock at that time but admitted that they had done so a little earlier, and it transpired that when Glastonbury had attempted to speak to Radstock John had likewise sent 'O.A.' to them. Stationmaster Jarrett and Telegraph Clerk John justified their sending on the up special without an order partly on this difficulty in contacting Glastonbury and partly on Percy having telegraphed earlier that he would arrange crossings, while Percy pleaded that he had lacked the necessary information from Wellow and Radstock to arrange the crossings. Captain Tyler considered that a most serious responsibility rested upon Percy for not arranging a crossing of the two trains which collided, but that at the same time the Radstock stationmaster was not justified in sending on the up special at a time when the down train was so long overdue. Crossing Agent Percy had undertaken to arrange crossings for the up special but had sent none to the stationmasters who therefore could not give them to the drivers. Driver Joshua Bishop of the up special had crossed six trains between Wimborne and Radstock but had received only one crossing order, and so he could hardly be blamed for leaving Radstock without one. There were admittedly too many stations on the Glastonbury – Bath telegraph circuit, preventing free communication.[675]

Summarising the causes of the disaster, Captain Tyler highlighted no less than thirteen defects in the operation and regulations:

1. The existence of Foxcote box constituted a breach of the regulations given by the Company under seal to the Board of Trade that only one engine in steam or two coupled together should occupy the Radstock – Wellow section at any one time.
2. Signalman Dando was comparatively inexperienced, could not use the single needle telegraph instrument in his box and was not supplied with sufficient oil to light his up distant signal.
3. The responsibility of working the single line was too often entrusted at Wellow to Hilliard, a boy of 15, paid 7s 6d a week and working 14 hours a day, who had to book passengers, collect tickets and attend to the accounts. On the day of the

[675] Ibid.

accident he had been in sole charge of the instruments from 6.30 p.m. to 11.15 p.m. in the absence of Stationmaster Sleep. When the collision occurred, he had been on duty 15¾hours.

4. The system of block signalling had been regularly neglected between Radstock and Midsomer Norton, also at Wellow and other stations.
5. The train advice book used by the crossing agent had no columns for Radstock or some other stations.
6. The superintendent had issued no timetable for the up relief train. The crossing agent arranged crossings with down trains at Bailey Gate and Templecombe, but it had crossed train No. 18 down without a crossing order at Shepton Mallet and he had failed to keep himself informed of its running to Radstock.
7. The crossing agent had improperly checked one or more messages from Radstock which would have informed him that the relief train was approaching Radstock.
8. The up relief train was allowed to cross six down trains with only one instruction.
9. The Wellow stationmaster and telegraph clerk had allowed the down special to proceed to Foxcote without informing the latter box.
10. Five or six minutes later they had accepted the up relief train from Foxcote before the down special had had time to reach Radstock.
11. The Radstock stationmaster had let the up relief train leave there without a crossing order, placing too much reliance on the previous assurance that the crossing agent would arrange crossings, and not persisting in attempts to contact Glastonbury.
12. The drivers of the two trains had not asked for crossing orders or telegraph passes before leaving Radstock and Wellow.
13. There were too many telegraphic speaking instruments on the Bath circuit.

Captain Tyler considered Stationmaster Sleep the person most directly responsible for the disaster. Telegraph Clerk Hilliard he considered less responsible because he was Sleep's subordinate, his age, his long hours and the extra duties he had to perform. To a lesser degree he blamed Crossing Agent Percy, Stationmaster Jarrett at Radstock and Superintendent Difford.

In conclusion, Captain Tyler wrote, 'The risk of working must be materially increased when, as in the present instance, the regulations do not apply to the mode or appliances of working; when the rules as regard the mode of telegraphing trains, or the use of crossing orders or telegraph passes, are not faithfully carried out, when there are too many telegraph-instruments on one wire; when signal lamps cannot be lighted for want of oil; when special trains are run without printed notices, or even proper telegraphic advices; when wrangling takes place, and important messages are improperly checked from head-quarters on the telegraph-instruments, and when the duties supposed to be performed by responsible station-masters are allowed, in practice, to devolve upon telegraph-clerks of immature age or experience, employed for long hours, and taken away constantly to duties incompatible with the proper attention to the simple details prescribed with a view to safety. Railway traffic worked under such conditions cannot, whatever the system employed, be expected to be carried on without serious accidents. It is by the avoidance of these defects in the first instance, rather than by seeking for any improvements in systems of working, that safety must be sought in the working of single lines of railway.'[676]

[676] Ibid.

The jury at the inquest on the victims of the disaster found a verdict of manslaughter against Stationmaster Sleep and censured Telegraph Clerk Hilliard, Stationmaster Jarrett and Crossing Agent Percy.[677] However, no report of Sleep's trial has been unearthed.

An accident near Guyastua, Pennsylvania, United States on 23rd July 1877 was an indirect result of the Great Railroad Strike of 1877, itself a reaction to very poor working conditions and a 10% reduction in wages. The strike had turned violent and resulted in an embargo on traffic over the Pennsylvania Railroad. Trains were diverted instead from Blairsville over the Western Pennsylvania Railroad in charge of a Western Pennsylvania engineer. On the evening of Monday 23rd, a westbound mail train was due at Sarpsburg at 5.20 p.m. and an eastbound express a few minutes earlier. The main line was single track here and the mail being a first-class train and the express only a second class train the express had to wait to cross the mail train. Instead the express left Sarpsburg on the single track and was running at thirty or thirty-five miles an hour when its engineer caught sight of the mail train, also running at full speed. The engineers whistled and applied their air brakes and the two trains shortly collided.

Killed in the collision were Pilot Engineer Jacob Feil of the Philadelphia Express, Fireman Harry Hill of the mail train, Express Messenger and Baggagemaster Fred Louther of the mail train, Mail Agent Al Price and passenger John McCarthy of Altoona. Fireman McCabe of the express was at first thought to have been killed but in fact escaped with slight injuries. Engineer William McElvey of the mail train jumped and escaped with a sprained ankle and Baggagemaster John Fox of the express lost a couple of teeth. Conductor J.D. Snyder was unfamiliar with the Western Pennsylvania line and had been relying on Pilot Engineer Feil, whose mistake caused the accident.[678] [679]

Accidents due to train drivers or other railwaymen being intoxicated have never thankfully been very common but they did occur from time to time, as in Wales on 28th September 1877. The accident occurred between Three Cocks Junction and Talgarth stations on the Mid-Wales Railway, and three-quarters of a mile from the former place. A down Mid-Wales Railway goods train from Llanidloes, consisting of an engine and tender, a loaded waggon, 25 empty waggons, a composite carriage and a brake van collided with an up Midland Railway goods train consisting of an engine and tender, 19 loaded waggons, 16 empty waggons and a brake van.

The Mid-Wales Railway was worked by the block telegraph, in addition to which the drivers of all trains were issued with a pass ticket by the signalman at each station before proceeding to the next one. The trains involved in the collision were scheduled to cross one another at Talgarth. The Midland train had only begun running on 17th September and since then had in fact only crossed the Mid-Wales train at Talgarth twice but at Three Cocks nine times.

On the 28th September, the Mid-Wales train was in charge of Guard Nathaniel Evans,

---

677 The Times, London, England, 9th, 12th, 15th, 18th, 21st,23rd,25th,26th and 28th August and 1st and 13th September 1876

678 Altoona Evening Mirror, Altoona, Pennsylvania, United States, 25th July 1877

679 The Democrat, Indiana, Pennsylvania, United States, 26th July 1877

with Driver George Hulse and Fireman Moses Williams. It left Llanidloes at 5.25 p.m. and was half an hour late by the time it arrived at Three Cocks Junction. Nine minutes were spent doing shunting there, during which time Driver Hulse left his engine and went into the station refreshment room, accompanied by Stationmaster William Thomas, and according to the stationmaster had a glass of beer, while Fireman Williams, Guard Evans and Porter Roger Powell got on with the shunting. Before the shunting had been completed, Driver Hulse went to the foot of the signal box steps and asked Signalman Thomas Williams, "Now then, where's that pass?" The signalman did not answer, and the driver then went into the box briefly, but only to chat about the war between Russia and Turkey. After two or three minutes the driver returned to his engine. The next thing Signalman Williams knew was that Hulse's train was setting off towards Talgarth. The signalman went out onto the platform of his box and called out to Stationmaster Thomas, Porter Powell and a wheeltapper named Rees Pritchard, "Where are they going?" Someone answered, "They're off." "Stop them, the other train's coming," the signalman had shouted, and he grabbed a hand lamp showing a red light and waved it violently at the departing train while they all shouted. They could not attract the attention of Guard Evans and the train did not stop. The driver had not given up the pass from Boughrood. He would normally have done so when he was handed the pass for Talgarth, which he had not been given.[680]

The Midland train was in the charge of Guard John Sherratt with Driver William Dowdeswell, and left Brecon at 8 p.m. on time, but was five minutes late at Talgarth. Signalman Thomas Davies there gave Driver Dowdeswell a ticket for the Talgarth-Three Cocks section, receiving the Talyllyn – Talgarth pass at the same time, and the train set off. Driver Dowdeswell said that he had gone a mile and three-quarters and was looking round the weatherboard of his engine for the Three Cocks distant signal when instead he saw the white light of the Mid-Wales engine coming towards him. He reversed his engine and whistled and as there appeared to be no response from the other engine he and his fireman jumped. In the collision, the tender and two waggons of the Mid-Wales train were thrown onto its engine, which was badly damaged. The Midland engine was less badly damaged. Six waggons of each train were also damaged. Driver Hulse was firing the engine at the moment of impact while Fireman Williams was occupied breaking up coal and neither saw the Midland train. Hulse was killed outright and Williams badly scalded. Guard Sherratt of the Midland train was also hurt.

The Board of Trade inspecting officer, Colonel F.H. Rich, had little to say about the accident, which was clearly the fault of Driver Hulse, who had left Three Cocks without giving up the pass from Boughrood, without receiving the pass on to Talgarth and against the Three Cocks signals. It seemed certain, Lt-Col Rich stated, that Driver Hulse had not been in a fit state to take charge of his train when he left Three Cocks, owing to drink, and had met his death in consequence of his own neglect. According to Guard Evans, Driver Hulse had had drinks not only at Three Cocks but previously on the journey at Llanidloes and Builth Wells, and it would seem likely that he was at least alcohol dependent if not an alcoholic. As might be expected, the inspecting officer considered that Stationmaster Thomas had also been much to blame for accompanying the driver to the refreshment room. It was contrary to the company rules for drivers to leave their engines or leave them in charge of their firemen. Although the method of working the

680 Board of Trade report

single line had had nothing to do with the accident Lt-Col Rich strongly advised that the train staff and ticket method be adopted in addition to the block telegraph, the staffs being delivered to the guards and given by them to the drivers; also, that starting signals be provided.[681]

The first serious railway accident in Japan took place in the aftermath of the Satsuma Rebellion, an uprising of disaffected samurai officers against the new imperial government. By 1st October 1877 the rebellion had come to an end and special troop trains were being run to bring the soldiers back from the front, with associated empty coaching stock workings in the opposite direction. The troop trains and empty stock workings were run as extra sections of timetabled trains and the Kōbe - Ōsaka – Kyōto line on which the collision occurred being single track trains had to wait at crossing places for all the sections of an opposing train. The stationmasters were Japanese, as were the engine firemen, but the engine drivers were Europeans. The stationmasters were notified of special trains by the traffic manager, Walter Page Smith, by telegraph, but the drivers sometimes had better knowledge of the traffic workings than the local stationmasters, and some of them tended to rely on their own perceived better knowledge over orders from the stationmasters. It was this which led to the accident.

On the evening of 1st October, a wet and stormy night with poor visibility, a train of empty coaching stock from Kyōto to Kōbe was waiting at Nishinomiya station for a train from Sumiyoshi. A troop train came off of the single line, running as the first section of an ordinary passenger train, and the European driver of the empty stock train, who had perhaps failed to realise than any troop specials were being run that night, mistook it for the ordinary train which was following it. Hearing someone blow a whistle he then set off without confirming that it was the right of way whistle from the stationmaster as per the rules, and against the semaphore signal at danger. He started off so suddenly that the guard of the train was left behind and covered about three miles before colliding with the ordinary train, which fortunately was only carrying a few passengers. The driver lost his life in the collision along with one of the firemen. Reports spoke of three deaths, the third possibly being the other driver, whom another report stated was seriously injured. The other fireman was badly cut, and two other persons were seriously injured. The wreck was so complete that the number of vehicles could only be determined by a count of the number of wheels and axles. The accident stressed the need for better communication between the Japanese stationmasters and the European engine drivers and for a general tightening up of the safety regulations.[682] [683]

It was a rather complex series of events which culminated in a head-on collision of trains between Northampton and Billing Road stations, England, on Thursday 18th October 1877, in an accident with similarities to that near Thetford in January 1854. The line was owned by the London and North Western Railway, but the Midland Railway had running powers over it, and a Midland passenger train was one of those involved in the accident, the other being a ballast train of the owning company.

---

681 Ibid.

682 Report on the Railways of Japan Dated April 10th, 1885 pg. 14

683 Early Japanese Railways 1853 – 1914 by Dan Free, pg. 100

The sequence of events which led to the accident began with the derailment of a goods train which was being shunted at Castle Ashby station, which blocked both running lines. The Castle Ashby stationmaster telegraphed to Northampton for the breakdown gang and the Northampton stationmaster, George Norris, in turn contacted District Inspector James Gibson, told him what had happened, and instructed him to accompany the breakdown train to Castle Ashby and arrange single line working as necessary. In accordance with the regulations of the London and North Western Company, printed notices were kept at the various stations and signal boxes which must be filled in before single line working was introduced. Inspector Gibson duly filled in the required notices and left copies of them at Northampton Station, Northampton Midland Junction and Hardingstone Junction signal boxes, the notices all stating that the up line (to Northampton) being blocked all traffic was to be worked over the down line (to Wellingborough) with himself acting as pilotman. Arriving at Castle Ashby the inspector gave the signalman there the same notice but then discovered that in fact the up line had been cleared, while the down line remained blocked by derailed waggons, and there were two trains waiting to proceed to Northampton. He therefore despatched the first, Midland passenger, train and followed it himself on a London and North Western train. While at Castle Ashby, Gibson did not inform the stationmaster there, named Henry Hafford, of the single line working arrangements, did not tell the signalman, Thomas Gardener, of the alteration to having single line working over the up line, nor say anything to Stationmaster John Briggs at the intermediate station at Billing Road.

Back at Northampton Gibson informed the signalmen at Hardingstone Junction and Midland Junction boxes of the alteration to using the up line for single line working and under his instructions a Midland passenger train was crossed over onto the up line and proceeded towards Castle Ashby and Wellingborough, going down the up line. The inspector then prepared to follow the Midland train with a London and North Western passenger train from Bridge Street station, but Stationmaster Norris asked him to produce the proper notice that he had made the proper arrangements with Castle Ashby to work the train over the up line, signed by the man there. As the inspector could not produce the notice the stationmaster refused to allow the train to proceed on the up line but allowed it to proceed on the down line, which the stationmaster knew was clear as far as the north-east end of Castle Ashby station, cautioning the inspector and the driver to go very carefully. Signalman Richard Smart at Hardingstone box was not happy with the North Western train going on the down line, the inspector having previously instructed him that the up line was to be used for single line working, but eventually permitted it to go on after Gibson had satisfied him that he could do so safely. The train set off again, but had one gone a short way when it was stopped by the guard of the Midland train which had preceded it, and which had collided with the ballast train. The Midland guard naturally expected the North Western train to becoming on the up line, like his own.[684]

Signalman Gardener at Castle Ashby had only been on the railway for three months. He had spent a fortnight at Willesden undergoing training, been examined at Euston and then been sent to Castle Ashby to learn the box under Signalman Taylor, being passed to work the box there by Inspector Gibson after another three weeks. He was able to work the box, which controlled a simple layout with up sidings and two mains crossovers, under ordinary circumstances, but had no knowledge of or experience in single line

[684] Board of Trade report

working. After Gibson had departed for Northampton with the second train as pilotman a ballast train had arrived from Wellingborough and Gardener had lowered his signals for the train to proceed, although the notice he had been given by Gibson prohibited him from letting any train proceed unless the inspector was there to start it. Gardener pleaded that he had not read the notice, did not understand the working and had assumed that he could let the ballast train proceed as the Billing Road signalman had cleared back the passenger train.

Guard William Finney was in charge of the Midland train involved in the collision, which should have departed from Northampton at 7.10 p.m. but did not get away until 8.10. The train consisted of an engine and tender, four coaches and a brake van. The driver of the train was Edward Goodman and its fireman Richard Yeoman. Driver Goodman told the Board of Trade inquiry into the accident that having finally got away an hour late his train was stopped by signals for another 45 minutes at the Midland Junction box. Finally, they were crossed over to the up line and they proceeded after he had checked with Signalman Smart at Hardingstone Junction that all was OK. They had gone two miles, and had passed the gates at Paper Mill Crossing, travelling at 10 or 15 miles an hour, when he saw the ballast train coming towards him. He shut off steam, reversed his engine, shouted to Fireman Yeoman to apply the tender brake and whistled for the guard's brake. He and his fireman had then both jumped, and neither was hurt.[685]

In charge of the ballast train was Permanent Way Inspector Grant, who was riding on the engine with Driver Arthur Islip and the fireman, Brakesman William Herne was riding in a van at the rear of the train. Driver Islip stated that his train had left Wellingborough about two and a half hours late because of the derailment at Castle Ashby. They were not told, either at Wellingborough or at Castle Ashby, that the line was blocked or that there was single line working. The signals at Castle Ashby were at danger but were cleared as they approached, as were the signals at Billing Road. Three-quarters of a mile beyond Billing Road, as his train was travelling at about 15 miles an hour, he saw the tops of some carriages about 30 yards away and called to the others. He remembered Inspector Grant shouting, "Oh dear! Oh dear!" There was a green light on the front of his engine; he did not see a light on the front of the Midland train. There had been no time for any of them to jump. He and his fireman were somewhat hurt, and the inspector was badly scalded.

The collision forced the passenger train back 100 yards when a coupling snapped, and four carriages and a brake van ran back a further 150 yards. The leading carriage of the passenger train was completely smashed, and its passengers killed or injured. It was occupied largely by members of the county police force who were returning from the County Quarter Sessions. Louis Poole, the Superintendent of Police at Wellingborough, was killed instantly, Police Constables Thomas Barratt of Rothwell and Isaac Wilson of Kettering, together with William Letherland, a Wellingborough newspaper proprietor, were fatally injured and about 18 other passengers more or less badly hurt.[686]

Lt-Col Rich ruled that the accident had been caused by the neglect of Inspector Gibson, firstly in appointing Signalman Gardener for duties which he was incapable of performing and secondly for neglecting to make the proper arrangements for single line

---

685 Ibid.

686 The Times, London, England, 20th October 1877

working, while Signalman Gardener was clearly at fault for permitting the ballast train to pass him when the pilotman was not present. It was clear, the inspecting officer continued, that Gardener had not been at all aware of the responsibilities placed upon him while the printed instructions given to him by Inspector Gibson, that single line working would be carried out using the down line, were the opposite of what was actually done. As for Gibson, he was an old and experienced servant of the company, but he had neglected his duty in many respects. He had not furnished notices to all the stations on the road, had neglected to get a duplicate notice countersigned by the man at Castle Ashby, had neglected to inform the Billing Road stationmaster of the proposed single line working and had not suspended block telegraph working. Furthermore, as the derailed waggons were on the crossover at the north-east (Wellingborough) end of Castle Ashby station there had been no need in any case for single line working over the eight miles between Northampton and Castle Ashby, it should instead have been arranged between Castle Ashby and Wellingborough. Finally, the inspecting officer commended Stationmaster Norris at Northampton and Guard Finney of the Midland train and noted that had the latter been fitted with continuous brakes the collision might have been averted.[687]

Two inquests were held because Police Constable Wilson only died of his injuries some days later; both returned verdicts of manslaughter against Inspector Gibson and Signalman Gardener.[688]

Shortly after 11 p.m. on the night of Wednesday, two goods trains collided head on at Emu Plains, New South Wales, killing three railwaymen outright and seriously injuring two or three more. The trains involved were the 9 p.m. down train (No. 16) from Sydney, consisting of an engine and tender, twenty-two loaded waggons, a brake van and a banking engine and an up special consisting of an engine and tender, nine waggons of coal from the Brownfels mines and a brake van. The two engines were almost destroyed in the collision, their boilers being crushed into one another, and the waggons piled up in a heap and caught fire. Driver John Egan of the up train and Fireman Wiggins and Assistant Guard Brady of the down train were killed outright, and Driver Michael Stephenson of the down train and Fireman John Larkins of the up train were seriously injured. Guards Healey, George Purdue and James Baker escaped injury, as did Driver Edward Cane and Fireman Edward Lang on the banker. The collision occurred at about 11.10 or 11.15 p.m.[689 690]

The origin of the accident lay in a change made two months earlier in the manner in which the single lines were worked. Until then the lines had been worked by the electric telegraph using so-called working orders, with the stationmasters arranging the crossing places between trains and issuing the guards of trains with written line clear reports instructing them to what point their trains could be run. This system had been abolished and decisions as to the crossings of trains had been left to the judgement of the guards. On the evening in question, the up special arrived at Blue Mountain station where it

[687] Board of Trade report

[688] The Times, London, England, 22nd, 26th and 30th October and 17th November 1877

[689] Australian Town and Country Journal, Sydney, New South Wales, 2nd February 1878

[690] Sydney Morning Herald, Sydney, New South Wales, 20th February 1878

should have stopped to obtain information as to the line being clear to Penrith. The night officer at Blue Mountain, John King, stated that he had telegraphed Penrith to obtain this information but had been unable to get a reply. Nor did the up special stop, but instead ran through the station at about five miles an hour, and the night officer jumped onto the step of the brake van and asked Guard Purdue "Where are you going to?" King testified that the guard had replied "Wascoe's," but Purdue claimed that he had said "Emu Plains," a siding on the Penrith side of Wascoe's. Rule 146 dictated that special trains were to be off the line fifteen minutes before the time of an ordinary train but to what point a train could run and be clear in time had been left to the discretion of the guards since the abolition of working orders. After the special had passed, and as soon as he could get the circuit, King had informed Penrith that the up special had passed him at 10.05 p.m. and would cross No 16 down at Wascoe's.[691]

Guard Purdue was found guilty at the Central Criminal Court in Sydney on 19th February of the manslaughter of Driver Egan and despite a strong recommendation to mercy from the jury was sentenced to two years' imprisonment.[692] [693] If the sentence was harsh, the verdict of the jury would appear to have been the correct one. It would seem likely that Guard Purdue, knowing that No. 16 goods was often half an hour or so late, had determined to press on to Emu Plains in the hope of getting home earlier (he lodged at Penrith) and in doing so had trespassed upon the time of the opposing train, which unfortunately for him was not running late that night. He almost made it, as his train got to within three-quarters of a mile of Emu Plains before the collision. Such accidents due to 'running on short time' were of course common at this time in the United States.

On the same day that Guard Purdue was sentenced it was announced that an improved system of single line working, using a train staff and tickets combined with the electric telegraph, would be introduced.[694]

The Pittsburg Cincinnati and St Louis (Panhandle) Railroad express (train No. 6) which had left Pittsburg, Pennsylvania at 11.47 p.m. on Tuesday 6th August 1878, met with disaster early the following morning one and a half miles west of Mingo Junction, Ohio, United States, when it collided with a freight train. The express consisted of an engine and tender, two postal cars, one for Cincinnati and one for St Louis; a baggage car, an emigrant car, two ordinary cars and a sleeper car and was in the charge of Conductor Edward Morrill of Columbus, Ohio, with Engineer John Dugan and Fireman Jed Fulton. The train had left Steubenville seven minutes late and was running fast when the collision occurred. Engineer Dugan applied the air brakes and thus somewhat reduced the magnitude of the collision, he attempted to jump at the last moment, as his fireman had already done, but was killed in doing so, though his fireman escaped. Engineer Charles Graham of the freight train, making his first trip, became confused on seeing the headlight of the express and failed to reverse his engine.

The trains met on a curve and in a fog at 1.50 a.m. Both engines were destroyed by the impact and thrown down a thirty-foot embankment. The Cincinnati postal car was

691 Sydney Morning Herald, Sydney, New South Wales, 20th and 21st February 1878
692 Sydney Mail and New South Wales Advertiser, Sydney, New South Wales, 23rd February 1878
693 Sydney Morning Herald, Sydney, New South Wales, 1st March 1878
694 Ibid, 21st February 1878

wrecked and three of the four mail clerks in it were killed, the other sustaining a broken leg. The St Louis car was more fortunate, being thrown down the embankment and ending up stood on its end; its occupants escaped with slight injuries. The baggage and emigrant cars were wrecked, and it was in the latter that the greatest loss of life occurred. The emigrants were of English, Irish, German, Swedish, Danish, Swiss and possibly other nationalities. The telegraph operator at Mingo Junction could not be found and it was two hours before help could be procured from Steubenville as one of the railwaymen, a Mr Morehead, had to run all the way there to bring the news of the disaster. Meanwhile uninjured passengers and a few local farmers did what they could to help the injured, who were eventually taken by a special train to Steubenville. The bodies of the dead were also taken to Steubenville, and because most of them were emigrants, great difficulty was found in trying to identify some of their bodies.[695]

A newspaper reporter wrote, 'Conductor Morrell and Mr J. Harley, a passenger from Edinburg, Pa, especially distinguished themselves by their noble efforts to rescue the sufferers. The maimed, and the dead outright, in many instances, were fastened in the debris, and the cries of those who looked death in the face harrowed up the souls of the listeners. A distressing case was that of Mrs Haley, who was found lying in the underbrush, 30 feet from the wreck, crushed and apparently dying, grasping closely in her arms her infant child, which was quite uninjured. An entire family of emigrants from Scotland, named Bromman, were more or less injured, and the members were scattered and ran about through the darkness hunting each other and uttering the most painful lamentations. Another peculiarly sad face was the Cullen family, of Connecticut, emigrating to Kansas. The grandmother was killed, the father and mother seriously injured, and the crippled daughter probably fatally hurt.'[696]

Those who lost their lives included Engineer Dugan, Head Postal Clerk Frank D. Graham, Assistant Postal Clerk W.R. Johnson, Local Mail Agent August Andrews, Mrs Anna Cullen, of Bridgeport, Connecticut; John Carry, of Tyrone, Ireland; Mary McDonald, of London, England; Bernard Miller, of Bremen, Germany and Fred Gross, of Mannheim, Germany.[697] [698] A few days after the accident the number of dead had reached 15 and several other persons were stated to be dying, so the final death toll could have reached 18 or19.

The second section of freight train No. 13, in charge of Conductor Sterling, was apparently at fault. According to one newspaper report the conductor claimed that his watch had been stopped for twenty minutes without his being aware of the fact, while another stated that he had looked at his watch just before the accident and it showed that his train had four minutes to cover the quarter mile to a siding at Mingo Junction. The newspaper considered that the story of the watch not working was highly improbable and pointed out that Engineer Graham had also had a watch and had not claimed that his had stopped.[699] It was said that the freight train should have taken the siding at Alexandria Roads, four miles short of Mingo Junction, for the passenger train.[700] The conductor was

---

695 Steubenville Weekly Herald, Steubenville, Ohio, United States, 9th August 1878

696 La Plata Home Press, La Plata, Missouri, United States,17th August 1878

697 The Sun, Baltimore, Maryland, United States, 8th August 1878

698 The Monmouth Review, Monmouth, Illinois, United States, 16th August 1878

699 La Plata Home Press, La Plata, Missouri, United States, 17th August 1878

700 The Standard, Greensburg, Indiana, United States, 16th August 1878

arrested after the accident on a charge of criminal carelessness[701] but it has not been possible to trace any report of a trial.

A minor accident on the Launceston – Hobart line of the Tasmanian Main Line Railway on 15th February 1879 was due to the gross disobedience of the regulations for train staff and ticket working by the stationmasters at two consecutive stations and by an engine driver as well. On the evening of Friday 14th February the axle box of engine No. 3, working the up Launceston – Hobart mail train, had broken and the engine was taken off at Antill Ponds, the mail train being worked forward by a spare engine, No. 2, which was kept there. To bring the disabled engine back to Hobart, Locomotive Superintendent Cundy sent Driver Joseph Rogers and Fireman James Allen to Antill Ponds to collect it on Saturday. The disabled engine duly left Antill Ponds for Hobart on the single line, but neither the stationmaster there nor his colleague at the next station at Oatlands gave Driver Rogers either the train staff or a ticket, and the driver seems to have been happy to set off without them. All went well as far as Oatlands but between there and the next station at Jerusalem (later Colebrook) the engine came into collision with another light engine proceeding towards Oatlands.[702]

Back at Hobart, the morning down express to Hobart proved to be more heavily loaded than expected. It was strengthened with extra carriages and it was decided that its engine, No. 8, was not powerful enough to work it alone over the steepest part of the line. Another engine, No. 4, was therefore got ready as quickly as possible and followed the express to Brighton, where it was coupled on to No. 8 and the train ran double-headed to Tea Tree. Here the lighter engine, No. 8, was detached and the express continued with No. 4, the heavier and more powerful locomotive. Mr Cundy, who had been driving engine No. 4, then returned towards Hobart with engine No. 8 but had only got as far as O'Brien's Bridge when he was handed a telegram from the general manager informing him that the fireman of engine No. 4 had accidentally damaged its firebox, and asking Mr Cundy if he could take engine No. 8 down to Oatlands in case the down express needed further assistance, or else leave it there to work the evening up express. Having gone down the line to Tea Tree and back up it again Mr Cundy thus set off back down it once more.

The Locomotive Superintendent reached Jerusalem in due course, and the stationmaster there having given him a ticket for the Jerusalem – Oatlands section, he set off again at 12.15 p.m. Fortunately, he was going very slowly because he thought that there might be platelayers working on the track, and Driver Rogers was also going slowly, perhaps having had misgivings realising that he had not got the staff, when the engines met on a curve near Coal Mine Creek. Mr Cundy suffered a slight head injury from being thrown against the look-out glass and a slight thigh injury from a falling toolbox. His fireman, John Shaw, suffered a cut finger and minor scalds from a broken gauge glass. Driver Rogers and Fireman Allen escaped with a slight shaking. The fronts of the engines were only a little damaged and the line was cleared in fifty minutes. One wonders what the locomotive superintendent said to Driver Rogers when he found that he had neither the staff nor a ticket – probably words that were not printable. Not surprisingly the Antill Ponds and Oatlands stationmasters were both summarily

---

701 The Indianapolis Journal, Indianapolis, Indiana, United States, 12th August 1878

702 The Mercury, Hobart, Tasmania, 17th February 1879

dismissed, though the newspaper report did not say that Driver Rogers had been.[703]

An accident near Clementon, New Jersey, United States on 14th August 1879 was another involving a special excursion train. The train in question left Philadelphia, Pennsylvania at 2.30 p.m. for Atlantic City over the narrow (3ft 6in) gauge Philadelphia and Atlantic City Railroad. It consisted of an engine and tender and nine cars, filled with three hundred men, women and children, and had been arranged by the Simier Yacht Club to take excursionists to see a firework display at Atlantic City. In the opposite direction, a freight train coming from Atlantic City and consisting of just an engine, a tender and one car was approaching Clementon station, expecting to meet the excursion train there, but instead the excursion train continued beyond that point and the trains collided. The two engineers saw one another almost simultaneously and both whistled for brakes, then Engineer Thomas Pennypacker and Fireman John Lafferty on the excursion engine and Engineer Ellwood Johnson and Fireman Enos Reed on the freight train had all jumped. The freight train had slowed up considerably when they met, the excursion train was travelling at about 15 miles an hour. There was an embankment about nine feet high at the accident point which the freight train partly fell down, while the excursion train remained on the track.[704]

Five people died in the accident; Conductor Carl Worthington, Brakemen Erskine Wood and Jacob De Ginther; and two passengers, Jacob Smith, employed by the Union News Agency and Charles Schall, a deaf mute. The conductor and the brakemen had rushed to the platforms to apply the brakes upon hearing the brake whistle. All those who lost their lives had been standing on the platforms of the cars, most of which were smashed, while of the passengers sitting in the cars only two were slightly injured. No one on the freight train was hurt. The accident naturally created great excitement among the passengers and the male passengers reportedly had to stop screaming women passengers from jumping out of the windows, though later the men had to help the women and children out through the same windows. Two physicians who had been riding on the train did all they could and were later assisted by doctors from Philadelphia and Camden. Traffic over the line resumed after two hours with passengers making their way on foot between trains on either side of the wreckage. The road was finally cleared at 11 p.m. Some of the excursionists later proceeded as planned to Atlantic City; others returned to Philadelphia. The bodies of the victims were taken to Camden at 8 p.m. to await the coroner's inquest, while as news of the accident spread, friends and relatives of the excursionists gathered at the railroad ferry to check that their loved ones were safe.

The inquest on the victims of the accident found that it had been due to the criminal negligence of Assistant Superintendent J.S. Verts. At the hearing of the case against the assistant superintendent under Justice Cassidy Superintendent Ellis Clark stated that he had given Mr Verts a written notice of the running of the 2.30 p.m. excursion train, Mr Verts acting as the train dispatcher at Camden. As such Mr Verts should have had telegraphic notice of the passing of the freight train from every station between Atlantic City and Clementon, which should have been recorded in the order books.

Telegraph operator Charles Redmond next gave evidence at the inquiry. He stated that

---

703 Ibid.

704 The Sun, Baltimore, Maryland, United States,15th August 1879

there had been no order book since he had started there on 11th June; instead the orders were written on scraps of paper. The freight train should have left Atlantic City at 10.45 but did not leave until 11.48. He had notified Mr Verts of the freight train being at Hammonton at 1.50, it should have been there at 12.30. Mr Verts told him to tell Hammonton that two or three extras would leave Camden at about 2.25. The operator said he had telegraphed this information to Hammonton but as it had not been an order Hammonton had simply replied 'O.K.' Mr Verts had seen the message and told him that it should have been addressed to the engineer and conductor and read 'Look out for them.' There was no telegraph station between Hammonton and Clementon. Mr Verts had advised him that the excursion train was leaving at 2.30 at 2.20. Half an hour after the collision the operator at Clementon had advised him that he thought that there had been a collision and he had told Mr Verts five minutes later.

Engineer Johnson testified that the first notification of the extras that he had had was at Hammonton at about 1.50 or 1.55 while he had been oiling his engine. He had not signed the telegram as it had only been a telegram, not an order. It would take the extra thirty-six minutes to run from Camden to Clementon. His conductor, John Ewing, suggested that they run to Cedar Brook, 4½ miles from Hammonton, which they could easily do safely, but he had said to the conductor, "John, this message is not an order which gives the passenger train any right over us."

Conductor Ewing stated that he was the boss of the train, but it was his first day as a conductor and as a young man he did not feel he could overrule an old railroad hand. He suggested to the engineer that they lay off at Dick's Switch, three miles short of Clementon, but the engineer had simply repeated that the despatch was not an order and he would not stop.[705]

On the next day of the inquiry, Mr Verts insisted that the message he had given to his operator had been an order and repeated that the operator had told him that the freight train would lay off at Williamstown Junction. If it had done so there would have been no accident. He blamed the accident on the engineer, Johnson, being headstrong.

Operator Charles Redmond denied having told Mr Verts that the freight would stop at Williamstown Junction.

Operator E. Redmond at Hammonton denied telling Camden that the freight would lay off at Williamstown Junction. He said that his son had asked him if Conductor Ewing had understood the message and had replied that he had read it to the crew of the freight train and advised them to go no further than Tansboro.[706]

At the end of a long inquiry, Superintendent Verts, Operator Charles Redmond, Engineer Johnson and Conductor Ewing were all committed for manslaughter.[707] At their trials in early 1880, the superintendent, operator and conductor were acquitted but the jury were unable to agree in the case of Engineer Johnson. It is unclear whether he was tried again or whether, as seems more likely, the case against him was finally dropped.

Coincidentally, another single line collision involving an excursion train and a goods

---

[705] The Philadelphia Inquirer, Philadelphia, Pennsylvania, United States, 19th August 1879

[706] Ibid,21st August1879

[707] Ibid, 26th August 1879

train occurred the day after the Clementon accident, 15th August 1879, this time on the Chemins de fer de l'Ouest in France. The excursion train was carrying passengers from Argentan and intermediate stations to the seaside watering place of Granville, on the Atlantic coast, when it collided with a goods train between Flers and Montsecret. Both trains were travelling at some speed when they met on a curve and in a cutting at about 7 a.m. The engines were driven into one another and some the leading carriages of the passenger train were described as having been 'smashed to atoms' and 'cut to pieces,' while fire reportedly broke out in the wreckage, though it does not seem to have played a significant role in the accident. Drivers Hamel and Bataille, Fireman Vautray of the passenger train, Conductor Quenion of the goods train and five passengers were killed outright or died shortly of their injuries. The passengers who lost their lives were M. LePrince, a Flers butter merchant; M. Dumont, of Brideville; Mesdames Gassion and Bretonelle, of Flers and Madame Durand of Oisellerie. Thirty-two other persons were injured, of whom no less than 14 were not expected to survive at the time of the report.

The accident was due to a mistake made by M Vapel, the sub-stationmaster at Flers. The goods train was running an hour and a half late and, being under the impression that it had passed, he had allowed the passenger train to proceed. Immediately as the train had gone out of sight, he had realised his mistake and fainted, since when it was stated that he had almost lost his mind, was weeping continuously and refusing to eat. He was arrested after the accident. There was a suggestion that thundery weather might have contributed to the accident, perhaps by interfering with the working of the telegraph instruments.[708] [709] [710]

708 The Times, London, England,18th August 1879

709 The Daily News, London, England,18th August 1879

710 The Magnet, London, England, 25th August 1879

CHAPTER 7

RUNAWAY TRAINS AND VEHICLES

A variety of situations can lead to a train or part of a train running away out of control. If there are railway staff in attendance, the train or engine can only run away out of control if there was a failure of the brakes or if the brakes are incapable of holding it on a steep gradient. In the case of unattended vehicles, the action of gravity alone is sufficient to precipitate a runaway. In contrast to the situation on a road, a train or vehicles can sometimes run for many miles without mishap. an accident occurring only if and when the runaway meets with another train, comes to the end of the line or gets to a curve too sharp for the runaway to negotiate at the speed at which it is travelling.

The earliest runaway that appears to have been recorded was a tragedy which occurred at Leckhampton, England, on the branch from Cheltenham to the stone quarries there, which was actually a 3 ft 6in gauge plateway. The branch included a rope-worked inclined plane, and on Wednesday 30th October 1827 a waggon containing upwards of a ton of stone was being lowered down the incline when the rope broke and the waggon ran away and crashed, striking and killing two boys aged 7 and 10 who were playing near the tramway as it did so. The father of one of the boys had been operating the windlass.[711]

Rope or cable worked inclines were common on early railways where there was a need to surmount severe gradients which were too steep for the primitive early locomotives, and accidents such as that at Leckhampton occurred from time to time when the load proved too much for the rope or cable hauling it and it broke. Thus, accidents due to waggons running away down such inclines occurred for example in England on 4th January 1830 (Etherley Incline, Stockton and Darlington Railway), 5th January 1830 (Brusselton Incline, Stockton and Darlington Railway), 24th December 1830 (Middleton Incline, Cromford and High Peak Railway), June 1831 (Between Bolton and Leigh, Bolton and Leigh Railway) and April 1832 (Law Incline, Dundee and Newtyle Railway).

[711] The Times, London, England, 5th November 1827

An early accident of a similar kind occurred in the United States on 25th July 1832 on the Quincy Granite Railway in Massachusetts, built like the Leckhampton branch to serve a quarry. It was not therefore a public railway but on the day of the accident Messrs Andrew E. Belknap and John G Gibson of Boston, Mr W G. Bend of New York and a Mr Backus of Santiago, Cuba, were riding up the incline on an inspection tour. They had almost reached the top of the incline when, as the newspaper report described it 'some of the machinery giving way' the car ran back down the incline at high speed, overturning at the bottom and falling with its occupants fifteen or twenty feet onto a mass of stones and rubbish. Mr Backus was killed outright and Mr Gibson very seriously and from other reports fatally injured. Mr Belknap suffered a broken leg, but Mr Bend escaped with bruises.[712]

A runaway on the Liverpool and Manchester Railway on 23rd February 1833 occurred on the Sutton Incline, England, east of Rainhill, when the brake of a goods train 'became unmanageable' and it ran away until its engine was derailed. Driver M'Cannis was knocked off the engine tender, the wheels of the train ran over his legs and he was fatally injured, dying before he could be brought to the Liverpool infirmary. Another man, presumably either the fireman or the guard, was slightly hurt but was expected to recover in a few days.[713] Another runaway on the same line a few days later ended more happily. A fireman briefly left an engine unattended at Manchester and it moved off and could not be stopped. It ran the whole 30 miles to Liverpool without mishap. As it approached the latter place an employee, fearful of what might happen should it run into the station, threw a piece of timber across the line and derailed it. No one was hurt though the locomotive was considerably damaged.[714]

The upper part of the Stanhope and Tyne Railway in the north of England was opened on 15th May 1834 and like the more famous opening of the Liverpool and Manchester Railway four years earlier the event was marred by an accident. Four waggons carrying forty or fifty people were being hauled up the 1 in 13 and 1 in 21 gradients of the Weatherhill Incline when a sudden jerk of the rope broke a shackle and the waggons began to run backwards down the bank. A railwayman at the foot of the incline was faced with the choice of letting the runaway waggons continue down the even steeper Crawley Incline, with gradients of 1 in 8 and 1 in 12 or diverting them in a siding. He chose the second option and the runaways collided with four loaded mineral waggons stabled in the siding, killing two men and a boy.[715]

A local newspaper found humour in an accident involving a runaway locomotive near Selby, England, on 15th January 1836, likening the event to the adventures of John Gilpin,

[712] Boston Centinel, Boston, Massachusetts, United States, 26th July 1832

[713] The Times, London, England, 6th March 1833, reproduced from the Liverpool Albion

[714] Ibid, 13th March 1833, reproduced from the Manchester Courier

[715] The Railways of Consett and North-West Durham by G Whittle, pgs. 205-6

the subject of an 18th century ballad about a man carried ten miles on a runaway horse.[716] The engine, the *North Star*, was proceeding towards Selby on the Leeds and Selby Railway and was about three miles from its destination when its driver, stooping down to adjust the controls, overbalanced and fell off it. He was not hurt but unfortunately his fireman, who perhaps had not been doing his job very long, had no idea how to stop the engine, and panicking jumped off it as well, being severely stunned in doing so. The now unmanned engine steamed on, various crossing keepers hurriedly closing their gates across the roads in front of it, until it reached the closed doors at the entrance to Selby station, which it smashed though. It was finally stopped when it collided with and severely damaged a number of carriages stabled in the station, though at least this prevented it from smashing through another set of doors and falling into the River Ouse.[717]

On 13th July 1837, a week prior to the official opening of the first section of the London and Birmingham Railway, from Euston Square, England, to Boxmoor, two special trains were arranged to make a trip over the line for the pleasure of selected guests, with a lunch between the outward and return journeys at Boxmoor. The trains left Euston Square at 1 p.m. and 2 p.m. and the first set off back to town at 4 p.m. Various difficulties were met on the return journey, with the boiler running low of water and also a derailment, so that the train did not reach Euston Square again until 5 50. On the approach to the terminus the engine was detached and run onto a siding, the carriages being allowed to run by gravity into the station. Due no doubt to an obvious lack of experience, the line not having yet been opened, and perhaps to an overconfidence in the ability of the brakes to slow up the carriages, the vehicles were not brought to a halt before they had collided violently with the 'barrier wall' at the end of the line, the impact throwing the passengers against one another and causing a number of injuries, two gentlemen and one lady breaking their noses and others losing front teeth. *The Times* commented that, 'We do not know that personal blame can be attributed to any one, but too much caution cannot be taken to avoid such occurrences in future, which might have produced most serious consequences.' The second train returned without incident.[718]

The procedure at termini of engines being detached from passenger trains and run into sidings, the carriages being then allowed to continue to the platform under the control of the guard, was adopted on a number of early railways before it was discarded as being too dangerous. Thus, the Camden and Amboy Railroad in the United States used it, at its South Amboy, New Jersey, terminal, and an accident there duly resulted on 11th September 1839. In this case the line terminated at a wharf on the Raritan River (from where there was a boat connection to Manhattan) so the procedure was more hazardous than usual. The engine was detached from its train 100 rods short of the wharf and run on to a siding, and the eight cars were allowed to run down to the wharf under the control of a single brakeman, who according to a New York newspaper sometimes had to control a dozen cars. On the day of the accident some of the brakes were faulty, or the rails were

[716] Wikipedia – John Gilpin

[717] The Times, London, England, 25th January 1836 –reproduced from the York Chronicle

[718] Ibid, 14th July 1837

extra slippery, or perhaps the brakeman not experienced enough. For whatever reason the cars were not halted before they collided with a baggage van being unloaded into the steamboat *Commerce*, and the baggage van and two cars ended up jammed into the wheelhouse of the steamer. Two passengers were badly hurt, and one later died.[719]

In another accident marring the opening of a new line celebrations at the opening of the Housatonic Railroad were somewhat spoilt by an accident at Bridgeport, Connecticut, United States on 12th February 1840. A train approaching the end of the line failed to stop in time due, according to a contemporary report, to a brake tender neglecting his duty. The report implied that there was no engine at the front of the train so possibly here again it had been detached. The railroad terminated at a dock on the Pequonnock River, but a pile of cordwood prevented the cars falling eight feet into the water (the tidal river being at low water) and a potential disaster. As it was no one was very seriously hurt. As at Euston Square the brake tender's mistake might have been due to inexperience, given that it was a new railroad.[720]

On 26th June 1840 the so far relatively minor runaway accidents were followed by what was reportedly a major disaster in the Kingdom of the Two Sicilies. The railway between Napoli (Naples) and Portici was still under construction, but notice was given that a train would be run over the completed portion of the line, and crowds flocked to buy tickets for the novel journey. King Ferdinando II and the remainder of the Royal Family were anxious to take part, but were persuaded, very wisely as it was to turn out, that the trip was too dangerous. The Prince of Salerno would also have gone but by chance was taken ill an hour before the train started and so had to drop out. The actual number of passengers who did make the journey was estimated at somewhere between 100 and 300. Shortly after the train had set off its driver was reportedly 'seized with apoplexy' – suffered a stroke or a heart attack perhaps but what was his fireman doing? The train being without its driver, and its fireman for some reason not taking over control of the engine, it ran away until it reached the end of the completed part of the line and fell from an unfinished bridge into a river. Twenty lives were said to have been lost on the spot and 'all the others dangerously wounded.'[721]

It was the negligence of an engine driver, rather than a medical condition, which led to an engine running away in England on Saturday 11th July 1840, in this case without anyone being hurt, though much damage was caused. The driver, on the little Sheffield and Rotherham Railway, went away leaving the engine in charge of a boy. Possibly the boy fiddled with its controls, at any rate the engine began to move off towards Sheffield, whereupon the boy, not knowing how to stop it, jumped off. The engine, supposedly travelling at a mile a minute, crashed into the station and buildings at Sheffield causing damage which it was estimated would cost close on £250 to repair. 'The damage would have been greater had not an old man thrown something on the rails and upset the

---

[719] A History of Railroad Accidents, Safety Precautions and Operating Practices by Robert B Shaw, pgs. 217-8}

[720] Freeman and Messenger, Lodi, New York, United States,5th March 1840

[721] The Times, London, England, 21st July 1840

engine.'[722] The engine driver presumably lost his job.

The Western Railroad depot at Springfield, Massachusetts, United States, situated at the foot of a short but steep descent into the Connecticut River valley, was an ideal spot for a runaway accident. The regulations decreed that heavy freight trains were to stop at the top of the incline, be divided, and worked down it in two portions, but on 18th December 1840 a long freight train got out of control before it could be brought to a halt and gathering speed ploughed through the brick wall of the depot engine house at the foot of the bank, killing the master mechanic, a Mr Willard, a carpenter and two brakemen on the train. The fact that the train had included a number of Worcester Railroad waggons, which unlike those of the Western Railroad were not provided with brakes, was suggested as one of the causes of the accident.[723]

Liverpool Lime Street station, England, the terminus from 1836 of the Liverpool and Manchester Railway, was another outside of which in the early years (and up until 1870) engines were detached and the carriages of passenger trains allowed to descend by gravity, in this case down an incline of about a mile at 1 in 93, mostly in tunnel, from Edge Hill. This procedure was apparently carried out successfully many times, but on 25th December 1841 there was an accident to the carriages of the York mail train, which arrived at Liverpool at 5 a.m. that morning. Guard William Davidson was for some unstated reason not at his post to apply the brakes, and the carriages emerged from the tunnel into the station at considerable speed and collided with two empty carriages in the station yard, causing considerable damage. Perhaps not surprisingly, given the early hour on Christmas Day, the incoming train consisted of only two carriages and there were only three passengers, one of whom was slightly injured, as was the mail guard. Guard Davidson was subsequently brought before the courts and sentenced to two months' hard labour in the Kirkdale House of Correction.[724]

An unusual runaway occurred on the Newcastle and North Shields Railway between Howdon and North Shields, England, on 18th April 1843. As a passenger train from Newcastle-upon-Tyne to North Shields was starting away from Howdon station at about 9 a.m. the bolt between the engine and its tender fractured. The poor engine driver, who at that moment had been standing with one foot on the engine and the other on the tender, fell in the gap between them and the whole train of carriages ran over his arm, which was so dreadfully lacerated that a forequarter amputation was necessary. Having left its tender and train behind, the locomotive, the *Wellington*, proceeded along the line at speed for the two miles to North Shields, where the railway terminated in a deep cutting, and its progress was halted when it buried itself almost four feet deep in a block of solid masonry. The damage caused was estimated at over £1,000. The newspaper report on the accident concluded with the statement that. 'Not a single passenger has been killed on

[722] Ibid, 17th July 1840

[723] A History of Railroad Accidents, Safety Precautions and Operating Practices by Robert B Shaw, pg. 218

[724] The Times, London, England, 1st January 1842

this line since it opened in 1838.'[725]

On 22nd April 1846 an accident happened at Montpelier station, France line of the Chemin de fer de Montpellier à Cette. An engine, the brakes of which were known to be faulty the previous day, ran away and collided with two goods waggons, killing a workman. Three months later the two drivers and the fireman held to be responsible were given sentences for manslaughter through negligence ranging from four months' imprisonment to 100 francs fine by the Montpelier Criminal Court.[726]

Incidents of runaway engines seem to have been quite common in the early days and another which caused considerable damage to property, though there was no loss of life or injury, occurred at Brighton, England, on the recently formed London Brighton and South Coast Railway on 8th May 1847. The waggons of an arriving goods train were left in a siding at 6 a.m. that morning and the engine, No. 54, returned to the down main line on the London side of the locomotive works. Its driver and fireman then got off the engine and went into the works for some reason. They alleged that they had left the engine secure, but in a few minutes, it was seen moving down the track towards the station. The driver and fireman ran to re-board it, but as it was moving at 12 to 15 miles an hour it was too quick for them. The station porters, seeing the engine advancing towards some carriages standing near the buffers beat a hasty retreat and the engine collided violently with the carriages, demolishing them and throwing parts of them against the parcel's office and refreshment rooms, which were also damaged. The buffer stops were broken and part of the end of the platform torn up but the engine itself escaped with damage to its front buffers. It was fortunate that the accident occurred at an early hour when the station was devoid of passengers. The company ordered the driver and fireman to be taken into custody for having contravened the regulation which stated that under no circumstances were engines to be left unattended on a running line, and thus having caused the accident.[727]

Brighton seems to have been an unlucky spot at this period for on Sunday 4th July 1847 there was another runaway engine under different circumstances. In this case the driver of the station pilot left the engine in charge of his fireman while he went to get his dinner. While the driver was away the fireman, William Harrison, ran the engine to a point where he could fill its tank with water, and having done so, instead of returning to the spot it had started from, got it into his head that it needed to be on another road. There being no pointsman at the points to change them the fireman jumped down from his engine to do so himself. As he then tried to regain the engine his foot slipped and the engine ran away towards the station without him, like its predecessor two months earlier, and collided with and smashed two luggage vans stood at the end of the London arrival platform. One of the vans was thrown against the parcel's office damaged in the earlier accident, knocking down its front wall and badly damaging it. Two clerks who had been in the office only a short time before were very lucky to have left it just before the

[725] Ibid, 20th April 1843

[726] Wikipedia – Liste des accidents ferroviaires en France au XIX siècle

[727] The Times, London, England, 11th May 1847

accident happened and by doing so probably saved their lives. The damage to the vans and office amounted to almost £400. Fireman Harrison duly found himself before the local magistrates and was fined £10, with the alternative of two months' hard labour, for neglect of duty. As if this was not enough there was a third accident, though not a runaway, at the terminus the very next day. A special train arrived from London carrying a party of the "Foresters" (possibly members of the Foresters Friendly Society). It was a long train of 23 carriages and its driver, misjudging the power necessary to bring the train to the platform, put on too much steam, so that his engine collided with a second-class carriage, forcing it through a hoarding erected in front of the workmen busy repairing the damage caused the day before. The momentum of the train was then exhausted, and no further damage was done.[728]

An accident to a Great Western Railway excursion train carrying 1,500 passengers from Gloucester, Stroud and Cheltenham to see the 1851 Great Exhibition in London, England, ended without serious consequences on 8th July 1851 but was to be the forerunner of several major disasters which occurred in similar circumstances in the years to come. Beyond Stroud the line climbed very steeply for five miles at gradients as steep in part as 1 in 60 and 1 in 75, to the long Sapperton tunnel. The double-headed train of 23 carriages surmounted the bank with some difficulty, but upon arriving at the London (far) end of the tunnel it was discovered that 11 carriages had become detached from the train. It was feared that the runaway carriages, which had only one guard to brake them, would go off the line of a curve and be thrown down a high embankment or alternatively be run into by the following mail train. The carriages must have only been running back at low speed as the pilot engine of the excursion train was able to cross over onto the down line at the next mains crossover, chase after the runaways, overtake them and meet the mail train slowly ascending the bank. The driver of the mail train having been thus warned reversed his engine and set back down the bank, allowing the runaway carriages to collide fairly gently with his train, and the combined train arrived safely back at the foot of the bank. The excursion train passengers were naturally much alarmed and two or three of them were severely bruised in jumping from the carriages. The train finally arrived at London Paddington about three hours late.[729]

An excursion train from Paris to London had a narrow escape at Folkestone Harbour, England, on the South Eastern Railway, on 12th August 1851.The branch from Folkestone Harbour climbed steeply to the main line at Folkestone Junction. The train was about halfway up the incline when the coupling behind the locomotive gave way and the whole train began to run back down the incline. Onlookers feared that the whole train might run off the end of the line and end up in the sea, but the Folkestone Harbour stationmaster gave orders for planks and beams of timber to be thrown across the line. The collisions with these obstructions shook up and injured several of the passengers but served their purpose in slowing the train down, so that the impact with the buffer stops was less severe, and only one vehicle, a luggage van, was forced over the bank into the water, while the penultimate carriage was considerably damaged. Seven or eight passengers were more or less hurt, but

[728] Ibid, 6th July 1847
[729] Ibid, 9th July 1851

no one was killed. The train reached London three hours late.[730]

The Great Western Railway Swindon-Gloucester main line was the scene of another runaway accident on 4th December 1851, this time on the westbound ascent to Sapperton Tunnel. A down goods train left Swindon at about noon with 32 waggons. One waggon was left at Minety station and five at Kemble, but it took seven on there and so continued with 33 waggons. It then proceeded as far as a point just before Tetbury-road station, where its driver found that the train was too heavy to climb the 1 in 150 gradients. He therefore brought seven waggons on to the station before returning for the remainder. He then asked his guard what waggons he could leave there as his engine was getting low on water and would not be able to get the whole train up the bank to Sapperton tunnel. Eleven waggons were duly left at Tetbury-road and the engine took the remainder of the train up the bank to Sapperton tunnel and down the far side to Brimscombe station. Here the guard told the stationmaster, Mr White, that some waggons had been left at Tetbury-road and asked him to send the banking engine to fetch them. The newly appointed stationmaster there was informed that the previous stationmaster, Mr Dunscombe, had frequently sent the banker to pick up waggons, and so after it had assisted the 10.10 a.m. passenger train from Cheltenham up the bank to Sapperton the banker followed it to Tetbury-road to collect the waggons. The banker was duly hooked onto the waggons by the Tetbury-road goods porter and set off, but the rails being very greasy it stalled at Sapperton Siding, just short of the top of the incline. Driver Cripps of the banking engine then decided to divide his train and having no scotches put a piece of two-inch planking under the wheels of the last waggon. He then got on his engine and told his fireman to unhook the last five waggons. The fireman did so, whereupon the detached waggons immediately began to run back down the 1 in 94 gradient. The policeman at Sapperton Siding tried to arrest them with a sleeper but failed, and he then arrempted to turn them across to the up line but could not get to the points in time. The policemen at Tetbury-road and Kemble stations did not have time to turn the waggons across to the up line either and a mile on the Swindon side of Kemble the waggons collided with the down Parliamentary train which had left London Paddington at 7 a.m. The collision badly damaged the engine of the passenger train and smashed five or six of the runaway waggons to pieces, but the driver and fireman of the train somehow escaped injury and its passengers were only cut and bruised. The line was blocked for some time but the passengers, some of whom had probably already endured a long journey from London, were able to continue their journey on the mail train after a delay of three hours.

Captain Douglas Galton investigated the accident for the Board of Trade and blamed it mainly on the carelessness of Driver Cripps, who had relied on a piece of board holding the wagons rather than taking the trouble to walk back the few hundred yards to the Sapperton Siding policeman's box, where scotches could have been obtained. He was also blameable for starting from Tetbury-road without a guard. The inspecting officer also pointed out that had the electric telegraph been installed the Tetbury-road stationmaster would have had time enough to warn Kemble of the approach of the runaways.[731] [732]

[730] Ibid, 15th August 1851

[731] Ibid, 6th December 1851

Due to a mistake by a night cleaner at Shrewsbury, England, an engine with the somewhat ironic name of *Mazeppa* ran away on the morning of 29th May 1852 and collided with a passenger train on the Shropshire Union Railway with fatal consequences. The engine had worked the 9.5 p.m. train from Stafford into Shrewsbury the evening before and the night cleaner, Joseph Thompson then took out its fire, packed its regulator and put in a new fire, and leaving the engine getting up steam had gone home to bed at 5.50 a.m. As far as he could recall he had left the engine's regulator closed. The locomotive was left unattended at this point until the day cleaner, David Tinsley, came to work at 6.10 a.m., ten minutes late. Thompson had gone off ten minutes early and the engine had thus been left unattended for 20 minutes. Cleaner Tinsley discovered that the *Mazeppa* was missing and was wondering what had become of it when a platelayer ran into the engine shed and said that he had seen the engine running away. Another engine was then sent off in pursuit of the runaway and its driver saw the steam left by the *Mazeppa* three miles from Shrewsbury but nothing more of it until he reached Donnington station, 14 miles from Shrewsbury, where it was found that *Mazeppa* had run into the back of the 6 a.m. train from Shrewsbury to Stafford, badly damaging several of its carriages and injuring a number of its passengers, one of whom died some days later. The Donnington stationmaster had seen the *Mazeppa* approaching but had assumed it was just the engine he had requested the day before coming to take away some coal waggons.

The inquest on the dead passenger found a verdict of manslaughter against Joseph Thompson. He had disobeyed the regulations in not checking that the engine regulator was closed, that it was thrown out of gear and that its brake was screwed down. His employers stated that he was a steady man, to which Captain George Wynne, the Board of Trade inspecting officer, sarcastically commented that neglecting to do those three things and leaving his work early was rather a lot of neglect for a steady man. The inspecting officer also criticised the locomotive department for not having detected that its cleaners were going off early and coming to work late and and punishing their conduct, though the locomotive department of the northern division of the London and North Western Railway (which worked the Shropshire Union line) did not share his views.[733]

The Dowlais Railway in South Wales was a two-mile line which ran from Dowlais Ironworks to the Taff Vale Railway, over its last 1,700 yards the line fell to the Taff Vale at 1 in 12, the incline being worked by a stationary engine. In 1852 the line carried a small amount of passenger traffic, with three daily trips using one carriage. The passenger trains were pulled by an engine from the ironworks to a point 400 yards from the start of the incline where a spring coupling allowed the engine to be detached from the carriage and run into a siding. The momentum of the carriage carried it a couple of hundred yards further and it was then drawn by horses to the top of the incline, where a rope was attached, and the carriage was lowered at low speed to the bottom.

On 27th December 1853 the engine was detached as usual, but the carriage was travelling too fast and running to the top of the incline under its own power mounted

---

[732] Board of Trade report

[733] Ibid.

scotches placed there, standing eight inches above the rails, and its wheels coming down upon the rails again it descended the incline without the rope having been attached. At the bottom of the incline the carriage left the rails, overturned, slid across the Taff Vale line and ended up bottom up and shattered to pieces on a river bank. A woman passenger lost her life and six other passengers were injured.

Captain George Wynne investigated the accident and wrote that when he had inspected the line in 1851, he had required that the carriages be fitted with brakes capable to stopping the carriages independently of the rope. It appeared at the inquest into the death of the woman passenger that these carriage brakes had not worked properly, and the guard in charge of the train stated that he had frequently reported problems with the brakes but that nothing effective had been done to remedy the problem. In addition, the engine pulling the carriage had been in charge of the fireman, who had only ever driven on the line once before and had probably got up too high a speed due to inexperience. The carriage had also been built not by a proper carriage builder but by the master carpenter at Dowlais Works, with no knowledge of carriage construction, a fact which explained its inadequate brake power.

The management of the line had suffered from a dispute between the management of the Taff Vale Railway Company and that of the Dowlais Works, with the result that both had repudiated the management of the line and rules and regulations issued by the Taff Vale manager had become a dead letter. Captain Wynne remarked that he was surprised that a serious accident had not happened before and noted that there had in fact previously been three "wild runs" with waggons loaded with mineral due to ropes breaking.

Passenger traffic was suspended on the railway following the accident pending the arrival of a proper carriage built by a railway carriage builder, the height of the scotch was increased to fourteen inches and another scotch of the same height was added twenty yards from the end of the level stretch of line.[734]

An accident near Mallow, Ireland, on the Great Southern and Western Railway on 20th October 1855 resulted in the death of a drover. The approach to Mallow station from Killarney was up a steep incline. A mixed train, consisting of 37 waggons, loaded with merchandise, turf and pigs, two carriages and a brake van, arrived at Mallow station at about 7 p.m. When it came to a stand it was found that a greater portion of the train had been left behind. The front portion was set back, the parts were re-coupled, and the train was about to set off towards the station when it was discovered that there was still part of it missing and it was set back again. When a porter went back to re-couple this third part, he found the body of the drover. A man who saw the accident happen from the roadside described how the train had divided twice and of seeing the second part run back and collide with the third part.

Captain George Wynne for the Board of Trade noted that the waggons which were carrying the pigs were ordinary goods waggons, the ends of which had been permanently roofed over with a tarpaulin cover in the centre. Confined close together in such waggons, the animals were liable to suffocate and to prevent this happening a drover had

---

[734] Ibid.

had to ride with them, but to avoid being bitten by the pigs the drovers tended to ride on top of the waggon doors, as this drover had been doing and from where he had been knocked off and killed in the collision between the parts of the divided train. Had proper cattle waggons been provided there would have been no likelihood of the animals suffocating and hence of the drovers needing to ride in the waggons with them, and the drovers could have ridden instead in third-class carriages, as they did in England.

The couplings had evidently broken from too great a strain having been placed upon them as the train climbed an incline and it was, Captain Wynne stated, highly improper to attach passenger carriages to such a train, composed of 40 vehicles and having inadequate brake power; to render it safe for passengers it should have had three additional brake vans, added to which situated at the end of the train the passengers must have had the full benefit of the smell of the pigs. It was also striking that the engine driver had not noticed his train becoming divided twice. If there had been communication between the driver and the guard notice could have been given to the driver of the first division and the accident would probably have been prevented.[735]

The first of several major disasters in different countries resulting from a collision with goods waggons which had run away occurred near La Louvière in Belgium on 31st May 1858. Two waggons of coke were detached from a goods train at La Louvière station and when the goods train moved off the waggons, which had clearly not been properly secured, began to move backwards down an incline. A railwayman attempted in vain to stop the waggons until rounding a curve an approaching excursion train from Mons appeared, when he had had to jump off to save his own life. Because of the curve the driver of the approaching train had little warning of the obstruction, but he and his fireman also managed to jump off and save themselves. The train consisted of 16 carriages but only the first two were smashed in the collision. The death toll was a heavy one nevertheless, seven being reportedly killed outright and the number of dead subsequently increasing to 21, with 52 others injured, though some of the passengers had time to jump out. A commercial traveller, M.E.D. van Mol, of Ghent, was the only man to escape unhurt from one of the two destroyed carriages. He had been sat next to a Madame Brouet, who had a two-and-a-half-year-old girl on her lap. At the moment of impact, he had seized the child and jumped, saving both himself and the child, but the mother was killed. A man named Marchiennes was picked up covered in blood but was unhurt, the blood was that of a fellow passenger. Two medical men on the train helped treat the injured.[736]

Less than three weeks after the Belgian disaster, runaway waggons caused a bad accident in England. The location was Springwood Junction, just outside of Huddersfield, where the Lancashire and Yorkshire Railway branch to Penistone left the London and North Western Railway Manchester – Leeds main line just outside of Huddersfield station. The Penistone line climbed almost all the way over the 13 miles from Springwood Junction, with a ruling gradient of 1 in 96 past Lockwood and Berry Brow stations to Honley, almost three miles from the junction. There was telegraphic

---

735 Ibid.

736 The Times, London, England 5th June 1858

communication on the main line, but not on the Penistone branch.

At 9.30 am on 17th June 1858 the down Holmfirth goods train left two waggons in the warehouse siding at Honley station, where there were already seven waggons, and Porter Sykes was sure that they had been scotched and their brakes put down. They were still all there and apparently secure at 12.10, when the stationmaster went for his dinner, but at 2.12 p. m. the porter told him that five waggons were missing, and the stationmaster found the brakes off and the scotches removed from the four remaining waggons. Meanwhile the five waggons had run down the bank at increasing speed to Springwood Junction. It is not clear if the runaways were noticed by the station staff at either Berry Brow or Lockwood stations but in any case, with no telegraph, they had no means of sending a warning forward. They were noticed by the London and North Western signalman at Gledholt on the main line and he telegraphed a warning to his colleague at Springwood "Look out for Lanky" but the latter had very little time to react to the warning, and the 12.30 p.m. passenger train from Leeds, which had left Huddersfield at 1.40 p.m., 10 minutes late, was approaching his box. The driver of the Manchester train saw the runaways at the last moment and put on steam to try and get clear, but they struck the last two third class carriages, fatally injuring three passengers and more or less hurting about ten others.

It emerged at the inquest on the victims of the accident that the yard was frequented by a number of carters and other non-railwaymen engaged in loading and unloading trucks and with no one having admitted to having done so it could not be discovered who had removed the chocks, which were not locked, and released the brakes of the waggons. The jury therefore returned a verdict of manslaughter against a person or persons unknown, recommended that the scotches should be provided with locks and that a rock at Springwood which interfered with the signalman's view of the Penistone branch should be removed.

Captain George Ross, the Board of Trade inspector, pointed out that if locked chock blocks were to be employed the congested goods yard needed to be enlarged and suggested the installation of blind sidings to prevent runaway waggons from reaching the main line. He also recommended the introduction of the electric telegraph on the line and the repositioning of Springwood Junction box.[737] [738]

1858 was a bad year for runaway accidents, for there was another major disaster on 23rd August, this time in England on the then still independent Oxford Worcester and Wolverhampton Railway, between Round Oak and Brettell Lane stations. The accident was described by the Board of Trade inspecting officer who investigated it, Captain H.W. Tyler, as 'decidedly the worst railway accident that has ever occurred in this country.' It involved two excursion trains, which had been advertised by the company as running between Wolverhampton and Worcester and back at very low fares. They were supposedly Sunday School excursions, being provided for the pupils and teachers of various schools along the line, but on the day of the accident were in fact carrying slightly more adults than children.

The excursionists left Wolverhampton at 9.21 a.m. in a single train of 24 carriages and

---

[737] Ibid,18th and 19th June 1858

[738] Board of Trade Report

two vans, drawn by one engine. As it was leaving Brettell Lane station the central couplings and side chains between two carriages gave way, the same thing happening as it left Hagley station, while at Droitwich a fractured screw coupling was discovered. In each case Guard Frederick Cook made temporary repairs. On the journey to Worcester Guard Cook had had half a dozen passengers with him in the brake van; they had been smoking and drinking together and Cook admitted to letting one or two passengers operate the brakes. The breakages not having occurred near the front of the train, where the strain on the couplings was greatest, and the couplings not being obviously defective, there was a strong suspicion that the van brakes had been mishandled. The train reached Worcester safely at 12.32.

At Worcester the rolling stock inspector repaired the damaged side chains but left the broken central couplings, both because they were difficult to access and because he considered the replacement goods couplings to be adequate. On the return journey, the gradients were mostly rising, and it was decided that the load would be too much for one train, so the decision was made to work two trains back to Wolverhampton. The first return train, of 28 carriages and two vans, duly left at 6.30 p.m., hauled by one engine, but an additional engine was attached at Stourbridge. The second train followed with 14 carriages and two vans hauled by a single engine.at 6.47. The first train reached Round Oak station at 8.10 p.m. and the second arrived at Brettell Lane at 8.11 and left there at 8.14.[739]

While the first train was standing at Round Oak a foreman platelayer heard a "snap" and saw a portion of the train running back towards Brettell Lane. The Round Oak booking clerk, who was collecting the tickets, also noticed the breakaway and tried to telegraph Brettell Lane, Kidderminster or Stourbridge to send a warning but could get no reply from any of them. Guard Cook alleged that he had heard a "snap" as he had eased off the brakes at Round Oak and he had then reapplied it, skidding the wheels, that the speed had first slackened and then increased, and that he had stayed in the van until he had seen the other train at the Moor Lane Bridge, two thirds of a mile from Round Oak, when he had jumped out urging the other passengers in the van to do the same. The driver of the second train observed the runaway carriages coming down the line towards him and braked and had almost brought his train to a stand when the collision took place, but even so the rear brake van and the carriage next to it were 'spilt into matchwood' and the next carriage badly damaged. The accident left 14 passengers dead and about 50 injured.[740]

Captain Tyler did not believe Cook's testimony and carried out a series of experiments at Round Oak with a train of 17 similar carriages, releasing them down the down the incline from Round Oak station, whereupon each time the carriages were stopped short of the point of collision. Captain Tyler stated that he could only conclude that the brakes, which Cook himself had stated had been in good condition, had not been applied at all, and that Cook had not been in the van when the carriages began to run away. In support of his conclusion, the inspecting officer pointed to the condition of the brake screw of the van, which clearly demonstrated that the brakes had been off at the moment of collision. Confronted with the inspecting officer's accusations Cook made the far-fetched statement at the inquest that he had applied the brakes but had released them

---

[739] Ibid.

[740] The Times, London, England, 25th August 1858

again immediately before the collision. Captain Tyler believed that Cook had been on the platform when the train became divided and failing to get back into the van had either ridden on one of the carriage footboards or had run after the carriages to the scene of the collision.

The immediate cause of the accident, Captain Tyler said, was a rebound of the carriages after the train had pulled up at Round Oak. Examination of the couplings after the accident showed many of them to have defective welds but had Cook's place been taken instead on the two journeys by a trustworthy man the accident would not have been likely to have taken place despite the inferior couplings. Cook had been a goods guard for eight years and had previously worked excursion trains. The inspecting officer remarked that, 'It cannot for a moment be supposed that a man habitually trustworthy should on this occasion only have so far forgotten himself as to invite the passengers into his van, to smoke and drink with them, to employ them at his break handle, and four times to fracture the couplings in one day by his carelessness; and if the company or their officers were not aware of his character previously, then it can only be said that they ought to have been aware of it, and that they ought to have used an amount of circumspection that would have prevented them from appointing a careless man, as he proves clearly to have been, to such important duties.'

The inspecting officer then went on to make some general remarks about the tendency of railway companies to run excursion trains carrying large numbers of more or less unruly passengers with less precautions than those adopted for ordinary passenger trains, added to which the trains involved in the accident had been much overcrowded. As for criticisms made that the time interval between the two trains had been insufficient, he pointed out that had the interval been greater the runaway carriages would have attained a higher speed before the collision. Nor had the electric telegraph been of any use in preventing the accident, though it was very desirable that it should be used to report the progress of trains from station to station.

The want of proper discipline in the administration of the company was indicated, the inspecting officer wrote, by the fact that although the excursion trains were advertised as being solely for schoolchildren and their teachers in practice all and sundry had been allowed to travel on them. Furthermore, the record book at Round Oak used to record the arrival and departure times of trains had fallen out of use, it having been filled up and no replacement ordered.[741]

After a lengthy inquest on 5th October, the coroner's jury returned a verdict of manslaughter against Guard Frederick Cook, at the same time condemning the management of the Oxford Worcester and Wolverhampton Railway for being unconcerned about grossly insubordinate conduct by stationmasters, taking insufficient care in selecting materials for shackles and chains and having insufficient staff at stations. The jury also strongly condemned the practice of allowing passengers to ride in unattended brake carriages where they could interfere with the brakes.[742]

Two passengers, who were amongst the injured, had testified that Cook had been in the van up to the moment of the collision, evidence which Captain Tyler had rejected, but

---

[741] Board of Trade Report

[742] The Times, London, England, 26th and 27th August 2nd, 8th, 14th, 15th, 16th and 17th September and 1st, 2nd and 6th October 1858

at Stafford on 29th November Baron Bramwell ruled that in view of the conflict of testimony Cook could not be rightfully convicted and the case against him was thrown out.[743]

A short atmospheric railway was opened in France in 1847 to run from Pecq on the left bank of the River Seine to Saint-Germain-en-Laye. The gradient of 1 in 28 between the two places as the railway climbed out of the river valley was too steep for the use of contemporary ordinary locomotives. Atmospheric power was instead used to haul carriages up the incline from Pecq to Saint-Germain-en-Laye and they descended by gravity to Pecq, from where a conventional steam locomotive took them forward. On 6th September 1858, by which time the line was part of the Chemin de fer de l'Ouest, there was a 'rupture of part of the machinery' of the carriage at the head of a descending train and it ran away at increasing speed until it collided at Pecq with the engine waiting to work it forward. The conductor of the train, M. Lacoste, and two women passengers, Mmes Michel and Rogier, were killed and 22 other passengers injured, some seriously. The power of steam locomotives having increased, atmospheric operation gave way to the use of ordinary steam traction in 1860.[744 745]

No one was hurt as a result of a runaway engine accident on the West Cornwall Railway on Tuesday 3rd April 1860, but considerable damage was caused all the same. On the arrival of the last down train at Penzance station, England, Monday evening, 2nd April it was taken to the engine shed for its nightly service and at 5 o'clock in the morning its fire was lit ready for it to be taken out by its driver and fireman at 6 a.m. In preparing the engine the cleaner had carelessly left its regulator open and its brakes off. He went away for a short while and when he returned the engine had gone, steam having been raised by the fire that he had lit much more quickly than he had expected. Trundling out of Penzance station with no one on it the locomotive passed over the wooden viaduct outside of the station, smashed through the level crossing gates at Marazion station, ran more slowly up the incline towards St Ives Road, then gained speed on a downhill stretch, continued over numerous unoccupied level crossings, broke through many pairs of level crossing gates and ran on past Hayle, Gwinear Road and Camborne. The second level crossing gate at the next station at Pool ended up on its cowcatcher and by now it had almost run out of puff. At Carn Brea a railwayman saw it coming along at reduced speed with a crossing gate attached to its front and was able to board it and end its wild journey. It had covered about 17 miles. Warning of the runaway had been telegraphed to Truro and the first down train had been held back at Newham station, thus averting a possible collision. Another engine was also sent from Penzance in pursuit of the runaway but was unable to catch it up and it then returned to Penzance to be ready to work the first up train.[746] What action was taken against the cleaner was not recorded.

Although its results were much less serious, an accident at London Kings Cross, England, on 30th May 1860 bore similarities to the Round Oak disaster of two years earlier, inasmuch as it was allegedly due to the mishandling of the brakes by a guard. *The*

---

[743] Ibid, 2nd December 1858

[744] Wikipedia – Atmospheric Railway

[745] The Times, London, England, 9th and 10th September 1858

[746] Ibid, c12th April 1860 – reproduced from the Plymouth Journal

Times carried the headline, 'Frightful accident on the Great Northern Railway' [747] which was something of an exaggeration as no one was killed and only a relatively small number of passengers injured, but the accident was admittedly dramatic. An excursion train had been run to the North of England the previous Saturday and the train involved in the accident was its return working, which had left Manchester at 10.21 a.m. with 23 carriages and two brake vans. Extra carriages had been attached at Penistone and Sheffield and the train had left Retford at 1.20 p.m. with 33 carriages and two vans. Stops were made at Carlton, Grantham, Little Bytham, Peterborough, Biggleswade and Hitchin, either for the engine to take water, for refreshments or due to adverse signals. Tickets were collected at Hitchin and there the train also had to wait for an up express to pass it, and it left there at 5.40 p.m., an hour late. It was subsequently checked by signals at Potters Bar because of the 11 a.m. pick-up goods train and given a caution signal by a platelayer at Southgate due to track work.

The line falls from Potters Bar for eight miles at 1 in 200, beyond which there is a brief climb, followed by a further fall of about three miles, including a final mile and a quarter at 1in 107, followed by a level or slightly rising stretch for 600 yards to the terminus. Driver Thomas Annis told the Board of Trade inquiry into the accident that he had shut off steam two and a half miles out of Kings Cross, and had told his fireman, Church, to apply the tender brake as they passed the signal hut at Holloway Bridge, at the start of the 1 in 107 descent.[748] The fireman said that he had already applied the brake and the head guard that he had applied *his* brake 200 yards further back. Evidence as to exactly where the engine was reversed varied between the driver, his fireman, the head guard, the signalman at the south end of the Canal {Gas Works} tunnel, and Inspectors Viger and Pigott at Kings Cross. Captain Henry Tyler concluded that the driver had in fact reversed his engine as it emerged from the last (Canal) tunnel.

The signalman at the south end of Copenhagen tunnel, over three quarters of a mile out, saw that the train was a heavy one requiring assistance in pulling up, and that the brake of the front van was on, but that of the rear van was "quite off" and he could not see the guard. The signalman at the south end of the Canal tunnel also saw that the brake of the rear van was off and shouted out to the guard, but there was no response. A carriage greaser standing near the station signal likewise noticed that the train was coming in too fast, saw that the brake in the rear van was off, and shouted to the guard to apply it, to no avail. With all its brakes not having been applied, the train entered the arrival platform at 8 or 10 miles an hour, collided with the buffers, forced them into the road, and was only finally stopped by a mound of earth in front of the works for the new Metropolitan Railway. Fifteen passengers were injured but the engine crew escaped, and the engine was little damaged.

The carriage greaser went to the rear brake van after the train had stopped and found the Manchester Sheffield and Lincolnshire Railway guard, James Warrender, lying on some luggage. He spoke to the guard, who did not answer, and who seemed confused and could barely stand up. Two passengers meanwhile called the attention of the signalman on the arrival platform to the brake of the rear van and the latter noticed particularly that the break blocks were not touching the wheels. The guard did not hang about after the accident but staggered off to his nearby lodgings after four or five minutes, the platform

---

[747] Ibid, 31st May 1860

[748] British Rail – Main Line Gradient Profiles – the Board of Trade Report gives the gradient as 1 in 110

signalman noting that he was walking very unsteadily. There could be no doubt, Captain Tyler stated, that the guard had been drunk. He had been perfectly sober at Manchester and Retford, but at Hitchin he had appeared "fresh" though not incapable of doing his job. There he was seen to drink two glasses of ale, bought for him by excursion train passengers.

Regardless of the rear guard's condition the inspecting officer pointed out that with the available brake power the train could only just have been pulled up, and the safety of its 800 passengers was at stake. The brake power of the train, while just sufficient for ordinary circumstances would have been totally inadequate in the case of an emergency requiring a sudden stop such as an obstruction on the line or a broken tyre or derailment, 'and I trust that the Directors will take care that no train shall ever approach their terminal station again with so small a proportion of brake power attached to it as was employed on this occasion, when there were only two working breaks, besides that on the tender, out of 35 vehicles.'[749]

Guard Warrender was sentenced to two months' imprisonment with hard labour at the Clerkenwell Police Court the next day for having been drunk and negligently omitting to perform his duty, by which the lives of several persons travelling on the railway were endangered.[750]

Having described the Kings Cross accident as a 'Frightful accident,' on 5th September 1860 *The Times* wrote of a 'Terrible Railway Collision' and this time the headline was more accurate. The accident happened near Helmshore station, England, on the Lancashire and Yorkshire Railway. This station was situated on the Bury-Accrington line, about three miles short of the summit near Baxenden of a ten-mile climb from Bury with a ruling gradient of 1 in 73.[751]

On 30th August the Lancashire and Yorkshire Railway advertised an excursion train to run from Colne to Manchester. This train was run at the request of the proprietor of the Bellevue Gardens in Manchester and hence was known as the Bellevue train. It was expected that 1,000 passengers would make the journey and an engine, and 24 carriages were provided. This train (henceforth train No. 1) left Colne at 9.45 a.m. on 3rd September 1860 to pick up passengers at all stations between there and Radcliffe Bridge (between Bury and Manchester) but before it had reached Burnley (between Colne and Accrington) the stationmaster there had wired Accrington to say that he had 350 passengers there waiting for it, with many more at Barracks station and hundreds at Nelson, and he asked for Accrington to send another engine and as many carriages as possible. The superintendent thereby arranged for another train consisting of an engine and 18 carriages to be provided (train No. 2). These two trains were able to accommodate all the excursionists as far as Accrington but were then quite full while a wire from Haslingden announced that there were about another 400 intending passengers waiting there. A third train was therefore arranged (train No. 3) starting from Accrington with 23 carriages, to which eight more carriages were added at Ramsbottom. The three trains, each with two guards, had all arrived in Manchester by 1 p.m.

---

[749] Board of Trade Report

[750] The Times, London, England, 1st June 1860

[751] British Rail- Main Line Gradient Profiles

Whereas on the journey to Manchester the trains had mostly had favourable gradients, the reverse was necessarily true on the return journey, and at Manchester it was arranged that two additional engines be provided. The first return excursion left at 10.45 p.m., consisting of 13 carriages pulled by one engine. This was train No. 2, its weight reduced by taking off five carriages because of the less favourable gradients. This train reached Accrington safely. What had been train No. 3 left next at 11.10 p.m., still with 31 carriages but now drawn by two engines. Finally, train No. 1 left at 11.31 p.m., still with 24 carriages and also drawn by two engines, though it was arranged for the pilot engine to be removed to the rear of the train at Ramsbottom and bank it from there. The superintendent accompanied the second train on its return journey and an inspector the third. Train No. 3 was checked by signals at Windsor Bridge and on restarting a cotter between two carriages broke, also the hook of the drawbar of the carriage next to the tender. Temporary repairs were made and the carriage with the broken cotter was removed at Ramsbottom on the orders of the superintendent, the train continuing with 30 carriages. Guard Tomlinson rode on a carriage fitted with a brake fourth from the end of the train and Guard Chippendale in a brake carriage about fifth from the engine. The two guards could see train No 1 held at the signal outside of the station as their train left Ramsbottom.[752]

Train No. 3 stopped at Helmshore station at 12.30 a.m. on 4th September 1860, and in accordance with the normal practice the guards released their brakes and got down to attend to the passengers. Superintendent James Shaw stated that after the train had come to a stand there was a slight rebound, he heard something snap and Guard Chippendale told him that some of the carriages were running away. With that he had hurried to the leading engine, got it detached and had set off back on the up line with a view of either overtaking the carriages and diverting them via a crossover to the up line or failing that of at least warning train No. 1, but before either could be done the accident had happened.

Guard Tomlinson said that when the train had stopped at Helmshore the rear four carriages were standing on the 1 in 78 gradient and the others on the 1 in 100. The train had been just about to start, and he had been standing on the platform, when he heard something snap. He had hurried back to the brake carriage and screwed on the brake as hard as he could, the speed of the runaways had not increased much after that and they had been travelling at five or six miles an hour when they struck train No. 1.

The collision between the 17 runaway carriages and the train took place on a curve and in a deep cutting 650 yards from Helmshore. Because of the curve the driver of train No. 1 had very little warning before the collision. Train No. 3 had left Ramsbottom at about 12.16 and train No. 1 at about 12.28 so the third train must have gained considerably on the second by Helmshore. The rear runaway carriage was broken to pieces, the next thrown against the side of the cutting but little damaged and the next again broken up, while the remaining carriages were less badly damaged. The engine of train No. 1 was thrown off the line and damaged and the tops of two of its carriages were knocked off. Ten passengers were killed outright, and an eleventh died later, about 30 other passengers were seriously injured and about 50 less badly hurt.

Colonel W Yolland for the Board of Trade concluded that the accident had not arisen from the misconduct or neglect of any of the railwaymen involved, but from an

[752] Board of Trade report

objectionable mode of working excursion trains, not confined to the Lancashire and Yorkshire Company. He noted that the accident would not have occurred but for the fracture of the coupling and side chains, and the fractures were the necessary result of having a train of 30 carriages. In addition, the drivers of such trains could not see the rear of them and could only tell if a part of the train had broken away by the reduced drag on his engine. The present accident could also have been much more serious had the driver of the banking engine not heard the whistle for brakes from the front engine and stopped pushing. The inspector had informed him that one of the reasons for having an engine at the rear was in case of a breakaway but he, Colonel Yolland, considered the practice of banking trains most objectionable and dangerous, and that it should be given up.

The accident would not have occurred, Colonel Yolland continued, had not Guard Tomlinson alighted from the brake carriage to attend to the passengers, but he could not be blamed for doing so, as he was merely following normal practice. Nor could the accident have occurred had Helmshore station not been situated on an incline.

The accident would probably not have happened, the inspecting officer considered, had the excursionists been taken home in five trains instead of three and might have in any case been prevented had train No. 3 been provided with more brake power. Had the same precautions been taken with it as with ordinary passenger trains it would have been supplied with two sets of Newall's brakes, two sets of ordinary brakes and four guards to work them. The inspecting officer also indicated that he had tried experiments with trains of 17 carriages on the incline at Helmshore which reinforced the views that he had stated, for example were the brakes left applied the carriages would not run away in the first place.

Referring to the recent similar accident at Round Oak, Colonel Yolland stated, in conclusion, that the means of avoiding further such accidents was firstly to cease running mammoth excursion trains and secondly to increase the brake power provided on such trains, to have a brake van at the rear of each and not to permit the guard to leave his van if the train was stopped on an incline. He ended his report by suggesting that since the railway companies seemed to take little notice of inspecting officer's recommendations on increasing brake power, brake vans at the rear of trains or communication between drivers and guards, the Board of Trade should refuse to sanction the opening of any new railway lines which included stations on inclines.[753]

The coroner's jury found a verdict of "Accidental death," adding that the amount of brake power and number of guards on so large a train were insufficient and that the company should follow Colonel Yolland's suggestions.[754]

At the end of 1863 a branch was being constructed from the Peebles branch of the North British Railway at Leadburn, Scotland, to Dolphinton. The new line climbed from Leadburn for three-quarters of a mile at 1 in 67 and 1 in 150 and the main line fell from Leadburn towards Edinburgh for about a mile and a half at 1 in 64, 1 in 80 and 1 in 173.

At 2.25 p.m. on 29th October 1863 five waggons, one loaded with iron girders and four with sleepers, arrived at Leadburn for the new line and were left on the branch to be taken away by a contractor's engine, the exit to the main line being protected by a chock

[753] Ibid.

[754] The Times, London, England, 6th, 7th, 10th, 11th, 12th, 13th, 15th and 17th September 1860

block and a spragg placed under the wheel of the last waggon, by order of the stationmaster. The contractor's engine came down to collect the waggons at 4 p.m. with a stonemason and three navvies riding on the engine but no brakeman, and all went to the Leadburn inn, where the navvies remained. Driver, fireman and stonemason rejoined the engine at 4.45 and the driver attempted to take the waggons along the new line to Linton. The rails were very slippery and there was a strong wind with the result that the train stalled 1,000 yards from Leadburn. There being no brakeman the driver got down and put some stones under the wheels of the front waggon, then seeing a problem with the coupling between that and the engine unhooked it, whereupon the waggons began to roll back. The driver applied a waggon brake and shouted to his fireman to "Come down steady". Instead, the stonemason, doubtless trying to help, released the engine brake and opened its regulator, whereupon the engine ran against the waggons, knocking the driver aside and injuring the fireman. The waggons ran back down the gradient and the driver ran after them and succeeded in climbing aboard them but jumped off as they approached the stop chock, fearing they would be derailed, fell on his head and was badly hurt. The waggons were not derailed by the stop chock but knocked it aside, ran through the points and continued down the main line towards the next station at Penicuik.

The Leadburn stationmaster saw the waggons running away and tried to telegraph a warning to Penicuik station but could get no reply, then the instrument went dead, from which he rightly concluded that the waggons had collided with the 4.20 passenger train from Edinburgh. The driver of the passenger train did not see the runaways at all before the impact and the fireman only saw them at the last moment; both were badly hurt. Three of the carriages of the passenger train were badly damaged, a boy passenger was killed, and other passengers were injured.

Captain H.W. Tyler attributed the accident to the absence of the brakeman, the action of the stonemason and the lack of better protection from runaways off the branch at Leadburn. The driver said that he had started off without the brakeman because the latter had made some remark that he, the driver, did not catch and had gone off somewhere and had not returned in ten minutes. The brakeman said that he had told the driver that he was going away to have his dinner. The stonemason said that he had only done what he had seen done before and denied that after the accident he had told the driver that the latter must keep quiet about his having interfered with the engine.

As with regards to the protection from runaways at Leadburn, Captain Tyler said that the stop chock had failed because it was set too high so that the waggon wheels had just knocked it out of its fastenings. A shallower stop chock had since been substituted, but it was a very serious oversight that a locked safety switch had not been provided.[755]

The North Eastern Railway's branch from Malton and Pickering to Whitby, in the north of England, in the early 1860s still included a by then rather anachronistic rope-worked incline, three-quarters of a mile long, with gradients of 1 in 14 and 1 in 15, called the Goathland or Beck Hole Incline. Soon to be done away with, the incline was the location of a serious accident on 10th February 1864. On the evening of that day the 6.45 p.m. Parliamentary train from York left Malton at 8.02 p.m. seven minutes late and arrived at the top of the incline at 9.08 p.m. nine minutes late. There the engine was run

[755] Board of Trade Report

around its train and pushed it against the incline brake van in the usual manner. Just as the whole of the train was beginning to descend the incline the rope broke. Incline Brakeman Legg tried to put down sand, but the sand pipes were choked up and having secured the brake handles he felt that he could do no more and jumped off 300 yards from the top of the incline. It had been snowing, the rails were very slippery and ice under the wedges of the incline brake van meant that it could not arrest the runaway vehicles. The whole train descended at increasing speed until it derailed on a curve beyond the foot of the incline. The carriages were badly damaged and two passengers in a first-class carriage were killed and another badly injured. Ten other passengers were less badly hurt.

The rope which had failed was provided in place of a previous one which had also failed. It was much worn and should have been replaced earlier, but the decision as to when the rope was to be changed was left to Engineman Goulden, the man in charge of the stationary engine, without any proper supervision from above, something which Colonel H.W. Tyler was not happy about. Nor was he satisfied with the fact that there was no communication, by electric telegraph or otherwise, between the top and bottom of the incline.[756]

The incline was abolished on the completion of a deviation line on 1st July 1865.

With American newspapers being fully occupied with covering the events which followed the conclusion of the civil war, a bad railroad accident, even at home, received little attention, thus there were only scant reports of one on the Central of New Jersey Railroad on 1st December 1865. The accident happened at White House, New Jersey, United States. A coal train was ascending an incline when a coupling broke and several cars ran back, colliding on a curve with a following express passenger train. The engineer and fireman of the latter just had time to jump off and were unhurt, but seven passengers were killed and twenty others more or less injured.[757]

From the west coast of India to the east of Bombay (now Mumbai) the land rises 2,000 ft in 30 miles to the edge of the Deccan plateau. The two main lines of the Great Indian Peninsula Railway had to surmount this barrier, which they did by means of the Thull Ghat (contemporary spelling in the newspaper, now spelt Thul Ghat or Thal Ghat) and Bhor Ghat inclines, on the lines to Igatpuri and thence to Calcutta and to Poonah (now Pune) and thence to Madras, respectively. The two main lines combined at Kalyan, 33 miles east of Bombay. Both inclines had ruling gradients of 1 in 37 and as might be expected posed considerable operating difficulties. The line over the Bhor Ghat Incline opened in 1863 and that over the Thull Ghat Incline in 1865.[758] The latter incline had been open less than a year when it suffered an accident on about 28th December 1865. A goods train got out of control on the bank, rushed through a station and was precipitated down a 70ft embankment. Six lives were lost, the engine and waggons destroyed and the goods 'utterly damaged.'[759]

---

[756] Ibid.

[757] The Alleganian, Cumberland, Maryland, United States, 6th December 1865

[758] The Railway Magazine 1982, pgs. 10-11

[759] The Times, London, England, 17th January 1866

A severe storm led to a railway accident in France on the Chemins de fer de l'Ouest on 10th February 1866, fortunately without serious consequences. Some empty carriages were stabled in a siding at Sotteville-les-Rouen station, and to guard against their running away were 'firmly fixed with bars and padlocks.' A gale force wind at 4 p.m. proved however more than a match for these safeguards, pushing the carriages up against the obstructions and breaking them, so that the carriages ran out onto the running line. They continued their course along the line as far as the Quatre Mars bridge, where they were struck by a train coming from Paris, the driver of which could not see them for more than 200 metres because of a sharp curve and thus did not have time to pull up, though he was able to reverse his engine and lessen the force of the impact. The eight empty carriages were 'knocked to pieces' and four carriages of the Paris train were slightly damaged but no one was hurt. There was some delay while the debris was cleared from the line, but the passengers arrived in Rouen only 29 minutes late.[760]

Runaway waggons were the cause of a disastrous accident near Abergele, North Wales, on the Chester and Holyhead section of the London and North Western Railway, on 20th August 1868.

On the morning of the day of the accident, a pick-up goods train left Crewe for Holyhead, and at Saltney wharf, about three miles west of Chester, it took on two waggons of paraffin oil from a siding of the Flintshire Oil Works, the waggons containing 50 casks of the inflammable liquid, about 7¾ tons worth. The two waggons were placed at the rear of the train, next to the guard's van. The train continued to Abergele station, shunting at different points en route, did more shunting at Abergele and left there at 12.15 p.m. consisting of engine and tender, 26 empty waggons, 17 loaded waggons and the van. The train arrived at Llandulas {Llandulas station was later renamed Llysfaen, and a separate station named Llandulas opened a mile or so nearer Abergele}, the next station, at 12.24, came to a stand opposite the station signal, and the stationmaster's son put that signal and the distant signal to danger.

At the east end of LLandulas station were Llysfaen lime sidings. The station was situated on a gradient of 1 in 660, falling towards Abergele, and the fall continued at varying gradients over the 3½miles to Abergele, including a mile at 1 in 100. There was insufficient room for the whole of the pick-up goods train in either of the Llysfaen sidings, but it could have been divided and split between the two, and as the down Irish Mail was due at 12.39 it should have been put away straight away, in accordance with the Company's rule, which stated, "goods trains, when likely to be overtaken by a passenger train, must be shunted at stations where there are fixed signals at least ten minutes before such passenger train is due." The stationmaster, Samuel Eaton, and goods train guard elected, however, to ignore this rule and do some shunting first. Six waggons and the guard's van were detached and left on the main line on gradients falling at 1 in 147 and 1 in 100, the senior brakesman (Richard Williams) having put the brake hard on. Some empty waggons were fly-shunted into the lime sidings, and a timber waggon with a check waggon on each side of it was fly-shunted back onto the waggons which had been left on the main line. The driver was then engaged shunting more waggons into the sidings when he saw the timber waggons strike the standing six waggons and brake van on the main

[760] Ibid, 16th February 1866

line, and all of them begin to move off down the incline, pursued by the senior brakesman, but with the waggons soon outdistancing the latter. The brakesman at first signalled for the driver to try and pursue the runaways but then signalled him to stop, as he had heard the sound of a collision.[761]

The down Irish Mail had left Chester four minutes late with no booked stops before Holyhead. It passed Abergele, travelling at 40 miles an hour, at 12.39 p.m., five minutes late. It was somewhat less than two miles further on when its driver, Arthur Thompson, saw the runaway waggons, but he was unsure for a moment which line they were on. When he realised that they were on his line he shut off steam and saw that his fireman had applied the brake. He then prepared to jump and shouted to his fireman, Joseph Holmes, "For God's sake, Joe, jump; we can do no more!" The fireman did not jump though and was one of the fatalities. The collision itself was not a violent one, but engine, tender, guard's van and the leading carriages were instantly enveloped in dense smoke and flames, such that it was impossible to reach any of the carriages at the front of the train or to uncouple the Post Office van from the front (Chester) portion. Both driver Thompson and a platelayer's wife, a Mrs Dicken, testified to having heard the voice of a woman passenger from one of the front carriages but the inspecting officer, Lt-Col F.H. Rich, decided that they were mistaken and that all the passengers at the front of the train must have been at once rendered unconscious either by the shock or by smoke inhalation.[762] He was doubtless anxious to rule out any idea that one or more of the passengers might have suffered the agony of burning to death. Thirty-one passengers, Fireman Holmes and Head Guard William Smith died in the accident and their bodies were all but consumed in the fierce fire, before it was eventually extinguished with buckets of water just before 8 in the evening. Lt-Col Rich wrote that the bodies of the victims could only be described as 'charred pieces of flesh and bone.' The up Irish Mail was shortly due but Rear Guard Hinton, who was stunned in the collision, sent two labourers down the line with fog signals and flags to stop it and a possible further disaster was prevented.[763] None of the surviving passengers or railwaymen was more than slightly hurt.

The Abergele accident created something of a sensation at the time for, no doubt, several reasons. With 33 dead it was of course a major disaster anyway but there was also the horrific nature of the accident plus of course the fact that it did not involve just ordinary working-class people, as for example the previous disasters involving excursion trains, but had numbered nobility amongst its victims, including Lord and Lady Farnham and the Rev Sir Nicholas Chinnery, a baronet.

Lt-Col Rich found that the immediate cause of the accident was the neglect of the senior brakesman in not properly securing the vehicles left on the main line. There was no doubt that he had screwed down the brake in his van, as he had said, as otherwise the waggons would not have remained where they were at all, but it seemed likely that he did not put down the brakes of the other six waggons, and the cast irons cogs of the van's brake wheel were probably broken when the timber waggons were fly-shunted into them. The senior brakesman should also have let down the brakes of the timber waggons so that they rolled back quietly onto the others. As well as the two brakesmen the stationmaster was very culpable for not having directed that the train be put into the sidings

---

761 Board of Trade Report

762 Ibid.

763 The Times, London, England, 22nd August 1868

immediately upon its arrival at Llandulas.

Moving on to other matters, the inspecting officer was critical of the working timetable which had down goods trains departing from Abergele at 12.5 and 12.15 in front of the Irish Mail, timed to pass at 12.34, and the second goods shunting out of the way of the express at Llandulas. Furthermore, Llandulas station and Llysfaen sidings had never been inspected by the Board of Trade and were quite unfit in extent and means of safety for the work usually done in them. An extra siding, long enough to hold the longest train ever sent there, should be installed so that no shunting need be done on the main line, and the block telegraph system should be introduced between Abergele and Llandulas, as well as on other parts of the line.

The London and North Western seemed to have a very lax system of supervision, with no inspectors to instruct and supervise guards to see that they performed their duties properly.

Special regulations appeared to be required for the conveyance of dangerous goods such as paraffin oil and gunpowder.

Locking of the doors of railway carriages should be prevented[764] (as noted earlier thus permitting passengers to quickly escape in the case of a fire, while incurring the risk of passengers accidentally opening the doors while the train was in motion and falling out).

The inquest into the deaths of the victims of the accident returned a verdict of manslaughter against Senior Brakesman Richard Williams and Junior Brakesman Robert Jones and strongly censured Stationmaster Samuel Eaton. They also recommended that more time should be allowed for the shunting of pick-up goods trains out of the way of expresses and that shunting should be done clear of the main lines. Like Lt-Col Rich the Assessor at the inquest was keen for it to be thought that all the passengers had died from instantaneous suffocation rather than from burning.[765] Stationmaster Eaton was later also charged with manslaughter but the case against him was eventually dropped as the two brakesmen refused to testify against him on the grounds that they might incriminate themselves.[766]

The jury took only 10 minutes to acquit the brakesmen at Ruthin Assizes in the Spring of 1869.[767]

Just over three years after the runaway on the Thall Ghat Incline on the Great Indian Peninsula Railway in western India, the company suffered a similar disaster on its nearby Bhor Ghat Incline, on the night of Monday 25th January 1869. The up night mail train from Poonah was descending the incline with just a single 0-8-0 saddle tank engine at its head when it got out of control and running away between Khundulla and the reversing station the engine, four brake vans and five carriages crashed into a wall at the end of the line at the reversing station. Eighteen passengers were killed and about 35 injured. The brake power on the train was stated to have been more than adequate but falling mist had

---

[764] Board of Trade report

[765] The Times, London, England, 22nd, 24th, 25th, 26th,27th, 28th, 29th and 31st August and 1st, 2nd, 3rd, 4th, 5th, 7th, 11th, 12th, 14th, 15th and 16th September 1868

[766] Ibid, 19th, 21st, 25th, 26th, and 28th September and 2nd· 6th, 9th and 14th October 1868

[767] Wikipedia – Abergele rail disaster

prevented the brakes from biting the rails.[768 769 770]

A down express from London left Carlisle at 9.25 p.m. on 10th February 1870 and was just passing Gartsherrie station, Scotland on the Caledonian Railway at 12.20 a.m. on 11th February when its fireman saw the red lights of an engine coming towards them, and his driver saw the white lights of another engine. The fireman applied the brakes, his driver reversed the engine and then a collision occurred in which both were seriously injured, but more especially the driver.

The red light seen by the fireman turned out to have been that on an engine on the other, up, line, which was chasing an engine, No.182, which was running away up the down line. The engine was a pilot engine which had been engaged in shunting the goods yard at Buchanan Street. Its driver had gone for his supper at about 11 p.m. and his fireman, David Wilson (according to the Board of Trade report, *The Times* calls him Robert Hay) recommenced shunting operations before his driver had returned. Just before midnight, when the foreman shunter signalled the fireman to stop after a shunting move the engine failed to stop, and instead ran away through Buchanan Street tunnel. Realizing that something must be wrong the foreman shunter set off in pursuit on another pilot engine, till he was stopped after about 1¾ miles by a goods train in front at St Rollox. Driver Alexander Anderson on the engine of this train had also realised something was wrong and set off in pursuit with his engine in turn. He overtook the runaway at Garnkirk but then had to reduce speed for signals. He overtook the runaway a second time near Gartsherrie and stopped so that his fireman could switch the runaway onto the up line via a mains crossover, but the fireman slipped on the snow and ice, the chance was lost, and the collision was unavoidable.

The runaway engine was completely broken up in the collision, the engine of the passenger train was damaged, the leading third-class carriage of the train was completely destroyed, and the front two compartments of the adjoining composite carriage broken up. One passenger was killed outright, a second died of her injuries and eight others were more or less hurt. The body of the fireman was found back along the line and it was clear that while looking out for a signal while shunting he must have slipped on the footplate and his foot having caught in some part of the engine he had been dragged along the line for some distance.

Lt-Col Rich could only comment that the driver of the pilot engine should not have been away for as long as he had. The Caledonian Railway Company awarded Driver Anderson £20 and his fireman £10 for their efforts to stop the runaway engine.[771 772]

A major disaster occurred in the little country of Luxembourg on 8th May 1870. A passenger train left Luxembourg City at 8.30 p.m. bound for Mersch, Ettlebruuck and beyond and stopped at Dommeldange station on the Wilhelm Luxemburg Railway.

---

768 The Railway Magazine 1982 pgs. 9-10

769 The Magnet, London, England, 1st February 1869

770 The Times, London, England,1st February 1869

771 Board of Trade Report

772 The Times, London, England, 12th February 1870

Following it was a goods train from Esch-sur-Alzette consisting of 40 ten-ton waggons loaded with minerals. The latter train had only just reached the start of a steep falling incline when it apparently suffered a brake failure and ran away at high speed towards Dommeldange. The station staff at Dommeldange, seeing the runaway train and realising the danger, opened the doors of the carriages of the passenger train and shouted to the occupants "Sauve qui peut" (Save yourselves who can). Some of the passengers succeeded in jumping out, while most of those who did not were killed as the engine of the goods train smashed the carriages to pieces. Some of the people standing on the platform were also hurt by flying debris. The driver and fireman of the goods train jumped from their engine and escaped. Initially it was reported that 20 or even 27 people had died but later, probably more reliable reports put the number of dead at 12 or possibly 13. There were also many injured, including M. Gesvais, Mersch burgomaster and brother of the Luxembourg Minister of State, and his wife. Workmen were engaged all night in clearing the line. Most of the victims were poor miners going home with their week's pay.

The Dommeldange stationmaster, the inspector of the line and the driver and brakeman of the goods train were arrested after the accident.[773] [774]

An accident occurred in the United States at the western end of the Kansas Pacific Railroad, at the time still under construction, on 9th August 1870, somewhere near Salina, Kansas. The circumstances of the accident are not entirely clear from the sketchy newspaper reports but it would appear that a heavily loaded 'iron train' – possibly a train loaded with iron rails – got out of control 'in consequence of the iron clogging the brakes' and ran into the back of what was described as a 'boarding train,' probably a train which housed the labourers working on the construction of the line. The two rear cars of this train were demolished, and the end of a sleeping car was crushed. Six people were killed and six or eight injured, the casualties being Mexicans who had been sleeping at the time.[775] [776]

A disaster reminiscent of that at Abergele two years earlier happened on the Manchester Sheffield and Lincolnshire Railway at Stairfoot, England, on 12th December 1870, although in this case the runaway waggons ran into the rear of the train and there was no fire.

This accident had its origins in shunting operations at Barnsley Low Level station, during the course of which Guard John Hathershaw left ten waggons standing on the up line, eight of them empty, with a wooden spragg pushed in the wheel of one of them. Two gas tanks were then fly shunted onto the standing waggons. The guard said that the fly-shunted waggons were moving so slowly that they had not reached the standing ones before he did. He took hold of the chain to couple them up but his hands being very cold he failed to do so. The gas tanks waggons collided with the standing ones, the spragg was knocked out or broken and the ten waggons started off down a 1 in 119 incline, followed

773 Idem, 10th, 11th and 13th May 1870

774 Freeman's Journal, Sydney, New South Wales, 6th August 1870

775 St Joseph Herald, St Joseph, Michigan, United States, 20th August 1870

776 Grand Traverse Herald, Grand Traverse, Michigan, 18th August 1870

by the tank waggons. The line fell for over seven miles from Barnsley towards Mexborough, with the initial short stretch of 1 in 119 being followed by gradients of 1 in 72 and 1 in 97 to Stairfoot station. A pointsman at a crossover road at the south end of the station managed to pin down the brake on the leading gas tank waggon but then slipped off. A second pointsman further on also got onto the foremost gas tank waggon but could not check the runaways and shortly struck his knees on the central iron girder of an underbridge and was knocked off. The runaway waggons passed signal boxes at Pinder Oaks Colliery Junction, Quarry Junction and Old Oaks Colliery Junction but of these only the man at Quarry Junction had control of any facing points by which to divert the runaways, and he could only have sent them onto the Midland Railway line which also had falling gradients. The distances between Barnsley station, the three signal boxes and Stairfoot station were all short so there was little time for warnings to be forwarded, Stairfoot station being only 250 yards from Old Oaks Colliery Junction and a mile and a half from Barnsley.[777]

By the time that the runaway waggons had reached Stairfoot station they were travelling at 30 or 40 miles an hour on the steep descent. Standing at the platform there was a local train from Barnsley which was just setting down passengers. They and the station staff had only a moments warning of the runaways as the latter rumbled over an iron underbridge 200 yards north of the station and the engine driver just had time to look to see what was coming when he was knocked down by the shock of the collision, while the guard and the stationmaster were thrown over a railing onto a line at the back of the platform. The impact broke the third-class brake carriage at the end of the train to pieces and half of the next carriage was broken up, as were nine of the runaway waggons, which were heaped on top of one another at the end of the passenger train. Twelve passengers were killed on the spot or died soon afterwards and three more died later. Another 59 were more or less injured. Some of the dead passengers were believed to have been standing on the platform.

Lt-Col F.H. Rich, investigating the accident for the Board of Trade, found that it was due to the gross neglect of Guard Hathershaw in relying on one spragg to hold ten waggons on the incline. The brake of the front gas tank waggon must also have been defective. Guard Hathershaw stated that he had done all that he had been taught to do since becoming a guard four years ago, to which the inspecting officer commented that, 'If the stationmasters, and persons who should be looking after the railway, allow the servants to do their duties in such a very dangerous and inefficient way, from day to day, they have much to answer for, and if the guard's statement is true the supervision of the brakesmen must be insufficient.'

The inspecting officer further commented that the backing down of the train from the Barnsley High Level station and the stabling of waggons on the main line on a falling gradient while the train was marshalled were 'most objectionable practices.' Proper protection with blind sidings should be provided and the east, west, north and south sidings at the Low-Level station should be rearranged and interlocked with the signals. He also found fault with the fact that the engine of the passenger train was running tender first and that its guard was travelling in the brake compartment next to the engine instead

---

[777] Board of Trade Report

of at the rear of the train.[778]

Because three passengers died later two coroner's inquests were held. Probably not surprisingly, both juries found Guard Hathershaw guilty of manslaughter. *The Times* noted that the accident had happened on the fourth anniversary of the great Oaks Colliery disaster, in which 361 miners had died.[779]

On the evening of 15th March 1871 the Cornwall Railway was lucky to escape what could easily have been a major disaster near Liskeard, England. Liskeard station stood in a deep cutting and was a crossing place on the single line. The goods yard was situated on the up side at the west end of the station but there was also a mileage siding on the up side at the east end. West of Liskeard the single line fell very sharply at 1 in 60 for half a mile to the tall Moorswater Viaduct before climbing again to the next station at Doublebois. There was no direct connection from the mileage siding to the up main line so in order to transfer a waggon from it to the main goods yard the vehicle had to be run via a through road to the down loop, thence be taken along that line out onto the single line at the Doublebois end of the station, before being brought back onto the up loop and thence into the goods yard.

On the late evening of the day of the accident it was raining hard, blowing a gale and pitch dark. At 9.28 p.m. the stationmaster instructed the policeman to move a waggon loaded with wheat from the mileage siding to the goods yard. The policeman in turn instructed the driver of an engine to pull the waggon out of the mileage siding and take it along the down loop. The shunt movement was stopped short of the end of the loop, the policeman put the waggon brakes down and the engine was detached, ran out onto the single line and was taken into the up loop. The policeman then released the waggon brakes to let it run out onto the single line by gravity, prior to the engine coming back to collect the waggon to draw it forward into the up loop. Between the operations of running the waggon out from the down loop onto the single line and of the engine backing from the up loop to collect it, the points at the end of the loop had of course to be reversed, so the policeman, having already partially reapplied the waggon brakes, enlisted the services of a porter to take over the operation of the waggon brakes while he (the policeman) managed the points. Having operated the points, the policeman seized a sprag to assist in stopping the truck, but at this point he accidentally knocked his handlamp against a wall and put it out, and in the resulting pitch darkness could not properly see what he was doing. Thus, after a couple of futile attempts to use the sprag to stop the truck the sprag flew out and knocking him to the ground rendered him *hors de combat.* The porter, meanwhile, thought that he had got the waggon brake pinned hard down, but despite this the waggon continued to gather speed on the 1 in 60 falling gradient until he was unable to keep up with it. The Board of Trade inspecting officer, Colonel W. Yolland, noting that the porter had only been at Liskeard two months, and had never performed such a shunting movement before, doubted if he had actually got the brake on as hard as he thought, as this would have been a difficult task for an inexperienced person on a dark night.

Notwithstanding the mishaps experienced by the policeman and the porter, no further accident should have occurred. The runaway truck should have just descended to the foot

---

778 Ibid.

779 The Times, London, England, 14th, 15th, 16th' 17th, 19th, and 23rd December 1870

of the incline at Moorswater Viaduct, run a short distance up the rising gradient beyond and then run back down again and stopped on the viaduct. The line was worked on the block system with the electric telegraph, and obviously no train should have been permitted to leave Doublebois until the runaway waggon had been retrieved and brought back to the station. Unfortunately, the Liskeard stationmaster, whose booking office was placed at the top of the cutting, from where there was a restricted view of the line, came out of his office and made just a cursory look down the line. Seeing nothing of the waggon he then proceeded to give "Line Clear" to Doublebois, at about 9.42, for the 6.30 p.m. up passenger train from Penzance. This train consisted of an engine (*The Sol*), a luggage van and three carriages, the rearmost of which was a brake second in which the guard was riding. In the darkness its driver had no warning of the obstruction until his engine collided with the waggon actually on Moorswater Viaduct. The collision derailed the leading wheels of the truck but not the trailing wheels, and the train then pushed the waggon in front of it until waggon and train were both clear of the viaduct, which at one point was 150 ft high. The engine, van and leading third-class carriage were also derailed but fortunately did not go down the steep slope on the approach to the viaduct either. Two passengers and the driver were injured, none of them seriously. The engine was damaged, the body of the truck was destroyed and there was considerable damage to the track.

As a result of the accident the stationmaster was dismissed for his negligence and Mr Margary, the company's engineer, agreed to install a direct connection from the mileage siding to the up loop line.[780]

An accident in Wales on the Hereford Hay and Brecon Railway on 24th October 1872 would have been less serious than it was had not an unauthorised passenger been riding with the guard in his van.

At 10.30 p.m. on 23rd October a Midland Railway train left Brecon, consisting only of two engines and two brake vans. It ran over the Brecon and Merthyr Railway to Tallylyn Junction, where it picked up a number of waggons, left there at 11.30 p.m. and proceeded via the Mid Wales line to Three Cocks Junction, arriving there at midnight. The train (rear) engine left Three Cocks Junction, passing onto the Hereford Hay and Brecon line, at 12.10 a.m. on 24th October with 21 loaded waggons, 14 empty waggons and a van as a special train, and was followed by the pilot (leading) engine with 18 loaded waggons, 15 empty waggons and a van at a time variously stated to have been 12.15 or 12.20. The first train left Hay at 12.25 a.m. and Eardisley at 12.50 a.m. The gradients of this line were severe, and between Kinnersley, the next station, and Moorhampton the train began first to slip and then stalled altogether near the top of Norton Cannon (or Moorhampton) Bank, which consisted of about a mile at 1 in 60. The driver then waited a minute, reversed or let his train drop back a few yards and had another go, whereby he managed to surmount the incline, but at the same time he realised that he had lost part of his train. He then proceeded to Moorhampton station and put the waggons which were still attached to his engine into the sidings.

Goods Guard John Gabb testified that his van had not remained stationary for more than a minute or two before it began to run back. He had shown his driver a red light,

780 Board of Trade Report

thinking that perhaps he was just setting back to get a start, screwed the brake down harder and then told his passenger, a young woman, named Charlotte Pincher, to jump out, the van and waggons then moving at eight or nine miles an hour. He himself had jumped out when the train was moving at about 15 miles an hour and was not hurt, but the train was going too fast then to pin any waggon brakes down. The girl had been lying on his great coat and as she had followed him to the door of the van, he had assumed that she would jump out after him. He then followed the runaway waggons and shortly heard the noise of them colliding with the second train. Arriving at the spot he had found the body of the girl, who had doubtless been too scared to jump. The runaways had covered about three-quarters of a mile.

The driver of the second, ordinary train said that he had seen the special train leaving Hay station as his had arrived there. He had not seen it again till he saw its lights as he neared the bottom of Norton Cannon Bank. A platelayer was stationed there because of a cutting slip and he (the driver) had asked the platelayer if the first train had stalled on the bank but the man did not appear to understand what he was asking. When he looked again the lights were closer. He had reversed his engine, called to his fireman and both had jumped off just before the collision, by which time their train had almost stopped. The collision was a violent one, the brake van and four of the runaway waggons were smashed up, the engine of the ordinary train was overturned and fell down an embankment and two waggons of that train were broken up. The young woman must have been killed instantaneously; it took hours to release her body.

Colonel W Yolland inquired into the accident for the Board of Trade. He pointed out that the line was worked by staff and ticket, which eliminated the possibility of head-on collisions on the single line, but not on the block system. According to the guards of the two trains the interval between them at Three Cocks and at Hay had been only five minutes, though the Mid Wales inspector at Three Cocks stated that it had been ten minutes. Again, the guards put the interval at Eardisley at ten minutes, while the Eardisley night porter put it at 18 or 20 minutes, though he kept no record of times. Colonel Yolland strongly recommended that the directors of the company introduce the block system

The death of the young woman was, the inspecting officer remarked, due to the gross misconduct of the guard in permitting her to travel in his van. Why she was making the late-night journey was not stated – had she missed the last train perhaps or was she the fiancée or mistress of the guard? It was also probable, Colonel Yolland stated, that if Guard Gabb had got out of his van when the waggons began to run back and pinned down some brakes, he could have prevented the accident. Finally, it was unclear if the weight of the train had exceeded the regulation load over that part of the line as it could not be determined if the loaded waggons had carried goods or minerals; if the latter the regulation load must have been exceeded. On lines with steep inclines a larger margin of safety should be provided as trains often became divided on restarting.[781]

At the inquest at the Moorhampton Hotel on the death of Charlotte Pincher, the coroner's jury returned a verdict of "Accidental death" with a recommendation that the company introduce the block system.[782]

---

[781] Ibid.

[782] The Times, London, England, 7th November 1872

Nine soldiers died and 22 were injured when a freight train became divided and some runaway waggons collided with a troop train at Montabliz on the Palencia-Santander line in Spain on 11th March 1874.[783] [784]

Another runaway train accident occurred in Spain on 11th September 1874 when part of a mixed train ran away from Vallada station and collided with the rear of a freight train which had left the station shortly before, killing four and injuring 75, in an accident between Vallada and Montesa on the La Encina – Valencia line.[785]

A derailment on an incline near Campania, Tasmania, on 13th April 1875 was apparently the third such accident there. A train consisting of an engine and six waggons loaded with sleepers left Bridgewater for Campania early on the morning of that day in charge of Guard William Tolson, with Driver James Woollens, Fireman George Andrews and Permanent Way Inspector Daniel Booker. The journey went normally until the train reached the top of a steep incline near Campania. Here Driver Woollens stopped, and the brakes of the waggons were all pinned down before the train began the descent. The train was then set in motion and with slippery wet rails the driver lost control of it despite the precautions taken and it came down the bank at increasing speed until it left the rails on a curve. The fourth waggon, in which Inspector Booker was riding, was thrown over an embankment and crushed him to death. The driver and fireman were unhurt and the guard only slightly so.[786]

The inquest into the death of the inspector was prolonged but ultimately returned a verdict of "Accidental death."[787]

The line from Teil to Alès, in France, had only been open a month on 7th June 1876 when a heavy freight train of 40 waggons loaded with rails and coal got out of control on the descent from Aubignas to Teil, despite the desperate efforts of its driver and fifteen(!) brakesmen to stop it. The train finally derailed on a curve on the outskirts of Teil. The engine and waggons were overturned, and five people were crushed to death and four more seriously injured.[788]

The passengers on a Central of New Jersey Railroad train from Philadelphia had a narrow escape on Saturday 11th November 1876 as it arrived at Jersey City, New Jersey, United States at 4.10 p.m. The train of nine cars arrived on time and was only travelling at 10 miles an hour as it entered the depot but due to a supposed failure of the brakes it failed to pull up, collided with the buffers at the end of the line, smashed through a

[783] Ibid, c16th March 1874

[784] Wikipedia – Accidentes ferroviarios en España

[785] Ibid.

[786] Cornwall Advertiser, Launceston, Tasmania, 16th April 1875

[787] Launceston Examiner, Launceston, Tasmania, 1st May 1875

[788] Wikipedia – Liste des accidents ferroviaires en France au XIX siècle

freight office beyond, and engine, tender, baggage and smoking cars went into the dock on the Hudson River, while a third car crashed into the smoking car and ended up on the edge of the dock. Engineer William Peebles and his fireman jumped from their engine. Passengers and railwaymen succeeded in rescuing all of the 18 passengers from the submerged smoking car and they escaped with bruises and cuts from broken glass, though 'the greatest consternation prevailed' among the large number of passengers for the Centennial Exhibition (the first American World's Fair, held in New York in 1876 to celebrate the centenary of the signing of the Declaration of Independence in 1776)[789] Engineer Peebles was seriously injured.[790] [791]

In 1868 Swiss engineer and railway pioneer, Kaspar Wetli, had developed his roller wheel system for use on mountain railways. On 30th November 1876 the system was given a trial on the Swiss Wädenswil-Einsiedeln-Bahn with engine No. 253 and a waggon loaded with rails. On the outward uphill journey from Wädenswil to Schindellegi things looked promising, but on the descent the train could not be braked either with the engine or waggon brakes or by reversing the engine, and the driver, two firemen and ten passengers riding on the waggon realised than an accident was inevitable. Some of the passengers jumped off, others were thrown off, and one of the firemen tried to screw down the waggon brake harder. A kilometre and a half from Wädenswil the waggon left the rails, the coupling between it and the engine snapped and it finally came to a halt, saving the remaining passengers, including Wetli himself. The locomotive continued its journey at increasing speed and was travelling at an estimated 120 kilometres an hour when it reached Wädenswil station, where the engine was derailed and overturned, and its boiler ruptured. One of the firemen was killed, and the driver and the other fireman were injured. A person on the station and one of the passengers who had either jumped or been thrown from the waggon also lost their lives, and all the others who had abandoned the waggon were injured.

An initial investigation of the accident blamed it on a graphite film on the rails as the result of excessive braking having charred the wooden brake blocks, but a second investigation blamed leaked oil from an undetermined source. Whatever the cause it meant the abandonment of the Wetli system. The Wädenswil-Einsiedeln-Bahn was later worked as a conventional adhesion line.[792]

A serious runaway occurred in Hungary, then part of the Austro-Hungarian empire, on 23rd August 1879. It involved a freight train of 48 waggons, eight of which, with all their contents, consisting of 600 centals (60,000 lb or about 27,240 kg) of oil and 24 centals (2,400 lb) (about 1089 kg) of manufactured goods caught fire and were totally consumed. The value of the destroyed goods, excluding the waggons, was put at 100.000fl (£10.000). Railwaymen saved the remaining 40 waggons by uncoupling them and pulling them clear with a breakdown train. The accident occurred at Steinburch, only just outside of Buda-Pesth (contemporary spelling) and most of the fire brigade of the city were soon

789 Wikipedia – Centennial Exposition

790 The Bucks County Gazette, Bristol, Pennsylvania, United States, 16th November 1876

791 The Elyria Constitution, Elyria, Ohio, United States, 16th November 1876

792 Wikipedia – Walzenradsystem Wetli

at hand to fight the fire. In less than an hour a crowd of several hundred spectators had also gathered, seriously impeding the salvage operation and also inviting a worse disaster should burning debris around the derailed steam engine of the train cause a boiler explosion. A deputation of Honvéds (territorial army soldiers), sent like the firemen by train from Buda-Pesth, were needed to preserve order and to keep the curious spectators at a safe distance.

The manner in which the accident had happened was explained afterwards by the freight train driver. His train had arrived at the start of a steep descent outside of Steinburch where he had whistled for the guard to apply the brakes prior to the descent, but for some reason his whistle signal had been misunderstood and the brakes had not been applied (or possibly they had, but not sufficiently). He had then begun the descent and finding that the speed of his train was increasing alarmingly he had reversed his engine, hoping thereby to control it, but without success. Anticipating a disaster, he had told his fireman, who was married, to jump, so as to save himself for his family, and the fireman had done so successfully. Shortly afterwards the leading waggon of the train had collided with his engine and the impact blew off or forced off a safety valve such that he too was thrown from his engine. This had probably saved his life, and he escaped with a few bruises. The train then became a total wreck, the waggons piling upon one another and the engine being forced off of the rails. The first crash had burst the barrels of petroleum in the waggons and as the train came to a standstill the oil flowed over the wreckage and along the ground and at once was set alight, the conflagration spreading 'like lightning.' Some people nearby witnessed what had occurred and at once ran to the nearest signal box from where news of the accident was sent to Steinburch and Buda and a rescue operation immediately begun. The newspaper report on the accident concluded with a reference to its similarity to the Abergele disaster, 11 years earlier, apart from the lack this time of casualties.[793]

In a repeat of the 1874 accident, two waggons loaded with rails became detached from a train, ran away and collided with a ballast train stopped on the line between Montabliz and Santiurde de Reinosa, Spain, on 15th October 1880, killing six Caminos de Hierro del Norte de España employees and injuring five.[794]

On the evening of Wednesday 27th October 1880, a goods train consisting of an engine and 14 coal waggons was waiting at Lier station, Belgium, to meet an opposing passenger train, the line onwards being to Herstal and Liège being single. While the train was waiting the fireman went away for an unstated reason, leaving the driver alone on the engine. The latter, who was only slightly hurt, was unable to explain how the train came to run away from Lier but thought that he might have fallen asleep and then, suddenly waking up, have inadvertently opened the regulator. The line fell steeply from Lier towards Liège and the heavy train was soon going at a considerable speed. About four kilometres before reaching Herstal station it came into head-on collision with the passenger train which had left Liège at 8.14 p.m. The collision was reported to have been terrific, with a report like that of artillery which could be heard a long way off. The

---

793 The Times, London, England, c1st September 1879

794 Wikipedia – Accidentes ferroviarios en España

engine, guards van and a carriage of the passenger train were piled into one another and smashed to atoms, forming one heap of debris. The couplings behind the third carriage broke and the rear vehicles of the train ran back down the incline to Herstal station, apparently without coming to any further harm. The bodies of the driver of the passenger train, two guards and three passengers, only one of whom could be quickly identified, were recovered from the wreckage, the corpse of the first guard being terribly mutilated. The fireman of the passenger train survived but had to submit to the amputation of his left leg. After the collision railwaymen and others did what they could to assist the injured pending the arrival of doctors. At the time that the report of the accident was written the work of clearing the line was still incomplete and it was feared that the number of those listed as injured would increase. A judicial inquiry into the accident was opened by the Liège Instructing Magistrates.[795]

[795] The Echo, London, England, 29th October 1880

CHAPTER 8

BRIDGE ACCIDENTS

Some of the most serious railway accidents of the nineteenth century resulted from the failure of bridges. Bridges can fail for a variety of reasons. The inherent design of the structure may be faulty, the materials used in its construction may be poor or inferior, the materials may deteriorate over time or the foundations of the bridge may be eroded. Often a variety of factors played a part, as is evident from the Tay Bridge disaster.

The earliest bridge accident which appears to have been recorded occurred on the Merthyr (Penydarren) Tramroad as early as 1815. On 15th February of that year the uppermost of two wooden bridges by which the line crossed and re-crossed the River Taff near Quakers Yard, South Wales, collapsed beneath a train; horses, trams, load of iron, haulier and four other people riding on the trams going down with the bridge. One of the horses was killed and the other much cut. The haulier and one of the passengers were badly hurt, the other people slightly bruised. The failure was blamed on the decayed state of the bridge timbers. It was noted that the lower bridge was in a similarly poor condition, and that their poor state had been known about for some time, without it seems anything having been done about it. It was proposed to replace the bridge using oak or iron but, in the end, both bridges were rebuilt as stone arches.[796]

The South Carolina Canal and Railroad was one of the early American railroads to be built on a continuous wooden trestle, a method of construction which found favour for a few years because it eliminated certain problems such as animals straying onto the line and collisions on level crossings. Sometime in March 1837 a spark from a locomotive set fire to the woods through which the line ran, including the brush at the base of the trestle, but the fire was put out, the trestle was checked and found to be apparently undamaged by the fire, and probably no more thought was given to the matter. A few days later, though, a 12 ft high section of the trestle, which obviously had been damaged after all, collapsed beneath a passenger train and an engine, a servant's car and a ladies' car fell to the ground. The passengers and servants escaped unhurt, but the engineer jumped on the

[796] Webpage – The Penydarren Tramroad: Industrial Railway Society

wrong side as he felt the trestle giving way and the engine fell on top of him, crushing him to death.[797] The accident happened at an unstated location between Charleston and Hamburg, South Carolina.

In 1837 the Baltimore and Ohio Railroad covered bridge across the Potomac River at Harpers Ferry, West Virginia was the longest in the United States. On 3rd September 1837 the whole structure collapsed beneath a light engine and tender. Given that the bridge had been only open a year it was rather surprising that its failure was attributed to rotting timbers. In the Spring of 1838, the central span of the structure gave away again beneath a heavy freight train. No one was very much hurt in either accident.[798]

The next bridge accident also occurred in the United States, this time on the Canajoharie and Catskill Railroad on 4th May 1840. A mixed train was crossing the 15 ft high High Rock Bridge over Catskill Creek at Durham, New York, and the engine had just reached solid ground when the structure collapsed, and all five cars of the westbound train went down with the bridge. One of the approximately 40 passengers on the train, a Mr Jehiel Tyler, of Durham, who had only got on the train at Stevens for a ride, was killed outright, while a coloured railway hand suffered two broken thighs and a broken leg, and his recovery was considered doubtful. Several other passengers were fairly seriously injured, and a quantity of merchandise was carried down the creek and lost.

The creek was much swollen by recent heavy rains and had doubtless damaged the foundations of the bridge and caused it to fail.[799]

The accident led to the financial collapse of the railroad company and the line was shortly abandoned.[800]

Heavy rains were also to blame for the next bridge failure, the collapse of an occupation bridge (Johnston's Bridge) which carried the South Eastern Railway over a road connecting two farms, owned by a Mr Johnston and a Mr Beckett, between Tonbridge and Penshurst, England. (It is often stated that it was a bridge over a tributary of the River Medway which collapsed but this is incorrect.) Adjoining the bridge on the up side of the line and to the south of it was a large ballast pit, six or eight feet deep, in which water tended to collect at times of flooding. Much rain had fallen in the days preceding 20th January 1846 and on the evening of the 19th the rain was exceptionally heavy. A torrent of water flowing under the occupation bridge from the Medway River on the down side of the line met floodwaters pouring from the meadows on the up side of the line at the corner of the ballast pit, and the eddy thus formed had in a short time undermined the embankment supporting the overbridge.[801]

The up goods train involved in the accident had left Dover at 8 p.m. on 19th January in the charge of Guard Charles Leonard, with Driver William Doyle and Fireman George Doyle (William's brother). The train reached Tonbridge at 12.30 a.m. on 20th, one

797 A History of Railroad Accidents Safety Precautions and Operating Practices by Robert B Shaw, pg. 19
798 Ibid, pg. 50
799 Boston Morning Post, Boston, Massachusetts, United States, 8th May 1840
800 Wikipedia – Canajoharie and Catskill Railroad
801 The Times, London, England, 22nd January 1846

waggon was detached, and it departed at about 12.45. The train had gone about a mile and was only travelling at about eight miles an hour, when Guard Leonard felt a sudden shock and it stopped. Going forward he found that the engine, tender and one waggon had disappeared into the water. He could not see anything of the engine crew but having sent a man back towards Tonbridge to protect the line he subsequently found the fireman, who had dragged his injured brother from the wreckage. Driver William Doyle was fatally hurt and died soon after they had brought him back to Tonbridge station.

The inquest jury returned a verdict that Driver Doyle had died through the collapse of Johnston's bridge caused by an unexpected flood of water, adding that in their opinion it was a dangerous practice to make drains under arches (one had been built to let water from the ballast hole drain into the Medway; the local engineer stated that it had nothing to do with the accident) and that the ballast hole was too near to the railway. The jury imposed a deodand of 1s on the engine and tender.

The accident was investigated by Major-General Charles Pasley for the Board of Trade, but his conclusions are not known as it has not proved possible to locate a copy of his report.[802] Possibly the report was never published, as in June of the same year the same inspecting officer approved some bridges and earthworks on the Edinburgh-Berwick line of the North British Railway. These works failed after heavy rain only three months later, as did temporary replacements to them, also approved by the Major-General Pasley. As a result, he came in for serious criticism and he was sacked in November, when the railway inspectorate was reorganised.[803]

Thus far we have dealt with bridge failures due to rotten wooden timbers, a wooden trestle catching fire and to collapses due to the bridge foundations having been washed away by floodwaters. The next accident, on 24th May 1847, was the first to be blamed on another cause, a design failure.

The Chester and Holyhead Railway was still being constructed in May 1847 (it opened from Chester to Bangor on 1st May 1848) but part of the Shrewsbury and Chester Railway, which ran over Chester and Holyhead metals for the last couple of miles into Chester, was already open. About a mile from Chester the line crossed the River Dee. Originally a five arch brick bridge was to have been constructed, but the design was then changed to three pairs of spans of cast iron girders, supported on piers of Runcorn sandstone, the bridge being designed by none other than the famous Robert Stephenson, son of George Stephenson.

The train involved in the accident was the 6.15 p.m. departure from Chester to Ruabon, in charge of Guard George Roberts, with Driver James Clayton and Fireman Thomas Anderson. It actually left Chester five minutes late, and consisted of an engine and tender, second-class carriage, first-class carriage, second-class carriage, luggage van and an empty second-class carriage, with about 25 passengers. Driver Clayton stated that he had been travelling at 15 to 20 miles an hour as he crossed the bridge, though the inspecting officer thought it was probably considerably more. Driver Clayton said that nothing unusual happened as he passed over the first two spans of the bridge, but as his engine reached the third span, he had felt the bridge sinking. He had immediately opened

802 Ibid, 27th January 1846

803 Wikipedia Her Majesty's Railway Inspectorate

the regulator wide and his engine had jumped forward and separated from the tender. The engine and tender reached the far bank, but the tender was derailed and his fireman, who had been standing on it, was thrown off and killed. The remainder of the train fell 36 feet to the water. In addition to Fireman Anderson, Guard Roberts and three passengers lost their lives. The passengers killed were a Mr John Matthews of Llangollen and Messrs Charles Nevitt and Isaac Jones of Wrexham.

Driver Clayton must have had nerves of steel for after the accident he went forward with his engine to Saltney to inform the staff there of the accident, then crossed over onto the down line and took his engine back to Chester to inform the people there of the accident, trusting to the integrity of the down line bridge span which was parallel to the one which had collapsed. It held.

The accident was, of course, the subject of an inquiry by the Railway Inspectorate of the Board of Trade, headed by Captain J.L.A. Simmons of the Royal Engineers, as well as of an inquest. Several theories were put forward to account for the failure of the structure. Robert Stephenson, being anxious to defend his design, proposed that the failure of the structure had been precipitated by the derailment of one of the carriages of the train. However, the evidence of several eyewitnesses to the accident, of several of the surviving passengers who had been on the train and of Driver Clayton, was that there had been no derailment prior to the failure of the structure. Stephenson's theory was supported by a number of other railway engineers such as Joseph Locke, engineer of the Grand Junction Railway, who proposed that the engine tender had derailed[804] though the fact that such a derailment could cause the bridge to collapse implied that the structure was inherently too weak, and Locke stated that he would not have designed such a bridge himself.

An alternative theory was put forward by Henry Robertson, engineer of the Shrewsbury and Chester Railway, who theorised that the top flange of the centre girder of the bridge had first failed due to compression from the weight of the locomotive, excessive vibration and the addition on the day of the accident of 25 tons of ballast to the bridge, a precaution adopted to prevent the oak beams of the structure from catching fire as had recently happened to a wooden bridge on the Great Western Railway.[805]

As a part of the official investigation into the accident, Captain Simmons and engineer James Walker conducted a series of experiments in which they ran an engine and tender over the bridge and they noted that there was a deflection of the girder with the engine at rest of almost 2½ inches, which lessened to about 1½ inches under the moving engine, but when there was also a considerable degree of oscillation. Captain Simmons concluded that both the deflection and the oscillation would weaken the cast-iron girders. The inspecting officer and the engineer were of the opinion that the wrought iron trusses fixed to the cast-iron girders to reinforce them did not in fact do so at all and that the bridge design was basically flawed.

The two investigators also examined the broken girders, but all of the pieces were not retrieved from the river, while lacking a knowledge of metallurgy and not therefore carrying out, for example, a determination of crack formation, they did not learn very

---

[804] Disaster on the Dee by Peter E Lewis, pg. 107

[805] Ibid, pg. 111

much, though they commented that there were no obvious defects in the castings.[806] Noting that much heavier trains than that involved in the accident had previously passed over the bridge successfully they decided that the repeated vibrations must have weakened the structure, an early recognition of the problem of metal fatigue.

The conclusions of the investigation were passed, before the publication of the official report, to the inquest jury. The coroner, in his summing up, explicitly ruled out a verdict of manslaughter or even negligence against the bridge designer, Robert Stephenson, but stressed that the verdict of the jury would be important for railway construction. The jury took only an hour to find verdicts of 'Accidental death' on the five victims of the accident. They added that: - 'We are further unanimously of opinion, that the aforesaid girder did not break from any lateral blow of the engine, tender, carriage or van, or from any fault or defect in the masonry of the piers or abutments; but from its being made of a strength insufficient to bear the pressure of quick trains passing over it.

'We feel that the 11 remaining girders, having been cast from the same pattern and of the same strength, are equally weak, and consequently equally dangerous for quick or passenger trains as was the broken one.

'We consider we should not be doing our duty towards the public if we separated without expressing out unanimous opinion that no girder bridge of so brittle and treacherous a metal as cast iron alone, even though trussed with wrought iron rods, is safe for quick or passenger trains, and we have it in evidence before us, that there are upwards of 100 bridges similar in principle and form to the late one over the river Dee, either in use or in the course of being constructed, on various lines of railways. We consider all of these unsafe, more or less, in proportion to the span; still all unsafe.

'We therefore call upon Her Majesty's Government to institute such an inquiry into the merits or demerits of these bridges, as shall either condemn the principle or establish their safety to such a degree, that passengers may rest fully satisfied there is no danger, although such bridges may deflect from 1½ to 5 inches.'[807]

A modern (2007) analysis of the accident concludes that the bridge collapse resulted from the catastrophic failure of the centre of a girder at a fatigue crack which had originated in a small blowhole or casting crack and had gradually grown over time until it reached a critical length.[808]

After the accident, the bridge was quickly rebuilt with shallow arches beneath the girders relieving the latter of much of their load. It was again rebuilt in wrought iron in 1870-71.[809] General Pasley, the inspecting officer who had approved the structure before it was brought into use, had already resigned his office, as noted above, in November 1846, probably because he was forced to, and as the inquest jury had desired a Royal Commission was set up to investigate the safety or otherwise of cast iron girder bridges. In July 1849 it condemned the use of trussed cast iron bridges. Nevertheless, untrussed cast iron bridges continued to fail in Great Britain up until 1891.

---

806 Ibid, pg. 114

807 The Times, London, England 18th June 1847

808 Disaster on the Dee by Peter E Lewis, pgs. 124-6

809 Ibid, pg. 132

On 27th October 1847 an evening work train was crossing the Miller River, near Athol, Massachusetts, United States, on the Vermont and Massachusetts Railroad, when the structure gave way, precipitating the 13 men on the train 30 or 40 feet into the river. Newspaper reports spoke variously of four or six men dead and of the others being seriously injured. The train had consisted of only an engine and two cars loaded with 'railroad iron,' presumably rails. The bridge was a new one, the line not yet being open to traffic, the train being reportedly the first to cross it; it was hoped that the structure could be repaired in time to avoid any delay in bringing the line into use. The accident was blamed on 'imperfections in the iron used for rods in the centre of the bridge.'[810][811][812]

Back in England, the South Eastern Railway had a fortunate escape from a more serious accident at 1 a.m. on the morning of 20th October 1848. An up fish and goods train which had left Margate just before midnight the previous day was crossing a bridge over the River Stour three miles from Canterbury when the structure gave way and engine, tender and train fell into the river. The contents of the train, destined for London, were strewn about the scene and the track was badly damaged but no one was hurt. Information of the accident was telegraphed to London and Superintendent Finnigan was despatched to begin an inquiry into the mishap. *The Times* reported that 'the mail train was detained a short time, while arrangements could be made for forwarding the passengers and letter-bags.' Recent flooding was supposed to have undermined the bridge.[813]

Just over a year later, on 6th December 1849, a similar sort of accident occurred between Newark and Collingham, England, on the Midland Railway branch to Lincoln. Again, it was the middle of the night, in fact about 4 a.m., when the brick abutments of a wooden bridge gave way under the weight of the engine of the down mail train. The shock caused the couplings between the engine and the leading carriage to snap and the momentum carried the engine to safety, but the carriages piled up upon one another and fell into the floodwaters of the River Trent. No one was killed, though the fireman suffered serious head injuries and the guard hurt his hip. The driver and Post Office guard escaped with shock and there were fortunately no passengers on the train. As at Canterbury the floodwaters had undermined the bridge.[814]

The next bridge accident, in the United States, was more serious. It occurred to the Westcolang Bridge of the New York and Erie Railroad, not far from Mast Hope, Pennsylvania, half an hour after midnight on 31st July 1850. As a freight train was passing over the bridge, its engineer felt it beginning to give way and opening his regulator just about managed to reach the stone abutment. His engine was derailed in the process and was thus saved from being pulled back into the gap by the following waggons. The bridge was about 60 feet in length and carried the line 25 feet over a

[810] Gettysburg Adams Centinel, Gettysburg, Pennsylvania, United States, 8th November 1847

[811] Fort Wayne Sentinel, Fort Wayne, Indiana, United States, 20th November 1847

[812] A History of Railroad Accidents Safety Precautions and Operating Practices by Robert B. Shaw, pg. 51

[813] The Times, London, England 21st October 1848

[814] Ibid, 8th December 1849

ravine, through which flowed a floom serving an adjoining sawmill. Fifteen freight cars fell into the said ravine and were piled up in the valley below. The train was carrying mainly livestock, comprising 113 head of cattle, 450 sheep and 300 pigs. Of these only 25 cattle and 100 sheep could be saved. Three humans also lost their lives; a brakesman whose name the newspaper reporter could not obtain and two drovers, Henry Clapp, of Menton, Ohio, in charge of the sheep and swine and Mr Bondall, of Corning, New York, who had been in charge of the cattle. Six others on the train survived, though Conductor Hatch was badly hurt. He had had to jump from the tender, while the owner of the cattle and two of his men had had to jump from waggon to waggon to save themselves. The property loss was considerable, each car being valued at $700 and each head of cattle at $50. The total loss was estimated at $30,000 to $35,000. The accident could though have been much worse as an express train had crossed the bridge only a short time before.

The line and bridge had been examined only a few days before and declared safe, but the newspaper reporter considered the wooden structure to have been too weak and thought that a stone arch should have been provided instead.[815]

Spain suffered a bridge accident on 27th September 1855. A light engine was en route from Madrid to rescue a train which had been stranded by floodwater at kilometre five of the Madrid – Alicante line, four workers riding on the engine together with the driver and fireman. The engine only managed to get as far as the Abroñigal Bridge at about kilometre 1 when the latter structure, which crossed a stream much swollen by torrential rain, collapsed beneath it. Five of the six men on the locomotive were drowned but one miraculously survived. Washed over half a kilometre downstream he was eventually able to catch hold of a tree, and after several hours was able to get out of the water.[816]

The first major disaster resulting from a bridge failure happened in the United States at Gasconade, Missouri, on the Pacific Railroad (later Missouri Pacific) on 1st November 1855. Like several others before and afterwards, the accident occurred on the opening day of the line, in this case the Pacific Railroad route from St Louis to Jefferson City Missouri. To celebrate the opening, a special excursion train was booked to leave St Louis at 8.30 a.m. for Jefferson City. The 12, or possibly 14, coach train, crowded with six hundred invited guests, according to one report did not actually depart from the Seventh street station until 9 a.m., but if so, it presumably had a fairly easy schedule, for by Hermann the half hour delay, if there was one, had been made up. All had gone well so far, with the guests being entertained by an accompanying brass band and the train being cheered with shouts, waved hats and handkerchiefs at stations and villages along the line. A ceremonial cannon greeted the train at Hermann. There were likewise no problems onwards to Gasconade, but here disaster struck. The Gasconade River, a tributary of the Missouri, was the largest waterway to be negotiated between St Louis and Jefferson City, and was crossed by a 750 feet six-span iron girder structure resting on six massive stone piers. The problem was that the bridge was still not complete on what was planned to be the opening day. Rather than the safe option being taken of postponing the opening until the bridge was finished the decision was made to run trains over the

---

815 Boston Post, Boston, Massachusetts, United States, 3rd August 1850

816 Webpage Accidentes Ferroviarios en España by Manuel J Marcos Montero pg. 88

incomplete structure, with the track laid on the false work erected during its construction. The approach embankment on the St Louis side of the bridge was also unfinished and here the track was laid on a wooden trestle. On the morning of the accident a heavily loaded gravel train was successfully run over the bridge to test it and it was therefore deemed safe for the excursion train to cross it at reduced speed.

A few minutes after 1 p.m. the excursion train reached Gasconade. The engine and leading baggage car had barely crossed the first 130 ft span between the east abutment and the first pier when the span collapsed, drawing the engine back into the gap. The weight of the engine, baggage car and two loaded passenger cars had proved too much for the span. The locomotive ended up reversed in direction and upside down in the shallow water. The following passenger cars piled up behind it. At least 29 passengers lost their lives in the disaster according to a local newspaper report a few days later, the victims included Thomas O'Sullivan, the chief engineer of the railroad, who had been riding on the engine. One of the survivors of the accident was the president of the railroad, the ironically named Mr H.E. Bridge. Immediately after the accident, a severe thunderstorm added to the discomfort of the survivors but probably helped to prevent fire breaking out in the wreckage.

Doctors McDowell and McPherson were fortunately among the guest on the train and did their best to treat the survivors with the limited resources to hand. Couriers hurried back to Hermann and reportedly within an hour another train had arrived at Gasconade to take the survivors back to St Louis, which was reached at about 11.30 p.m. that night.[817]

The disaster was investigated by a committee of engineers appointed by the legislature. The engineer of the train, William Tucker, who survived the accident, stated that his speed on entering the bridge had been six miles an hour but it seemed more likely to have been nearer 15 miles an hour. The majority of the investigating committee attributed the accident solely to the train's excessive speed over the temporary structure, but one committee member, whose dissenting report was probably nearer the truth, declared that the structure had been unsafe for trains at any speed. He stated, 'The attempt of running a locomotive and cars, upwards carrying human life over a river like the Gasconade, upwards of thirty-five feet above its bed, upon such frail structures, can only be ascribed to incompetency, recklessness or infatuation.'[818]

The Hudson River Railroad suffered a most unusual bridge accident at Spuyten Duyvil, New York, United States on 14th January 1856. The train involved was the Sunday afternoon mail train from Albany to New York. It had left Albany at 4.15 p.m., hauled by two engines. Progress was very slow due to accumulated snow and ice on the track, and it was several hours late when it reached Spuyten Duyvil, 18 miles from New York, at 1.30 a.m. on Monday morning. Here the line crossed Spuyten Duyvil creek by a wooden trestle bridge with a draw span and it was at this bridge that the train's journey ended abruptly. The leading engine crossed the bridge safely, but it went down under the train engine before its engineer had time to shut off steam. A milk car, freight and baggage cars followed the second engine off the bridge, but the baggage car did not go into the water and three following passenger cars did not go off the bridge, though the

[817] Hannibal Messenger, Hannibal, Missouri, United States, 6th November 1855

[818] A History of Railroad Accidents Safety Precautions and Operating Practices by Robert B. Shaw, pgs. 51-2

front car was derailed. The passengers thus escaped but Engineer George Parsons and Fireman Howard Root on the train engine lost their lives and Engineer Henry Dawson and Fireman Peter Banta on the pilot engine were severely scalded.[819]

What had happened was that on the previous day a strong easterly wind had blown large masses of ice against the trestle, around and beneath the timbers. The creek was tidal, and the ice, rising with the tide, had lifted the bridge structure above its supports. Unfortunately, when the tide fell again the sills did not come down properly on top of the piles so that the track was left completely unsupported, held in place only by the tension of the rails. Not surprisingly, as soon as the train ran onto the bridge, the unsupported track gave way. The leading engine got to the far side but did not escape unharmed as its boiler was pierced by a broken rail, which was how its driver and fireman came to be scalded.[820]

The eastbound New York Central Railroad Cincinnati express passed Rome, New York, United States at 6 a.m. on the morning of 11th May 1858, running a few minutes late. A few miles further on, a mile below Whitesboro it met the westbound Utica accommodation train on the double track as both were crossing the Sauquoit Creek Bridge, a minor structure about 20 feet in length and only 10 feet above the water. The two engines passed one another but just as the leading baggage car of the express got onto the bridge its middle stringer gave way and the two trains were thrown in towards one another. The accommodation train struck and shattered the top of the leading passenger car of the express but the momentum of the latter carried it to the east end of the bridge, the front of the second express passenger coach went down into the stream and its end struck the east abutment of the bridge, the third passenger car ran under the second and the fourth half telescoped the third.

The accommodation train consisted mainly of freight cars with a passenger car at the rear.[821] This coach did not reach the bridge and all the casualties were in the eastbound express. At least eight passengers lost their lives in the accident, including a Mr William Perkins of Rochester, who succumbed to his injuries the following day. A number of other passengers on the express were more or less injured.

The *Auburn Weekly American* gave a detailed account of the accident from one of the express passengers, a Mr William Shube, a former editor of a Mansfield, Ohio, newspaper: 'He says it is the third time he has been through a similar scene within nine months. He was in a car which rolled down an embankment near Hornellsville, on the Erie Road, and again on the same Road when two trains came in collision near what is called Tip-top Summit. On the latter occasion he was carried by the force of the collision through the flying splinters of the baggage car and landed on top of the wood in the tender, and yet escaped without any injury other than painful bruises. He says he has learned to take calamities coolly.

'The first intimation Mr Shube had that all was not right on the train, he heard a shriek apparently from a female voice, immediately followed by a cracking sound. Having made up his mind that the surest way of escaping injuries upon such occasions, was to make

819 New York Daily Times, New York, United States, 15th January 1856

820 A History of Railroad Accidents Safety Precautions and Operating Practices by Robert B. Shaw, pgs. 53-4

821 New Albany Daily Ledger, New Albany, Indiana, United States, 12th May 1858

fast to something, he grasped the seat firmly, and was immediately carried, seat and all, over the head of the man in front of him. He felt the fire flash from his eyes by coming in collision with the head of the unknown man, who was lying dead in the baggage room. Mr Shube says he distinctly heard the crushing sound of breaking bones as the old man was thrown against the side of the car; he said it sounded in his ears like that produced by a dog "crouching the bones of a partride" and was to him the most horrible noise he ever heard. He thinks the old man died instantly.

'Mr Shube was in the second car; he saw a shower of splinters flying and got a glimpse of daylight shining through the end car, as he went down, following the engine, tender, baggage and first passenger car. The third car struck the second in end and dashed it in pieces, a heap of ruins. Mr Shube was assisted out of the midst of the wreck by friendly hands; his injuries are a sprained hip and one of the small bones of his leg broken. He says the hip was injured in the same place where its firmness was severely tried on the Erie Road, but he speaks confidently of being on his way in a few days. Few men live to pass through such scenes as Mr Shube has experienced. His own escape as well as that of many others, stop little short of the miraculous.' [822]

The inquest which was held on the victims of the accident heard that the bridge had been renewed three years earlier. A witness, Erastus Crandall, stated that at that time he had noticed that one of the bridge timbers was rotten and had mentioned it to Track Master Everts. Everts denied this. The said track master had examined the bridge a week before the accident by striking it with a hammer. Mr Priest, the superintendent of the line, contended that the accident had followed the breaking of an axle of the baggage car of the express, though all agreed that post-accident examinations had shown the bridge to have been unsafe, and Mr Priest testified that 'knowing what he does now, he should not think the bridge would sustain two engines with loaded trains passing at the same time in opposite directions.' In addition, it emerged that most or all of the bridge cords were composed of what was variously termed white elm or, by its detractors, bastard elm. Every witness at the inquest other than Messrs Evert and Priest agreed that the wood was unsafe for railroad purposes.

Various witnesses at the inquest; Messrs Park, Crundall, Taylor and others testified that the unsoundness of the bridge might have been discovered by test borings. Leach, the baggageman, who had initially stated that the axle of the baggage car had broken, later admitted that he had only noticed the broken axle as the car was leaving the bridge, and that it might have been a consequence rather than a cause of the accident. There was no evidence that the railroad company had suspected that there any problem with the bridge before the accident.

The jury recommended that every bridge on the line should be thoroughly examined and put in perfect order. Furthermore, every wooden bridge on the line should be replaced by one of iron or stone, preferably stone. The *Auburn Weekly American* wrote on 26th May 'Hereafter, everyone passing over the road will feel a sense of insecurity, until iron or stone is substituted for these frail wood structures. We conjure the company, as they value the lives of their passengers, their own good name as the safest road in the country, and their dividends, which must be sadly diminished by frequent accidents, like that at Whitesboro, not to delay a single day in commencing a reform in the particulars

[822] Auburn Weekly American, Auburn, New York, United States, 19th May 1858

indicated.' [823]

Flooding following heavy rain in the American state of Georgia caused two accidents on the Columbus and Macon Railroad on the last day of 1858, one of them a major disaster. The more minor accident occurred to the morning passenger train from Macon two miles from Columbus, where a culvert was washed away, derailing the train and killing the fireman and a wood passer. The engineer, G.W. Smith, suffered a broken jaw but the passengers escaped.

The major disaster occurred to the corresponding morning train from Columbus to Macon, at the bridge over Randall's Creek at Adams (or Odum's) Mills, about 15 miles out of Columbus. Undermined by the floodwaters the bridge collapsed under the train and 17 of the 29 people on it lost their lives. Only seven bodies had been recovered at the time of the third report into the accident. The dead included Engineer J.H. Miller, the fireman, the wood passer and Conductor W.H. Snell.[824] [825]

The Albany Vermont and Canada Railroad suffered a bad accident on 2nd August 1859 when the mail train due in Albany at 7.25 p.m. was crossing a wooden trestle near Schaghticoke, New York, United States, 17 miles north of Albany. The trestle over the Tomhannock Creek collapsed under the train. As seems to have often been the case the engine managed to get clear of the trestle, but the coupling between it and the tender broke and the tender, baggage car and two passenger cars fell 38 feet and ended up in eight feet of water. The baggage car was wrecked, and Express Messenger David Russell, Mail Agent Charles Pympton, Brakeman Charles Bethelon and an unnamed baggageman in it were killed. The first passenger car came down endwise onto the tender and the second passenger car on top of it. Nine people in all died in the accident and about 20 were injured. The newspaper report stated that only one passenger was among those killed, Howard Wright, an Albany merchant, though two of the others were Mrs Lucina Cooley and Mrs Cuyler, the wife and mother-in-law of Conductor Cooley. The other two dead were machinists Patrick Connolly and Dennis Cahill.

The trestle involved in the accident had been regarded as insecure for some time and the company had commenced to rebuild it a fortnight before the accident. Meanwhile trains were ordered to pass over it at reduced speed and it would seem that the engineer had whistled for brakes as the train was approaching it, but Brakeman Bethelon had not had time to apply them, either because he was slow to react or because the engineer had whistled too late.

Coroner Madden of Troy held an inquest on the death of Charles Plympton. The jury brought in a verdict that, 'the said Charles Plimpton came to his death from injuries received by the falling of a bridge on the Albany Vermont and Canada Railroad, over the Tomhannock Creek, while engaged in a regular train of said railroad as Mail Agent; and as said bridge was in a rotten and unsafe condition, the Jury believe that those having

[823] Ibid, 26th May 1858

[824] Fort Wayne Weekly Republican, Fort Wayne, Indiana, United States,5th January 1859

[825] Atlanta Weekly Intelligencer, Atlanta, Georgia, United States, 6th January 1859

charge of said road should be highly censured.' [826]

At 2.15 a.m. on 26th September 1860, a dark and very foggy morning, a Midland Railway goods train left Derby for Normanton, England. It consisted of an engine and tender, 27 loaded salt waggons, two loaded goods waggons, two empty waggons and a brake van. After a half hour stop at the city wharf in Derby the train proceeded, slipping badly with the heavy load and bad rails, so that as it passed Ambergate station the train was only travelling at 14 miles an hour. Half a mile further on the driver realised that the trailing wheels of his engine had left the rails and bringing his engine to a stand went back to investigate. He found that the tender and the four waggons behind it were also derailed, and behind these waggons the next nine waggons were piled up in a heap in the turnpike road which had been crossed by the cast iron Bull Bridge, one of the girders of which had failed. The driver had had no idea that the bridge had failed, nor had his guard, who merely found that his van had suddenly came to a stand. Neither the van nor the rearmost 18 waggons were derailed.

By the time that Captain Tyler, the Board of Trade inspecting officer, had arrived at the scene of the accident, the fracture in the girder involved had rusted over, but the evidence of those who had examined it shortly after the accident made it very clear what had happened. The cast iron girder had failed because of a massive flaw, which it was strange had not been detected before, either at the foundry where it was cast or when the bridge was being erected in 1837. The inspecting officer could only suggest that perhaps it had been comparatively small in the first instance and had increased over time. He pointed out that it was very fortunate that the girder had finally failed beneath a goods train and not a passenger train but suggested that the driving wheels of the slipping engine might have put an extra strain on the girder causing its final failure.

Following the accident, the Midland Railway engineer determined to have the lower surfaces of all other cast iron girders on the line scraped and examined to detect any further defects. Captain Tyler also suggested that any other girders that were not bolted together should be so secured. [827]

On 11th June 1861 the 6 a.m. London and North Western Railway mineral train from Kenilworth to Leamington made its usual run, and subsequently set off on the return journey with empty coal waggons at about 7 a.m. It had covered half of the return journey to Kenilworth and was travelling at seven or eight miles an hour, when a cast iron underbridge over the junction of two roads near Wootton, England, suddenly failed beneath the engine. The luck which had attended the Bull Bridge accident on the Midland Railway the year before ran out this time and the driver and fireman were crushed to death by the tender, though the guard and a porter riding in the van with him were unhurt.

The bridge had been constructed in 1844. It was a composite structure with two timber trusses connected together below by five cast iron girders. One of the cast iron girders had been patched in 1853 to cover a defect. Captain H.W. Tyler found that all five cast iron girders had failed; three of them had flaws in the sections of fracture. The first girder

[826] Banner of Liberty, Middletown, New York, United States, 10th August 1859
[827] Board of Trade report

from the Leamington end had a blow hole, the third had two similar defects and the second had clear signs of a previous crack. There could be no reasonable doubt that this second girder was the first to fail under the engine. The strain of the failure of the second girder would have caused the first and third girders to fail in turn, the fourth and fifth girders finally giving way as well. The second girder had been much weakened by the patching up job done in 1853, described by Captain Tyler as 'an improper proceeding' which was likely to lead to the ultimate failure of the girder. It would have been better if all five girders had been removed in 1853 and replaced by new ones of better and more suitable material.

The inspecting officer concluded that it was too common a practice on the railways to neglect the strengthening of bridges to cope with the increased weight of the engines using them, 'so as to ensure the margin of strength that it is desirable to interpose between a reasonable degree of safety and actual failure.' [828]

A brief report from Lisbon, Portugal, on 12th December 1862 detailed an accident on the Eastern Railway. A work train loaded with materials and also carrying about 200 workmen was crossing the Puente de Sor (Sor bridge) when the structure suddenly collapsed and the whole train was precipitated into the river. About 100 people were reportedly killed and many others injured.[829]

The state of Mississippi., United States, was the scene of the next bridge disaster on 19th February 1863, an accident which bore similarities to that at Spuyten Duyvil, just over seven years earlier. The American Civil War was in full swing and the city of Vicksburg was under attack by the Union army led by General Sherman. Reinforcements were urgently needed and, at 3 a.m. on the morning of the 19th a Southern Railroad passenger train left Meridien, Mississippi, bound for Vicksburg. Most of the passengers on the train were therefore soldiers but there were a few civilians and a Confederate government paymaster.

Thirty-five miles west of Meridien floodwaters due to winter rains had significantly weakened a bridge over Chunky Creek. No train had passed over the structure since the previous day, when a passenger train had been permitted to negotiate it only after all its passengers had alighted from it. More debris borne by the floodwater had piled up against the bridge since then and the weight of the debris had shifted the structure, the track being six inches out of alignment at the point where it ran onto the structure. Efforts were made to repair the damage but there were just insufficient men to do the job before the next train was due. Clearly in these circumstances prompt measures were necessary to stop any approaching trains and hold them until matters could be put right, but the subsequent investigation found that the precautions taken to stop traffic had been seriously inadequate. Sectionmaster Absalom F. Temple had been warned by the conductor of another train of the 3 a.m. departure from Meridian and had asked a railwayman named Hardy to stop the train at a water tank short of the bridge. When Hardy protested that he lived too far away from the line to do so the Sectionmaster told Hardy to get a negro to warn the train instead and an elderly negro man was ordered to

---

828 Ibid.

829 The Times, London, England, 20th December 1862

stop the train with a lantern. The sectionmaster also erected a thick pole in the middle of the track at, he thought, a distance of 150 yards from the trestle, this being the usual means employed to warn trains of danger. Feeling that there was no more to be done he had then gone home to bed and knew nothing of the accident until told of it the next morning.[830]

The engineer of the train either completely failed to observe the warnings in the dark night or else observed them only too late to pull up. His engine left the track as it entered the bridge and plunged into the cold icy water, which was about 15 feet deep, followed by the string of four or five boxcars carrying both passengers and freight. The engine, named *Hercules*, finished up out of sight in the deep water and the boxcars piled up on top of it. Over 40 passengers died in the accident, most of them by drowning. Engineer Isaac P. Beauchamp and the unnamed fireman also lost their lives, but the conductor swam to safety. Most of the dead passengers were soldiers, many from the Mississippi and Tennessee Infantry regiments and including Major William H. Lilly.[831]

Rescue came quickly as there was a Confederate military training camp of the First Battalion of Choctaw Indians close to the accident scene. Many of the passengers who survived were rescued by Indians who had stripped and jumped into the flooded stream to save them. Over the next few days most of the bodies of the victims were recovered from the stream and they were buried along the railroad right of way or on the section master's farm. A large amount of money, reportedly $80,000, was also recovered from the possessions of a William P. Grayson, a lawyer who was a Confederate purchasing agent responsible for buying cotton. He was one of those killed.

On 27th March 1863 a freight train broke through a bridge seven miles below Gallatin, Tennessee, United States. Three men and 120 head of cattle lost their lives.[832]

October 7th, 1863 saw another bridge accident in Spain. It occurred on the Gerona Railway to a train carrying passengers who had come from France by diligences plying between Perpignan and Gerona. Severe thunderstorms overnight had been accompanied by torrential rain throughout the district, leading to massively swollen rivers and streams. The floodwaters had swept through the town of Vich, leaving it in ruins and drowning a great number of its inhabitants. Meanwhile the train from Gerona had left Sirvac for Barcelona at 4.30 p.m. The storm was still raging, and the floodwaters were at their highest when the train of nine carriages arrived at the Riera Seca (also known as the Alabern or Vert) bridge at Hostalrich at about 6 p.m. The engine and leading two carriages managed to cross the temporary wooden bridge successfully, but the flood-weakened structure then gave way and the remaining seven carriages were precipitated into the torrent. Efforts were begun at once to rescue the passengers, but many were drowned. There had been reportedly 175 or 180 people in all on the train and there was talk of 135 or even 150 dead, but only twenty bodies were later recovered from the carriages with 30 or more other passengers more or less injured. The final death toll

[830] Wikipedia – Chunky Creek Train Wreck

[831] Webpage – The Chunky Creek Train Wreck by Greg S. Boggan

[832] Philadelphia Inquirer, Philadelphia, Pennsylvania, United States,30th March 1863

appears to have been 21.[833] [834]

By a strange coincidence two major disasters occurred in England in two days in 1865, on the 7th and 9th of June, and both resulted from the inadequate protection of work being carried out on the track. The first was at Rednal, on the Great Western Railway, and is described in the chapter dealing with derailments due to track defects. The second disaster, on the South Eastern Railway, at Staplehurst, involved a bridge, and will be described here.

The Staplehurst accident is probably better remembered than most accidents of that period because it involved a boat train, the Tidal Mail, running from Folkestone Harbour to London Bridge, and whose passengers on the fateful day chanced to include the famous novelist Charles Dickens, who was returning from a holiday in northern France with his mistress, the young actress Ellen Terman, and her mother.[835]

During the eight or ten weeks preceding the 9th June, work had been in progress replacing the longitudinal timber baulks of three low viaducts to the east of Staplehurst station. The work was being undertaken by a gang of four carpenters, three platelayers and a labourer in charge of Foreman Platelayer (Ganger) Henry Benge. Rather than one line being taken out of use while the baulks were being exchanged and single line working being brought into operation, the work was being done between trains. Foreman Platelayer Benge, who worked on the line every day, was of course perfectly familiar with the times of all the booked trains and so could easily choose suitable gaps to do the work in. The only problem was the Tidal Mail, carrying passengers and mail from the cross-Channel ferry, which because of the limitations of the harbour at Folkestone could only berth at high tide, and whose times of running therefore varied from day to day according to the times of the tides. The times of this train were given on a separate page in the working timetable. Unfortunately, Foreman Benge misread the timetable and thought the train was due at nearby Headcorn at 5.20 p.m., its time on the 10th, rather than the 3.15 p.m. that it was due on 9th. The leading carpenter had also been supplied with a working timetable, but it had been mangled by the wheel of a passing train and rather than requesting a replacement he had simply relied on his foreman.

Although the times of all the ordinary trains, including the irregular times of the boat train, were therefore available to the gang, there was always the possibility of an untimetabled train appearing, such as a light engine sent out to assist a disabled train. To cater for such a possibility the company regulations decreed that before the road was opened up or disturbed in any way a signalman was to be sent back the line in the direction from which any such train might come, equipped with five detonators and a red flag. He was to place one fog signal on the rails every 250 yards, and two ten yards apart at 1,000 yards (or further if there was a curve or a steep falling gradient), and there remain to exhibit a red flag to any approaching train. Had this regulation been carried out the disaster would probably not have happened. Instead labourer John Wiles went back only 10 telegraph poles or about 554 yards (the driver, fireman and guards of the boat train thought he was much closer in than that) and he was only supplied with two

833 The Times, London, England, 12th and 17th October 1863

834 Webpage - Accidentes Ferroviarios en España by Manuel J Marcos Montero, pg. 34

835 The Dover Mails by Dan Garrett, pg. 23

detonators and told not to use them unless it became foggy. It was a bright sunny day. The similarities to the causes of the 1863 Chunky Creek disaster in the United States are obvious. The flagman had gone out the same distance before and had been far enough out to stop two ballast trains and a light engine, but these had all been travelling much more slowly than the boat train, which was booked to run at 50 miles an hour over this dead straight stretch of line.

When "protracted repairs" on the line were being undertaken it was also the custom to inform the train crews by printed notices as an additional precaution, but the company engineer pleaded that as the track was restored to its normal condition between the replacement of each baulk the repairs were not protracted, a view not shared by Captain F.H. Rich, the Board of Trade inspector. At any rate no printed notices to enginemen were sent out.[836]

The up Tidal Mail departed from Folkestone Harbour station at 2.38 p.m. and Folkestone Junction, where it had to reverse, at 2.48 p.m. Meanwhile, the work of replacing the timbers of the three viaducts was almost complete, with the last baulk being replaced on the low viaduct over the River Beult, 170 feet long, and carrying the railway only 10 feet above what at the time of the accident was only a small muddy stream.

The boat train passed through Headcorn at 3.11 p.m. and reached Beult Viaduct at about 3.13. It consisted of an engine (No. 199) and tender, a guard's van, a second-class, seven first-class, and two second-class carriages and three vans bringing up the rear and was carrying 80 first-class and 35 second-class passengers, most of whom had arrived at Folkestone Harbour by steamer from Boulogne. Driver George Crombie stated that he only became aware of the danger about 500 yards from the viaduct, which would have therefore been about when he was passing Platelayer Wiles, and that he had been able to reduce his speed from 45 or 50 miles an hour to 10 or 12 miles an hour by the time he reached the viaduct. Captain Rich thought that he was mistaken and that the boat train had been still travelling at not less than 30 miles an hour when it left the rails. Five of the fourteen carriages of the train were supplied with brakes, the front guard being able to operate Cremer's patent brakes on the two adjoining carriages and two of the rear vans having brakes. None of the three guards had however realised the danger before Driver Crombie whistled for brakes and Captain Rich thought that the train had only been about 250 yards from the bridge when the brakes were applied. The momentum of the train carried the engine across the viaduct despite two lengths of rail having been removed and it came to rest on the ballast 28 yards beyond the end of the viaduct. The tender, leading van and second-class carriage also remained on the viaduct, The leading first-class carriage ended up hanging from the viaduct with its rear end resting on the dry river bed, the next first-class carriage fell upside down into the river and the remaining five first-class carriages, the next second-class carriage and the front of the rear second-class carriage also went into the river. The rear three vans remained on the embankment.[837]

Ten passengers, comprising seven women and three men, lost their lives in the accident according to the Board of Trade report and 40 were more or less badly hurt. Other sources state that there were eleven dead, possibly one victim died of their injuries some time later. Driver Crombie escaped unhurt and Fireman William Beattie was only

---

[836] Board of Trade report

[837] Ibid.

very slightly injured. Two of the guards were also hurt, one rather badly. Charles Dickens and his companions were unhurt. He managed to get one of the guards to unlock the carriage door (the carriage doors had been locked before departure in accordance with the usual practice) and fetching a plank he had climbed carefully down from the coach and made for the river. There he had filled his top hat with water and washed blood from some of the badly injured victims, also administering brandy from his hip flask, only to become aware that several were too badly hurt to be helped, when they died shortly afterwards. All the survivors had been removed from the wreckage by 6 p.m. and the daughter of a member of the local gentry arranged carriages to take them to nearby houses, while a group of surgeons arrived from Guy's (situated next to London Bridge station) and other London hospitals on a special train to carry out amputations and apply splints to broken limbs.[838] Describing the disaster, Dickens wrote, "No imagination can conceive the ruin of the carriages, or the extraordinary weights under which the people were lying, or the complications into which they were twisted up among iron and wood, and mud and water." His later descriptions of the accident omitted any reference to Miss Terman, he being anxious to keep the affair secret and thus avoid any scandal. It was said that Dickens never fully recovered from the disaster, and he died on its fifth anniversary.[839]

Captain Rich was of the opinion that the boat train would have reached London safely, despite the rails having been removed on the Beult viaduct, had the rules of the South Eastern Railway been observed. He noted approvingly that the regulation about laying detonators was an additional precaution not generally adopted by other companies. It appeared, however, that the regulations had been daily disregarded for the previous ten weeks between Headcorn and Staplehurst. For this he blamed not only Foreman Platelayer Benge but also Permanent Way Inspector Joseph Gallimore, who was responsible for 36 miles of track and who was supposed to visit every part of the line if not daily at least several times a week, and whom it seemed had taken no notice of the fact that rules were being broken, something which he could hardly have remained unaware of. Captain Rich added that he feared that the regulations for flagmen to go out certain distances and guards to go back to protect their trains were too often disregarded. The inspecting officer also considered that such prolonged works should not be carried out without notice of them having been issued to train crews. He suggested that in future permanent way inspectors should be required to submit periodical returns of works in progress and the dates at which they had visited them, which returns could be checked by stationmasters or other officials, and that the inspectors might be requested to state that they had checked that their orders were being properly carried out.[840]

The coroner's inquest upon the victims of the disaster, held in the Railway Hotel at Staplehurst, returned a verdict of manslaughter against Foreman Platelayer Benge and Permanent Way Inspector Gallimore.[841] Commenting on their subsequent fates, a recent (2017) writer states, 'The well-spoken and well-represented Gallimore walked free {from the manslaughter trial} but working-class Benge, represented by a poorly-prepared clerk, was found guilty and served six months hard labour. His S.E.R. career over, he returned

---

[838] The Times, London, England, 10th June 1865

[839] The Dover Mails by Dan Garrett, pg. 26

[840] Board of Trade report

[841] The Times, London, England, 12th, 13th, 17th, 20th and 22nd June 1865

to agricultural labouring. By 1902 his brooding over the words, "Did feloniously kill…eleven people", turned his mind and he spent the last three years of his life in an asylum.'[842]

A bridge over Long Creek at Ainsworth station, Iowa, United States, on the Washington branch of the Milwaukee and Mississippi Railroad, gave way under the weight of the engine of a train crossing it on 10th August 1865. The bridge was about 100 feet long and thirty feet high. It did not collapse instantly, and the engine and tender made it across the structure. The coupling between the tender and the train snapped and the baggage and freight cars went down to the creek below. The last two cars remained on the bridge, but the passenger car telescoped the car in front of it. Two passengers, a soldier named Jack Iron, of Company H., 33rd Iowa infantry regiment, and a man named William Stephens, of Washington, Iowa, were killed, as was Baggagemaster Charles Sullivan, of Muscatine, Iowa, who had been standing on the platform of the passenger car.

A passenger on the train described the accident to a reporter, 'The car was well filled with passengers, and the fright and consternation can be much more easily imagined than described. The moment I saw the forward car coming into ours I immediately sprang for the back door and was on the point of leaping into the creek. The car rocked and crashed as though it would capsize every moment, but it finally became stationary. The bridge was so broken and inclining to one side that it was very dangerous to get from the car onto the bridge. We lifted the women and children from the rear and had to carry them off. They were almost paralyzed with fright.'

As to the cause of the accident, the passenger stated that 'The railroad company are most criminal for gross neglect. The inhabitants of Ainsworth were perfectly exasperated. They regarded the bridge unsafe for a long time, and the company were aware of its insecurity. I went under the bridge to examine the woodwork. The props supporting it were completely rotten at the bottom and were like rotten stumps broken off. I secured some of the wood and have it with me as a sample.' [843]

Just over a fortnight after the Ainsworth accident, on 25th August 1865 a considerably worse one occurred in the state of Tennessee, United States, on the Richmond and Danville Railroad. The morning passenger train from Nashville to Huntsville reached Reynolds station, 67 miles from Nashville, on time at about noon, and there crossed an opposing freight train. Departing from there it was crossing a trestle over Richmond Creek, three-quarters of a mile south of the station, at about 12.20 p.m., when the structure failed under its weight immediately behind the engine and nine cars went down into the creek with a tremendous crash, pulling the locomotive with them. Immediate efforts were made to extricate the injured passengers and recover the bodies of the dead and a relief train was sent to the scene, presumably from Nashville. Up to the time that an informant to the *Nashville Dispatch* newspaper left the scene, 30 injured had been taken from the wreckage, together with twelve bodies. However, a car which had contained about 30 passengers remained at the bottom of the creek and all its occupants had of course perished. Engineer D.W. Magee was seriously injured and Fireman James

[842] The Dover Mails by Dan Garrett, pg. 26

[843] Philadelphia Enquirer, Philadelphia, Pennsylvania, United States, 24th August 1865

Deremer, of Oswego, New York, was scalded to death. Express messengers Mosely and Hazlett, of Pulaski, Tennessee, were killed and a third express messenger was injured. The injured were taken to Athens and Huntsville. The train had been well filled, and it was regarded as a miracle that anyone had escaped. The report mentioned imperfect timber being used to construct the trestle, implying that this was the cause of the disaster. Robert B. Shaw does not describe the disaster but gives the number of fatalities as 35.[844] [845]

A bridge collapsed beneath a train between Pedaso and Marrano, on the Tronto-Ancona line, in Ascoli province, Italy, on about 31st August 1865. Four railwaymen lost their lives. 'An official investigation has been opened upon the subject.' [846]

On 26th June 1867 India suffered the most serious railway accident to have so far occurred in that country. The scene was a point between Bhosawal and Khundwah, and ten miles from the former station, on the Great Indian Peninsula Railway. The railway here crossed a tributary of the River Taptee called the Sookee Nuddee or River Suttee, which for much of the year was a dry river but which during the monsoon season was fed by streams running from the Santpoora Hills and became a powerful river. On the day of the accident, it had swollen to a torrent 230 feet wide and 15 feet deep. A bend in the river increased the probability of it overflowing its banks when it was in flood and to deal with this possibility a separate flood stream had been constructed, 60 feet wide, to divert a proportion of the water from the main river channel. A larger bridge of seven 30 ft arches crossed the main river and a smaller bridge of two 30 ft arches crossed the flood channel, with a substantial embankment of loose shingle between the two structures. On the night of the accident the raging torrent tore away the embankment behind the abutment of the smaller bridge, leaving the rails suspended over a gap 50 ft long, 10 ft wide and 20 ft deep. Into this trap ran a passenger train consisting of an engine and tender and seven carriages, carrying a number of passengers variously estimated at from 57 to 112. It was dark, and the train driver had no warning of danger. The engine and tender somehow made it across the gap but the remainder of the train, with the exception of the rear van, fell into the river. It was officially stated that there were 13 dead or missing and 20 injured but the number of dead was generally believed to be at least five times as high. *The Times* in London was, however, able to reassure its readers that, 'With the exception of three Europeans in charge of the train all the passengers were natives.'

Sir Seymour Fitzgerald, the Governor of Bombay, ordered a thorough investigation into the accident. *The Times* correspondent added that, 'It is not yet known whether any blame attaches to the company in respect to this accident; the bridges are said to be sound enough now, though one of them is reported to have been very unsound lately. Several other of the bridges on the Great Indian Peninsular Railway are said to be in an unsafe state. The bridges across the Taptee viaduct on the Nagpore line are more especially commented upon; and on two separate occasions the Government consulting engineer

[844] Nashville Dispatch, Nashville, Tennessee, United States, 27th August 1865

[845] A History of Railroad Accidents Safety Precautions and Operating Practices by Robert B. Shaw, Appendix

[846] The Times, London, England, 4th and 7th September 1865

reported several arches to be giving way.'[847]

Thomaston, Connecticut, United States, was the location of a bridge accident on 5th September 1867. A Naugatuck Railroad passenger train which had left Waterbury at 7 p.m. for Winstead was crossing a bridge over the Branch River (referred to as the Branch Brook on a 2018 map) during a severe storm when the bridge gave way. The bridge must have been only a short one for the engine made it across the structure and only the following combined baggage and smoking car, which had become detached from the tender, went down, the remainder of the train stopping on the line, the coupling in the rear of the combined car having also given way. The car was swept down the river with its 16 or 17 occupants and into the Naugatuck River, being carried about three miles.[848] [849] As usual, accounts of the number of victims varied but a local newspaper reported the following day that six bodies had been recovered and four others were still missing. All the mail and express matter in the baggage section of the car was also lost. The survivors had clung to the rim of the car, which had lost its roof in the fall, and were rescued by citizens of Thomaston who had heard their shouts, had seized ropes and had been able to secure the car just in time to prevent it going over a dam at an old woollen mill below the village.

Among those who escaped from the car was Baggagemaster "Larry" McDermott, who years later related his experiences, 'The car was the first smoker ever run on the Naugatuck road. One end was partitioned off for express and baggage. There had been a number of passengers in the car earlier in the evening, but we had taken on a gang of workmen, and their strong pipes, not mixing well with cigars, had caused most of the passengers to go back into the other cars. I was sitting on a canvas trunk with Egbert Sperry of Wolcottville (now Torrington) beside me when the car dropped. Mr Skinner, the express messenger, was the only other person in the baggage compartment. We went down so suddenly that not a word was spoken by any of us. When I felt myself in the water, I made a spring for the side door, but I could not have opened it. Somehow the roof went off and I must have got out that way.

'I was more or less dazed and when I came fully to myself, I was struggling in the water some ways below the bridge and grasping at everything that I could find. I had just got hold of a timber when I heard a man behind me choking and gasping. I cried out 'Who is it?' and caught the word 'Skinner.' I said, 'Hold on a little longer,' or something like that, but there was no reply and next day, Skinner's body was found, not far from this place, and Sperry's near it, both caught in the alders. A moment or two after I spoke there was a lightning flash, and I saw the shore close to me. I reached it, and as I pulled myself out, I heard going past what I knew afterward to be the car with the men holding to it. They had not been thrown out at first when the roof went off as we three had been, because they were in the seats in the smoker and probably held on.' [850]

The Philadelphia and Erie Railroad suffered a serious accident on the night of Wednesday 8th July 1868 near Union or Union Mills, Pennsylvania, United States. The

[847] Ibid, 5th and 7th August 1867

[848] Sandusky Daily Commercial Register, Sandusky, Ohio, United States, 7th September 1867

[849] Philadelphia Inquirer, Philadelphia, Pennsylvania, United States, 7th September 1867

[850] Torrington Evening Register, Torrington, Connecticut, United States, 24th July 1900

company's express passenger train due at Warren, Pennsylvania, at 11 p.m. consisted of an engine and tender, a baggage car, two passenger cars, a sleeping car and 'several' emigrant cars. The bridge over French Creek gave way after the passage of the engine and tender and the following cars plunged into the creek, 'piling up in an inextricable wreck.' Five passengers lost their lives and a number of others were injured. Those passengers killed were Miss A.S. Brainard, of Warren, a milliner; Miss Priscilla Feese, of Flat Rock, Ohio, and her two children aged nearly 2½ and fourteen months; and the three-year-old daughter of Scott Patton of Erie City. In addition, Conductor John Boyce suffered serious internal injuries and Brakeman William Fury and Baggagemaster Charles Greedy were also hurt.

The failure of the bridge does not seem to have been due to the washing away of its supports but rather due to its inherent weakness. There was a suggestion of sabotage, as some of the victims had allegedly been robbed, but nothing was found to substantiate this hypothesis. The local newspaper, reporting on the accident, commented, 'We believe the bridge was one of the kind known as "Deck" bridges, the roadway being carried over the top, and a kind in which, with proper lateral bracing, the full strength and benefit of the materials used in construction are obtained. Notwithstanding the assertations made by the R.R. authorities, that a recent searching examination proved the bridge to be in a safe condition; the plain conclusion must point to some defect in the structure, which existed, in spite of their scrutiny.'

The newspaper concluded that, 'This casualty is one of the necessary attendants upon an imperfect and irresponsible system of railroading. As long as perishing materials are used in bridge construction and railroad officials are exempt from criminal accountability, for negligence and want of vigilance, these monster monopolies will claim their annual list of victims.' [851]

A westbound passenger train of the Chicago Burlington and Quincy Railroad left Ottumwa, Iowa, United States at 1.30 a.m. on Wednesday 14th July 1869, in charge of Conductor Miller, Engineer Ericsson and Fireman Davenport. It consisted of 15 cars, including baggage, passenger and sleeping cars and freight cars loaded with lumber. The fifth car had been chartered by a Mr Herriott to Russell station; he was moving with his family from Princeton, Illinois, to Taylor County, Iowa, and the car was carrying his five horses and personal effects. Mr Herriott was also riding in the car with four of his children, three girls and a boy, ranging in age from seven to thirteen. Mrs Herriott was coming by the next train with their other two children. There were about 15 other passengers on the train.

Much rain had fallen in the previous 48 hours, and it had been raining very hard in the hours before the train's departure. One of the other passengers on the train, who had joined the train at Ottumwa, bound for Chariton, had considerable reservations about making the trip, fearful that the floodwater might have damaged the track, but in the end reluctantly decided to do so.

The train left Ottumwa on time and was due at Albia, Iowa, at 3.47 a.m. but did not leave there until 4.49 a.m., having run cautiously because of possible problems with the

[851] Warren Mail, Warren, Pennsylvania, United States,14th July 1868

track. Two miles west of Albia the line crossed Coal Creek by a pile bridge, (Bridge No. 75), 45 feet long, the piles having been driven 20 feet into the ground. The previous summer the railroad company had dumped 14 car loads of stone into the creek around the piles to guard against the foundations being washed away. The morning before the day of the accident the section boss had informed Mr Potter, the Albia station agent, that the track was not in good condition. Mr Potter duly went out and inspected the track and found that the sand ballast had been washed away but the bridge appeared to be safe. He flagged an eastbound train, which passed over the bridge without any problem, as did two further trains, one of them with eighteen cars of livestock. The bridge still appeared to be safe and the station agent no doubt retired for the night with no misgivings.

Unfortunately, during the night the creek rose to an unusual level and the foundations of the centre piles of the bridge were undermined. When the westbound train passed onto it at about 5 a.m. it collapsed and the engine, tender and seven cars went into the creek. Had the train not been going very slowly the remaining cars would also have gone into the creek. Because the passenger cars were at the rear of the train the other passengers escaped, but Mr Herriott and his four children went down with the bridge, as did the engineer and fireman. Mr Herriott and his 11-year-old daughter chanced to be awake and were able to prize open a door; they were eventually rescued by local residents, who secured the car to the shore with ropes. The rescue had taken several hours as Conductor Miller and others had first had to travel three miles to Albia town for ropes and assistance. Before they had returned those left at the scene had tried to construct a raft but lacking ropes to tie it together it had swept away almost at once by the strong current. The other three of the Herriott children, who had been asleep, drowned inside the car. The engineer and fireman survived, the fireman managing to cling to a tree and the engineer being washed some distance downstream. Fireman Davenport suffered exposure due to his prolonged stay in the water, Engineer Ericsson was more seriously injured. Mr Herriott and his surviving daughter were little hurt, but the father was naturally extremely distressed at the loss of three of his children. He also lost much of his property, though three of his horses were saved. Shortly after the accident the Burlington company paid him $1,000 in agreed compensation and subsequently the company reportedly paid him an additional $700 as well as paying him for the expenses of searching for his children and their burial.[852]

Another bridge accident in India on about 1st November 1870 resulted in an unknown number of casualties. About 130 miles from Madras (now Chennai) the north west branch of the Madras Railway bridged the Cheyair (now Cheyyeru) River by a long iron girder bridge of 30 spans of 60 feet each, the girders resting on solid masonry piers. The bridge was situated between Rajampett (130th mile) and Nundaloor (137th mile) stations. On the evening of the day of the accident torrential rain had swollen the river to the extent that it had completely washed away several of the piers and the girders supported by them had fallen into the river.

The night train departed from Madras at 7 p.m. for Cuddapah (now Kadapa) and Gooty, its first and second-class carriages nearly full of passengers. It arrived at the bridge with no one aware there was anything wrong and fell into the torrent. Fortunately,

---

[852] Burlington Daily Hawk Eye Gazette, Burlington, Iowa, United States, 15th and 16th July 1869

it had not been travelling very fast and it seems likely that the rear of the train remained on the track, though the newspaper report does not say. The engine driver, three firemen, an under-guard and a number of passengers drowned. The passengers who lost their lives included the Rev. Mr Seymour, Mrs Seymour and their child, who had been on their way to Secunderabad, and a barrister of the Madras High Court.[853]

Floodwaters following torrential rains undermined a bridge at Aidin (now Aydin) in Turkey causing a bridge to fail beneath a train of 15 carriages on about 29th January 1871. Four people lost their lives in the accident.[854]

A passenger train consisting of an engine and tender, a baggage car, two passenger coaches and a sleeping car left Nashville, Tennessee, United States, on the night of 3rd July 1871, on the Nashville and Northwestern Railroad, in charge of Conductor Landis. There were 14 passengers in the first passenger coach, about thirty in the second, ladies, car and three in the sleeping car. Running at its usual speed the train travelled as far as a bridge over the Harpeth River, sixteen miles from Nashville and close to Kingston Springs. The engine, tender and baggage car negotiated the bridge safely, but the centre pier of the structure gave way beneath the leading passenger car, which fell twenty feet to the river below, landing on its side in eight feet of water. The following ladies' car fell end up onto the passenger car, its passengers all falling to the front end of it and the seats falling onto them. The sleeping car lodged itself against the upper end of the ladies' car. In the front car a Mr John Campbell of Nashville and another unnamed passenger were the only two who were free to move; between them they were able to extricate another eight passengers. The other four passengers were either killed in the fall or drowned. In the ladies' car, thirteen of the occupants were killed outright, their bodies piled up with those of the injured passengers. The three passengers in the sleeping car were only bruised and assisted in the rescue work.

The engine and baggage car ran forward to Kingston Springs station where the operator was woken up and telegraphed news of the disaster to Colonel Thomas, the superintendent of the line, who received it at about half an hour after midnight. The superintendent immediately set off for the scene with the wrecking train and a body of physicians.[855]

As always, with such disasters, there were tragic stories to tell, 'Mr Charles Campbell, of Edgefield, was an invalid with consumption; had spent last winter in Florida and was on his way with his wife and child to spend the summer at the Springs. This reporter called to see him and family last evening, and conversing with Mrs Campbell, while their beautiful little child was playing on the grass in the yard, heard her speak hopefully of the future. They had not then fully determined to go on the evening train. But soon after, Mr Campbell thought he was strong enough, even after the labors of the day, to go. In two hours afterwards they were dead. Mr Campbell's sisters and brothers were going to visit them at the Springs to-day and enjoy a sort of picnic. They went to the depot to take the train, at 4 o'clock, but were horrified at meeting there the dead bodies of the dear brother

853 The Times, London, England, 22nd November 1870

854 Ibid, 3rd February 1871

855 The Daily Sun, Atlanta, Georgia, United States, 12th July 1871

and sister they were going to surprise.'

The newspaper report added that the Nashville and Northwestern had for years been known to be unsafe to travel on but since its lease to the Nashville and Chattanooga Railroad it had been undertaking repairs and had for some time been considered as safe as any road in the country. The bridge in question had been carefully examined and was thought by all to be perfectly safe.[856]

A rather unusual accident occurred at Navasota, Texas, on 29th July 1871. A construction train, which included a wrecking car, was passing over the Navasota river bridge at about 4 p.m. when the derrick of the wrecking car, which could not have been properly secured, struck the lateral braces and rods of the bridge, causing it to collapse, and the engine and nine cars to fall 40 feet into the river. Eighteen labourers had been riding on the train and five of them were killed outright and a white man and a negro were fatally injured. None of the names of those killed was known at the time of the newspaper report. Engineer Patrick Lanigan, Fireman R.A., Murphy and Conductor Rom Lynch were dangerously injured, and several other men were hurt.[857]

On 24th January 1872 floodwaters undermined a four-arch viaduct across the Brague valley two kilometres from Antibes, France, on the Chemins de fer de Paris à Lyon à la Méditerranée. At 6 p.m. it collapsed beneath omnibus train No. 492 from Menton to Grasse. Six people died and ten were more or less seriously injured.[858] [859]

Spain suffered another bridge disaster on 9th September 1872. The accident involved the Barcelona – Valencia mail train which met with disaster at the Riera de San Jorge Bridge between Hospitalet and L'Ametlla stations and three kilometres from the latter. Floodwaters from a storm which had broken some hours earlier had undermined one of the piers of the bridge which collapsed under the train. The engine and seven of the nine carriages of the train fell about 25 feet into the water. A broken coupling saved a third-class carriage and a brake van at the rear of the train or possibly just the rear van (accounts vary). About 17 people lost their lives and 25 were injured. Some of the bodies were washed out to sea.[860] [861] [862]

A correspondent in Santander described the disaster for *The Times* in London, and wrote, "All that remained on the line was the luggage van, with five fortunate persons in it – an official and four gendarmes, who behaved with singular self-devotion and courage in trying to save the lives of the passengers during the three hours they remained alone and unassisted. Sixteen corpses, including those of General Smith (a Spaniard) and the Military Governor of Tortosa, with his entire family, have been recovered. Several of the

---

[856] Ibid.

[857] Galveston Daily News, Galveston, Texas, United States, 30th July 1871

[858] The London Times 24th and 29th January 1872

[859] Wikipedia - Liste des accidents ferroviaires en France au XIX siècle

[860] The Times, London, England, 10th, 11th and 17th September 1872

[861] Webpage - Accidentes Ferroviarios en España by Manuel J Marcos Montero, pg. 41

[862] Wikipedia – List of Rail Accidents in Spain

corpses were terribly crushed; in others death appeared to have been caused by drowning only. The bodies of several passengers remain in the water and are difficult of recovery. A first and a second-class carriage also remain submerged. In one of the first-class carriages two ladies – mother and daughter – were found dead, locked in a close embrace, their eyes full of tears." [863]

The bridge across the Kelani River in Ceylon was a substantial affair, 800 feet long with eight spans of 60 feet built on screw piles and twelve spans of 25 feet on brick piers. Following monsoon rains it collapsed beneath an engine on 20th September 1872. The report did not state if the engine crew survived the accident.[864] The structure was rebuilt with plate girder spans on cast iron piers.

At 3.15 a.m. on the morning of 19th April 1873 there were two passenger trains waiting to proceed from Stonington Junction, Connecticut, United States, eastwards over the single line of the New York Providence and Boston Railroad to Providence, Rhode Island and Boston, Massachusetts. One of them was the slightly delayed 2.30 a.m. Stonington steamboat train, which carried passengers who had arrived by the steamer *Stonington* from New York to Stonington onwards by rail to Providence and Boston. The other was the 3.15 a.m. Shore Line express passenger and mail train from New York which allowed passengers to travel directly between New York and Boston by rail apart from a ferry over the yet to be bridged Thames River between New London and Groton. Conductors Thomas Sprague of the express and Orrin Gardiner of the boat train were engaged in a heated argument as to which train was to go first. Eventually Switchman J.W. Thayer let the boat train, which was in any case scheduled to leave first, have priority, and it departed, with the Shore Line express following ten minutes behind.

Fourteen miles east of Stonington was Richmond Switch, Rhode Island. In the following year this would become Wood River Junction, with the opening of the 5-mile-long branch to that place, but at the time of the accident it just a siding where trains could cross one another. In the vicinity of the siding was an insignificant wooden bridge, 20 feet in length, over Meadow Creek, a tributary of the Pawcatuck River, itself a tributary of the Wood River. Built between 1833 and 1837 the bridge had piles going six feet below the bottom of the creek and was considered as solid a structure as any on the line. About 100 yards up-stream of the railway bridge was a dam across the creek, constructed to provide waterpower for a grist mill. On previous occasions when heavy rain had turned the stream into a torrent the mill dam had overflown. On the evening before the accident, a Mr Kenyon, who had been running the mill during the day, seeing the mill pond getting too high, had left the gates of the mill open to drain some of the excess water before going home. This, apparently, had not been sufficient, and the build-up of water had eventually caused the dam of boulders and dirt to fail. The mass of water, timbers and boulders had then rushed downstream and the small railway bridge not being large enough to carry the flood a second dam had formed just above the bridge. Water overflowing this second dam had in turn undermined the supports of the bridge and it had

[863] The Times, London, England, 17th September 1872

[864] Webpage - An Historic Journey in 1864 – The Sunday Times, Sri Lanka, 3rd January 1999

collapsed, leaving the track hanging over the gap.

The boat train was hauled by an engine named *N.F. Dixon*, in the charge of Engineer William Guile or Guild and Fireman George Eldred. It consisted behind the engine and tender of three platform cars loaded with baggage, a second-class car, four first-class cars, of which one was a smoking car, and a caboose. Its occupants had of course no knowledge of the bridge collapse which had happened probably an hour earlier. The engine crossed the gap before striking the far embankment and falling back into the gap, followed by the three platform cars, the second-class car and two first-class cars. The third first class car, the smoking car and the caboose remained on the line. The engineer and fireman died either from being crushed between engine and tender or from severe scalds. Shortly after the accident fire broke out in the wreckage, probably started from an overturned stove in the second-class car, which was mainly carrying sailors and emigrants to Boston. The fact that the passenger cars fell on top if the three platform cars did however save many of the passengers from drowning in the flooded creek and the provision of the cars with Miller platforms meant that none of them was telescoped.

Further danger threatened because the Shore Line express was following only ten minutes behind the boat train. Conductor Gardiner was reportedly trapped between two seats in the wreckage, but Brakeman Walter Monroe was able to run back down the line with a red lantern and stop the following train. The express was then able to move slowly up to the scene of the accident and pull the rear cars of the boat train back clear of the fire. The remainder of the train was destroyed by the fire and there was little left of the bodies of some of those who had lost their lives. It was thought that in addition to Engineer Guile of Providence and Fireman Eldred of Kingston five passengers lost their lives; Albert Allen of Providence, Jerry Creamer of Boston, John Callahan of New York and two men who could not be identified, who were believed to have probably been emigrants. About 25 other passengers were seriously injured and another 20 slightly hurt.

Rescue operations were quickly begun after the accident and by 11 a.m. the same morning the first train was passing over a new bridge. The rest of the day was spent retrieving the bodies of the dead, clearing the wreckage and dragging the stream. The injured were removed to Providence and the bodies left where they were until later in the day for the inspection of the Kingston coroner, Mr Clark, who decided that no inquest was necessary. Had the well-filled Shore Line express preceded the boat train instead of following it the disaster would probably have been much worse.[865] [866] [867]

At dawn on Thursday 24th September 1874 a southbound passenger train on the Selma Rome and Dalton Railroad went through Waxahatchie Creek Bridge, Alabama, United States, falling sixty feet and being completely wrecked. The engineer, fireman and several passengers were killed. and the remaining passengers were injured, many seriously. The dead included William Boyd, a former judge of the Alabama Supreme Court.[868]

---

[865] Boston Globe. Boston, Massachusetts, United States, 21st April 1873

[866] Philadelphia Inquirer, Philadelphia, Pennsylvania, United States, 21st April 1873

[867] Webpage – The View from Swamptown: George Eldred and the Wreck at the Richmond Switch

[868] Boston Post, Boston, Massachusetts, United States, 25th September 1874

The wrecking of a Memphis and Charleston Railroad express train due to the collapse of a bridge over Spring Creek, near Tuscumbia, Alabama, United States on 22nd November 1874 was but part of a wider natural disaster, for the town was struck by a powerful tornado. The train was going very slowly as it approached the bridge but due to a curve the engineer barely had time to reverse his steed before it left the structure, turning a complete somersault in its 50 feet fall. The mail and baggage cars also fell from the bridge but the breaking of the coupling at the rear of the baggage car saved the passenger coaches behind it from also going into the creek. The engineer, who was severely scalded, was the only casualty. The solid rock bed of the creek meant that the engine was a complete loss. At least twelve people in the town were killed by the tornado and the town was almost entirely destroyed.[869]

At 10.30 p.m. on the evening of 6th July 1875 a mixed train departed from Valparaiso, Chile, bound for the capital, Santiago, where it was due to arrive at 6 a.m. the next morning. It was in charge of Conductor Manuel Valdivieso Araos and Driver Francis MacCabe, an Irish immigrant, and consisted of an engine and 14 cars. There were six palanqueros (descendants of escaped slaves who spoke a Spanish-based creole language) acting as brakesman under Head Brakeman Severo Padilla and twenty-three second and third-class passengers, mostly market vendors. There was also just one first-class passenger, a French artist named Cheri Labrocaire. The passenger cars were marshalled at the rear of the train with the third-class car in front, followed by the second-class car and with the first-class car at the very rear of the train. The goods carried on the train included 979 bundles of valuable merchandise; mirrors, pianos, silk goods and so on. There was also a freight car loaded with containers of paint and paraffin together with a second car loaded with coke for iron ore smelting.

Disaster struck the train at 11.45 p.m. between Peñablanca and Limache stations as it was crossing the Limache Bridge, an apparently strong 60-metre-long iron girder structure supported by three stone piers. The centre pier of the bridge failed and the whole of the train was precipitated into the river except the engine and one freight car. The cars piled up in a pyramid on the floor of the stream below with the passenger cars on top of the freight cars and the first-class car at the top. Broken oil lamps from the passenger cars then set fire to the wreckage and fed by the paint, paraffin and coke in two of the freight cars the fire burnt fiercely. M. Labrocaire, the French artist, managed to get out of his carriage and assisted by an uninjured brakeman was able to rescue six of the passengers, including two women. Meanwhile, Driver MacCabe hurried with his undamaged engine to Limache for help and at 1 a.m. on the Wednesday morning the Limache governor, José Orrego, returned with a rescue party which included a doctor and a priest. The governor wrote later that there had been little that they could do.

At dawn, Driver MacCabe telegraphed his superiors from Limache with news of the disaster, "I have just returned from the scene of the accident. The whole bridge is completely burned. Only one complete car is saved plus some odd items. The goods are almost all burned.

"The conductor, the head brakeman and two palanqueros are missing.

[869] Madison Wisconsin Journal, Madison, Wisconsin, United States, 27th November 1874

"I do not know how many of the passengers are missing. Only one has been found drowned. Everything else is a pile of ashes. F MacCabe."

It was eventually determined that five passengers, three brakemen and Conductor Araos had died in the accident. Two of the surviving passengers were slightly injured. Another passenger who escaped death had bought a third-class ticket to Limache at Peñablanca station but had then fallen asleep and missed the train. The accident was also very expensive financially with a loss of 1,00,000 pesos of merchandise and about 35,000 pesos worth of damage to equipment.

Crowds of spectators went to the scene the day after the accident and at times could only be controlled by prodding them with rifle butts.

Several questions were raised by the disaster, most obviously why had one of the apparently solid granite piers suddenly failed? There were also questions regarding how one of the oil lamps had set fire to the wreckage and as to why two cars of potentially inflammable materials had been marshalled in the centre of the train.[870] [871]

Between 11 p.m. and midnight on 6th March 1876 a Baltimore and Ohio Railroad freight train with a passenger car attached was crossing a bridge 'over a narrow passage' at a point stated to have been 65 miles from Harpers Ferry, West Virginia, United States, on the Valley branch, which must have been somewhere near Bowman, Virginia. The middle span of the bridge gave way and the whole train fell through it. Engineer F Donovan, of Sandy Hook, Maryland; Brakemen D.J.D. Chapman and F. Jefferson of Winchester, Virginia and eight or nine passengers lost their lives. Half a dozen other passengers were injured.[872]

The next accident to be considered, on 29th December 1876, was the worst railroad accident to occur in the United States in the whole of the nineteenth century. It happened on the Lake Shore and Michigan Southern Railroad, one of the constituent companies of which, the Buffalo and Erie Railroad, had already suffered 'the Angola Horror' nine years earlier. The scene of the catastrophe was Ashtabula, Ohio (by analogy with the Angola accident the event would later be termed 'the Ashtabula Horror) and the train involved was passenger train No. 5, the westbound *Pacific Express*.

The bridge at Ashtabula consisted of a single 150 feet long wrought iron Howe Truss span, carrying two tracks 76 feet above the shallow Ashtabula River and suspended between two solid masonry abutments. The design had been developed in the 1840s by William Howe, hence its name, together with his three brothers in law, Amasa Stone, Azaria Boody and Daniel Harris. The design was popular with railroad companies because it was very strong over long spans and easier to construct than other designs. Such bridges were usually constructed in wood or a combination of wood and iron but the Ashtabula one was constructed entirely of wrought iron. Amasa Stone was the president of the Cleveland Painsville and Ashtabula Railroad company, the Lake Shore predecessor which originally owned the line, and chose to have an entirely iron design, a

---

870 Webpage – La Ruta Fatal by Julio Hurtado (in Spanish)

871 The Times, London, England, 30th August 1875

872 Fort Wayne Weekly Sentinel, Fort Wayne, Indiana, United States, 15th March 1876

choice probably not unconnected with the fact that his brother was a partner in the company which provided the bridge beams. The detailed drawings for the bridge were done by a Joseph Tomlinson, who later resigned or was fired after a disagreement with Stone over the stress allowable on the compression elements of the bridge, Tomlinson believing a smaller amount of stress should be allowed. Tomlinson also later testified that the iron used in the bridge had been of a smaller size than that specified but it had been used anyway.

The express consisted of two engines and eleven coaches consisting of two express cars, two baggage cars, three passenger coaches, one of them a smoking car, a drawing room coach and three sleeping cars. The train was in charge of Conductor B. Henn. Driving the pilot locomotive, the *Socrates*, were Engineer Dan McGuire and his fireman (name not recorded) while the train engine, the *Columbia*, was in the charge of Engineer G.D. Folsom and Fireman Peter Levenbroe.[873] A heavy snowstorm was raging in from Lake Erie as the train made its way westwards, and with a twenty-inch snowfall and visibility limited to 50 feet it was only travelling at 12 or 14 miles an hour, and was running two and a half hours late.

Engineer McGuire and his fireman heard a cracking sound just as the pilot wheels of the *Socrates* were approaching the far abutment of the bridge and they could feel their engine beginning to sink. McGuire instinctively opened his throttle wide and the *Socrates* shot forward onto dry land, dragging its derailed tender with it. The coupling behind the tender broke and glancing backwards the engineer glimpsed the second engine and the cars of the train plunging into the ravine one after the other as they reached the bridge, the whole span of which had collapsed. Almost at once upset stoves in the coaches set them ablaze in several places.

A local newspaper related events after the accident, 'The scene which followed could have no description that would portray its awful horrors. The crash of the falling bridge was heard for over a mile, and the lurid light that reflected on the sky a moment later brought men with ropes and buckets to the rescue. Those who could be picked from the wreck, about fifty in number, many of whom were fatally injured, were rescued before the flames reached them, and among these were the conductor and train boy. Heartrending shrieks from victims that could not be reached rent the air and froze the blood of strong men, who would gladly have risked their own lives but were powerless to save.'

'Of those on board a large majority perished. One man relates that he was in a double seat with three others over a game of cards, and when the train went down two were instantly crushed to death and the third terribly injured. He alone was saved. Two little babies were rescued, but one in an almost lifeless condition. In another place fire and smoke came in such volumes and so suddenly that the men were driven back just as they were on the point of reaching a woman and child, who were wedged in a mass of timbers, but still alive. One man was fastened in near the engine, and there, for full thirty minutes, heroic endeavors were made to loosen the weight above him but in vain. Of those taken out many have already died.'

'Citizens have thrown open their houses, and the sufferers are receiving all the care that human interest can bestow. Two were at once were conveyed to the residence of the Mayor, whose efforts to save and care for the injured were untiring. The latest estimate

---

[873] Webpage -Ashtabula Railway Historical Foundation: Ashtabula Bridge Disaster Archives

places the number killed at 100.' [874]

The fire engine *Lake Erie* soon arrived from the village of Ashtabula, but in the chaos prevailing after the accident confusion prevailed as to the priorities, and some of the railwaymen told the firemen to help with the rescue of the victims rather than fight the fire. Thus, although there was plenty of water available the hose of the fire engine was not attached to the steam pump in the pumping engine, the fire was soon completely out of control and lives were lost. This led to unfounded rumours and talk later that the railwaymen had been keen that all the bodies should be completely consumed by the fire so that the number of victims could not be determined. Adding credence to the story was that the mayor and many of the inhabitants of Ashtabula worked on the railroad.[875] A second fire engine, *Protection*, had to be hauled a mile through a snowstorm over almost impassable roads and arrived too late to save lives, but was not used to try and subdue the fire.[876]

The final number of dead in the catastrophe according to what appear to be the most reliable figures was 98, comprising 90 passengers, two express messengers, three porters, a baggageman, Sleeping Car Conductor Harry Wagner and Fireman Levenbroe on the *Columbia* engine.[877] The toll would not be exceeded in the United States until the bridge accident at Eden, Colorado in August 1904. Among the fatalities, was the evangelist and Gospel hymn composer Philip P. Bliss, who had initially escaped the wreck but died when he returned to try and save his wife. No trace of either of their bodies could be found.

The catastrophe shocked the nation and, in addition to the usual inquest, a special committee of the Ohio Legislature was set up to examine it while there was also a third investigation. The coroner's jury sat for 68 days and returned a verdict that 'the bridge had fallen as a result of defects and errors in its design, construction and erection, and that a major defect in the design was that every member depended for its efficient action upon the probability that all or nearly all of the others would retain their position and do the duty for which they were designed, instead of giving to each member a positive connection with the rest, which nothing but a direct rupture could sever. The members of each truss were, instead of being fastened together, rested one upon the other, as illustrated by the following particulars: the deficient cross-section of portions of the top chords and some of the main braces, and insufficient lugs or flanges to keep the ends of the main or counter braces from slipping out of place, in the construction of the packing and yokes used in binding together the main and counter braces at the points where they crossed each other in the shimming of the top chords to compensate deficient length of some of their members; in the placing, during the process of erection, of thick beams where the plan required thin ones, and thin ones where it required thick ones.' The jury also blamed the railway company for failing to conduct an examination of the bridge by a competent bridge engineer during the eleven years it had been in use, which said examination would have revealed its defects, and for failing to use a heating apparatus constructed such that its fire would be immediately extinguished should the car overturn.

---

874 Cambridge News, Cambridge, Ohio, United States, 11th January 1877

875 A History of Railroad Accidents Safety Precautions and Operating Practices by Robert B. Shaw, pgs. 55-6

876 Webpage – Ashtabula Railway Historical Foundation: Ashtabula Bridge Disaster Archives

877 Ibid.

The fire brigade was criticised for failing to use either of its engines on the blaze.[878]

The findings of the Ohio investigating committee paralled those of the jury. Three eminent engineers advised the committee, very likely they also advised the coroner's jury. The committee found that the failure had occurred in the second or third panels of the south truss, with either the top chord or the compressive brace initiating the failure. The load had been appropriate for the bridge, but its design had been faulty in several respects. It was excessively heavy. The compressive diagonal braces were of different sizes mixed together. The separate beams were not tied together continuously, the end bearings were faulty, and there were no positive mechanical connections between the braces and the angle blocks to prevent movement of the ends of the braces. The compressive top chords were also of mixed sizes and not tied together correctly. They were bent by train weight and were only braced laterally at every other panel point. The x-bracing in several areas was inadequate. The angle block castings should have been continuous.

In conclusion, the engineers found that the bridge failure was not due to defective iron or sudden cold weather but to a construction which was not in line with established engineering principles. The specific cause of the collapse was a fatigue crack which had originated in a small air hole end had grown with repeated stress over eleven years. The strength of that component had been reduced by the cold weather and the weight of the train had caused a brittle crack to form and the structure to fail.[879]

The third investigation by the American Society of Civil Engineers came to similar conclusions as the previous two, though engineer Charles MacDonald concluded that an air hole in a faulty iron lug had reduced the strength of the bridge by half and that otherwise it might not have failed.

A few days after the accident Charles Collins, the Lake Shore chief engineer, who should have inspected the bridge but never had, committed suicide by blowing his brains out. Cornelius Vanderbilt, the aging Lake Shore president, also died four days after the disaster, his death probably hastened by it. Amasa Stone, on whom the chief responsibility for the catastrophe was placed, was defiant of public opinion and was never formally charged or prosecuted with negligence in connection with the affair, but also did away with himself several years after the accident, very likely because it had weighed upon his mind. The legislative committee made a number of recommendations relating to design codes, design loads, allowable stresses, minimum strengths, design review by experts, supervision of bridge construction and periodical expert examinations of bridges, but none of them became law. The catastrophe cost the Lake Shore and Michigan Southern company $495,722. A wooden replacement bridge was brought into use on 17th January 1877.[880] Ten years after the catastrophe the Lake Shore finally introduced steam heating on its passenger trains and by 1900 many of the long bridges on its line had been replaced by embankments and culverts.[881]

A goods train of 14 loaded waggons was crossing a bridge over the River Aluta, on

---

878 Ibid.

879 Webpage – I-35 Bridge Collapse/Ashtabula River Railroad Disaster

880 Ibid.

881 Webpage – Ashtabula Railway Historical Foundation: Ashtabula Bridge Disaster Archives

the line from Bucharest and Crajova, Romania, on the evening of 20th May 1877 when the structure failed and ten of the waggons fell into the river. *The Times* correspondent in Bucharest wrote that 'Several persons are killed, but the details are not yet known here.'[882]

The Ashtabula catastrophe was rather unusual as regards its cause. The next American bridge disaster, in Iowa, on 29th August 1877, had a much more typical cause, a bridge was washed away. The disaster occurred about eight miles east of Des Moines at a point where the Chicago and Rock Island Railroad crossed the Little Four Mile Creek by a stone arch bridge resting on five feet thick abutments and twelve feet above the water. The creek was normally a placid little brook but unprecedented rainfall in the hours before the accident had transformed it into a raging torrent. It was thought that the swollen stream had washed away the bridge some time before the train reached the spot, leaving the track hanging over the gap. The bridge was approached around a curve and Engineer Billy Rakestraw probably had no warning of the disaster.[883]

The disaster occurred at about 4 a.m. The engine ended up on the far side of the gap, half buried in the earth. A Barnum's Circus car was behind the engine and went into the stream. Next came the baggage and mail car, which passed over the circus car, smashing it to pieces. The first passenger car pitched head down into fifteen feet of water, the next passenger car telescoped it and the third passenger car telescoped both of them. The gap was now completely filled with wreckage and there was no room left for the rear sleeping car, which remained on the line.

A newspaper report of the disaster stated, 'The scene at the time of the accident is described as having been terrific. The rain was falling in torrents, accompanied by wind, lightning and violent thunder. The crash put out the lights and the scene of terror ensuing may well be imagined. Men who were not injured and could get out went at once to rescue the living and the wounded. They had to go a mile to a farmhouse to get axes to chop them out. There were many pitiful scenes and tender incidents. One mother was killed sitting between two children who escaped unhurt. One little girl, who had lain in the water for four hours, with a heavy man lying dead beneath her body, was discovered to be breathing, and was rescued and restored and now shows no signs of injury. The dead were brought here {Des Moines} on a train, reaching here at eleven o'clock. Some twenty-five of the wounded came with them. The officers of the road were at the wreck all day caring for the dead and wounded.' [884]

Eighteen people died in the disaster, among them Engineer Rakestraw. The dead included eight circus employees. Another fatality was a Thomas Dunaway of East Des Moines, who was possibly riding on the train as a trespasser, as he had reportedly recently run away from the local lunatic asylum.[885]

The year 1879 began with what was possibly the worst railway catastrophe in the

---

[882] The Times, London, England 22th May 1877

[883] Burlington Daily Hawkeye, Burlington, Iowa, United States 30th August 1877

[884] Ibid.

[885] Madison Wisconsin State Journal, Madison, Wisconsin, United States, 30th August 1877

world up to that date. It occurred in Turkey, on 11th January, on the line between Philippopolis and Adrianople. The previous year, Turkey had been defeated in a war with a coalition of Russia, Bulgaria, Romania, Serbia and Montenegro[886] and a Russian military train was involved in the accident. A bridge across the River Arda (also spelt Orda) collapsed under the train and a Russian general, several officers and about 200 men were drowned when the first and second-class carriages were precipitated into the river. The engine and three third-class carriages escaped. 'Communications with Constantinople have been interrupted for some days in consequence.'[887]

At 6 a.m. on the morning of 18th February 1879 a northbound train on the Selma Rome and Dalton Railroad went through a bridge over Mulberry Creek, thirteen miles from Selma, Alabama, United States. The train consisted of an engine, one freight waggon loaded with 60 bales of compressed cotton, mail and express cars and two coaches. The engine passed over the bridge safely, but the remainder of the train went down. The rear end of the sleeper car overhung the bridge abutment, allowing its occupants to escape. The wreckage caught fire and all the cars other than the rear sleeper were totally consumed. The creek was much swollen, but the bridge was also described as having been 'perfectly rotten.'

Killed outright in the accident were George R. Evans, tax assessor of Dallas county, and two or three unnamed negro passengers. Conductor White, General Superintendent M. Stanton and Mail Agent N.Y. Hunter were believed fatally injured and two brakemen, a baggagemaster, express messenger, porter and 14 passengers less badly hurt. It was estimated that it would take a month to repair the bridge.[888] [889] [890]

A severe storm which affected Sicily and southern Italy on 28th September 1879 washed away a bridge on the railway between Ciro and Cruceli. A passenger train fell into the torrent, killing the engine driver, while all the passengers were more or less injured.[891]

An eastbound train consisting of an engine, 19 cars of cattle and hogs (pigs) and a caboose, on the St Louis Kansas City and Northern Railroad, was crossing a bridge over the Missouri River at St Charles, Missouri, United States, at about 8.30 p.m. on 8th November 1879 when it met with disaster. The train was an extra and was in charge of Conductor Frank Kinley with Engineer John Ploughman and Fireman William Teneyeck. As the engine neared the end of the bridge the engineer heard the structure collapsing behind him and pulled the throttle wide open. The jerk broke the coupling pin behind the second car of the train so that the locomotive and two cars got clear of the bridge, but the remainder of the train fell 80 feet to the river. The men on the locomotive could hear the

---

886 Wikipedia – Russo-Turkish War (1877-1878)

887 Lloyds Weekly Newspaper, London, England, 26th January 1879

888 Columbus Daily Enquirer Sun, Columbus, Georgia, United States,19th February 1879

889 Birmingham Iron Age, Birmingham, Alabama, United States, 19th February 1879

890 Boston Post, Boston, Massachusetts, United States, 19th February 1879

891 Stevens Point Daily Journal, Stevens Point, Wisconsin, United States, 4th October 1879

cries of the men and animals in the water but realising that there was nothing that they could do to help them hastened to St Charles for assistance. Conductor Kinley chanced to be riding on the engine and so escaped but two passengers, Joseph Weaver, a wealthy stock-raiser and president of the First National Bank of Malvern and his friend, a Mr Hyde, who was accompanying Mr Weaver to St Louis, went into the river and were killed. Also killed was Brakeman George (or Joseph) Barndhart, while Brakeman Charles Irwin was dreadfully injured and died the next day. Drover John Summers suffered a broken leg or thigh, drover Fred Davis sustained only slight injuries and stock raiser J.M. Strahn, of Malvern, jumped from the caboose onto the No. 2 pier of the bridge at the very last moment when he heard the crashing of iron and timbers and the splashing of the water. 'His escape was one of the most wonderful on record.' All the stock except the animals in the first two cars was drowned.

The bridge was an iron Howe Truss structure like that at Ashtabula, but with three spans. It was the west span which failed, both ends apparently giving way at the same moment so that the span ended up in the river in an almost horizontal position, still filled with cars. It was fortunate that the bridge failed when it did, for the St Charles accommodation train had crossed it not long before. Why it had failed was unclear. C. Shallersmith, of the Baltimore Bridge Company, who had personally supervised the construction of the bridge, suggested that it had been damaged by a derailed vehicle of the train. However, Conductor Kinley had testified before the coroner's jury that there had been no prior derailment of his train. There was also a suggestion that the structure had been struck by lightning, there having been a severe storm an hour before the accident. The conductor also told the coroner's jury that he had taken hundreds of heavier trains over the bridge with no problem and that it had been recently inspected by a competent person and declared sound. He could give no explanation for its failure.

The cost of the damage to the bridge was estimated at $75,000 and the loss of cars and livestock at $25,000. A temporary track was laid to a ferry across the river for the transfer of passengers and freight while a temporary wooden truss bridge was erected.[892]

On Monday 29th December 1879, *The Times* in London advised its readers of a 'Dreadful Accident on the Tay Bridge' and the 'Loss of a Passenger Train' and proceeded to give the first account of what probably remains the most famous railway accident in world railway history, a sort of railway equivalent of the sinking of the *Titanic* ocean liner in 1912.

The direct railway route northwards from Edinburgh, Scotland, to Dundee, Montrose, Arbroath and Aberdeen had to cross two broad river estuaries or firths, the Firth of Forth and the Firth of Tay. Neither estuary was bridged for many years after the opening of the lines between the rivers, the river crossings being provided by ferries, one across the Firth of Forth from Granton to Burntisland and the other across the Firth of Tay from Ferry Port on Craig to Broughty Ferry. These two ferry crossings made the journey between Edinburgh and Dundee slow and inconvenient and there was an alternative, if considerably longer, all rail route via Larbert, Stirling and Perth. Furthermore, the direct route was owned by the North British Railway company and the indirect route via Perth by the Caledonian Railway company. There was therefore considerable interest in bridges

---

[892] Indianapolis Journal, Indianapolis, Indiana, United States, 10th November 1879

over the two rivers and a Tay Bridge had been proposed as far back as 1854. The bridging of the two estuaries was, however, neither a simple task nor a cheap undertaking and it was not until July 1871 that the Act of Parliament for the construction of the Tay Bridge received the Royal Assent. The proposed Forth Bridge was then still at the planning stage. The Tay Bridge was designed by engineer Thomas Bouch, who was granted a knighthood following its completion.

The Tay Bridge was a lattice girder design, using both cast and wrought iron. The girders were to be all supported on brick piers, which were originally intended to rest on bedrock which trial borings had wrongly implied existed at no great depth below the surface of the river. At each end of the structure the single-track railway line ran on the tops of the bridge girders, most of which therefore lay below the level of the tops of the brick piers. To accommodate sailing ships heading up-river to Perth extra clearance was required in the centre of the bridge, and thus here the track was placed inside the girders which were placed above the tops of the piers. This central section became known as the "high girders." Later it became evident that the supposed solid bedrock lay much deeper than originally thought and Bouch had to redesign the bridge. Brick-lined wrought iron caissons were constructed on the river bed. Sand was removed until the caissons rested on the consolidated gravel layer which the trial borings had wrongly interpreted as solid bedrock and the caissons were filled with concrete. To reduce the load on the ground beneath the caissons open lattice iron skeleton piers replaced the brick piers. Each pier had six cast iron columns arranged in a hexagon to take the weight of the bridging girders, with wrought iron braces and diagonal tiebars linking the columns for rigidity and stability. Crumlin viaduct in South Wales, designed by a different engineer, and Belah viaduct, designed by Bouch himself, were of similar design, but the engineering details of the Tay Bridge were simpler, lighter and cheaper.[893]

At the same time as Bouch was redesigning his bridge, having discovered the lack of solid bedrock the original contractors, Messrs de Bergue, of Cardiff, went bankrupt. The work was taken over by Hopkin Gilkes and Company of Middlesborough, successors to the company which had constructed Belah viaduct. Further delay then occurred in February 1877 when two of the "high girders" fell while being fitted into place. Nevertheless, an engine crossed the structure in September 1877 and by early 1878 the bridge was complete. At the time of its opening it was the longest bridge in the world. United States Civil War general and later president Ulysses S. Grant, on a visit to Dundee, commented that it was "a big bridge for a small city." At the end of February 1878 Major General C.S. Hutchinson examined the bridge on behalf of the Railway Inspectorate and passed it, subject to a 25-m.p.h. speed limit to trains passing over it. Doubtless very fortunately for his own reputation he added a comment to his report that, 'When again visiting the spot I should wish, if possible, to have an opportunity of observing the effects of high wind when a train of carriages is passing over the bridge.' He did not, however, impose any restrictions upon trains using the bridge during times of high winds. Bouch had, in fact, enquired about wind pressure and been advised by the Astronomer Royal, Sir George Airy, to allow for a pressure of 10 lbs per square foot. Few in Britain seemed aware that in France allowance was made for a pressure of 55lbs, and in the United States 50 lbs.[894]

---

[893] Wikipedia – Tay Rail Bridge

[894] The Fall of the Tay Bridge by David Swinfen pg 112

The bridge was brought into use on 1st June 1878 and just over a year later Queen Victoria travelled over it in the Royal Train on her return from Balmoral to London.

The 28th December 1879 was a Sunday. The day had begun peacefully enough as far as the weather was concerned, but by 5.15 p.m. the wind was rising fast and an hour later it had become a gale. At about 6.05 p.m., the 5.50 p.m. local train from Newport crossed the bridge on its way into Dundee. Meanwhile, the down express which Dundee railwaymen knew as 'the Edinburgh,' which had of course actually started from Burntisland, was at Thornton Junction, 29 miles to the south. It was in charge of Guard David McBeth, with Driver David Mitchell and Fireman John Marshall and consisted of engine No. 224, a third-class carriage, a first-class carriage, two third-class carriages, a second-class carriage and a luggage van.

At 7 p.m. the train stopped at Leuchars Junction station. There was no branch train connection for St Andrews on Sundays. A first-class passenger on the train, Mr Linskill, was expecting a coach to meet the train to pick him up but there was no coach waiting for him when the train pulled into the station. Leuchars Junction station stood in an isolated position and the wind was by now blowing very hard. Mr Linskill decided that a better option was to go on to Dundee and spend the night there, presumably in a hotel. Then, just as the train was about to depart, Stationmaster Thomas Robertson saw the lights of Mr Linskill's carriage coming along the road. He helped his passenger out of the first-class carriage once more and in doing so unwittingly saved Mr Linskill's life.[895]

Beyond Leuchars Junction the train made its last stop before Dundee at St Fort station, then it had to stop at Tay Bridge South signal cabin, where Signalman Thomas Barclay handed Driver Mitchell the train staff for the single line section over the bridge and the train proceeded on to the structure. Signalman Barclay gave the 'Train entering Section' signal to the North cabin at 7.13 p.m. according to his train register book.[896] With the signalman in cabin at this moment was a surfaceman named John Watt. While Barclay was writing in his train register book and attending to his stove Watt was watching the progress of the train through the window in the door of the cabin. There was a full moon, but it was obscured by clouds, and a powerful gale was blowing across the bridge. Watt stated that he saw sparks flying from the wheels of the train as it was 200 yards from the cabin and continuing for about three minutes. Then there was a sudden flash of light followed by total darkness, sparks, tail lamp and flash of light all disappearing. Watt feared that something had happened to the train and said so to his friend. Barclay reassured the surfaceman that the train's lights had simply disappeared over the hump of the bridge, as they always did. They waited, but the lights of the train did not reappear. Nor did the 'Train out of section' signal come from the North cabin. Barclay tried first the telephone and then the block instrument. Both were dead. Now alarmed, the two men tried to make their way out onto the bridge, but the force of the wind was too great. They made their way down to the river bank, and when the moon shone through a gap in the clouds they could see that a large portion of the bridge was down.[897] Watt made his way to St Fort station and finding the stationmaster there out continued the four miles to Leuchars Junction where he blurted out to Stationmaster Robertson. 'Oh man, the brig's doon.' The stationmaster could not believe it and alarmed by Watt's appearance and

895 The Tay Bridge Disaster: New Light on the 1879 Tragedy by John Thomas pg. 83

896 Report of the Committee of Inquiry

897 The Times, London, England,30th December 1879

demeanour sent for the village doctor to examine him. The doctor thought that Watt was telling the truth, whereupon the stationmaster borrowed the doctor's gig and drove to St Andrews station, where Watt's story was confirmed.[898]

In Dundee a handful of people who were not at church or sheltering from the storm witnessed the collapse of the bridge, though none was certain at that moment what they had seen. George Clark and his visiting brother William were watching the train. George had sailed in the China Seas and had encountered a typhoon; he thought the wind that Sunday night was a hurricane. He had turned away the moment that the bridge collapsed and did not see it, but his brother William did and cried, 'Look at the fire, the train is over the bridge, look at the fire!'[899] James Black Lawson had wrapped up well and was watching the bridge with a friend, a Mr Smart. They saw the lights of the train on the bridge, then they suddenly vanished. 'There is the train in the river,' Smart shouted.[900] [901] George Clark and James Lawson made their way to Tay Bridge North cabin, stood high up on the north end of the bridge. They shouted to the signalman, but he could not hear them above the noise of the gale. They were halfway up the steps to the cabin when Signalman Henry Somerville came down to meet them. Lawson asked Somerville where the train was, but the signalman could not make out what he was saying. The three men sheltered in a railway arch at the bottom of the steps where it was easier to be heard. Somerville told the other two that the train had been sent on to him at 7.14 but it had not reached him (at 7.30). He had tried to contact the South cabin at 7.17 and found that communication with this colleague had failed. That was enough for Lawson and Clark and they hurried to Dundee Tay Bridge to inform Stationmaster James Smith.[902]

Stationmaster Smith had already become uneasy about the non-arrival of the train, for which the North cabin signals had been standing off ten minutes. He asked Guard Shand of the waiting Newport local train if he could telegraph the South cabin. The guard attempted to do so, but the instrument was dead. Even when he heard Lawson and Clark's story the stationmaster was not convinced the train was in the river but suggested (hoped?) that it might have been stopped in time and retreated towards St Fort. Meanwhile, Lawson had gone on to the harbour office and Clark to the office of the ferry superintendent. Stationmaster Smith went to the Dundee locomotive superintendent, James Roberts, who was in the engine shed. Roberts told the stationmaster that Signalman Somerville had come to him and said that two members of the public had alerted him that the bridge was down. Stationmaster and superintendent went to the North cabin where Somerville repeated his story and tried contacting the South cabin again, without success. Smith and Roberts decided that their only option was to make their way out onto the bridge on foot and find what had happened. It was not a task to be undertaken lightly. Battered by the gale and drenched by spray they edged their way along the bridge, clinging to the handrail. After a while the ordeal proved too much for the stationmaster and he had to stop. The superintendent forged onwards alone, half walking and half crawling, until he reached the last of the low spans. The "high girders" should have been next, but they had gone. He recounted later that, 'I did not stay many

---

[898] The Tay Bridge Disaster: New Light on the 1879 Tragedy by John Thomas pg. 100

[899] The Bridge is Down by André Gren, pg. 25

[900] The Tay Bridge Disaster: New Light on the 1879 Tragedy by John Thomas pg. 92

[901] The Times, London, England,30th December 1879

[902] Ibid.

seconds. Directly I saw it I turned around and came back and had not time to see anything.'[903] There could now be no doubt that the bridge was down.

News of the fall of the bridge spread like wildfire in Dundee.[904] For a while the people in Dundee clung to the hope that the train had somehow been able to return to the south bank, but a boat from the training ship *Mars*, moored in the firth, brought the news that a deck watch on that ship had seen the train fall from the bridge. Then at 10 p.m. the *Dundee* ferry struggled across the estuary from Newport and confirmed the worse. The *Dundee* was put out again in the vain hope of picking up survivors, but it was no use. Everyone on the train had died. A disaster with no survivors was quite a common feature of a shipwreck on the high seas but was unprecedented in a train accident. The tickets of the passengers had been collected at St Fort and from these it was concluded at the time that there had been 69 or 70 passengers on the ill-fated train, plus the driver, fireman, guard, a mail guard and a GPO employee, a total of 74 or 75 in all.[905] A modern review of the evidence suggests however that this figure is too high, being based on an overestimate of the number of children on the train, and that the true number of dead was 59.[906] Only 46 of the victim's bodies were ever recovered.

With no welfare state yet in existence the loss of a breadwinner in an accident could mean great distress for the surviving members of a family. The North British company set up a relief fund in Dundee to try in some measure to alleviate the suffering of surviving dependents and letters were written to the chairman of the fund by local ministers, doctors and schoolteachers on behalf of deserving parishioners, patients and former pupils. Typical was a letter from Dr Boyd of St Andrews:

'Dear Sir, there is a poor young creature here, a girl of twenty-two with a child ten weeks old, left a widow by the awful disaster. Mrs Ness is her name. Her husband, a fine young fellow, walked away that evening to join the train at Leuchars. He was a cleaner in the employ of the railway company. Her father, Brand, has been thirty-eight years a driver in the company's service. They are all particularly decent. She is left with her child a burden on her parents. Her mother told me they should never want while the parents could support them, and they had not dropped a hint to me, tho' seeing them constantly, of any aid from the fund. But I have taken upon myself to bring the case before you, and I think a little help to provide mourning and the like, and generally to give a lift to the poor girl, would be well bestowed.'[907]

By 30th December it had already been announced that instead of the normal Board of Trade investigation held into British railway accidents there would be a formal inquiry conducted under the provisions of the 1871 Regulation of Railways Act, and conducted jointly by Mr H.C. Rothery, Wreck Commissioner, Inspector General William Yolland of the Board of Trade and William Henry Barlow, President of the Institute of Civil Engineers.[908] The inquiry opened in Dundee on 3rd January 1880 with the taking of evidence from witnesses and Henry Law, a qualified civil engineer, undertook technical

---

903 The Tay Bridge Disaster: New Light on the 1879 Tragedy by John Thomas pg. 95

904 The Times, London, England, 30th December 1879

905 Report of the Court of Inquiry

906 The Fall of the Tay Bridge by David Swinfen pgs. 139-45}

907 The Tay Bridge Disaster: New Light on the 1879 Tragedy by John Thomas pgs. 108-9

908 The Times, London, England,30th December 1879

investigations. The inquiry later shifted to Westminster, London.

The preliminary inquiry took the evidence of railwaymen and eyewitnesses as to the events on the night of the disaster and of divers sent down to examine the remains of the bridge. After the taking of the diver's evidence, the inquiry was adjourned, and Henry Law began his investigations on 26th January. The inquiry resumed on 26th February when ex-Provost Robertson, who had been a season ticket holder between Newport and Dundee, gave evidence as to the excessive speeds of northbound trains through the high girders causing perceptible vibrations. After vainly complaining to the Dundee stationmaster he testified that he had gone back to using the ferry for northbound journeys, using the train only for southbound ones. Other passengers supported the provost's evidence and bridge painters testified as to the bridge shaking when a train entered the high girders, both east-west and more strongly up-down, especially when the trains were going fast. The evidence of the passengers and painters occupied two days and another three days was taken up with the testimonies of the moulders and ironfounders at the Wormit foundry where the bridge columns had been built on site.[909] The foundry workers complained that the columns had been made using 'Cleveland iron' which was less easy to cast than 'good Scotch metal' and more likely to result in defective castings. Salt water was used to dampen moulds, cores were inadequately fastened and moved, resulting in columns with uneven wall thicknesses. Foundry foreman Fergus Ferguson testified that when lugs had been imperfectly cast the missing metal was added by 'burning on' – forming a mould around the lug and adding molten metal which would hopefully fuse to the column.[910] [911] Blowholes and other casting defects regarded as minor faults were filled with a concoction known as Beaumont Egg (a paste of beeswax, fiddler's rosin, fine iron filings and lampblack melted together) and the casting used rather than being rejected. Several local train drivers were also questioned as to their have driven trains over the bridge at excessive speeds. They all denied having done so, though Locomotive Foreman Roberts admitted having cautioned them and having issued a notice regarding excessive speed after ex-Provost Robertson's complaints.[912] The inquiry was then again adjourned, to resume in London on 19th April.

The third phase of the inquiry began with the testimony of Henry Noble, the man appointed by Bouch as bridge inspector. Noble was a conscientious inspector of brickwork and had made a study of the tides and current around the bridge, but he was not an engineer, and knew nothing of ironwork.[913] Noble told the inquiry that while checking the river bed for scouring he had noticed that some of the tiebars were 'chattering' and had remedied the problem by packing them with pieces of bar iron, paid for out of his own pocket. Not being an engineer, he had not realised that he was thus fixing the tiebars in a permanently distorted position. Nor had he told Bouch of the problem.[914] He did tell Bouch about cracks in four of the columns spotted by the painters commissioned to work on the bridge but having satisfied himself that the cracks were not

---

[909] The Tay Bridge Disaster: New Light on the 1879 Tragedy by John Thomas pg. 126

[910] Ibid, pg. 127

[911] Wikipedia – The Tay Bridge Disaster

[912] The Tay Bridge Disaster: New Light on the1879 Tragedy by John Thomas, pgs. 131-5

[913] Ibid,pg. 72

[914] Ibid, pg. 74

serious had merely bound the columns with wrought iron hoops.[915] [916]

The Evidence of Henry Law shortly followed and occupied several days. He listed a whole catalogue of defects in the design and workmanship of the bridge:

-The bridge piers had not shifted or settled but the masonry showed poor adhesion between the stone and cement.

-The connecting flanges on the column sections were not fully faced.

-The bodies of the columns were of unequal thickness – Bouch blamed poor workmanship of which he had been unaware.

-The channel iron horizontal braces did not butt up against the body of the column

-The gibs and cotters of the diagonal braces were roughly forged, unfaced and too small.

-Every tie bar to the base of one of the columns of the southernmost fallen pier had been packed.

-The bolt holes for the lugs had been cast tapered rather than cylindrical, so that with off-centre loading the lugs would fail at a much lower loading.

-The bracing had failed by the lugs giving way; possibly the lugs had been burnt on.[917]

Bouch himself was examined on the 30th April. By then, the fallen girders and train had been lifted from the river. He theorised that the second-class carriage and the adjoining van at the rear of the train had been blown against the bridge girders by an exceptionally strong gust of wind and had thus caused it to fail. He was basically admitting that the failure of just two struts of the bridge had been sufficient to cause its collapse, and thus condemning himself out of his own mouth.[918] Dugald Drummond, responsible for North British rolling stock, subsequently demolished Bouch's theory by pointing out that the wooden carriage was too flimsy to have shifted the bridge girder and if it had struck it would have simply been reduced to matchwood.

The remainder of the inquiry was devoted to exchanges on the strength of metals and wind pressures. Bouch admitted that he had made no special allowance for wind loading, nor had one been required by the Board of Trade.

Two reports were issued by the inquiry, one by Yolland and Barlow and the other by Rothery. The reports had much common ground. Both found that:

Neither the foundations nor the girders were at fault.

The wrought iron was not of the best quality but was not a factor.

The cast iron was fairly good but had presented problems in casting.

The workmanship and fitting of the piers was poor in many respects.

The cross-bracing of the piers and its fastenings were too weak to resist severe gales.

Strict supervision at Wormit foundry had been lacking.

---

915 Ibid, pg.76

916 Wikipedia – The Tay Bridge Disaster

917 Ibid.

918 The Tay Bridge Disaster: New Light on the1879 Tragedy by John Thomas pg. 164-8

Supervision of the bridge after it had been brought into use was inadequate. Furthermore, Noble had no definite instructions to report problems with the ironwork and in any case no knowledge of what to look for. He should however have known to report the loose tie bars.

The 25-mph speed limit over the bridge had not been enforced and had been often exceeded.

Rothery added that Bouch should have done more when the supposed bedrock was found not to exist. Young and Barlow were of the opinion that the bridge had failed at the south bend first and did not exclude the possibility of the train having become derailed, as per Bouch's theory. Rothery thought that the bridge might have failed first at the north end, explicitly dismissed the idea of the train have capsized and that the bridge had been previously strained by gales and high train speeds. Yolland and Barlow suggested the possibility of the fracture of a leeward column and that the Board of Trade should establish rules regarding permitted wind pressures. Rothery thought that the question of wind pressures should be left to the engineers.

Finally, Yolland and Barlow refused to blame specific individuals and were not happy with Rothery writing of 'we' and 'our opinion' as opposed to 'I' and 'my opinion' in his report when he placed the main responsibility for the disaster on Sir Thomas Bouch. Rothery wrote, 'The conclusion, then, to which we have come, is that this bridge was badly designed, badly constructed and badly maintained, and that its downfall was due to inherent defects in the structure, which must sooner or later have brought it down. For these defects both in the design, the construction, and the maintenance, Sir Thomas Bouch is, in our opinion, mainly to blame. For the faults in design he is entirely responsible. For those of construction he is principally to blame in not having exercised that supervision over the work, which would have enabled him to detect and apply a remedy to them. And for the faults of maintenance he is also principally, if not entirely, to blame in having neglected to maintain such an inspection over the structure as its character imperatively demanded.' [919]

Bouch described Rothery's report in a letter as 'unmerciful'.[920] Following the publication of the report he was forced to resign as engineer for a new Tay Bridge and the catastrophe meant an end to his ideas of designing a Forth Bridge. Less than a year later he was dead. Bouch was not the only person to be blamed; some blame was assigned to the overall manager of the project. Albert Grothe, to Gerard Camphuis, who managed the foundry, to foundry foreman Fergus Ferguson and even to honest Henry Noble, for taking on more than he was qualified to do.[921] But others, such as the directors of the North British Railway, and Major General Hutchinson of the Board of Trade, who had passed the bridge as safe in 1878, escaped blame. Given that the Board of Trade was part of the team conducting the inquiry it was probably not surprising that it was reluctant to blame itself.

Only ten days after the collapse of the bridge it was the subject of what became a famous poem by William McGonagall, widely regarded as the worst poet in history. The first two lines of his poem read, 'Beautiful railway bridge of the silv'ry Tay, Alas! I am very sorry to say.'

---

[919] Report of the Committee of Inquiry

[920] The Tay Bridge Disaster: New Light on the1879 Tragedy by John Thomas pg. 177

[921] The Fall of the Tay Bridge by David Swinfen pg. 112

A second Tay Bridge opened in June 1887 and a complementary Forth Bridge in March 1890.

Modern analyses of the accident have come to two different conclusions. One suggests that the principal cause of the collapse was metal fatigue fatally weakening the cast iron lugs, the other theory attributes the collapse to a fundamental flaw in the design of the bridge inasmuch as it did not make adequate allowance for wind pressure. The two theories seem irreconcilable.[922]

On Thursday 23rd December 1880 a Carolina Central Railroad passenger train from Shelby to Charlotte, consisting of an engine and tender, three box cars, a combination mail, express and baggage car and a passenger coach was crossing Indian Creek trestle, near Lincolnton, North Carolina, United States when the structure gave way. The whole framework of the trestle gave way and the complete train other than the engine and tender fell seventy feet into the ravine, at the bottom of which was only a small creek. The cars became uncoupled as they fell and overturned, the passenger car ending up upside down and the combination car on its side. The remainder of the trestlework then collapsed onto the cars, completing their destruction. Overturned stoves in the wooden cars set them alight.

Engineer Hall stopped his engine and with Fireman Smith hurried down to the wreckage with an axe. He found one of the passengers, F.N. Smith, pinned beneath a seat by an immense beam. Smith was perfectly conscious but the engineer and fireman, even working together, were unable to lift the massive beam pinning Smith down above his knees. Engineer Hall then realised that the only solution was to cut Smith out and began to hack away at the beam. The flames came closer and closer and soon Smith was roasting alive and he pleaded with the engineer to try and free him, 'For God's sake try and save my life!' Hall replied that he would do the best he could and did not abandon his efforts until his clothes were literally burned from his body. Smith remained conscious a long time and with his last breath thanked the engineer for his efforts. The other five passengers on the train also burned to death, and all that remained of them was charred bones. It seemed at first as if the conductor of the train had also been a victim, but he was later found in the woods, 'a raving maniac, and blood oozing from a ghastly wound in the head.' [923]

[922] Ibid, pgs. 107-11

[923] Bloomington Progress, Bloomington, Indiana, United States, 5th January 1881

CHAPTER 9

OPENING BRIDGE ACCIDENTS

Accidents involving opening bridges appear to have occurred in only a handful of countries with the majority occurring in the United States. It seems that most countries managed to a large degree to avoid building such bridges, which from a railway point of view are a constant source of expense and worry. Shaw states that the early railroads on the eastern seaboard of the United States were forced by opponents of the new form of transport to build such bridges over even insignificant waterways where they were never really needed.[924]

Strangely enough the earliest accident to a train involving an opening bridge occurred in a country hardly well-known for such features, namely Belgium. On 20th August 1838 the King and Queen of Belgium travelled by royal train from Brussels to Ostend, and the accident occurred to the return working of that train. which set off at about 20:30, accompanied by six officials. Approaching the swing bridge (pont tournant) at Snep the train driver whistled, and a signal light indicated that the bridge was closed and everything in order but when the train got onto the structure it was derailed and the engine, tender and leading carriage of the train fell into the river, killing two railwaymen, presumably the unfortunate driver and fireman. One of them was crushed when the carriage fell onto the tender.

Accounts of how the accident happened were conflicting, partly at least because it was almost dark at the time that it occurred, but it would seem likely that the bridge was closed but not properly secured, the bridge keeper having failed to properly secure it, and that before the accident he had left his post and gone drinking in a neighbouring public house. Several persons who had noticed that the bridge was not properly secured had asked him for the keys to it, but he had refused to give them up. When the engine had entered the bridge, it had become derailed and in doing so had caused the bridge to open, thus causing the accident. A little time later the regular train from Bruges reached the scene but its driver was alerted by shouts from people at the bridge and managed to pull

[924] A History of Railroad Accidents, Safety Precautions and Operating Practices by Robert B. Shaw, pg. 187

up his train in time to avoid a further accident.[925] [926]

The next accident, in The Netherlands on 10th March 1843, claimed the life of Adrien Desiré Teyler van Hall, a trainee engineer who was also the cousin of C.C. van Hall., the engineer director of the Hollandsche Ijzeren Spoorweg Maatschappij. This was also the first time that anyone had lost their life in a train accident in the Netherlands. Van Hall had decided to himself drive the locomotive *Vesta* at the head of the 11:15 a.m. train from Leiden to Amsterdam. As the train reached the bridge at Warmonder Leede the engine was derailed and overturned, and van Hall was fatally crushed beneath it. Driver Beyers was seriously injured. He also ended up beneath the engine and buried in sand but was saved from certain death by Chief Conductor Oberson. Trainee Driver Schmetterling suffered a slight head injury. The *Vesta* was badly damaged, and the leading goods waggon and third-class carriage behind it were also damaged, but no passengers were hurt.

The derailment was due to the bridge tender, after closing the span, having neglected to turn the eccentric. The outer end of the span had therefore been insufficiently supported. Engineer director F.W. Conrad told a meeting on 13th March that the two bridgetenders, Konings and De Jong, had been dismissed by the company for their negligence and their places taken temporarily by a carpenter from the waggon works and a watchman. As a further measure to prevent a recurrence of the accident it was proposed that a protecting signal should be erected which was interlocked with the bridge so that it could not be cleared unless the span were properly closed and secured, a rather farsighted idea in 1843. The bridge was converted to a fixed span in 1863.[927]

The first of a number of opening bridge accidents in the United States occurred on 1st October 1846.A bag of mails, bound from Boston to Europe via the steamship *Cambria*, was mistakenly picked up from the Post Office by the mail messenger for Sandwich, Massachusetts, in mistake for his own. By the time that the mistake had been discovered the Old Colony Railroad train for Sandwich had left, but the chief clerk at the post office, Thomas Pomeroy, undeterred, arranged to have the departure of the *Cambria* delayed by an hour, and requisitioned an engine, the *Quincy* to follow the regular train. As the Sandwich train had a number of stops it was easily overtaken in a dozen miles, the missing mailbag recovered, and the engine set off back to Boston to catch the waiting steamer. Unfortunately, the Old Colony, which at this time had been open less than a year, was one of the American railroads on the eastern seaboard mentioned by Shaw as having been forced to provide opening bridges for river traffic, and where such were provided to give the boats priority. Thus, it was that the bridge over the Neponset River, a couple of miles north of Quincy, Massachusetts, had been re-opened for river traffic by the bridge tender, no regular train being expected. Presumably the *Quincy* must have crossed the bridge before it was re-opened on its outward journey, perhaps because it was at that point right behind the ordinary train. To add to the danger there was a fog lying over the Neponset Meadows that afternoon.

The predictable result was that the *Quincy* plunged from the open bridge into the river and sank out of sight. Everyone on the engine survived and the mails were subsequently recovered 'very much dampened.' They obviously failed to catch the steamship, giving

---

[925] Le guide indispensable sur les chemins de fer de belgique de Duplessy by Joseph Duplessy pg. 157

[926] The Times, London, England. 25th and 27th August 1838

[927] Spoorwegongevallen in Nederland by R.T.Jongerius pgs. 21-22

New York people, who had not been happy before about having the trans-Atlantic mail sent via Boston, a new source of complaint. Pomeroy was awarded $500 by the railroad for his injuries.[928]

An accident occurred in England upon the Midland Railway on 31st October 1846, near Burton-on-Trent. A swing bridge over a canal had been opened for the passage of a barge when a passenger train for Birmingham approached, a porter having forgotten that it was due.[929] The bridge could not be closed in time and due to the foggy weather prevailing the fixed signal which was provided could not be seen.[930] A man ran 50 yards down the line to warn the engine driver and the brakes were then applied but it was too late to stop. The momentum of the engine and tender carried them across to the far bank which they struck and embedded themselves in. The leading second-class carriage was 'smashed to atoms' but was fortunately empty, the other carriages 'escaped with a violent concussion, which alarmed the passengers, but not one sustained any injury. The driver and fireman also escaped unhurt, but the guard was seriously hurt and the bridgetender was also injured. Other trains were delayed for several hours. 'Mr Commissioner Balguy was in the train in which the casualty occurred, on his way to the Bankruptcy Court, and the cause of his non-arrival was communicated by means of the electric telegraph.'[931] The bridge was subsequently converted to a fixed span.[932]

There was a minor accident at Lincoln, England, on 13th October 1848, when a few contractor's waggons ended up in the River Brayford due to carelessness on the part of a navvy who failed to notice that the bridge was open, though he managed to save his horse from falling into the river.[933]

Rather more serious was an accident in the United States on Saturday 23rd April 1853 at a bridge across Rancocas Creek, 10 miles from Camden, on the Camden and Amboy Railroad. It was due to the gross thoughtlessness of the bridgetender, who hearing the blast from an approaching boat immediately opened the swing bridge without having first displayed any warning signals along the railway line and disregarding the 2 p.m. passenger train from Philadelphia, which was also approaching the bridge. The engine and two coaches fell into the water and a third coach was left hanging from the bridge. Somehow no one was seriously hurt.[934]

As can be seen all the opening bridge accidents so far dealt with were relatively minor, involving at the most one or two fatalities, but less than a fortnight after the Rancocas Creek mishap a major disaster occurred, also in the United States, at Norwalk, Connecticut.

---

[928] A History of Railroad Accidents Safety Precautions and Operating Practices by Robert B Shaw pgs. 187-8

[929] Webpage – Burton-on-Trent Local History

[930] Board of Trade Report

[931] The Times, London, England, 10th November 1846 – reproduced from the Birmingham Journal

[932] Webpage – Burton-on- Trent Local History

[933] The Times, London, England, 23rd October 1848 –reproduced from the Lincolnshire Chronicle

[934] A History of Railroad Accidents Safety Precautions and Operating Practices by Robert B Shaw pgs. 188-9

The Norwalk bridge was one of half a dozen opening bridges provided in Connecticut to enable the New York and New Haven Railroad to cross various navigable rivers emptying into Long Island Sound. It consisted of a wooden span pivoted on a central pier to provide two 60ft openings for ships. The track was about 15 feet above highwater level, when the water was eight to ten feet deep with about 25 feet of mud beneath it. Three hundred yards west of the bridge, around a fairly sharp curve, was South Norwalk station. This bridge stood normally closed to shipping, being opened by the bridge tender upon the approach of a ship only if no trains were approaching. Protection for trains was provided by a two-foot diameter red ball, suspended from the top of a 40ft pole, which the bridge tender would lower before opening the span. Subsequent examination showed that the driver of an eastbound train could first see the ball when his train was 3,312 feet from the span.[935]

On the morning of the disaster (6th May 1853) Captain Blythe of the *Pacific* steamer whistled for the bridge to be opened at 10.15. William Harford, the bridge tender, checked by listening that no trains were close at hand, lowered the signal ball, waited several more minutes to be on the safe side, then opened the span and the *Pacific* sailed through. The span had been open 14 or 15 minutes, and Harford was in the process of closing it again, when he heard the train coming around the curve towards the bridge and its driver whistle twice. He could not close the span in time, it took a minute to a minute and a half to do so. He told the coroner's jury that he had been much frightened.[936]

The train involved was the 8 a.m. express from New York to Boston, the only train not booked to call at South Norwalk. It was running eight minutes late when it ran off the bridge at about 25 mph, the momentum carrying the engine across the western channel. The leading baggage car, two mail cars and two passenger coaches followed the engine off the bridge and a third coach ended up hanging off of the span and broke in two. The two coaches which fell into the river were not heavily damaged and most of the fatalities were from drowning. The crew of the Pacific and others pulled a number of passengers from the river but the death toll of 46 made it probably the second worst railway accident in the world up to this date, surpassed only by the Versailles derailment. Seven physicians returning to New England from the Sixth Annual Meeting of the American Medical Association in New York were among the dead but there were some lucky escapes, such as that of express messenger Fuller, who had been riding in the leading baggage car, and escaped almost unhurt. A Miss Griswold, of a prominent Connecticut family, survived because Dr John C. Warren continued to try and resuscitate her long after she had been declared dead. The engineer, Edward Tucker, fireman Ellman and conductor Charles H. Comstock all survived by jumping, the fireman escaping unhurt, the conductor suffering numerous bruises and the engineer a broken leg.[937]

Immediately after the accident there were calls for the engineer to be lynched, either hanged or shot, but he was spared to face the coroner's inquest, where he stated that he had been on the road two months and as to the signal ball declared, 'Am certain the ball was up – *cannot* be deceived. Did not look out of the window through the glass but entirely out {probably meaning he put his head out. Did not think he could have mistaken anything for the signal. Can see signal just before I get to the bridge. Am certain I saw

[935] Ibid, pgs. 189-193

[936] Bangor Daily Whig and Courier, Bangor, Maine 11th May 1853

[937] A History of Railroad Accidents Safety Precautions and Operating Practices by Robert B Shaw pg. 190

signal – Am very careful. The brakeman usually sees the signal and supposed signal was seen by him and let up the brakes. I received no other directions other than to look out for the ball for the draw. Sounded whistle the moment I saw the end of the draw. Brakes not applied – thinks train could have been stopped before going off. Gave signal before getting to the depot. Engine might have gone in, but not the cars… Norwalk is the worst bridge on the road to stop the train before coming to the draw, because the distance is much shorter than any other.'[938]

The problem was that too many other witnesses, including the bridge tender but also the master and engineer of the *Pacific* and various passers-by, all agreed that in the fact the ball had been lowered long before the train arrived. In the end the engineer had to concede that in the fact of such overwhelming evidence the signal ball must have been down, though he *thought* he had seen it showing.

There was a speed restriction of ten miles an hour over the bridge which for every other train, which stopped at nearby South Norwalk station anyway, was a formality, but the practice had grown up, and been allowed to continue unchecked, of the express passing over the bridge at twice that speed, and the superintendent of motive power admitted that engineers who lost time were frequently discharged, giving them a strong motive to run faster than was permitted to make up time. There was also the possibility that injuries sustained by Engineer Tucker in a previous accident (which was not his fault) had affected his physical condition in some way and there were allegations made of engineers reading newspapers while driving, though it was not suggested that Tucker had been doing so.[939]

Two months after Norwalk, on 8th July 1853, there was another opening bridge accident near Wilmington, Delaware, United States, on the Philadelphia Wilmington and Baltimore Railroad, though this time the deaths were limited to the engineer and fireman of a freight train. The rules of the railroad were for the engineers of trains to approach the bridge slowly, ready to stop, but if they were shown a white light by the bridge tender then the line was clear to proceed. Bridge Tender James Darley told the inquest on the bodies of Engineer Charles Girvin and Fireman Reuben Titus that he had closed the bridge at midnight for the down train to pass, then opened it for ships again. Feeling tired he had then stepped into his house a few feet, sat down and fallen asleep without intending to. Awoken by the whistle of the 2 a.m. Wilmington to Pennsylvania freight train he had rushed out and shown the "All clear" light to the engineer of the train, forgetting his half-asleep state that he had left the bridge open. He had then realised what he had done, but too late to stop the train falling through the open span into Brandywine Creek (also referred to in the report as Red River), killing the engine crew. Since the accident Darley had been 'almost crazy' and had attempted to commit suicide by drowning himself.[940] Clearly the accident was the bridge tender's fault, but one is left to wonder what hours he was expected to work. Were he and his wife expected to look after the bridge day and night in the manner of a gatekeeper at a rural level crossing?

An engine on the St Helens and Runcorn Gap Railway which had conveyed the night

---

[938] Bangor Daily Whig and Courier, Bangor, Maine 11th May 1853

[939] A History of Railroad Accidents Safety Precautions and Operating Practices by Robert B Shaw pg. 191

[940] Bangor Daily Whig and Courier, Bangor, Maine, United States, 12th July 1853

express to Garston returned towards Warrington at about 2 a.m. on 17th November 1858 (the date of the accident is unclear, it was possibly 24th November) This was an irregular working, running after the time of the last booked train, and the stationmaster at Sankey Bridges, England, just short of Warrington, had retired to bed, leaving a swing bridge across a canal open for the passage of flats. The *Actæon* engine, 'a very powerful one,' approached the bridge at speed on a foggy night and the driver being unable to see any signals which had been exhibited the engine crashed into the canal, although its tender, which remained partially attached to it, did not leave the bridge. The driver and fireman miraculously escaped unscathed. The engine had been put onto a Kings Cross – Garston goods train at Guide Bridge in place of an engine which had failed and was required to return to Guide Bridge the same night, a fact which no one thought to convey to the Sankey Bridges stationmaster.[941]

Goods traffic over the line had to be abandoned until the mess had been cleared away and the bridge repaired, while passenger trains stopped on either side of the bridge and the passengers made their way between them via a temporary footbridge over the canal. 'About 10 o'clock a large body of men arrived from St Helens, and, under the superintendence of Mr Cross, engineer to the company, immediately commenced the arduous work of getting the engine upon the line again, and until that is accomplished the canal traffic will be stopped.'[942]

Seven people lost their lives in the next opening bridge accident, near St Georges station, Delaware, United States, on 15th January 1862. A train on the Delaware Railroad set off through an ice, sleet and rain storm and in the early morning darkness of that day from New Castle to collect a load of wood. It consisted of an engine, thirteen empty flat waggons and an old passenger coach in which the conductor, Mr Albert Butler, and about 20 labourers were riding. It passed the deserted St Georges station, shortly beyond which was a high opening bridge over the Chesapeake and Delaware Canal. The bridge was normally kept open for canal traffic, and only closed when a train was due, a signal light being exhibited when the bridge was closed, as at Wilmington nine years earlier. On this morning no light was showing, indicating danger, but the engineer was slow to react, and when he did finally apply the brakes the wheels skidded due to a layer of ice on the rails. The train could not be pulled up quickly enough, and the engine plunged into the canal, followed by car after car, and finally by the coach. The conductor, standing on the rear platform of the coach, had time to jump off and save himself, but no time to warn the workmen inside the coach, who went down with it and were killed and injured in the mass of iron, wood and human bodies. Engineer Josiah Anderson, said to be a careful and reliable man, was killed together with his fireman, Edward Menam, and workmen William McElwee, William Menam, Joseph Dolan, Nathaniel Shay and Joseph Griffins. A number of the other workmen were injured, some seriously. It was speculated that Engineer Anderson, who rarely drove over the line, had forgotten that he was so close to the bridge.[943] [944]

---

[941] A Regional History of the Railways of Great Britain: Volume 10: The North West by Geoffrey O. Holt, pg. 92

[942] The Times, London, England, 26th November 1858 – reproduced from the Warrington Guardian

[943] Philadelphia Ledger, Philadelphia, Pennsylvania, United States, 16th and 17th January 1862

[944] Webpage – Terrible Railroad Calamity at the C & D Drawbridge

The worst opening bridge accident of the nineteenth century, and indeed the most serious at any time up to the present (2019), occurred at Beloeil (St Hilaire), Canada, on 29th June 1864. The train involved in the catastrophe was carrying about 458 immigrants from Germany and Eastern Europe who had landed at Quebec City the day before from the sailing ship *Necker* and were on their way via the Grand Trunk Railroad to Montreal. The immigrants were treated very poorly by the railway company, riding in coaches which were actually windowless grain cars, fitted with wooden benches and without even proper lanterns for lighting, not much better in fact than the trucks in which the Nazis transported Jews to the extermination camps 80 years later. Because of the number of them some also travelled in the rear brake car. In charge of the train was conductor Thomas Finn, with Engineer William Burney, Fireman Nicholas Flynn and Brakeman Gideon Giroux. Engineer Burney had held that position only eleven days and did not know the Montreal line. He had asked for a replacement but been told that none was available. Instead Brakeman Giroux rode with him on the engine as a pilot but returned to the brake van when the train loaded more wood at St Hilaire at 1.5 a.m.[945]

From St Hilaire the line first descended parallel to the Richelieu River, then curved to the right to cross the river by a swing bridge to arrive at Beloeil station. The opening bridge was needed because the river was busy with traffic and the 50 ft clearance between bridge and water was insufficient for steamers and towed sailing ships. Unlike a number of other locations railroad traffic was here given priority and ships were to be stopped 15 minutes before the scheduled arrival time of a train. The immigrant train was, however, not a regular scheduled train but a special, the running of which Bridge Tender Nicholas Griffin knew nothing about, and he opened the span for a southbound steam tug, the *Champlain*, which was towing eight barges and sailing ships. According to Rule No 24 in the Grand Trunk rulebook all trains were required to stop dead short of the bridge and check that it was closed, but this rule was routinely disregarded, like the 10 miles an hour speed restriction at Norwalk, and trains tended to just slow down instead. There was a semaphore signal protecting the bridge, but it was hard to see and would not be likely to be spotted in the darkness by an engineer who was not likely to be looking out for it. There was also a row of red lights along the side of the swing span itself which could be seen by an approaching train when the span was open.[946]

The immigrant special left St Hilaire at 1.15 a.m. and gained speed as it descended the incline to the bridge, sight of which was hidden by a row of trees. Then as the train rounded the curve Engineer Burney caught sight of the row of red warning lights on the open span. He knew his speed was too high and whistled for brakes but Brakeman Giroux, busy trimming a lamp, was slow to respond. Before he had really done so the train had reached the bridge and disaster was inevitable. The engine landed on a barge laden with grain and then bounced into the water and the coaches fell from the bridge one after another, the first ones being crushed by those behind.[947]

The first help came from the crews of the ships which had been passing through the span at the time, and volunteers from nearby villages also soon joined in the rescue work. The language barrier between the immigrants and the rescuers proved a problem until the arrival of a special train from Monrtreal, carrying German and Polish speaking

[945] Wreck! Hugh H. Halliday pg. 32

[946] Ibid.

[947] Ibid, pg. 33

representatives who could act as translators.[948]

No less than 97 passengers lost their lives in the disaster, along with Conductor Finn and Fireman Flynn. Engineer Burney, however, escaped with bruises. The coroner's inquest into the accident was held the same day. Engineer Burney, who was clearly the person directly responsible for the accident, was allowed to question witnesses but not to give evidence on his own behalf. Thomas King, locomotive foreman at Richmond, denied that Burney had asked for someone else to drive the train because he did not know the line, whereas Brakeman Giroux told of having to identify many features between Richmond and St Hilaire which Burney was unfamiliar with. There again although he had only been an engineer eleven days Burney had worked over the line as a fireman for several years, and crossed the Beloeil bridge about 30 times, and should have learnt the road as part of his training towards becoming an engineer.[949]

In the end the jury censured Thomas King, 'The locomotive foreman at Richmond displayed a want of judgement and caution in putting a driver in charge of a train without previously ascertaining by personal investigation whether such driver knew the road and signals,' but it was engineer Burney whom they found guilty of "gross negligence." Burney was sacked by the company but not prosecuted. It was stated that he remained in Montreal. 'broken mentally and physically, scorned and taunted by children and adults as the culprit in the Beloeil bridge disaster.'[950]

An accident at Troy, New York, United States, on Saturday 23rd September 1865 could have been a lot worse. A Rensselaer & Saratoga Railroad train was backing towards Troy over a long bridge across the Hudson River. The swing span was open at the time, but the engineer somehow misunderstood them and only noticed the open bridge when it was too late. The engine *Jay Gage* which was running tender first, plunged 40 feet into the deep water, followed by a passenger car. The accident ended fortunately; Engineer Broughton and Fireman Meser jumped from the engine as it went down and clung to the bridge and the three passengers in the submerged car were able to escape from it. The newspaper report goes on to speak rather mysteriously of a 'burning locomotive 'which was 'threatening to set fire to the bridge' implying that some or all of the engine must have ended up out of the water, so that it could catch fire.[951]

A serious accident occurred in the Vale of Neath Railway lines serving the Swansea Docks in South Wales on 29th November 1865 when an engine and no less than 30 loaded coal waggons fell into the docks, killing the driver and fireman of the train. The 581 yard long Swansea Eastern Viaduct in the docks included in its length two drawbridges, the New Cut bridge, with a span of 62 feet and the North Dock bridge, slightly longer with a span of 73 feet. Both were telescopic bridges, opened and shut by hydraulic presses controlled from adjacent valve houses. Rail traffic over the bridges was controlled by telegraphic signal stations and fixed signals, No. 1 signal station being at the pump house east of both drawbridges, No. 2 station at the New Cut valve house, and No. 3 station at the North Dock valve house, which worked with Wind Street Junction

---

948 Ibid, Pg. 35

949 Ibid, pgs. 36-7

950 Ibid, pg. 38

951 New York Times, New York New York, United States, 24th September 1865

signal box west of the North Dock Bridge. The outdoor signals at the two bridges were worked in connection with the hydraulic presses working the bridges and could not be cleared unless the bridges were closed. The signals were gas lit.

Before a train was allowed to proceed from either end it had to be offered and accepted by each of the four signal stations and the signalmen at the two valve houses were especially ordered not to give "Line Clear" until they were sure that their bridge was closed and secure, while drivers were ordered not to proceed unless the bridge signals had been cleared. In clear weather, with the whole distance between No. 1 signal station and Wind Street box being less than 600 yards, it was possible for the signalmen at each end to see that the intermediate bridge signals had been cleared, but this was not possible if it was foggy or there was too much 'copper smoke' (from nearby copper works).

On the early morning of the day of the accident a down coal train was ready to depart and the regular signalman not being there the foreman at No 1. signal station asked "line clear" from the New Cut valve house for the train at 6.18 a.m., received "line clear" at 6.19 a.m. and let the train proceed. The foreman said that he could see the bridge signals at the New Cut were clear but could not see those at the North Dock because of the fog and smoke. The signalmen at Nos 2 and 3 stations were arrested after the accident and did not give statements at the subsequent Board of Trade inquiry but Charles Bartlett, the signalman at No. 2 station, made a statement to the police, in which he said that on being asked for the coal train by No. 1 station he had asked for "Line Clear" from John Howells at No. 3 station and received it in half a minute. As the coal train had passed him, travelling at about six miles an hour, Driver William Cole had been working the regulator in and out and had complained of having a heavy train again and no sand to apply to the slippery rails. He had given the driver a white light to indicate all was clear. When the guards van reached him, the guard had asked him about a red light and he had noticed, apparently for the first time, that the signal protecting the North Dock bridge was at danger. Shortly afterwards there was a crash as the engine and waggons fell from the open bridge. Some time later he had met John Howells, who was his brother in law and lodged with him, and had asked Howells why the latter had given him back "Line Clear" when the bridge was open. Howells had replied,' Oh dear, dear, it's my fault.'

Porter Signalman James Evans at Wind Street Junction said that Howells had also told him that it was all his (Howells') fault. Evans admitted that he had also failed to check that the protecting signals at North Dock were off before giving "line clear" to Howells.

Guard George Gerrish of the coal train agreed with what Signalman Bartlett had said about his asking Bartlett about the red light and said that Bartlett had replied 'Oh Lord, it's too late!' The guard had shown a red light to his driver and had time to pin down the hand brakes of five or six waggons before the engine fell from the bridge.

The accident was investigated by Colonel Yolland for the Board of Trade. He stated in his report that, I do not recollect any instance in which many servants of a railway company appear to have been so remiss and guilty of such gross negligence; and as it appears to me, those who lost their lives were about the least to blame.' Thus, the signalmen at Wind Street and No 1 signal stations were specially enjoined, the one not to give "Line Clear" and the other not to start the train until the bridges were closed, and the signalmen at Nos 2 and 3 stations not to give "Line Clear" till they had closed the bridges by their own actions. It was true that Bartlett had received "Line Clear" from No. 3 signal station but he had failed to see the signal there at red and had misled the driver with a white hand signal only 200 yards from an open bridge. The driver of the engine had been preoccupied with its wheels slipping and because the engine was running bunker first therefore had his back to the line ahead, Colonel Yolland thought that the bunker running of engines should be prohibited. He was also unhappy with the fact that the signalmen had been working for 16 or 17 hours continuously for several days prior to the accident. He recommended that it should be made impossible for a telegraphic message to be sent between Nos 1 and 4 signal stations unless the drawbridges were closed, the opening of either bridge to break the circuit, and the closing to restore it, also improvements to the Tyer's telegraphic instrument in use.[952 953]

A swing bridge accident from an unusual cause happened in Chicago, Illinois, United States on 8th November 1871. As the Chicago and North Western Railroad engine *Wheeler* was approaching the Kinzie street bridge with a passenger train during the evening the tug *McClellan* collided with the structure, knocking it open, and the locomotive and tender fell into the river, the brakes saving the rest of the train. Engineer Ed Pearce was rescued seriously injured and Fireman Charles Hall escaped with slight injuries, but a William A. Clark, from the purchasing agent's office, who had also been riding on the engine, went down with it and drowned. His body had not been recovered at the time of the report.

Captain Duane Ryder stated that he had been proceeding up the North Branch of the river and when near the bridge had noticed a train approaching and had at signalled "two bells" for the tug to back up. Then seeing that this was not enough he had signalled his engineer to back the tug with all the steam at his command. Notwithstanding these efforts his tug had hit the bridge and knocked it open. After the accident he had at once turned his tug round and had gone down river to the offices of Messts Fox & Howard, where he had reported the accident, and where he was arrested by Police Officer McCabe. The tug's engineer was also subsequently arrested.

Caprain Ryder 'was considered to be one of the most careful tugmen on the river. 'He blamed the bridge not having been bolted in time for the accident, implying presumably that had it been firmly secured the impact of his tug would not have been sufficient to knock it open. His engineer stated that he had obeyed his captain's signals but had not had enough steam. Not surprisingly a large crowd of people who had gathered at the scene blamed Captain Ryder for the accident.[954]

A very brief report mentioned a train on the Canada Southern Railroad going through an open bridge in to the Welland Canal at Welland, Ontario, Canada on 24th April 1876, killing the engineer and fireman. 'It is supposed that the engineer was asleep.'[955]

At 8 a.m. on Thursday morning, 9th August 1877, the 7.45 a.m. passenger train from Jersey City on the New York and Long Branch Railroad fell from an opening bridge near Oceanport into the Shrewsbury River at a point midway between Long Branch and Red Bank stations. This accident was one of those caused by an opening bridge being closed but not properly secured. The span should have been locked with two heavy iron latches

952 Board of Trade Report

953 The Times, London, England, 30th November and 7th December 1865

954 The Lee County Democrat, Dixon, Illinois, United States, 15th November 1871

955 The Atlanta Constitution, Atlanta Georgia, United States, 27th April 1876

which fed into iron slots let into caps on the last bents of piles, worked by an iron lever in the middle of the bridge. The tender of engine No. 129 pulling the train left the track directly it passed onto the bridge, followed by the mail car and the four passenger coaches, all of which ran over the 500 ft bridge on the sleepers. When the locomotive and mail car reached the end of the bridge they telescoped into one another and rolled from the bridge to the bank of the creek, 10 feet below, followed by three coaches which turned over onto their sides in the water. The very crowded rear car was derailed but remained on the bridge. The engine and tender were so crushed together as to be barely recognisable. The air was filled with the cries and screams of women and children who feared that they would drown in the river, but all were pulled from the cars through the windows and rescued. Calls for assistance were telegraphed for in each direction from Oceanport and a special train was despatched from Long Branch, only three miles away. The most seriously injured were taken in carriages and row boats to Oceanport Village, half a mile away, others were taken to Long Branch and Ocean Grove. About seventy people were injured in the accident, though no deaths had been reported up to Friday evening. Most of those injured had been in the third car of the train, which fell into the deepest water.[956] [957]

During the afternoon of 14th October 1879 a Chicago Milwaukee and St Paul Railroad yard engine and trucks were driven onto the Burnham opening bridge at Milwaukee, Wisconsin, United States, in a thick fog, the engineer being unaware that the bridge had been opened for a passing vessel. Engine, tender and one car fell into the river. The engineer and fireman jumped, and Engineer Len Rutter was picked out of the water probably fatally hurt, though his fireman escaped with severe bruises. Harry Hill, a telegraph operator who had been riding on the bumper of the engine, was buried beneath the wreckage and crushed so as to be unrecognisable. Also riding on the engine was Switchman Mike Toohey, but he jumped off in time and escaped unhurt. The loss to the railway company was estimated at $25,000.[958]

A New York and Greenwood Lake Railroad train consisting of an engine, baggage car and one coach carrying about twenty passengers, mostly businessmen from Arlington, Monticello and Greenwood lake going to New York, plunged off the Hackensack drawbridge, New Jersey, United States, on 4th November 1879. Engineer Ira Dexter jumped from his locomotive and was killed and one merchant, a J.B. Puttle, was fatally injured. Fireman John Masker, Conductor William Stork and Brakeman Jesse Williams were badly bruised and Baggagemaster J.B. Wood injured his back. Wood, sat at the front end of the passenger car, had seen the fireman jump onto the footboard of the car and thence into the river and rushed to the door in time to see the engine falling from the bridge, when he himself had jumped. J. Harriman (or Harrison), presumably a passenger, suffered a cut head and internal injuries and swallowed a quantity of salt water. The engineer left a wife and family. All the passengers were rescued from the coach before it sank by men in small boats; several of the passengers were badly bruised, and all were

[956] The Evening Gazette, Port Jervis, New York, United States,11th August 1877

[957] The Bucks County Gazette, Bristol, Pennsylvania, United States, 16th August 1877

[958] Brooklyn Daily Eagle, Brooklyn, New York, United States, 14th October 1879

drenched.[959] [960]

An accident which was expensive in terms of property loss rather than loss of human lives occurred near Woodbridge, New Jersey, United States at about 2 a.m. on Tuesday 6th January 1880, when a night freight train from Long Branch fell through a drawbridge over the Rahway River. The train was wrecked, and its cars consumed by fire. Two railwaymen were slightly hurt. Property loss was estimated at $40,000.[961] [962]

959 Bangor Daily Whig and Courier, Bangor, Maine, United States, 5th November 1879
960 The Weekly Miner, Butte, Montana, United States, 11th November 1879
961 Chester Daily Times, Chester Pennsylvania, United States, 7th January 1880
962 Galveston Daily News, Galveston, Texas, United States, 7th January 1880

CHAPTER 10

TRAIN ACCIDENTS INVOLVING BRIDGES

Bridges could feature in train accidents in one of two ways. Either the bridge itself could collapse for a number of reasons such as faulty design, faulty construction or because its foundations had been damaged; or a train could fall from a bridge, the structure itself remaining intact. Bridge failures have been covered in the previous chapter; this chapter will concern itself solely with accidents where the bridge itself was not at fault. It may seem perverse to list such accidents in a separate chapter, instead of in the chapters dealing with primary causes of the accidents, the procedure adopted elsewhere in this book, but the fortuitous presence of a bridge has such a potential for transforming what might have been only a minor accident into a major disaster that it has been decided to treat such accidents separately. For the same reason, accidents occurring in tunnels have been given a chapter of their own.

The earliest railways seem to have avoided any accidents due to trains falling from bridges and the first accident of this type does not appear to have happened until 6th April 1843, when the locomotive and two cars of an Erie Railroad freight train were derailed and fell 15 feet from the high trestle over the Hackensack River, five miles from Piedmont, New York. Conductor Henry W. Watson, who was riding on the engine, was killed instantly, but the engineer and fireman, although knocked unconscious, were apparently not badly hurt. One of the cars was loaded with pig iron and calves, and 19 of the animals were crushed to death by the pig iron. The accident could have been much worse, as 14 passengers had also been riding on the train until a little before the trestle, at which point they had transferred to a passenger train which overtook the freight. The cause of the derailment was not stated.[963]

A bridge featured in the accident which happened between Falmer and Lewes, England, on the London Brighton and South Coast Railway, on 6th June 1851 but this accident involved a derailment possibly due to vandalism and is covered in the

---

[963] Webpage – First conductor killed on the Erie: Catskill Archive

Vandalism chapter.

On 7th February 1856 a train on the New York and Harlem Railroad was derailed near Dover Plains, New York, United States and fell 30 feet into deep water. The engineer and fireman were severely injured. The bridge was totally wrecked.[964] It is unclear from the newspaper reports if the bridge collapsed following the derailment of the train on it or if the bridge failed as the primary event.

The accidents of this type so far had also been minor, but on 12th March 1857 there was a really bad accident in Canada, near Hamilton (also referred to as the Desjardins Canal wreck) on the Great Western Railway. The *Hamilton Spectator* newspaper described the accident in graphic detail, 'The train from Toronto, due in this city at 5.34 p.m. yesterday evening, had come from Toronto as usual, and was proceeding at a moderate rate to pass the swing-bridge over the Desjardins Canal. Those residing at a distance may not know that a cutting has been made as an outlet to this canal through the Burlington Heights, and that the railway crosses it by a swing bridge, 60 feet at least above the level of the canal. At this time, of course, the water below was covered with ice about two feet thick.

'Just before the train reached the bridge, the engine ran off the track, owing, it is supposed, to some defect in the axle. This, however, is a mere surmise, founded only on the observation of some marks on the road for some distance on the other side of the spot where the accident occurred. The immense weight of the engine, cutting through the timber of the bridge, produced the effect naturally to be expected. The whole structure gave way, and with one frightful crash the engine, tender, luggage car, and two first-class passenger cars broke through the severed frame work and leaped headlong into the yawning abyss below. The engine and tender crashed at once through the ice. The baggage car, striking the corner of the tender in the act of falling, was thrown to one side and fell some ten yards from the engine. The first passenger car rushed after and turning as it descended, fell on its roof, breaking partly through the ice, and was crushed to atoms, while the last car fell endways on the ice, and strange to say, remained in that position.

'The loss of life was of course frightful – there were 90 passengers on the train, and the list of those who have escaped only number about 20. As far as we can learn, everyone in the first car was killed; those who were not crushed being drowned by the water, which nearly filled the car. About thirty were in the last car, of whom ten were taken out dead, and most of the others were fearfully mutilated. The conductor, Mr Barrett, the Deputy Superintendent of the Line Mr Muir, and Mr Jessop, one of the Auditors, who were on the hind platform, jumped off and escaped. The Express Messenger, Mr Richardson, a conductor on the road, and the mail conductor, were with the baggageman. The latter jumped over the baggage he had piled up ready for delivery and escaped with but slight injury while the three others went down, but miraculously enough were not much hurt. The engineer and fireman went under the ice with the

[964] New Albany Daily Ledger, New Albany, New York, 9th February 1856

locomotive, and their bodies have not yet been recovered.'[965]

The train involved was a local accommodation working which had left Toronto at 4 p.m. About 59 or 60 people died in the accident, apart from Engineer Alex Burnfield and Fireman George Knight all those killed appear to have been passengers, mostly ordinary people but also including Isaac Buchanan, Vice President of the line, Samuel Zimmerman, a prominent businessman and railway promoter, and Captain Tooley, 'a popular commander on Lake Ontario.' At least one passenger kept his wits and saved himself, 'A German lad in the hind car, seeing the Conductor hastily retreating to the back door of one of the cars, took alarm and followed him, and he, the Conductor, and two others were the only ones who escaped with trifling injuries.'[966]

The investigation into the cause of the accident determined that the front axle of the locomotive, the *Oxford*, had broken at or near the point where the double track line became single just before the bridge, derailing the engine just as it came onto the structure, which as the newspaper article had indicated had not been strong enough to withstand the shock and had collapsed under the engine. The engine had been checked at Toronto on the morning of the accident but the only tests available at the time to check the soundness of an axle were of course naked eye examinations and tapping with a hammer. No rust was found at the point where the axle had fractured so it is very unlikely that the technology of the day could have prevented the accident.

Questions were naturally raised about whether the bridge had been strong enough, but a majority of the engineers consulted expressed the view that it had been strong enough to carry 500 tons at a time when most trains were not heavier than 125 tons. Frederick P Rubridge of the Department of Public Works dissented and claimed that the structure had inherent weak spots and was built of poor materials.

The jury at the inquest on the victims of the disaster found that the fracture of the axle could not have been foreseen and that the bridge had been strong enough to carry the traffic passing over it provided that no derailment occurred. They recommended the replacement of the wooden bridge with a double-tracked steel one to eliminate the set of points on its approach and the revival of a law dictating that all trains come to a stand before crossing the structure. The canal was in fact already falling into disuse and would eventually be filled in.[967]

The Cornwall Railway, from Plymouth to Truro, England, had only been open two days when it had a serious accident on 6th May 1859. The 7.25 p.m. down ordinary passenger train from Plymouth consisted of a tank engine, second, first and second-class carriages and a waggon. It had stopped at Saltash station and was running on an embankment between Saltash and St Germans, and approaching the timber Grove Lake Viaduct when the engine mounted the right-hand rail. The derailed engine then collided with the right parapet of the viaduct, carrying away a considerable portion of it, and fell nearly 30 feet to the creek below and was embedded in the mud, taking the first two carriages with it. Driver Henry Briscombe, Fireman David Hannaford and Head Guard

---

965 Auburn Weekly American, Auburn, New York, United States,18th March 1857 reproduced from Hamilton Spectator, 12th March 1857

966 Ibid.

967 Wreck! By Hugh H. Halliday, pgs. 24-30

William Hoskin, who had been riding in the leading compartment of the second-class carriage, all lost their lives, but the passengers, who numbered about 14, escaped, though some of them suffered cuts and bruises. The driver and fireman were suffocated or drowned in the mud, but the guard would probably have survived had he not been struck by the following first-class carriage falling upon the roof of his compartment. The passengers were rescued by Under Guard Richard Paddon and a St Germans waterman named Henry Spencer. Three women who were injured were taken to nearby Orchard Farm, where two remained overnight; the only others hurt were a man with a bruised face who walked to St Germans and a little girl who suffered an injured arm.[968]

Captain George Ross investigated the accident on behalf of the Board of Trade. There was no obvious problem with the engine, a new one of almost identical design to those used successfully on the South Devon Railway line, which had similar characteristics to the Cornwall Railway, for the last eight years. Nor was there anything obviously wrong with the permanent way, which was of the type used on the Great Western Railway with horizontal timbers and transoms. The train was due at Saltash at 7.42 and St Germans, 5¼, miles from Saltash, at 7.56, which allowing five minutes for starting and stopping gave an average speed of 35 miles an hour. He attributed the accident to the use of a 'heavy' (39 tons) tank engine over a new road which had possibly also been affected by heavy rain on the previous couple of days and said that he was glad to learn that the Cornwall Railway Company was to reduce the speed that trains needed to run at between Saltash and St Germans.[969] The jury at the inquest into the deaths of the three victims passed a verdict of "Accidental death" and recommended that the speed of the trains should be reduced, especially while the line was new.[970] The viaduct fell out of use with the opening of a deviation line between Saltash and St Germans in 1908.

Some accidents involving bridges were fairly minor affairs. Thus, on 10th October 1859 in the United States a westbound freight train ran into some horses on Scioto Bridge, just west of Circleville, Ohio, throwing the engine off the track, whereby one span of the bridge collapsed, and the engine and six freight cars fell into the river. Engineer James Downey and Fireman Michael Crowly survived, though they were injured.[971]

On 29th of the same month, what *The Times* described as 'a frightful accident' but which the Board of Trade did not consider serious enough to warrant an investigation, a mishap which occurred just north of Belper, England, on the Midland Railway. An up coal train from Clay Cross to Peterborough was crossing a wooden pile-bridge at Messrs Strutt's dam, about a mile north of Belper, when a waggon was derailed from the breaking of an axle. The derailed waggon ripped up the cross beams of the bridge and no less than 25 waggons laden with coal fell into the swollen stream below. The engine and tender were already over the bridge and the shock broke the coupling to the guard's van at the rear of the train, temporarily knocking the guard unconscious but saving his van

---

968 The Times, London, England, 9th May 1859

969 Board of Trade Report

970 The Times, London, England, 11th May 1859

971 Washington Herald, Washington, Ohio, United States, 13th October 1859

from falling from the bridge, it instead rebounding and moving back along the up line. The guard, as soon as he had recovered himself, knowing that another coal train was following, ran back down the line but he had only covered 50 yards before the second train came around the corner and it was unable to stop before it ran into the brake van. The driver and fireman of this train jumped off and were unhurt, but the engine and van were both damaged. Single line working over the down line was thought likely to be necessary for several weeks and goods and mineral traffic was interfered with.[972]

Rather more serious than the previous two accidents was one on the West Cornwall Railway at Penwithers Bridge, near Penwithers Junction, England, on Sunday 30th August 1863. At this date a new goods siding was being constructed between Penwithers Junction, just west of Truro and the point where the Falmouth Branch left the main line, and the aforementioned bridge, which was about 15 chains west of the junction on the Penzance line. When work on the siding had finished on the Friday evening a tipping waggon was left on it, supposedly properly scotched with stones. Only a few trains ran on Sundays on the West Cornwall line, and Richard Tregoweth, the porter on duty at Penwithers Junction, went home after the arrival of the up train due from Penzance at 10.52 a.m., not returning until 6 p.m. in time for the 6.15 p.m. down train from Truro, at which time he lit the signal lamps, except that of the up main distant, which he deferred until after the 6.15 down train had gone. He said that he had looked towards the tipping waggon in the siding, which appeared to be in its proper place, but added that as the siding was not yet finished, he did not consider it his responsibility.

The following 7.15 p.m. train to Penzance eventually got away at 7.27. It had started at 7.15 but was found to be too heavy for its engine to haul up the sharp climb out of Truro on the greasy rails so had to be set back into the station for three empty carriages to be taken off to reduce the load. When it started again a brakesman was riding on the front of the locomotive to sand the rails. The train passed Penwithers Junction all right, and was approaching the western end of the siding, when its fireman saw the tipping waggon, which had run down the siding and presumably had become derailed foul of the main line. Hearing the shout of his fireman, the driver of the train, which was travelling at about 15 miles an hour, immediately whistled for brakes, reversed his engine and applied the tender brake but with a falling gradient and greasy rails the distance was too short, and the engine struck the waggon, pushed it some distance forward and then struck it a second time. The second collision derailed the engine, which slid along the sleepers and ballast until it struck the truss and platform of a wooden underbridge, carrying them away. Reaching the opposite stone abutment of the bridge the engine then fell into the road below, followed by the tender and two leading carriages, both fortunately empty. The other carriages of the train remained on the embankment. The poor driver died of his injuries shortly after being taken from the wreckage and the brakesman was fatally scalded by hot water and steam from broken valves and cocks and died in the infirmary two days later.

Captain F.H. Rich, the Board of Trade inspecting officer, remarked that both the majority of the coroner's jury and the officers of the Company believed that the waggons had been maliciously placed foul of the main line, especially as there had been previous

---

972 The Times, London, England, 31st October 1859

attempts to derail a train, but he considered that the waggons had probably been driven there by a squally north-easterly wind. Since the accident a chock block had been provided, but he recommended the provision of a second chock block at the east end of the siding, with spring padlocks needed to open but not to lock them, indicators at the points and a catch siding at the western end. The gradients of the sidings could also be improved.[973]

An accident 'highly significant of the perils of the modern highway' in the words of *The Times* occurred on Monday 29th April 1867 near Larbert, Scotland, on the Caledonian Railway. A long goods train of 52 waggons and a guard's van was en route from Perth to Edinburgh when the twentieth waggon suffered a broken axle as it was crossing Carron viaduct. As a result, 22 waggons loaded with sheep and cattle for the Edinburgh markets, three vans for dressed meat and the brake van fell from the viaduct to the lade of the river and an adjoining meadow. Guard Esson was quick enough to jump out of his van before it fell, but two cattle drovers named Hastings and Cartwright went down with it. The guard was slightly hurt, the two drovers rather more badly so. A large number of animals in the waggons which fell from the viaduct were killed or injured, those injured being slaughtered on the spot, and the cattle suffered more than the sheep. About 300 sheep were rescued alive, some of them in the waggons which remained on the line, and about the same number of animals were killed or afterwards slaughtered. About 100 sides of meat were retrieved from the vans which had also fallen from the viaduct. It was estimated that it would take two days to clear the line and the total cost in loss of plant, damage to track and viaduct and loss of stock was put at about £15,000.[974]

An accident in Ireland on Friday 9th August 1867 happened at a location about a mile south of Bray station which was likely to cause anxiety in the minds of nervous passengers, where the Dublin Wicklow and Wexford Railway was carried via a Brunellian wooden viaduct across a ravine between two rocky headlands, with the Irish Sea close at hand on one side of the line. As the 6.30 a.m. up passenger train from Enniscorthy entered the viaduct at Ramscalp (or Brabazon Corner) the engine became derailed and breaking through the handrail on the left or landward side of the viaduct fell into the ravine, 33 ft below, pulling with it the tender and three carriages. The leading brake third was smashed to bits, though a lad riding in it in the service of the company somehow escaped unhurt. The body of the second third-class carriage was also smashed and fell on its end, thus supporting the third third-class carriage which ended up hanging from the viaduct. The remaining three carriages and brake van of the train remained on the line. One passenger, Mr Murphy, a shopkeeper of Tinahely, was killed outright and a Mrs Hackman or Haikman died later from her injuries. Twenty-three other passengers were more or less badly hurt, and the driver and fireman were also seriously injured but were probably lucky to have survived at all. Had the train fallen on the seaward side of the viaduct the accident would doubtless have been a lot worse than it was.

By the time that Colonel W. Yolland, the Board of Trade inspecting officer, had arrived at the scene on 14th August the line had been re-laid so that traffic could be

[973] Board of Trade Report
[974] The Times, London, England, 2nd May 1867

resumed but he had the advantage of an inspection made by a Mr J. Fogerty, a civil engineer, a couple of hours after the accident. From his own inspection, Colonel Yolland stated that there appeared to be three different types of rails laid on the viaduct; an old 80 lbs bridge rail on the northern part of the viaduct, an old 80 lbs flanged or flat bottom rail on the left rail south of this point and partly on the right rail, and a much worn old 56 lbs bridge rail on part of the right rail. Mr Fogerty wrote that "The cause of the accident was, in my opinion, solely due to wretched manner in which the line was being re-laid; and, having considerable experience in laying permanent way, I unhesitatingly say that the line was not in a fit state to run over at this point."[975]

Belated work on relaying the line had been in the process when the accident occurred and the track of the company was apparently in a generally poor state, as they had not been able to pay any half-year dividend to their shareholders due to the cost of replacing rails and sleepers. The first train over the line after the accident, due in Bray at 5 p.m., was almost derailed at the same spot and the next up train was slightly derailed, its passengers considering themselves fortunate to be able to walk the last mile to Bray.

After the Friday evening incidents, Mr Murphy told the platelayers to leave the track alone until he had examined it on the Saturday morning and sent a note of his findings to the Board of Trade inspecting officer, but Permanent Way Inspector William Cosgrave and Ganger Simon Carty looked at the line before dawn on Saturday with hand lamps and concluding that the left rail was too low instructed their gang to lift it 'so that they could make the road all right, unknown to anyone.' The inspector and the ganger then left the scene, leaving their men to get on with the job, so that they could say at the coroner's inquest 'that they had not altered the road.' Colonel Yolland wrongly states in his report that these facts did not come out at the two inquests, and that if they had done the jury might not have returned a verdict of "Accidental death," but according to *The Times* report, they did come out in cross-examination in at least one of the inquests, the ganger adding that he and the permanent way inspector had been suspended as a result. The jury nevertheless returned a verdict of "Accidental death."[976]

Colonel Yolland concluded that the accident was entirely due to the permanent way, which 'had been left in a very bad condition on the evening of Thursday 8th- much worse than was necessary if proper care had been taken' and added that, 'Certain portions of the line between Bray station and this viaduct have been re-laid with new rails on cross sleepers, and, in my opinion, a great mistake was made when it was determined, certainly not too soon, to put down new rails on the other parts on the old longitudinal timbers, many of which, as I have already stated, are seriously decayed, instead of entirely renewing the road on new rails on new cross sleepers. If the remainder of the line is like the portion between the viaduct and Bray station, the sooner it is re-laid the better, as far as the safety of the travelling public is concerned.' He also criticised the lack of a guard rail on the viaduct, as recommended by his colleague Colonel Rich after an 1865 derailment, which would probably have prevented the accident. He recommended that the Company acquire a weighing machine to adjust the weights on the wheels of their various engines, noted that the ganger did not have a level to adjust the cant of the rails he was to maintain, and added that 'from the statements which were made, I should not consider that Inspector Cosgrave and Ganger Carty were proper men to be entrusted with the

[975] Board of Trade report
[976] The Times, London, England, 10th,12th,13th and 15th August 1867

looking after and relaying of the permanent way, on whose condition the safety of the public travelling on the line must entirely depend.'[977]

Four months after the Bray accident it was easily surpassed in terms of casualties by what became known as 'The Angola Horror' in the United States, described in contemporary newspapers as 'Another Railroad Murder' and 'The Railroad Slaughter.'

Angola was a small station in New York state on the Buffalo and Erie Railroad, situated about 15 miles west of Buffalo close to the shore of Lake Erie. About a quarter mile east of the station the railroad crossed the shallow Big Sister Creek by a truss bridge, with high ground leading to it from the station side and an embankment on its far side. The bridge was about 100 feet long and carried the line 30 feet above the stream. Directly below the bridge a mill dam made a pond out of the stream. The train involved in the accident was the New York express, due in Buffalo at 1.30 p.m. but already running over two hours late on Wednesday 18th December 1867 when it left Erie, Pennsylvania, at 12.43, due to a broken bridge on the Cleveland and Toledo Railroad. It consisted of four baggage cars, an emigrant car and three first-class coaches. The train was in charge of Conductor Benjamin Sherman with Engineer Charles Carscadin, Fireman Charles Newton, Front Brakeman Gilbert Smith, Rear Brakeman John Vanderburg and Baggageman James Woods.[978] Formerly it had stopped at Angola, but the stop had been taken off and the train was running at considerable speed when it reached the station at 3.10 p.m. Just as the train was passing through the station the rear car became derailed at a set of points, and it was then dragged along over the ties (sleepers) for a quarter of a mile to the bridge itself. A few, but not many, of the passengers at its front end who had sufficient presence of mind managed to make it forward to the car ahead, while the train crew were making desperate efforts to apply the brakes. The penultimate car was pulled off the track in turn, but it was the rear car which became uncoupled and fell over on its left side against the bridge abutment and then slide down the ice-covered slope to the level of the creek. Its terrified passengers were all thrown into a heap at its bottom end and before they could recover themselves the stove crashed down on top of them from the top end. Its coals set the wooden car ablaze at once and only three of the 44 passengers in the car got out of it alive, theatre agent Isadore Mayer and Amos H. Thomas and his wife, Mary. In addition, Buffalo and Erie engineer Robert Dickson, a passenger in the rear car, jumped from the back of it just before it fell, urging his friend Frank Walker to do the same, but Walker did not follow and was killed. Another lucky survivor in a different way was a young John. D. Rockefeller, later to become the richest American of all time, who managed to miss the train at Cleveland. In contrast to the last coach, only one passenger died in the penultimate car, which eventually also became uncoupled and fell down the embankment about 270 feet further on. He was caught in an opening in the roof which closed upon him again. This car was badly damaged, and coals were scattered from the stove, but here there was no significant fire.[979]

A newspaper reporter described what happened next, 'It was some time moments before any one could reach the scene to attempt assistance. The second car in the

---

977 Board of Trade report

978 The Angola Horror by Charity Vogel, pg. 69

979 Boston Post, Boston Massachusetts, United States, 23rd December 1867

meantime – the first car thrown from the track {this is incorrect, the last car was the first to be derailed} - had plunged over the opposite side of the embankment, falling a less height, but suffering almost as complete a wreck, and its few uninjured passengers had enough to occupy their attention. The train, which had run some distance before being stopped, was backed to the scene as speedily as possible, and those on board set to work with promptitude and energy. But the steep and ice-covered slope to the car, most horribly situated, was not easily descended, and when helping hands reached it, they were without weapons for fighting the flames or for breaking into the wreck. As soon, of course, as they could be summoned, the people of Angola and nearer inhabitants ran to the spot, and it was not long before many buckets were employed in pouring water upon the burning wreck; but a raging fire is not to be overcome with buckets, and those desperately labouring at the work had to suffer the unspeakable anguish of seeing their efforts made utterly in vain. The hideous, remorseless flames cracked on; the shrieks died into moans and moans into silence more terrible as the pall of death drew over the scene. Except one little fragment of the side of the car nothing but a heap of smoking cinders was left for those gathered around to search into with sickened hearts. How many of the bodies were utterly burned up to a handful of undistinguishable cinder it is impossible to tell, nor can it be known in all cases, probably for many days, who they were. Only as distant friends enquire for missing ones likely to have been upon the ill-fated train can the terrible list of mortality be made up.' [980]

News of the accident and a call for assistance were telegraphed from Angola station to Buffalo while Dr Frederick Hoyer, who had been a passenger on the train, was the first physician to treat the injured. He was soon joined by Angola doctor Romaine Curtiss and local villagers, some of whom worked for many hours tending the wounded and taking them into their homes, especially those of Josiah and Huldah Southwick and Frank and Thankful Griffith.[981] Later an infant's cry would alert rescuers to a baby swaddled in blankets wrapped in the snow, which people had repeatedly walked by earlier thinking it was just a cushion or piece of baggage. The child was unharmed.

When news of the disaster had reached Buffalo, a number of physicians had gone to the station, from where a special relief train left at 5 o'clock for the scene, accompanied by William Williams, the President of the line, and Robert. N. Brown, the Superintendent,[982] together with police officers, workmen and reporters as well as doctors. When the relief train arrived at Angola, work was begun to retrieve the bodies of the dead and they were removed to the freight house. Meanwhile, the second westbound through passenger train of the day, the Cincinnati express, arrived, carrying among other passengers J.D. Rockefeller, who telegraphed to his wife from Angola that he was safe, but that the train that he had missed had met with a bad accident. With the line cleared of the front of the Cleveland train and the relief train, the Cincinnati express continued its journey, taking with it the uninjured and slightly hurt passengers from the Cleveland

[980] Ibid.

[981] The Angola Horror by Charity Vogel, pgs. 117-124

[982] New Albany Daily Ledger, New Albany., New York, 19th December 1867

train.[983] The more seriously injured followed with the coffins of the dead in another train the next day.

The derailment was due to a rather obscure and uncommon fault. Not something obvious such as a broken rail or an obstruction on the track but a slight bend in the forward axle of the rear truck of the rear car, a deviation of only three-quarters of an inch, barely sufficient to be visible to the naked eye. It was sufficient however to eventually make the car jump as it met a set of facing points and it so happened that it had done so at the points on the approach to the bridge.[984] In this respect, it was a very unlucky accident. Other factors of course had made the accident worse; the lack of adequate brake power to pull the train up quickly and perhaps especially the combination of stoves for heating and easily flammable wooden coaches. This combination would cause many more accidents before better alternatives in the way of continuous automatic brakes and steam heating replaced them.

The Angola tragedy was followed in the United States by another, less well known one, on 25th February 1870. This accident happened on the Mississippi Central Railroad at Buckner's trestle, forty feet high and three miles from Oxford, Mississippi, at about 4 p.m. The victims involved were reportedly mainly emigrants en route to Texas. The engine of the train crossed the trestle successfully but the remainder of it left the track, crushing the trestle work and completely wrecking the baggage, mail, express and three passenger cars. The cause of the derailment was believed to have been the unsound trestle timbers which had permitted a rail to slip out of place, throwing the cars through the trestle work or into the ditch.

A newspaper report stated, 'Every car was utterly destroyed; the luggage, mail and express freight broke loose and were scattered on the sides of the ravine in which the wreck plunged. The first and second passenger cars were scattered into pieces. The remaining passenger car kept the track, while its forward end rested on the wreck of the second car in the ravine, at an inclination of fifty degrees. Two women, four children, twelve white and three coloured persons are known to have been killed. And it is believed some others.' Among those killed were a Colonel Spears, a planter residing fourteen miles from Jackson, who was returning from Chicago with thirty labourers; Mrs Duncan Garrett, who had just arrived from Scotland via New York, and J McDonald, Supervisor of the southern end of the line.[985]

The accident could have been more serious still, for 'The excursion train from Louisville, with delegates of General Freight Agents, failed to connect at Humboldt, else the loss of life would have been fearful.' The excursion apparently continued instead as a separate train, arriving at the scene half an hour after the accident had occurred.

At about 11 a.m. on the morning of Thursday 16th June 1870 the Vermont and Massachusetts Railroad passenger train which had left Boston at 7.30 a.m. in charge of Conductor Holden was rounding a curve near the Long Bridge, between Athol and

[983] The Angola Horror by Charity Vogel, pgs. 138-40

[984] A History of Railroad Accidents, Safety Precautions and Operating Practices by Robert B. Shaw pg. 322

[985] Philadelphia Inquirer, Philadelphia, Pennsylvania, United States, 28th February 1870

Royalston stations, in Massachusetts, United States, when its engineer saw a hand car standing on the bridge. He immediately did everything that he could to stop but the distance was too short, and the engine struck the hand car, was derailed and thrown from the bridge into the river. The leading baggage car and a passenger car were dragged after the engine, the forward passenger car ending up beneath the baggage car while the following passenger car piled up on top of it. The engine, baggage car and one passenger car were completely demolished, and it was thought a miracle that all the passengers in it were not killed. As it was two passengers; George Johnson of Fitchburg and James O'Connell of Indian Orchard were killed outright, and a Mrs Almond Brewer died after undergoing a leg amputation. About 15 other passengers were injured, as well as Engineer William Foster, Fireman H.M. Jones, Brakeman George Howard, Baggagemaster William Elliott, Roadmaster E.O. Britton, Mail Agent A.S. Whitcomb and Express Messenger Albert Backwell.

Surgeons from Fitchburg arrived soon after the accident and were joined by surgeons and citizens from Athol and Orange. Another train in charge of Conductor Bangs was also soon at the spot and took the injured to Athol.

The accident was apparently due solely to a mistake by Sectionmaster P.B. Morse, who admitted that the hand car should not have been on the bridge at the time and said that he had made a mistake in looking at his watch. Bridge, track, engine and cars were said to have all been in perfect order. No passenger had been killed or even had a bone broken on the line in the previous 19 years.'[986]

What was described in the newspaper reports as a Chesapeake and Ohio Railroad freight train, but which was really a mixed train as it included a single accommodation car, had just passed Jerrys Run, Virginia, United States, and was making its final ascent of the climb into the Allegheny Mountains on 27th October 1870 when it became divided. The rear portion of the train, complete with the accommodation car, began to run back down the gradient and travelled as far as the trestle over the run, from which the vehicles fell 80 feet and were smashed to pieces. Two men, Mr Gay and a man named Woodson, were killed outright and two other men, Mr Shepard and Mr Heppard, were injured. 'It will take several days to repair the broken trestle work.'[987]

On 10th November 1870 it was the breakage of a wheel which threw part of a train from the track as it was crossing a trestle near Columbia, South Carolina, United States, on the Greenville and Columbia Railroad. The derailment caused the trestle to fail and the mail car, second class and two other passenger cars fell 25 feet from the structure, the engine and tender at one end and the last passenger car at the other remaining on the track. The last car was supported by the wreckage below it. The accident occurred about 12 miles above Columbia, between Frost's Mill and Littleton. The train had been due in Columbia at 3.45 p.m. A relief train was sent to the scene from there and returned later with the dead and injured.

Mrs Phillip Fogarty of Charleston, Charles Joy of Newberry, both South Carolina,

---

[986] Boston Post, Boston, Massachusetts, United States,17th June 1870

[987] Madison State Journal, Madison, Wisconsin, United States, 29th October 1870

coloured stewardess Grace Montague and Minerva Parr, coloured, were all killed outright and Stephen Smith, of Newberry; Hatsona Lomex, coloured, State Senator for Abbeyville, South Carolina; and ex-Confederate Colonel James Haywood were fatally injured. A number of other passengers, Conductor Stephen Smith and Express Messenger Isaacs were also hurt.[988]

The complex disaster at New Hamburg, New York on 6th February 1871, which included a derailment, a collision, a fire and a trestle failure is detailed in the Derailment, Broken Axles and Defective Vehicles chapter.

An accident occurred on the Cincinnati and Indianapolis Junction Railroad east of Morristown, Indiana, United States at about 12 noon on Friday 22nd December 1871. A short westbound mail train was within twenty or thirty yards of a bridge over Beaver Meadow Creek when it was derailed by a broken rail. The momentum carried the train the short distance onto the structure and the engine and cars fell fourteen feet from the trestle. Engineer James Hamilton was killed instantly and Conductor Martin Flenner, mortally injured, staggered to his feet, asked "Are the boys safe?" then collapsed and died within a few minutes. Fireman John Williams was badly scalded and suffered a broken arm and an unnamed baggagemaster sustained internal injuries. Pat Kilkenny, a section foreman travelling as a passenger, was also fatally hurt and died later. None of the other passengers in the single car they were occupying was seriously hurt, though most were bruised and shaken.[989]

At 11 a.m. on 20th July 1872 the engine and tender of the Mulhouse-Paris postal train were derailed on Dannemarie Viaduct, between Altkirch and Belfort in France, and fell from the structure, killing the driver and fireman. The remainder of the train remained on the viaduct, though two carriages overturned. No passengers were hurt. The cause of the derailment was not stated.[990] [991]

Christmas Eve 1872 saw what the Philadelphia Inquirer referred to as 'The Railroad Holocaust;' another major disaster in the United States, at Prospect, Pennsylvania. The eastbound Buffalo Corry and Pittsburg Railroad express train involved in this accident, train No. 6, due to leave Corry at 1.15 p.m., was a very short one, consisting apart from the engine and tender of only a baggage car and one passenger car. Engineer Joseph Hare had shut off steam on the falling gradient leading to Prospect station, but heavy snow on the ground had impeded the progress of the train, and he went to re-apply it. As he was about to do so, about midway along a 200 ft long low (20 ft high) trestle, he realised that the tender was off the line. He blew the whistle at once for brakes and reversed his engine and at that moment the bell cord snapped. He had looked around and saw the two cars falling from the trestle. The engine and tender made it across to the other side of the

---

[988] Bloomington Weekly Leader, Bloomington, Illinois, United States, 17th November 1870

[989] Indianapolis Journal, Indianapolis, Indiana, 23rd December 1871

[990] The Times, London, England, 22nd July 1872

[991] Wikipedia - Liste des accidents ferroviaires en France au XIX siècle

structure where he brought his locomotive to a stand and ran back to assist.

Brakeman Claude Hanson told a reporter afterwards that he had been railroading for fifteen years, but with the Buffalo Corry and Pittsburg only five weeks. He had been the only brakeman on the train but had been assisted by Baggagemaster Frank Taylor. At the time of the accident they had been in their proper positions, he at the forward end of the passenger car facing the baggagemaster at the rear end of the baggage car, and he (Hanson) had taken up the slack. The train was only running at five or six miles an hour. The first warning that he had had was when the cars parted about midway along the trestle and at the same moment the forward part of the passenger car began to fall from the trestle and the baggage car with it. He had shouted to the Baggagemaster to jump but the latter said "No," and clung to the rail. He had jumped when the car was about 10 ft from the ground and was able to leap clear of the car. When he got up, he went to Taylor and asked if he was hurt to which Taylor replied that he was dying. He shouted to Engineer Hare for a screw jack to release the baggageman and then, seeing that nothing could be done to save his life, for some water for him. They managed to release the dying baggageman and then, seeing that they needed more help, he ran to Prospect station, a quarter mile away, and finding two busses there, sent them off for doctors and assistance. Returning to the scene he found that fire had broken out at each end of the passenger car, the cars having fallen upside down and the big trucks (bogies) crashed through them. Aided by the enginemen and a trackman, he had scooped up snow to try and extinguish the flames. Three other trackmen had soon arrived and also began to pile snow on the flames and work began to release the trapped passengers. They managed to get five out, but one, Edward Bell, of Sherman, New York, died soon after being released. By the time that all five were freed about twenty men, trackmen and farmers, were at the scene and were able to keep the flames in check for a time. They found Conductor Fayette Flanders in the centre of the car with his feet trapped and slowly roasting. He begged them to get him out if it meant cutting his feet off. They put a chain under his arms and three of them pulled him out, the chain breaking some of his ribs. A lady near the conductor was jammed between the seats but they could do nothing to help her. A New York woman was burned to a crisp, nothing but the steel hoops of her skirt and a few small body parts remaining. The others were charred and blackened trunks.[992]

The rescuers were greatly hampered by having no water and very little equipment. The snow which was scooped up and thrown on the flames did not have that much effect and the two axes available made little progress on the mass of interlocked wreckage. The woodwork of the cars had been thickly coated with varnish which made the fire burn fiercely. It was presumably started by overturned stoves.

A newspaper report recounted, 'The flames were steadily gaining, and the shrieks of the imprisoned victims were appalling. Now and then a rope was put through a crevice, fastened to a form and pulled out by main strength. Sometimes it would be a lifeless body, and charred arms, legs or head would drop in the struggle. In three or four instances the head was thus separated from the trunk, and in as many more cases the body was bereft of its limbs. The body of a woman was first secured, the head and one arm snapping off. The other arm was untouched, a kid glove covering the hand, and on the fingers jewelled ornaments.

[992] Oil City Daily Derrick, Oil City, Pennsylvania, United States, 27th December 1872

'The bodies of three other women were taken out, none of which, it is believed, can be recognised. In all, it is estimated that nineteen bodies, removed from the burning car, are unrecognizable. It is supposed that in addition to the ladies recovered a number of others were consumed in the burning car, and nothing but their blackened bodies will be discovered. Adding these to the number of injured that will die, it is estimated that the total number of lives lost will not fall short of thirty. The whole number of persons in the two-cars was about forty-five.

'After the accident, Frank Taylor, of Corry, one of the rescued passengers, was leaning against some object, and seeing an acquaintance passing called him by name. The friend turned and saw Taylor, apparently uninjured, but suddenly he said, "Tell my wife," and fell over dead.

'But the inmates of the burning cars were shrieking for aid, and Mr Taylor's friend did not see him again until his corpse was brought in and placed beside others. With no means of extinguishing the flames, and the heat being too intense to admit of all the bodies being extricated, nothing could be done by the spectators but to stand quietly by and see the bodies slowly consumed.'[993]

Especially sad was the fate of Mrs and Mrs Wilbur Rice, newlyweds on their way to spend Christmas at Rochester, both only children and who had only been married on the 5th of the month. Their bodies were amongst those unrecognizable after the accident, but it was hoped that friends would be able to identify them from their clothes. As for the other victims, the newspaper noted that, 'Judging from the clothing, baggage and appearance of those coming to identify them, a number of dead belonged to the lower classes.'

In addition to the passengers who were killed, along with Baggagemaster Taylor, about 20 were injured, many of them badly, as of course was Conductor Flanders. it was hoped that the conductor might recover if his internal injuries did not prove too severe. Another report, however, stated that his condition was critical. Mail Agent Earl Bacon, at first reported to have been killed, was in fact only slightly injured, 'having secured the registered letters and extricated himself by a jump.' The engineer and fireman also escaped unhurt of course as the locomotive did not go down with the cars.[994]

The accident was caused by a broken wheel on the rear truck of the tender. The tender had kept track until got to the trestle, when it dragged and tore up a rail, derailing the following cars. It was rumoured after the accident that the timbers of the trestle had been rotten, but this was untrue, the trestle having been rebuilt only the previous season. The engine and trestle were also new.[995]

For the next accident, we return to England and another viaduct on the Cornwall Railway, this time nearer Plymouth. Early on the morning of 15th April 1873 the 2 a.m. down goods train from Plymouth, consisting of an engine, 15 waggons and two brake vans, did not leave until 4.39. It had passed through Devonport station and the short Devonport tunnel beyond and had reached the southern end of Camel's Head or Weston Mill Creek Viaduct, travelling at 18 or 20 miles an hour, when its driver felt a small jerk

993 Philadelphia Inquirer, Philadelphia, Pennsylvania, United States, 27th December 1872

994 Ibid.

995 Oil City Daily Derrick, Oil City, Pennsylvania, United States, 27th December 1872

behind him, and looking around found that he had lost the greater part of his train, only three waggons remaining attached to his engine. He then shut off steam, sounded the brake whistle and brought what was left of his train to a stand at the northern end of the 396-yard-long viaduct. He then heard the head guard, who had been in the van at the rear of the train, calling him back, and leaving his fireman in charge of the engine he walked back over the viaduct to find the rest of his train at the southern end of it, some waggons and the rear van on the rails, some on the top of the embankment, down the side of the embankment and some in the mud of the creek below.

The head guard told the Board of Trade inquiry that he had not realised that anything was wrong until his train reached the branch to Keyham dockyard, which left the single track main line by a set of points which were trailing for down trains, 70 yards short of the viaduct. Here he felt the train slow up and applied his brake, assuming that the viaduct must be still under repair, then he had heard the noise of the waggons derailing, had applied his brake harder and his van had quickly stopped. He was worried about his mate and could not see his van amongst the derailed waggons on the viaduct, but then heard him groaning below the viaduct, and the second guard asked if someone could move a large cask which was on his legs and was crushing the life out of him. He went down to his mate but could not lift the cask alone, but with the assistance of the driver and three other men who had arrived at the scene the barrel was lifted off the poor guard and he was carried into an adjoining field. As this was being done, the men heard the whistle of the down mail train, which proceeded to collide with the guard's van of the goods train at a speed of about eight or nine miles an hour.

The Board of Trade inspecting officer, Colonel W. Yolland, thus effectively had two different accidents to investigate, the derailment and the subsequent collision. The officers of the Cornwall Railway attributed the derailment to the axle of a waggon breaking in Devonport tunnel and running over the ballast and transoms for 1½ miles before coming into contact with the end of Camel's Head viaduct and finally throwing off the rest of the train. Colonel Yolland was not happy with this explanation, pointing out that had the axle broken that far back it was unlikely that the waggon would have run as far as it did. That a waggon had derailed in the tunnel was shown by the marks on the permanent way but Colonel Yolland was unprepared to give a definite cause for the derailment, though he noted that the track was not in very good condition. He thought that the axle had broken in the final derailment. Mr Margary, the company engineer, said that they had recently had difficulty in obtaining a supply of new rails.[996]

As noted in the description of the Liskeard accident two years earlier, the single line was worked with the assistance of the block telegraph, and the collision which followed the derailment was the result of a mistake by the Devonport signalman. The latter was dismissed after the accident but had made a statement of what he had done to the company superintendent, Mr Compton. He stated that Saltash had given him "Line clear" for the goods train at 4.45 a.m. The Cornwall Junction signalman had then asked for the down mail train at 4.49, he had given "Line clear" and gone out to put his down distant signal at "All right" to permit the mail train to approach his box, leaving the inner signal at "Danger." Returning to his box at about 4.54 he had seen the needle of the Saltash block instrument vertical. At 4.59 Cornwall Junction had given "Mail Train Off" and he had then tried to contact Saltash but could get no reply. After five or six minutes of vain

[996] Board of Trade Report

attempts to contact Saltash he had assumed that the Saltash man must have forgotten about the mail train and gone out, had given the driver of the mail an "All right" signal with his arm and cleared his inner signal for it. He had considered cautioning the driver of the mail but then decided that there was no need to do so and so did not bother. The entry that he had made in his book as to having received the "Train in" signal for the goods train from Saltash at 5 a.m. was, he said, a mistake.

The needle of the Saltash block instrument had of course gone vertical because the derailed waggons of the goods train had brought down the telegraph wires. The signalman had entirely disregarded the 3rd rule of the company signalling regulations, which stated that under no circumstances must a train be allowed to leave a station without the "Line clear" signal having been obtained, also the 20th rule, which was to the effect that when the telegraph was out of order the drivers of trains must be instructed to proceed with great caution.[997]

The guard who was the sole casualty of the accident seems to have survived it, but he was still in hospital at the time of the Board of Trade inquiry.

On 23rd November 1874 a westbound morning passenger train on the Memphis and Little Rock Railroad jumped the track on a trestle a short distance from Madison, Arkansas, United States, and plunged into the mud eight feet below. Contradictory reports stated that no one was hurt and that the engineer, a new employee, was fatally injured.[998]

At the end of 1874 there was a very lucky escape on Harbuck (or Knitsley) Viaduct, on the Consett branch of the North Eastern Railway in the north of England. On the evening of 21st December, the 7.55 p.m. passenger train from Durham to Newcastle via Lanchester was approaching the viaduct when a vehicle became derailed and a first-class carriage was thrown 65 feet from the viaduct to the valley below. Most fortunately there was no one in it and only two of the 20 passengers on the train complained of injury.

Although this accident did not, in the end, prove very serious it clearly needed investigating and Captain H.W. Tyler looked into it. After a careful examination of the permanent way and rolling stock and interviewing all the railwaymen involved, he came to the conclusion that it was due to a loose wheel on third-class carriage No. 1027. He theorised that as the gauge had widened between the loose wheel and the other wheel on the same axle it had left the rails. Unable to get away from the end of the axle it had been dragged along, and the axle and both wheels had remained under the carriage, until in consequence of the bumping of the loose wheel against the deep transoms between the longitudinal sleepers on the viaduct the horn-plates had given way and were torn from the carriage. As the axle and wheel were torn from the carriage, they had formed a serious obstruction to the following vehicles, and first caused the following second-class carriage to leave the line and then the first-class carriage behind the second-class to fall from the viaduct.[999]

[997] Ibid.
[998] Steubenville Daily Herald, Steubenville, Ohio, 24th November 1874
[999] Board of Trade Report

The luck which had attended the above accident on the North Eastern Railway in England was absent in the United States on 15th January 1878. On the evening of that day a crowded eight car excursion train, drawn by two engines, was carrying passengers back from a Moody and Sankey Protestant religious revival meeting at Hartford, Connecticut to various points via the Connecticut Western Railroad. The train had left Hartford at 9.20 p.m. and was crossing a bridge over the Farmington River near Tariffville when the structure collapsed. The two engines, named *Salisbury* and *Tariffville*, went down the river bank on the west side of the bridge and three cars followed, two of them being partly submerged and the third capsizing and also ending up partly in the water. A fourth car stopped with its leading bogie overhanging the edge of the bridge. The roofs of the first two cars were broken in during the fall and some of passengers were able to escape through them.

After the accident, Jonathan Jones, the superintendent of the company, went to Tariffville station and telegraphed for assistance to Hartford, both to his own Company's office and to the office of the Hartford Providence and Fishkill Railroad, who also sent help,[1000] and a relief train left Hartford for the scene at 12.45 a.m., while local hospitals were notified to be ready for patients. Also, a passenger on the train was the president of the line, Caleb Camp, and he sought help locally. A newspaper reported that, 'The citizens of Tariffville had been aroused by the ringing of the alarm bell, and promptly hurried to the scene and rendered good service. It was necessary to approach the wrecked cars over a treacherous reach of ice. The people went out on the ice to the edge of the water and did their best to get out the struggling wounded and the dead. The bodies as fast as recovered were drawn on improvised sleds and placed on the Hartford special train.'[1001] There were 13 deaths in the accident, including Engineer George Hatch, and 70 others were injured. 'Five young men from New Hartford, among the killed, were of a party of six who were on the platform, enjoying a moonlight ride and whistling in chorus. The only survivor of the six was inside at the time of the accident.'[1002]

Conductor T.M. Elmer described his experiences to a reporter, 'I had been in the mail compartment in the baggage car, where I left Mr Jones, the superintendent, assorting tickets. I had just started to go through the train and had got just by the front door of the baggage car and was about to open it, when I felt the rear end settling, and heard a loud crash. The whole car dropped down with but a very little pitching on one side and struck. Just as it struck a heavy beam fell down upon the roof and came through, striking me on the head and knocking me down (there is a cut near his left temple, but not a serious one). I was stunned by the blow, but soon saw what the situation was. There were three other men in the car, whom I did not know. It was very dark in the car, so that I could see nothing but a streak of light which came through the hole in the roof made by the beam. This beam had come end first and had crushed in the roof, so that there was plenty of room for a man to crawl through. The other men saw the opening and we tried to get out but found that we could not reach the roof. Meanwhile, the water was coming into the car through the floor and windows and was rapidly rising. It got up to a point above our waists and then stopped, when we knew that the car had struck bottom, and then we felt safer. We then went to work to rescue ourselves, which was accomplished by lifting one

[1000] Webpage – Connecticut Histtory.org – The Tariffville Disaster

[1001] Middletown Daily Press, Middletown, New York, 16th January 1878

[1002] Indianapolis Journal, Indianapolis, Indiana, United States, 17th January 1878

man up to the roof, and, after he got through, he assisted another, and then the others were helped out. After getting out we went to a car behind, which was lying corner-ways, out of which the passengers were crawling through the windows. The car was pitched down, and it appeared as if more of the passengers must be jammed on the lower end. The cries were heartrending of men and women, many of whom seemed to be suffering from extreme pain. Some were crying piteously, "Oh, don't step on me!" "Get off, for my legs are broke!" "Help us; do!" and all such cries and appeals. We assisted several ladies out, and got out one young man, who was up to his neck in the water. After doing this, I was so thoroughly chilled through that I became numb and had to give up.'[1003]

The bridge was a Howe Truss structure with two 163 ft spans and was said to be in good condition. There was controversy over the cause of the accident. One theory put forward was that it was due to the excessive weight of the two engines, one weighing 32 and the other 28 tons without coal and water. It was stated that, 'One man who stood in the depot at Tariffville as the train left there noticed its make-up and reported yesterday that he waited in suspense to see whether it would pass over the bridge safely, and, while he was waiting, heard the crash.'[1004] A director of the Company, ex-Governor Holley, said later that he had thought the trestlework weak, and had intended to raise the matter shortly at a board meeting. There were calls in the press for a law prohibiting the working of locomotives in multiple over bridges. Later investigation disclosed that the bridge had not merely been overloaded but had been unable to withstand the extra stress imposed upon it when the train engine had become derailed from an undetermined cause.[1005]

A passenger train was crossing a trestle three miles west of Jacobsburg, Ohio, United States, on the Bellaire and Southwestern Railroad narrow-gauge line, on Thursday 19th February 1880, when its engine jumped the rails, dragging the baggage car and several passenger cars after it. At one point the trestle was 80 ft high but the derailment occurred shortly after the train had come onto it and before it reached the highest point. Even so the cars fell 35 feet and shrieks of the passengers as they did so were described as heartrending. The cars fell bottom up and the victims were only released with difficulty. About half a dozen passengers were very seriously injured, though no one was killed outright, and Conductor Burkhead, Mail Agent Booth and Express Messenger Hayden were also quite badly hurt. Mrs A.B. Caldwell, the wife of a director, and Mr A.P. Wiley, a farmer living near Barr's Mills, died from their injuries the following day, at which time the condition of ex-Congressman Lorenzo Danford was stated to be critical. There had been upwards of 40 passengers on the train. A third death was that of a man named Jones, who had been working at the wreck site and fell from the trestle.[1006] Fireman John Abbey reportedly managed to uncouple the engine from the baggage car as the derailment occurred to prevent it too falling into the ravine and Brakeman Burchfield 'had time to extinguish the fire in one coach.'[1007] It is not clear whether this means a fire that had accidentally broken out or whether he put out the fire in the stove to prevent a blaze breaking out.

1003 Philadelphia Inquirer, Philadelphia, Pennsylvania, United States, 17th January 1878

1004 Ibid.

1005 A History of Railroad Accidents, Safety [Precautions and Operating Practices by Robert B Shaw pg. 57

1006 Philadelphia Inquirer, Philadelphia, Pennsylvania, United States, 21st February 1880

1007 Boston Weekly Globe, Boston, Massachusetts, United States, 24th February 1880

The next American accident involved not a derailment but a collision on a trestle. When the Pittsburgh Cincinnati Chicago and St Louis (Panhandle) Railroad train from the east was very late reaching Indianapolis the regular Terre Haute and Indianapolis (Vandalia) Railroad train with which it was meant to connect was dispatched on time and an extra train was run to carry through the passengers westward from Indianapolis. On 24th September 1880 the Panhandle train was much delayed, and an extra train was therefore duly put on, leaving Indianapolis almost three hours after the booked time of the regular train and two hours and twenty-one minutes after the latter was due to arrive at Terre Haute. The line being single a crossing therefore had to be arranged with an eastbound freight train and an order was intended to be sent to the crew of the freight train advising them that the extra passenger train was running 2 hours and 15 minutes behind the schedule of the regular train. Due to a mistake in the transmission of the order by either the dispatcher or the operator it reached the freight train as indicating that the passenger extra was running 2 hours and 50 minutes late, thus leading them to believe that they had 35 minutes more on the extra than they actually had. Both trains were running at considerable speed when they met on a curve somewhere near Bridgeport, Indiana, nine miles from Indianapolis, and by chance on a trestle. Having done what they could to stop their train, the engineer and fireman of the freight train jumped from their engine and escaped with slight injuries but Engineer Oscar Rankin and Fireman William Saunders on the passenger engine did not jump and were killed when their engine plunged from the forty foot high trestle. In the postal car, Chief Mail Clerk Benjamin Dillehunt was probably fatally injured and Mail Agents Ellis Miller and William Lewis were badly hurt. The remainder of the train was not much damaged and the passengers, of whom there were not that many, escaped with a shaking up.[1008]

[1008] Boston Daily Globe, Boston, Massachusetts, United States, 25th September 1880

CHAPTER 11

TUNNEL ACCIDENTS

In his classic work on British railway accidents, *Red for Danger*, L. T. C. Rolt remarks on the acute trepidation felt by Victorian travellers when they were passing through a tunnel, and their oft repeated pleas to the railway authorities that all tunnels should be artificially illuminated, pleas which, probably for practical reasons if nothing else, fell upon death ears. He goes on to point out that these fears were by no means unjustified, especially in the period before the introduction of the absolute block system, when train were merely spaced by the time interval system. If a train should chance to break down in the open, there was a good chance that the crew of a following train might be able to see it in time to pull up, even if there was insufficient time for the guard of the first train to go back and protect its rear. In the smoke and steam-filled darkness of a tunnel there was no such hope. Furthermore, should a serious collision occur in a tunnel the confined space would be likely to result in the wreckage being hideously compressed, and should a fire break out in a long tunnel provided with ventilating shafts, the latter could act as chimneys. In addition, the rescue work would be hampered by the darkness and the difficulty of working in a confined space where the use of breakdown cranes was impossible, and the only resort was to drag the wreckage bodily out of the tunnel.

Because of the special characteristics of tunnel accidents, this book will devote a separate chapter to them, although in fact, despite the obvious dangers of accidents occurring within them, only one accident serious enough to be termed a major disaster occurred within one during the whole of the period covered by this book. The reasons for this fortunate situation are not entirely clear. In part it may be due, paradoxically, to the obvious dangers of a collision in particular in a tunnel led to many of them being provided with some form of signalling from an early date, for example in England Clay Cross tunnel on the North Midland Railway was reportedly protected by the electric telegraph as early as 1841, Clayton tunnel on the London and Brighton at about the same date and Box tunnel on the Great Western in 1847.[1009] Another and perhaps more important factor may simply be that in many countries tunnels themselves were rare, as for example in the United States, where most lines were built as cheaply as possible, and

---

[1009] The Victorian Railway and how it Evolved by P.J.G. Ransom, pgs. 151-5

tunnels, being expensive to build, were therefore avoided as much as possible. Additionally, up until 1901 there were only a few urban underground railways, or subways as the Americans call them.

The world's first railway tunnel was a short one at Frichley, Derbyshire, England, on the Butterley Gangroad Railway, opened in 1793.[1010] No accidents have come to light within its brief length.

The first accident to occur within a tunnel seems to have been a rear collision in Primrose Hill tunnel, England, on the pioneer London and Birmingham Railway, on 11th February 1839. Between 8 and 9 p.m. on the evening of that day, a luggage train belonging to Messrs Pickford and Co (who at this date were carrying goods for the railway company) and consisting of an engine and 36 waggons, 26 of them loaded. The train was running five hours late because its engine was not powerful enough to pull its load. When it was within a few yards of the Chalk Farm, or south end of the tunnel, one of the couplings snapped and the rear portion of the train was left behind in the tunnel. A few minutes later a mail train approached the tunnel from the north, or Kilburn side. The guard of the goods train must have had sufficient time to run back and warn the policeman stationed there, and the latter showed a red light to the driver of the mail train, but his warning signal was obscured by the smoke and steam emanating from the tunnel, and it entered the tunnel unchecked, travelling at between 20 and 30 miles an hour. The collision was a violent one, severely damaging the engine and one or two of the carriages of the mail train. The two guards, named Peoford and Roots, were somewhat badly hurt and a lady and two gentlemen passengers were cut by flying glass from broken carriage windows. The driver and fireman, and the remaining passengers escaped with minor injuries, the escape of the engine crew being described as miraculous.[1011]

Another two years elapsed before the next tunnel accident, this time in Knighton Tunnel, England, on the Midland Counties railway. This accident, on 16th January 1841, was neither a collision nor a derailment, but rather a failure of the tunnel roof. The up mail train from Derby had just left Leicester when, as it entered the tunnel, it ran into a mass of brickwork and earth which had fallen from the roof of the structure, possibly as the result of a recent thaw. The train was only travelling at low speed and only one passenger was hurt, his face badly cut by broken glass from a smashed window, as in the Primrose Hill accident. It took two hours to clear the line, and the train was apparently drawn back into Leicester station, where it was kept waiting for another two hours. Upon arriving at Rugby, the passengers for London by the London and Birmingham Railway had to endure a wait of a further five hours for the arrival of an engine to take them forward, so that they finally reached London Euston station at 2 a.m. the next (Sunday) morning rather than at 5 p.m.[1012]

[1010] Historic England webpage: Where is the World's Oldest Surviving Railway Tunnel?

[1011] The Times, London, England,19th February 1839

[1012] Ibid, 23rd January 1841

A similar accident to the Knighton Tunnel one occurred in Box Tunnel, England, on the Great Western Railway in March 1845. Rumours as to the insecurity of Isambard K. Brunel's famous tunnel had circulated since its opening and in August 1842 Brunel had assured the directors of the company that the tunnel was completely safe. Nevertheless, following a severe frost nearly three years later a piece of stone weighing about three hundredweight fell from the roof of the tunnel in front of a light engine, derailing it.[1013]

An accident in Summit Tunnel, England, on the then Manchester and Leeds Railway on 21st June 1843 was typical of those that occurred in the early days. The mail train which had left Leeds at 11 a.m. was allowed to enter the tunnel without its engine driver having been warned that a luggage train had not long preceded it. As usual smoke and steam in the tunnel restricted visibility and the mail train overtook the luggage train about halfway through the two mile bore beneath the Pennines, smashing two or three waggons but without anyone being hurt.[1014]

The first tunnel accident to result in a fatality seems to have been one which occurred in Clay Cross Tunnel, England, on the 29th March 1844. What was described as a mineral train, which usually meant a coal train, was passing through the tunnel when its driver had a problem with his engine, its weigh-bar shaft being out of order. He got down from his engine to attend to the problem and while he was underneath his machine a following passenger train ran into the back of his train and he was killed.[1015]

On 25th June 1847 two trains collided head-on in the single track Couzon Tunnel, near Rive-de-Gier, in France, on the Chemin de fer de Saint Étienne à Lyon, due to a lack of vigilance from one of the railwaymen charged with ensuring the safety of the line. Two people died, and another was injured, while a number of waggons and an engine tender were smashed.[1016] A newspaper report on the accident stated that an ascending locomotive hauled train collided with a descending train of waggons, implying that waggons were allowed to descend the tunnel by gravity. After the accident, the Sub-Prefect of St Ėtienne, who chanced to be on the line in the course of his duties, stopped a passenger train from Lyons and proceeded by special train to Rive-de-Gier, where he ordered the arrest of the tunnel keeper for having permitted the locomotive hauled train to enter the bore although he had been informed that a train was descending through it. He also organized the clearing up operation.[1017]

Tunnel accidents began to occur in Great Britain with increasing frequency from about 1850 as more railways were opened, most of the accidents being rear collisions, with the occasional derailment or collision with rubble which had fallen from the tunnel

---

1013 History of the Great Western Railway Vol 1 by E.T. MacDermot, pgs. 68-9

1014 The Times, London, England, 28th June 1843

1015 Board of Trade report

1016 Wikipedia - La Presse du 28 juin 1847, p. 2

1017 Courier de Lyon, Lyon, France, 26th June 1847 via Morning Post, London, 30th June 1847

roof. What might be described as the first bad accident happened on 30th April 1851, in Sutton (sometimes incorrectly called Frodsham) Tunnel, on the Birkenhead Lancashire and Cheshire Junction Railway. The accident occurred on the evening of the principal day of the Chester races, when large numbers of racegoers were making their way homewards to Warrington and Manchester over the railway in question. Five trains carrying returning punters were involved, the first of which left Chester station at 5.45 p.m., carrying 430 passengers, and the second at 6.30 p.m., carrying 470 passengers. As the second train was about to set off, a vast crowd of people arrived at Chester station, the races having just finished, and soon there were about 5,000 people at the station. Many of them crossed the lines and climbed into some empty carriages labelled 'Manchester via Warrington.' The stationmaster decided that it would be useless to draw these carriages into the station platform and the third train thus departed from the sidings at 6.50 p.m. This was the first of the trains directly involved in the accident, the first and second trains completing their journeys without incident. The third train consisted of an engine named *Druid* weighing 15 tons, two first-class, six second-class and 10 third-class carriages, the latter being open waggons in which the passengers were forced to stand. An estimated 912 passengers travelled by this train. The fourth train was considerably lighter; it consisted of two first-class, nine second and third-class covered carriages, a cattle waggon pressed into use to carry passengers(!) and a flat waggon loaded with a private carriage. This train left Chester at 7.15 carrying about 430 passengers. As well as being lighter that the train which preceded it, this train was drawn by No. 16, a more powerful engine than the *Druid.* The fifth train started from Chester at 7.30 p.m. and with two first-class and 12 second and third-class covered carriages carrying an estimated 470 passengers was again lighter than the third train, while its engine, the *Albert*, weighed three tons more than the *Druid.*[1018]

It had been raining since the *Druid* left Chester with its train and with slippery wet rails and a heavy load it took 40 minutes to reach Frodsham station, after one intermediate stop, as opposed to the 26 minutes allowed for ordinary trains. About 100 passengers left the train either at the previous station or at Frodsham and it departed from the latter station at 7.33, facing a climb at 1 in 240 from there, followed by a short level stretch and then a long climb at 1 in 264, partly in Sutton Tunnel. As the train set off the engine's driving wheels immediately began to slip. The fireman jumped off and assisted by a platelayer began to sand the rails and the speed of the train increased somewhat on the level stretch but then fell as it reached the 1 in 264 gradient, and the train was travelling at little more than a mile an hour as it entered the tunnel. The fourth train meanwhile, which had left Chester at 7.15, was doing much better, and reached Frodsham at 7.41. The Frodsham stationmaster held this train there for two minutes as the previous one was still in sight, but then fearing holding up the trains behind, he then allowed it to proceed. The driver of No. 16 engine could still see the third train ahead of him and having a lighter load and a more powerful engine soon closed up on it. The guard of the third train signalled to the driver of the fourth to help push them through the tunnel, which the driver of No 16 engine did, but he had to regulate his speed to avoid a collision and it was 7.55 before the fourth train entered the tunnel. The third train had taken 22 minutes to cover less than two miles.

The combined third and fourth trains moved slowly through the tunnel, but the *Druid*

[1018] Board of Trade report

engine's driving wheels began to slip again the moment the trains joined up and about halfway through the 1 mile 180-yard tunnel the combined train stalled. Full of steam the tunnel was also pitch dark; the carriages had no lighting and the guard of the third train did not even have a hand lamp, though the guard of the fourth did have one, and that train also carried a tail lamp.

With the line being worked on the time interval system and the certainty of another train following protection of the stalled train was clearly a matter of great urgency, but here the railwaymen concerned seem to have dithered. The fireman of the third train was sent back by his driver and making his way cautiously between the carriages and the tunnel wall in the darkness and past several obstructions had got about halfway along the combined train when he met the guard of the fourth coming forwards. Meeting the fireman, the guard then belatedly went back and getting to the end of his train, where he was able to get onto the four foot, he began to run, but it was now too late. Seeing the engine of another train entering the tunnel the guard completely lost his presence of mind and fell over in the four foot, the train passing over him as he lay on the ballast without hurting him. The fifth train, hauled the *Albert*, had made as good a progress as the fourth, and had reached Frodsham station at 7.55, just as the first and second trains were entering the tunnel. The Frodsham stationmaster held it for two minutes, but then let it go without telling its driver of the two trains ahead. The rules of the company dictated an interval of only five minutes between trains, and as the third train had been gone 24 minutes and the fourth 14 the stationmaster decided that no caution was required. No special directions had ever been issued as to the intervals of time to be allowed between trains entering the tunnel, nor any special instructions regarding the handling of the abnormal traffic during race week.[1019]

Travelling at considerable speed, the fifth train entered the tunnel. Its driver found the tunnel full of smoke, but that was to be expected, and he had received no caution at Frodsham. He was also aware that if he reduced speed too much there was a danger of his train being run into by the following one and he therefore pressed on. The *Albert* thus collided at high speed with the rear of the fourth train, destroying several carriages and derailing both them and the engine. The wreckage completely filled the tunnel and the body of one of the passengers was found crushed between the remains of a carriage and the tunnel roof. Needless to say, there was a scene of fearful confusion as 1,600 passengers found themselves packed together in the darkness, especially when another train was seen approaching the tunnel, the sixth departure from Chester. Fortunately, a second collision was avoided as a man had run back and was able to stop it. A Manchester police superintendent who was a passenger felt the time by his watch and put it at 8.3. In all, six passengers lost their lives in the accident and a great many more were injured.[1020]

Captain Laffan, the inspecting officer of the Board of Trade who investigated the accident, found the all the servants of the railway company in any way involved in the collision had acted 'with a singular absence of all ordinary prudence and discretion.'

Rather than arranging additional protection during race week, the company secretary had withdrawn the two policemen normally stationed near the Sutton tunnel who might

---

[1019] Ibid.

[1020] The Times, London, England, 2nd,3rd, 5th,6th,7th,8th,9th, 10th, and 12th May 1851

have warned the driver of the *Albert* of the danger.

The locomotive superintendent had attempted to work the extraordinary traffic of race week with a stock of engines insufficient for even ordinary traffic. He had classed the *Druid* engine as weighing 21 tons instead of only 15 and had thus allowed an underpowered engine to haul the train and he had allowed the open carriages to be overloaded.

The drivers of the *Druid* and No. 16 engines had been extremely imprudent in attempting to take the combined train through the tunnel plus had failed to send someone back to the tunnel mouth to warn following trains, should the combined train stall.

The guards of the third and fourth trains had failed to protect them in the rear when they had entered the tunnel at low speed and the guards were aware that another train was likely to be following close behind. Furthermore, the guard of the fourth train, when he did finally go back, had showed, in Captain Laffan's words, 'a want of self-possession and firmness which should prevent his ever being employed in a similar capacity again.'

The driver of the *Albert* had entered a tunnel which he knew to be unprotected by a policeman or signals and which was full of steam at headlong speed regardless of the fact that the tail lamp of a preceding train could not be seen for more than a few yards.

The Frodsham stationmaster, despite his having seen the difficulty which the *Druid* engine was having with its train, had failed to foresee the possibility of a delay or stoppage in Sutton tunnel and had let the following train proceed without cautioning its driver.

The inspecting officer nevertheless felt that the main responsibility for the accident lay with the management of the railway company who had resolved to try and work the exceptional race traffic without any extra engines, even though the stock of engines was insufficient even for the ordinary traffic. They had also taken no special measures to protect the line, even when trains were to follow one another as quickly as they could be filled, and despite special precautions having been taken on the Chester to Birkenhead line. In contrast, on the London and North Western Railway Chester to Crewe line platelayers had been employed along the whole length of the railway for the race traffic, acting as policeman and stationed within sight of one another.

The inspecting officer concluded his report with a series of recommendations to prevent a recurrence of the accident: -

The two stations which the company already proposed to build at each end of the tunnel should be completed and linked by electric telegraph so that no train was permitted to enter the tunnel until the previous one had left it (Halton and Norton stations subsequently opened at the Chester and Warrington ends respectively of the tunnel).

Each passenger train should carry two guards.

The stock of engines and carriages should be increased.

The company should engage an efficient staff of officers and men

All carriages passing through Sutton tunnel by day or night should be lit. The lights could be added and removed at the new stations on either side of the tunnel.

The interval of time between following trains should be considerably increased, and the drivers told how long ago the previous train had passed.[1021]

The coroner's inquest into the accident returned a verdict of 'Accidental death', the jury adding a rider echoing the observations and recommendations of the Board of Trade's inspecting officer.[1022]

Some tunnels were the scene of more than one accident, thus there were rear collisions in Stoke Tunnel, England, on the Great Northern Railway main line, on 14th December 1853 and again less than three months later on 7th March 1854, to the exasperation of the inspecting officer of the Board of Trade. The first accident involved three trains, a goods train which left Corby station for Grantham at 12.30 p.m. a coal train which followed from that station at 12.35, and a second coal train leaving the same place at 12.45. As the driver of the first coal train approached Stoke tunnel, he found it full of smoke and steam and so proceeded cautiously to avoid running into the train in front. The driver of the following coal train was unfortunately less cautious and ran into the preceding one two thirds of the way through the tunnel, seriously injuring its guard. The line was still worked by the time interval system, the regulations providing that a danger signal be shown for five minutes and a caution signal for eight minutes more after the passage of a train. Lieut Tyler commented that these intervals were clearly insufficient for coal trains following one another, let alone for an express train following a coal or goods train. However, he stated that he was glad to learn that the company planned Corby and Great Ponton stations, at the south and north ends of the tunnel, to prevent two trains being in the tunnel at the same time.[1023]

Nothing had been done as regards the installation of the electric telegraph by March of the following year and the second accident was almost a repetition of the first. Again, three down train were involved. The first, a goods train, passed through Corby station at 9.36 a.m. The second was a ballast train which had been shunted at Corby for a Doncaster race special and other trains, and left there at 9.45, and the third, a coal train, passed according to the stationmaster at 10 a.m. The driver of the ballast train was cautioned both by some platelayers and by the driver of an up coal train that he was catching up the train in front and hence entered the tunnel cautiously, only to be run into from behind by the following coal train in a re-run of the previous collision. The driver of the third train said that the signal at Corby had been lowered from danger to caution as he had approached, and the signal porter there had told him that the ballast train had been gone some time and would be out of his way. He had proceeded, travelling at about seventeen miles an hour, but on receiving, like the ballast train, caution signals from the platelayers and having also been warned by the driver of an up coal train had shut off steam and entered the tunnel at little more than walking pace. He had been going only slightly faster when his engine ran into the guard's van of the ballast train. A labourer riding in the van and one of thirty-two accompanying the train were seriously injured.

Lieutenant Henry Tyler, the inspecting officer of the Board of Trade, did not believe the times quoted by the Corby stationmaster, pointing out that the collision must have occurred some minutes before 10 and that according to the stationmaster's times the coal

---

1021 Board of Trade report

1022 The Times, London, England, 13th and 14th May 1851

1023 Board of Trade report

train had run into the ballast train four miles beyond Corby some minutes before the former train had even left Corby, which was clearly impossible. The times given by the signal porter were equally unreliable, and Lieut. Tyler believed it likely that while the ballast train did leave at 9.45 the coal train had followed it less than five minutes afterwards. The stationmaster was meant to keep a time book, but the times of none of the three trains were recorded; the goods because it had passed without stopping, the ballast train because those were not recoded, and the coal train because it was not booked to stop. Had the times of all trains been recorded, the inspector pointed out, it would tend to ensure the proper time intervals being kept.

Concluding his report, the inspecting officer stated that he hoped that having now had a second accident in Stoke tunnel the Great Northern company would no lose no time in adopting measures to prevent a third.[1024]

By May 1859, the 1,645 yard long Bowling Tunnel, not far from Bradford, England, on the Lancashire and Yorkshire Railway, was being worked on the block system using the electric telegraph, a fact which failed to prevent a collision in it on the 6th of that month. At the east (Bradford) end of the tunnel a Great Northern Railway branch joined the Lancashire and Yorkshire line from Bradford and this end of the tunnel was controlled by a Great Northern signalman, who worked with a Lancashire and Yorkshire signalman at Low Moor station at the other end of the tunnel.

At 3.32 p.m. on the day of the accident, a passenger train from Leeds entered the tunnel, and as it did so a train from Bradford arrived at the junction. The Bowling Junction signalman subsequently learning by telegraph from Low Moor that the Leeds train had arrived there the Bradford train was allowed to pass into the tunnel in turn. As it did so, a second train from Bradford arrived at the junction, and after some short delay was permitted to enter the tunnel in turn, whereupon it overtook and ran into the rear of the first Bradford train about 200 yards from the Low Moor end of the tunnel, more or less injuring six passengers and two railwaymen.

Clearly one of the two signalmen involved had made a bad mistake; but which one? The Bowling Junction signalman told the inquiry into the accident that he had telegraphed the first (Leeds) train "Train in" to Low Moor, had received an acknowledgement and let the train go. The first Bradford train having shortly arrived at his hut, he had telegraphed "Train waiting" to his colleague, received the reply "Line blocked" and had held the train until he received the message "Line clear", when he had sent "Train in" and let the second train go. Upon the arrival of the second Bradford train he had again telegraphed "Train waiting" to Low Moor, whereupon the latter signalman had telegraphed "Train out." He had then sent the "Train in" signal to Low Moor for a third time and let the second Bradford train into the tunnel. Following the collision, the Low Moor signalman had asked him why he had let the third train in, and he had replied "You gave me it out", referring of course to the second train.

The testimony of the Low Moor signalman agreed with that of the signalman at Bowling Junction up to the time of the arrival of the second Bradford train at the entrance to the tunnel, but he stated positively that he had exchanged no signals regarding this

---

1024 Ibid.

train until after the collision. Upon hearing the noise of the collision near his end of the tunnel he had telegraphed Bowling box and asked, "What did you let that train come in for?" but had received no reply other than the ringing of his bell to acknowledge the message.

On the face of it, the Board of Trade inspecting officer wrote it was hard to tell which of the two signalmen had made the mistake. There had, however, been a marked difference in the manner in which the two signalmen had given their evidence. The Low Moor signalman, a much older and less intelligent man, had told a plain story without difficulty, whereas the younger Bowling Junction man, had, 'using superior sentences' contradicted himself in a way which implied that he was telling a made-up story. There was also the matter of a complaint made by the Low Moor signalman to his stationmaster regarding an incident a week later, on 13th May, when a goods train had appeared at the Low Moor end of the tunnel at 2.40 a.m. without its having been signalled from Bowling Junction. Looking into the matter the Low Moor stationmaster had found that the goods train was recorded in the Bowling Junction train register book as having been signalled into the tunnel at 2.03 and out at 2.06. These were close to the usual times that the train passed, but half an hour earlier than the time it had actually passed on this day. The Bowling Junction signalman had offered an ingenious if improbable excuse to explain this discrepancy about his watch stopping at 2.03 and his having had to guess the times.

Captain Henry Tyler stated that, having reviewed all the circumstances, he was sure that it was the Great Northern signalman who was to blame for both the accident and the later incident. He considered, however, that the telegraph arrangements, which were those in general use on the Great Northern Railway, were in need of improvement. At the time of the accident the telegraph needles were used both to signal the trains and to indicate their nature and stood vertical except when messages were actually being sent. The inspector thought that a better system would be to have one needle reserved for each line, with the needle being fastened over to "Line blocked" by the signalman at the far end of the tunnel upon a train entering it until it emerged from the other end, so that the signalmen had a constant indication of the state of the line, whether blocked or clear, and did not need to rely on their memories. He also recommended the installation of proper signals at the Low Moor end of the tunnel, supplying the Low Moor signalmen with a train register book, and was of the opinion that the Great Northern signalman was not a fit person to do the job.[1025]

An accident in Tottenham (later Wood Green) Tunnel, England on 27th July 1860, on the Great Northern Railway main line, involved a derailment rather than a collision and could easily have been much more serious than it was. It involved an up express train from Edinburgh to Kings Cross, which had left the Scottish capital at 10 a.m. and was due in London at 9.30 p.m. The train had just entered the tunnel, travelling at close to 60 miles an hour, when its engine left the track, crossed over to the down line, threw off a wheel, returned to the up line and scraping along the wall of the tunnel came to a stand in 400 yards. The vehicles of the train were not greatly damaged but both lines were blocked and a corresponding down express, due to leave Kings Cross station at 9.15 p.m., usually passed the up train at about this point. A disastrous collision in the tunnel

[1025] Ibid.

between the down express and the derailed vehicles of the up train was therefore a distinct possibility. It was prevented by the signalman, John Tarver, in Wood Green box, at the south (Kings Cross) end of the tunnel. He had received the "Train in" message from the Southgate signalman at the north end of the tunnel and a bell had rung in his box shortly afterwards, indicating that the up train had passed over a treadle working in connection with a signal at the north portal of the tunnel. Expecting the train to emerge from the tunnel shortly he had then left his hut to replace his distant signal at the south end of the tunnel behind the train. The train had failed to appear, but he could hear a lot of whistling from it. Returning to his box he had asked the Southgate signalman via the telegraph instrument if the latter could see anything of the train but could get no reply. Hornsey box, to the south of him, had then telegraphed notice of the down express and he had again tried to contact Southgate box but without success, the telegraph needles just quivering and shaking. Realising that there must be something wrong with the up express he had stopped the down train and sent a porter to fetch the stationmaster. The down express was just pulling up at Wood Green as the fireman and guard of the up train emerged from the tunnel on foot and explained what had happened. The down line was not clear for the express to proceed until 5.30 the next morning. None of the 40 or 50 passengers on the up express was hurt but its fireman suffered injuries to his hand and arm. The derailment had resulted from the fracture of the leading axle of the engine, No. 206, which had contained a flaw.

Following the accident, the locomotive superintendent of the company undertook to replace all the axles of the same type used under similar engines.[1026]

The Tottenham Tunnel accident was something of an exception inasmuch as it involved a derailment. Rear collisions continued to be the main class of tunnel accidents and two more occurred in Great Britain before the end of 1859, both on the Caledonian Railway in Scotland; a passenger died in an accident in Bishopton Tunnel on 16th July [1027] and another collision occurred near Glasgow Buchanan Street terminus on 27th August.[1028]

On 18th January 1860 a similar accident took place in the United States, on the Hudson River Railroad in New York state; a Harlem Railroad train came to a stand in Yorkville Tunnel and was run into by a New Haven Railroad train, considerably injuring a Mr Palmer of White Plains. A man had been sent back with a flag to warn the following train, but his flag was not seen through the smoke and steam.[1029] In October 1860 (exact date unknown), an accident of the same type seems to have occurred in Rilly La Montagne Tunnel in France. Then on 25th August 1861 came the Clayton Tunnel disaster, the worst accident to occur in a tunnel anywhere in the world up until the end of the nineteenth century.

Aware of the dangers posed by the 1 mile 499 yard long tunnel under the South Downs the London Brighton and South Coast Railway had already provided a signal box

---

1026 Ibid.

1027 Ibid.

1028 The Times, London, England, 1st September 1859

1029 Berkshire County Eagle, Pittsfield, Massachusetts, United States, 26th January 1860

at each end of it in about 1853, each provided with a single-needle instrument, a clock and a telegraph bell. The line up to the tunnel from the Brighton terminus however, and beyond the north box, was still worked by the time interval system with a stipulated minimum interval of five minutes between following trains. Clayton Tunnel South box was provided with only one signal, an up distant 300 yards from the hut. This was one of a type used extensively by the Brighton Company, known as a Whitworth signal. The signal was designed to be semi-automatic; the signalman could place it at "Clear" or "Danger" by operating a wheel in his hut, but a treadle was provided which was intended to put the signal automatically back to "Danger" when it was depressed by the wheel of a passing train.

Like the Sutton tunnel accident just over ten years earlier, the Clayton tunnel accident occurred at a time of heavy traffic, the ordinary trains being supplemented in this instance by excursion specials. On the day of the accident, a Sunday, three passenger trains were despatched from Brighton to London in quick succession by the assistant stationmaster at Brighton, Charles Legg. The latter told the inquiry into the disaster that he had let them go at 8.22, 8.27 and 8.36 by his watch, but the evidence of guards, drivers, signalmen and others indicated that their actual departure times had been about 8.28, 8.30 or 8.31, and 8.35. Their advertised departure times were 8.05, 8.15 and 8.30.

The first train was a fortnightly excursion from Portsmouth, consisting of an engine and tender and 16 carriages. Delayed in calling at sixteen stations between Portsmouth and Brighton, it had not arrived at the latter station until 8.16 and had left there 23 minutes late. It reached the tunnel in safety, but the automatic signal failed to return to "Danger" behind it. Passing through the tunnel it continued non-stop to Norwood Junction, eight and a half miles from London. Behind this train was a regular excursion from Brighton, consisting of an engine and tender and 17 carriages. Delayed in starting by the late running Portsmouth train it set off at 8.31 according to its driver, or about 8.30 according to the station foreman and at 8.30 exactly according to the under-guard. When it approached the south end of Clayton tunnel the Portsmouth train had not yet been telegraphed as having left the bore, but the distant signal stood at "Clear and the train passed it with steam on" The signal had, as noted above, failed to go back automatically as it should have down and Signalman Henry Killick in the south box had failed to restore it manually. Seeing the second train about to enter the still occupied tunnel he did, however, seize his red flag and wave it violently at the driver as the second train passed him. John Scott, the driver of the Brighton excursion glimpsed the red flag as his engine entered the tunnel, but it took him some time to pull his train up, and it was clear (although the driver later denied it) that he had thereupon commenced to set his train back through the tunnel to the south portal to find what was wrong.

Meanwhile, the third train, an ordinary passenger working from Brighton to London, consisting of an engine and tender and 12 carriages, had left the terminus at 8.34½, 8.35, 8.36 or 8.37 according to different credible witnesses. The Brighton Station Yard signalman gave the passing times of the three trains as having been 8.29, 8.32 and 8.36 and the evidence of other witnesses suggested that they might have been even closer together. The driver of the third train, William Gregory, also found the Clayton Tunnel South distant signal showing "All right" and Signalman Killick holding out a white flag for him to proceed. Entering the tunnel at, according to his estimate, 25 miles an hour, but probably more, he soon saw two red lights ahead. There was little time to do more than shut off steam and reverse his engine, while his fireman applied the tender brake, before the collision occurred. The engine of the colliding train mounted the brake carriage at the end of the excursion train, pushed it to one side, pushed forward the next carriage and finally came to rest almost vertical, its funnel striking the tunnel roof, 24 feet above the track. The force of the collision pushed the first train forward about 50 yards. It was in the carriage beneath the colliding engine that most of the fatalities occurred, many of the poor victims being scalded to death by escaping steam from the engine. Twenty-one bodies were recovered from beneath the engine and 23 passengers in all died in the disaster. The guard in the rear brake carriage had time to jump out before the collision and so escaped unhurt.[1030]

Like the Bowling tunnel accident, this disaster clearly arose from a mistake made by one or other of the signalmen. Signalman Henry Killick in the south box and Signalman James Brown in the north box agreed that the "Train in" signal had been given for the Portsmouth train when it entered the tunnel and the "Train out" given when it left it. Killick stated that he had given the "Train in" signal to Brown (after he had waved a red flag to its driver) for the Brighton excursion immediately Brown had given him the "Train out" for the Portsmouth train and had then gone out to show a white flag to the guard. Brown stated however that he did not get the "Train in" for the second train until a minute after he had given "Train out" for the first. Killick stated that on the approach of the third train he had asked Brown if the second train was out and had only let the third train go when Brown replied "Yes". Brown, however, denied receiving any message from Killick after the second "Train in" until he received one which began with "The train was not out," to which he had answered "Yes" and ended "no, they are run into one another." From other evidence, the inspecting officer, Captain Henry Tyler, thought Killick's account more likely than Brown's as to these messages. However, if Killick had been supplied with a stop signal outside of his hut as well as a distant signal (as already provided at the north box) and he had been obliged to place this signal to "Danger" behind every train entering the tunnel, the accident would probably not have occurred.[1031]

Captain Tyler went on to describe exactly how he thought that the accident had occurred:

The three trains had left Brighton within an interval of seven minutes, say three minutes between the first two and four minutes between the second and third, and had maintained these intervals as they reached Clayton tunnel. Signalman Killick had failed to notice, or noticed too late, that the automatic signal had failed to work behind the Portsmouth train. He had then very properly waved a red flag at the second train as it entered the tunnel before the first had been signalled back as having left it. The second train had restored the automatic signal to "Danger" or Killick had done so. Thus far, no accident was inevitable, as the second train had been stopped by Killick's red flag and the ordinary train would be stopped behind it by the automatic signal. The accident had happened because Killick, little imagining that the second train was backing towards him and thinking that its driver had not seen his red flag, had then too impatiently enquired of Brown if the second train was out. Brown, believing that Killick was referring to the first, Portsmouth, train had answered "Yes" and Killick, failing to realise that it was too soon for the second train to have passed through the tunnel, had cleared his signal and allowed the third train to enter the bore. The crash was then unavoidable and the third train,

---

[1030] The Times, London, England, 26th August 1861

[1031] Board of Trade report

travelling at perhaps 20 or 25 miles an hour, had collided with the second backing towards it at about 5 miles an hour, the collision occurring 200 yards within the tunnel.

Continuing, Captain Tyler stated that the Hastings Junction signalman had not regarded the short intervals between the three trains as being anything unusual, and that the Brighton stationmaster had apparently only done what had often been done before in reducing the intervals between the trains to less than five minutes. The stationmaster was, however, occasioning greater risk to the trains should one materially slacken speed, break down or be brought to a stand, especially as the line between Brighton and Clayton tunnel was obstructed by curves, cuttings, overbridges and the 488-yard Patcham tunnel.

Signalman Killick might have overlooked the failure of the automatic signal after the passing of the first train but had it at "Danger" when the second approached. Given the system under which he was working he could not be blamed for allowing the third train to proceed having been informed that the second was clear, though a more prudent man might have stopped to consider whether the second train could have cleared so quickly and have waited to ascertain if the second train had responded to his red flag.

Signalman Brown in the north box had made 'a simple, though momentous mistake' in supposing that Killick's telegraph query had applied to the first train rather than the second.

Driver Scott of the Brighton excursion, the inspecting officer considered, had committed a grievous error, as it turned out, in backing his train, but he had been placed in a difficult position. If he had acted strictly according to the regulations, he would have stopped his train where it was and sent his guard back to find what was wrong, but this would have subjected his passengers to a disagreeable detention in the tunnel. A less cautious man might have proceeded cautiously to the north end of the tunnel, but would have disobeyed the regulations in doing so, and have been subject to severe punishment should any accident have occurred.[1032]

The inspecting officer said that he was somewhat surprised to find that record books recording the times of telegraph messages were kept by neither the Brighton or Hassock's Gate stationmasters, nor by the Hastings Junction or Clayton tunnel signalmen, although such books were employed on many lines of railway, particularly where signalmen with one telegraph needle for two lines of rails were required to trust to their memories. Such books, keeping a permanent record and liable to periodical or irregular inspections, were likely also to discourage irregularities such as the despatch of trains at less than the authorised time intervals. The inspector then embarked upon a lengthy discussion as to the merits of the absolute block system over the time interval system. He conceded that the line through Clayton tunnel upon which the accident had occurred was already worked using the electric telegraph, but it was worked, he stressed, under a most unsatisfactory system. He basically made the same recommendation as he had made with regard to the Bowling tunnel accident two years earlier; that there should a separate telegraph needle used for each line of rails, and that the signalman (A) receiving notice of the approach of a train from his colleague at (B) should at once by means of a pin inserted in the handle of his instrument block over the needle at (B) to "danger". Both signalmen would then have a visual reminder that there was a train in the section between them, the needle not being released by the signalman at (A) until the train reached him.

[1032] Ibid.

At the time of the Clayton tunnel accident, in contrast, there was only one telegraph needle between the two boxes for both up and down lines, which could not therefore be kept at any particular indication but had to stand vertical between the sending of messages, the signalmen trusting to their memories of what messages had been sent. Had this defective system not existed Brown's mistake would, the inspector thought, have been much less likely. The main cause of the accident was, therefore, not the mistakes made by the signalmen but the defects in the system which had permitted them to happen. In order for the system to work properly, a third needle should also be provided for the signalmen to describe trains or exchange any other messages, the provision of a separate single needle instrument preventing them from being tempted to use the first two needles other than to indicate either "Line clear" or "Line blocked."

Another point raised by the inspector was the length of time that the signalmen were expected to remain on duty. As it happened, Killick and Brown had only just come on duty, but they were anticipating working for 24 hours without a break. The traffic manager had intended them to work shifts of only 18 hours on alternate Sundays, so that they might switch from working days to nights and vice versa, but the men themselves had elected to work 24 hour Sundays so as to get one whole day a fortnight with their families. Common humanity, as well as common prudence, Captain Tyler stated, should mean that the signalmen should at least have alternate Sundays free without the need to resort to such measures. Relief signalmen should instead be provided to take their place. The captain was also distrustful of automatic signals, stating that he had not recommended their general adoption because they could not be guaranteed not to fail to work, and the men coming to rely on them might then not be sufficiently alive to their duties to work them manually should it prove necessary. He also considered that the signal in question had not been properly maintained.[1033]

The inquest on the victims of the disaster began the day after it had occurred and lasted for nine days. At its conclusion, the jury announced that they had decided that signalmen Killick and Brown had made a mistake or had a misunderstanding in working the telegraph but there had been no negligence on their part. Likewise, they ruled that Driver Scott of the leading train had made a mistake in reversing his train back through the tunnel but had not been careless. The person whom they felt most to blame, and against whom they returned a verdict of manslaughter, was Assistant Stationmaster Legg at Brighton, who had despatched the trains at such short intervals.[1034] It appears that this charge was, however, not proceeded with.

Another tunnel collision occurred in the United States on 17th September 1862 when the 6.30 train for Newburyport (spelt Newbury Port in the newspaper account) on the Eastern Railroad collided with an excursion train as it was passing through Western Tunnel in Massachusetts state. Two firemen lost their lives and 30 or 40 passengers were more or less injured. [1035]

1033 Ibid.

1034 The Times, London, England, 27th,28th,29th,30th and 31st August and 2nd,3rd,4th, 5th, 6th and 11th September 1861

1035 Burlington Weekly Hawkeye, Burlington, Iowa, United States, 20th September 1862

An accident with unusual features occurred in New Street North Tunnel, adjoining Birmingham New Street station, England, on the London and North Western Railway, on 28th June 1865. The 760-yard tunnel had been under repair for three years, with defective portions of its brick arch being taken down and replaced. The work was being carried out between 11 p.m. and 5 a.m. each night, when no trains were running. A timber platform was provided for the men to work from, standing rather more than 14 feet clear of the rails. On the morning of the accident, a goods train left Adderley Park sidings, a mile and a half south of New Street station, ten minutes late at 10.10 a.m. It consisted of an engine and tender, six waggons carrying pig iron, one loaded with bran, one carrying a disabled waggon, eighteen empty trucks and a guard's van. After having been stopped ten minutes at the ticket platform south of the station the train entered the north tunnel at 10.25. It had travelled less than 400 yards when its driver felt a jerk and the brakeman was thrown down in his van. The driver stopped his train, but before he could do anything a train ran past in the opposite direction and collided with the derailed waggons of his train and other debris obstructing the tunnel. The brakeman was hurt in the leg and briefly trapped in his van by items thrown against the door but was then able to make his way back through the tunnel to the station to warn the signalman what had happened.

The second train involved in the accident was a passenger train which had left Dudley for Birmingham at 10.01 a.m. and had entered the north end of the tunnel at 10.28. In the smoke and darkness, its driver knew nothing of the obstructions before he ran into them. None of the passenger train was derailed and its coaches were little damaged, but 21 passengers were injured by the violence of the sudden stop, three of them seriously, and a guard was thrown down and knocked unconscious for some time.

It transpired that what had happened was that the disabled waggon, which was loaded onto another waggon of the goods train, had fouled the scaffolding at two points, waggon and scaffolding then falling onto the up line where they were struck by the passenger train. Both waggons were high sided coal trucks, No. 304 and 703. Hired by Messrs W. and T. Phillips, London coal merchants, they had been loaded on 24th June at Willesden Junction, and had travelled from there to Rugby, and from Rugby to Adderley Park sidings. There was no loading gauge at either Willesden or Adderley Park for the waggon examiners to determine whether or not the waggons were within the maximum loading gauge. Experimenting with similar waggons the Board of Trade inspector, Captain Henry Tyler, found that when loaded they were in fact five or six inches out of gauge. He said that ordinarily the blame for the accident would rest with the chief foreman at the initial dispatching stations, as succeeding stations would assume that the waggons had been properly loaded. He could not, however, blame the foreman at Willesden in the absence of a loading gauge there and he recommended that gauges should be forthwith provided at all stations at which waggons were loaded or goods trains despatched. In conclusion he remarked that even then there should have been a small amount of clearance and there must have been some irregularity in the shape of the tunnel arch or the waggon which had combined with an oscillation of the truck to cause the accident.[1036]

The 7.45 a.m. London and South Western Railway down passenger train from Yeovil to Exeter, England, had almost reached its destination at about 9.20 a.m. on Saturday 25th

[1036] Board of Trade report

November 1865 when it ran into trouble in Blackboy Tunnel, less than a mile from Exeter Queen Street station. Steam had already been shut off as the train entered the tunnel, preparatory to stopping at the ticket platform, and when the driver saw a heap of rubble on the track in front of him he braked hard, but on the falling gradient he could not stop the train before it ran into the obstruction. The train consisted of an engine and tender, guard's van, three third-class, one second-class and one first-class carriage. The engine and tender, van and third-class carriages were thrown off the track by the shock but beyond a slight injury to the guard there were no casualties. The passengers then abandoned the train and made their way on foot along the line to the station. A short time later more earth, bricks and rock fell and crushed the roof of one of the third-class carriages. The tunnel roof must have failed a short time before the arrival of the train as the up trains to London at 7.30 and to Exmouth at 8 a.m. had passed through the bore without any problem, nor had the ganger who had walked through it at 8 a.m. noticed anything wrong.

By the time that Colonel W Yolland of the Board of Trade had arrived to inspect the tunnel, the company were in the process of building a new brick arch, so that he could not see the state of the tunnel as it had been immediately after the accident, but he was informed that there had been an arched space about 20 feet high above the arch of the tunnel which the workmen must have failed to pack when the tunnel was built in about 1860. There was no appearance of any quantity of water, though a newspaper report spoke of heavy rain on the days preceding the accident. As well as repairing the arch the company proposed to line the tunnel with a brick wall for 20 ft on either side of the opening. The tunnel was brought back into use, with single-line working, on 3rd December; in the meantime, passengers were conveyed in omnibuses and coaches between Exmouth Junction and the city centre. There was a strong feeling among local residents that given the difficulty of making the tunnel perfectly secure it should be converted into an open cutting – they were to be disappointed, 154 years later it remains a tunnel.[1037] [1038]

The Great Northern Railway main line out of London Kings Cross, England, was the location of a spectacular accident on 9th June 1866 in Welwyn North Tunnel which involved three trains. The events which led to the accident began at 11.20 p.m. that day when a down goods train consisting of an engine and tender, 38 empty coal waggons and a guard's van entered the 1,066-yard tunnel, which was the second of two between Welwyn and Knebworth stations. The train had got about a quarter mile into the tunnel when one of the boiler tubes of its engine burst, water put out the fire and it stalled in the tunnel. Sent back by his driver the fireman of the train asked the guard, Joseph Wray, what was to be done. Wray suggested letting the train run back down the falling gradient to Welwyn station, but this procedure was strictly against the regulations, and the driver refused to do so. Wray then failed to go back to protect the rear of his train. At 11.36 a Midland Railway goods train approached Welwyn (between 1857 and the opening of its own line to St Pancras in 1868 the Midland reached London over the Great Northern line from Hitchin). It was checked by signals, then allowed to proceed into the tunnel, and in the usual scenario the smoke and steam in the tunnel prevented its driver seeing the tail

1037 Ibid.

1038 The Times, London, England, 27th November 1865

lights of the stalled train ahead, which his engine collided with at a speed of 20 or 25 miles an hour. Guard Wray paid dearly for his failure to protect his train when he was killed in the collision. Travelling irregularly with him in the van had been a former Great Northern employee named John Rawlins, who had been latterly employed on the Metropolitan Railway. Rawlins was fatally injured in the collision and died three days later. Shortly after the first collision, a Great Northern up fast goods train (engine and tender, 31 waggons and two vans) entered the tunnel from the north or Knebworth end and a second collision ensued, completely blocking the tunnel with the wrecked waggons of three trains. To complete the work of destruction, about four hours later men engaged in beginning the clearance operation dropped some naphtha from one or more of their lamps amongst a waggon load of furniture packed in shavings and a fire broke out which burnt until the afternoon of the 11th, destroying thirty waggons and badly damaging the engine of the up train. The tunnel was not re-opened for traffic until 3 a.m. on the 12th. In addition to the two railwaymen who lost their lives, two others were injured.

Captain F.H. Rich, who investigated the accident on behalf of the Board of Trade, placed the principal blame for it on the deceased Guard Wray, who had apparently remained quietly in his van for about 16 minutes after his train had come to a stand, in disobedience of the company regulations, which directed him to go back a mile to protect his train, and not to trust to the protection of the block system. Secondly, of course, there had been a mistake made in the signalling between the Welwyn and Knebworth boxes.[1039]

When the Midland train arrived at Welwyn at 11.36 the signalman there, James Bradford, had not received notice from his colleague at Knebworth, Joseph Harding, that the Great Northern train had cleared the section. He had therefore telegraphed Knebworth to ask if the Great Northern train had arrived. The Signalman Harding was emphatic that he had replied "No," Signalman Bradford equally positive that the answer had been "Yes." The Great Northern train, had it not stalled, would have cleared the section at about 11.30, at a time that the Welwyn signalman was busy telegraphing an up train, and he assumed that he must have thereby missed observing it. The telegraphing of trains between Welwyn and Knebworth was done with a single speaking instrument with "In" and "Out" marked on it. The needle was moved to the same side for "Out" and "No", with the number of beats being the difference between them, and Captain Rich considered that the Welwyn signalman must have misread "No" as "Out". The captain believed that both signalmen were speaking truthfully, and that each genuinely believed that it was the other who had made the mistake. The Welwyn signalman should however not have telegraphed asking if the first train was clear but merely signalled "Train waiting." It was very desirable, the inspecting officer added, that the Great Northern company change their system of telegraphing, as they had stated they intended to do, and to have instruments capable of being blocked over to "Line clear" or "Line blocked" with separate speaking instruments.[1040]

At the inquest into the deaths of Rawlins and Wray the jury took only ten minutes to return a verdict of "Accidental Death". Their foreman observed that 'with regard to the important question of "Yes" or "No" given by Harding to Bradford there was no evidence beyond that of the signalmen themselves, who directly contradicted each other. Under these circumstances, the jury would not take upon themselves the responsibility of

---

[1039] Board of Trade report

[1040] Ibid.

attributing blame to either.'[1041]

A serious accident which involved a collision in the Midland Railway's Dove Holes Tunnel, England, resulted in five deaths on 9th September 1867. It began with the departure from Chapel-en-le-Frith station at 5.20 or 5.25 p.m. of a ballast train, consisting of an engine and tender, nine waggons and a guard's van. Riding on the train, mostly in the van, were 36 contractor's labourers, also a little girl named Martha Vaines, the daughter of a blacksmith at Dove Holes, who had wished to get to Peak Forest and had missed her train. Guard Tooth, on the other hand, was riding against the rules on the engine, perhaps because there was no room in the van. One of the contractor's men, who was meant to be acting as flagman, was also riding on the engine, while the fireman had left the train at Chapel-en-le-Frith and gone into the village for drink, only returning, after the accident, intoxicated. Before the train had left Chapel-en-le-Frith the ganger in charge of the men and the guard had been told distinctly to run through Dove Holes tunnel to the next station at Peak Forest, to cross over there onto the down line, and return thence to the tunnel, inside of which the ballast was to be unloaded. This was necessary because an up cattle train was shortly due to follow them. Disregarding these instructions, the guard of the ballast train arranged with his driver to stop and unload the ballast from the up line. Compounding this irregularity was the failure of the guard to either go back himself or to instruct the man deputed to act as flagman to go back and protect the train according to the regulations, which decreed that the guard or a competent person should go back 800 yards, putting down detonators at 400 and 800 yds. Instead, no one left the engine and after the train had been standing in the tunnel about five minutes it was run into from behind by the cattle train, which had left Chapel-en-le-Frith at 5.33. The guard's van and two waggons of the ballast train were smashed to pieces, and 11 or 12-year-old Martha Vaines, who naturally had remained in the van while the labourers got down to unload the ballast, was killed outright.

The collision in the tunnel was not the end of the accident though. The two engines pulling the cattle train were thrown off the line by the impact and the coupling between the second engine and the leading cattle truck was broken, so that the 27 cattle waggons, third-class carriage and guard's van which made up the train at once began to run away down the 1 in 90 gradient on which the tunnel was situated. This steep gradient continued in the direction in which the cattle train was running away through Chapel-en-le-Frith station to New Mills, eight miles from the tunnel, beyond which it eased slightly to 1 in 100 and 1 in 140 to Romiley station, five miles further on. The guard of the cattle train, having no hope of stopping it on the steep incline with just the hand brake in his van, wisely jumped out, and four cattle drovers followed suit, all escaping with little injury. The runaway passed through Chapel-en-le-Frith station at about 30 miles an hour and had probably reached 50 or 60 miles an hour by the time it got to New Mills. The New Mills signalman had been advised of the runaway by Chapel-en-le-Frith and had stopped an up passenger train from Manchester. The driver of the latter saw the cattle train coming towards him and had time to reverse his engine and jump off before the runaway struck it. Thirteen or fourteen of the cattle waggons were broken up in this second collision, three cattle drovers were killed outright and a fourth fatally injured. Thirteen cattle and 120 sheep were killed or had to be put down. The driverless passenger train ran back in

[1041] The Times, London, England, 11th, 12th and 13th June 1866 its turn down the bank from New Mills, but the New Mills signalman telegraphed the fact to his colleague at Marple and the latter diverted the train from the up to the down line. Its engine subsequently out of steam and the gradient having changed to a rising one the second runaway came to a stand about five miles beyond New Mills without further mishap.[1042]

The line concerned in the accident was worked by the block system using double needle speaking instruments, which the signalman could pin over to indicate there was a train in the section, but without separate speaking instruments. The Chapel-en-le-Frith signalman had telegraphed the departure of the ballast train to the Peak Forest signalman, and the latter had pinned his instrument to "Train on line." He had then left his cabin to take a note from the stationmaster to the adjoining lime works, before returning to tea with the stationmaster. Coming back to his cabin after about fifteen minutes, he said he had found the pinned up line needle shaking and had understood that the Chapel-en-le-Frith signalman wanted him to release it, so he had unpinned it for about a minute. The Chapel-en-le-Frith signalman told the inquiry into the accident that he had been trying to call Peak Forest using the down line needle, and when he had found the up line needle free, he had understood that the line was clear. He had then started the cattle train away without bothering to get "Line clear" for it.

Lt-Col Rich remarked that there was no excuse for the Peak Forest signalman releasing the up line needle, and that he had shown a total ignorance of his duties. He had been less than a month in the railway service, having previously been a gentleman's servant, and had been placed at Monsaldale (later written Monsal Dale) and Matlock Bridge to learn the telegraph. In his one examination by an inspector prior to taking over at Peak Forest he had made two mistakes, and, the inspecting officer commented, 'had shown himself quite unfit for the responsible position in which he was placed.' Surprisingly the Chapel-en-le-Frith signalman, whose actions had also contributed to the accident, was not criticised in the report, but the inspecting officer stressed the need for separate speaking instruments, plus the need for a record of the passing times of trains at every station, and not just here and there. Guard Tooth of the ballast train, Lt-Col Rich stated, had disobeyed his instructions by riding on the engine, had arranged with his driver to unload ballast on the up line instead of the down line and had failed to protect his train. The evidence of the driver, guard and fireman appeared to be entirely false, many other railwaymen had kept back information which might reveal the guilty parties and one or other of them had disregarded Regulations 7, 8, 19, 21, 22, 37, 41, 47, 50, 51 and 59! Concluding his report, Lt-Col Rich stated that he thought it inadvisable that contractors should be used to maintain a line used by passenger trains, that the brake power of the cattle train was defective, that the Midland Company should adopt an improved method of telegraph signalling and that signalmen should be trained at places set aside for the purpose (signalling schools?) and not allowed to work at permanent stations until they were competent.[1043]

Separate inquests were held upon Martha Vaines and the four drovers. Both juries returned verdicts of "Accidental Death," the first-named criticised the Peak Forest stationmaster for sending the signalman away from his cabin and proclaimed that the

---

[1042] Board of Trade report

[1043] Ibid.

signalman, being insufficiently trained, had not been capable of doing his job.[1044] [1045]

Upholland Tunnel, England, on the Lancashire and Yorkshire Railway Liverpool-Wigan-Manchester main line, was the scene of an accident on 13th September 1870. The morning of the 13th was a very wet one, and before breakfast Foreman Platelayer John Ball told his gang of three men that if the heavy rain continued, they would go into the tunnel and turn a rail, so that they could stay where it was relatively dry. Two of the platelayers had then gone off to a nearby public house and did not return. Despite his gang thus having been reduced by half Ball went into the tunnel about 9 a.m. with the remaining platelayer, a man named Melling, they taking their tools with them plus two naphtha lamps, and the two at once began to work on the rail, loosening the wedges, removing most of the fishplate bolts and chipping off as much of the top edge as they thought necessary for it to be replaced after turning. The foreman ganger then left Melling in the tunnel and went back to Pimbo Lane station (renamed Upholland 1900)[1046] close to the west end of the tunnel, to complain to the permanent way inspector, whom he expected to arrive there by train from Wigan at 10.23, about the absent platelayers. The inspector failed to arrive as expected, however, and Ball returned to the tunnel, with the intention, he later explained, of turning the rail after the passage of the 11 a.m. express from Liverpool at about 11.30. His plan changed when he saw a special up goods train approaching. Knowing, he said, that goods trains usually stopped for water at Upholland signal station, at the east (Wigan) end of the tunnel and that the line through the tunnel was protected by the electric telegraph, so that the express would not be allowed to leave Pimbo Lane until the special goods had left Upholland signal station, he judged that he had a full ten minutes to do the job. He had not checked his watch, but also knew that the down express from Manchester, due at 11.20, had not yet appeared. At about 11.20 or 11.21, therefore (according to the signalman's train register books), as soon as the up goods had passed, Ball and Melling removed the remaining fishplate bolts, lifted the rail out and turned it over ready to replace on the chairs. Here they ran into difficulty, as they found that a sufficient amount of the crushed surface of the rail at the Upholland end had not been removed to enable it to drop into the chairs at that end. They were hammering away at the rail at that end to try and force it in when the up express appeared in sight as it approached Pimbo Lane station. Ball at one began to run back through the tunnel, waving his lamp until the latter went out after he had run 200 yards, but he continued to run, shouting to the engine crew as the train passed. Melling continued to hammer away at the rail until he saw Ball's lamp go out when he also began to run back, but his lamp went out too after he had covered about 40 yards, and all he could do as well was shout.

The Pimbo Lane signalman knew nothing, of course, of the plans to remove a rail, and the express having been accepted by Upholland all his signals were clear. The express entered the tunnel at about 30 or 35 miles an hour. Its fireman saw the platelayers, as he thought, getting out of the way and the driver stated that he saw the platelayer's lamps, but neither would admit to seeing the lights being waved or hearing the platelayers shouting. Reaching the gap in the outside rail the engine gave a jump, broke off several pieces of a rail and was thrown against the left tunnel wall, coming to a stand 105 yards

1044 The Times, 11th and 13th September 1867
1045 The Magnet, London, England, 16th September 1867
1046 Clinkers Register of Closed Stations

beyond the gap. The driver and fireman were unhurt. The passengers in the leading carriage, which was crushed, were less lucky, and two of them, described in *The Times* as a workman with a ticket for Rochdale and a gentleman bound for Oldham, were killed.[1047] Several other passengers were more or less injured.

The Board of Trade inspecting officer, Lt-Col C.S. Hutchinson stated that accident was clearly the fault of Foreman Platelayer Ball, (who was subsequently taken into custody, charged with manslaughter). The Lancashire and Yorkshire Railway regulations stated that such an operation was not to be done within 15 minutes of a train being due, nor without a man being sent back half a mile with a red flag and two detonators, and that the fixed signals were not to be relied upon to provide protection. All of these rules had been disregarded. Clearly with only half his gang Ball could not send a man back leaving only himself to do the job, but there was no great urgency for him to turn the rail and he had not asked for any assistance.[1048]

The earliest accident which the author has been able to trace to have happened in any of the major tunnels through the Alps occurred in what is commonly known as the Mont Cenis Tunnel (its more correct name is the Fréjus Tunnel) just over a year after its opening. A lengthy letter in *The Times*, written by a passenger on one of the trains involved, described what had happened. A passenger train bound for Italy left Modane, France, at 4 p.m. on Sunday 6th October 1872. The train had travelled about half way through the bore when there was a terrific noise and an awful crash, which threw the writer from his seat. No-one knew what had happened of course, all was total darkness and within five minutes the coupé began to fill with smoke. It transpired that the passenger train had run into a goods train in the middle of the 8½ mile long tunnel. The driver of the passenger train lost his life and several of the passengers were reportedly mortally wounded, three of the carriages being 'smashed to splinters.' The writer and her fellow passengers had something of a nightmare experience in the hours that followed. The heat in the tunnel was intense and the passengers were trapped there in the darkness, in what resembled an oven, and almost with air, for some five hours. At last, six engines were brought into the tunnel to drag the passenger train back out, and their efforts to do so created further alarm among the passengers, many of whom wrote their names on pieces of paper lest they should die in the tunnel. Eventually the carriages were removed from the tunnel and the passengers taken back to Modane station. Here, the writer continued, some of the women had been in convulsions, men were crying, and a number of the passengers had had to be carried from the coupé. She and her sister had held up wonderfully in the tunnel, but they had both fainted when they put their feet onto safe ground. Her nerves now were weaker than a baby's and she thought that it would be months before she regained her health. Accounts of the accident in several (Italian? French?) newspapers had altogether misrepresented the accident, and she considered the mismanagement that night shameful. A touch of Victorian melodrama from the writer perhaps, but it was doubtless a very frightening experience for those involved.[1049]

1047 The Times, London, England, 14th September 1870

1048 Board of Trade report

1049 The Times, London, England, 18th October 1872

Two express trains collided in Blea Moor Tunnel, England, on the Midland Railway main line from Leeds to Carlisle, on 19th August 1880. The 3 p.m. Leeds to Carlisle express came to a stand near the north end of the tunnel because the Westinghouse brake pipe blew off between the engine tender and the front guard's van, the brakes thus applying automatically and stopping the train. Before the problem could be fixed and the train could continue its journey, it was run into from behind by the 10.35 a.m. express from London St Pancras to the North. The main damage was caused, not surprisingly, to the rear van of the first train, which was broken up, while the end of the next carriage and the engine of the colliding train were also damaged. Only three passengers complained of injury.

The accident was the result of a signalling error. The signal boxes involved were those at Blea Moor Sidings, to the south of the tunnel, and Dent Head, a simple block post north of the 1 mile 869-yard-long tunnel. Signalman Daniel Hewitt at Blea Moor told the Board of Trade inquiry into the accident that he had received the "Be ready" signal for the Leeds train from Ribblehead box at 4.37p.m., the "Train on line "signal at 4.42 and it had passed him at 4.45. When he had cleared it back to Ribblehead he had been offered and had accepted the London express. "Train on line" for the latter train had been received from Ribblehead at 4.47 and it had arrived at his box at 4.50, whereupon its pilot engine (which would have assisted the train up the long climb to Blea Moor, probably all the way from Leeds) was detached. The usual time for an express to clear Dent Head was four or five minutes but at 4.53 it had still not been cleared back. He had given Dent Head one beat on the bell to attract his attention and then, the up needle hanging vertical at that moment as there was no train in the section, he had manipulated that needle to indicate to the Dent Head signalman that the latter had not cleared out the Leeds express, whereupon the latter had done so, and had accepted the London express. This use of the block instrument to attract the attention of the other signalman was of course completely irregular, and also unnecessary as there was a speaking telegraph instrument between the two cabins. It was in fact via the speaking telegraph that the Dent Head signalman had come back shortly afterwards and asked where the first express was. Signalman Hewitt said that he had tried to ask the Dent Head signalman what express he meant but could get no reply from him. This was about 4.54 or 4.55. Subsequently he could get no communication with either Dent Head or Dent cabins and did not know what had happened until a platelayer came to his box to tell him at about 5.40. His train register had an entry made and erased in the "Line clear" column for the second express which he explained was the time at which Ribblehead had cleared back the pilot engine which he had accidentally written on the down line page instead of the up line.[1050]

Signalman Arthur Samuel at Dent Head stated that he had received the "Be ready" for the Leeds express at 4.44 and "Train on line" at 4.47. The last up train had passed at 2.56. Several minutes after he had been given the "Train on line" Blea Moor had given him one beat on the bell and then begun to shake the needle for the up line violently. Upon this he had unpegged the needle of his instrument for the down line, for reasons which he did not seem to be able to explain, before pegging it again over to "Train on line". He had then asked Blea Moor via the speaking instrument where the first train was but had received no reply. He explained that a leaf of the train register book for the 18th and 19th had been torn out because it had been dirtied by the guard of the Leeds express coming into his

[1050] Board of Trade report

cabin and writing a telegraph message on it. He had copied the times onto another page and then thrown the leaf into the fire. Unfortunately for him Guard Hardy of the express denied writing any such message and inspecting officer Hutchinson did not believe this unconvincing if imaginative excuse.

Major-General C.S. Hutchinson found that the cause of the stopping of the Leeds express was the blowing off of the brake pipe due to it being furnished with an inferior brass fastening. The Westinghouse Brake Company had alerted the Midland Railway to the problem and of the advisability of substituting improved malleable iron clamps in July 1878 and had reminded them of the problem three times since, but the replacement with improved fittings had not been completed by the time of the accident. No accident need have occurred however but for a serious mistake in the block working between Blea Moor and Dent Head cabins, which the inspecting officer decided was due to an extraordinary error on the part of the Dent Head signalman in clearing out the Leeds train and accepting the London one, subsequently destroying the page of his train register book to try and conceal his error. Even then the collision would probably have been prevented if the rear guard of the Leeds train had gone back more speedily to protect it, as the evidence showed that the train must have been stopped in the tunnel for 11 or 12 minutes before it was run into at 4.59. Guard George Beevers had wasted valuable time, the inspecting officer concluded, in going forward 200 yards to find what was wrong before going back to protect his train. Reading the evidence of the guard it is impossible not to have sympathy with him; running back through a dark tunnel was not much fun for a middle-aged man and Guard Beevers stated that at one point he had fallen over, hurting his knees, that his lamp had gone out as he fell, and he had had to relight it. Hearing the second express coming he had put two fog signals down, shown a red light to the driver and shouted. Colonel Hutchinson stressed that the duty of going back at once should a stoppage occur could not be impressed too strongly upon rear guards, who were too inclined to rely upon the block system and were afraid of being left behind.[1051]

[1051] Ibid.

CHAPTER 12

LEVEL CROSSING COLLISIONS

Wherever roads and railways intersect, they must do so either by means of an over or under bridge or alternatively on the level. Bridges have the advantage of allowing road and rail traffic to operate entirely without interference from one another, but they are much more expensive to provide, especially in flat countryside, due to the need to provide extensive earthworks for approach embankments or cuttings as well as the cost of providing the span itself. Where underbridges are provided there is often also the need to build retaining walls and there can be problems with drainage.[1052] The provision of level crossings, especially if, as in many parts of the world, they are unstaffed, is the much cheaper alternative, but where they exist, they introduce an additional source of danger to both road and rail users. Robert B. Shaw in his outstanding book on American railway accidents, '*A History of Railroad Accidents, Safety Precautions and Operating Practices*', writes that he has excluded ordinary grade (i.e. level) crossing collisions from consideration on the grounds that they are not proper train accidents, inasmuch as the trains usually suffer much less harm from such accidents than the road vehicles involved. This book, on the other hand, will attempt to include, amongst others, a selection of such accidents that have features of interest even if none of the casualties in them, if any, occurred to train users.

The internal combustion engine was only invented in 1884 and the first motor car in 1885 or 1886.[1053] As far as the author has been able to discover, the first level crossing collision involving a motor vehicle did not occur until 1st May 1904, in France[1054] and thus all the accidents that we will be concerned with here will involve the collisions of trains with vehicles drawn by animals (usually horses).

It is difficult to believe that no level crossing accidents with, for example, horses and carts or waggons occurred on lines such as the Stockton and Darlington Railway in

---

1052 Backtrack Volume 24 pg. 88

1053 Wikipedia – History of the Internal Combustion Engine

1054 Wikipedia – Liste des accidents ferroviaires en France au XX siècle

England in the earliest days but no reports of any have been found. The first level crossing collision to be recorded appears to be one which happened at Thornton Lane crossing, England, on the Leicester and Swannington C Railway on 4th May 1833[1055] when the engine Samson collided with a horse and cart. The accident reportedly led to the fitting of 'steam trumpets' to the company's locomotives by a musical instrument maker of King Street, Leicester, and thus the invention of the locomotive whistle. However, the author C.R. Clinker in his book on the Leicester and Swannington Railway disputes the authenticity of this occurrence as there is no mention of it in the detailed minutes of the Leicester and Swannington Company directors which list numerous minor accidents, no report of it in the locomotive superintendent's monthly report for May 1833 and indeed no record in the Leicester trade directories of any musical instrument manufacturer in King Street. The story is therefore possibly apocryphal, and it is not clear if such an accident ever happened.

There is no such uncertainty regarding the next level crossing accident as news of it was reported in the London *Times*. On Friday 11th April 1834, as the 2 p.m. first-class coach train from Manchester on the Liverpool and Manchester Railway had proceeded about half way down the Whiston Incline, a man in charge of two horses and a cart rashly attempted to cross the line when the train was only 40 or 50 yards away. A collision was unavoidable. The two cart horses were killed, the carter suffered a broken arm and serious leg injuries, and the engine of the train was overturned and its tender badly damaged. The coaches of the train fared better as only one carriage was slightly damaged and none of the passengers was hurt. The train was delayed three quarters of an hour until another engine could take the coaches forward. The railway line at that point was dead straight for half a mile in each direction, with nothing to interrupt the view of an approaching train, and like so many level crossings to follow in coming years, the accident was clearly entirely the carter's fault. Possibly, in this case, the unfamiliarity of the new form of transport was a factor. *The Times* commented that it hoped that before long, railway operations would be so well understood, and the necessity of caution so obvious, that a man would no more think of crossing a railway before looking right and left than he would of crossing Cheapside in London without doing the same.[1056]

A fatal level crossing collision occurred near Daubhill, England, on the Bolton and Leigh Railway on 21st February 1839. The accident happened at the level crossing with the turnpike road. A train from Bolton had reached the Daubhill stationary engine (at this date the line included two inclined planes) and a watchman was occupied with dealing with this train when a carter named Samuel Evans, employed by a Mr William Green of Little Hulton, came along the road in charge of two horses and carts, on his way home to Bolton. The leading horse had stepped onto the railway line, with the carter at its head, when a luggage train from Leigh appeared and struck them. Man and horse were thrown beneath the engine and the cart was driven over an adjoining wall. Both the carter and the horse were killed.[1057]

[1055] History of the Midland Railway by Clement E. Stretton, pgs. 26-7

[1056] The Times, London, England, 14th April 1834

[1057] Ibid, 28th February 1839

Also in 1839, the British government passed the first legislation concerning level crossings, a clause in the Turnpike Act of that year ruling that where the railways were crossed by public roads gates must be provided, attended by 'good and proper persons' and kept shut across the road except when road users needed to cross. Gates had been provided already on some of the early railways to prevent animals from straying on the line while being herded over a level crossing, now they were to be made compulsory. Public road level crossings were now distinguished from occupation level crossings, where the railway crossed a private road, and accommodation level crossings, where a new railway divided a piece of private land. At both of these types of crossing, the users of the crossings were themselves responsible for their safe use.[1058]

Level crossing collisions continue to be recorded sporadically in the early years. On 17th January 1840 a train collided with a horse and sleigh on a crossing about a mile beyond the West Brookfield, Massachusetts, United States, depot. A Mr Hale Young of North Brookfield was killed and his horse so badly injured that it had to be put down. On his way to a place named Warren, Mr Young had turned onto the crossing and stopped, apparently to give way to a number of sleighs coming in the opposite direction. His face had been muffled up against the cold and he had failed to hear either the bell of the train's locomotive or the warning shouts of its conductor, presumably he had also failed to see the imminent danger he was in.[1059]

On 7th and 11th September 1841 trains collided with a stage coaches on level crossings, not in the Wild West of the United states but at Euxton and Bridgwater, England.

The Euxton accident happened at the station of the same name on the North Union Railway main line from Parkside, on the Liverpool and Manchester Railway to Preston. A parish road from Chorley to Eccleston and Southport crossed the railway at this point and the accident involved the Chorley and Southport stagecoach, which was struck by a train of empty coal waggons drawn by the *Asa* locomotive, on its way from Preston to Springfield colliery, Ince, near Wigan. The accident was a serious one; one of the passengers on the coach (the Rev Mr Ivy, a Dissenting minister) was killed outright and six other passengers and two employees of the coach were injured, as was Thomas Nickson, the crossing keeper, whose left leg had to be amputated.

The accident occurred between 12 noon and 1 p.m. As the train approached the station its driver received no warning of danger and he did not therefore reduce speed. He only saw the coach entering the crossing at the last moment, when it was too late to avoid the accident. The level crossing was provided with gates, but the gatekeeper had left them open and gone into his lodge. The coach driver, seeing the gates open as he neared the crossing, naturally did not anticipate any problem and neither saw nor heard the train, while on the approach to the station from Preston the railway line was in a cutting, hiding the coach from the engine driver and the train from the coach driver.

1058 Webpage - Network Rail, History of Level Crossings

1059 Steamboat disasters and Railroad Accidents in the United States by Southworth Allen Howland, pg. 304}

Clearly the accident was directly due to the negligence of Nickson, the gatekeeper. Hearing both train and coach nearing the crossing and anticipating an accident, because he had left the gates open, he had rushed out of his house at the last moment and had sustained his injury in trying to stop the coach. It appeared that he regularly left the gates open, and in his defence pleaded that he had had complaints from crossing users that they were kept waiting while the gates were opened, and on one occasion someone had actually 'pulled up the gates by main force.' The problem of impatient road users was probably exacerbated by the fact that before the aforementioned 1839 Turnpike Act the gates had stood normally open across the road, only being closed for the passage of trains. The instructions issued to the gatekeeper were also out of date, referring to the time when the gates were normally left open to the road: 'Gatemen and policemen are to be constantly on the look out, and to open the proper gate where the engines are to pass through.'

Lt-Col Frederic Smith, who inquired into the accident, stressed that the crossing keepers' instructions needed urgently to be updated, and that an order should be issued that trains should whistle when approaching turnpike or parish road level crossings. He was also not happy that the engine of the coal train had been travelling tender first, due to the lack of a turntable at the Springfield colliery. He added that he was satisfied to learn that the directors of the North Union Railway proposed to substitute bridges for five of the most important of the seven turnpike or public road level crossings between Preston and Parkside, that at Euxton being among those to be done away with.[1060] In 1842, perhaps partly as a result of the Euxton accident, the Board of Trade was given the power to authorise the normal position of the gates to be across the railway at specified level crossings, thus reverting to the situation before the 1839 Act.

At the time of the Bridgwater accident, that place was the temporary terminus of the Bristol and Exeter Railway's main line from Bristol; the further opening of the line to Taunton having to await the completion of the Somerset Bridge over the River Parrett just west of Bridgwater. Bridgwater station had separate arrival and departure platforms, beyond which the rails continued for a short distance towards Taunton, crossing a road leading from Bridgwater to a farm on the level. On the arrival of a down passenger train from Bristol, its engine had to draw the coaches of the train out over this level crossing prior to pushing them through a set of points and back over the crossing to the up or departure platform. Meanwhile, the horse-drawn coaches and omnibuses which had been waiting in the station yard for the arrival of the train needed to pass over the level crossing to get to Bridgwater town centre and thence via the turnpike road to Taunton and Exeter. On Saturday the 11th September 1841 the down mail train arrived from London and Bristol at 4 o'clock in the afternoon. Waiting outside the station to carry passengers on to Taunton, Exeter and beyond by road were five coaches and two omnibuses. While the passengers' luggage and the mails were being transferred from the station platform to the road vehicles the train had been uncoupled behind the second class or open carriages and the engine had drawn these forward over the level crossing and backed them into the up platform. The engine had then returned over the crossing and was stopped on the Taunton side of it, its driver awaiting a signal from the switchman to come back over the level crossing, couple up to the first-class coaches and transfer them to the up side.

[1060] Board of Trade report

While the engine was thus stationed beyond the crossing it seems likely (witness statements were conflicting) that the *Brilliant* coach for Exeter had already left the station yard, had passed over the level crossing and was off on its way. The switchman then gave the "All right" signal to the engine driver to set back onto the first-class carriages and at the same moment the *Exquisite* coach, also bound for Exeter, set off from the station yard and entered the level crossing. The front wheel of the coach was struck by the tender of the engine and the coach overturned. It had been carrying eleven passengers, only one of whom, a Mr Bruford, was sufficiently seriously hurt to be unable to continue his journey. He was taken to the Clarence Hotel in the town and attended by two surgeons. By the Monday he was well enough to be able to continue to Taunton. Less fortunate was an elderly woman standing by the line, described as a very poor lady of 73, who suffered a broken leg and severe bruising when the coach fell onto her, but Mr Toogood, one of the surgeons, said that he had great hopes for her recovery.

The switchman, Benjamin Cox, stated that he had had no written orders as to what to do before the accident, though he had been issued with some since. He had not seen the *Exquisite* coach when he had signalled to the engine driver to come back but the coach had come out of the station yard immediately after. He had held up his arm to the coachman and shouted to him to stop but Johns (the coachman) had ignored him. Johns had appeared to hesitate before crossing the line but then proceeded when the engine was no more than five yards away. The coachman himself admitted stopping momentarily at the crossing and said that he had then 'thought his best chance was to drive on.' Johns denied having heard the engine whistle before it began to set back though other evidence suggested that it had done.

Professor Barlow, who inquired into the accident, stated that all the railwaymen blamed the coachman for the accident, and Mr Ashbee, the station superintendent, quoted a Mr Slocombe, a passenger on the *Exquisite* coach, as having said that 'no blame was attributable to the railway people; it was all Johns' fault, he would whip on.' The professor was of the opinion that the engine and the coach had started off at the same moment, and that coachman Johns, intent on overtaking the *Brilliant* coach in front of him, had either not heard the engine whistle or had disregarded it, hoping to cross the line in front of the engine. But, added the professor, however heedless or reckless the coachman might have been, or however careless the switchman might have been, there could be no excuse for 'the dangerous and reprehensible' practice of having engines crossing and re-crossing the road at the very moment that coaches were departing from the station. Since the accident an order had been issued that after transferring the second-class carriages the engine was not to again pass over the crossing until all the coaches had gone and the station yard gates had been locked. The professor considered that this was a half measure, and it would be better not to transfer any of the carriages of an arriving train until all the coaches had departed from the station yard or the gates had been locked. He noted that the operation of moving a rake of carriages only took three minutes, so little time would be lost in delaying the operation.[1061]

The line onwards to Taunton was opened on 1st July 1842[1062] The level crossing was subsequently abolished.

[1061] Ibid.

[1062] History of the Great Western Railway by E.T. MacDermot, Vol1. Appendix 1

A narrow escape from a serious accident occurred at Eguisheim, in France, on 25th October 1845. The level crossing concerned was provided with gates, though the newspaper report did not state whether a crossing keeper was provided or if they were user-worked. At any rate there was a thick fog on the day of the accident when a Lyons mail coach entered the crossing only for the postilion (the driver of the coach, who rode one of the horses, in contrast to a coachman, who rode on the coach itself) to find that the gate on the far side of the line was closed. A train was at that point heard to be close at hand, and there was no time to open the gate. With great presence of mind, the postilion positioned the coach and horses parallel to the railway line in an attempt to avoid a collision. The manoeuvre could not be done quite quickly enough though and the engine of the train caught the hind wheel of the coach and smashed it to pieces, throwing the mail guard 25 yards into a ditch. The horses and a passenger in the coach nevertheless escaped with some severe bruises, and neither the postilion nor the mail guard was seriously hurt.[1063]

An early example of the mechanisation of English agriculture was provided by a brief report on an accident which occurred at Tortington Crossing (near Ford) on the London Brighton and South Coast Railway on 25th April 1847. Owing, the report said, to the negligence of the crossing keeper a ballast train collided with a man driving a horse and a thrashing machine, killing the horse and injuring the man.[1064]

The crossing keeper at Buttevant Crossing, Ireland, was also considered to be at fault for an accident which occurred there on 6th July 1850, though in this case blaming him seems harsh. Between 5 and 6 p.m. that day, two horses pulling cars loaded with culm (a type of anthracite-like coal) approached the level crossing with the Great Southern and Western Railway line. The first passed over the railway safely, but a train was then close at hand and the crossing keeper, Michael Noonan, told John Callaghan, in charge of the second horse, to wait. Instead of doing so, Callaghan whipped up his horse and attempted to cross. He had fatally misjudged the speed of the train and both he and the horse were killed. Captain George Wynne of the Royal Engineers, the inspecting officer, considered that Noonan had done his job very inadequately in only warning Callaghan and not *resisting* him. This seems rather unfair, surely Callaghan himself was at fault and if Noonan had attempted to stop him too forcefully, he might have been assaulted for his trouble. Interestingly, though, the report added that Noonan himself had since been dismissed for an assault on the police and had emigrated to the United States. The inspecting officer was also critical of the unique type of level crossing gates employed on the line, which when closed across the road left a gap of three feet for the passage of pedestrians. This Captain Wynne thought was ‘an objectionable sort of barrier in a part of the country where the people are much inclined to resist authority.’ Instead ‘the gates should be made to shut entirely across, with a good horizontal bar, to enable the gate-keeper to oppose a proper resistance to violence’ as ‘from the account given to me, the

1063 The Times, London, England, 5th November 1845
1064 Appendix to the Report of the Commissioners of Railways 1848

practice of persons forcing themselves through is not uncommon.'[1065]

All the level crossing collisions mentioned so far have been relatively minor affairs, involving at worst the death or injury of a handful of road users, but on 29th August 1855 a much more serious accident occurred in the United States, near the town on Burlington, New Jersey, and in an exception to the usual rule this time the victims were train passengers rather than road users.

On the morning of that day, a Dr Hannigan or Heineken was on his way from his home in Columbus, New Jersey, to visit several patients in nearby Burlington. Coming along for the ride in his two-horse carriage were his wife and child and a neighbour, a Mr Thomas Antrim. The doctor's journey took him over a level crossing with the Camden and Amboy Railroad two miles outside of Burlington. As he reached the crossing the doctor took out his watch, saw that it was about 11.15, and remarked to his companions that it should be a good time to cross the line, as the up train to New York and the down train to Philadelphia should both have passed. Thus, reassured the doctor seems to have entered the crossing without looking very carefully along the line, and he failed to see the rear end of a train which was being set back towards the crossing. The side curtains of the carriage had been pulled shut in an effort to keep out some of the dust being kicked up by the horses from the road and this no doubt impeded his vision and explained why he had also failed to see Mrs Mary Cook, who lived by the crossing, standing at her front gate waving her arms violently to attract his attention.[1066] Finally, when he was only about 25 feet from the crossing, the doctor did see the train and tried to rein in his horses, but the animals panicked and dashed directly in front of the oncoming train.

The train which hit the doctor's carriage was the 10 a.m. departure from Philadelphia, which, the line being single track, was scheduled to cross the down train from New York at 10.52 at Burlington. It had arrived there on time, but the New York train was not there. After waiting ten minutes it had thus proceeded cautiously eastwards, as per the rules, its engineer, Israel Adams, looking out for the late-running down train. It had got about three miles when the New York train was sighted. Adams had whistled for brakes, the train had been brought to a stand, he had whistled again and then begun reversing his train to the nearest siding at Mount Holly Junction. It had already been set back about two miles when the collision occurred.

Situated at the wrong end of the eight-car train, busy, as he had said afterwards, with his gauges, and with his view ahead obscured by lineside vegetation, Engineer Adams had been unaware of the horses and carriage crossing the railway before the impact. Conductor Isaac Van Nostrand was also unaware of the danger. Brakeman Maurice Maxwell, perched on his "gig" at what was then the front of the reversing train, could see what was about to happen but could not communicate quickly enough with the engineer and conductor. The then leading car off the train struck the team of two horses and rolled or pushed one of the animal's bodies 153 feet down the track, left the rails, jolted along the line for another quarter mile and then overturned into a ditch. The second car, in which most of the casualties occurred, was thrown broadside across the track and rammed by the third car. No less than twenty-three train passengers died in the accident,

---

[1065] Board of Trade report

[1066] A History of Railroad accidents Safety Precautions and Operating Practices Robert B Shaw, pg. 281

including Major Boyce, of the U.S. Coast survey, and his daughter; Baron De St Andre, French Consul at Philadelphia; and Messrs Meredith and Dallam, Baltimore merchants.[1067] Many other passengers were injured. The occupants of the doctor's carriage, on the other hand, escaped with a few bruises.

As is often the case in disasters, there were individual acts of bravery and self-sacrifice. The Hon W. B, McClay, of New York, a former member of Congress, was badly injured, but begged those who came to his aid to give their attention to others, as did another passenger whose foot was badly crushed. Meanwhile, feeling against the train crew ran high, and there was talk among the surviving passengers of lynching Conductor Van Nostrand, but in the end wiser counsel prevailed. Engineer Adams reportedly became suddenly deranged upon witnessing what had happened.[1068]

At the inquest into the accident there were allegations that the train from New York and been approaching too fast, so that Engineer Adams had had to hurriedly reverse his train, but the two engine crews denied this, and nothing was proved. The bell fitted to the front of Adam's engine was not working as the bell - cord was broken but with the engine being at the far end of its train it could probably not have been heard by the road users anyway. The coroner's jury declared that the engine whistle had not been sounded either, but the railway company insisted that it had been blown repeatedly, and that there were a number of witnesses to that effect. Dr Hannigan was, in any case, quite deaf, though a whistle, if blown, might presumably have been heard by his wife or neighbour.

There was no independent board investigating railway accidents in the United States at this time but a report on it was published by the railway company, in which the railway was entirely cleared of any responsibility for the accident, all the blame being placed on the unfortunate Dr Hannigan.[1069] This report provoked a storm of protests against the joint companies, the most outspoken critic being the Rev Cortlandt Van Rensselaer of Burlington. As Shaw points out the practice of trains proceeding beyond their crossing places on a single line until they met one another, and one of them then setting back to the nearest siding, was outdated even in 1855. Ironically there was a telegraph line along the railway and Conductor Van Nostrand had used it while his train was stopped at Burlington to inquire as to the running of the opposing train and had been told that it had left Bordentown at 10.56. No thought was given, though, it seems, to use the telegraph to actually arrange the crossing of the two trains.[1070]

More typical than the Burlington tragedy was a level crossing accident near North Bridgewater, Massachusetts, United States, on Saturday 30th July 1859. This time the victims were confined to the road users and the accident was clearly not one which could be blamed on the railway, except perhaps inasmuch as it happened at a level crossing, like that near Burlington, lacking gates or a crossing keeper, but outside of the British Isles such level crossings tended to be the norm. The accident occurred at what was

---

[1067] The Adams Sentinel, Gettysburg, Pennsylvania, United States, 3rd September 1855

[1068] Ibid.

[1069] Report of Executive Committee to the Directors of the Delaware and Raritan Canal and Camden and Amboy Railroad Transportation Companies

[1070] A History of Railroad accidents Safety Precautions and Operating Practices by Robert B. Shaw, pgs. 282-3

known as the 'Mill Crossing' and involved a fast train of the Old Colony and Fall River Railroad, known as the 'Lightning Train' because it made few intermediate stops in its journey from New Bedford to Boston. At the time of the accident the train was not, in fact, travelling particularly quickly, and as it approached the crossing its engineer whistled vigorously as he saw a covered waggon (referred to in some reports as a carryall) coming along the road and nearing the crossing. Witnesses who were on the train, at the adjoining depot, and working at a nearby mill thought at first that the waggon driver would stop and wait for the train, but at the last moment he whipped up his horse in an attempt to cross in front of it. The horse got across in time, but the engine of the train struck the waggon, smashing it and carrying pieces of it some way down the line. The inmates of the waggon were thrown out by the impact and gravely injured. The train stopped in a short distance and its crew and some of the passengers went back to the accident scene. They found a mother and her daughter covered in blood and insensible; the two died within a few minutes. The waggon driver, a Mr Stetson, a widower of Hanson, and the fourth victim, a young woman from Hanson, were also initially unconscious and it was feared that Mr Stetson would not survive, but the young lady soon regained consciousness and was apparently not badly hurt. The party had been on a shopping expedition from Hanson (or South Hanson) to Bridgewater. The several witnesses to the accident all agreed that the railwaymen involved were entirely blameless in the matter.[1071] [1072]

Sometimes careless level crossing users were luckier. Thus, on 20th December 1868 an accident occurred at a level crossing near Bowden station, South Australia. A hawker of vegetables named Batty tried to cross the railway in front of a train in his horse and cart. He perhaps misjudged the speed of the train, which smashed his cart to pieces; he was knocked down and stunned and his horse thrown into a ditch. He soon recovered however, as did his horse, which 'quickly regained its legs, and made good use of them.' A buffer of the engine was somewhat damaged.[1073]

A mistake by a gatekeeper was the cause of an accident to the family of a Lyons, France, ironfounder near that place on 5th October 1879. Failing, he said, to hear the whistle of an approaching express the gatekeeper allowed the ironfounder's horse and carriage over the line, and the express, coming around a curve, dashed into them. The train cut the horse in two and Madame Peysson, the wife of the ironfounder, and the coachman were both killed, while a servant maid had both her feet cut off and was not expected to live. M. Peysson and his two-year-old son suffered head injuries; two other children were unhurt.[1074]

[1071] The Portsmouth Times, Portsmouth, Ohio, United States, 9th August 1859

[1072] The Burlington Weekly Hawk Eye and Telegraph, Burlington, Iowa, United States, 6th August 1859

[1073] The Leader, Melbourne, Victoria 9th January 1869

[1074] The Times, London, England, 7th October 1879

CHAPTER 13

RAIL/RAIL CROSSING COLLISIONS

As noted at the beginning of the previous chapter, at locations where railways and roads intersected, the idea of having them cross on the level had obvious advantages in terms of economics over the cost of building an over or underbridge, especially if the level crossing was unmanned and the construction of a bridge would require the building of substantial approach embankments or possibly approach cuttings leading to an underbridge. With road users being limited in the early days to pedestrians, cyclists and horse-drawn vehicles, the dangers to a train from a collision on such a level crossing were also likely to minimal. For railways crossing other railways on the level though the situation was very different. A collision with another train on a crossing was likely to be far more serious than a collision with, say, a horse and cart, and, furthermore, a train driver approaching a level crossing and seeing another train nearing or on it could not pull up in anything like the distance that a road user could. It is no surprise, therefore, that in most countries, where the railway system was either built directly by the state or by private companies under state supervision that rail/rail crossings were not provided. It was only in countries where the railways were built by competing private companies without state interference that such crossings tended to exist, most especially in the United States, where lines were often built as cheaply as possible and the economy of not needing to construct a bridge was doubtless an important consideration. Thus, the United States had many rail/rail level crossings, Great Britain at one time had about twenty or so and its colonies, such as Ireland and Canada, had a handful. One or two did also exist elsewhere, such as at Istravoz (later Hilal), Turkey, where the Oriental Railway crossed the Smyrna – Cassaba Railway.[1075] As has been pointed out by Robert Shaw, the American companies also seemed to have a perverse proclivity for building rail/rail crossings where they could easily have been avoided, perhaps because in the early days train speeds were low, trains were infrequent and night operation was unusual, so that the dangers were not fully appreciated.[1076]

[1075] Wikipedia - Hilal railway

[1076] A History of Railroad Accidents, Safety Precautions and Operating Practices by Robert B. Shaw pg. 99

The earliest rail/rail crossing accident appears to have occurred on the 11th September 1839 at the intersection of the Glasgow Paisley Kilmarnock and Ayr Railway and the Kilmarnock and Troon waggonway at Barassie in Scotland. A policeman was stationed at the crossing but, despite his warnings, a young man named Orr employed on the waggonway persisted in attempting to cross the Paisley Kilmarnock and Ayr line with two horse-drawn trucks of coal in front of the 12-noon train from Irvine. The latter passenger train, drawn by the engine *Marmion*, cut through the coal waggons but in doing so the fireman of the *Marmion* was knocked off the engine and was found lying on the ground with an ear torn off.[1077] What happened to Orr is not recorded but it seems likely that he was prosecuted for his negligence.

The next rail/rail crossing was the first of many to occur in the United States. On Tuesday 1st November 1842 a Western Railroad train left Worcester, Massachusetts, at 9.30 a.m. and a Norwich and Worcester Railroad train left its Worcester station at the same time. The two trains collided at the flat crossing of the two roads only half a mile away. The engine and baggage car of the Western train were thrown from the track in the collision and slightly damaged, but no passengers were hurt (the report did not indicate if only one of both trains were carrying passengers).[1078]

In 1845 it was the practice of the Preston and Wyre Railway to run excursion trains from Preston, England, to Fleetwood at 8 a.m. on Sundays and on alternate Sundays, when the tide at Fleetwood was favourable for bathing, large numbers of people used them. August 17th was one such day and a train left with at least 1,000 passengers in 21 carriages, getting away five minutes late because of their number. The Preston and Wyre station at the former place was situated on the east side of the Lancaster and Preston Junction Railway, which was crossed on the level at an angle of 60 degrees. The excursion train had just got under way and was on the crossing when a train from Lancaster ran into it between the eighth and ninth waggons. Eighty or ninety passengers were thrown out, but they all got up again and clambered into other carriages, the upset waggons were righted, and the train proceeded. It was only after arrival at Fleetwood that it was discovered that one man had a broken thigh, the only apparent casualty. The engine of the Lancaster train was derailed and its crank axle damaged but none of its 270 passengers was hurt.

Captain J Coddington investigated the accident on behalf of the Board of Trade. He stated that the Lancaster train had left there at 6.50 a.m. and was due at 7.50 a.m. but was late. He considered it gross mismanagement for a heavy excursion train to be started when it was known that a train crossing its line was due at any moment. In extenuation, it was pleaded that as the Preston and Wyre was the older line its trains had right of way over the crossing and Lancaster and Preston trains were meant to give way to them.

Furthermore, the crossing itself was dangerous, being situated in a cutting with limited visibility, though a signalman was provided there, furnished with red and white flags and lamps. Should trains approach the crossing simultaneously he was expected to show one a red flag and the other a white. The drivers of approaching trains were instructed to

1077 The Glasgow and South Western Railway by Campbell Highet, pg. 45

1078 Boston Post, Boston, Massachusetts, 3rd November 1842

whistle and to approach the crossing cautiously. The signalman was on duty every day from 6 a.m. to 8.30 p.m. As well as the four lines he had two sets of points to attend to.

On the day of the accident, the Lancaster train had whistled, and the signalman had heard it, but he was watching the Fleetwood train and had not time to get across the line and stop the Lancaster train before the collision.

Captain Coddington remarked that it was absolutely necessary for public safety that additional precautions be taken at the crossing. A coal store and waggon siding obstructing the view northwards should be removed. Better communication between the two companies in arranging the departure times of their trains was also desirable.[1079]

One can imagine that Captain Coddington was not best pleased to have to report on another collision on the same spot on 16th March 1847. In this case, a train of 34 waggons of cattle and sheep, driven by a Charles Hodges, belonging to the Lancaster and Carlisle Railway and proceeding southwards to Penrith, had got to within about 70 yards of the crossing when an engine driven by cleaner Robert Dunderdale, employed by the Preston and Wyre Company, came out of their station and stopped exactly on the crossing prior to backing into a siding. He could not get his train clear of the crossing in time and there was a collision which damaged both engines, though no one was hurt.

Since the previous collision, improvements had been made by the provision of a fixed signal and of a separate pointsman, so that the signalman at the crossing had no other work to do. The occurrence of another collision showed, however, the inspecting officer pointed out, that further measures were needed. Firstly, the Preston and Wyre engine shed was placed so near the crossing that if an engine needed to be removed to make way for another it had to foul the crossing.

Secondly the signalman, Thomas Carter, had to remain within a few yards of the signal post, even if no trains were due for several hours, as otherwise he lost sight of approaching trains. It was very difficult to expect a servant to remain rooted to the spot for hours at a time with nothing to do. Instead, an arrangement should be employed whereby there were signal arms and lamps for both lines and the arms remained at danger until altered for one particular line, drivers being required to approach the crossing prepared to stop. This arrangement was employed at the junction of the Brighton and Croydon lines near Croydon, at the junction of the Ely and Peterborough and Northern and Eastern lines near Ely and elsewhere with success.[1080]

The crossing was abolished in 1885.

The next crossing accident happened near Newark, England, on 4th October 1852, at the crossing of the Great Northern Railway main line from London Kings Cross to Doncaster and the Midland Railway branch from Nottingham to Lincoln. The up Great Northern express train from Edinburgh had left Doncaster 53 minutes late and was an hour late by the time it reached Newark. The crossing was protected by distant and stop signals and the driver of the express found the distant signal, 600 yards from the crossing,

---

[1079] Board of Trade Report

[1080] Ibid.

off for him, but the stop signal at the crossing was hidden for some distance by the girder bridge carrying the railway over the Newark dyke. When it came into view, he found it to be at danger and saw a Midland goods train on the crossing. He did everything in his power to pull up, but his train cut clean through the Midland train two or three waggons from its rear, coming to a stand 130 yards from the crossing.

The Midland Railway driver said that his train had left Lincoln ten or fifteen minutes late and had been pulling about 30 waggons and a van. The distant signal, 612 yards from the crossing, had been at danger, but had been cleared when he whistled. The stop signal at the crossing was off when he first saw it. He had been close to the crossing when he first saw the Great Northern train and, as it was too late to stop, he had put on steam in the hope of clearing the crossing and had almost done so.

The signalman at the crossing stated that he had replaced the up Great Northern signals to danger after the passage of a cattle train half an hour previously and had not cleared it since. He had also thought it safer to keep the Midland distant signal at danger until it was whistled off because it was a dark and wet night.

The fireman and guard of the express backed up their driver in saying that the distant signal had been 'off' and it was also found to be 'off' after the accident, but this was attributed to damage caused by the collision. Subsequent to the inquiry a Midland engineer wrote to Captain George Wynne, the investigating officer of the Board of Trade, suggesting that the Great Northern distant signal had been clear due to a faulty repair to it, but the inspecting officer was not convinced and thought it more likely that the Newark Crossing signalman had cleared his signals for the up express, which was well overdue, and had not replaced them.

Captain Wynne thought that it would be much better if all the signals at the crossing were kept at "Danger" and only cleared for the passage of a train, being replaced to "Danger" immediately afterwards. This method of working had been suggested before but objected to by the Midland Company. The inspecting officer said that the Midland ran no expresses over the crossing, and he could not see why they objected, except as a matter of punctilio; he rejected Midland suggestions that their trains might be delayed or that Midland and Great Northern drivers might be tempted to race one another to the crossing. He concurred with the Midland engineer's objection to the design of the night distant signals, whereby a coloured glass was raised in front of the lamp to show danger rather than the lamp rotating, and he considered that a third signalman should be provided at the crossing to reduce the hours that the signalmen had to work. The accident would not have occurred, he pointed out, had the express not been late.[1081]

What was probably the second rail/rail crossing to occur in the United States was far more serious than any of those which had preceded it there or in Great Britain and was also, briefly, the most serious railroad accident in North America up to that time. The Michigan Central and Michigan Southern Railroads crossed and then re-crossed one another on the level on the approaches to Chicago, Illinois. On 25th April 1853 trains of the two companies collided at Grand Crossing, the one nearer to Chicago, and situated on level ground amidst a swampy area. The 9 p.m. Michigan Southern express from Chicago

[1081] Ibid.

to Toledo was running half an hour late, having had to wait half an hour at a point not far from its starting point to make a connection with a late running eastbound Chicago and Rock Island train. The Michigan Central train had left Michigan City no less than seven hours late.

Engineer Edward Davis of the Michigan Southern train had only been with the company for a month and had only made three journeys with his conductor, H.L. Whiting. His train ran into the side of the Michigan Central train on the crossing while travelling at about 25 miles an hour, striking its sixth car. The Michigan Southern train should have given way to the Michigan Central one, but the rules were badly written and difficult to understand. They read: 'Regulations to be observed at the railroad crossing between Ainsworth and the Rock Island Junction – trains going from Chicago on the Northern Indiana Railroad (Michigan Southern) must avoid trains going either way on the Illinois Central Railroad (also used by the Michigan Central), trains on the Illinois Central Railroad going either way will avoid trains going towards Chicago on the Indiana Railroad.' Which company's trains had priority depended, therefore, on whether they were going to or from Chicago. Engineer Davis admitted that he had never understood the regulations. He also had some excuse inasmuch as the Michigan Central train was running without a headlight. The Michigan Central locomotive had also made two previous trips without a headlight, the light having failed due to a defective wick and the problem not having been corrected despite Engineer Thomas Rockham having reported it. Instead his fireman had placed a hand lamp at the front of the engine.

Engineer Rockham had approached the crossing at about 20 miles an hour but had whistled for brakes and slowed to five miles an hour when he reached it. He had seen the Michigan Southern train also approaching but could not estimate its speed and relying on the fact that according to the rules his train had priority over the crossing he had kept going. Ironically, therefore, if he had either not reduced speed to begin with or having done so had pulled up there would have been no collision.

Engineer Davis stated later that he had not seen the Michigan Central train until he was within a quarter of a mile of the crossing, when he had reversed his engine and whistled for brakes, but too late. The Michigan Central train was a mixed one and consisted largely of 22 freight cars, but with two emigrant cars placed sixth and seventh from the engine, as chance would have it just where it was hit by the Michigan Southern train. The two cars were smashed in the collision and thrown from the embankment into the adjoining swamp, where some of those who had escaped death from the impact were drowned in the shallow water. The two engineers and their fireman all escaped unhurt, the *Chicago Tribune* commenting 'as usual.' Engineer Davis and his fireman had jumped from their locomotive, which together with the leading baggage car and a second-class car of the Michigan Southern train was wrecked.[1082]

After the accident, Conductor H.L. Whiting ran back to Rock Island Junction, from where he commandeered an engine and proceeded to Chicago with news of the disaster. Likewise Engineer Rockham detached his engine and drove into Chicago on the Michigan Central line. Upon the arrival of Rockham's engine with the news messengers were despatched into the city and a locomotive and car was sent from the city with doctors Palmer and Clark and arrived at the accident scene at about midnight (the

[1082] Dixon Telegraph, Dixon, Illinois, United States, 7th May 1853 collision had occurred at 10.15 p.m.). This relief train returned to Chicago at 2.30 a.m. with passengers from a first-class carriage of the Michigan Southern train which had been thrown onto its side on the northern side of the Michigan Central track; none of these passengers was killed or even seriously hurt. A *Chicago Tribune* reporter commented that at this point, four and a half hours after the accident, no Michigan Southern relief train had yet appeared to rescue their injured passengers, who could only be reached via their line, as the crossing was surrounded on all sides by water so that getting from one line to the other was difficult if not impossible.[1083]

Heart-rending stories were told of the scene after the accident, 'Here sat a poor woman with a broken limb, and her little daughter stood by her side, weeping and begging for assistance.' Most of the victims were German immigrants, and 'There lay a young German, dead, his sister by his side, wringing her hands, and crying, "mein Bruder, mein Bruder!" Here a woman crying, "Oh, my father," there a woman wailing for the loss of her infant; another dead, her mangled features but partially concealed by a cloak, and at her side only a faithful dog.'[1084] About 20 passengers died and about 60 were injured. The scene of death and destruction did not, though, prevent an opportunist thief from stepping over the body of the express messenger to steal the Boston and Albany mail bags.

The inquest on the accident placed the blame for it on the crews of both trains, and on Jerraut, the Michigan Central machine shop superintendent, the latter presumably because of the unlit headlight of the Michigan Central train.[1085]

Another rail/rail crossing collision occurred in England at the end of 1855, on 28th November, this time not at Preston but on the Great Western Railway main line, not very far from London Paddington. At West London Crossing the main line was crossed by the single line of the West London Railway, which having left the London and North Western Railway at Willesden and passed under the Grand Junction Canal, passed over the Great Western line to terminate at Kensington Basin. The West London line fell at 1 in 50 and 1 in 39 but a catch siding was provided for runaways and a barrier at the crossing, manned by a policeman. A timetable was provided at the crossing giving the times at which West London trains were allowed to pass over it, for example after the passing of the 8.15 a.m. down train from Paddington until 8.40 a.m., together with a set of regulations relating to the fixed signals, procedures in the event of delays or a train breaking down on the crossing and so on. Day and night signals were provided on the Great Western and West London lines, and as an additional precaution a second up distant signal was provided in Acton Cutting, 2,519 yards from the crossing, worked by a signalman stationed one mile from the crossing with instructions to repeat with the Acton Cutting distant signal the indication of the crossing distant signal. If during foggy weather the Acton Crossing signalman was unable to see the indication of the crossing signal, he was instructed to maintain his signal at danger, place two detonators on the line and walk forward to a point where he could see them. The stop signals at the crossing, but not the distant signals, were provided with lamps. The crossing was not to be used before 6 a.m.

[1083] Ibid.

[1084] Ibid.

[1085] A History of Railroad Accidents, Safety Precautions and Operating Practices by Robert B. Shaw pg. 102

or after 6.15 p.m.

Lt-Col W. Yolland, the Board of Trade investigating officer, pointed out that at the time of year when the accident happened, when it did not get light until 7 a.m., the two up distant signals were worse than useless, as in the hours of darkness the Acton Crossing signalman could not see the signal he was expected to repeat, nor in any case had he the proper means of repeating them. Electric telegraph working had been established between Paddington and Ealing but was not used to protect the crossing.

At about 6.15 a.m. on the morning of the accident, Henry Hill, the signalman at the crossing, had raised the barriers and cleared his distant signal on the West London line for one of their trains from Willesden to pass, having placed his main line signals at "Danger." At about 6.20 a.m. he became aware of an approaching up Great Western goods train, and first walked towards it and then hurried back to the crossing to show a red light to the West London train, which was then rapidly descending the incline to the crossing, its engine propelling its train. The Great Western train had just entered the crossing when its guard's van, behind the second engine of the train, was struck by the guard's van of the West London train. London and North Western guard John Moore on the West London train was killed, his van being completely destroyed, and the Great Western guard very seriously injured. Seven or eight coal trucks of the West London train and the van and five or six trucks of the Great Western train were damaged, and the line was blocked for three hours.

The Great Western train was the 4 p.m. up goods from Bristol, running 5¼ hours late due to heavy traffic, extra stops, an engine problem and having to shunt for other trains. Driver John Brown, in charge of its pilot engine, said that he could not see the Acton Cutting distant signal because it was dark but that he received a white "All clear" light from the Acton Cutting signalman. Driver Manasseh Collins on the train engine also saw the white light. Driver Brown said that he could not see the up crossing distant signal either but saw the signal at the crossing at danger when he was 150 yards inside the crossing distant signal. He had then reversed his engine, whistled for his guard's brakes and done everything in his power to stop.[1086]

The Acton Cutting signalman, named Bennett (his Christian name was not stated), was 21 years old. He testified that he had been in charge of the box since 5th the month after five weeks working as a signalman at Ealing. He did not think that the West London line opened until 6.55 a.m. and it had never been explained to him that he must walk up the line to see the crossing stop signals. No up goods train had passed him between 6 a.m. and 7 a.m. until the day of the accident. No explanation was really given as to why he had shown the driver of the goods train a white light, Lt-Col Yolland attributed his mistake to ignorance, forgetfulness or neglect of his duties.

The Acton Cutting Signalman's mistake apart, the inspecting officer stated, the arrangements for protecting the crossing were faulty. Mechanical means were required for the indication of the crossing signals to be repeated at the Acton Cutting box, or alternatively the use of the electric telegraph. Under the existing arrangement, the Acton signalman had during the night, or when it was foggy, to walk nearly half a mile to find the position of the crossing signal before returning to set his own distant signal, during which time the indication of the crossing signal might have changed. Continuing, the

[1086] Board of Trade Report

inspecting officer wrote that, 'The placing of an inexperienced young man but two months in the employ of the Company, in charge of so responsible a duty as the repetition of signals to cover a crossing so dangerous as this was known and admitted to be, and the omitting to exhibit lamps on the distant signals during the winter months, so as to make them available before it is daylight, are circumstances of such an extraordinary character as to be difficult of belief; but these facts are undeniable, and the death of one man, and injury to another, has directly resulted from such unfortunate arrangements.'[1087]

Lt-Col Yolland was not impressed with a proposal to move the Acton Cutting box 300 or 400 yards further east, with the lack of a brake van at the rear of the Great Western train, with its great unpunctuality, or with its length (90 waggons after leaving Twyford according to the Board of Trade report). Nor did he like the fact that the West London train was being propelled, a fact dictated by the lack of any run-round facilities at Kensington Basin.

As far as recommendations were concerned the inspecting officer said that Acton Cutting box should be repositioned at the crossing up distant signal and its signalman furnished with the times of up main line trains passing his post. The distant signals should be supplied with lamps. The repeating distant signal should be kept at "Danger" and the crossing signalman informed before it was cleared for an up train and there should be communication between the two signalmen by means of a back-signal or a bell.

In conclusion, LT-Col Yolland remarked that as there was no exchange of traffic between the two companies at the crossing, and given the danger of a loaded coal train running down the steep incline to the crossing, it would be much better if the latter was abolished, by carrying the West London line over the Grand Junction Canal and the Great Western, and he believed the matter was being discussed by the directors of the two companies.[1088]

The coroner at the inquest into the death of the guard showed considerable sympathy for the Acton Crossing signalman who had directly caused it, pointing out his inexperience, the arduousness of his duty and the fact that he had to work daily from 6 a.m. to 7 p.m. in all weathers, for only 15s a week. At the conclusion of the inquest, at the coroner's suggestion, the jury immediately returned a verdict which stated simply that the guard had died as a result of a collision between the two trains on the crossing and did not fix blame on anyone.[1089]

After several years delay an overbridge was duly built and the crossing abolished in October 1860.[1090]

On 8th December 1856 a bad accident occurred at Alliance, Ohio, when a Cleveland and Pittsburgh Railroad train collided with the rear part of a Pittsburgh Fort Wayne and Chicago one at the crossing of the two lines, two cars being thrown off the line and into the station buildings. Eight people died and several were injured. It was believed that all

---

1087 Ibid.

1088 Ibid.

1089 The Times, London, England, 1st, 3rd, 5th and 13th December 1855

1090 History of the Great Western Railway Vol 1 by E.T. Macdermot, pg. 232

those who died had been standing on the station platform at the time.[1091]

A coroner's jury who inquired into the accident found that it was due to the northbound Cleveland and Pittsburgh train entering the crossing when the westbound Pittsburgh Fort Wayne and Chicago train was already on it and that it was the fault of Engineer John Cherry of the Cleveland and Pittsburgh train. Engineer Cherry was at that time reportedly still 'at large' at the headquarters of the Cleveland and Pittsburgh Company but said that he was willing to come forward to stand trial as long as he could 'protected from the violence of an excited public,' suggesting that he feared that he might be beaten up or even lynched.[1092]

Engineer Cherry later surrendered himself and on 17th December made a long affidavit regarding the accident. He wrote that his train had taken wood and water at Bayard, 12 miles from Alliance, and had approached the latter place on time. He had whistled for the station and crossing, shut off steam, and finding that his speed was not dropping had whistled for brakes. He had seen the engine of the Fort Wayne train east of the crossing and it appeared to be stationary. Finding that his train was not going to stop at the usual place, he had whistled for brakes a second time and then also found that the Fort Wayne train was on the crossing. He had reversed his engine and given it steam. He had never previously passed over the crossing without stopping with a passenger train, and only once at low speed with a freight train after his conductor had told him that the line was clear. His train was travelling at not more than 12 miles an hour at the moment of collision. After the accident, his life had been threatened a number of times and after about twenty minutes, declining the offer of a Cleveland and Pittsburgh watchman to hide him in his house, he had walked to the next station at Limaville and caught a train to Cleveland.[1093]

It will be seen that Engineer Cherry's affidavit did not really explain how the accident happened, but it seems possible that he had just run into Alliance a little too fast and braked too late, or perhaps he did not get the assistance that he had expected from his conductor in braking the train, though he made no attempt in his statement to blame the latter.

Just north of Darlington station, England, the North Eastern Railway main line crossed the main line of the pioneer Stockton and Darlington Railway on the level at S and D Crossing. On 7th September 1863, less than two months after the Stockton and Darlington had been absorbed into the North Eastern, there was a collision on the crossing.

S and D crossing was protected by fixed signals worked from a hut in the south-eastern angle of the crossing, with a signalman on duty there throughout the 24 hours and with communication by telegraph bell with Aycliffe, five miles to the north, and Darlington to the south. There were also sidings on the Stockton and Darlington line to the north-west of the crossing, worked by a pointsman there, but he went off duty after 7.40 p.m. until 6.40 a.m. next morning.

[1091] The Ohio Repository, Canton, Ohio, United States, 10th December 1856

[1092] Ibid, 17th December 1856

[1093] The Independent Democrat, Elyria, Ohio, 31st December 1856

On the evening of the accident, a train of 25 waggons loaded with ironstone which had left Ingleby at 4 p.m. for Lady Londonderry's works at Seaham arrived at the crossing on the Stockton and Darlington line at 8 p.m. Its driver whistled for the crossing at the same time that Aycliffe announced on the telegraph bell that the up (southbound) mail train had left there. The signalman let the ironstone train over the crossing from east to west, then cleared his signals for the up mail train which had just come in sight. As he was doing so, the Darlington station telegraph bell announced that the northbound 6.10 p.m. York – Newcastle Parliamentary train had arrived there. As the mail train reached the ticket platform south of the crossing Darlington signalled that the Parliamentary train was leaving and the signalman, seeing that the crossing was clear, lowered his signals for it and was just about to send it on to Aycliffe as it passed him when he heard a crash.

The night was dark and there was a strong wind. The driver of the ironstone train had taken it forward almost as far as the former Stockton and Darlington station, then commenced to shunt it into the sidings in the north-west angle of the crossing, where the waggons were to be stabled overnight before being taken northwards to Seaham in the morning. Unfortunately, a set of points had been left in the wrong position so that the waggons were pushed back onto the crossing. The driver of the ironstone train saw what was happening but as the guard had uncoupled the waggons from his engine, he could do nothing about it. The driver and fireman of the Parliamentary train were naturally not expecting to find waggons in their path on the crossing. Both were much hurt, and several passengers were also injured.

Guard Chaytor of the ironstone train had asked a number taker named Gale to check that the points were set right for the shunt movement, but the latter objected to doing so as he was unfamiliar with the sidings and the location of the points and only had a hand lamp and not a signal lamp. Guard Chaytor heard the number taker's protest but seeing him walk in the direction of the points rashly assumed that he had checked them anyway. He then uncoupled the waggons from the engine, leaving his driver with no control over them when it as seen that they were going the wrong way.

Captain H.W. Tyler for the Board of Trade said that since the accident a set of safety points had been installed to prevent a recurrence. They needed however to be connected to the crossing box. He recommended various other improvements at the crossing, such as the interlocking of the main signals for the two lines so that they could not be cleared simultaneously, the supplementing of the telegraph bells with instruments, uniforms for the crossing signalmen and paying them higher wages than the 17s a week which they currently received.[1094] [1095]

A faulty shunting move was also responsible for a collision at Clifton Crossing, England, on the London and North Western Railway, on 17th February 1866.

The London and North Western line from Eccles via Tyldesley to Wigan crossed its branch from Patricroft to Molyneux Junction on the Lancashire and Yorkshire Railway at an angle of about 75°, and fixed signals were provided which were interlocked, so that were those for the Clifton Branch taken off those for the Eccles - Wigan main line were necessarily at danger. The Wigan line was worked by the telegraph and the Eccles

1094 Board of Trade Report

1095 The Times, London, England, 8th and 9th September 1863

Junction signalman communicated with the Clifton Crossing signalman by means of a gong when a down train was telegraphed to Eccles from Weaste, thee miles nearer Manchester. The distance from Clifton Crossing to Patricroft on the Manchester-Liverpool line was so short that in moving waggons into or out of the Patricroft sidings at the start of the branch the shunting move had frequently to foul the Wigan line at Clifton Crossing. There were no special regulations relating to these shunting moves.

The Clifton Crossing signalman stated that at 3.40 p.m. on the day of the accident a goods train was backed up the wrong line from Patricroft along the branch and came to a stand over the crossing at 3.50 p.m. The 3.40 passenger train from Manchester to Tyldesley was then due so the signalman spragged the waggons which stood on the Molyneux Junction side of the crossing, uncoupled them and the engine drew the front waggons forward onto the Patricroft side of the crossing and stopped there, thus leaving an opening at the crossing itself. The signalman then had the Manchester train signalled to him from Eccles Crossing, accepted it and cleared his main line signals. He said that he then heard a noise and going to the door saw the front portion of the goods train being backed towards the crossing. He waved his arms at the guard to indicate danger and then threw back his down signals just as the passenger train was about to pass them. He distinctly denied having given any signal to the goods train guard to re-join the two portions of the train.

The goods guard on the other hand stated that after his train had been divided his engine had taken some waggons out of the Patricroft sidings and he had then exchanged hand signals with the signalman, whom he understood to have agreed to the train being reunited. He had got as far as recoupling the two portions of the train and his driver was just about to move ahead when the passenger train ran into them.

Driver Edward Sackfield and Fireman John Daniels, of the Manchester passenger train were apparently not keeping a good look out and the fireman said that they had been only 20 yards from the crossing when his driver had called out, "See that, Jack," and had then seen the home signal at "Danger" and had jumped from the engine. The Manchester train cut clean through the waggons but in doing so its engine and tender were derailed and went down an embankment. Driver Sackfield, who had stuck to his engine, was very seriously hurt and died before he could be brought to Manchester.

Colonel W. Yolland looked into the accident for the Board of Trade and stated that he was unable to decide whether the signalman or the guard had given a true account of the events leading up to the accident, and who therefore was to blame. It was perfectly clear, however, that a most objectionable practice of shunting trains over the crossing without all the signals being kept at danger had been adopted, as if it were a station yard. It was, he said, difficult to believe that such a practice had gone on so long. He thought it unfortunate that the level crossing had ever been sanctioned and recommended that all shunting over the level crossing should cease.[1096] [1097]

Dundalk, Ireland, was the scene of the next accident of this type. At about 9 a.m. on 18th July 1868 the 5.45 a.m. up mixed train from Portadown to Drogheda of the Dublin and Belfast Junction Railway collided with a goods train of the Irish North Western

---

[1096] Board of Trade Report

[1097] The Times, London, England, 20th February 1866

Railway on the level crossing just south of Dundalk station. There were only three passengers on the mixed train and none of them was hurt, nor were any of the railwaymen.

The crossing was protected by a four-armed stop signal placed in the north-east angle of the crossing and applying to all four directions. Because of an intervening overbridge, which obstructed the view, the down and up signals for the Irish North Western line were placed on the 'wrong' side of the signal post. A signalman in a hut at the crossing had orders not to let an Irish North Western train over the crossing within 15 minutes of a Dublin and Belfast Junction train being due.

On the day of the accident, the Portadown train was 25 minutes late, having been delayed at Goragh Wood, and did not arrive at Dundalk until 8.55, with an up express being due at 9 a.m. Its driver did not therefore stop at the up platform to put down the passengers but continued, whistling for the crossing signal, so as to draw forward past a mains crossover and set back into the down platform out of the way of the express. The crossing signal was cleared and as he moved forward, he saw the crossing signalman walking up the Irish North Western line towards the ticket platform with a red flag, then as soon as his engine was clear of some buildings, he saw the tail of an Irish North Western train coming towards him. He whistled to attract the attention of the other driver and put on steam, but the Irish North Western train struck the 16th waggon from his engine.

The Irish North Western train was a portion of the 4 a.m. up mixed train from Enniskillen to Dundalk. A banking engine was attached to its rear at Castleblaney, on which were riding the regular Driver Woods and Driver Evans. The train reached Dundalk ticket platform a few minutes late and here the front engine took part of the train around the curve to the passenger station, leaving the rear engine to push 18 waggons over the crossing to the quay. At this point Driver Woods left the engine to get his coat from the van and to alter the points for the train to go towards the crossing. Driver Evans started the engine and whistled for the crossing. The Irish North Western men all agreed that the crossing signal was lowered for them, but thrown back as the train neared the crossing, the signalman at the same time showing a red flag. By this time Driver Woods had re-joined the engine and by reversing it managed to almost stop before the train reached the crossing. One waggon of the North Western train and some horse-boxes and waggons of the Junction train were derailed and damaged.

The signalman at the crossing denied having ever lowered his signal for the North Western train at all but said that the engines of the two trains had whistled at about the same time and having lowered the signal for the Junction one he had taken a red flag to exhibit to the North Western one. Lt-Col C.S. Hutchinson, the Board of Trade inspecting officer, was inclined to think however that the signalman, whom he described as an old man hardly fit for the job, had in fact pulled off for the North Western train and then, seeing the Junction train coming in fast, had thrown it back again and shown the North Western train a red flag, hoping that it might pull up short of the crossing, which if the engine had been at the front and its driver had been keeping a proper look out it probably would have done. The speed of the Junction train appeared to have been greater than was justified in approaching such a dangerous crossing.

Lt-Col Hutchinson urged the repositioning of the signal at the crossing, the provision of a lower arm, interlocking to prevent the display of conflicting signals and the provision of safety points leading to blind sidings on the North Western line on either side of the crossing.[1098]

A mile from the depot at Buffalo, New York, United States, the lines of the New York and Erie and Buffalo and Erie Railroads crossed one another at right angles. At about midnight on 12th /13th November 1868 the crossing was the site of a collision which could have been a lot more serious. A New York and Erie passenger train was passing over the crossing when it was run into by a Buffalo and Erie cattle train, which cut through the baggage car and derailed the engine and other cars. It was said that 'Had the collision occurred one second later there would have been fearful loss of life' doubtless because the cattle train would then have struck the passenger cars rather than the baggage car. As it was the only human casualty was Baggagemaster Robert Decatin, who suffered a broken leg. A large number of cattle and hogs were killed.[1099]

The worst rail/rail crossing collision in England occurred at St Nicholas Crossing, just south of Carlisle Citadel station, early on the morning of Sunday 10th July 1870, when London and North Western Railway up passenger and mail train collided with a North Eastern Railway goods train on the crossing. Five passengers were killed and 30 more, and one of the guards of the mail train, were injured.

The main line from Carlisle south to Penrith was crossed on the level 520 yards south of Carlisle Citadel station by a goods line of the North Eastern Railway, known as the Canal branch, which ran from its Newcastle and Carlisle line to connect with both the Maryport and Carlisle Railway and the North British Railway. The crossing was protected by fixed signals worked from a box situated in the north-east angle of the crossing, the stop signals at the crossing itself being all placed on one post, which because of the angle of the crossing made those relating to the North Eastern line but imperfectly visible. Distant signals were also provided, except on the Citadel station side, and at the suggestion of the Board of Trade a notice board had been erected 30 yards west of the crossing on the Canal branch reading, "No engine to pass this board until the crossing signal is lowered."

On the morning of the accident the up mail train from the North departed from Carlisle Citadel station on time at 12.47 a.m., consisting of an engine and tender, a brake van, 10 carriages, a Post Office van and a second brake van. Driver Hugh Norris told the subsequent inquest and the Board of Trade inquiry that he had had clear signals on leaving the station and at the crossing, which he had approached travelling at about six miles an hour (but which was probably really at least double that). Looking out on the left side, he had seen the signalman doing something with his hand lamp which he did not understand and this made him look back his train. At the same moment he felt a drag and immediately saw that the rear coaches of the train were off the line. He had stopped in about 150 yards and going back had found the sixth to eleventh coaches derailed and the fifth and six badly damaged.

The goods train involved in the collision was on its way back from the North British

[1098] Board of Trade Report

[1099] Davenport Daily Gazette, Davenport, Iowa, United States, 14th November 1868

Railway Canal station to the North Eastern Railway's London Road goods yard. It did not appear in the timetables but ran daily, between 3 a.m. and 4 a.m. except on Sunday nights when it ran three or four hours earlier. The engine of the goods train, No. 36, should have been in charge of Driver Robert Pattinson and Fireman Joseph Rowell. The driver and fireman came on duty at 7 p.m. on the evening of Saturday 9th and shunted the London Road yard until 9 p.m. when together with their guard, Michael Shields, they adjourned to a pub. The driver and guard returned about 11 p.m. and shortly began shunting a train which had just arrived from Newcastle. By 11.30 p.m. when Fireman Rowell returned, 13 waggons were ready to be taken to the North British Canal station. At this point Driver Pattinson told his fireman to take charge for two or three minutes and disappeared to the rear of the train for some reason, perhaps just to relieve himself. When he returned a few minutes later the night yardsman, named Henderson, told the driver that his fireman had taken the engine and waggons to the North British yard without him. Driver Pattinson did not seem unduly worried and went to a pointsman's cabin where he remained until he heard of the accident.[1100]

Fireman Rowell was advised by his solicitor not to give any evidence either before the coroner or the Board of Trade investigating officer, Lt-Col C.S. Hutchinson.

Guard Shields described setting off on the engine with Fireman Rowell, having collected the staff for the single line branch. He said that he had ridden on the engine in case he needed to rouse the keeper at a crossing near the canal, who might be asleep, and he stated that it was not uncommon for trains over the branch to be run with just the driver or fireman, and without brake vans. They had reached the Canal station without any problem at about 12.20 a.m., and a train of 19 loaded and 8 empty waggons having been made up started on the return journey at 12.40. This time the guard was riding in his proper place on the last waggon of the train. Guard Shields said that the light of the back (distant) signal had been out but that of the signal at the crossing itself was showing a dim red light. The top speed of his train had been 12 or 14 miles an hour, and it was reduced to seven or eight miles an hour before the collision. He had seen the lights of the mail train when it was about 400 yards from the crossing and had felt soon afterwards that a collision was inevitable.

Fireman Rowell apparently spoke to his locomotive superintendent shortly after the accident and told him that the distant signal had been out, he had not seen the home signal and had only seen the mail train when it was 200 yards away. He had then done everything that he could to stop and had reversed his engine and put on steam the back way. He could have stopped in another five yards.

Signalman John Dalton, who had worked the crossing box for the last 18 years, said that all his signal lamps, which were gas lit, had been burning, except the up Canal branch distant. The previous night the lamp of this signal had not been lit when he came on duty at 10 p.m. and he had asked the signalman whom he was relieving to light it on the way home. The next morning his mate had told him that he had wasted three matches trying to do so. The lampman had been informed and thought that he had fixed it, but it went out again later and given the previous problems and the night not being very dark Dalton had not bothered with it further, though he had an oil lamp in the box which he could have used instead. The North Eastern goods train had gone down the branch at

---

[1100] Board of Trade Report

about midnight and he had recognised Guard Shields as one of the two men on the footplate of the engine. At 12.45 Crown Street box had gonged him that the mail was ready to depart, and the road being cleared he had lowered his signals. At that moment he had heard the engine of the North Eastern train whistling some way off, but not giving the three distinct whistles for the crossing. As the mail train was going over the Crown Street bridge, he caught sight of the goods train approaching, apparently rather quickly. Thinking that it was coming rather too fast he had grabbed his hand lamp, turned it to red and waved it up and down at the North Eastern engine. One of the derailed carriages of the mail train had struck his box following the collision and he was a good deal shaken.[1101]

Lt-Col Hutchinson said that Fireman Rowell was clearly the person directly responsible for the collision. Having returned from the pub he had voluntarily taken control of the train and had lost control of it on the return journey. Even though the light of the up distant had been out there were plenty of landmarks to show him where he was, and had he been driving with proper caution he could have stopped at the stop board. He had also broken the rules regarding treating the absence of a signal as a danger signal and restricting the speed of his train at the crossing to four miles an hour. Driver Pattinson was greatly to blame for not having made any effort to re-join his engine and Guard Shields for allowing the fireman to drive the engine alone and for riding on the engine on the outward journey. Yardsman Henderson was to blame to a lesser extent. The culpability of the four men was however much mitigated by the fact that the irregular methods of working were known to and permitted by higher officials. No copies of the rules relating to the crossing appeared to have been distributed to the North Eastern Railway staff.

The practice of working goods trains over two important passenger lines without a brake van was, the inspecting officer considered, reprehensible. One of the causes of the accident appeared to be that the working of trains over the Canal branch was treated as a kind of shunting operation in a goods yard. Signalman Dalton could not be acquitted of all blame for he had just accepted the defective condition of his distant signal lamp, despite having had an oil lamp he could have substituted in his box. The rule restricting the speed of trains between the north end of the Citadel station and St Nicholas Crossing to four miles an hour appeared to be systematically broken and should be either strictly enforced or expunged. The signals at the crossing should be repositioned and those at the junction with the Newcastle-Carlisle line interlocked.

The most obvious means of preventing a repetition of the accident, Lt-Col Hutchinson concluded, was of course to eliminate the crossing, and its abolition could become a practical in connection with the alterations to the station which would be needed in connection with the opening of the new Midland Railway line. He earnestly recommended the directors of the companies to consider it. Meanwhile as a temporary remedy he recommended the installation of safety points leading to blind sidings on each side of the crossing on the Canal line, with similar points at the crossing of the branch and the Maryport and Carlisle line.[1102] [1103]

[1101] Ibid.

[1102] Ibid.

[1103] The Times, London, England, 11th, 12th, 13th, 14th, 20th and 21st July 1870

The coroner's jury at the inquest not surprisingly found Fireman Rowell guilty of manslaughter, while severely censuring Driver Pattinson and Guard Shields, and 'trusted that the result of this inquiry would be to stir up the North Eastern Company to a sense of their responsibility, so that for the future the lives of the public should not be placed at the mercy of reckless and incompetent men.'[1104]

St Nicholas Crossing was abolished, replaced by an overbridge carrying the Canal branch over the London and North Western main line, in the summer of 1877.[1105]

Plymouth, Indiana, United States was the scene of the next accident of this type on 14th December 1871. An Indianapolis Peru and Chicago Railroad local mixed train was on the crossing when it was struck by a Pittsburgh Fort Wayne and Chicago Railroad freight train at about noon. Two freight cars of the Peru train and the engine of the Fort Wayne train were damaged, but no one was hurt. Both trains had just restarted after having stopped. Both lines were blocked until about 5 p.m.[1106]

An accident on Byers Green Crossing, England, was unlucky in that within a very short time the crossing was to be abolished anyway, with work on a new line with an overbridge in place of the level crossing well advanced. At the time of the accident the North Eastern Railway main line was crossed on the level at an oblique angle by the branch from West Hartlepool to Spennymoor, which ran alongside the main line from Ferry Hill (later Ferryhill) station. The crossing was protected by home and distant signals in each direction, worked from Byers Green Crossing cabin, though with the home signals for the main line on the same signal post, 90 yards north of the actual crossing.

At about 10.10 p.m. on the evening of the accident Signalman Thompson in the crossing box lowered his signals for a down main passenger train. As he replaced his down main home signal behind that train, he heard a passenger train from Ferry Hill to Spennymoor whistle for the crossing while it was about half a mile off. As the Spennymoor train got within 200 yards of his home signal he saw an engine and van, which had been slowly approaching on the down main, and waved his red hand lamp at both trains. The trains however collided and the passenger engine becoming separated from its carriages the latter ran a short distance up the 1 in 100 gradient, stopped and began to run back to the crossing, where they would have collided with the derailed van of the other train, but Thompson managed to turn them into a siding where the fireman of another train was able to stop them. Eight passengers were slightly injured, and the guard of the passenger train and the fireman of the engine and van were badly hurt. Signalman Thompson declared most positively that he had never lowered his signals for the engine and van.

The passenger train, from Stockton, was due to leave Ferry Hill at 10.05, the same time as the main line passenger train, with which it connected there, and as usual left a minute or so later, at 10.08. Its driver said that all the signals at Ferry Hill and Byers Green Crossing had been off for him and he had seen nothing of the engine and van until he had seen the reflection of the headlights on the side of the van at the last moment,

---

[1104] Ibid, 3rd August 1870

[1105] Rail Centres: Carlisle by Peter W. Robinson

[1106] Fort Wayne Daily Sentinel, Fort Wayne, Indiana, United States,16th December 1871

when it had been too late to do anything. The buffer of his engine had struck the van and his engine had become detached from its train. The fireman of the Stockton train confirmed that he had seen the main line signals at danger as their train had left Ferry Hill and as it was nearing the crossing.[1107]

The engine and van were returning from West Hartlepool to Gateshead, having taken a coal train to West Hartlepool. Its driver and fireman were comparatively ignorant of the line between Ferry Hill and West Hartlepool, the driver not having been over it for four years, and they were given as a pilot a driver named Parker, notwithstanding the latter having told the Gateshead foreman that he had not been over the line for two years, the foreman saying that their guard knew the line. The trip to West Hartlepool and back to Ferry Hill had been without incident and at the latter place the pilot, Parker, had given over charge of the engine to the regular driver, Heron, who said that he was acquainted with Byers Green Crossing. Heron alleged that he had approached the crossing at about four miles an hour, and that the crossing signal had been lowered to "Caution" when his train was 40 yards south of the crossing, then raised to "Danger" again as he reached the crossing and a red flag waved from the signal cabin. Parker, at the Board of Trade inquiry into the accident, confirmed what Driver Heron had said about the down main signal having been lowered to "Caution" and then replaced to "Danger" and a hand signal shown from the cabin. The stationmaster, assistant stationmaster and a goods inspector at Ferry Hill stated, however, that immediately after the accident, Parker had told them that he knew nothing as to the position of the signals as he had been sat down having his supper. Parker predictably denied having made this statement. The guard, Buckle, who because he was the only one who knew the road had exchanged places with the fireman, backed up the story of the two drivers.

Considering the conflicting evidence Lt-Col C.S. Hutchinson thought it exceedingly improbable that the signalman, 'an experienced and intelligent man,' would have cleared the main line signals at the moment the branch train was going over the crossing. It was, he thought, more likely that Driver Heron and unintentionally encroached on the crossing, especially as the signal applying to it was placed 90 yards on the wrong side of it. The signalman and the fireman of an East Hartlepool train deserved credit for diverting and stopping the runaway carriages. It was unfortunate that what was apparently the first collision at this crossing should occur so soon before it was removed but in the interim the down home signal should be moved to a point south of the crossing.[1108]

Another minor crossing collision occurred on 21st July 1873 at Llandilo Crossing, South Wales. Here the Great Western Railway main line from Swansea to Milford was crossed by the single track Llanelly Railway. The crossing was protected by stop and distant signals on the Great Western line, with just stop signals at the crossing on the Llanelly Railway.

On the day of the accident, the 7.45 a.m. passenger train from Gloucester, due at the crossing at about 1 p.m., was 13 minutes late and a coal train arrived at the crossing at the time the passenger train was due. The signalman, not knowing the whereabouts of the passenger train, began to let the coal train across, his main line signals being at "Danger".

1107 Board of Trade Report

1108 Ibid.

He then heard the whistle of the passenger train and then saw it, but its driver had shut off steam and whistled for brakes before he had passed the down distant signal, 567 yards from the crossing, and the signalman was confident that he would pull up. When it got closer and he could see that it would not stop in time he ran towards it with a red flag but the driver of the passenger train, who had reversed his engine, put it back into forward gear and put on steam in the hope of beating the coal train over the crossing.

The driver of the Llanelly train only realised the danger when he was 270 yards from the crossing. He likewise reversed his engine and whistled for brakes and had reduced his speed from about 18 to four miles an hour when his engine struck the brake van at the rear of the passenger train, knocking it and two carriages off the rails. The coal train was not derailed but the buffer-plank of its engine was broken. Eight passengers and the guard of the passenger train were slightly hurt.

Colonel F.H. Rich blamed the accident on the driver of the passenger train. The latter had claimed that the down main distant signal was off, but this appeared not to have been the case, and he could have seen the home signal, which he admitted had been at danger, more than 1,000 yards from the crossing. His carelessness could only be attributed to a complete absence of mind.

Since the accident the Great Western had provided new signals at the crossing, but distant and home signals should also be provided on the Llanelly Railway and the signals interlocked. It was very desirable to do away with crossing altogether.[1109]

The crossing was finally removed about 1912.[1110]

The two relatively minor British accidents at Byers Green and Llandilo Crossings were followed on 23rd August 1873 by a bad accident at Retford, England, in which three passengers died and about 35 were injured. The Great Northern Railway main line from London to Doncaster was crossed on the level at Retford by the Manchester Sheffield and Lincolnshire Railway main line from Sheffield to Grimsby. The crossing was protected by home and distant signals in each direction except that an up distant signal to the north of the crossing was dispensed with, the rule being that all up (southbound) trains were to stop short of the crossing. The crossing signals were controlled by a Manchester Sheffield and Lincolnshire signalman in the crossing box there and there was a separate box, the "tower signal cabin" 80 yards to the north of the crossing controlling the station signals, the station being situated north of the crossing.

In June 1861 a regulation was issued that before the tower signalman cleared his up signal for a train to proceed over the crossing, he ring a bell to the signalman at the crossing to indicate that he had a train waiting to cross, and must keep his signal at "Danger", only lowering it after the crossing signalman had lowered his signal. This regulation was only meant to apply to trains which had stopped at the station and did not preclude the tower signalman lowering his distant signal if the station itself was clear.

The rule as to all trains stopping at the station was occasionally broken for the running of special horse trains and by the date of the accident was daily broken in the case of the up Scotch express unofficially known as the *Flying Scotchman* which left Edinburgh at

---

1109 Ibid.

1110 Track Layout Diagrams of the GWR and BR(W) Vol 57: Llanelly by R.A. Cooke

10.25 a.m. In the case of these trains, the signalmen were informed of their running and the tower signalman did not lower any of his signals until he had had the train accepted by the crossing signalman and the latter had lowered his signal.

On the day of the accident, the up Scotch express was very long when it reached York and the decision was made to divide it there. The main train left with the passengers at 4.50 p.m., half an hour late, and was followed at 5.25 by an engine pulling eight waggons of fish and a brake van. The main train reached London Kings Cross safely. The fish train arrived at Doncaster at 7.2 p. and left there at 7.10 p.m. in charge of Driver George Bryant, who had been booked to work a light engine from Doncaster to Grantham to pick up a goods train but instead was detained at Doncaster to work the fish special.[1111]

Driver Bryant ran the 17 miles from Doncaster to Retford in 28 minutes at a speed of about 40 miles an hour. The up distant worked from the tower box was off and the tower box home signal, which could only display "Caution" or "Danger" was a "Caution", but this merely denoted that the station itself was clear. The signal at the crossing was at "Danger" and an inspector in the goods yard 400 yards north of the crossing signalled to the driver to pull up. The fish train's guard, Thomas Patrick, saw the inspector's signals and screwed down his brake but Driver Bryant did not see the warning, nor apparently the crossing signal at "Danger," while his fireman, John Greatorex, was busy oiling the engine and did not return to the footplate until the train was halfway through the station. By this point in time, Driver Bryant had realised the danger but it was too late.

The crossing signals were off for a Manchester Sheffield and Lincolnshire excursion train from Deepcar and Sheffield to Grimsby and Cleethorpes, carrying workmen of Messrs Grayson and Lowood on their annual outing, along with their wives, children and sweethearts.[1112] The engine of the excursion train, which was travelling at 10 miles an hour, reached the crossing as the fish train was 50 yards from it. The excursion train driver, Jonas Smith, put on steam to try and get across but the engine of the fish train caught the rear of the fifth carriage and the front of the sixth, cutting the train in two. These two carriages and four others were derailed, as was the engine of the fish train, which struck and demolished the signal box, though the signalman, George Whelpton, was not much hurt. The three passengers who were killed were in the fifth carriage, which also struck the signal box. Three fish trucks were derailed and damaged.[1113]

Colonel F.H. Rich investigated the accident. He said that Signalman Nat Prass at the Retford North box should have shown Driver Bryant a red flag, but the signalman said that he had thought that the train was going to stop in the station. Signalman Charles Whitehead had debated turning the fish train into the curve which connected the two lines but feared that the resulting sidelong collision would be worse than a direct one. While he was wondering what to do, he forgot to throw his signals back, though the trains were probably too close together anyway then for it to have made a difference.

The view of the Great Northern train obtained by Driver Smith of the excursion train was restricted by the station buildings.

The accident was clearly due, Colonel Rich wrote, to the unaccountable conduct of Driver Bryant. He was well acquainted with the road, and although the absence of a

---

[1111] Board of Trade Report

[1112] The Times, London, England, 25th August 1873

[1113] Board of Trade Report

distant signal for the crossing was likely to mislead an engine driver unfamiliar with the station the signal of the crossing box standing at "Danger" could be clearly seen by the driver of an up train from 900 yards away and there was no doubt as to its showing "Danger." Had he reacted to the caution signal given by the inspector in the goods yard he could still have pulled up. The fireman should have been keeping a lookout for signals instead of oiling the engine, but he was acting under the authority of the driver. The guard of the fish train was apparently not as alert as he might have been. He did not blame the Retford North or tower signalmen for not showing danger signals to the fish train as they were expecting it to stop and were taken by surprise and were then too confused to act.

It could not be safe under any circumstances to allow trains to approach under clear signals to within 80 yards of a train crossing the line at right angles, especially on a line where the trains travelled at high speed and the traffic was continually increasing. So long as the crossing existed distant signals should be provided in all directions and no train should be allowed to run up to the station while another was crossing on either line. The station needed resignalling and interlocking. There would be no difficulty in passing one line over or under the other so as to eliminate the crossing altogether.[1114]

At the inquest Driver Bryant was committed to trial for manslaughter.[1115]

Signalman Whitehead committed suicide on 22nd December. He had been in a depressed state since the accident.[1116]

Driver Bryant did not come to trial until 10th March 1874 when after a lengthy hearing he was acquitted.[1117]

The level crossing was finally replaced when the Manchester Sheffield and Lincolnshire line was carried under the Great Northern line in 1965, 92 years after the accident. Colonel Rich would not have been very happy that its abolition had taken so long.

At about 11.30 a.m. on 16th October 1873 a Boston Clinton and Fitchburg Railroad local freight train was butted by the engine of a Worcester and Nashua Railroad way train on the crossing at Clinton, Massachusetts, United States. Two cars were derailed and one smashed and both engines were derailed. No one was hurt.[1118] The crossing was replaced by a bridge at an unknown date.

The Pennsylvania Railroad express which had left Jersey City at 8.45 a.m. on the morning of Monday 23rd November 1874 ran into a Central of New Jersey Railroad coal train at the crossing of the two lines at Elizabeth, New Jersey, United States. This crossing was apparently controlled from a signal tower and the towerman had decided to give the coal train priority in what was presumably an operating error, as had no accident occurred the express would have been delayed. The air brakes of the express allegedly

---

1114 Ibid.

1115 The Times 27th, 28th and 30th August and 6th and 17th September 1873

1116 Ibid, 23rd December 1873

1117 Lloyd's Weekly Newspaper, London, England, 15th March 1874

1118 Daily Kennebec Journal, Augusta, Maine, United States, 17th October 1873

failed to work. Engineer Garrabrandt whistled for his two brakesmen to apply theirs, which they did, but the ten-car express could not be pulled up in time. The sixth truck from the rear of the coal train was knocked off the track and the coal thrown high into the air. Three other coal trucks were piled on top of it and demolished. The express train engine was thrown onto the up line and collided with the engine of a Philadelphia train which had just stopped at Elizabeth station, wrecking both locomotives. Some of the frightened passengers of the express reportedly jumped but there was no report of injuries. Engineer Garrabrandt saved himself by jumping.[1119] The crossing was replaced by a bridge at an unknown date.

What was to prove the last accident of this type of any seriousness in Great Britain occurred at Bedford, England on 12th March 1875. At Bedford, the London and North Western Railway's Bletchley – Oxford line was crossed on the level by the Midland Railway's branch to Hitchin at the west of the North Western station and to the south of the Midland station. The crossing was controlled by a signal box with fixed signals, but some of the signals, such as that for an up train starting from the North Western station, were placed beyond the crossing they were intended to protect, as at Byers Green Crossing (qv).

On the day of the collision the 6.45 a.m. up Midland passenger train from Hitchin to Bedford was going over the crossing when it was run into by the 7.15 a.m. North Western passenger train from Bedford to Bletchley. One passenger, the Rev W. Sprott, a Presbyterian minister returning home to Glasgow, was fatally injured and died two days later in the Bedford Infirmary; four other passengers, all on the North Western train, were injured as were the fireman and guard latter. One carriage of the North Western carriage was broken to pieces, another was badly damaged, and the signalman's hut was demolished.[1120]

North Western signalman Richard Bishop told the Board of Trade inquiry that he had been at the crossing since the Midland branch had opened in 1857. There had never been an accident or near miss at the crossing previously. He had heard two whistles from the engine of the Midland train, had seen it approaching and had lowered his signals for it, those for the North Western line being at "Danger." When the Midland train was near the Elstow Road bridge he had heard the whistle of the North Western train as it left the station. He had not anticipated any problem as North Western trains normally came forward to about 50 yards of the crossing while the signal was at "Danger" It was his practice to show a red flag to a North Western train approaching the crossing if a Midland train was crossing. The driver of the North Western train had then applied steam and he (Bishop) had grabbed a red flag and run towards the train, waving his flag at the driver from the four-foot space. The driver had not appeared to see him, and he had had to jump out of the way into the six-foot, from where he continued to wave the flag.[1121]

Driver John Perkins of the Midland said that just after he had come under the Elstow Road bridge he had seen the signalman running towards the North Western train waving his red flag and had said to his fireman, "There is someone coming who has not seen the

[1119] The Bucks County Gazette, Bristol, Pennsylvania, United States, 26th November 1874

[1120] The Times, London, England, 13th, 15th and 16th March 1875

[1121] Board of Trade Report

signal." As soon as he saw the North Western train, he had reversed his engine. His signal at the crossing was off for him. He saw the North Western train going over the crossing and realising that he could not stop had jumped off his engine, which left the rails. Either it or the derailed North Western carriages struck the pointsman's hut.

The driver of the North Western train, who coincidentally was also named John Perkins, declined to give evidence on legal advice. His fireman, Edwin Leadbeater, said that they had set off from the station in the usual way, smoke and steam preventing him at that moment from seeing the crossing signals. He had then been firing when his driver opened the whistle violently and looking up had seen the Midland train bearing down on them. He had gone for the brake, but his driver said, "Let her come," and opened the regulator to try and get over the crossing in front of the Midland train. He did not see Signalman Bishop until after the collision.

Captain H.W. Tyler said that there was no doubt as to the cause of the collision. The servants of the Midland train were clearly not at fault as the signals had been clear for them. The driver of the North Western train could not be blamed for having started from the station as this appeared to be the prevailing practice, but there could be no doubt that the signal at the crossing had been at "Danger" against him and furthermore the signalman had attempted to stop him with a red flag. However, the driver's decision once he found he could not stop to try and get across in front of the Midland train was doubtless the correct one.

With a view to reducing the risk of a further accident at the crossing, the inspecting officer continued. the signalling arrangements should be improved. A raised signal box should be constructed, and signals erected on all sides of the crossing clear of the fouling points of the crossing and a separate crossing with some goods lines, a starting signal should be provided at the North Western station, worked from the crossing box, and the signals interlocked with one another in accordance with modern Board of Trade requirements.[1122]

The jury at the inquest at Bedford on the death of the Rev Sprott found a verdict of manslaughter against the North Western driver, John Perkins.[1123] At the Bedford Assizes the driver was found guilty of manslaughter, but with a strong recommendation to mercy, the jury there commenting that they considered the signal and general arrangements at the crossing highly dangerous and unsatisfactory. Driver Perkins escaped with a fine of £20.[1124]

William Campbell, a train hand, was probably fatally injured and two others were severely hurt when a Camden and Atlantic Railroad express ran into a gravel train of the 3ft gauge Camden Gloucester and Mt Ephraim Railroad at the right-angle crossing of the two roads in Camden, New Jersey, United States on Saturday 18th December 1880. 'An investigation resulted in the dismissal of the engineer of the express train, the yard master and the flagman.'[1125]

[1122] Ibid.

[1123] The Times, London, England, 20th March 1875

[1124] The Magnet, London, England, 2nd August 1875

[1125] Daily Gazette and Bulletin, Williamsport, Pennsylvania, United States, 20th December 1880

CHAPTER 14

FIRES

Fires often broke out on trains following derailments or collisions, especially perhaps in the more northerly areas of North America where trains were commonly heated by wood or coal stoves, especially in the earlier years, and a stove which overturned in an accident would more likely than not set the wooden coach ablaze. The substitution of gas for oil for train lighting proved to be another hazard, and a number of accidents were made much more severe by fires started by leaking gas. This chapter will confine itself, however, to what might be termed primary fires, that is accidents where there was no precipitating derailment or collision but merely a fire.

Steam locomotives, of course, employed fires to produce the steam which powered them, and the throwing off of sparks or small pieces of burning coke which could set alight goods carried in open waggons was an early hazard. The earliest accident due a fire appears to have happened on the Liverpool and Manchester Railway in June 1833, when three waggons carrying American cotton caught fire on Chat Moss, England; waggons and cargo suffering considerable damage. Tarpaulins were already in use to try and protect the loads but in this case hot ashes dropped by the locomotive had been thrown up by the waggon wheels and got under the tarpaulin. As preventative measures ash boxes were added to engines or wire screens placed behind their tenders to try and catch the hot cinders.[1126]

Another Liverpool and Manchester Railway fire on a goods train in 1834 continued to burn until it was noticed by a passer-by, resulting in a directive that a brakeman ride on the last waggon to detect such fires earlier.[1127] Despite this measure, on 22nd July 1835 £2,000 worth of damage was caused when three waggons containing one hundred packages of goods caught fire as a Manchester – Liverpool train was approaching the Whiston Incline. 'The accident originated, it is supposed, from friction, owing to the great speed at which the train was travelling.'[1128]

[1126] Liverpool and Manchester Railway Operations by Thomas Donaghy pg. 106

[1127] Ibid, pgs. 106-7

[1128] The Times, London, England, 25th July 1835

Preparations for an excursion over the railway from the Russian imperial capital of St Petersburg to the imperial palace at Czarskojeselo (spelt Zarskojeselo in the 1838 newspaper report) involved the movement of the offices of the imperial household from St Petersburg to Czarskoyeselo on 23rd April 1838, the day before the planned excursion, with various items included a quantity of silver plate. During the journey of the train carrying the household, one of its family carriages was set on fire by a spark from the engine. The engine driver failed to hear the cries of the frightened passengers, or perhaps, as the correspondent wrote, he was too confounded by the fire to stop the train. The passengers were forced to jump for their lives and fifteen of them were hurt, though only one very seriously. The carriage was more or less burnt out together with its contents, which included silver plate worth 30,000 roubles, which was melted. The planned excursion was postponed a day, but Tsar Nicholas I made an excursion on the railway on the day it should have taken place as an encouragement to the railway company, whose officers were said to be in consternation over the affair.[1129]

A similar accident to the Russian mishap occurred on the Great Western Railway between London and Twyford, England, on Saturday evening, 11th April 1840. Some burning cinders from the engine landed in the front boot {sic} of the down Bristol mail and set it alight. The guards saw the fire and attempted to get the engine driver to pull up by blowing their horns and shouting but could not attract his attention. By the time the train reached Twyford, the whole of the boot was on fire, and nearly the whole of its contents, which consisted of valuable parcels and passengers' luggage worth from £400 to £500, were consumed. There was some small consolation in the fact that the fire did not reach the letter or newspaper bags.[1130]

On Thursday 30th April 1840 it was the turn of the London and Birmingham Railway and the 8.30 p.m. down mail train from London Euston. At this date, passenger luggage was carried on top of the carriages, and 'Thursday being magazine night' a portion of the magazines accompanied the luggage on top of a first-class North Union Railway carriage. When the train was between Leighton Buzzard and Wolverton, the guards saw flames on the top of the carriage and applied their brakes, and this time the engine driver did realise the problem, and stopped the train, though with the train running fast downhill, and the wind blowing from the north-east, the fire had progressed somewhat before the train could be halted. The ten passengers in the affected vehicle escaped unhurt and some of the property was saved but the majority of it was destroyed. The fire was presumed to have originated in a spark from the engine which had penetrated the tarpaulin covering the luggage or lodged underneath it.[1131]

In May 1840 the Great Western Railway down mail train was in trouble again. It was found to be on fire on Tuesday 26th, when it was between Twyford and Reading, England, and was stopped as soon as possible. The fire was in one of the Bath mail bags, containing newspapers. Water was fortunately available close at hand, and the company's

---

[1129] Ibid, 8th May 1838
[1130] Ibid, 15th April 1840
[1131] Ibid, 4th May 1840

servants and some of the passengers were able to put out the fire before any serious damage was done, carrying the water in their hats. Some newspapers were lost, and some were partially burnt, but the letter bags entirely escaped. A spark from the engine was put down as the cause.[1132]

The driver of a train on the Birmingham and Gloucester Railway on 2nd July 1842 became aware of a fire which had broken out in one of the goods waggons of his mixed train as it was between Defford and Eckington stations, England, between 8 and 9 a.m. that Saturday morning. The reporter for the *Worcester Chronicle,* who described the accident, was critical of the driver for increasing the speed of the train and continuing to Eckington rather than stopping in mid-section and detaching the burning waggon 'As the increased velocity of the train caused the atmosphere to act upon the fire in the manner of a blast-engine, the flames rapidly extended, and in a few minutes the train presented the appearance of a moving mountain of fire.' First, second and third-class carriages were behind the goods waggons, and the passengers were described as having been doubtless in great terror at the fire. There was water available at Eckington but by the time the train reached the station the blaze was too far advanced for four waggons to be saved and they were 'consumed down to the very axles' along with their contents, though the passengers escaped. The reporter pointed out that had the fire broken out at a greater distance from a station the tragedy of the Versailles accident might have been repeated on a smaller scale. The loss of property was estimated at between £600 and £800. The waggons had belonged to the Grand Junction Railway Company and were laden with goods destined for Manchester. As usual, the fire was blamed on a spark from the engine falling among the goods.[1133]

When a minor fire broke out in a waggon of a mixed train in Belgium on 2nd May 1843 a panic which broke out among both the passengers and the railwaymen riding on the train made what should have been just a minor accident into a bad one. The 7.30 a.m. departure from Liège had reached Gingelom when two guards riding in the second waggon, which was carrying small packages of goods, noticed smoke issuing from the middle of the waggon. On moving some of the packages to find the cause of the smoke they burst into flames, which enveloped the guards. The guards at once signalled the train to stop and the conductor also blew his whistle to alert the driver. Unfortunately, this alerted the passengers who looked out and saw the fire, whereupon many of them panicked and jumped from the moving train, though a few more level-headed passengers and two guards attempted to stop them. In this way, three passengers were killed outright and 15 or 16 injured, either directly from the fall or from ending up under the wheels of the train. One of the injured passengers, a German artisan, suffered a fractured skull and there was thought to be little hope of his recovery. Two guards also lost their lives, one of them from smoke inhalation, the other possibly from jumping from the train, and two other guards were injured. Had the passengers not panicked and simply kept their seats they would have been in no danger as the train was quickly stopped and the burning waggon detached and drawn a safe distance away. None of the carriages of the train were damaged and even the waggon involved was only partially burnt. In this case, the fire was not caused by a spark from the engine but by the breaking of two bottles of varnish

[1132] Ibid, 30th May 1840

[1133] Ibid, 9th July 1842

carried in the waggon.[1134]

Another minor fire, probably caused by burning cinders, occurred near Duffel in Belgium on 5th May. This time the burning waggon could be detached from the train and it continued its journey in a few minutes without any serious consequences other than the destruction by fire of the waggon concerned.[1135]

Also on 5th May 1843 a fire occurred between Leicester and Rugby, England, on the Midland Counties Railway, from a rather unusual cause. It involved a flat waggon loaded with a horse-drawn carriage, a fairly common practice on the early railways, in Great Britain and elsewhere. To protect the carriage from excessive friction from the rope securing it to the waggon about half a pound of cotton waste had been placed on its wheels, and it was this that became ignited, in turn setting the waggon itself alight. On seeing the fire, the guard attempted to communicate with his driver, but a horse-box marshalled at the front of the train prevented him doing so, and the train travelled three miles before it stopped. Most at risk, as pointed out by the writer of a letter to *The Times*, who also referred to the Belgian railway fire a couple of days earlier, were the passengers riding in an open carriage behind the affected waggon. Once the train had stopped the fire was quickly extinguished with water from the engine. The writer, and another who wrote to *The Times* two days later, were keen to stress the need for the adoption of some means of instant communication between drivers and guards.[1136]

Flax, used for making linen, was an inflammable cargo which caught fire on more than one occasion on the Dundee and Arbroath Railway in Scotland. Two waggons loaded with flax had caught fire near Dundee on 19th May 1841, but water was available close at hand and the fire was extinguished before much damage had been done.[1137] The second accident on 17th July 1843 was more destructive. Some labourers working on the line saw that the 2 p.m. mixed train from Dundee was on fire when it was midway between Dundee and Broughty Ferry. The train consisted of four waggons of flax immediately behind the engine, one loaded with miscellaneous goods (flannels, cloths, lace, and other valuable items), more waggons loaded with flax and the passenger carriages at the rear. Hot ashes carried from the engine by the wind on a very stormy day set the flax in the leading two trucks alight and the blaze rapidly spread to the waggons behind. The train was stopped and with difficulty the rear waggons of flax and the passenger carriages behind them were detached from the burning part of the train and taken out of danger. This time, though, no water could be obtained for a considerable time and the railwaymen assisted by the passengers had to resort to braking up the burning waggons and scattering any unburnt goods about the lineside, by which action any articles which had not already caught fire were saved. The fire could then be extinguished, but the leading two waggons were almost completely consumed and the next two and that containing the miscellaneous goods were badly damaged, the financial

1134 Ibid, 8th May 1843

1135 Ibid,11th May 1843

1136 Ibid, 10th and 11th May 1843

1137 Ibid, 27th May 1841

loss amounting to £300 or £400. The newspaper report, originally published in the *Glasgow Courier*, added that 'Considerable alarm was for a short time manifested by the female passengers, whose fears, however, were soon allayed when they saw themselves detached from the fiery train.'[1138]

A prize bull, en route from the property of a Mr Houghton, of Upton, near Aylesbury, to the Royal Agricultural Society's show at Southampton, was the only victim of a fire on the London and Birmingham Railway, two miles from Tring, England. A spark from the engine set fire to straw in the truck in which the bull was riding, which was attached to the 7 a.m. train from Aylesbury on 23rd July 1844. The fire reduced the interior of the waggon to charcoal and the poor bull was badly burnt and had one of its eyes completely destroyed. The animal was returned to Aylesbury by the 10 a.m. train and taken to Mr Lepper, the local veterinary surgeon, 'where he remains without any chance of his recovery.' It seems a little surprising that in view of its injuries the animal was not immediately put down, but presumably it was too valuable a beast for this option to be considered, at least while there was any hope of its recovery. *The Times* considered that 'surely some better plan might be devised than littering beasts sent by railway with dry straw, would not sawdust be better.'[1139]

Animals were again the victims of a train fire on 26th May 1845. This time it was pigs which were involved. The 7 p.m. Midland Railway mixed train from Birmingham to Derby consisted mainly of waggons of pigs and cattle destined for the Derby market the next day. It was within five miles of its destination, and near Willington, England, station when a waggon was discovered to be on fire, with 20 pigs, each weighing upwards of 18 stones, in the midst of the flames. The guard at once stopped the train but much confusion arose for a while from the cries of the pigs and the alarm of the passengers who were enveloped in smoke from the burning waggon. Where the rain stopped a stream ran close to the line and with plentiful water thus available the fire was put out, but not before the whole cargo of beasts had been rendered unfit for sale and the waggons much burnt. The dealer in charge of the pigs had the five most badly burnt put down on the spot and their owner, Mr Brewitt, of Nottingham, found when the remainder of the animals were forwarded to him that they were so badly burnt and scalded that he had them put down as well. 'The damage done cannot be less than 100l,' the report added. Like the Tring accident the fire was blamed on a spark from the engine setting fire to loose straw in the waggon carrying the pigs.[1140]

What was described as a cattle box, though it was carrying household furniture rather than animals, caught fire on the 7 a.m. train from Glasgow on the Edinburgh and Glasgow Railway on about 25th August 1845 as the train was nearing Castlecary, Scotland. Again, a mixed train was involved, with the cattle box and a luggage van immediately being placed behind the engine and followed by open carriages for the passengers. The furniture in the cattle box was covered only by a sheet of coarse

[1138] Ibid, 26th July 1843
[1139] Ibid,24th July 1844
[1140] Ibid, 29th May 1845

tarpaulin, and the passengers could see red hot cinders thrown out by the engine landing upon it. When the fire broke out the passengers in the open carriages naturally became alarmed and they began to shout to the driver to stop, while the guard waved his flag and the passengers waved their hats and umbrellas. The driver continued, however, to Castlecary station, either because he was unaware of what was happening behind him or because he thought it best to continue to a place where he knew he could get a plentiful supply of water to extinguish the fire. Before the train reached the station, the fire had spread to the van containing the passenger's luggage and a burning fragment of the cattle box was blown into one of the open carriages, causing great alarm to the passengers, and one, in an adjoining carriage, jumped out and suffered facial injuries. Castlecary station was, however, close enough at hand for the train to reach it before the fire spread to the passenger portion of the train, but the leading van and its furniture were completely destroyed with a loss of about £150 or £200.[1141]

It cannot have been every day that news of a railway accident was sufficient to affect the stock market but the misrepresentation of a fire affecting a train on the Paris and Rouen Railway (Chemin de fer de Paris à Rouen) apparently caused a considerable if short-lived depression in the French Bourse. The accident was presumably initially reported as having been much more serious than it actually was, and memories of the Versailles catastrophe again doubtless played a part. What actually happened on 22nd July 1846 was that there was a mishap with a lamp in the mail van of a train at an unstated location between Paris Saint-Germain and Rouen, with the result that the spirits of wine (alcohol) that it contained were spilt onto a table and set fire to some papers there. A Post Office clerk tried and failed to extinguish the fire and then opened a window and cried out for help. His shouts seeming to be unheeded and the fire increasing he then jumped from the van, breaking a leg in doing so. The train was pulled up shortly afterwards and the fire easily dealt with.[1142]

On 8th December 1847, what *The Times* headlined as an 'Extraordinary Railway Accident', involved the private carriage of the Earl Zetland, loaded onto a flat waggon as in the May 1843 accident and incorporated into a train on the same stretch of line between Leicester and Rugby, now a part of the Midland, Railway. Travelling in the carriage were the Countess Zetland and her maid. As the 1.15 p.m. train from Leeds was about four miles from Rugby and travelling at some 40 miles an hour a spark from the engine set fire to the imperial {the top of the carriage} and presently the interior of the carriage was ablaze. In order to escape the blaze, the countess and her maid climbed out of the carriage onto the flat and ignoring the caution to hold on given by a male passenger in an adjoining carriage the maid shortly threw herself from the train. None of the passengers succeeded in alerting the guard to the danger and the latter remained unaware of the fire until informed by a pointsman at Rugby, 150 yards before the ticket platform where the train stopped anyway. A light engine was then despatched to search for the missing maid with two men on its front buffer beam carrying lights as well as a large signal lamp with a reflector. Unfortunately, the crew of the engine were not given any

---

[1141] Ibid, 30th August 1845

[1142] Ibid, 25th July 1846

instructions as to where the maid was likely to have fallen from the train and some of the passengers thought she had jumped off five, seven or even fifteen miles back. As a result, when they came up to her lying across the rails less than four miles from Rugby, they were unable to stop before running her over, amputating three of her fingers. She also sustained a fractured skull, either from the blow from the engine or from the initial jump, and at the time of the last newspaper report was in Leicester Infirmary, where there were fears that lockjaw (tetanus) might set in.[1143]

The fire was investigated by Captain J.L.A.Simmons of the Royal Engineers who did not find fault with the guard of the train for failing to realise that the train was on fire, as he stated that the guard could hardly be expected to see, from his low van at the rear of the train, without standing on his seat, even the roof of the second carriage from him, let alone the Earl Zetland's carriage five vehicles ahead. Moreover, he could not, in his position unprotected from the weather, smoke or dust be expected to keep a constant lookout. The inspecting officer also thought that the fire had probably spread mostly downwards as it had not been noticed either by a gang of platelayers five miles from Rugby or by a policeman a mile outside the station. The inspector was critical, however, of the fact that there had been only one guard with a train of 13 carriages and thought that there be always at least two. To deal with the difficulty of guards seeing along the length of a train he suggested the equipping of guard's vans with projecting boxes, in the proportion of one guard's van and guard for every four carriages of the train.[1144]

A train fire in France in June 1849 was not without its humorous side. A train from Calais to Paris, which included a waggon filled with pigs, came to a stand between Pérenchies and Lille stations and its passengers were alarmed by cries of 'Fire!' They jumped out of their carriages and then found that the pig waggon was enveloped in fire and smoke. The frightened pigs meanwhile had escaped from their burning prison and rushed helter-skelter in all directions. 'Then was witnessed one of the most ludicrous scenes that it is possible to imagine. Several of the passengers, with great good will, hastened in pursuit of the truant swine, and at length, after a hot chase, they were all recovered and re-incarcerated. The fire had been extinguished in the interim.'[1145]

A brief report in *The Times* mentioned a fatal accident to a Madame Brueure, daughter-in-law of Baron St James, who was burnt to death while travelling in the coupé of her diligence (presumably a carriage loaded onto a flat waggon) near Melun, France, on the Paris – Lyons line (Chemins de fer de Paris à Lyon), on the 10th June 1850. Some Lucifer matches (self-igniting matches which would ignite when struck against anything, not just the material on the side of a matchbox) left in the carriage were blamed for the accident.[1146]

The London and North Western Railway was still carrying passengers luggage on the

---

[1143] Ibid, 9th and 14th December 1847

[1144] Board of Trade report

[1145] The Times, London, England, 29th June 1849

[1146] Ibid, c13th June 1850

roofs of its carriages in mid-1851 and on 21st June of that year some of it loaded onto the 10.15 a.m. train from London Euston-square caught fire when the train was two miles beyond Wolverton, England. Travelling in a compartment of the carriage behind that which was on fire, on their way to a dinner given to a Mr Fox at Rugby (possibly Henry Fox Talbot, the Victorian scientist and inventor) were Mr Paxton, of Bass, M.P.; Mr Cochrane, the active superintendent of the building of the Crystal Palace; and Mr Crampton, the engineer of the Dover-Calais Submarine Telegraph Company. Mr Crampton realised what was the matter and as at this date there was still no means of communication with the engine driver, he opened the carriage door and made his way by means of the footboards past the carriage which was on fire towards the engine. He was almost there when he was observed by the driver and guard and the train was stopped. His companions, meanwhile, had a narrow escape as the fire penetrated their compartment via the door which had been left open and set their clothes alight in several places. They put the fire out with difficulty and suffered some singes and blisters. Several of the passengers also lost their luggage.[1147]

Another fire on Sunday 10th August 1851 broke out on the 9.30 a.m. up train from Liverpool, on the London and North Western Railway line, nine miles north of Crewe, England. Again, the passengers in the massive (reportedly nearly 40 carriages) and crowded train were unable to communicate the danger to the guard or driver, and this time there was no brave soul willing to make their way along the footboards to the engine as Mr Crampton had done. Instead, the passengers resigned themselves to waiting until the train crew became aware of the blaze and the train was stopped. The passengers then assisted the train crew to separate the burning carriage from the remainder of the train and it was left to burn itself out, passenger's luggage and all. After half an hour its charred remains were reattached to the train, which proceeded with it to Crewe, where it was left behind.[1148]

A carriage belonging to the Messagieres Nationales (a French shipping company) caught fire between Chatellerault and Ingraudes, France, on the line between Tours and Poitiers on 23rd March 1852. The cries of the passengers alerted the train driver and the train was stopped. With difficulty the burning carriage was separated from the remainder of the train and no further damage resulted, but the coach and its contents were destroyed except for 40,000f in bullion. Carelessness caused the fire; some smokers having allowed the hot ash from their cigars to drop amongst the straw in the carriage.[1149]

A failure to properly protect goods from sparks from the engine caused an expensive fire on a goods train in Egypt on about 1st April 1856. Cotton was being carried by the railway opened from Kafr el-Zayat to Alexandria in 1854 because the Mahmoudieh Canal was closed to navigation due to its being silted up and there was thus no alternative to the railway, although the fees charged to merchants by the railway company were described as being enormous, and it was said that no care was taken of their property. As

---

[1147] Ibid, 23rd June 1851

[1148] Ibid, 13th August 1851

[1149] Ibid, 27th March 1852

if to emphasise the lack of care being shown, a train consisting of 20 trucks bringing about 1,100 bales of cotton to Alexandria caught fire and the whole consignment was destroyed.[1150]

A bad accident in Spain on 16th September 1856 involved a diligence from Seville loaded onto a flat waggon of an Albacte – Madrid mail train on the Aranjuez Railway. The fire broke out between Huerta de Valdecárabanos and Villa sequilla Yepes stations and the train covered two miles before the blaze was spotted by the crew and it was stopped. By then a number of passengers had jumped off the moving train and in doing so two sisters, a woman and a child were fatally injured, while a fifth passenger burnt to death. One of the remaining eight passengers was badly injured but the others escaped with contusions, including a woman who was so overweight that she had to pay for two seats, and whose escape was considered miraculous. After the accident some of the passengers were so infuriated (perhaps because the train did not pull up more quickly) that they assaulted the engine crew and guards and even menaced a government inspector. One of the assaulted guards was badly wounded. The diligence and the waggon which had been carrying it were reduced to cinders. The fire was believed to have been caused by the carelessness of a cigar smoking passenger in the diligence.[1151] [1152]

France suffered another serious train fire on Tuesday 28th July 1857. Train No. 24 from Strasbourg to Paris was running fast near Oiry station when its baggage waggon caught fire. The engine driver was not immediately aware of the problem and the speed of the train fanned the flames and increased the intensity of the fire. When he did become aware of the blaze he immediately whistled for the guard's brakes, the train was stopped, the burning vehicle detached from the rest of the train and the fire extinguished without too much difficulty. In all, it should have only been a minor accident but the carriage immediately in the rear of the baggage van, which had come from Metz, was occupied almost entirely by women and children. Seeing the flames coming from the baggage waggon they panicked and despite the efforts of another passenger to prevent them a number of the women jumped from the moving train with their children in their arms. In this way two women and two children were killed and eight others slightly injured. Those killed in jumping from the train were believed to include a Madame Boulichoff, a cousin of Prince Balabine, an attaché to the Russian Embassy in Paris. The railway directors commented that 'This misfortune is a fresh example of the danger arising from the public not observing the railway regulations. If, as those regulations command, the travellers had not attempted to alight from the train while in motion, but had waited for it to stop, which it did in a few minutes afterwards, we should not have had to deplore any accident.' Very true, no doubt, but one suspects that faced with the apparent prospect of being burnt to death the directors might well have done the same thing. The fire was believed to have originated in the ignition of some combustible material contained in a travelling trunk, the contents of which were unknown as they were sealed when loaded at

[1150] Ibid, c7th April 1856

[1151] Ibid, 24th September 1856

[1152] Wikipedia – Accidentes Ferroviarios en España

the Forbach custom house.[1153]

A fire in the carriage of a passenger train on a Great Western Railway train between Slough and London Paddington, England, on 9th October 1857 and another between Plumpton and Penrith, England, on the Lancaster and Carlisle Railway on 12th of the same month led to a letter and then an editorial in *The Times* newspaper. In both cases, the problem was the usual one of a lack of a means of communication between the passengers, the guard and the driver. In the Great Western case, according to the letter to *The Times*, the passengers, on becoming aware of the fire, had attempted to attract the attention of the engine driver for 20 minutes with the waving out of carriage windows of handkerchiefs attached to umbrellas, without success, until the attention of a policeman was finally attracted as the train was approaching the ticket platform outside of Paddington station.[1154] In the other case after luggage on the roof of a carriage had caught fire and (according to *The Times* editorial, but see below) a passenger had, like Mr Crampton back in 1851, made his way along the outside of the carriages until he reached the guard at the rear of the train, only to find that the guard had no way of communicating with the driver. The passenger had then attempted to make his way forward to the engine the same way but before he had done so his actions had been observed by some platelayers who put down detonators and stopped the train. *The Times* editorial commented that 'Now, this really will not do. Burning alive is not a thing to be risked or tolerated. We have put it down even in India, and we can't have it in England....'[1155]

Perhaps because of the editorial in the lofty *Times,* both accidents were the subject of full Board of Trade reports. In the case of the Great Western accident, to the 6.50 a.m. train from Plymouth, the train driver apparently did see the passengers waving from the carriages just after passing West London Junction and about a mile short of the ticket platform. A switchman at West London Junction had also seen the passenger waving the handkerchief on the end of his umbrella but being unable to see any fire had assumed that he was just messing about. The train was stopped, and the passengers and their luggage helped out of the carriage, which shortly began to burn fiercely, and a lack of water meant that it was burnt out. The fire was attributed to the peculiar design of the first-class carriage involved, in which the carriage wheels protruded through the carriage floor, screened by sheet-iron and wooden boards and covered by carpet. The strap of an axle-box and the axle–box itself having broken the frame of the carriage had dropped down and the sheet-iron casing had come into contact with the tire of a wheel, and the resulting friction had heated up the sheet-iron and ignited the wooden boarding and carpet.[1156]

The Great Western Company, Lt-Col W. Yolland continued, contended that the best means of dealing with such emergencies was to have a travelling porter seated on the rear of the engine tender, in a very exposed position, with no other duty than to keep an eye on the train for possible problems. The travelling porter on this train had, however, been discontinued on the grounds that it was a slow train. The company objected to the establishment of a means of communication between the passengers and the guard on the

[1153] The Times, London, England, 31st July 1857

[1154] Ibid, 12th October 1857

[1155] Ibid, 16th October 1857

[1156] Board of Trade report

grounds that it would be misused by nervous passengers who would stop trains unnecessarily. Lt-Col Yolland pointed out that in the accident under consideration it was not likely that the travelling porter would have been able to see the fire in any case. He did not think that passengers should be able to directly stop trans, but that they should be able to communicate with guards.[1157]

In the case of the Lancaster and Carlisle accident, to the 2.15 p.m. up parliamentary train from Carlisle, it was luggage carried on a carriage roof which had caught fire, probably by a spark or cinder from the engine. The facts of the case as detailed in the Board of Trade report were somewhat different from those stated in *The Times*. There were two guards on the train, and the attention of the under guard had been attracted by a platelayer waving a red flag, whereupon he had applied his brake, looked out, seen the fire and whistled to attract the attention of his driver. He had failed to do so and had then walked along the tops of the carriages to the second-class carriage which was on fire and at the same time some more platelayers, who had seen that the train was on fire, had laid fog signals and so stopped the train. The under guard had seen one passenger who had got out onto the step of the second-class carriage, but it was the platelayer's flag that had attracted his notice, and not the passenger coming back along the train to him. The head guard said that he had been occupied sorting parcels and had not noticed the fire or known that anything was wrong until the detonators exploded.

LT-Col Yolland also investigated this accident and pointed out that even two guards had been insufficient to protect the passengers from the fire as the under guard, when he became aware of the fire, could not communicate with the driver, and the train was only stopped thanks to the actions of the platelayers, whom he considered deserved a reward. The train would have taken four minutes or more to reach Penrith and with (as reported) six ft high flames anything could have happened by then. The inspecting officer stressed, as he had in his report on the Great Western fire, the need for a means of communication between the guards and the drivers. This subject had been repeatedly brought before the directors of British railway companies before and in March 1853 a committee of railway general managers had reported in favour of such a system, but little had been done. A system enabling passengers to communicate with guards but not drivers, was also he thought desirable, and abuse of it could be provided for by the company's by-laws.[1158]

Eight or ten valuable rams, worth from £10 to £15 each, and riding in the rear van of a train on the South Wales Railway were the victims of a train fire near Cardiff, Wales, on Saturday 25th September 1859. By the time the train was stopped the van was well alight, with flames rising to a considerable height, and the poor animals inside were burnt alive, their fat falling onto the wooden sleepers and setting them alight. As soon as possible, the van was detached, and it was then pushed slowly along with iron bars so that the burning fat falling from it onto the sleepers could be put out by throwing soil onto it. The whole of the woodwork of the van was consumed and the ironwork of its frame was too hot to touch for some time. The remains of the animals when pulled out of the van 'presented a horrible sight.' The origin of the fire was not explained but it was strongly suspected that it had been started by the man in charge of the sheep, who clearly must

[1157] Ibid.

[1158] Ibid.

have escaped the fire, smoking his pipe; either because of a spark from the pipe or from a Lucifer match carelessly discarded onto the floor of the van.[1159]

In France on 17th September 1860 it was three horses belonging to Emperor Napoleon III, and riding in a horse-box which formed part of a special train from Chambéry, that were killed in a fire on the Chemins de fer de Paris à Lyon et à la Méditerranée. The driver and conductor of the train saw smoke coming from the horse-box at a place named La Tuvière, near La Burbanche station, and the train was stopped, and the burning horse-box detached, but not before the horses had burnt to death. The loss was estimated at 20,000 francs, the horses being worth 12,000 francs and the vehicle 8,000 francs. The cause of the fire was uncertain. The axle-boxes were full of grease, so it was not due to an overheated axle box, and it was theorised that a spark from the engine had set fire to straw in the horse-box.[1160]

Some horses belonging to New South Wales politician George Oakes were luckier than those belonging to the French emperor. On Christmas Eve 1860 Mr Oakes, a former member of the New South Wales Legislative Council,[1161] was travelling on a train to Campbelltown with his travelling brake loaded onto a flat waggon and his horses in adjoining horse-boxes. Between Liverpool and Campbelltown, New South Wales, the brake, which was covered by a tarpaulin, was discovered to be on fire, presumably due to a spark from the engine. The train was stopped but the brake was destroyed before the fire could be extinguished, and the horse were only saved from the fire with some difficulty.[1162]

A special train carrying Archduke Ferdinand Maximilian Joseph, Lieutenant Field Marshal de Schmerling, Count de Bombelles and other dignitaries from Berlin to Vienna caught fire on the evening of Wednesday 9th January 1861, half an hour after setting off at an unstated location in Prussia. The overheating of a hot air stove set fire to the saloon carriage in which the Archduke and party were riding and the fire spread rapidly. At the risk of his life, Count de Bombelles climbed onto the roof of the carriage and his cries were eventually heard by the engine driver, who stopped the train. The velvet hangings in the carriage were with difficulty torn down, its furniture was thrown out of the window and water and snow were used to put out the fire. The damaged carriage was then detached, and the Archduke, having taken a seat in another part of the train, the journey was resumed.[1163] The Archduke had had a fortunate escape, but he would not be so lucky six years later when after a failed attempt to establish an empire in Mexico he was shot by a firing squad.[1164]

[1159] The Times, London, England, 28th September 1858

[1160] Ibid, 17th September 1860

[1161] Wikipedia – George Oakes – Australian politician

[1162] The Maitland Mercury and Hunter River General Adviser, New South Wales, 3rd January 1861

[1163] The Times, London, England, 19th January 1861

[1164] Wikipedia – Maximilian I of Mexico

The hazards of transporting valuable items were demonstrated by a fire on a Great Western Railway train between Reading and Slough, England, on 30th July 1863. Sir E. Erdley Wilmot had just been appointed judge of the Marylebone County Court, and three of Mr Fish's large vans were bringing his furniture and effects from Bristol to London. As a result, it was presumed, of a live coal from the engine chimney falling onto its tarpaulin, the middle van, carrying the most valuable parts of the judge's property, caught alight. On the fire being discovered the train was taken as quickly as possible to the first station at which water was available, but this was seven miles away, and the very rapidity of the train, combined with the unusually hot and dry weather, made the fire burn all the more fiercely, and the van and its contents were completely consumed. Ironically the van had been chosen on account of its being considered the safest and most commodious, for three of Sir Edwards largest and most valuable paintings, one a portrait of the judge's grandfather, the commissioner selected by William Pitt the Elder to settle the claims of the American loyalists. The other two vans were slightly crushed in the efforts to remove the burning middle one but escaped comparatively unscathed.[1165]

The sleeping car of the Pennsylvania Railroad Erie express caught fire from an overheated stove on Monday 27th February 1865 near Lancaster, Pennsylvania, United States. Before long the car was burnt out. The passengers just had time to escape but lost all of their clothes and luggage.[1166]

An accident on the Paris – Strasbourg line (Chemins de Fer de l'Est) in France on about 5th May 1865 had features identical to several already described. At about 4.30 p.m. a woman and her 18-month-old child boarded the train at Luneville and took seats in the otherwise empty ladies' compartment. Twenty minutes after leaving Luneville the woman noticed that the carpet near the door felt hot and discovered a small hole through which fire was entering the carriage. As the train continued its progress towards the next stop at Sarrebourg, she tried covering the hole first with her foot, then with a cushion. The fire was soon growing rapidly, and she tried screaming from the window, but to no avail. As the train passed through Avricourt station flames were issuing from either side of the carriage, but there was no one at the station to see them. In great terror, she then leant out of the window and was able to shake a man who was asleep in the next carriage. Awakened, he looked out of the window, saw the fire, and he and a fellow passenger screamed as loudly as they could, but still to no effect. Finally, the train reached Rechicourt, where some men working on the line saw the fire and alerted the train driver. The train was stopped, the fire was put out with buckets of water, and men with water travelled in the carriage for the remainder of the journey to Sarrebourg to extinguish the fire should it break out again. At Sarrebourg the coach was taken off the train. Another woman passenger, travelling in another carriage, had heard the cries of 'Fire' and jumped out of the still moving train. She escaped with severe bruises.[1167]

Considerable carelessness on the part of the railway officials concerned would seem

---

1165 The Times, London, England, 1st August 1863

1166 The Adams Sentinel, Gettysburg, Pennsylvania, United States, 7th March 1865

1167 The Times, London, England, 11th May 1865

to have been the cause of a fire on a train near Mittagong, New South Wales, on Christmas Day, 1867. What was termed a mail train included waggons of hay and coal, and the waggon loaded with hay was, according to a letter in a local newspaper, marshalled immediately behind the locomotive but not even provided with a tarpaulin or other covering. It was no surprise, therefore, that as the train was labouring up an incline a spark from the engine set the hay alight. Cries of alarm from the train passengers alerted the train driver and the train was stopped, No one was hurt, though several passengers had jumped from the moving train.[1168]

As a Bristol and Exeter Railway down goods train loaded with Esparto grass for local paper mills was approaching Hele & Bradnich station, England, on the evening of February 14th 1868 it was seen to be on fire. The train was shunted into a siding and it was found that a large portion of the dry grass had been burnt and the waggon was badly damaged. The grass had probably been set alight by a spark from the engine. Up traffic was considerably delayed by the accident.[1169]

A fire on a North Eastern Railway train on Thursday 30th July 1868 caused a major problem locally. A goods train from Malton to Whitby was on the Saltersgate Moors, between Levisham and Goathland, England, when two of three waggons of "pickled" sleepers (sleepers treated with creosote to preserve them) were found to be on fire and the blaze spread rapidly, such sleepers being highly inflammable. The sleepers of the track were soon involved, and the rails ended up twisted out of shape, such that a section of track had to be re-laid before traffic could resume. The fire then communicated itself to the surrounding moorland, where the ling and turf were bone dry, and it spread rapidly. The fire burnt all day but by the evening it was thought that its progress had been arrested. On the Friday, though, it was still burning strongly, and it was thought that nothing, but heavy rain would extinguish it, though men were hard at work trying to thrash it out with tree branches. The origin of the fire was unclear, as a gang of platelayers 200 yards back from where it was discovered had seen nothing amiss.[1170]

The accidental ignition of a package of Lucifer matches caused a fire in a goods train bound for Italy as it was ascending the Termignon Incline, 37 kilometres from St Michel de Maurienne in France on about 6th September 1868. The matches were a dangerous cargo but had not been declared as per the company regulations. No one was hurt.[1171]

Famous opera singer Adelina Patti, the Marchioness de Caux, found herself in danger when she as returning by train to France from a tour to St Petersburg on about 16th March 1870. At the Prussian frontier, her party engaged a special saloon, fitted with sleeping compartments, and she had retired to rest when in the middle of the night she was awoken with a sense of suffocation and venturing to the saloon of the carriage found it

---

[1168] The Goulburn Herald and Chronicle, New South Wales, 1st January 1868

[1169] The Times, London, England, 17th February 1868

[1170] Ibid, 1st August 1868

[1171] Ibid, 14th September 1868

filled with smoke. The stove used for heating the carriage had got too hot and set fire to the flooring of the carriage. The whole party were soon aroused and attempted to extinguish the fire, which had spread to a sofa. With little means of quelling the fire, and with the train being an express, their situation looked dangerous as there was no means of communicating with the driver or guard. They were fortunate that the train soon slackened speed as it entered Bromberg, Prussia, station and they were able to alert someone to the fire, stop the train and get it extinguished.[1172]

By 25th October 1871 a communication cord was available on a Midland Railway train from Manchester to London St Pancras, unfortunately it failed to work when it was needed, as detailed in a letter to the Times newspaper from a passenger on the train. According to his account as the train due in St Pancras was between Bakewell and Matlock a spark from the engine set fire to some 'greasy rag or stuff' on the roof of a carriage, and the blaze was fanned by the wind. The occupants of the carriage pulled the communication cord only to find that it did not work, and a passenger got out of the carriage, climbed onto the top of it and proceeded over several carriages to inform the engine driver. The train was duly stopped and the fire, which had nearly burnt through the roof, was extinguished with a few buckets of water. If the account was true, the intrepid passenger's exploit must have been a very hazardous one on a line with frequent overbridges and tunnels.[1173]

Communication cords might have been in use on the Midland Railway in England by 1871 but their use does not seem to have extended to New South Wales by late 1873. On 6th September of that year a luggage van loaded with furniture towards the rear of a train from Windsor to Sydney caught fire as the train was a few miles from Blacktown, and the passengers in the adjoining carriage on seeing the smoke and flames were obliged to resort to shouting to attract the attention of the driver and guard. One male passenger was sufficiently frightened to jump from the moving train, escaping more or less unhurt, and a woman passenger tried to follow suit but was restrained by a fellow traveller. Eventually, the driver heard the passenger's cries and the train was stopped, the burning van was detached from the carriages on either side of it and the fire was extinguished. The origin of the fire could not be determined.[1174]

On the same day as the New South Wales fire, 6th September 1873, careless and negligent passengers caused a fire in the carriage of a train in Scotland. The 8.20 p.m. train from Helensburgh to Glasgow Queen Street on the North British Railway had only covered a mile and a half when a third class carriage was seen to be on fire. Panicking passengers jumped out of the windows and six – five men and a young woman – were found lying along the line with some of them badly hurt, one suffering a broken leg. Their alarm was understandable as the carriage was completely destroyed in a short time. As it was a stopping train, it was not provided with a communication cord and it was fortunate that the fire was spotted as quickly as it was by the driver and guard. Lt-Col

---

[1172] Ibid, 28th March 1870

[1173] Ibid, 26th October 1871

[1174] The Tasmanian Tribune, Hobart, Tasmania, 10th September 1873

C.S. Hutchinson, who investigated the accident for the Board of Trade, pointed out that accidents such as this one provided strong arguments for fitting all trains with communication cords. The passengers on the train had included ship's carpenters carrying with them naptha used for caulking operations. Some of the naptha from one of the cans carrying it had been spilt onto the floor of the carriage and a smoker had thoughtlessly thrown a lighted match onto it. Posters at Helensburgh and other stations warned passengers of the illegality of carrying inflammable materials with them onto trains, but the inspecting officer thought that it was difficult to prevent them from doing so. In view of the circumstances, the liability of the railway company for the injuries caused to the passengers was to be subject of a legal action.[1175] [1176]

The most serious primary train fire up to that date anywhere in the world occurred near Komoka, Ontario, Canada on 28th February 1874 on the Great Western Railway. At this date the area where the accident occurred was a major oil drilling area, and the train involved in the accident was an accommodation one, a mixed train consisting of an engine and tender, three oil tanker waggons, a baggage car, a second-class coach, a smoker and a first-class coach. It left London at 6.28 p.m. with about 60 passengers, most of them travelling in the rear carriage. As it was descending an incline towards Komoka at about 25 miles an hour, Conductor John Mitchell thought he saw a fight among the passengers around the toilet of the first-class car. In fact, they were attempting to fight a fire which had begun in the water closet due to a coal oil lamp either falling from the wall to which it was secured with a spring-clamp or exploding. Efforts to extinguish the fire with cushions were unsuccessful and the conductor closed the toilet door, and in an effort to calm the passengers shouted that everything was under control. It was not, as there was no means to stop the train. Had the toilet door been kept closed the fire might have been confined there but other passengers opened it to fight the fire and a strong draught stimulated the fire, which soon began to spread rapidly down the coach. A few passengers attempted to get through the fire to the smoker car ahead but most of them crowded to the rear platform of the first-class coach, staying there until smoke, fear of being burnt or the crush of other passengers forced them to jump off. Other passengers jumped through windows.

The train was provided with a communication cord but on mixed trains such as this one the cords caused problems by becoming caught in or broken by the tanker waggons, and the train crews, with the knowledge and connivance of the local stationmasters, had solved the problem by disconnecting the cord until the tanker waggons had been dropped off. Conductor Mitchell thus had no means of getting his driver to stop the train. He ordered the brakeman, William Burke, to make his way over the roofs of the carriages and the tanker cars to the engine. Burke started off but soon returned, afraid of falling off the train. Conductor Mitchell was now desperate, all the more so as one of his own daughters was a passenger on the train, and he himself undertook the hazardous journey. He reached the locomotive in about two minutes and Engineer George Williams stopped the train. A Windsor train was following close behind but the brakeman, followed by the conductor, ran back down the line and it was able to pull up just short of the

[1175] The Times, London, England, 8th September 1873

[1176] Board of Trade Report

accommodation train, which was now burning from end to end.[1177] The bodies of seven or eight (newspaper accounts vary) victims who could be identified were afterwards taken from the first-class coach.[1178] [1179] [1180] Several other corpses reportedly remained unidentified and three more victims died later. A number of other passengers were more or less seriously injured. Among the wreckage were found two gold watches, which could be identified, and a number of gold nuggets, supposedly from melted jewellery.[1181]

At the inquest into the victims of the accident, attention was paid to the lamp which had caused the fire – whether it had been properly secured or interfered with by a drunken passenger and the type of oil it used. The couplings of the cars, which had been hard to separate after the accident, received less consideration. Most of the jury's attention was devoted to the bell ropes and the fact that the one on the accident train had been disconnected (as was the one on the following Windsor train). The jury recommended legislation banning kerosene oil lamps on trains, censured the railway company higher management for its unawareness that bell ropes were routinely disconnected and recommended that the conductor, brakeman and engineer of the train be charged with manslaughter for their disconnection of the communication cord. At the Middlesex Assizes in May, however, the Grand Jury returned a "no bill" on the manslaughter charges which had therefore to be dropped.[1182]

When a fire occurred in the mail and express car of a train nine miles east of Petersburg, Virginia, United States, on the Atlantic, Mississippi and Ohio Railroad, on 9th July 1874, the bell rope had not been disconnected but the fire itself had destroyed it. The car contained an unusually heavy Northern and Southern mail, all of which was consumed. Route Agent Jones and J.L. Jennings, the only occupants of the car, did their best to save the most valuable part of it, being badly burned about the face and arms in their efforts, and eventually had to jump from the train to save their lives, escaping with bruises from the fall. The train ran on for several miles before the fire was discovered and the car was burnt down to its wheels before the fire could be put out. The cause of the fire was unknown.[1183]

Mysterious circumstances surrounded a fire on the mail train from Clones, due in Belfast at 3.30 a.m. at Lisburn, Ireland, on the Ulster Railway, on 28th August 1875. The fireman of the train lost his life in the fire and the driver of the train, at the subsequent inquest, made a statement, which he adhered to under a strict examination, that having observed sparks coming from a waggon just behind his engine he had whistled and stopped his train. He and his fireman, a man named Lechford, had then gone back to ascertain the cause. On their opening the door of a van containing six hogsheads of whisky, the spirits had exploded; he had been thrown across the platform and had lain

---

1177 Wreck! By Hugh A. Halliday, pgs. 46-7

1178 Bangor Daily Whig and Courier, Bangor, Maine, United States, 3rd March 1874

1179 Hornellsville Triune, Hornellsville, New York, United States, 6th March 1874

1180 The Times, London, England, 13th March 1874

1181 Hornellsville Triune, Hornellsville, New York, United States, 6th March 1874

1182 Wreck! By Hugh A. Halliday, pgs. 50-2

1183 The Coshocton Age, Coshocton, Ohio, United States, 17th July 1874

there for a while insensible. When he had recovered, he had seen his fireman lying on his back in the van, surrounded by blazing whisky, and the fireman had died half an hour later. Other evidence produced at the inquest proved that the whisky could not have caught fire due to an overheated axle-box and must have been ignited by some flame or light introduced into the van. It was conjectured that the fireman had gone from the engine to the van during the journey from Lurgan to Lisburn, obviously with the knowledge and connivance of the driver, to steal the whisky and in trying to do so had accidentally set it alight. The inquest jury returned a verdict of "Accidental death."[1184]

On about 26th June 1876 fire broke out on a train of the Punjab Northern State Railway as it was crossing the newly opened Alexandra Bridge over the Chenab River, India.[1185] Great damage was caused and three waggons laden with tea and cloth were burned.[1186]

1184 Board of Trade report

1185 Wikipedia – Punjab Northern State Railway

1186 The Times, London, England, c27th July 1876

CHAPTER 15

CHEMICAL EXPLOSIONS

Nineteenth century railway accidents involving explosions were of two types; those which might be defined as chemical explosion and locomotive boiler explosions. Accidental chemical explosions, such as those due mishaps with blasting powder or munitions, were rare, whereas locomotive boiler explosions were much more common. This very short chapter will deal with chemical explosions.

At 7 a.m. on the morning of Thursday 27th June 1850 a goods train on the Glasgow Paisley Kilmarnock and Ayr Railway left Mauchline, Scotland, station for Glasgow. Its 23 waggons included nine carrier's carts of general goods, covered with thick woollen wrappers and tied down with ropes, which were being carried two and two on flat waggons. The train had covered a mile and a half when there was a small explosion in a cart two or three waggons back from the engine. This separated the train into two, and the rear portion ran back downhill for 300 yards, when there was a second much more powerful explosion, followed by a third lesser one. Every house and building within a couple of miles was shaken by the blast and afterwards the ground was strewn with the smoking and burning remains of carts, waggons and goods. A large quantity of burning spirits ran into a watercourse and damaged vegetation, bales of cloth were on fire by the lineside and some items had been blown a considerable distance. By great good fortune no one was hurt.[1187]

The explosions were investigated by Captain Laffan of the Royal Engineers on behalf of the Board of Trade. He stated that it could not be determined how the accident had occurred. A spark or a loose coal from the engine could have come into contact with some of the gunpowder being carried in the carts, or the explosions might have been due to friction or an overheated axle. The matter was not in any case of importance, what was more to the point was that the gunpowder was being carried clandestinely without the knowledge or consent of any of the railway servants, and without any extra precautions having been taken to guard against the risk of an explosion. The railway company had the

---

[1187] The Times, London, England, 1st July 1850

power, he added, to fine the carrier £10 for a breach of the company's bye-laws in failing to declare the contents of the carts, but were reluctant to enforce the penalty as if they did so it might interfere with an impending action for damages for the loss of the trucks and goods.[1188]

Later, in 1850, Captain Laffan found himself investigating an explosion the London and North Western Railway at Birmingham, England. A goods train was moving slowly over the viaduct on the approach to the station at 3 a.m. on 7th November 1850 when there was a loud report, the engine was thrown a foot or so into the and derailed, and part of the viaduct abutment was destroyed. Again no one was hurt. It was clear at once that there had been an explosion, but its origin was nowhere near as obvious as that of the one at Mauchline just over four months earlier. At first it was attributed to the release of inflammable gas from the marshy ground surrounding the viaduct, but Captain Laffan discounted this theory and after much careful investigation deduced that it had resulted from a broken gas main in the vicinity of the viaduct which had been leaking for some months before the accident.[1189]

The next accidental explosion on a train did not occur until 1862, in the early stages of the American Civil War. It took place on the afternoon of 7th November 1862 a quarter of a mile distant from Lafourche station, Louisiana, on what the newspaper report described as the Algiers Railroad, presumably a reference to the New Orleans Opelousas and Great Western. An ammunition waggon behind the tender of the engine blew up, 'and the limbs of some six or seven persons, supposed to have been in the ammunition car at the time of the explosion, were scattered in all directions.' Eleven people were killed outright, and 17 or 20 officers and privates were injured. Lafourche is just outside of New Orleans, which had been occupied by the Union on 1st May, so the train was presumably carrying Union soldiers. Two waggons were completely destroyed by the explosion, which was blamed on a spark from the cigar of a careless smoker.[1190]

Another explosion of ammunition occurred in Nashville, Tennessee, United States on 7th September 1865, a few months after the end of the American Civil War. The train consisted of eight waggons and a single passenger car and was carrying 200 people. For some reason, two waggons loaded with powder, shot and shell were marshalled immediately behind the locomotive, and it was supposed that sparks from the engine had reached the cargo via cracks in the car. Seven persons were killed by the explosion, including the fireman; his driver was badly injured and the conductor slightly so. Those slightly hurt included Mr Vanvalkenburg, Assistant Superintendent of the Telegraph. All but the two cars at the rear of the train were wrecked. The explosion reportedly shook the buildings of the town like an earthquake.[1191]

[1188] Board of Trade report

[1189] Ibid.

[1190] Burlington Weekly Hawkeye, Burlington, Iowa, United States, 29th November 1862

[1191] Janesville Gazette, Janesville, Wisconsin, United States, 9th September 1865

Three Bridges station, England, on the London Brighton and South Coast Railway, was the scene of the next explosion on Sunday 13th December 1868. At about 1.50 a.m. that morning a down goods train from London to Brighton had dropped off a number of waggons, among which was what was referred to as a closed iron box truck, labelled "Newhaven" About 2.30 a.m. the driver of the 4.5 a.m. Three Bridges to Tunbridge Wells goods began to make up his train, assisted by Guards F. French and J. Harris. While he was coupling some trucks, guard Harris heard something leaking and went with a lighted lamp to investigate. As soon as he got into the iron box truck there was an explosion which knocked him down and set him alight. Driver Philcox of the Tunbridge Wells train ran to assist the guard and removed the guard's burning clothes as the only means of helping him.

Guard French, who had been standing near the iron box waggon at the moment of the explosion, was also enveloped in flames, and was assisted by Shunter Colwell, who removed most of his burning clothes. Driver Philcox then took the two guards in a waggon to Brighton and they were admitted to Brighton hospital, where Guard French died the same evening and Harris about a week later.

The waggon in which the explosion occurred had come from Liverpool, via the London and North Western Railway and the West London Railway, and was destined for Dieppe, France, via Newhaven. It had contained nine casks of what was described as "refined naphtha of British manufacture." Immediately after the explosion, the naphtha had begun to burn furiously and had destroyed the woodwork of the truck it was in, set fire to a waggon of coals, burnt a waggon of hay and partly burnt a waggon of goods.

Colonel W. Yolland investigated the accident on behalf of the Board of Trade. He noted that revised regulations for the transport of dangerous goods had been drawn up after the Abergele disaster the previous August, and was distinctly unhappy that Mr Kay, the district goods manager of the London and North Western Railway, and Mr Hawkins, general manager of the London Brighton and South Coast Railway, could nevertheless state at the inquest upon the two guards that refined naphtha was not recognised as a dangerous substance and could be sent to the Brighton company without special notification, in opposition of the 1st, 6th, 7th and 11th of the modified regulations. He was not sure, though, that even these would have prevented the accident, and recommended that the naphtha should be sent in special trains, as recommended by Colonel Rich after the Abergele disaster, though he was not sure that the quantities sent would justify such trains. He noted that the Belgian State Railways only received such articles such items for forwarding at a limited number of stations, though they did not seem to use special trains.[1192]

As the last train of the day was close to departure from Liverpool Lime Street, England, station on, the London and North Western Railway on 17th February 1870 there was an explosion in the guard's van. The accident happened just as the passengers were taking their seats, but no one was hurt though the guard's van was 'blown to pieces.' The gas, used for lighting, was stored in a metre {sic} in the van.[1193]

[1192] Board of Trade report

[1193] The Times, London, England, 19th February 1870

By the beginning of February 1871, the Franco-Prussian War was almost at an end, with the French continuing to suffer heavy defeats despite a partial truce on 28th January. At 7.45 a.m. on the morning of 5th February 1871 train No. 481 left Marseilles for Toulon. It was carrying about 500 passengers, many of them soldiers from the Eastern Army returning to their depots or going home. Also attached to the rear of the train were four waggons laden with 20 tonnes of gunpowder. The train left Bandol station normally and was about three kilometres from Saint-Nazaire (now Sanary) when the ammunition waggons exploded. The blast was powerful enough to destroy the whole train of 11 carriages and 13 waggons, with the exception of the three leading vehicles. The sound of the explosion was audible about ten kilometres away, houses near the scene were demolished and trees uprooted. Of those on the train eighty were killed outright, many badly mutilated, and over 200 more or less injured. Nothing remained of the bodies of the two gendarmes who had been escorting the waggon. The engine and tender of the train, farthest from the explosion, remained on the track, and its driver continued toward Saint-Nazaire to summon assistance. Many of those hurt were cared for at local farms. An ambulance was improvised in the chapel of Notre-Dame-de-Consolation in Saint-Nazaire and the bodies of the dead were taken to Ollioules village. The cause of the explosion was never established; possibly it was due to the ignition of the gunpowder by an overheated axle-box.[1194]

The catastrophe was hushed up by the press, which was subject to wartime censorship. On 6th July 1871 the Aix-en-Provence Court of Appeal exempted the Chemins de fer de Paris à Lyon et à la Méditerranée (PLM) company from prosecution given the special circumstances of the disaster but on 6th May 1872 ruled that they must still pay compensation to the victims.[1195]

[1194] Webpage – La Catastrophe dite 'de Bandol.'

[1195] Wikipedia – Liste des accidentes ferroviaires de France au XXIX sIècle

CHAPTER 16

BOILER EXPLOSIONS

The railways of the nineteenth century were powered almost entirely using steam engines, and locomotive boiler explosions were therefore much more common than chemical explosions. Although the explosion of a locomotive boiler could often prove fatal to the engine crew, it was rare for train passengers to suffer and probably because of this information regarding boiler explosions in countries outside of Britain and the United States has been hard to obtain.

Boiler explosions could occur in three different ways. Firstly, the boiler barrel itself could fail under normal working steam pressure due to inherent weakness or corrosion. Secondly, the safety valves provided to release steam to avoid an excessive pressure, could fail, either because they had been deliberately interfered with to obtain a higher pressure to get the engine to work harder or because they had been incorrectly assembled. Finally, the engine firebox could collapse, usually due to the water having been allowed to get too low so that the firebox crown sheet overheated and weakened.[1196]

The very first known boiler explosion was also strangely probably the one which resulted in the most fatalities of any before 1901, though those who died were not train passengers. The engine involved, named the *Mechanical Traveller*, was of a very unorthodox design with a horizontal boiler and cylinders which powered mechanical 'feet' which pushed the engine along the Newbottle Waggonway in County Durham, England. The accident happened on 31st July 1815 at a place called Philadelphia. The 13 or 16 people killed were curious spectators who had been stood watching the engine working.[1197] The cause of this boiler explosion is not stated (possibly, not known) but conceivably was due to its poor design and or its construction by a blacksmith who had no theoretical knowledge of such matters.

1196 Locomotive Boiler Explosions by Christian Hewison pg 11

1197 Ibid. pg 26

The next locomotive boiler explosion arose through its driver fiddling with the safety valves, a primitive design of which had existed as long ago as 1679[1198] and thus well before the invention of the first steam locomotive. On 28th February 1818 the boiler of the *Salamanca* locomotive exploded on the short Middleton Railway between Leeds and Middleton, England. Its driver paid for his tampering by being 'carried, with great violence, into an adjoining field the distance of one hundred yards'[1199] and lost his life.

Two boiler explosions occurred in England on the Stockton and Darlington Railway in 1828. On 19th March engine No. 5 blew up at Simpasture, a mile or so from Shildon, severely scalding its two firemen, one of whom, John Gillespie, died, but leaving the driver, engineer George Stephenson's older brother, unhurt. Then, on 1st July, engine No. 1, the *Locomotion*, blew up at Aycliffe Lane (now Heighington) killing its driver, John Cree. It remained working on the railway, after repairs, until 1841.[1200] Both accidents arose from the fact that the safety valves, which had been weighed down during the journeys (presumably to make the locomotives work harder), had not been released when the trains were at a stand, so the boiler pressure just built up until the boiler gave way.[1201]

On 17th June 1831 the boiler of a locomotive unofficially named *The Best Friend of Charleston* exploded at Charleston, South Caroline, United States on the South Carolina Canal and Railroad Company. While it was standing at the 'upper revolving platform' and its engineer, Nicholas Darrell, was away, its black slave fireman, described only as 'a negro belonging to Mrs Surr' either lay on or sat on a stout piece of wood which he had placed over its safety because he was fed up with its whistling noise. He died as a result of the explosion, having suffered a broken thigh and other injuries. In the typical racist language of the period, the *American Railroad* Journal wrote that 'a negro placed himself on the top of a safety valve of a locomotive engine during the absence of the engineer; it proved no seat of safety to him, and resenting the indignity blew poor Sambo sky-high.' The explosion took place as soon as the engineer returned and set the locomotive in motion. The engineer was scalded but survived and several others were hurt.[1202] [1203] [1204] Salvageable parts of the engine were used to build another one – the *Phoenix*.

What should have been a very minor accident on the Liverpool and Manchester Railway on the afternoon of Friday 1st February 1833 ended up as a much more serious affair due to the lack of caution of some of the passengers, together no doubt with their lack of awareness of the dangers of the new mode of transport. The 3 p.m. second-class passenger train from Liverpool had travelled as far as Parr's Moss, near Newton, when its progress was halted by a goods train in front, which had stopped because its engine, the *Bee*, had a problem with its eccentric strap. Guard Edward Lewis of the passenger train

---

[1198] Wikipedia – Safety Valve

[1199] Leeds Mercury, Leeds, England, 7th March 1818

[1200] Locomotive Boiler Explosions by Christian Hewison pg 26

[1201] Tomlinson's North Eastern Railway by William Tomlinson, pg 148

[1202] A History of Railroad Accidents, Safety Precautions and Operating Practices by Robert B. Shaw pg 16

[1203] Wikipedia – Best Friend of Charleston

[1204] Webpage – South Carolina Encyclopaedia – Best Friend of Charleston

went forward to see what was wrong with the goods train and shortly afterwards the engine of his own train, the *Ajax*, suffered a burst boiler tube. As a result, a quantity of water fell into the fire and much steam was generated, while the *Bee* was also generating some steam. Curious as to the reason for the stoppage several passengers got out of their carriages to investigate and unwisely placed themselves on the opposite running line. Just at that moment a train of waggons from Bolton to Liverpool came up, and engine pulling 12 or 16 waggons, travelling at about 15 miles an hour, and its driver being unable to see the passengers because of the cloud of steam they were run down, three being killed outright. A fourth passenger, Robert Hilton, aged sixteen, on his way from Belfast to Halifax, was so badly injured that it was thought impossible for him to recover. At the inquest on the victims Driver John Wakefield testified that he had seen some passengers alighting from his train, but he and his fireman had been busy working on their engine and they had had no time to worry about the passengers before the other train arrived. Guard Lewis stated that previously there had been policemen stated at half mile intervals along the line but now they were only stationed where there were points to operate and at level crossings.[1205] The initial day of the inquest was fully reported but the national press of the time seems to have then lost interest in the case and the verdict of the inquest is unknown.

The little Middleton Railway had a second boiler explosion between Leeds and Middleton, England, on 12th February 1834, and Driver James Hewitt, the world's first regular engine driver, lost his life. This time the cause was different, it was not the driver's fault but rather that the boiler was in poor condition, repairs to it not having been properly done after the death of engineer John Blenkinsop. The railway reverted to mainly horse haulage the following year.[1206]

Towards the end of 1834, on 12th November, a second boiler explosion occurred in the United States, this time on the Baltimore and Ohio Railroad, presumably at or near Baltimore, Maryland, killing the engineer, named Neff. The fireman was of train of what were described as 'burden cars' was considerably, but not it was thought dangerously, injured and the conductor escaped. The newspaper account added that at the time of the accident the engine had been in the possession and under the management of its owner or of his immediate agent, and was being tested to determine its capabilities, previous to it being accepted by the railroad company. The locomotive obviously failed its test![1207] [1208]

A locomotive which had only been recently imported into Belgium from Mr Robert Stephenson's works in England, and which 'was admired by all competent judges,' nevertheless blew up as it was about to leave Ghent with the 3.15 p.m. passenger train on 18th January 1839, killing its driver and fireman and injuring a third railwayman. As to the cause of the explosion the report stated that 'the only conjecture that can be formed is that the engineer, who was remarkably zealous, forced the pressure, and became the first

---

1205 The Times, London, England, 4th and 10th February 1833

1206 Wikipedia – Middleton Railway

1207 The Adams Sentinel, Gettysburg, Pennsylvania, United States, 17th November 1834

1208 A History of Railroad Accidents Safety Precautions and Operating Practices by Robert B. Shaw pg. 47

victim of his misfortune.'[1209] The possibility of the boiler itself having been defective does not seem to have been considered.

An explosion at Bromsgrove, England, the Birmingham and Gloucester Railway, on 10th November 1840, rather ironically involved a locomotive named *Surprise*. The engine had been designed by an American, Dr William Church, and built by a Samuel Goddard of Birmingham. It had already seen three years' service on other lines before it was submitted to the Birmingham and Gloucester company for evaluation. It had made a seven-mile test run without incident and was parked over an ash pit at Bromsgrove station when it exploded without warning at 5.50 p.m. killing Driver Thomas Scaife instantly and injuring Engine House Foreman Joseph Rutherford so severely that he died the next day. Several other people were injured.

The inquest on the bodies of Scaife and Rutherford was reported on in great depth in *The Times* newspaper. The jury seem to have accurately identified the main cause of the explosion as the insufficient thickness of the boiler plate, reduced from an original three eighths of an inch to only three sixteenths of an inch, partly due to corrosion while the engine had been standing unused at Birmingham. A contributory factor, according to a later writer, was the presence of sand from the local water supply, but (the writer said) the coroner and the foreman of the jury had steered the jury away from any verdict which might have embarrassed the railway company. The jury returned a verdict of "Accidental death."[1210] [1211]

A very brief report by the Railway Inspectorate of the Board of Trade mentions a boiler explosion at an unstated location on the Hartlepool Dock and Railway (to give the company its full title, the report simply refers to the Hartlepool Railway). The accident, at about 1.30 p.m. on 28th December 1843, caused the death of a Jane Davison, described as an old woman who was not a company servant but merely a relative of one of the workmen employed on the railway. The cause of the explosion was not stated.[1212]

At about 9 a.m. on 21st March 1844 a locomotive left Petersburg, Virginia, United States, on the Petersburg Railroad, with a train of thirty-two burden cars. It had just passed the first curve in the line when it was discovered that a car had left the track. The train was stopped, the car replaced on the rails, and the engineer set his engine into motion once more, when its boiler exploded without warning. The newspaper accounts describe the engineer as having been blown into three pieces, while a fireman was so fatally injured and died within a few minutes and a second fireman was badly scalded. 'It is supposed by some… that the steam was permitted to accumulate at too great a "head," so that when the Engine was put in motion, the explosion took place.'[1213]

1209 The Times, London, England, 22nd January 1839

1210 Webpage -WR Magazine: The Bromsgrove Rail Accident

1211 The Times, London, England, 10th and 25th November 1840

1212 Board of Trade report

1213 The Adams Sentinel, Gettysburg, Pennsylvania, United States, 1st April 1844

The locomotive *Richmond* exploded at 8.40 p.m. on 2nd September 1844 at Reading, Pennsylvania, United States, on the Philadelphia and Reading Railroad. There had been a severe thunderstorm raging at the time and the theory was advanced that the explosion had resulted from the locomotive being struck by lightning. Keen to promote this theory as an alternative to the possibility of any inherent defect in the engine that they had built, the Norris Locomotive Works engaged Professor Dionysius Lardner, a British natural scientist, to write a pamphlet in support of it.[1214]

The explosion of the boiler of the engine of a goods train on the Saint Étienne – Lyon line (Chemin de fer de Saint Étienne à Lyon) in France on 11th September 1845 killed the driver and fireman of an engine standing on an adjoining line and injured three other people.[1215]

The boiler of the locomotive *Windsbraut* exploded as it stood at the departure platform of Leipzig, Saxony, station on 21st May 1846. No one seems to have been hurt in the accident though a possibly inaccurate engraving of the accident shows at least one person on the engine at the time.[1216]

An engine named *Neversink* blew up on the evening of Thursday 14th January 1847 (date also given as Friday 15th) near Mill Creek, Pennsylvania, United States, on the Philadelphia and Reading Railroad, while it was pulling a train of empty burden cars, killing its engineer and five of his assistants, and dangerously injuring a sixth man. Another report stated that all seven died. Why the empty train needed a crew of seven was not explained.[1217] [1218]

The boiler of one of two engines of a construction train hauling soil on the line being built from Paris to Chartes exploded at Versailles, France, on 4th March 1849. A driver was killed, and a fireman and three workmen injured.[1219]

At about 7.30 p.m. on the evening of 27th June 1849 an up goods train left Plymouth Millbay station on the South Devon Railway and got as far as Plympton, England, having run without steam over the preceding two miles of easy gradients. Beyond Plympton is the very steep Hemerdon bank, a two-and-a-half-mile ascent at 1 in 42, and Driver George Thompson opened the regulator of the engine *Goliath* wide, as was customary. He immediately heard a loud crack and was thrown from the engine. The boiler of the engine had exploded. It made several bounds and ended up overturned, while Fireman Henry Evans, who was also thrown from the locomotive, suffered a fractured skull and

[1214] A History of Railroad Accidents Safety Precautions and Operating Practices by Robert B. Shaw pgs 47-8

[1215] Wikipedia - Liste des accidentes ferroviaires en France au XIX siècle

[1216] Eisenbahnunfälle in Sachsen pg. 41

[1217] Republican Courier Gettysburg Pennsylvania, United States, 18th January 1847

[1218] The Sandusky Clarion, Sandusky, Ohio, United States, 26th January 1847

[1219] Wikipedia - Liste des accidentes ferroviaires en France au XIX siècle

died. Driver Thompson himself was seriously injured but survived.[1220]

The accident was the subject of a detailed investigation by Captain J.L.A. Simmons of the railway inspectorate who accepted Driver Thompson's statement that he had not interfered with the safety valve to get more steam pressure to climb the bank but he decided that the gauge glass, showing the amount of water in the boiler, could not have been full as the driver had stated and that although the single fusible plug of tin had not melted the explosion must have resulted from the water level being allowed to get too low, baring the crown-sheet of the firebox and allowing it to weaken through overheating. The driver might have been deceived by a choked gauge glass and trusted entirely to it rather than trying the gauge cocks.[1221] A recent writer has suggested as other possibilities that scale on the crown-sheet might have deprived it of water and that the firebox might have developed cracks after too rapid day-to-day cooling after use.[1222]

The locomotive *De Hoops* left the Amsterdam engine shed, in The Netherlands, at 12.30 p.m. on 5th July 1849 to work the 1 p.m. train to Haarlem. A couple of minutes before departure time there was a very loud bang and a violent escape of steam from the engine. The driver, fireman and a baggagemaster who was in the vicinity at the time suffered more or less serious scalds. The front of the fire-door was ruptured, and the top plate pushed in.

A locomotive foreman immediately tried to determine what had happened. The fusible lead plug had not melted, and the safety valve appeared to be working properly. The engine was towed away to the workshops and examined, and the thickness of the boiler plates was measured. The conclusion of F.W. Conrad, the engineer director of the Hollandsche Ijzeren Spoorweg Maatschappij railway company, was that the failure of the firebox was not due to it being faulty and must be ascribed to a sudden overheating of a part of it which water had access to. His conclusions are described by a modern Dutch writer as being pure speculation and proof that in 1849, in the Netherlands as in England, the causes of boiler explosions were the grounds for wild theories.[1223]

A boiler explosion at Wolverton, England, on 26th March 1850 on the London and North Western Railway bore similarities to that involving *The Best Friend of Charleston* in the United States 19 years earlier. Engine No. 157 had been in the workshops at Wolverton and was then left on a siding overnight and a labourer instructed to get up steam so that it could go out as a pilot engine at 6 a.m. This he duly did and by 5 a.m. it was blowing off steam. A fitter and another labourer appeared at this hour and the second labourer, fed up with the continual noise of the escaping steam, screwed the safety valve down as hard as he could to shut it up. By 5.40 it was blowing off again and when its driver arrived at 5.55 it was blowing off harder, but the driver apparently did not notice the steam pressure. As he opened the regulator at 6.10 the engine blew up. The main casualty of the accident, ironically enough, was the labourer who had screwed down the safety valves, who was badly scalded and had one of his perhaps over-sensitive ears

[1220] The Times, London, England, 29th June 1849

[1221] Board of Trade report

[1222] Locomotive Boiler Explosions by Christian Hewison pg 31

[1223] Spoorwegongevallen in Nederland 1839-1993 by R.T.Jongerius pgs 24-5

blown off.

Captain R.M. Laffan investigated the explosion on behalf of the Board of Trade and naturally blamed the labourer for screwing down the safety valve, but he also found fault with the material used for the boiler plates and angle irons. He recommended the substitution of a fusible plug made of an alloy which melted at 330 Fahrenheit for one of lead which only melted at 612 Fahrenheit, so as to to better guard against explosions resulting from excessive boiler pressure.[1224]

On about 20th October 1850 (exact date not determinable from newspaper account) at 7 a.m. an accommodation train left Vicksburg, Mississippi, United States, drawn by the locomotive *America*. It had proceeded about a mile when the boiler of the engine exploded violently. There were five people on the engine and tender at the time; Dennis McConnel, the engineer, Thomas Adrian, a carpenter employed by the company since the opening of the line, Mr McMurray, a coppersmith going out to repair some pumps and two railwaymen not otherwise identified, one or both of whom was presumably the fireman. Mr Adrian and the two railwaymen were killed instantly, the engineer died about two hours later and Mr Murray was badly hurt and scalded but it was thought he would recover. The locomotive and tender were wrecked but the passenger and freight cars undamaged.

The newspaper report on the accident added that 'The explosion was undoubtedly caused by a deficiency of water in the boiler. The train was running a heavy "up-grade," and it is probable the engineer delayed letting water into the boiler until he should have ascended it. The locomotive was an old one but was considered perfectly safe. Such accidents are very rare and can be scarcely produced by another cause than we have mentioned.'[1225]

The year 1853 was barely more than a week old when the Midland Railway company had a boiler explosion at Fishponds, England, just outside of Bristol on its main line to Gloucester. Three quarters of a mile from its starting point at Bristol the line climbed a steep bank with about two miles at 1 in 75. The 20-waggon goods train for Gloucester which left Bristol at 8.10 p.m. on 8th January 1853 was therefore provided with two engines, but even so, it stalled on the incline. The decision was therefore taken to divide the train. Steam was shut off and the driver was handing blocks to the fireman to chock the wheels prior to uncoupling the train when the boiler gave way. The driver only realised what had happened when his fireman shouted that something had hit him but the noise seemed very loud to people living in cottages some distance away, and the force of the blast was sufficient to carry part of the boiler and steam dome 14 or 15 ft over the roofs of some lineside cottages to land a quarter mile away.

Captain Galton of the Board of Trade was unable to find anything wrong with the boiler and after conducting a series of experiments with safety valves and concluded that given that three quarters of an hour had elapsed since the engine setting off and the

---

1224 Board of Trade report

1225 The Daily Sanduskian, Sandusky, Ohio, United States, 23rd October 1850 – reproduced from The Vicksburg Sentinel

explosion, during which time the driver stated that he had had a very good fire and plenty of steam, that he was 'led to the conclusion that this explosion was caused by a deficiency of water in the boiler in consequence of which the top of the fire-box became heated, and caused a too rapid generation of steam.'[1226] A recent writer has pointed out that the captain failed to explain how steam could be generated from water that was not there; that he might have least have checked that the pumps had been working properly and the water gauges had been in order, and that the explosion was probably due to defects in the iron barrel of the boiler which had escaped Galton's attention.[1227]

The next boiler explosion of consequence occurred at Longsight, near Manchester, England, on the London and North Western Railway on 6th March 1853 when an explosion in the roundhouse there killed six men and blew away part of the roundhouse roof. The cylindrical part of the engine boiler remained intact, but the copper firebox was destroyed, its outer casing blown off and the axle of a driving wheel broken. Captain George Wynne conducted the investigation into this accident for the Board of Trade. He found that the engine had been built in Manchester in 1840 and had been relegated to shunting duties since 1846 because it was not powerful enough to pull trains. The firebox had not once been renewed and the iron casts forming the firebox castings were 'much honeycombed' by corrosion and eaten away in rings around the stays. Captain Wynne described the stays themselves as 'extraordinary specimens of corrosion, being thickly encrusted in some parts with an oxide, and in other parts greatly reduced in thickness.' Eleven of the boiler tubes he had found wholly or partially choked with sulphate of lime (calcium sulphate) and one plugged up, and 'this is the working condition in which it had just been turned out of the repairing shop.'[1228]

At the inquest into the accident one witness, a George Scholes, stated that he had seen the driver screwing down the safety-valve on the dome of the engine at 8.35.[1229] Other witnesses however said that they had seen steam blowing off at this safety-valve after the time that it had supposedly been screwed down and the inspecting officer did not believe Scholes' story. The said safety-valve was found screwed down after the explosion, but Captain Wynne did not hesitate to suggest that it had been screwed down afterwards by the shed staff so that they could escape the responsibility for their having let an engine go out in such an unsafe condition, while the other safety valve had 'disappeared' so that the inspector could not see it.

At the conclusion of his report, the captain wrote that: 'The condition in which the engine was allowed to remain for 13 years, without ever subjecting the part of the boiler where the most corroding action goes on, to inspection, cannot be too strongly condemned. Such a practice, I believe, to be altogether unusual in any carefully conducted establishment. The neglected state of the engine arose, no doubt, from the consideration that being employed only as a pilot engine, and that the duty being light, it did not require the same attention as a goods or passenger engine. This, no doubt, was true to a certain extent; but in this instance the principal has been carried beyond its limit,

---

[1226] Board of Trade report

[1227] Locomotive Boiler Explosions by Christian Hewison pgs 35-6

[1228] Board of Trade report

[1229] The Times, London, England, 10th March 1853

and the consequence has been a large sacrifice of life.'[1230]

Before the end of March 1853 another boiler explosion had occurred at Brighton, England, this one happening on 17th of that month. London Brighton and South Coast Railway Engine No. 10, which was about to work a train to Littlehampton, blew up at 7.14 a.m., killing Driver John Young, Fireman John Elliot and a fitter named Richard Baker.[1231] In this case, another driver who had previously worked the same engine on the Eastbourne branch saw Driver Young screwing the safety-valve on the dome down to a setting of over 100 lb/sq. in and walked over to tell him that the normal setting was only 801 lb. Driver Young did not respond, and about five minutes later the dome valve was blowing off very strongly, and a minute after that the boiler gave way.

Captain Galton investigated the accident and discovered that the boiler plates had lost some thickness due to corrosion and that the boiler manufacturers stated that the boiler should only be worked at a maximum pressure of 60 lb/sq. in. He attributed the explosion to the driver's action in screwing down the safety valves combined with the weakened state of the boiler and recommended that the safety valves should be repositioned at a point where the driver could not manipulate them. He was also critical of the policy of the Brighton company in offering a premium to the drivers who used the least amount of coke, as this encouraged them to try and work their boilers at a higher pressure, stating, 'this premium on a saving of coke is a premium on working at a high pressure.'[1232]

Strangely a boiler explosion at Aylesbury, England, on 2nd June 1853 does not seem to have been the subject of a Board of Trade report, although it involved the death of a passenger, a Joseph Dendrip (surname illegible in newspaper report), a gamekeeper. The latter had charge of some cattle destined for London and was in a brake van immediately behind the engine when the explosion occurred. He was severely scalded and died in Aylesbury Hospital two days later. Driver Fleming, in charge of the engine, was also badly scalded but survived. Captain Wynne of the Board of Trade apparently examined the remains of the boiler and could find nothing wrong with it, while Mr M. Connell, the London and North Western Railway locomotive superintendent at Wolverton stated that it was 'a question whether or not the safety-valve had become locked' by which he presumably meant accidentally jammed rather than deliberately interfered with. At the inquest the jury heard 'a great number of witnesses' but were unable to determine the cause of the explosion. The jury strongly criticised the practice of drovers being carried in brake vans next to the engine and felt that the supervision at Aylesbury station was insufficient.[1233]

On 16th June 1853 a New York and Erie Railroad engine attached to a freight train suffered a disastrous boiler explosion about a mile east of Susquehanna, Pennsylvania, United States, killing 11 people instantly and injuring many others. The engineer was

---

[1230] Board of Trade report

[1231] The Times, London, England, 18th March 1853

[1232] Board of Trade report

[1233] The Times, London, England, 6th and 7th July 1853

among the fatalities while the fireman was seriously hurt. The brief newspaper report did not state who all the other victims were, possibly they were lineside residents.[1234]

The 5th April 1855 was a bad day for boiler explosions, with one at Rugby, England and a second at Greenock in Scotland. In the Rugby explosion, engine No. 164 was shunting some waggons at 5.50 a.m. when it blew up. At the time it was in charge of a man known as a "turner" who was supposed to be quite competent to be in charge of the engine. He survived the explosion but someone else apparently did not as there was an inquest held, He said that before the accident the boiler had been full of water and the safety valves blowing off. He had not examined the safety valves to see what pressure they indicated and a steam gauge on the engine was out of order. George Wynne, now Lt-Col Wynne, examined what was left of the boiler after the explosion and could find nothing wrong with it but the safety valves were found to have been screwed down and Wynne was sure that this was the cause of the accident, and as the "turner" would have gained little by interfering with them, that they must have been screwed down before the engine left the shed. He criticised the lack of supervision at the shed and stated that 'the foreman in charge, having neglected to submit the engine to a proper examination, before allowing it to go out, is chargeable with great neglect.'[1235]

The Greenock explosion was also investigated by Lt-Col Wynne who found that in this case the explosion was the result of severe corrosion. Engine No. 51 of the Caledonian Railway had blown up at 6.25 a.m., killing two cleaners and injuring four more men. Its boiler had a deep groove about an inch from and above the line of rivets which had weakened it until it had suddenly given way. The boiler had not had an internal examination for seven years.[1236]

The down mail train leaving Carlisle for Glasgow at 5.56 a.m. on 7th April 1856 was only three miles north of Carlisle, England, when the boiler of its engine exploded violently, killing its driver and fireman and injuring two passengers. The whole train was derailed, the engine performed six somersaults and landed in a field and the Perth carriage of the train overturned. A messenger on horseback brought news of the accident to Carlisle station at 6.30 a.m. and five doctors went to the scene of the accident but could do nothing for Driver Thomas Musgrave or Fireman Isaac Sessford.[1237]

Lt Col Wynne began his report on the explosion by saying that as both enginemen had lost their lives and the collateral evidence was very defective the reason for the accident must be pure conjecture. After the accident the Caledonian Railway employed a Professor Rankin of Glasgow to conduct experiments on the causes of boiler explosions, the results of which were accepted by the inspecting officer. Lt-Col Wynne concluded that the explosion was due to the faulty design of the firebox roof supports and 'an evolution of a volume of steam which there are reasonable grounds for supposing may

---

1234 Democratic Banner, Davenport, Iowa, United States, 24th June 1853

1235 Board of Trade report

1236 Ibid.

1237 The Times, London, England, 8th April 1856

have been generated from the overheating of the firebox, caused by a deficiency of water.'[1238] A recent author points out that the report does not mention the state of the safety valves or water gauges or when the boiler was last washed out and suggests that 'In all probability, the roof girders had so much loose scale around them that they and the crown sheet's upper surface were deprived of water, becoming so hot as a result that the crown sheet softened and then gave way, breaking and pulling down the girders with it.'[1239]

On 11th November 1856 an engine described as being 'of a very primitive construction' and about 20 years old blew up at Percy Main, England, on the Blyth and Tyne Railway, killing its fireman. The fireman worked at the smokebox end of the engine attending to the fire and the driver at the opposite end. The report did not say where the water gauge was located or whether the driver or fireman was responsible for replenishing the boiler. Two safety valves were designed to keep the boiler at its working pressure of 60 lb/sq in. At the time of the explosion the engine had been stationary 20 minutes and the driver was away having his dinner.

Lt-Col Wynne concluded that while the engine had been at rest for 20 minutes the pressure in the boiler had kept increasing due to the inability of the safety valves to carry off the surplus steam but his observations have been labelled as 'dreadful nonsense' by author Christian Hewison, who points out that it would have been better if the inspecting officer had examined and described the company's boiler inspection arrangements and had stated the thickness of the boiler plates and their factor of safety.[1240]

Three men were killed in a boiler explosion near Montezuma, New York, United States, on 20th July 1857. They were the engineer, whose name was not given in the newspaper report, the fireman, Ambrose Christian and a Mr Ostrander. No railroad company was mentioned in the report and Mr Ostrander was described as the 'contractor in charge of the work' so presumably this was a locomotive employed on a construction train. The force of the explosion blew Mr Ostrander seventy to eighty feet down a bank and into a swamp while the engineer and fireman were blown nine rods (about 150 feet).[1241]

One of several boiler explosions in England in 1857 occurred on the London and South Western Railway at Basingstoke on 10th October. The 9.p.m. up goods train from Southampton arrived there at 1.5 a.m. and was immediately taken to the ash pit. A porter rode on the engine to the ashpit and asked the driver if he might take on more waggons, to which the driver replied that he could hardly get along with the load he already had. Ten minutes later the boiler exploded, killing the driver and fireman.

A locomotive foreman, who reached the scene of the explosion five hours later, stated that he had found the safety valves 'locked', that is screwed down, and his evidence was

---

[1238] Board of Trade report

[1239] Locomotive Boiler Explosions by Christian Hewison pg 44

[1240] Ibid. pgs 45-6

[1241] The Adams Sentinel, Gettysburg, Pennsylvania, United States, 20th July 1857

confirmed by the locomotive superintendent, but the guard of the train said that the safety valves had been blowing off as the train came into Basingstoke, and a porter agreed. Lt-Col Wynne pointed out that if the driver had screwed down the safety valves after his engine reached the ashpit so much pressure as to cause the explosion could hardly have built up in ten minutes. The inspecting officer commented that 'Among locomotive superintendents, I have found an almost universal opinion to prevail that nine tenths of the boiler explosions are caused by the drivers tampering with the safety valves. With such a prevailing belief, it would be naturally supposed that an endeavour would be made to render any such tampering as difficult as possible; but this has not generally been the case, and in the present instance the valves were left so completely exposed, and the power of locking them was so easy, that it was almost a temptation for the driver to do so if he found himself in difficulties with the load.'

Lt-Col Wynne concluded that the accident had resulted from a flaw in the combustion chamber part of the firebox on an engine designed by Locomotive Superintendent Joseph Beattie to burn coal rather than the more expensive coke, and to consume the resulting smoke. Two distinct laminae which formed the plate had failed to properly unite when the plate was forged. 'It was therefore a weak point from which the effects of corrosion must sooner or later yielded before the other parts of the boiler were worn out, and therefore the question of whether or not the valves were tampered with is not a matter of such serious importance.'[1242]

In early February 1858 *The Times* reported on a 'Fearful Explosion' on a minor railway in South Wales. The accident had actually occurred the previous month, on 28th January, at Pantyffynnon, Wales on the Llanelly Railway, a mineral line. The engine involved, named *Victoria,* dated from 1841, and was similar to that which had been involved in the Blyth and Tyne accident, with the driver working at the front end of the engine where the water gauge, test cocks and steam pressure gauges were situated and the fireman working at the other end. For the seven years prior to the explosion the engine had been used only to work ballast trains, not being powerful enough to haul mineral trains, and for two and a half years of that time it had been laid up out of use.

On the day of the accident the engine had brought down six waggons of ballast, each weighing five tons, from the Amman branch. Stopping to take water at Clun-y-Moch, where the driver noted the steam pressure to be 35 lb/sq. in and that the water glass was half full. On arriving at Pantyffynnon the ballast waggons were dropped off and the driver left the engine at the passenger platform and went into the telegraph office to ask the locomotive superintendent for permission to use another engine to take the ballast train up the 1 in 100 bank to the Limestone branch at Cil-y-rychan, as he doubted if *Victoria* could manage the job. While he was in the office the engine blew up.[1243] The driver thus probably saved his life, and the fireman was only injured, but the station was crowded, and David Davis, who was standing near the fireman, was fatally injured, as were two 15-year-old boys, John Harris and John Morgan, Harris being, in the words of *The Times*, 'fearfully mangled.'[1244]

[1242] Board of Trade report

[1243] Ibid.

[1244] The Times, London, England, 8th February 1858

Lt-Col Yolland investigated the explosion and concluded that it had been caused by 'the ordinary pressure of steam acting against the enfeebled and worn out part of the old portions of the plate on the top of the boiler.'[1245]

Only ten days after the Pantyffynnon accident the London Brighton and South Coast Railway suffered a boiler explosion at Caterham Junction (later renamed Purley), England, when the boiler of the 6 a.m. passenger train from London Bridge to Brighton blew up. The engine had been delivered to the company in November 1849, when its working boiler pressure had been 120 to 130 lb/sq. in, but this had been reduced to 90 lb/sq. in by the time of the accident. The damage to the engine was limited to the bulging inward of the copper firebox just below the firebox door but the accident nevertheless resulted in the death of the fireman. The engine had been working the down Parliamentary train, and an assisting engine had been attached to climb the New Cross bank. The train reached Caterham Junction on time and had standing there about a minute when the explosion occurred.[1246]

Lt-Col Wynne reported on the explosion and stated that the explosion had occurred at ordinary working pressure. He added that this was the last report he was going to make on a boiler explosion and he was therefore anxious to point out that the majority of such accidents had taken place at ordinary steam pressure and could be traced to the boiler being worn out or some defect in its construction and not to wilful tampering with the safety valves, which was the favourite theory of locomotive superintendents to relieve themselves from blame. In the case of this explosion the inspecting officer found that the thickness of the copper below the firebox door had been reduced from the friction of filling the firebox with coke from $\frac{7}{8}$ in to $\frac{1}{4}$ in. The locomotive foreman in charge of the engine was of the opinion that such a thickness was perfectly safe, and said that he had known engines work perfectly safely with copper worn down to only $\frac{1}{8}$ in thickness, but Lt-Col Wynne stated that this explosion, and two others that have recently occurred, owing to the wearing of the copper, show that it is not safe to trust to the tenacity of the material, when from long use it has been worn down to half its original thickness.'[1247]

Another boiler explosion with fatalities occurred in Scotland on 24th July 1858. The accident occurred at Burnmouth on the North British Railway main line from Edinburgh to Berwick. Engine No. 66 was working the 10.10 a.m. train from Edinburgh, which was a mixed train as far as Dunbar, thence purely a goods train. A light load of 21 waggons had been taken up the 1 in 96 bank from Innerwick to Grantshouse (spelt Grant's House in the Board of Trade report) and after Reston station only eight waggons remained to be taken the 11-odd miles to Berwick, on a continuous falling gradient and stopping only at Burnmouth. The train arrived at Burnmouth at 2.19 p.m., 14 minutes late, dropped off a waggon and was in the process of attaching another when the boiler exploded, the train having been there no more than 10 minutes. The engine was shattered by the explosion

[1245] Board of Trade report

[1246] Ibid.

[1247] Ibid.

and the boiler thrown 140 ft. Driver James Ritchie was trapped in the wrecked machinery and was extricated with difficulty; he supposedly died from internal injuries, while fireman James Caril was severely scalded.

Lt-Col Yolland investigated the explosion. He found that the thickness of the copper of the firebox had been reduced from $\frac{7}{16}$ in to $\frac{5}{16}$ in but this was not sufficient to account for the accident and there was no appearance of the firebox having overheated. What was much more significant was the evidence of a telegraph foreman and former driver who said that he had previously ridden on the same engine with Driver Ritchie and had observed that one of the levers of the safety valves was pinned down by a small wooden wedge and that Driver Richie had told him that he (Ritchie) could go anywhere with a pressure of 150 lb/sq. in. The telegraph foreman said that he had seen Driver Ritchie use the same wedge both before and after that time to enable the engine to climb a gradient of 1 in 75, and another driver to do the same thing. Lt-Col Yolland considered that 'It is greatly to be regretted that this foreman did not report these circumstances before the accident; and as he had been an engine driver on the line, he could not be ignorant of the consequences that might at any time ensue from a recourse to such practices. I consider that he is very much to blame.' The other driver denied using any such wedge, but then 'he would say that, wouldn't he' to use a well-known quotation from the Profumo trial. The inspecting officer accepted that the explosion was due to the safety valve being wedge shut but did not explain very clearly why the other safety valve did not prevent the explosion, a more recent writer suggests that in fact both safety valves had been wedged closed to ascend the previous inclines and that the driver had then forgotten to remove them.[1248] [1249]

It also appeared that the maximum working pressure for the boiler had never been specified and no one knew what it was, including neither the North British Railway locomotive department nor the manufacturers, Messrs R and W Hawthorn, so that the locomotives could have been working for years at a higher pressure than was intended or that they had ever been tested to bear. The inspecting officer recommended that the makers of engines should provide them with a brass plate specifying the maximum boiler pressure at which they should be worked and the pressure at which it had been tested to, also that the second safety valve should be placed where the driver could not tamper with it.[1250] In view of Lt-Col Wynne's recent comments about most boiler explosions not being the result of drivers tampering with the safety valves it was of course ironic that an accident soon afterwards apparently was due to this very cause.

On the evening of 3rd October 1859 London Brighton and South Coast Railway Engine No. 108 was pulling the Newhaven to Brighton goods train from Lewes up Falmer bank, 4 miles at 1 in 88, with the Brighton yard engine banking the train of 46 loaded waggons. The train had travelled about a mile up the incline, and Engine No. 108 was observed to be labouring hard and slipping, with steam blowing off at the safety valves, when an explosion caused the top of the firebox to collapse, throwing the driver and fireman off the engine. The fireman was killed but the driver survived.

[1248] Ibid

[1249] Locomotive Boiler Explosions by Christian Hewison pgs 53-5

[1250] Board of Trade report

Captain George Ross investigated the accident and found that, as in the case of a recent boiler explosion at Lewisham, there was a defect in the design of the firebox, 'the top stays of which were so short as to not obtain direct support from the sides,' and he added that, 'No time should be lost in examining and making more efficient the fire-box crown stays of other engines of this class.' The inspecting officer also recommended that the levers of the safety valves should be repositioned so as to remove any suspicion of the driver having tampered with them and (as recommended by the jury at the inquest) that the maximum loads for trains on steep inclines should be regulated.[1251 1252]

South Devon Railway engine *Tornado* arrived at Totnes, England, station at 12.10 p.m. on 13th March 1860 from Newton (later Newton Abbot) with a goods train of 22 waggons and a van. Its driver did some shunting, took on some water and was engaged in more shunting when the engine boiler exploded. The driver, looking back for a signal from the guard, was killed outright and the fireman, who had been stooping down to pick up his coffee, was blown back over the coke box and so seriously injured as to be still confined to bed when the Board of Trade inspecting officer, Captain HenryTyler, made his investigation.

The inspecting officer found that the effect of the explosion had been to detach three portions from the outer shell of the firebox, one of which was blown through the side of a goods shed, the second against a brick wall and the third onto the permanent way. He discovered severe corrosion in the boiler at the line of fracture, with the metal having been eaten away by an average of $\frac{3}{16}$ in, in some places more, and blamed the corrosion on careless use of the caulking tool when the boiler was made.[1253] Christian Hewison comments that he may well have been right, but a more likely explanation was the tendency of boilers with lap joints which were not quite truly circular to become distorted by steam pressure, causing bending, localised stress and cracks which developed into grooves.[1254]

Captain Tyler mentioned as a factor accelerating the corrosion of the boiler the poor quality of water available at Newton up until November 1858, before that date it had been obtained from a well and often contaminated with salt water. Since that date the water quality had improved. There had also been a second failure with a sister engine, *Ixion*. In this case a crack in a firebox plate had fortunately been discovered before there was an explosion.

Captain Tyler had five recommendations; additional care in caulking, the placing of the lower plate outside instead of inside the upper one, the employment of good water, the use of sufficient stays and the periodical testing of boilers using hydraulic pressure. In addition, he thought it very desirable that every engine should have a steam gauge; the *Tornado* did not have one.[1255]

[1251] Ibid.

[1252] The Times, London, England, 8th October 1859

[1253] Board of Trade report

[1254] Locomotive Boiler Explosions by Christian Hewison pgs 56-7 and 59-60

[1255] Board of Trade report

The down Irish Mail of the London and North Western Railway was involved in boiler explosion four miles north of Rugby, England, on 4th July 1861. The train departed from London Euston at 8. 25 p.m. that day and left Rugby on time at 10.28 p.m. It was approaching an overbridge known as Easenhall Bridge when the boiler exploded without warning while the train was travelling at about 35 miles an hour over moderate gradients. The explosion was a very violent one and the barrel of the boiler was blown to pieces, parts of it striking the overbridge, while one of the pieces broke off the engine's right-hand driving wheel and the engine and all eight vehicles of the train were derailed. The driver was thrown off the engine against a telegraph pole and injured, his fireman was carried forward with the engine, dreadfully crushed and killed. A Post Office clerk and a guard were injured but the passengers escaped.

Captain Henry Tyler investigated and concluded that the accident was due to corrosion in the boiler of the *Bloomer* class engine, which had caused it to explode under ordinary working pressure. He believed that the corrosion must have already been present when the engine had been re-tubed in October 1857 but had not been noticed. He concluded that the only method by which such accidents could be prevented was by more frequent boiler examinations.[1256]

On Thursday 17th October 1861 the Erie Railroad suffered a boiler explosion at Hornellsville (later Hornell), New York, United States. Engine No. 186, named *Wisconsin*, which was used as a switching engine, blew up as it was standing near the water tank, a piece of boiler weighing several hundred pounds being thrown clean over several buildings to strike a barn. Other fragments were also thrown considerable distances. Engineer Charles Willour and fireman James Ogden were sitting in the engine cab at the time but although considerably confused were unhurt. The damage done was estimated at about $2,000. The boiler pressure had not been over 100 lb/sq. in at the moment of the explosion but the boiler was an old one and 'had doubtless become corroded and weakened by age.'[1257]

Engine No. 878 of the London and North Western Railway left Camden goods station at 1 a.m. on the morning of 5th May 1862 with a train for Rugby. Starting with only three or four waggons the train picked up eight more at Kilburn and had 12 more to take on at Harrow, England. These 12 had just been attached when the explosion occurred as the driver was opening his regulator to proceed. The explosion killed the fireman and seriously injured the driver.

Captain Tyler found severe corrosion along one of the middle lap joints of the boiler which was clearly the cause of the explosion. He found it impossible to believe that the plate had been sound when the boiler was examined and re-tubed a year and eight months previously, or that it had then been in a safe condition to remain in use. He alluded to the Rugby (Easenhall Bridge) explosion and another more recent one at Stella Gill on the North Eastern Railway, and to the fact that in both of those as in the Harrow explosion the boilers had been inspected not that long previously, but the corrosion which must have been present then was either missed or else ignored. He added 'These and other

---

1256 Ibid.

1257 Hornellsville Tribune, Hornellsville, New York, United States, 24th October 1861 instances which have been brought before their Lordships prove sufficiently that additional care and attention are much required in the maintenance of the boilers of steam locomotive engines. The common practice that exists of allowing them to run for six or seven years without internal examination, and without applying to them any test which would discover weaknesses arising from hidden defects, is evidently a most dangerous one.'[1258]

A brief report in an American newspaper described a serious boiler explosion near the Long Dock in Jersey City, New Jersey, United States, on 25th October 1862. The boiler of an Erie Railroad locomotive blew up at about 7.30 p.m. on that Saturday evening, the explosion being 'the most terrific one of the kind ever known, and the shock, as if of an earthquake, was felt throughout the city. Houses nearby were shaken, window-glass was shattered, and at the distance of a mile or more the windows and doors rattled violently.' The explosion killed five railwaymen and injured two more. Its cause was not stated.[1259]

The Great Western Railway suffered a boiler explosion just outside of London Paddington, England, station on 8th November 1862. The broad-gauge engine *Perseus* blew up in the engine shed at Westbourne Park at about 5.45 a.m., killing the lighter-up, Christopher Thompson, and two cleaners, James Wilson and John Elridge; a fourth man named Bishop was seriously injured.

Captain Tyler investigated the accident and found that once again the explosion had resulted from corrosion, in this case 'Extensive corrosion had taken place near the line of rivets at the bottom of the middle ring, and the weakness along this line has no doubt caused the explosion. There were also marks of pitting at the bottom of the firebox ring.' He was forced to repeat his comments made after the Harrow explosion about the need for additional care and attention in the maintenance of boilers and the need for boiler examinations at say every three years or 75,000 miles run. 'It is, in any case, in my opinion, most unsafe to allow an engine to run on without any examination of the interior of the boiler, as the *Perseus* has done, at full pressure, for 7 years and 5 months, and for a duty of 175,000 miles.' The Great Western locomotive superintendent at the time was the famous Daniel Gooch.[1260] [1261]

An accident with similarities to that near Rugby, England, the previous year occurred between Mills and Villeneuve-sur-Allier in France on 10th November 1862, when the boiler of an engine pulling a freight train exploded beneath an overbridge. This time the force of the explosion was sufficient to make the bridge collapse altogether onto the waggons behind the engine. The fireman was killed outright, and the driver was fatally injured. The head guard was also badly hurt but two other railwaymen on the train escaped injury.[1262]

1258 Board of Trade report

1259 Davenport Daily Gazette, Davenport, Iowa, United States, 30th October 1862

1260 Board of Trade report

1261 The Times, London, London, 10th, 12th and 21st November 1862

1262 Wikipedia - Liste des accidentes ferroviaires en France au XIX siècle

On 4th January 1864 a heavy freight train of the Chicago and North Western Railroad, drawn by two engines, one named *Gladiator*, arrived at Milton Junction, Wisconsin, United States, and took the siding at about 5 p.m. there to cross an opposing passenger train, which was running about an hour late, and the two engines of the freight train came to a stand opposite the two rear coaches of the passenger train, which was presumably stopped at the station platform. While they were standing thus, the boiler of the *Gladiator*, described as 'one of the heaviest and best engines belonging to the Northwestern Road,' exploded, overturning the penultimate passenger coach and sending fragments of iron and wood 50 feet into the air. A piece of engine boiler also tore a ragged hole three or four feet wide in the overturned car and all its windows, together with some of those in the ladies' car behind, were smashed. Two of the passengers were fatally injured and half a dozen others were hurt, some seriously.

Engineer Choate, in charge of the *Gladiator* stated that he had had two full gauges of water and only 90 lb/sq. in of steam pressure. 'If such is the case,' the newspaper report commented. 'of course, not the slightest blame can attach to him, and there must have been some unknown flaw in the boiler.'[1263]

The Midland Railway had a boiler explosion at Colne, England, on 5th May 1864. Engine No. 356 had arrived at Colne, England, at 2 a.m. with a goods train from Leeds. Having waited for about half an hour for a goods train from Manchester, it then did the shunting required to make up its train for the return journey to Leeds and was just about to set off when the explosion occurred. After the explosion, Driver George Parker was found under a waggon dead while his fireman was blown nearly one hundred yards, lost consciousness and was badly enough injured not to be able to return to work at the time of the investigation. Different pieces of the boiler were blown different distances and a spring balance pierced the roof of a cottage a quarter of a mile away in Bridge-end and injured an old woman named Mary Hartley who had been in bed at the time.[1264]

Captain Tyler investigated the accident and found yet again that the boiler was seriously corroded. He noted that this accident was one of three boiler explosions that he was investigating that had occurred that May, the others involving Great Northern and London and North Western engines; while he had investigated four other cases, two on the London and North Western and on each on the Great Western and North Eastern Railways, during the previous three years, all of which were due to seriously corroded boiler barrels. 'And these seven cases represent a more serious risk than would appear at first sight, that is daily incurred by the officers and servants of railway companies as well as by the public. Of the 6,500 locomotive engines and upwards which are in use on the railways of the United Kingdom, a large proportion are affected by corrosion to an extent which is more or less dangerous. For every engine that explodes there are a great number of others which have been much weakened from this cause, and which are constantly working with a less margin of safety than ought always to be preserved between ordinary pressure and bursting pressure.'

The inspecting officer advised that 'As regards construction,

---

[1263] Janesville Daily Gazette, Janesville, Wisconsin, United States, 5th January 1864

[1264] The Times, London, England, 6th and 10th May 1864

(1) the barrels should be made more perfectly cylindrical by the use of butt-joints and cover-strips in place of the lap-joints more commonly used

(2) The longitudinal joints should be placed in all cases above the water-line instead of below it, so as to prevent the risk of corrosion from the different actions previously referred to; and there is no necessity for having more than one longitudinal joint in each ring.

(3) The boiler should be firmly attached to the framing at one end only, the other end being allowed to slide backwards and forwards, to allow for expansion and contraction from changes in temperature, as is now frequently, but by no means always, done.

(4) The barrel should be strengthened at the vertical (or transverse) joints, and at intermediate intervals, either by the addition of belts or by the use of plates rolled thicker in the middle as well as at their edges.'

The other two May explosions referred to in Captain Tyler's report on the one at Colne took place on the 9th May at Bishop's Road station (renamed Paddington 1933) on the Metropolitan Railway on 9th May (which involved a Great Northern engine) and at Overton (later Orton Waterville) on the London and North Western Railway on 30th May. Both resulted in injuries but not fatalities.[1265]

Three people died in boiler explosion four miles east of Syracuse, New York, United States, on the morning of 8th June 1864. Those who lost their lives were the engineer and fireman of a westbound steamboat express train and a third unnamed person. Twenty of thirty other people were injured, several seriously, and the engine and three or four coaches of the express were wrecked.[1266]

The boiler explosions in England continued with one on 16th August 1864 at Camden Road station on the North London Railway. Driver John Ellis left Chalk Farm station on the London and North Western Railway at 9.32 a.m. on the day of the accident and reached Camden Road station on the North London Railway at 9.34. Here he stopped with his engine on the bridge over Randolph Street to await the arrival of the 9.5 a.m. train from Kew. This arrived at 9.39 and its engine was detached, ran around its carriages and pushed them gently forward to be coupled up to those of the Chalk Farm train, the whole train to then proceed to Fenchurch Street hauled by driver Ellis' engine. Just as the train was about to start there was a terrific explosion; the driver was thrown, badly scalded, onto the station platform, the fireman, William Davis, was thrown, with fatal scalds and head injuries, into the six-foot, and the engine thrown against the bridge parapet, which gave way and the engine fell 20 ft into Randolph Street, but landed on its wheels.

Captain Tyler found that the explosion had resulted from the failure of a number of copper stays on the left of the firebox and that the engine 'had been working for some

---

[1265] Board of Trade report

[1266] Burlington Weekly Hawk-Eye, Burlington, Iowa, United States, 11th June 1864

time in a critical condition.'[1267] [1268]

What was described as 'A Shocking Accident' occurred on the Philadelphia and Erie Railroad three miles from Erie Union station, Pennsylvania, United States on the morning of 31st August 1864. The engine concerned had left Erie to bring up and extra train when it blew up, killing the engineer, the fireman, the brakeman and a stranger named Parker from Chatauque County, whose reason for riding on the light engine was not stated; possibly he was a passenger who had missed his train or perhaps just a friend of one of the other three coming along for the ride. The explosion tore the engine to pieces and demolished a stretch of track.[1269]

The inhabitants of Salem, Illinois, United States had a rude awakening from their sleep at 1 a.m. on the morning of Wednesday 11th January 1865 when the boiler of an Ohio and Mississippi Railroad engine exploded. The engine was in charge of a freight train and was taking on wood between the mill and the depot when the explosion occurred. Its engineer and fireman were killed instantly, being thrown twenty or thirty feet, one on each side of the line. One brakeman was severely and possibly fatally injured but the other brakeman, who had been stood by the tender, was unhurt. Portions of the wrecked boiler were blown everywhere: 'A flue struck Seth Andrews' fine dwelling near the top of a window, knocking a hole through the siding, shivering glass, and peeling off a large amount of plastering... One piece of metal, two feet in diameter by six inches thick, was carried one hundred yards and must have passed directly over Mr Andrew's two storey house and came down some twenty yards from Rev John Williams' dwelling. Its weight could not be under three hundred pounds. Another piece was carried clear through a cooper shop, also, smashing glass and sash in two large windows of the large steam-mill.'[1270]

North Eastern Railway Engine No. 122 left Hartlepool, England, at 6 a.m. on 20th March 1865 with 22 waggons of ballast for Thinford. There the ballast was dumped to form a new siding to serve an ironworks and the engine ran around its now empty waggons to return to Hartlepool. Having got as far as Coxhoe, the driver of No. 122 found a coal train in front and began to set back his train onto the other running line as a passenger train was due to follow shortly from the Thinford direction. Then, however, the coal train moved off, and the driver of No. 122 decided to follow it and not wait for the passenger train. The fireman then began to get down some coal and the driver had left the engine for some reason when the fireman suddenly heard a tremendous noise but felt very little shock. His coat was burnt, and he was somewhat scalded but not otherwise hurt. He then found the body of his driver beside the engine, where he must have been when the explosion occurred and where he had taken the full force of the blast.

By the time that Captain Tyler had been asked to investigate the explosion and had got

[1267] Board of Trade report

[1268] The Times, London, 17th and 19th August 1864

[1269] Janesville Daily Gazette, Janesville, Wisconsin 2nd September 1864

[1270] Centralia Sentinel, Centralia, Illinois, United States, 12th January 1865

to Coxhoe the remains of the boiler had been cut up so that he could not examine it, but he was told that a seam had given way through a line of rivet holes rather than because of grooving. A modern writer, however, believes that grooving was in fact the cause and that the North Eastern Railway locomotive superintendent at the time, Edward Fletcher, had ordered that the boiler be cut up to prevent Tyler from seeing the evidence.[1271] [1272]

New York Central and Hudson River Railroad Engine No. 104, formerly known, the newspaper report stated, as the *Nathaniel Thayer*, left Syracuse, New York, United States, station just after midnight on 1st April 1865 with the westbound Night Express. It got as far as the approach to the Oswego Canal Bridge, a short distance west of the city, when its boiler exploded, scattering fragments in every direction but for once without causing any death or injury other than a slight cut on the head sustained by the engineer when he and the fireman were both thrown backwards by the blast. The front part of the boiler was blown completely out, and the engine whistle was blown high in the air and came down and penetrated the roof of a passenger coach. The engine and leading baggage car were derailed and wrecked.

The engineer, William Perry, was described as one of the most careful on the road. The cause of the explosion was something of a mystery. Lack of water was ruled out as the engine had only just started out, while an examination of the flues afterwards showed them to be clean and well looked after, while 'weakness in the boiler is not the cause assigned for the accident, as it has been in use a long time, but underwent several repairs recently which it was thought had made it as good as new.'[1273] So, an undetected area of corrosion or a faulty repair to the boiler probably caused the accident.

Another locomotive took the train forward to its destination.

A boiler explosion at Frankfort, Kentucky, United States on 12th June 1865 killed four persons and severely injured eight or ten more. A piece of the boiler weighing supposedly three hundred pounds was thrown a hundred yards, penetrated the roof of a house and slightly injured a young woman inside.[1274]

Driver James Whitehead and Fireman Robson Heppel arrived at the Blyth and Tyne Railway station at Newcastle New Bridge Street, England, at 7.20 a.m. on the morning of 15th December 1865 with engine No 25 and a train of 17 loaded chaldron waggons. They shunted the yard until the 7.30 a.m. passenger train had gone, then placed the waggons over the coal drops and the engine had been stationary for a minute and a half when its boiler exploded. The driver and fireman, who had been standing by the engine at the time were severely scalded and both died the same evening. Guard Hudspeth, who had been standing a few feet in front of them, was thrown into a manure heap and slightly scalded in the face.

1271 Board of Trade report

1272 Locomotive Boiler Explosions by Christian Hewison pgs 71-2

1273 Syracuse Daily Courier and Union, Syracuse, New York, United States, 3rd April 1865

1274 Ibid, 22nd June 1865

Captain Henry Tyler was not impressed with the state of the engine. Hr found that 'The heads of the stays had been almost entirely burnt away on the left side of the firebox, and there was hardly anything left to resist the pressure of the steam against the plate on that side but the threads of the screws. The thinnest portion of the plate, barely $\frac{3}{16}$ths of an inch, was not in the portion that first yielded, but at the back of the firebox. If the stays had been removed, as they undoubtedly ought to have been, when the engine was brought to the shops for the examination and repair of the firebox, the explosion would not have occurred. It appears that the fuel which is used on this line contains a large proportion of sulphur, which causes the heads of the stays to burn off and the copper plates to burn away more rapidly than they otherwise would do. The foreman boiler smith considered, when he sent the engine out with a partial removal of stays, that it would work for another month or so without a more extensive renewal; but he had no good excuse for doing so. The burning away of the heads of the stays is so simple and obvious a symptom of weakness, that no explosion from this cause ought, with common care and attention, be permitted to occur.'

Captain Tyler concluded his report by criticising the lack of written records at the Blyth and Tyne works at Percy Main, so that details of previous repairs had to rely on the memory of the foreman boilersmith and actual observation of the firebox.[1275] [1276]

A brief report in *The Times* referred to a letter from Alexandria detailing a boiler explosion at Kafres-Kaiat, Egypt, on about 2nd September 1866 which killed five people and injured many more.[1277]

A westbound lightning express freight train on the Baltimore and Ohio Railroad was climbing a bank near Littleton, West Virginia, United States, on 19th March 1867, when a violent boiler explosion occurred which overturned the engine and killed Engineer George Artis, Fireman E. Johnson and Train Supervisor Otto Burrell. Johnson and Burrell were killed outright; Artis was blown forward and aside about a hundred yards and was still alive when found but died in a few minutes. The tender was wrecked, and the leading car of the train overturned but the remainder of the train was undamaged. The cause of the explosion was not known.[1278]

The 5.30 p.m. passenger train from Manchester Victoria to Bolton on the Lancashire and Yorkshire Railway had just reached Halshaw Moor & Farnworth (later Farnworth & Halshaw Moor), England, station on 31st January 1868 when its boiler exploded. Driver Richard Partington was seriously injured, with scalds and a broken leg, but the fireman, who was down on the track beside the engine, escaped injury.

Colonel W. Yolland found that the explosion had occurred in the copper firebox, and that the thickness of the firebox plate, originally ½ inch, had been reduced by wear to less

---

[1275] Board of Trade report

[1276] The Times, London, 18th December 1865

[1277] Ibid, 27th September 1866

[1278] Petersburg Index, Petersburg, Virginia, United States, 27th March 1867

than $\frac{1}{8}$ of an inch along the line of first fracture. Colonel Yolland stated that 'it appears desirable that a more careful examination should be made in order to ascertain the thickness of the copper plates when engines have been running for such a length of time, and for such long distances as No. 115 engine, where the plate had been reduced at one part to less than one fourth of the original thickness. The Locomotive Superintendent, at my request, caused a hole to be drilled in the pate at the right side of the fire-box, and here the thickness had been reduced to somewhat under a quarter of an inch; so that it is apparent that the engine must have been running for some considerable time with an insufficient margin as regards the strength of the fire-box.'[1279]

Maatschappij tot Exploitatie van Staatsspoorwegen (SS) company locomotive No. 8 had spent much of the previous year in the workshops when on 31st March 1868 it was fired up ready to work the first passenger train of the day from Harlingen (The Netherlands) to Leeuwarden with Passed Fireman Hulscher and Fireman Leman. An hour before the departure time Hulscher complained of chest pains and went off sick, his place being taken by Driver Stonner.

Two minutes before the train was ready to leave, with the passengers already seated, Fireman Leman on the footplate and Driver Stonner nearer the tender, there was a very loud bang and thick clouds of steam followed by a shower of pieces of iron and lumps of coal. When the steam cleared it could be seen that the boiler had been blown to pieces, with parts of the locomotive blown considerable distances and the engine wrecked, while the tender and one axle of the leading baggage van were derailed, and both were damaged. Driver Stonner was only slightly injured, but Fireman Leman was killed.

Examination of the engine after the explosion showed that the firebox had suffered the most damage. The maximum working pressure of the locomotive was meant to be 6.1 atmospheres (about 87 lb/sq. in) and the safety valves were meant to open at that pressure, but it transpired that on the engine's two previous runs on 27th and 28th March two wooden wedges had been used to prevent the safety valves from opening, so that while the pressure gauge was indicating 6.1 atmospheres the actual pressure was in the order of 8 to 9.3 atmospheres (about 120 to 140 lb/ sq. in). The Raad van Toezicht (accident investigation board) which inquired into the accident reported on 8th May that they were very surprised that a locomotive could be operated in such a manner as No. 8 was on the day of the accident without the knowledge of more superior officers and that those officers really should exercise more supervision over their drivers and fireman. The board also recommended the use of stays which if broken would indicate this by releasing steam or water and of pressure gauges that would indicate the highest steam pressure attained rather than just the maximum permitted.

The accident was also investigated by the SS mechanical engineer who came to the conclusion that the explosion could not be ascribed to faulty boiler construction or material nor to insufficient water and he asked the opinion of J.G.W. Fijnje, the chief executive of the SS, who opined that the explosion must be due to an exceptional natural phenomenon.

Faced with these differing conclusions, the government minister appointed on 10th

[1279] Board of Trade report

June a commission composed of the Raad van Toezicht, Dr C.H.C. Grinwis, an Utrecht professor and J.G.W. Fijnje to carry out a further investigation. In the report of this body, J.G.W. Fijnje with great verbosity took the chance to agree with the essential points made by and the conclusion of the Raad van Toezicht without conceding that he was. Fijnje admitted that the firebox must have been in a less than perfect condition on 31st March and for this to lead to an explosion that the maximum permitted steam pressure must have been considerably exceeded (as the Raad had found) and basically the findings of the Raad van Toezicht were confirmed. No further boiler explosions occurred in the Netherlands.[1280]

The engine of a Le Mans – Caen mixed train suffered a boiler explosion at about 12.30 p.m. on 19th March 1870 between Fresnay-sur-Sarthe and La Hutte Coulombiers stations on the Le Mans – Mèzidon line in northern France. The engine driver was killed, and the fireman and conductor seriously injured but the passengers escaped injury.[1281]

The 4.45 p.m. Manchester Sheffield and Lincolnshire Railway passenger train from Sheffield to Penistone started five minutes late on 25th October 1870. The train proceeded without a problem to Deepcar station, where it stopped three or four minutes, but just as it was setting off again the boiler of its engine, No. 83, exploded. The driver was knocked back onto the tender and died in about ten minutes from a head injury, while the fireman was knocked unconscious for a short time and had post-traumatic amnesia, remembering nothing of the accident, but was not badly hurt. By chance, the same engine had suffered a broken connecting rod between Deepcar and Oughty Bridge stations only eight days earlier.

The explosion was investigated by Lt-Col F.H. Rich. By the time that he was able to examine the remains of the engine, the parts were so rusty as to make it impossible to detect any flaws, but he was told by Mr Charles Sacré, the Manchester Sheffield and Lincolnshire engineer, that a flaw had been plainly visible immediately after the accident. In the absence of any other evidence as to a cause for the accident, the inspecting officer accepted this flaw as the cause of the explosion and said that he thought that it had been present since the boiler was manufactured.

Told that the boiler had been tested hydraulically to a pressure of 130 lb/sq. in Lt-Col Rich said that he thought that this was insufficient when the boiler was to be worked up to 115 lb. He considered that boilers should be tested to double their intended working pressure, He also did not think that hydraulic testing was satisfactory and thought that boilers should be tested with steam to double the working pressure, the testing being done in an area shielded by earth or masonry to prevent danger to life and limb should a boiler perchance explode under the strain. This rather destructive idea was never adopted by any company.[1282]

The boiler of Lehigh Valley Railroad engine *Vulcan* exploded near Mauch Chunk, Pennsylvania, United States on 17th July 1871. it was pulling a northbound train of

---

[1280] Spoorwegongevallen in Nederland 1839-1993 by R.T. Jongerius pgs 26-28

[1281] Wikipedia - Liste des accidentes ferroviaires en France au XIX siècle

[1282] Board of Trade report

express cars. No less than six men were riding on the engine and five of them were killed outright while the sixth was said to be horribly mangled and could not live. Of the railwaymen with the train only the flagman, riding out of harm's way on the rear car, escaped. The bodies of the victims were thrown in all directions and made a heart-rending scene. Some had been blown 200 and 300 ft away. All of the men had families and three of them came from Weatherly, Pennsylvania. Three section men (platelayers) working on the line several hundred feet distant were badly shaken and injured.[1283]

On the 4th January 1872 a boiler explosion occurred on the Highland Railway at Fochabers (Orbliston Junction from 1893), Scotland. A goods train had left Inverness for Keith, and after doing work at several intermediate stations left Fochabers four minutes late with eleven waggons and a van pulled by a tender engine. It had travelled nearly a mile from Fochabers and was running down a falling gradient of 1 in 350 when the outer shell of the firebox exploded in a cutting. The blast separated the engine from its tender and it ran away for 600 yards before coming to a halt, while seven waggons were derailed and two of them damaged. Brakeman John Gerrie, who had been riding on the engine because it was more convenient at the various stops, was killed outright while Driver William Ross and Stoker D Bruce were injured.[1284]

Lt-Col C.S. Hutchinson decided that there was 'no reason to suspect and negligence or want of caution on the part of the driver or fireman, and that sufficient cause could be found for the explosion in the defective condition of the stays where a bulge occurred on the left side of the copper firebox, in the crystalline state of the edges of the adjoining iron plates, and in the comparatively weak single riveted joint at this same part of the firebox, at which part it is, from all the circumstances, most probable that the explosion commenced. It is in my opinion to be regretted that after the repairs of November 1871, the boiler was not admitted to a hydraulic pressure test, which would have been almost certain to discover the existence of the defective stays. I also think that on the occasion of these repairs the condition of the whole of the stays could not have been investigated as completely as it ought to have been.'[1285]

The little Whitehaven Cleator & Egremont Railway suffered a boiler explosion at Moor Row on 5th February 1872. Engine No. 8 with a train of 29 loaded coal waggons had just drawn up at a signal when its boiler exploded, killing the driver. The fireman had left the engine to collect his breakfast and neither he, nor the guard who was busy pinning down brakes, was hurt.

Lt-Col C.S. Hutchinson said that there was no suspicion that the safety valves had been tampered with. He found severe corrosion at the point where the boiler plates met, with the thickness reduced from $\frac{7}{16}$ in to $\frac{1}{16}$ or $\frac{1}{8}$ in and a good deal of honey-combing and stated that 'The reason, therefore, of this boiler bursting is very evident; the only wonder is, that considering the state of the bottom plate along the left joint, it should not have burst long since. Flaws of the description which led to this explosion are no doubt

[1283] Evening Gazette Port Jervis New York, United States, 24th July 1871

[1284] The Times, London, 6th January 1872

[1285] Board of Trade report

difficult to detect by inspection, and this makes it all the more important that boilers should be periodically subjected to a hydraulic pressure of nearly double the amount at which it is intended to work then, a test which had not hitherto been applied by this Company.'[1286]

The boiler of a Baltimore and Ohio Railroad engine, hauling an eastbound freight train, exploded on 22nd April 1872 near Parkersburg, West Virginia, United States, killing Engineer Milton Parker and Fireman Daniel Nelson outright and injuring Brakeman Michael Eagan so badly that he died within a few hours. The boiler was thrown 350 feet and the running gear and tender of the engine were thrown down the embankment. Three cars were thrown down the embankment into a creek on the opposite side to the running gear and tender.[1287]

The boiler of the engine of a La Rochelle – Nantes mail train exploded at 3:30 a.m. on 22nd June 1872 near Angerville on the Paris - Orléans (Chemin de fer de Paris à Orléans) main line and the train was derailed. The fireman was killed outright, the driver died the next day and three passengers were injured.[1288]

No. 4 engine of the Dublin Wicklow and Wexford Railway had taken water and was standing in Bray station, Ireland ready to set off with the 9 a.m. train to Dublin on 16th September 1872 when its boiler exploded, killing the driver and fireman. Neither the stationmaster and guard, who were stood on the station platform a short distance away, nor any of the passengers were injured. After the accident, twenty-one pieces of the boiler were picked up at different locations inside and outside the station.

The explosion was investigated jointly by Lt-Col Rich of the Board of Trade and L.E. Fletcher, chief inspector of the Manchester Steam Users Association, who failed to agree on its cause. Fletcher considered that the boiler had first given way at the manhole cover, torn away by the failure of the wrought-iron bolts securing it to the firebox, but Rich thought that the boiler had initially failed at the bottom end next to the firebox where it had been grooved by corrosion. The two agreed that the explosion had been brought about by excessive steam pressure, and the inspecting officer believed he knew why the pressure had been so high: 'From the evidence given to me there appears to be no doubt as to the cause of the explosion, viz., that the lever of the safety valve had been pressed or wedged down, where it passed through the hole in the weather-board of the engine. It appears that some of the drivers on the Dublin Wicklow and Wexford have been corrected for such practices more than once, and that the unfortunate engine-driver who was killed had been amongst the number and was known to do it. Another man had seen the safety valve lever wedged down and covered with waste on a previous occasion, when the pressure gauge was showing 160 lbs. No steam was escaping from the safety valves when the engine was waiting at Bray station attached to the 9 a.m. train on the morning of the accident. No complaint had ever been made of the engine. The driver that

[1286] Ibid.

[1287] Gazette Bulletin, Williamsport, Pennsylvania, United States, 24 April 1872

[1288] Wikipedia - Liste des accidentes ferroviaires en France au XIX siècle

was killed had been heard to say that it was a good engine, but that it required to be worked with about 20 lbs more pressure than the 105 lbs to the inch.'

Lt-Col Rich concluded his report by suggesting, again, that boilers should be periodically tested to twice their working pressure, that locomotive superintendents should examine all boiler manhole covers for lack of strength, and that the levers of safety valves should not pass through the weatherboard covers and the valves should be arranged so that drivers could not fasten them down.[1289]

The 4 p.m. freight train from Mestre, Italy, had just left that station for Padua on about 11th February 1873 when the boiler of its engine exploded, tearing it to pieces and sending its forward half with four wheels 500 metres along the line, while the tender and waggons were sent in the opposite direction and overturned. The fireman was blown up into the air, striking and snapping the telegraph wires, described a semi-circle and fell to the ground dead. What became of the driver was not stated in the newspaper report.[1290]

The Taff Vale Railway had a boiler explosion at its Cardiff Dock, Wales terminus on 15th July 1873. Engine No 14 blew up at 3.35 a.m. injuring the driver and fireman, a brakeman and a night time-keeper, all of whom were on the engine at the time.

Colonel Rich of the Board of Trade found that the engine, which dated from 1864, had had a number of repairs at different times, and in October 1871 the under plate of the carrel of the boiler next to the firebox had been found to be pitted, and a small patch, 21 in X 9 in, called a liner plate, had been rivetted to the bottom of the boiler plate to arrest the further progress of the pitting. The inspecting officer commented that: 'The method adopted in putting on the patch or liner plate, reduced the strength of the bottom plate of the boiler, which was already defective, about a half, and there can be no doubt that it exploded from the weakness created by this work. It is a mistake to patch locomotive boilers.'[1291]

The 11 a.m. goods train from Geelong to Ballarat, consisting of an engine and seven waggons had just left the station at Geelong, Victoria, at 10.45 a.m. on 20th August 1873 when the boiler of its engine gave way, and Driver Augustus Pazanan, described by the newspaper as an elderly foreigner, and Fireman T. Macnamara were fortunate not to have been killed outright, though they were injured. Engine No. 23 was badly damaged, and the leading waggon was overturned whereupon the following waggons piled up against it.[1292]

The Victorian Railways locomotive overseer, William Meikle, examined the engine the following day and found that an upper part of the outer casing of the firebox had blown out 'and is torn through the first or top row of stayholes on the one side and through the rivet holes on the other. The fracture through the stayholes shows in the most

[1289] Board of Trade report

[1290] The Times, London, England, 15th February 1873, reproduced from the Swiss Times

[1291] Board of Trade report

[1292] The Ballarat Star, Ballarat, Victoria, 21st August 1873

unmistakeable manner that the plate at that part was never sound, being laminated, or in two leaves, instead of one solid plate. This defect would arise from cinder or dirt being in the iron at time of manufacture, thereby preventing it from being welded into one piece, and most probably affected the quality of the other portions of that part. Such a defect it would be difficult to detect after the plate was finally rolled; and when the boiler was last repaired no sign of weakness was manifest at that place. Whether the ordinary working pressure of 130 lb to the inch was sufficient to rupture this plate is a matter of doubt, and it will not be easy to get evidence on the point as the safety valves were blown out of their seats, and could not be found last night, the seats themselves and the portion of the boiler in which they are being uninjured.'[1293]

Two days after the Victoria accident, on 22nd August 1873, the broad-gauge saddle tank engine *Hebe* of the South Devon Railway arrived at Brent, England, station at 4.29 p.m. with a down goods train of 17 waggons and two vans. This was the 4 a.m. stopping goods from Exeter, which had actually not started until 6.27 a.m. and calling at all stations had taken 10 hours to cover the 35 miles to Brent! The train had been standing five or six minutes when the engine blew up. The driver, fireman and a boy standing on a bank beside the engine were scalded but there were no fatalities.

Colonel Rich found that the engine had a water gauge, pump and injector but no pressure gauge. The driver and fireman said that the injector was not working due to the split pin of the wheel having broken, but that they had used the pump to keep the boiler filled with water. The inspecting officer also noted that the hole cut in the barrel of the boiler for the dome had not been strengthened by an angle iron ring and recommended that when a hole was cut in a boiler for a dome or manhole cover it should be, but nevertheless attributed the explosion to the failure of the driver and fireman to keep the boiler filled with water while climbing Rattery bank, between Totnes and Brent, and the firebox consequently overheating, causing the boiler to explode when the injector was got to work again and water was added at Brent.[1294] Christian Hewison states in his book that Colonel Rich was probably wrong, and that it was a weakness in the boiler caused by the rivets securing the dome being pitched too close together which had caused the explosion.[1295]

An accident occurred on the Great Western Railway's Cwmannan colliery branch near Aberdare, South Wales, on 22nd May 1874. A train of 66 loaded coal waggons was descending the branch behind engine No. 1072 when it became divided on a short level stretch due to the fracture of a "D" shackle and 15 waggons were left behind. The head guard signalled for the driver to proceed with the leading 51 waggons to Dare Junction, with the idea of leaving them there and then returning for the other fifteen, but in the end the alternative plan of the front of the train setting back to re-couple to the detached portion was adopted. The front of the train was duly set back. Stopped close to the detached waggons and the guard and brakesman were walking back from the engine with a new shackle when its boiler exploded, killing the driver and fireman and injuring two

[1293] The Argus, Melbourne, Victoria 22nd August 1873

[1294] Board of Trade report

[1295] Locomotive Boiler Explosions by Christian Hewison pgs 82-3

other people, one of them a young telegraph clerk who lost an eye, the other the engineman of a stationary engine who was only slightly hurt.

Colonel W. Yolland was somewhat at a loss in finding a cause for the explosion but concluded that the engine crew had let the water in the boiler get too low and uncovered the crown sheet, which then overheated, and the explosion happened when there was a sudden surge of water as the train was set back. 'I think this is far more likely to have happened than that the driver and fireman had either of them tampered with the safety valves.' He did, however, recommend that the boilers should either be strengthened or worked at a lower pressure and that the safety valves should be moved so as to be inaccessible to the driver and fireman so that they could not tamper with them.[1296] Christian Hewison points out that William Dean, the Great Western assistant locomotive superintendent, had considered that the firebox crown had collapsed due to over-pressure and over-heat, that the safety valves and levers connected with them had been blown off the boiler and that what really happened was probably that the driver had wedged the safety vales shut before pushing the 51 waggons up a 1 in 300 gradient and had then forgotten to remove the wedge when the train had come to a stand.[1297]

The boiler of a Lehigh Valley Railroad engine attached to a gravel train exploded three miles west of Bound Brook, New Jersey, United States, at 7 a.m. on the morning of 1st November 1875, killing Engineer Thomas Capering, and Brakemen Abraham King and William Thompson, and injuring 'about twenty others, many fatally.' All of the waggons of the train were thrown off the track by the force of the blast.[1298]

On the afternoon of 11th February 1876 the boiler of the locomotive of a Görlitz-Zittau passenger train exploded between Ostritz and Hirschfelde in Saxony, Germany.[1299]

A train carrying Glasgow and South Western Railway workmen from Kilmarnock to the permanent way workshops at Irvine was in a bad accident at Springhill Junction, Scotland, on 28th March 1876, when the boiler of its engine exploded at about 5.45 a.m. The results of the explosion were very serious; the driver, fireman and a fitter in the van behind the engine were killed outright, the guard in the van died the same afternoon six other men in the van were injured, three of them seriously, and all four occupants of a carriage behind the van, the only other vehicle in the train, were also hurt. The casualties in the van and carriage were caused by the engine having been lifted into the air by the explosion and then falling back onto them.

Lt-Col C.S. Hutchinson found that boiler plates were in good condition and concluded that the accident was due to a shortage of water and overheating of the crown sheet,[1300] a conclusion which Christian Hewison finds does not correspond with the evidence as detailed in the Board of Trade report, and Hewison writes that 'In fact, the crown sheet

---

1296 Board of Trade report

1297 Locomotive Boiler Explosions by Christian Hewison pgs 83-5

1298 The Burlington Weekly Hawk-Eye, Burlington, Iowa, United States, 4th November 1875

1299 Eisenbahnunfälle in Sachsen pg 53

1300 Board of Trade report

broke away from the firebox doorplate along the riveted seam and was then wrenched off the firebox sides as, undistorted and with all the girders above it still attached, it was bent down until it almost reached the tubeplate. It is most likely that a crack had developed along the top horizontal flange of the doorplate and that the steam pressure broke the copper at this place; such cracks do not as a rule appear overnight and regular periodical inspection would have detected signs of the fault in time.'[1301]

The officials of the Olean Bradford and Warren Railroad decided to have a trial run with a new engine just delivered from Gibbs and Sterrett's shop at Titusville and the locomotive had reached Babcock's Mills, Pennsylvania, United States, at about 7 p.m. on the evening of Monday 27th January 1879 when its boiler exploded. The blast killed engineer John Vaughan and brakeman Jackson Addis outright and by midnight that day fireman Michael Halloran, conductor George Graham (or Groshen), assistant superintendent Charles Shepperd and a man named Thomas Luby (or Lutry) had all died as well. Superintendent N.N. Allen and the engine manufacturer, Mr Sterrett, were also riding on the engine but escaped injury while three other men, who presumably were not riding on the engine, George Petersen, Al Gartside and Oscar Schett were all seriously injured.[1302 1303]

The *Ohio Democrat* wrote that 'It is presumed that the bottom of the exploded boiler leaked, and that the leak was out of sight, and as the water fell below the flues it was thrown into a spheroidal condition which, by sudden evaporation, caused the explosion.'[1304] The *Marion Daily Star* theorised that, 'It is supposed the explosion was caused by water being low in the boiler, and the fireman turning in cold water while the boiler was heated.'[1305]

On Saturday 27th September 1879 the 2.05 p.m. London Brighton and South Coast Railway up passenger train from Hastings to London started six minutes late and arrived at Lewes, England, at 3.04 p.m. It had been stopped there about ten minutes, had taken water and was just moving off again when the boiler blew up, killing Driver William Rookwood and seriously injuring Fireman John Wheatley and Guard Alexander Frazer. None of the passengers was injured.[1306]

Colonel F.H. Rich found that the explosion had resulted from the breaking of the left side of the firebox, which was made from copper plates $\frac{7}{16}$ in thick. At the point where the plate had given way, its thickness had been reduced by wear to less than $\frac{1}{16}$ in. It had broken under the pressure of the steam and been torn away from the rivets. Thomas Chilton, foreman boiler maker at New Cross, stated that he had examined the copper firebox 12 days before the accident and had found the plate to be 'a bare eighth in

1301 Locomotive Boiler Explosions by Christian Hewison pgs 87-9
1302 Gazette Bulletin, Williamsport, Pennsylvania, United States, 28th January 1879
1303 Evening Gazette, Port Jervis, New York, United States, 30th January 1879
1304 The Ohio Democrat, New Philadelphia, Ohio, United States, 6th February 1879
1305 The Marion Daily Star, Marion, Ohio, United States, 31st January 1879
1306 The Times, London, England, 29th September 1879

thickness' but had declared the firebox to be in good condition. Colonel Rich wrote at the conclusion to his report that, 'The engine and fire-box were examined at New Cross 12 days before it exploded, when it was sent into the shops in consequence of the tubes leaking. The examination was made by an experienced boiler-maker, who had been a considerable time in the Company's service. This man stated that when he examined the fire-box he thought that the plate of the fire-box at the left side was "a bare eighth" in thickness. I do not think it can have been $\frac{1}{16}$ thick at this time along the line where it gave way. He thought that the plate at the right side of the fire-box was a full eighth thick. There can be no doubt that under these circumstances he made a serious mistake in not reporting the state of the engine to his superiors, as no engine should be allowed to work when the fire-box plates are in any part less than $\frac{3}{16}$ in thick'[1307]

A passenger train from St Louis on the Ohio and Mississippi Railroad was approaching Caseyville, Illinois, United States on 17th September 1880 drawn by an engine which had just come out of the shops after a complete overhaul and partial rebuild. Everything went normally until the engineer attempted to shut off steam and slow down for the station when he was suddenly thrown back into the tender and partially stunned. He recovered himself and sprang forward to stop the engine and was astonished to find nothing but the firebox, the front of the cab and the boiler tubes. The train was still running, but soon stopped. The force of a boiler explosion had been all directed forward and upward so that the wheels had remained on the track and the momentum of the train had kept it going. Neither the engineer nor anyone else on the train heard the noise of the explosion. The conductor had applied the brakes when the train had failed to stop at the station.[1308]

A heavy Chicago and Alton Railroad freight train suffered a boiler explosion at Joliet, Illinois, United States on 30th December 1880. The engineer and fireman escaped but the head brakeman was scalded, probably fatally so. The explosion was blamed on low water in the boiler due to the pipes between the tank and the engine having frozen, a possibly unique explanation for a boiler explosion.[1309]

[1307] Board of Trade report

[1308] The Locomotive, November 1880

[1309] Ibid, January 1881

CHAPTER 17

VANDALISM

'Accidents' due to vandalism and sabotage have been a problem for the railways from the beginning. They are obviously very closely related, inasmuch as both represent deliberate attempts to cause trains to crash, as opposed to accidents caused through non-deliberate carelessness or negligence or purely accidentally. Here they will, however, be treated separately, as they are slightly different. This separation is perhaps a little artificial, and if the perpetrator of a given act is not caught or his or her motives not sufficiently obvious it may be difficult to decide which it was, but it has nevertheless been felt that the distinction to be worth making.

Vandalism is usually, though not invariably, carried out by either children or young adults and its results are most commonly, though not always, less serious than those following from acts of sabotage. The motives for train accidents resulting from vandalism are usually either simple curiosity, in the case of younger children – 'I wanted to see what would happen' or in the case of older children and teenagers a reflection of general anti-social behaviour; so that such individuals who attempt to derail trains also carry out other acts of 'mindless vandalism,' and perhaps also commit burglaries or carry out arson attacks.

Sabotage, by contrast, is usually undertaken by adults, who employ methods more likely to cause death and injury, so that while a vandal might for example just place an object upon the track or jam a stone in a set of points, a saboteur would be more likely to employ methods such as blowing up a section of track using explosives.

Unrecorded acts of vandalism were doubtless carried out by mischievous children from time to time on the early waggonways. The earliest case which has come to light is strangely enough one of those where it is difficult to be certain whether it was a case of vandalism or sabotage, but it will be treated, in the absence of evidence to the contrary, as the former. It happened on the pioneer Liverpool and Manchester Railway, a fortnight before the line was opened to the public, on 1st September 1830, at Broad Green, England. A temporary crossover had been here laid between the running lines and an unknown person had changed and fastened one of the tongues of the points the wrong way. A contractor's train coming from Liverpool came off the track at the points and a contractor who was riding on the engine jumped off as it became derailed, only to be run down and killed by the following waggons. His surname chanced to be Stephenson, and an erroneous report circulated that the famous George Stephenson, the engineer of the railway, had lost his life, but this turned out of course not to be the case. The Liverpool and Manchester directors immediately offered a reward of 200 guineas – a very large sum in 1830 – for the conviction of the perpetrator.[1310]

The *Liverpool Albion newspaper* was keen to point out that it was unjust to infer, from an accident due to malignity or inattention, that the new form of transport was peculiarly unsafe and, furthermore, that accidents were perpetually occurring on the public roads, not to mention the fact that the railway would carry its passengers a given distance in half the time, and charge considerably less for doing so, than the stagecoaches. It also informed its readers that once the line was open persons 'would be stationed at different points of the road, to watch it during the day; to keep the rails free from obstructions of every kind; and to exercise a vigilant inspection over every inch of road committed their care.'[1311]

Also slightly difficult to categorise as vandalism or sabotage but which, although it was the act of an adult, seems to come closer to vandalism, was what a local newspaper described as 'a most shameful outrage' on the line of the Eastern Counties Railway at Brentwood, England, on 22nd September 1840. Shortly after midnight a ballast engine had drawn up at Brentwood station, and the driver and stoker (fireman) had left it unattended for a short period while they went to get their supper, prior to raking its fire out for the night. A local man names Thorogood, who had been drinking with his companions, and was heavily intoxicated, decided, as he put it, "on having a bit of a lark." He got into the station by climbing over a fence, mounted the locomotive, turned on the steam, then jumped off the engine and ran away. At this time, Brentwood was the temporary terminus of the line, which only continued for a quarter mile beyond the station. The engine fortunately ran away towards the end of the line, rather than towards London, in which direction it might have done more harm, especially as there were large numbers of men working on the line between Brentwood and Romford during the night. It collided with a horse-box and a waggon loaded with coke stabled at the end of the track, wrecking them, and ran on until it ran off the end of the rails and embedded itself in the earth and quicksand there. The drunken Thorogood was found in the neighbourhood, admitted what he had done, and was given into the custody of the police, to be doubtless prosecuted by the railway company. The damage he had caused was estimated at between £40 and £50.[1312]

Vandalism has often involved the placing of obstructions on the line, which on occasions would derail a train but were more often swept aside or crushed by the weight of the engine. Thus on 19th September 1841 a 'diabolical attempt' was made to interfere with the mail train from Arbroath to Dundee on the Dundee and Arbroath Railway at Barry, Scotland. A wooden stile or pailing was placed on the rails, together with two large stones slightly further northwards. The day was very foggy, and the engine driver

[1310] The Times, London, England, 4th September 1830

[1311] Ibid, 8th September 1830

[1312] Ibid, 26th September 1840

did not see the obstruction until the last moment, when it was too late to pull up. The train hit the obstructions but was not derailed. Three persons were subsequently apprehended (whether they were children or adults, the report does not say) and were under examination by the Procurator Fiscal.[1313]

Again, on 29th December 1841 'A most diabolical attempt was made…to cause a further accident (to that in the Sonning cutting) on the Great Western line.' The down mail train from London was near Bathampton, England, three miles short of Bath, when the incident occurred. 'Some monster' had placed two large stones on the track, one on each rail. The train hit them. and the broad-gauge engine crushed one of them, while the other was forced to one side by the swords or guards on the front of the locomotive., both of which were bent. The train was stopped and the driver picked up the intact stone and took it with him to show his superiors in Bristol. It weighed about 40lb.[1314]

Between 6 p.m. and 8 p.m. the engine of a train was thrown off the line and overturned near Holmes station, England, on the Midland Railway, because someone had wedged an iron chain and a piece of stone into a set of facing points. The driver and fireman were slightly hurt. The Midland Company immediately offered a £100 guineas reward for the conviction of the offender.[1315]

On 15th February 1850 a large stone was placed across the rails at Chalons, France, 'on the Chalons and Dijon line' (of the Chemins de fer de Paris à Lyon) but the engine just passed over it and continued its journey without mishap.[1316]

What was allegedly a very serious act of vandalism, which resulted in the death of five people, occurred between Falmer and Lewes, England, on the Brighton and Lewes branch of the London Brighton and South Coast Railway. On 6th June 1851, the 12.05 p.m. departure from Brighton was running down the 1 in 88 incline between there and Lewes and passing through a chalk cutting around a left-hand curve, when it struck a wooden sleeper resting upon the near rail. The impact threw the engine off the line and it ran onwards, crossing the opposite running line and striking and demolishing the parapet wall of a road underbridge. Engine, tender and two of the four carriages of the train then fell into the road. One of these two carriages was an empty second-class one, but three of the four passengers in the other third-class carriage were killed, as was Fireman Chase, and Driver Jackson died from his injuries two days later. Some had fortunate escapes; the other passenger in the third-class carriage scrambled under the seat as the accident occurred and escaped with minor injuries while some Lewes tradesmen missed the train by a couple of minutes and were greatly upset at the time because they had to hire a fly (a

1313 Ibid, 24th September 1841

1314 Ibid, c1st January 1842

1315 Board of Trade report

1316 The Times, London, England, c18th February 1850

carriage put out to hire) to take them home.[1317]

The sleeper which had caused the derailment had been left by the side of the line by the platelayers, along with two others, about two months previously, to be used when they were needed. The sleepers had been stacked in their proper place when a platelayer passed at 8 that morning, and the 11.15 a.m. train from Brighton had passed the spot unhindered, so it was clear that someone must have moved it onto the line after that time.[1318] Suspicion fell on a ten year old boy named James Boakes, the son of a labourer, who lived in a house adjoining the line and close to the point of derailment. The boy was interrogated at his home by Mr William Acton, railway police superintendent, accompanied by Detective Sergeant Langley, of the Metropolitan police, who it appeared tried to force as confession out of the child until the child's father intervened and told the men to leave the house.[1319] Subsequently, Jimmy Boakes had to give evidence at two inquests, one on the passengers and the fireman at Lewes and, because he had died later and elsewhere, a second inquest on the driver. He admitted not wishing to go to Ashcombe on an errand for his mother until the train involved in the accident had passed as he wanted to see its new engine, and to having been in the potato garden adjoining the line before the crash, but stoutly denied throughout having put the sleeper on the line or having been on the line at all. In the end, both juries returned verdicts that the accident had been caused by the sleeper being placed on the line by a person or persons unknown. The Lewes jury criticised the speed of the train down the bank, and both juries were critical of the fact that the train had been running tender first. In an editorial published after the verdict of the first jury, *The Times* was of the opinion that Boakes was a convenient scapegoat for the accident.[1320]

It has been suggested more recently that Jimmy Boakes was in fact the guilty party, and that he had been playing see-saw on the line with the daughter of a neighbour (Guard Erle of the derailed train had spoken of having previously seen children playing see-saw by the line). If two children were involved it would explain more easily how they had been strong enough to lift a 9 ft sleeper weighing a hundredweight. Mrs Boakes and her neighbour, Mrs Weeks, the mother of the little girl involved, might have conspired to conceal the truth to prevent their children from suffering the harsh penalties meted out to juvenile offenders at the time. By a strange coincidence, Jimmy Boakes died from unclear causes just a year after the accident. There was a local rumour that he was 'not quite right in the head,' though he had acquitted himself well enough in court, apart from being frightened.[1321]

The child was not mentioned in the subsequent Board of Trade inquiry, but Captain Laffan was critical of sleepers being left in close proximity to the line in secluded spots giving an opportunity to 'persons mischievously disposed,' of the imprudent speed of the train and of the fact that the engine had been running tender first. He considered that the running of engines thus was 'most objectionable.' The engine driver could, while running tender first, not keep a proper look out without neglecting his other duties, fine dust blown up from the coke in the tender at any speed make his job even more difficult and,

1317 Ibid, 9th June 1851

1318 Board of Trade report

1319 The Times, London, England, 14th June 1851

1320 Ibid. 16th 17th 18th and 19th June 1851

1321 The Railway Magazine 1952, pgs. 799-801 and 846

with regard to the engine involved in the accident, the weight on the trailing wheels was less than half of that on the leading wheels, making a derailment all the more likely.[1322]

On about 27th June 1857 the night mail train on the North Devon Railway collided with a wheelbarrow which had been placed on the line at Coleford, England without any serious consequences other than the gardening equipment being smashed to pieces and its bits flying about the head of the engine driver, without causing him any injury. Sir J Duckworth took a dim view of the affair at the Devon Quarter Sessions a few days later and James Balsdon, aged 12, who had pleaded "Guilty," was sentenced to two months' imprisonment, presumably to be served in Exeter Prison.[1323] This was hardly, one would have thought, a suitable place for a 12-year-old boy, but then 'The past is a foreign country, they do things differently there.'[1324]

The death of an engine driver at Kincraig Siding, Scotland, on the Brechin branch of the Scottish North Eastern Railway, on Monday 27th January 1862, was the unfortunate result of children playing around with a set of points. Kincraig was a little used farm siding, which was sometimes unused for weeks or even months. On Saturday 25th January, however, a pointsman came from Brechin station to remove two empty waggons from it at 5 p.m. He said that he had padlocked the points again after removing the waggons and they must have lain for the main line that evening for a passenger train passed over them in each direction. No trains used the branch on the Sunday, but two 11-year-old boys and an 8-year-old girl admitted to the Board of Trade inspector, Captain Tyler, that they had played with the points. The first train on Monday morning left Brechin along the single-track branch at 7.45 a.m. As it neared the siding, a goods clerk who was riding on the engine observed that the siding points were wrong. He shouted a warning and he and the fireman jumped and survived, though the fireman was much hurt. The driver stuck to his post and was killed when the engine rammed the buffers at the end of the siding and finished up in a field. Only four or five of the 60 or 70 passengers on the train were injured.

Captain Henry Tyler was of the opinion that the Brechin porter must have failed to secure the padlock properly on the Saturday evening and he also found fault with the foreman platelayer, who should he said have walked his length before the first train. Had he done so that morning he would have noticed the points standing wrong, He suggested that the best remedy to prevent a repetition of the accident was that which the company were intending to adopt anyway, namely, to do away with the siding and its points altogether.[1325]

Two 'diabolical outrages' were committed in three days in January and February 1863. In the first Joseph Buckland, aged 16, and Robert Churchill, aged not stated, were charged before Mr Elliott, at Dorchester, England, with having placed three chairs on the

---

[1322] Board of Trade report

[1323] The Times, London, England, 3rd July 1857

[1324] Opening sentence of 'The Go Between' by L.P. Hartley

[1325] Board of Trade report

line two miles from Dorchester London and South Western Railway station. A train driver, seeing the obstructions on the rails, put on full steam and the weight of his engine smashed them. The prisoners admitted the offence and were committed to the assizes for trial.[1326] Then on 3rd February the driver of an up train, arriving at London Kings Cross station, reported that his train had run into some obstructions on the line between Seven Sisters Road and Holloway Road stations, England. A telegraph message led to a porter examining the line and finding a piece of wood and several iron chairs placed on the up line and a fire bar on the down line. The Great Northern Railway offered a £50 reward for the apprehension of the culprits.[1327]

On 23rd August 1863 a Sète to Tarascon express struck an obstruction on the line close to the start of a viaduct over the River Rhone at Beaucaire, France, at 12.30 and was derailed. The engine, baggage car and three coaches of the train collided with the parapet of the viaduct and fell over into a ditch. The engine driver, brakesman and four passengers lost their lives and ten others were injured.[1328]

A short time afterwards some children, apparently undeterred by the serious accident that they had just caused, were caught placing new obstructions on the line. One of them, a nine-year-old, was sent to a reformatory.[1329]

On the arrival on Wednesday 30th November 1864 of the 11.15 p.m. (also reported as having been the 10.20 p.m.) train from Melbourne to Sandhurst, Victoria (Sandhurst was named Bendigo until 1854 and again from 1891) at its end station the guard told the stationmaster that his train had run into some obstructions on the line as it was running downhill at some speed on the Sandhurst side of Elphinstone. On examination of the train, one of the guard irons at the front of the engine was found to have been completely torn away and some of the passengers spoke of a shock that had thrown them out of their seats.[1330] A suggestion that the iron chair which had apparently caused the near derailment had accidentally fallen from a train was dismissed by a local newspaper which pointed out that no train laden with material had passed the spot where the incident had occurred, near the proposed Chewton station, between the passing of the previous passenger train and that which had hit the obstruction. To offer sufficient resistance to cause such damage to the guard iron the chair must also have been fixed to the track.[1331]

On Sunday 1st October 1865 two boys placed stones and other obstructions on the track near Garland, Pennsylvania, United States. A Pittsburgh and Erie Railroad freight train was derailed, throwing its engine and several waggons down the embankment. The engineer was killed outright, Conductor Breckentidge was fatally injured and the fireman

---

[1326] The Times, London, England, 2nd February 1863

[1327] Ibid. 4th February 1863

[1328] Wikipedia - 'Liste des accidents ferroviaires en France au XIX siècle'- La Presse 26th August 1863

[1329] Ibid. La Presse 15th November 1863

[1330] The Age, Melbourne, Victoria, 2nd December 1864

[1331] Mount Alexander Mail, Mount Alexander, Victoria, 1st and 3rd December 1864

maimed.[1332]

In 1867 a banking engine was stabled at Wigan, England, to assist northbound trains up the hill towards Preston on the London and North Western Railway main line. Late on the night of Saturday 15th June it banked a coal train to Coppull and returned to Wigan. Finding that his engine's boiler was getting low on water, its driver, William Green, then ran the it up the line to Springs Branch, a couple of miles to the south, to refill it and get back to Wigan before the down limited mail, due to pass Wigan just before 1 a.m. on Sunday morning, 16th June. He had left Springs Branch on the return journey when just outside of Wigan the engine encountered an obstruction on the line, left the rails and continued along the six foot (the space between the tracks) where some new rails had been placed ready for relaying the track. The driver and fireman stuck to their posts and applied steam and managed to get the engine back on the line. Steps were taken to warn the down postal train and going back along the track the driver and fireman of the engine found that two of the new rails had been laid across the track. The down limited mail then slowly approached them, its driver having been cautioned at Springs Branch. The driver of the down mail having been told by Green what had happened said that his train too had run over something on the track. It was discovered that this was two chair bottoms had been placed on the track after the banker had gone by in a second attempt to derail the mail train. The banking engine was badly damaged, and the limited mail was delayed 45 minutes. Two men were seen near the scene but escaped. *The Times* added that it was expected that 'a large reward will be offered for the perpetrators of this outrage.'[1333]

A third attempt to wreck a train was made on Sunday 22nd of the same month which led to the arrest of one of the offenders. Between 10 and 11 on the evening of that day, a navvy named Michael Flatly was walking around the same area with a second man and a workman named Herne. Flatly and the unknown man then decided to have a "good jump" and placed a heavy sleeper across the up line and laid two more across it at right angles to keep them in place. Herne remonstrated with his two companions and removed the obstructions but the other two threatened him with personal violence if he interfered and he left the scene as they were putting the sleepers back onto the line, but was shortly overtaken by Flatly, who informed him that they had not only replaced the three sleepers but had added a fourth one for good measure. At the first opportunity that he had, Herne wrote a message to the Wigan stationmaster informing him that an obstruction had been placed on the line between Wigan and St Helens, but the note was too vague to be acted upon. The up limited mail, due through Wigan at about 11.30 p.m., passed without stopping at its normal time, and must have hit the obstructions, but it passed on unharmed. The broken sleepers were discovered by the lengthman when he inspected the line between Wigan and Springs Branch in the morning. Flatly was apprehended by the police on the following Thursday but his assistant had left the neighbourhood.[1334] This series of attempts possibly came closer to sabotage than vandalism.

A North British Railway goods train left Stirling, Scotland, at 4 a.m. on 5th October

[1332] Titusville Morning Herald, Pennsylvania, United States, 3rd October 1865

[1333] The Times, London, England, 18th June 1867

[1334] Ibid, 28th September 1867

1867 for Larbert, using that company's running powers over the Caledonian Railway. It travelled as far as a place named Torwood, when it ran into an obstruction which threw 15 of the waggons off of the line, and nine of them, one empty and eight loaded with baled goods, paper and coal, fell down a 20ft embankment. The engine crew and guard escaped unhurt. The obstruction was found to have been a screw jack which the surfacemen had failed to lock away in a box provided for the purpose. Various investigations were made afterwards but no clue could be found as to the person or persons responsible.[1335] Another attempt on a train was made on 13th April 1868 in the same area, this time near Larbert on the Caledonian's Denny branch. A number of chairs were placed on the rails and a set of facing points was left half open. The driver of a northbound train saw the obstructions and was able to reduce speed but the tender of his engine, which was running tender first, was derailed and its leading axle broken, two cattle trucks were derailed, and considerable damage was done to the permanent way. No one was hurt but the line was blocked for several hours, during which time passengers between Larbert and Denny had to walk the 3½ miles between the places.[1336]

An act of vandalism near Hanwell, England, on about 18th June 1873 prompted an angry letter to *The Times*, its writer demanding that Parliament permit judges to order the flogging of offenders in such instances, in addition to any other punishments they might see fit to impose. The accident was not directly reported in *The Times*, though it got into a local newspaper, but a Great Western Railway up fast train apparently struck a rail which had been laid across the track near Hanwell at a point where the railway ran on a high embankment. Fortunately, the guard of the engine struck the rail with sufficient force to break it in two and throw it clear of the line. The writer of the letter, one Edmund Lane, suggested that had the attempt succeeded 'every soul in the train would have been destroyed' and describer the perpetrators as 'demons in human shape.'[1337]

In an incident reminiscent of that at Brentwood, England, in September 1840 on about 29th March 1874 two girls from Detroit, Michigan, United States were arrested after they had climbed aboard a locomotive, opened its regulator to set it in motion and then jumped off. The engine ran away and collided with a freight train, causing $700 worth of damage. The newspaper report did not indicate if the girls had been drunk at the time.[1338]

A very serious act of vandalism occurred in Quebec, Canada, on 28th September 1875. The track of the Richelieu Drummond and Arthabaska part of the South Eastern Railway was being upgraded at the end of September 1875 with new rails being put down in place of the old ones, and one or two work trains would be sent out from Sorel in the morning, returning there in the evening. The trains consisted of flatcars on which up to 100 labourers would ride. One the outward journey the engine would pull the cars but, presumably because there was no turntable or run-round loop at the far end of the line, on the return journey the locomotive propelled the flatcars. At about 5.50 p.m. on the day of

[1335] Ibid, 8th October 1867
[1336] Ibid, 15th April 1868
[1337] Ibid, 24th June 1873
[1338] Elyria Independent Democrat, Elyria, Ohio, United States, 1st April 1874

the accident, a work train set off from St David for the return trip to Sorel with about 75 labourers riding on six flatcars. Riding on the engine were Engineer Henri Gaudette and his fireman while Conductor Nathan Swan, Superintendent Oscar Phelps, and Louis Senénecal, the contractor in charge of the relaying operations, were on the tender. Riding on the leading flatcar, with a lantern to warn the engineer of any hazards, was Brakeman Edouard Lachapelle. The train was running at between 8 and 12 miles an hour.[1339]

The train passed through Yamaska, 10 miles south-east of Sorel. A couple of miles further on it crossed the Rivière David. Shortly afterwards the brakeman glimpsed something on the track. He shouted to his engineer, waved his lantern violently and then jumped, followed by most of the workmen on the front car. The engineer, reacting to the warning from the brakeman, shut off steam and applied his brake, assuming that there was a cow on the line. Instead, the leading flatcar struck a pair of timbers which had been laid across the rails, overturned and became a barrier against which the next three flatcars slammed and overturned as well. The fifth car was derailed but did not overturn and the last car, tender and engine remained on the rails. The labourers on the overturned cars were described as having been 'fearfully crushed and mangled.' Ten or eleven of them were killed outright and another 25 more or less injured. The uninjured labourers tore up fences and lit bonfires to provide some light, the sun having set by the time of the accident, which was about 6.45 p.m. They then set about releasing their less fortunate comrades from the capsized cars. Assessing the situation, Engineer Gaudette steamed back to Yamaska and soon returned with a doctor, a priest, and some tools. The track was quickly cleared, the injured were loaded onto one of the flatcars and they were taken on to Sorel, arriving there at 11.15 p.m.

An inquest into the accident conducted by Ernest Turcotte heard evidence as to the speed and make-up of the train and the condition of the track. An inspector testified that he had passed over the line at 5.30 p.m. on a hand car and all had been in order. More to the point was evidence that two brothers, Denis and Basil Pepin, aged 12 and 14, had admitted placing coins, nails and wooden wedges upon the track on both the day of the accident and the previous day. They refused to admit having placed the timbers on the line but the suspicion against them was strong. They were arrested but as nothing could be proved against them, they were subsequently discharged. The coroner's jury returned a verdict that the railway company was in no way to blame for the accident, which had been caused by the action of parties unknown. No comment was apparently made as to the wisdom or otherwise of propelling the flatcars rather than pulling them, presumably because it was accepted that this had been unavoidable.[1340] [1341]

In one of several such incidents around this time in England on 11th April 1877 a London and South Western Railway train on what was described as the Salisbury branch involved a mail train which had left Exeter at 5 p.m. for Portsmouth. It ran into an obstruction while travelling at full speed a mile and a half short of its stop at Botley, at a point where a footpath crossed the line. A sleeper was found to have been laid across the

---

[1339] Wreck! By Hugh A. Halliday, pgs. 52-57

[1340] Ibid.

[1341] Lincoln Daily State Journal, Lincoln, Nebraska, United States, 30th September 1875

track. The engine guards were smashed but it kept to the rails.[1342]

The last down passenger train on the West Cornwall line was the target on the evening of 16th April 1878 when it struck a stone placed on the line near Avers Crossing, England, about half a mile east of Chacewater station. The lifeguard of the engine knocked the stone away, getting bent in the process, and the train passengers distinctly felt the shock. A porter sent back up the line from Chacewater discovered a large stone, about 36lb in weight, 20 inches long and 16 inches wide, placed on the line about 500 yards from a viaduct (probably Chacewater viaduct or possibly Blackwater viaduct).[1343] A couple of days later the Great Western Railway announced that it was offering a reward of £50 for the conviction of the offenders.[1344]

Many of the perpetrators of vandalism do not seem to have been caught but 10 year old James Gibbons, who lived at or near Bury St Edmunds, England, was less fortunate. At Ixworth Petty Sessions he was convicted of having attempted to upset a Great Eastern Railway train on 22nd May 1878 by placing a fishplate on the metals (the train broke the obstruction in two and escaped harm). The boy was sentenced to 14 days' imprisonment to be followed by five years in a reformatory.[1345]

Another attempt to upset a train near Wigan, England, was made on 8th September 1878, this time on the Lancashire and Yorkshire Railway line. As a cattle train was passing over a canal bridge half a mile from Wigan station its driver felt the engine pass over something, and it was found that a new iron chair, weighing nearly 30 lb, had been placed on the track. The same day two lads of about 10 years of age named John Kelly and John Clarke were arrested by a Detective Gorman, to appear before the magistrates the next day charged with committing the offence.[1346]

William Appleman, a young man of 22, committed an act of vandalism near Shohola station, Pennsylvania, United States, on the Erie Railroad, on 1st December 1878. He fastened a length of rail to the track, and then hid in bushes nearby for the arrival of train No. 5, an express. This train shortly appeared, and it was lucky that the leading wheel of the locomotive smashed the length of rail in two and the train was not derailed. Appleman was subsequently arrested and lodged in the county jail at Milford, Pennsylvania to await his appearance in court. This incident is of special interest inasmuch as after his arrest Appleman gave a detailed interview to a reporter from a local newspaper, the *Port Jervis Evening Gazette*. He told the reporter that he had had no intention whatever of doing anybody harm, but that as he was passing on his way home, he thought that it would be funny to see an engine throw an obstacle. He had been glad that the train had passed over safely, and 'would not have repeated the experiment for all

---

[1342] The Times, London, England, 13th April 1877

[1343] Ibid, 18th April 1878

[1344] Ibid, 20th April 1878

[1345] Ibid, 5th June 1878

[1346] Ibid, 9th September 1878

the wealth of the universe.' Asked if he had considered what could have happened, he replied that 'Yes, many times I have given the matter earnest thought and consideration and meditated on how easily a frightful accident might have occurred. But I did not commit the deed with any evil intention, I assure you.' Asked what he had done after the accident, he replied that he had gone home and spent a restless night. He thought that the court would probably send him to gaol for 10 years, but he might get a shorter sentence with good conduct.[1347] His motive sounds to be that one would expect from a boy of 8, not a man of 22. Clearly, he was very immature and probably of low intelligence.

[1347] Port Jervis Evening Gazette, Port Jervis, New York, United States, 7th December 1878

CHAPTER 18

SABOTAGE

Four distinct motives for sabotaging trains may be distinguished (after Robert B. Shaw):

(a) As an aid to intended robbery, either of the train passengers or to steal for example bullion being carried by the train;

(b) To satisfy a personal grudge against the railway company, such as by an ex-employee who feels that he has been wrongly dismissed;

(c) To hinder the operation of the railway, as for example during a strike;

(d) As an act of guerrilla warfare or terrorism, perhaps against an occupying enemy power. Sabotage against the railways was a feature of most of the nineteenth century conflicts from the American Civil War onwards.

The earliest cases of sabotage to have been recorded were on the London and Birmingham Railway during September 1838. Wooden logs were placed upon the line on several occasions in the neighbourhood of Wolverton, England, and one such obstruction derailed a train, though without causing any great damage to it. The railway company offered a reward of £10 (figure illegible in newspaper copy and possibly incorrect) for the apprehension of the offenders. It was supposed that the 'disgraceful species of outrage' was the work of former bargemen on the Grand Union Canal, many of whom had lost their jobs since Pickfords and the other great transport contractors had switched their business from the canal to the railway.[1348]

On 8th April 1839, the opening day of the Leipzig-Dresden line in Saxony, the *Saxonia,* the first locomotive to be constructed in the state, was diverted through a wrongly positioned set of facing points at Priestewitz station into the English-built *Adle*. This was anecdotally stated to have been an act of sabotage on the part of resentful

---

[1348] The Times, London, England, 1st October 1838

competitors of Johann Andreas Schubert, the engineer who had built the *Saxonia.*[1349] [1350]

It was eight years before the next act of sabotage. This took place on the Taff Vale Railway somewhere between Cardiff and Merthyr, Wales, at a location which is not entirely clear. There were in fact two attempts; in the first a set of points was interfered with, but this attempt was discovered by the railway servants before any accident happened. The second attempt, on 8th December 1847, was much more serious; a set of rails was removed at a point where the line made a sharp turn while running along the side of a steep valley. Fortunately, the driver of the last up passenger train saw that there was something wrong with the track and reduced speed, but his engine, tender and three carriages were derailed, overturning to the right against the side of the bank rather than to the left where they might have tumbled down the slope into the river. No motive for this act seems to have been discovered.[1351]

A gross over-reaction by an engine driver to a seemingly trivial wrong was the cause of the next act of sabotage. On the evening of 4th August 1848 the driver and fireman of the *Medusa* locomotive of the East Lancashire Railway were drinking in a beer-shop at Burnley, England. Cleaner John Rhodes, meanwhile, was ready to take the fire out of the engine and the fireman left the beer-shop and went to supervise him. Before the fire could be taken out, however, some person in authority ordered the two to go with the engine to Rose-green-tip, and they set off there with the fireman acting as driver and the cleaner as fireman. Driver Samuel Whittle then also returned from the beer-shop and finding that his engine had gone without him apparently became very angry (alcohol fuelled aggression perhaps), swore violently and determined to extract his revenge, and he fastened three waggon tail-boards, three iron furnace bars and eight iron picks to the track. When the engine returned, it was carrying, in addition to the fireman and cleaner, the Rev T.G. James of Burnley, the latter's brother from Liverpool and Mr Donaldson, the engineer of the line. Someone by the lineside having noticed the obstructions, the engine was signalled to stop but before it could do so it ran into them and was derailed. Having run some distance the locomotive regained the rails, but Cleaner Rhodes was thrown off the engine when it hit the obstructions and was then run over by it, suffering fractures to both thighs and having his right arm so badly injured that it had to be amputated. Driver Whittle was taken into custody and committed for trial by the magistrates.[1352]

The first act of politically motivated railway sabotage took place on 23rd September 1848 between Darmstadt, Hesse, and Heidelberg, Baden, on the Main-Neckar line. Sleeper fastenings were loosened and a section of track undermined. Two engines of an empty train were derailed and fell down an embankment and a railwayman was killed. Responsible for the derailment were participants in the Baden revolution, which attempted to form a republic in southern Germany under the sovereignty of the people.

1349 Eisenbahnunfälle in Sachsen by Claus Schlegel, pg. 41

1350 Wikipedia - Liste von Anschlägen im Schienenverkehr

1351 The Times, London, England, 13th December 1847

1352 Ibid, 15th August 1848

The revolution was finally crushed by the Prussians in July 1849.[1353] [1354]

The first of a number of acts of sabotage to occur in Ireland at various times during the nineteenth century took place near Trillick station on the Londonderry and Enniskillen Railway on 15th September 1854. The train involved consisted of nineteen carriages and a van, drawn by two engines. Travelling at very low speed, because the second engine was providing very little assistance, the train ran into three large stones which had been laid across the line. The two engines were thrown down an embankment and the leading carriage ended up diagonally across the rails, but a broken coupling meant that the remainder of the train did not leave the track. According to *The Times*, but not mentioned in the Board of Trade report, the fireman on the leading engine, named Mitchell was killed outright and Fireman Griffin on the second engine was fatally injured. The accident happened to what the Board of Trade report referred to as an excursion train, the evening return working of a train from Enniskillen to Derry, which was carrying, according to the initial report in *The Times,* a large party of Orangemen who had been on a reciprocal visit to their brethren at the latter place. A letter to the same newspaper some days later from the Earl of Enniskillen, who had himself been a passenger on the train, denied that all the people on the train had been Orangemen, stated that there had been neither ribands nor music and that the passengers had been Protestant clergy, gentry and local Enniskillen people celebrating the opening of the railway between the two places and the historical connection between the towns. That be as it may, the sabotage was clearly political in nature. *The Times* mentioned a rumour that the excursionists had burnt down the houses of several suspected parties in the neighbourhood and that the general opinion in the county of Tyrone was that the act was a 'deeply laid Popish conspiracy to send 700 or 800 Protestants to an untimely grave.'[1355] [1356] Seven men were arrested soon afterwards and accused of having caused the derailment but when their case came before the Tyrone Assizes in the summer of 1855 the Grand Jury found "No bill" and they were discharged.[1357]

At the end of April 1857, 150 conductors on the Baltimore and Ohio Railroad came out on strike over an apparently minor matter relating to the sealing of cars. As often seemed to be the case in the United States, the strike soon escalated into violence. Already on the 2nd May the Governor of Maryland had called out the military and the troops fired on strikers attempting to stop the special train carrying them, killing one striker and wounding others. The soldiers were then deployed escorting a freight train into Baltimore but at Ellicott's Mills, Maryland, just outside the city, the strikers had spiked a short 'T' rail to the track and the engine, tender and five cars were precipitated down the embankment, wounding two soldiers. The military then marched the remainder of the way into the city, leaving the disabled train in the charge of the Sheriff and his

---

[1353] Wikipedia – Liste von Anschlägen im Schienenverkehr

[1354] Wikipedia – Baden Revolution

[1355] The Times, London, England, 19th, 20th, 21st, 22nd, 23rd, 25th, 26th, and 27th September and 2nd, 6th,13th and 16th October 1854

[1356] Board of Trade report

[1357] Webpage - The Trillick Derailment 1854 by Desmond Fitzgerald

posse.[1358]

The American Civil War began on 12th April 1861 with the bombardment of Fort Sumter and the first act of sabotage against a train was committed only a fortnight later. Secessionists (Confederates) removed several sleepers near Annapolis, Maryland, United States, on 26th April and a train consisting of only a locomotive and one car left the tracks. A Colonel Vosburg was slightly hurt.[1359] On 2nd June the Confederates destroyed the trestle over Opequon Creek, just east of Martinsburg, West Virginia and then set 50 waggons alight and ran them off the destroyed structure.[1360] This action was not helpful to their cause as it left a large number of locomotives which they had captured, and which were sorely needed to work Rebel trains, stranded on the wrong side of the destroyed trestle.

One significant and two major acts of Civil War sabotage or probable sabotage occurred within just over a fortnight in September 1861. The first was near Cockeysville, Maryland, on 1st September 1861 when two or three cars of a train carrying a detachment of the Ira Harris Cavalry of New York, under Colonel Kilpatrick, were thrown from the track of the Northern Central Railroad. The derailment killed four soldiers outright and mortally wounded three more. It was ostensibly due to excessive speed, the implication being that the engineer of the train was a Rebel sympathiser. Following the accident, the engineer detached his engine and hurried into Baltimore, where he was arrested.[1361]

The above incident might perhaps have been the result of driver error but in the case the second there was no doubt that it was an act of sabotage. On 3rd September Confederate guerrillas known as bushwhackers set fire to the lower timbers of a 160 ft wooden bridge over the Platte River east of St Joseph, Missouri on the Hannibal and St Joseph Railroad. They then extinguished the fire, leaving the bridge apparently intact but incapable of supporting much more than its own weight. Just after 11 p.m. a train carrying between 85 and 100 passengers, including women and children, passed onto the bridge, which collapsed under the additional weight of the locomotive and the train fell thirty-five feet to the ground.[1362] The passengers were buried among broken timbers and thrown out of the windows, some landing in the river, here only a foot and a half deep. Seventeen bodies were recovered from the wreckage and an initial report said that two of the injured were not expected to live until the morning. Only three of the passengers were in a state afterwards to help the injured and one of them, a Mr Hager, then left the site at midnight, walked five miles and then procured a hand car to get to St Joseph for assistance. The Rebels had ensured that no help could come from the opposite direction by destroying another trestle.[1363] Union soldiers were ordered to track down and execute any bushwhackers they found, prompting Confederate Major General Price to protest to

---

1358 Alton Weekly Courier, Alton, Illinois, United States, 14th May 1857, reproduced from Baltimore Sun, 4th May 1857

1359 Bangor Daily Whig and Courier, Bangor, Maine, United States, 29th April 1861

1360 Wikipedia - Jacksons Operations against the B&O (1861

1361 The Boston Post, Boston, Massachusetts, United States, 10th September 1861

1362 Janesville Daily Gazette, Janesville, Wisconsin, United States, 7th September 1861

1363 Zanesville Daily Courier, Zanesville, Ohio, United States, 7th September 1861

his Union counterpart that the sabotage had been "lawful and proper" and that any bushwhackers captured should be treated as prisoners of war. Union general Halleck replied that the bushwhackers were "spies, marauders, robbers, incendiaries, guerrilla bands…in the garb of peaceful civilians." [1364]

On 17th September 1861 at about 8.30 p.m. a bridge collapsed beneath a train carrying a portion of Colonel Torebon's 19th Illinois Regiment near Huron, Indiana, United States, on the Ohio and Mississippi Railroad. Four passenger cars fell into a creek and a box car and a baggage car fell on top of them. The engine and leading car had passed over the structure safely. By the time of the second report on the accident, thirty bodies had been recovered and the train was on its way again with ninety-two wounded. More bodies were said to still lie in the wreckage. Bridge No. 48 was 60 feet long and 10 feet high. It had only recently been inspected and it was believed that its collapse was due to sabotage.[1365] [1366]

With the Huron episode significant sabotage against trains in the American Civil War seems to have largely ceased, or at least not been reported, though there was much destruction of railroad property, such as for example during General Sherman's March to the Sea in late 1864.[1367]

An act of sabotage on 5th June 1863 resulted in one of the largest numbers of deaths of any train accident in the nineteenth century. It happened to a Russian troop train at Malkinia, in Russian Poland, a state which although theoretically semi-autonomous was in reality a puppet state of the Russian Empire.[1368] Following the defeat of Russia in the Crimean War, the January Uprising of Poles against the Russian Empire began in late January 1863[1369] and it was during this rebellion that the attack on the train was carried out. A correspondent writing for *The Times* wrote that if the train was being operated by drivers and conductors in the service of the National Government it was taken quietly to a place where the insurgents were waiting to fall upon the soldiers. If not, the case was rather worse, as the train would run at full speed to a place where the rails had been removed and soldiers and civilian passengers alike would suffer. The train at Malkinia had it seemed only been carrying soldiers. A Pole told the correspondent of the constant anxiety of passengers on the Warsaw and Cracow Railway as to whether their train would be abruptly stopped, thrown down an embankment, fired upon or blown up before the end of its journey. On the train the Pole had ridden on, the passengers had asked the railway staff if it was a "safe train" they were travelling by, that is one not carrying any troops. They had been assured that it was and had made themselves comfortable for the trip to Warsaw, then at the last moment four luggage vans containing a company of infantry had been hooked up to it and they had a very anxious journey until the soldiers had alighted a few stations down the line. The Malkinia accident was clearly just the most serious of a

[1364] Wikipedia – Platte Bridge Railroad Tragedy
[1365] Zanesville Daily Courier, Zanesville, Ohio, United States, 18th September 1861
[1366] Dawson's Daily Times and Union, Fort Wayne, Indiana, United States, 19th September 1861
[1367] Wikipedia - Sherman's March to the Sea
[1368] Wikipedia – Congress Poland
[1369] Wikipedia –January Uprising

number of acts of railway sabotage in Russian Poland at this date. According to the *Gazeta Narodowa* in Lemberg the 'railway accident with a purpose' killed 13 officers and 140 other ranks and wounded 240 soldiers.[1370]

News of two further attacks on trains in Russian Poland reached *The Times* in February 1864. On the 18th insurgents blew up a goods train near Wloclawek and destroyed 16 waggons. The following day rails were removed, and a passenger train was derailed near Lazy.[1371] The January Uprising was crushed by Russia in late 1864 with harsh reprisals.[1372]

Robbery, not revolutionary zeal, was the motive for an act of sabotage directed against a Louisville and Nashville Railroad train at Bristow, near Bowling Green, Kentucky, United States, on 11th October 1866. The robbers loosened a rail but left it in place, with wires attached leading into the undergrowth where they were concealed. Engineers pulling trains carrying pay cars tended to drive cautiously, wary of robbery attempts, but the rails being pulled out of place in front of the locomotive were too much for the engineer's caution and the whole train was derailed. Paymaster G.W. Craig reportedly saved $6,220, presumably by hiding it, but the robbers escaped with $8.265, eluding a posse sent after them. Damage to the train cost the company several hundred dollars more and Trainmaster Church was injured when his car overturned.[1373] [1374] Less than a month later the train robbers tried again, this time at Salmon, near Franklin, Kentucky, also on the Louisville and Nashville Railroad, on 9th November 1866. This time they piled a heap of sleepers across the track, at 2 a.m. in the morning, which the engineer, from the light of an oil headlight, could not pick out, in time to stop, and the train was derailed, the engine overturning twice as it fell down a 30 ft embankment. Twelve robbers with blackened faces then plundered the 100 or so passengers on the train, accumulating £2,500 in money and watches before disappearing in an easterly direction. The railroad company had offered a reward of $2,000 for their capture after the earlier robbery and the culprits were subsequently caught and some of them hanged.[1375] [1376]

On 6th March 1867 the up Cork mail train ran into four or five telegraph poles which had been curt down and laid across both running lines near Knocklong, Ireland, on the Great Southern and Western Railway. The engine and tender kept to the rails after striking the obstructions, but a coupling broke behind the tender and the remainder of the train was derailed. None of the passengers was hurt beyond a few bruises. There were a number of military gentlemen travelling on the train, but also six women and 14 young children, the families of the domestics of the Protestant Bishop of Tuam, whence they were bound. The engine of the mail proceeded to Limerick Junction to raise the alarm and 50 troops, a few police and all the railwaymen that could be spared were sent to the

---

[1370] The Times, London, England, 9th July 1863

[1371] Ibid, 20th February 1864

[1372] Wikipedia – January Uprising

[1373] Bangor Daily Whig and Courier, Bangor, Maine, United States, 13th October 1866

[1374] History of the Louisville and Nashville Railroad 1850 – 1963 by Kincaid Herr

[1375] Bangor Daily Whig and Courier, Bangor, Maine, United States, 19th November 1866

[1376] History of the Louisville and Nashville Railroad 1850 – 1963 by Kincaid Herr

stranded train. The passengers, Post Office staff and mail bags were brought to Limerick Junction on this relief train, and about an hour and a half after arriving there were sent forward to Dublin, 'against the wish of the passengers and Post Office and railway officials,' in the words of *The Times*, as they wished to wait until dawn – so who demanded they go at once? The train proceeded cautiously, at about 12 miles an hour, and got as far as a deep cutting two and a half miles from Thurles, where it ran into further telegraph poles which had been sawn down and laid across the line and was again derailed. The rails had also been shifted. The two attacks were connected with a brief and poorly planned Fenian uprising against the British occupation of Ireland.[1377] A derailment near Castleblaney on the Irish North Western Railway on 16th March, in which a passenger died, was possibly also due to Fenian sabotage, but the Board of Trade inspecting officer could not be certain as to its cause.[1378]

The coming of the railroads to the central and western parts of the United States in the 1860s had a profound effect on the Red Indians (Native Americans) who lived there. The railroads brought troops to conquer the Indian tribes, brought settlers to take over their land and enabled the transport of buffalo hides eastwards, resulting in the decimation of the vast buffalo herds upon which the Indians depended for so many aspects of their lives. Breaking the power and independence of the Indian tribes east of the Mississippi river had taken two centuries. With the railroad's help the achieving of the same object west of the Mississippi took a mere thirty years. Surprisingly, therefore, the Indians seem to have made only one attempt at sabotaging the railroad which was devastating their lives.[1379]

At 8 a.m. on 6th August 1867 a westbound Union Pacific Railroad freight train left Omaha, Nebraska. At 12.25 a.m. on 7th August it reached Plum Creek station, Nebraska, 232 miles from Omaha. It stopped there ten minutes, then continued westwards, and had travelled another six miles when the engine and all seventeen cars were thrown off the line, the cars piled up in a heap and their contents scattered. Immediately after the derailment, a party of about a hundred Cheyenne warriors appeared out of the darkness and surrounded the train. The engineer, fireman and two brakemen were shot and scalped by the Indians, who then took the burning coals from the engine and set the wreckage ablaze. The conductor in his caboose managed to escape in the confusion and running back towards Plum Creek was able to flag down a following train, which, after hearing from him what had happened, reversed whence it had come. The fierce yells of the Indians dancing around the blazing wreckage echoed in the ears of the crew of the second train as they beat a hasty retreat. In a related incident, six men who had gone out, well-armed, to repair the telegraph wires, which had been cut by the Indians the previous evening, were attacked by them and three of the six were killed. The *Omaha Republican* of 8th August ranted about 'Indian murders' and 'wholesale slaughter' and urged the Government to take hold of the Indian question.[1380]

[1377] Wikipedia – Fenian Uprising

[1378] Board of Trade Report

[1379] Webpage – American Indians and the Transcontinental Railroad

[1380] Daily Journal, Indianapolis, Indiana, 13th August 1867, reproduced from Omaha Republican, 8th August 1867

At about 3 a.m. on the morning of 14th April 1868 the official in the postal car of an Erie Railroad train noticed the bell rope straighten out and break as the train was just west of Carr's Point or Carr's Rock, near Port Jervis, New York, United States. Conductor Judd realised at once that this meant that the rear cars of the train had become detached and going to the platform of his car, he watched the sleeping cars at the back of the train roll down a high embankment. The stove in the last car set it alight shortly after the derailment. The train involved was the Cincinnati express, due in New York at 8 a.m. The accident was a major disaster, twenty-four people died and 80 were injured, many of the victims were unrecognizable and some were reportedly burnt to death in the ensuing fire. The accident was initially ascribed to a broken rail.[1381]

The coroner's jury at Port Jervis returned a verdict that the accident was thus caused and that they thought trains should be run at lower speeds given the poor condition of the track. The Erie superintendent at the same time issued an order that no first-class passenger train was to run at more than twenty-five miles an hour, and no second-class passenger train or freight train at more than twelve miles an hour.[1382]

Rumours persisted though afterwards that the disaster had not been the result of a broken rail but of a deliberate act. On 28th May of the next year a man named John Bowen reported to an Erie watchman in the area of the accident that a farmer named James Knight was tampering with the track. Knight was duly arrested, tried and acquitted and suspicion then fell instead on Bowen himself. Bowen was arrested in turn and on 16th August in gaol at Milford, Pennsylvania, he made a full written confession of having committed sabotage at Carr's Rock and of perjury against James Knight. His motive for the perjury had been to secure the $2,000 reward offered by the company for the detection of any persons interfering with the track. Bowen was described as an Englishman, about 64 years old, and it was mentioned that he had previously been an Erie employee, but had been jailed by them on a charge of forgery.[1383] Bowen was convicted on 23rd September 1869 and sentenced to 10 years solitary confinement with hard labour in the state prison and a heavy fine. This seems a very lenient sentence for someone who had killed 24 people, though a local newspaper described him as a miserable man of probable weak intellect and opined that he was probably 'a more fit candidate for an insane or idiot asylum.'[1384] [1385]

An insurrection had broken out in Saragossa, Spain by October 1869 and on 7th October the insurgents derailed a troop train 'on the Andalusian Railway', killing three ordinary soldiers and a captain and wounding a large number of other troops.[1386] A fortnight later, *The Times* correspondent in Madrid reported that four men had been executed in Seville for carrying out the attack.[1387]

1381 The Daily Alton Telegraph, Alton, Illinois, United States, 16th April 1868

1382 The Alton Weekly Telegraph, Alton, Illinois, United States, 24th April 1868

1383 Hornellsville Tribune, Hornellsville, New York, United States, 26th August 1869

1384 Morning Herald, Titusville, Pennsylvania, United States, 30th September 1869

1385 A History of Railroad Accidents, Safety Precautions and Operating Practices by Robert B. Shaw, pgs.201 and 272-3

1386 The Times, London, England, c9th October 1869

1387 Ibid, 21st October 1869

War had broken out between France and Prussia in July 1870 and in a couple of months was already going badly for the French. The Prussian army was advancing towards Paris and the French destroyed railway structures in a bid to delay its advance, while the Prussians of course worked hard to repair the damage. Thus, *The Times* correspondent in Berlin reported on 7th October that 'the railway bridges between La Ferte and Meaux having been restored, the Germans can now bring up their artillery park, weighing 5,000 tons. Trains run day and night.'[1388] The French also carried out acts of sabotage on Prussian trains, mostly if not all undertaken by francs-tireurs, guerrillas who were an outgrowth in eastern France of rifle clubs and unofficial military societies formed after the 1867 Luxemburg Crisis, when France and Prussia had almost come to war.[1389] Thus, on 29th September 1870 saboteurs loosened the rails near Vitry-la-Ville, derailing a train. A letter from Epernay published in a German newspaper detailed the accident and stated that the engine driver had had both legs crushed, though the soldiers on the train had escaped with slight injuries. It added that the nearby town of Chalons had been ordered to pay (presumably the French equivalent of in francs) £700 as a punishment, and that 'Secret injuries to the railways are happening everywhere, so that trains can only travel very slowly, and with the utmost care.'[1390]

Another attack on a Prussian train was reported towards the end of October. A correspondent to *The Times* reported from Mézières on 29th October that, 'A band of volunteers, calling themselves the "Wild Board of the Ardennes" joined to another entitled the "Railway Destroyers," have caused a German train to run off the line between Launois and Saulces. A great many Germans were killed.'[1391] Another French report from Lille on 31st October stated that, "some Francs-Tireurs stopped a Prussian train on the railroad from Mézières to Rethel. This bold act of war cost the enemy 465 men."[1392] However, a few days later *The Times* quoted details of the attack from the German *Elberfeld Gazette*, which stated that the francs-tireurs had loosened the rails but left them in place, so the damage was not noticeable. 'The result was that 13 or 14 carriages had plunged down a declivity but fortunately most of them were unoccupied and only one soldier was severely injured. In the midst of the confusion, twenty or thirty francs-tireurs opened fire. After a couple of them, however, had been shot, they took to flight, seven being captured.' The truth of the matter no doubt lay somewhere between the conflicting accounts. The derailment must have occurred on about 27th October.

The attacks continued, including the derailment of a double-headed German troop train on 28th November. The troops had dismounted at Dornach and the train was proceeding to nearby Mulhouse when it was derailed, three sleepers having been loosened and put back in their place. An engine driver and several other persons were hurt. The Governor of Alsace directed as a result that the inhabitants of Dornach must

---

1388 Ibid, 8th October 1870

1389 Wikipedia Francs- Tireurs

1390 The Times, London, England, 10th October 1870

1391 Ibid, 31st October 1870

1392 Ibid, 1st November 1870

accompany every train between Dornach and Belleviller, and the town was fined (the equivalent of) £1,000.[1393] Captured francs-tireurs were executed by the Germans as armed non-combatants and vilified by the German army and press as murderers and highwaymen. Their attacks led to harsh reprisals against French civilians in nearby villages and towns, and "pacifying actions" involving whole regiments or divisions in areas where francs-tireurs were active, leading to a lasting enmity between the occupying soldiers and French civilians.[1394]

Reports of francs-tireur sabotage seemed to come at intervals of about a month, the next being on 24th December, Christmas Eve, 1870 when an attack was carried out by the usual tactic of tearing up the rails and putting them back in place to conceal the damage. The line between Chaumont and Troyes was targeted for the third time and a troop train carrying 500 men derailed. The francs-tireurs then sallied out from a nearby wood expecting to capture the party but the soldiers {or Landwehr as the presumably German source of this account called them) were unhurt and collecting their senses fired several volleys at the francs-tireurs, who fled leaving dead and injured behind.[1395] This is the last reported attack and the war ended with France's defeat on 10th May 1871.

On 2nd January 1871 a mail train was derailed on the Sevilla – Cádiz line between Jerez de la Frontera and El Cuervo stations in Spain. Seven died and 22 were injured. It was believed that the derailment was an act of sabotage aimed at General Caballero de Rodas, a prominent soldier who had been appointed general director of infantry in 1869, and who was a passenger on the train.[1396] [1397] [1398]

Another act of sabotage in Spain on 30th March 1872 was probably linked to the Carlists (supporters of the pretender Carlos V who wanted a return to autocratic monarchy rather than the more liberal regime established after the overthrow of Queen Isabella II in 1868) though the Third Carlist War did not begin until three weeks later.[1399] *The Times* correspondent in Madrid submitted a very detailed account of the attack on the train to his paper on 1st April. The Ferrocarril de Mediodia railway company followed the practice of having the takings from its various stations conveyed to Madrid each day by the evening mail train. This was a tempting target for the saboteurs, who chose an isolated location between Manzanares and Valdepenas which they knew the mail from Cordova to Madrid would pass about midnight. Between 9 and 10 p.m. they presented themselves to two *guardas del paso* (crossing keepers) and indicated that they wanted the mail stopped. Upon the latter protesting they were threatened with being shot if they did not comply and were forced to "assist" in removing the rails. The saboteurs explained

1393 Ibid, 13th December 1870
1394 Wikipedia Francs- Tireurs
1395 The Times, London, England, 3rd January 1871
1396 Wikipedia - Anexa – Accidentes ferroviarios en España
1397 Webpage - Federación Castellano Manchega de Amigos del Ferrocarril - Accidentes ferroviarios en España order de gravedad
1398 Wikipedia – Antonio Cabarello and Fernández de Rodas
1399 Wikipedia – Carlist Wars

that they had no desire to hurt anyone, their motive was robbery. Danger and caution signals (presumably lights) were placed in the rear of the obstruction to stop the train. The mail approached at too high a speed to pull up however, despite the driver braking, and the engine and eight carriages were derailed. The bandits also fired at the driver and fireman, but without hurting them. Once the train had stopped some of its passengers jumped out, but the saboteurs forced them to re-board 'with terrible oaths and threats' and they were told that if they remained quiet, they would not be hurt, but if they stirred, they would be shot. Alerted by the driver two Civil Guards on the train fired at the bandits and an infantry lieutenant, armed only with a sword, did likewise, but there were too many bandits and one of the Civil Guards and the lieutenant were wounded, while a passenger who had alighted from the train and was too slow to get back on was shot and fatally wounded. The bandits then rifled the wooden boxes on money in the guard's van and escaped to the mountains with an estimated equivalent of $2,000.[1400]

On 29th April 1872 a train passing through the Andalusia region of Spain was stopped by an armed band, who forced its driver and fireman to get off their engine and then started the train towards a point where they had removed the rails, so that it left the track.[1401] Again on 22nd September, saboteurs tore up the rails and derailed a mail train in Andalusia. The train was carrying 5,000,000 francs in specie but the appearance of the Civil Guard escorting the train resulted in the attackers taking flight. The passengers escaped with a few bruises.[1402] In another attack on about 17th November 1872 'thirty-five masked men destroyed the telegraph wires between Almanza and Albacete and upset a train.'[1403]

The Carlist rebels were very active in Spain at the beginning of March 1873. On the 9th they had destroyed the viaduct at Ormaiztegui on the Norte line, cutting communication with France, and they also about the same date interrupted the line at Garinchurqueta tunnel, cut the line between Zumarraga and Beasain, and burnt down Santa Ottalla station, while on the 8th they derailed a train between Tolosa and Beasin, taking its conductor and driver (who was French) prisoners. There were reports that the two railwaymen had been shot, but these were contradicted. 'The departure of the Irun trains was suspended.'[1404]

On 12th March 1873 passenger train No. 590 was derailed and thrown against the masonry wall of the cutting near Ycaztigueta Tunnel,[1405] killing the driver and two conductors. The Carlists then fired at it, killing the fireman, and a fight with the military escort of thirty soldiers riding on the train ensued. The escort was outnumbered and forced to retreat to a house where they fortified themselves in. The timely arrival of reinforcements saved them and the Carlists withdrew with their wounded, leaving one

[1400] The Times, London, England, 1st and 2nd April 1872

[1401] Ibid, 1st May 1872

[1402] Ibid, c23rd September 1872

[1403] Ibid, c20th November 1872

[1404] Ibid, 11th and 12th March 1873

[1405] Ibid, 14th March 1873

dead behind.[1406] According to a report from Hendaye on 13th March traffic on the Norte line was suspended after this attack. A report from Bayonne on 9th March had stated that the Carlists were determined to cut communications on the whole of the Norte line[1407] and with this attack they clearly almost succeeded in doing so, but a report from Madrid on 13th March[1408] contradicted that from Hendaye and stated that the railway company had been induced not to suspend the service, with the promise of more troops for their protection.

On 13th March 1873 the evening passenger train from Madrid was derailed by the Carlists between Zummaraga and Armaiztegin, without casualties.[1409]

On about 16th April 1873 three hundred Carlists commanded by Quito Constanti derailed a Barcelona train at Arbos. They made the passengers show their passports and took a sum variously stated as 3,000 pesetas and 647 douros (5 pesetas = 1 douro) from a lieutenant of the Civil Guard who (probably fortunately for him) was in private clothes. They destroyed the telegraph line and threatened the officials but 'the traffic, however, has not been stopped.'[1410]

On 29th May a train was derailed and plundered between Guero and Villacanas, Spain. The engine driver was killed and the fireman severely injured but none of the passengers was hurt,[1411] and on 12th June Carlists derailed the mail train from Madrid near Miranda.[1412]

The next act of railway sabotage worthy of mention takes us away from Spain to South America. Paraguay had suffered a severe defeat in 1870 in the War of the Triple Alliance against Argentina, Brazil and Uruguay. After the war, at the end of which Solano López, the then president, had been killed, a new more liberal government was formed with Salvador Jovellanos as president, and there were repeated attempts to overthrow it by López loyalists, with armed revolts in March and June 1873.[1413] About 7th June 1873 Colonel Germán Serrano and 800 men cut the railway between Cerro León and Pirayú, derailed two locomotives and tore up some track before they were driven off by government troops.[1414] Meanwhile a part of General Bernadino Canbellero's rebel troops had reached the railway at Patiño-cué and were heading for the capital, Asunción. It was possibly these rebels who sabotaged a bridge, causing a train to fall through it. The government then despatched another train with soldiers to bring in the casualties from the

1406 Ibid, 18th March 1873

1407 Ibid, 12th March 1873

1408 Ibid, 18th March 1873

1409 Ibid, c15th March 1873

1410 Ibid, c19th and c21st April 1873

1411 Ibid, 31st May 1873

1412 Ibid, c14th June 1873

1413 Wikipedia History of Paraguay

1414 Paraguay and the Triple Alliance: The Post-war Decade 1869-1878 by Harris Daylord Warren

bridge sabotage, but 'the engine driver forgot to put on the brakes coming down the Yuquery-hill. In the collision there were 5 killed and 16 wounded.'[1415]

Next, we go to Adair, Iowa, United States, where on 21st July 1873 a Chicago Rock Island and Pacific Railroad train was derailed. The rear sleeper of the eastbound train contained 'a company of aristocratic Chinese on their way to the Eastern colleges,' but the saboteurs were more interested in the fact that the train was carrying the California mails and express matter and three tons of bullion. Rock Island train No. 2 had left Council Bluffs at 5 p.m. and was about four miles west of Adair when as it was rounding a sharp curve at an isolated spot Engineer John Rafferty suddenly saw a rail shifted out of place. The train was travelling at about twenty-five miles an hour and although the engineer reversed his engine and applied the air brakes, the engine was derailed, thrown against the side of a cutting and overturned. Shots were also fired at the locomotive and Rafferty was killed in the derailment, but no bullet wounds could be found on his body and it was supposed he had died from concussion. Behind the engine the two baggage cars were badly smashed up. As soon as the train came to a stand six large men dressed in Ku Klux Klan uniforms entered the express car and threatened Conductor William Smith, Superintendent Boyce, Express Messenger John Burgess and three others, shouting "Get out of there, damn ye! Get out of there!" The gang also fired off a series of shots which slightly injured the conductor and the express messenger. Two of the attackers marched up and down the train threatening the passengers and telling them to get down out of sight. The robbers then took about $2,000 from the safe and cut open the mail sacks but took no letters. In less than ten minutes it was all over, and the robbers rode away southwards across the prairie. None of the passengers was hurt, their lives probably saved by the promptness of the engineer in trying to pull up his train. John Rafferty left a widow and three children. The gang had interfered with the track using equipment stolen from a hand car house at the nearest station.[1416] Those who had derailed and robbed the train were later identified as the James Younger gang, which included the legendary Jesse James and his brother Frank.

In August 1873 Carlist insurgents in Cartagena, anticipating defeat, were determined to attack the capital, Madrid. The advance was learnt of by the government, who sent 1,000 troops to the railway junction at Chinchilla, Spain, where an encounter with the Carlists took place. The defeated rebel leader Contreras retreated with his men by train, but the line had been cut, and the train was derailed on about 8th August, whereupon the insurgents continued on foot. It is unclear if the railway had been sabotaged by the government troops or by the Carlists themselves who had not anticipated needing to use it later.[1417]

Two accidents occurred in Spain on 11th September 1873. One of them was a major disaster. An Irun – Madrid express of the Norte Company (Caminos de Hierro del Norte de España) left the track at the exit from a bridge over the River Douro near Viana de

1415 The Times, London, England, 31st August 1873

1416 Sedalia Daily Democrat, Sedalia, Missouri, United States, 23rd July 1873

1417 The Times, London, England, 13th August 1873

Cega. Twenty-one people died and 58 were injured, the latter including 'a General and several persons of distinction.' *The Times* correspondent stated on 12th September that whether it was an accident or had been deliberately planned had not been ascertained.[1418] [1419] As to the other accident where a train ran off the rails near Tafalla 'on the Pamplona line' there was no doubt that Carlists had caused it by removing the rails.

Acts of railway sabotage at Indianapolis, Indiana, United States on 26th December 1873 and near Cincinnati, Ohio, United States on 27th December 1873 were due to a cause that we have not met before. There were carried out during a strike by Pittsburgh Cincinnati and St Louis (Panhandle) Railroad engineers, protesting against a reduction in their wages. At Indianapolis a train which left there during the evening was thrown off the line by an open switch and was then abandoned.[1420] At Columbia, on the outskirts of Cincinnati, at midnight, a Pennsylvania Railroad Pacific express from New York, carrying a sheriff and posse for protection, was diverted by an open switch into a siding and collided with some cars, injuring the engineer and damaging the engine. In other acts of sabotage the New York night express ran into an open switch at Cadiz, Ohio on the evening of Saturday, 27th and a watchman at Logansport, Indiana, found a misplaced switch just before a passenger train arrived at the spot. [1421] The strike soon fizzled out; it was one a number on no less than 18 different railroads between November 1873 and June 1874.[1422]

About 5th February 1874 Carlists 'threw a goods train off the line' at Pont Guadalmes, Spain.[1423] This was the last reported Carlist sabotage of a train. The Third Carlist War ended on 27th February 1876.[1424]

A serious act of sabotage was carried out in Austria half an hour after midnight on Thursday 4th November 1875. As passenger train No. 9 from Vienna to Prague was nearing Schanzenau it was thrown from the track on an embankment and the engine tender and whole train except the rear one or three (accounts vary) vehicles fell partly down the embankment and partly from a road overbridge. The engine ended up in the road and three parcels vans, a mail van and four passenger carriages piled up on top of it. Five people were initially reported as killed and nine wounded. The final number of dead was at least eight, comprising four railwaymen, a postal clerk and three passengers; some sources suggest that in all there were eleven deaths. It was discovered that a rail had been carefully removed from the sleepers, the nails and screws being left undamaged by the side of the line. No motive for the attack could be suggested, though the police suggested

---

[1418] Wikipedia – Accidentes ferroviarios en España

[1419] The Times, London, England, 13th September 1873

[1420] Bangor Daily Whig and Courier, Bangor, Maine, United States, 29th December 1873

[1421] Galveston Daily News, Galveston, Texas, United States, 30th December 1873

[1422] Trouble on the Railroads in 1873-1874: Prelude to the 1877 Crisis? By Herbert Gutman

[1423] The Times, London, England, 7th February 1874

[1424] Wikipedia - The Third Carlist War

that it was the work of anarchists.[1425]

The real motive for the attack only came to light thirty-five years later when the man who had been the local signalman (Bahnwärter) at the place at the time of the derailment made a death-bed confession. He related that having had a large family and being poorly paid he had been desperately short of money. He had thereby conceived the plan of 'discovering' an intended act of sabotage but being able to stop the train and so prevent an accident, whereupon he had hoped to receive a substantial reward for his actions. In furtherance of his plan he had loosened a rail and then positioned himself at a point some distance in the rear of the disturbed track with a red lamp to warn the crew of the approaching train. His scheme had come unstuck when a thick fog had descended just before the arrival of the train. The engine crew had failed to see his warning signal through the fog, had run onto the damaged track at undiminished speed and had been derailed.

The signalman was later transferred to another line and then retired prematurely because of mental problems. He received an increased pension due to the stress that it was supposed that he had suffered through witnessing the accident, no one knowing, of course, that the stress that he was suffering was from the knowledge that he himself had caused it.[1426]

An identical motive to that which lay behind the Schanzenau derailment was responsible for a series of incidents the following year in the United States. On 11th August 1876 a train on the Lake Shore and Michigan Southern Railroad was derailed at an unstated location due to interference with a switch. The locomotive of the train was badly damaged and its driver and fireman badly scalded. Another Lake Shore train was wrecked on 16th August and thrown down an embankment, spikes having been drawn from the sleepers and bolts removed from the fishplates. One passenger, W.R. Clemens, was badly hurt and later died from his wounds. Finally on 20th August another attempt was made on a train at Fairport, New York, but this time disaster was averted thanks to the train being stopped after a warning had been given by a George Adams. Unfortunately, it soon turned out that the 'hero' who had apparently prevented yet another wreck was in fact the person responsible for the attacks. After his arrest at Buffalo, New York, a few days later, Adams confessed to having been behind all three incidents, pleading that he had hoped each time to notify the railway authorities in time to prevent the accident and thereby 'place the company under obligation to furnish him with a situation.'[1427]

The Tucumán Ferrocarriles (Railway) of Argentina opened on 31th October 1876. Returning (presumably to the capital, Buenos Aries, though the brief newspaper report does not say) on the railway following its inauguration the train carrying the Argentinian president, Nicolás Avellaneda, 'and some 250 ladies and gentlemen,' was derailed. Rails had been spiked down across the line and the engine and a waggon left the track and were

---

1425 The Daily News, London, England, 5th November 1875

1426 Wikipedia - Eisenbahnunfall von Schwarzenau

1427 Daily Gazette and Bulletin, Williamsport, Pennsylvania, United States, 29th August 1876

badly damaged. 'but no loss of life occurred.'[1428] The motive for the sabotage was presumably political.

1876 was presidential election year in the United States and in the Southern states there was fraud and threats of violence against Republican party voters. In Florida, Republican Governor Stearns telegraphed from Tallahassee to the national Republican committee that he was sure of a majority in his state if they could get an honest canvass. 'The indications,' he added, were 'that violence ids to be freely resorted to, so as to prevent any returns from remote points in the interior.' A special train organized by the Republicans to run from Tallahassee to the Chattahooches, to collect returns from the western counties, had been 'Ku-Kluxed,' a few miles from Tallahassee the night before and 'thrown from the track, which was torn up and blockaded in several places.'[1429] Here then we have a new and perhaps unique motive for railway sabotage, to prevent the collection of votes during an election. Despite the best efforts of the Ku Klux Klan, the Republicans did indeed win Florida and Republican Rutherford B. Hayes became president.[1430]

The American railroad strikes of 1873 -1874 were followed by the Great Railroad Strike, also known as the Great Upheaval, of 1877, which began in July when the Baltimore and Ohio Railroad cut railwaymen's wages for the third time. The strike was a very bitter affair with railroad property being destroyed and trains derailed, while more importantly about 100 people lost their lives across the country. There were no trade unions at this date and city and state governments countered the strike with the use of armed militias, the National Guard, federal troops and private militias organized by the railroad companies. The strike was supressed after 45 days.[1431]

The first derailment of a train by the strikers appears to have been at 3 a.m. on 17th July, when one of 18 cars bound for Locust Point, Baltimore, Maryland, was thrown off the track just south of that place. The engine and several cars were derailed, and fire broke out in the wreckage. The engineer and fireman were slightly hurt.[1432] On 25th July an Erie Railroad train consisting of a baggage car, two passenger cars and a sleeper was derailed and partially overturned near Oswego, New York.[1433] On the morning of the next day, strikers ran a passenger train off the line at Twelfth street, Chicago, Illinois and having allowed the passengers to escape smashed the windows and furniture in the cars until troops appeared.[1434] A more serious incident occurred at Johnstown, Pennsylvania, on 27th July. Several trains of troops had been sent from the East to Pittsburgh via the Pennsylvania Railroad. As one of them passed through Johnstown depot it was attacked by a crowd who threw volleys of stones and pieces of metal at it, smashing carriage windows and injuring a number of the soldiers. Eventually the patience of Colonel

[1428] The Times, London, England, 18th December 1876

[1429] The Atlanta Daily Constitution, Atlanta, Georgia, United States, 10th November 1876

[1430] Wikipedia – United States Presidential Election 1876

[1431] Wikipedia - Great Railroad Strike of 1877

[1432] Daily Gazette and Bulletin, Williamsport, Pennsylvania, United States, 18th July 1877

[1433] Galveston Daily News, Galveston, Texas, United States, 26th July 1877

[1434] Weekly Nevada State Journal, Reno, Nevada, United States, 28th July 1877

Hamilton of the 1st U.S. Artillery, in charge of the troops, gave out and he pulled the bell rope to stop the train, doubtless intending to order them to confront and probably shoot at the crowd. The scab driver and fireman, who were guarded by soldiers, reversed the engine and slowed the train just in time, for a set of points lay open and a waggon laden with firebricks was waiting for the train at the switch. Engine, tender and five cars were derailed, struck the waggon and were piled up in a mass. Colonel Hamilton was seriously injured in the side and several soldiers badly bruised. The following third section of the troop train was held up behind the accident and several times attacked by the crowd. The soldiers from the derailed train arrested 100 of what the newspaper referred to as 'rioters,' 70 of whom were to be brought to Pittsburgh.[1435] Finally, on 31st July the engine, tender and one car of a troop train were thrown from the track two miles west of Keyser, West Virginia, on the Baltimore and Ohio line and ten soldiers were injured. This appears to have been the last derailment of the strike, which was by now being defeated.[1436]

A train carrying passengers returning from fêtes at Nancy met with disaster at 2 a.m. on 4th August 1879 at Xeuilley, France, on the Nancy – Vézelise line (Chemins de fer de l'Est). A wrongly placed set of facing points diverted it from the main line into a siding serving some old lime kilns. Five passengers were killed outright and another seven were to die of their injuries later. An investigation revealed that two railwaymen who had been dismissed were responsible for wrecking the train.[1437]

An attempt to assassinate Tsar Alexander II of Russia on about 2nd December 1879 near Moscow received extensive coverage in *The Times* newspaper in London (whether the attacked occurred late on 1st or early on 2nd December is unclear, one account states that it occurred at 11 p.m. on 1st, another implies that it happened early on 2nd). The Tsar had abolished serfdom in Russia in 1861 but this had not suddenly ended the grim rural poverty which existed in the country, and which so contrasted with the lives of the ruling autocracy. The result was the growth of a revolutionary socialist movement and the formation in 1879 of Narodnaya Volya (People's Will)[1438] which planned to assassinate the Tsar by blowing up his train. The imperial train, bound for Moscow from Livadia, reportedly left Simferopol station at midnight, following a luggage train consisting of 14 vehicles hauled by two engines which had left half an hour earlier. At some point along the route, however, the Tsar's train overtook the luggage train. This clearly confused the revolutionaries, who exploded a bomb under the luggage train at a point two and a half or three versts (about three kilometres) from Moscow. The blast destroyed a luggage van, blew seven other vehicles off of the line, and left a hole 5ft deep, 16ft long and 18 or 19ft broad. The Nihilists (as they were termed at the time) had procured a house near the railway line from where they had sunk a shaft and then laboriously bored a tunnel for about 50 yards to a point beneath the track. The Tsar's train had also had as passengers Count Alderberg, Minister of the Imperial Household; Count Milutin, Minister of War;

1435 Oshkosh Daily Northwestern, Oshkosh, Wisconsin, United States, 28th July 1877

1436 Ibid, 31st July 1877

1437 Wikipedia - Liste des accidents ferroviaires en France au XIX siècle

1438 Wikipedia - Narodnaya Volya

M. Possjet, Secretary of State; and various other high officials. The Tsar, who had survived several previous assassination attempts, "thanked the Almighty for having again saved him from the remorseless hand of the assassin."[1439] The fact that the order of the luggage train and the Tsar's was change en route possibly stemmed from an anonymous letter which had been received warning of the attempt.[1440]

A further attempt had been made at an unstated date in 1879 to blow up the Tsar's train near Alexandrovsk station on the Lozovaya – Sebastopol line. As in the Moscow attack, a tunnel had been dug beneath the railway line and here dynamite was placed in two chambers connected by an electric wire beneath the turf to the adjoining highway. As the Tsar's train passed a three-horse carriage came along the highway, stopped momentarily and then was driven away at great speed. It was supposed that the intention was to set off the charge from the carriage, but the scheme failed because unknown to the saboteurs the wire had been accidentally cut.[1441]

Narodnaya Volya finally succeeded in assassinating the Tsar with a bomb thrown at his carriage in St Petersburg on 13th March 1881.[1442]

Robbers who must have hoped for a much more profitable outcome were frustrated when they derailed the wrong train on the 19th December 1880 at a point between Elkhart Lake and Plymouth, Wisconsin, United States, on the Wisconsin Central Railroad. A number of loose rails placed across the track derailed an early morning wild (extra) freight train instead of the train carrying a pay car or the express due in Milwaukee at 3 a.m. The locomotive left the rails but remained upright and a brakesman went back to protect the obstruction from the Milwaukee express, which was following. He had only got back a short distance when he was accosted by a man who asked him if the pay car had passed. Told that it had the man pulled out a revolver and demanded the brakeman's money. He was given $40 as two other men appeared from the undergrowth on either side of the line. All three then made their escape.[1443]

[1439] The Times, London, England, 4th, 5th, 8th and 11th December 1879

[1440] Ibid, 14th January 1880

[1441] Ibid, 12th and 14th October 1880

[1442] Wikipedia – Alexandra II of Russia

[1443] Fort Wayne Weekly Sentinel, Fort Wayne, Indiana, United States, 22nd December 1880

CHAPTER 19

MISCELLANEOUS ACCIDENTS

The vast majority of railway accidents fall into one of a small number of categories such as derailments, collisions, bridge failures and so on. A few, though, do not easily fall into any obvious group, hence this final chapter on miscellaneous accidents. Many of the accidents listed in this chapter were fairly minor ones; they have been included because of their unusual features.

The earliest accident which might be considered to fall into the miscellaneous category is the famous death of Mr William Huskisson, MP on 15th September 1830. The story is well known. It was the opening day of the Liverpool and Manchester Railway. Eight trains were to be run from Liverpool to Manchester, one carrying the Prime Minister (the Duke of Wellington) and other dignitaries, including Huskisson, was on the more southerly track and the other seven formed a procession on the more northerly track. The departure from Liverpool was on time and although there was a minor mishap near Parr, when one of the trains on the northerly track derailed and another ran into it, matters were soon righted, and the trains continued to Parkside, England, midway along the line. Here the locomotives stopped to change water and, although they had been advised by the railwaymen not to do so, about 50 dignitaries alighted from the Duke of Wellington's train to stretch their legs, among them Huskisson. Huskisson and Wellington had been on bad terms since May 1828 but now Huskisson saw an opportunity for a possible reconciliation and made his way to the Duke's carriage. They shook hands and exchanged a few words, when a shout went up, "An engine is approaching. Take care gentlemen!" The engine was none other than the *Rocket*, the winner of the Rainhill trials the previous year. There should have been plenty of time, and most of the men got back into their carriages or climbed up the embankment till just two, Huskisson and a William Holmes, remained on the track. Just sufficient room was available for the men to stand safely between the side of the carriage and the advancing engine but at the last moment Huskisson had his doubts and attempted to climb into the carriage but failed to do so. Holmes cried out, "For God's sake, Huskisson, be firm!" but Huskisson seized the door of the carriage, which swung wide open, into the path of the *Rocket,* carrying Huskisson with it. The engine, driven by Joseph Locke, the renowned civil engineer, struck door and man, and Huskisson was fatally injured, the *Rocket* passing over his leg and thigh. Huskisson was attended by doctors at the scene and rushed up the line to Eccles where he was taken to a vicarage. His leg was amputated to no avail and he died at 9 p.m. The opening of the first railway in the modern sense had been marred by tragedy.[1444] [1445] At Huskisson's inquest the jury returned a verdict of 'Accidental death' without any deodand on the *Rocket*, implying that they viewed the accident as entirely Huskisson's own fault.[1446] Although this might be regarded as primarily a movement accident it was also a train accident given that the *Rocket* struck the carriage door as well as Huskisson, and it has therefore been felt that its inclusion here is justified.

The opening of the Leicester and Swannington Railway on 17th July 1832 was also marred by an accident, this time only a very minor one. It arose from what might be termed a problem with the loading gauge. The inaugural train left West Bridge, Leicester, England, at 10 a.m., with a covered-in open waggon for the directors, a second-class carriage and ten coal waggons fitted with seats, filled with about 400 passengers and hauled by the *Comet* engine. The accident, which only caused slight delay to the train and no injuries, occurred because the platelayers had previously packed a low spot in the track to the extent that the chimney of the engine fouled the roof of the 1,796-yard-long Glenfield Tunnel. Reducing the height of the engine chimney by six inches prevented a repetition of the accident.[1447]

Poorly laid track led to a potentially serious accident in January 1842 to a passenger train from Haswell to Hartlepool near Castle Eden, England, when a rail sprang up from the track and passed right through the coach, lodging in its roof. Despite the coach being full of passengers, no one was hurt, though one man was only saved by being thrown from his seat by the shock. The rail touched another man's hat without hurting him.[1448] Such snakeheads, as they were termed, were slightly more common on early American railroads, where a number of the early railroads used strap rails consisting of thin iron bars laid on longitudinal timbers. The weight of the train passing over them tended to loosen such straps so that they bent upwards and could penetrate the floor of a carriage. Early passengers much feared such accidents, though few seem to have actually occurred. Shaw was able to identify only two such accidents causing personal injury in the early years: on 12th August 1843 a snakehead pierced the floor of a carriage near Plainfield, New Jersey, on the Elizabethtown and Somerville Railroad, killing a mechanic whose throat it pierced, and on 10th June 1848 a loose rail penetrated the thigh of a "stout young lady" on the Auburn and Rochester Railroad, without apparently seriously injuring her.[1449] To these two accidents may be added one which occurred on 4th September 1850 near Albany, New York, on the Saratoga and Schenectady Railroad. A snakehead passed through the coat sleeves and scraped the flesh of a passenger, while also carrying him to the ceiling of the carriage and pinning him there. Even so he was not much hurt, but the

---

1444 The Last Journey of William Huskisson, by Simon Garfield

1445 Wikipedia – Opening of the Liverpool and Manchester Railway

1446 The Times, London, England, 18th September 1830

1447 The Leicester and Swannington Railway by C.R. Clinker, pg. 13

1448 The Times, London, England, 31st January 1842

1449 A History of Railroad Accidents, Safety Precautions and Operating Practices by Robert B Shaw, pg. 20

newspaper report commented that the accident showed the necessity of repairing the road.[1450] Strap rails did not in any case prove satisfactory and were soon replaced by conventional iron rails on all railroads.

Royal trains in Belgium were involved in two nineteenth century accidents, the first as early as August 1838 at Snep, when a swing bridge was involved (see chapter on Opening Bridge Accidents). In the second accident Louise of Orléans, the Queen of the Belgians, had a narrow escape on Tuesday 11th May 1847 when the train she was riding in was involved in a collision at Ans. She had accompanied her husband, King Leopold I, to Verviers and was returning alone to Brussels. Having been delayed in starting the Royal train was running fast when it was in collision at a junction with the 11 a.m. down train from Brussels, the locomotive of which struck the side of the royal carriage. Two officers of the Royal suite; General Chazal and General d'Hane were quite badly injured, and an unnamed servant suffered a broken arm. The Royal train had been preceded by a pilot engine, but this of course gave no protection against a sidelong collision. 'A strict inquiry has been instituted into the cause of the accident.'[1451]

Queen Victoria of Great Britain had made her first railway journey in June 1842, travelling from Slough (at that date the nearest point to Windsor palace) to London Paddington over the Great Western Railway. The Versailles tragedy had happened only just over a month earlier and there were reservations about her travelling by train, but she had dismissed them. In France at about the same time, the Council of Ministers were more successful in dissuading King Louis Phillipe from travelling by train from Paris to Rouen, en route to his château at Bizy, and he went by road instead.[1452] However, with trains the only easy means of long-distance travel in the nineteenth century and royalty needing to travel a lot practical considerations usually overrode theoretical fears regarding safety, the example of King Louise Philippe notwithstanding, and despite every precaution Royal trains were occasionally involved in accidents, most of which seem to have occurred in the 1890s, when there were several in a few years. Well before then, a train carrying the King of Naples was involved in a minor accident on the line between Naples and Capua, in the Kingdom of the Two Sicilies, on or about 23rd August 1859. The king had been on a visit to the town of Capua and had reviewed the troops there. His train was returning to Naples when it passed a goods train on the opposite line. The door of one of the waggons of the goods train was open and struck the carriage in which the king was riding, damaging it somewhat. No one was hurt but his Majesty and his two brothers were naturally considerably alarmed.[1453]

On 20th November 1848 it was the forces of Nature which caused an accident. A down goods train on the North British Railway was on its way to Edinburgh and running along a high embankment close to the sea east of Cockburnspath, Scotland. A gale force onshore wind was strong enough to blow of the roof of one of the carriages and took the

---

[1450] Weekly Wisconsin, Milwaukee, Wisconsin, United States, 18th September 1850

[1451] The Express, London, England, 14th May 1847

[1452] The Railway Age, by Cyril Bruyn Andrews, pgs. 46-7

[1453] The Express, London, England, 5th September 1859

guard who had been sitting in the carriage with it, depositing him at the foot of the embankment. One of the doors of the luggage van was blown off at the same time. The accident was only discovered at the next station. A search was made for the guard, who it was feared might have drowned given the proximity of the line to the sea; he was found seriously but hopefully not fatally injured lying down the embankment.[1454]

Reading passenger station, England, on the Great Western Railway, was in 1853, was one of Isambard Kingdom Brunel's one-sided stations, with platforms for both up and down lines on the south side of the railway. Both platforms had roofs supported by a row of cast iron columns. On the evening of Wednesday 12th October, the up goods train from Basingstoke, due at 10 p.m., arrived on time. As it passed through the platforms, travelling at some speed, a loose side flap hanging down from one of its waggons stuck each iron column in turn with great violence, knocking them out of the perpendicular, until the roof collapsed, partly onto the goods train and partly onto some empty carriages. No one was badly hurt and the down mail train, due at 10.12 p.m., was stopped outside the station, but costly repairs to the roof were necessary.[1455]

An accident which caused the death of a third-class passenger of a North Eastern Railway train at Felling, England, on 7th October 1854 had similarities to that to Mr Huskisson 24 years earlier. As a train known as the market train, which left Newcastle for Darlington at 4.25 p.m., was passing Felling station, its guard heard someone calling from towards the rear of the train and seeing a passenger without his hat assumed that was the reason. The train duly stopped as normal at Pelaw Main station, when the guard found that a passenger had suffered a head injury. Meanwhile, the driver of the market train had seen the door of one of the waggons of his own train fly open as they were passing Felling and strike the van of a passing down express. When they arrived at Pelaw, the driver asked two men in the van why they had opened the waggon door and they strenuously denied having done so. Only one man had been in the waggon when it left Newcastle, so the second man must had got into the waggon at Gateshead, opening the door from the offside. The door could not be fastened from the inside and had, the Board of Trade inspecting officer, Captain Henry Tyler, surmised, probably been opened while the train was passing Felling so that the two men could see where they were, it being a windowless covered van.[1456]

The guard of the 2.40 p.m. down express from York stated that his van had received a severe blow near Felling station, knocking brass and iron handles from the van. Both trains had been travelling at 30 miles an hour at the time. The fatal wound to the forehead of the passenger corresponded in shape to the end of the iron handle. The market train had 16 carriages and wagons. The van, the door of which had opened, was the leading vehicle, the guard's van was tenth in the train and the third-class carriage (open waggon) in which the deceased passenger had been riding was fifteenth.

Captain Tyler noted that the space between the two tracks was only 4ft 9 in, instead of the usual 6ft. The iron handle of the Great Northern Railway van of the down express

[1454] The Times, London, England, 25th November 1848
[1455] Railways, January 1943, pgs. 26-7
[1456] Board of Trade report

extended 1ft 2¼ into this space and the open door of the waggon of the market train extended 3ft 6¾ into the space, in total 4ft 9 in, the exact distance between the two tracks. With an oscillation of the train, and the door being swung open hard, the accident could thus be explained.

The passenger's life would probably have been saved, the inspecting officer remarked, had he been riding in a roofed carriage, or had there been sufficient clearance between the two lines. Any recommendation that the distance be increased would probably be of little avail because of the difficulties involved so, Captain Tyler said, he could only recommend that the Company stop using open carriages, that they prohibit the use of wide carriages owned by other companies and that they prevent passengers riding in non-passenger vehicles.[1457]

On Monday 12th November 1855 the New York and Harlem Railroad express from Albany to New York, United States, departed at 4.30 a.m. It consisted of an engine and tender, baggage car and three passenger coaches and was in charge of Conductor Robert White. It arrived at Chatham Four Corners, New York, at 5.38 a.m. and beyond there had to pass over a very exposed part of the line, where the railroad ran through a narrow valley between two peaks of the Berkshire Mountains. Severe gales were common here, especially when the wind was blowing in an easterly direction, and on the morning in question the wind was very high and accompanied by heavy rain. As the train was passing through the valley and near Boston Corners a sudden exceptional gust blew in the door of the baggage car, the vehicle was thrown from the track, and it rolled down a 40 ft embankment. The coupling between the baggage car and the engine snapped and the latter escaped but the wind blew the passenger coaches after the baggage car down the embankment. The coaches rolled over three times before finally stopping with a heavy crash. 'At this time of the morning it was pitch dark; the rain was falling heavily, and the groans and shrieks of the mutilated passengers were heard with dreadful distinctness even above the noise of the tempest.'

Conductor White extricated himself from the wreckage and despatched the engine to Millertown, the next station, for assistance. There had been thirty passengers on the train, of whom twelve escaped unhurt, but one, Francis Rathbone, a paper manufacturer, of Chatham Four Corners, was killed outright and Brakeman Harry Gaylord died of his injuries some hours later. The remaining passengers, Conductor White, Conductor Hart and Brakemen Duncan and Barnes were injured to a greater or lesser degree. A newspaper report added, 'No blame, of course, can be attached to any person in this casualty.'[1458] [1459]

The first recorded railway accident in Egypt occurred on 14th May 1858. It was notable both for its unusual character and because it cost the life of the prince who was the successor of the current Viceroy of the country. His Highness Ahmet Pacha had very reluctantly given in to the wishes of his relatives and travelled from Cairo to pay his respects to his uncle, the Viceroy, at Alexandria. The ceremony having been completed;

[1457] Ibid.

[1458] New Albany Daily Tribune, New Albany, Indiana, United States, 17th November 1855

[1459] Milwaukee Daily American, Milwaukee, Wisconsin, United States, 19th November 1855

he had hurried to the station to catch the up train back to Cairo. At this date, the railway to Cairo was still not continuous, as a bridge across the River Nile at Kafr-el-Zayat (the location is given as Kafir Lais in the contemporary newspaper report) had yet to be constructed and in the meantime ordinary passengers alighted from their train, took a steam ferry across the river and boarded another train on the far bank. To spare princes the inconvenience of having to disembark and re-embark they were conveyed across the river in their carriages, possibly road carriages loaded onto flat waggons. Disaster struck when the Arabs pushing the waggons onto the steamer omitted to put on the 'shappens' {presumably a word meaning scotches but unable to find in dictionary} and four waggons rolled off the river bank into the water. The alarm was given at once but Prince Ahmet Pacha, his suite, Rifaat Bey, the Minister of Commerce and Kernidin Pacha, formerly director of the transit administration, were all drowned. Halim Pacha, the brother of the Viceroy, was rescued from the river by some English railwaymen and survived. The newspaper report added that, 'The disastrous news was received shortly after the accident occurred, and both the European colony and the Arab population could not help manifesting how deeply they felt the sudden loss of the man in whom the hope of Egypt rested.'[1460]

Another accident which resulted from an act of nature occurred on the London and North Western Railway Stour Valley line at Deepfields, England, on 8th July 1858. The 4 p.m. passenger train from Birmingham to Wolverhampton set off during a heavy storm of rain and hail. After a while it outpaced the storm but getting to Deepfields it was again overtaken by the thunderstorm and approaching the station there it was struck by lightning, with a sound reportedly resembling the crack of a hundred rifles. The driver and fireman found themselves enveloped in a sheet of blue flame, the stoker experiencing a blow to the back of his head and temporarily losing consciousness. Every passenger on the train felt the shock, especially those sat nearest to the offside doors, the lightning apparently running along the rail on that side. Mr Benjamin Hicklin, commissioner of the peace for the borough of Wolverhampton, was struck on the right arm and knocked to the opposite side of the first-class carriage he was occupying, but the most badly hurt was Guard Henry Rixam, who had been standing up in his van applying the brake. He was momentarily paralysed but when his train stopped at Deepfields station he staggered out of his van, holding his head and reeling as if he were drunk. Had it not been for the stationmaster holding him he would have fallen over. The stationmaster put a porter in the brake van with Guard Rixam for the remainder of the journey to Wolverhampton and there the guard was taken to the inspector's office and given brandy and water, and after a while was able with difficulty to make his way home. Mr Hicklin was also found to be suffering from shock and complained of a benumbed arm.[1461]

The next event to be described was not an accident at all, but rather a deliberate attack on a train. In March 1862 the American Civil War had been in progress for almost a year. The state of Florida did not see any major fighting but on 2nd March 1862 some Union gunboats were near Fernandina Beach, Florida, United States, when they saw a train

[1460] The Daily News, London, England, 29th May 1858

[1461] Morning Chronicle, London, England, 10th July 1858

leaving the place. Supposing that the train was carrying escaping Confederate soldiers, the Union ships fired several shots at the two locomotives hauling the train. They did not succeed in preventing the train getting away, but afterwards two bodies were found by the track, one of them clearly that of a soldier.[1462]

A passenger train on the Madrid – Alicante line was near Villarobledo, Spain, on 18th August 1862, when a portion of the line was destroyed by a waterspout just as the train was passing the spot. The engine and several carriages of the train were precipitated into the hole nine or ten feet deep which had been formed. Over fifty people were injured, four of whom subsequently died, and several passengers had disappeared, supposedly carried off by the waters.[1463]

In December 1866 the Smithfield Meat Market in London, England, was under construction, with part of the Dead Meat Market being built over the lines of the Metropolitan Railway, and girders were being erected 100 yards west of Aldersgate Street station by the Thames Ironworks Company. By Wednesday 19th December four main girders had been erected transversely across the railway on brick piers and iron columns and seven cross girders had been laid parallel to the railway lines. The eight and last cross girder, 40 ft long and weighing 3¼ tons, was being put on place on that day. The girder was dragged partly into position by hand but the foreman of the gang, named Clancy, was then told by the "walking foreman," John Wilmot, to employ a portable steam engine, and proceeded to do so. Wilmot had intended only for the girder to be moved so far by the steam engine, but he then went away, it was moved further, and going too far it overbalanced and fell to the ground just as the down train which had left Moorgate Street at 12.40 p.m. for Kensington was passing. Most of the train had passed the spot but one end of the girder smashed into the second-class carriage at the rear of the train. An elderly woman passenger named Sarah Johnson was killed outright and two men, Charles Passmore, a clerk, and Henry Lakey, a silk mercer, were fatally injured and died at St Bartholomew's Hospital.[1464]

The Board of Trade inspector, Captain Henry Tyler, remarked that the method of moving the girder employed by Foreman Clancy would in any case have been imprudent and was attended with great risk when there were passenger trains running below. A responsible person should have been employed to stop all trains while the work was in progress. In practice no special arrangement had been made for ensuring the safety of the railway but the work had been carried on for two months on the understanding that no work would be done which would interfere with the safety of the trains. The inspecting officer noted that much work remained to be done in connection with the construction of the market and that it was desirable that some more stringent measures should be adopted to ensure public safety. The railway company were reluctant to station one of their men at the spot because they conceived that he would have no control over the contractor's workmen, plus the responsibility for public safety would then shift from the contractors to them. Captain Tyler felt, however, that instead of a mere understanding, a responsible

[1462] Philadelphia Inquirer, Philadelphia, Pennsylvania, United States, 21st March 1862

[1463] The Daily News, London, England, 26th August 1862

[1464] The Times, London, England, 20th and 22nd December 1866

railwayman should always be provided at the spot, in communication with the contractor's foreman and provided with distant signals to prevent the approach of any train should it appear not to be safe for them to pass.[1465]

All trains, and in the case of goods trains, the loads carried upon them, obviously normally had to fit within the loading gauge of the lines over which they travelled, lest they collide with lineside structures such as bridges, tunnels, signal posts, station platforms and so on. Great care was usually taken to ensure that trains were within the loading gauge, but just occasionally things went wrong, as was demonstrated by the 1865 accident in New Street North Tunnel, Birmingham described in the Tunnels Accidents chapter. Two years later there was another accident of a similar character. At 1 p.m. on 3rd June 1867 a Caledonian Railway passenger train left Edinburgh for Glasgow, Scotland, via Carstairs. It consisted of an engine and tender, a third-class carriage, a third-class carriage, a brake van, a cattle waggon, a carriage truck, a third-class carriage, a first-class carriage and a third-class brake composite carriage. Loaded upon the carriage truck was a van containing panoramic views, the property of a Mr Hamilton, bound from Coatbridge to Paisley. The van had actually come from Airdrie but as there was no loading ramp at Airdrie it had been taken by road to Coatbridge. There was a ramp but no loading gauge at Coatbridge, but the stationmaster there anticipated no problems with its size. The carriage truck was transferred to the Edinburgh train at Motherwell, where the engine driver had reservations as to its dimensions and asked a permanent way inspector to measure it, which the latter did. The inspector thought that there would be no problem at Uddingston but that it would not pass under a temporary bridge under the Barrhead Railway. Hearing the inspector say this the engineer of the line, who chanced to be present, remarked that in that case the load would not pass through the Shields or West Street tunnel either and it was directed that the load must be instead be taken to Gushetfaulds Junction and thence to Glasgow Southside by an alternative route. It was considered by the superintendent of the line that as the load had safely covered the 5½ miles from Coatbridge to Motherwell it should be able to pass over the remaining 12 miles to the Southside station via the alternative route. But the Motherwell stationmaster had his doubts and the engine driver "did not like its appearance."[1466]

As the train departed from Uddingston station, the driver stood on the hand rail of his engine and looked to see if the load would clear an overbridge 20 yards beyond the station. Seeing that it was not likely to he jumped down, reversed his engine and applied the brakes. The next moment the van struck the arch of the overbridge and was smashed to pieces, part of it damaging the end of the following third-class carriage and more or less injuring five passengers in the end compartment of that coach.

The Board of Trade inspecting officer, Captain Henty Tyler, stated that, 'In this case then, a van which ought not to have travelled at all with a passenger train, and the height of which, when loaded, exceeded the height prescribed by the company's printed regulations, was loaded and despatched without suspicion by the station-master at Coatbridge, and was allowed by his superiors to proceed from Motherwell, after the question had been raised and discussed, whether it would travel safely through the

---

[1465] Board of Trade report

[1466] Ibid.

bridges. There were no loading gauges or other convenient means of ascertaining whether the van came within the limit of the regulations or of safety, either at Coatbridge or at Motherwell, nor are there such means in connection with the passenger or the goods yard of the principal (Buchanan Street) station of the Company at Glasgow.' Captain Tyler was clearly not impressed.[1467]

The New York and Harlem Railroad had another accident near Boston Corners, New York, United States on Wednesday 17th November 1869 and again it was the result of a train being blown over by gale force winds. This time it was the mail train which had left Chatham at 9 a.m. for New York which was involved. The entire train apart from the locomotive was blown down the embankment and to make matters worse the express, mail and baggage cars caught fire and were destroyed along with their contents. Thirteen-year-old Peter McBride was killed outright, and several passengers were injured; one of them, Birdsay Curtis, died the net day. Brakesmen William Van Vlick, Reuben Brellis and Hiram Jacobin, Express Messenger John Hurd and Baggagemaster Frank O'Niel were also injured, the latter from inhaling scalding steam while attempting to save the baggage. As with the previous accident the inquest jury returned a verdict exonerating the railroad company and its employees from all blame.[1468] [1469]

An act of nature of a different sort caused an accident on the railway between Geneva and Aunumak in Switzerland on or about 3rd June 1870. A large mass of snow and ice became detached from the Monthoux Glacier as a passenger train was passing beneath it, its fall possibly triggered by the vibration of the train. The rear three carriages of the train were 'crushed to fragments' and three passengers killed and five others more or less injured. The accident would have been worse had the engine driver not applied full steam when he saw the avalanche coming down the mountainside and so managed to get a portion of the train clear.[1470]

One accident led to another at Maryhill, Scotland, on the North British Railway on 16th October 1871. The problems began with the mistake of a porter, who about 7.15 a.m., without consulting anyone, had moved a goods waggon from a siding so that it could be shunted to another siding for unloading. In doing so, he put the waggon foul of the down main line and into the path of a light engine for which the signalman had cleared his signals. Seeing the obstruction, the signalman threw his signals to danger, but the engine was too close at hand and it collided with the waggon at 7.18 a.m.

Mr Kirkman, the Maryhill stationmaster was going to leave the waggon where it was until the next day, as it had ended up standing clear of the main lines, but he was advised that its position would prevent any trains from getting into the goods yard, and so he belatedly ordered the travelling crane from Cowlairs, which arrived at 3.37 p.m. Matters were further delayed by a hunt for a hammer and chisel to remove a waggon buffer and

---

[1467] Ibid.

[1468] Philadelphia Age, Philadelphia, Pennsylvania, United States, 19th November 1869

[1469] Buffalo Age, Buffalo, New York, United States, 20th November 1869

[1470] The Times, London, England, c8th June 1870

then by having to shunt the crane to the up line out of the way of the 2.55 p.m. down mineral train from Sighthill Junction. Work then began but the crane had again to be shunted onto the up line for two more down trains. It having been brought back onto the down line, work was recommenced and had been in progress for ten minutes when the 3.45 p.m. up express train from Helensburgh arrived and collided with the balance weight of the crane. The collision damaged the engine and the right-hand side of all the vehicles of the train, killing a young woman and a boy and injuring 15 other passengers and the driver and fireman. The fatalities and most of the injuries occurred in a third-class carriage marshalled second in the train.

John Mayberry, the crane operator, told the Board of Trade inquiry into the accident that on being crossed back to the down line for the second time he had asked how long he had before another train was due and someone had said 'three-quarters of an hour.' He had not pointed out to anyone that the tail of his crane would foul the up line while it was in operation as he had thought it was obvious, it being evident that the crane could not lift the waggon without doing so. At other stations where he had worked the stationmaster had always supervised the operation (the stationmaster was away having his tea when the accident occurred).

Signalman Robert Stewart stated that there had been nothing foul of the main line when he cleared his up distant signal upon the Helensburgh express whistling for it (though Crane Operator Mayberry stated that the rear of his crane had fouled the up line for the whole ten minutes it had been working). Signalman Stewart said that he had then seen the obstruction caused by the crane, had thrown back his distant signal before the express had passed it, not having pulled any other signals, and had shouted out to the people with the crane and left his box with a red flag. He had never seen a travelling crane used before, had not realised that it would foul the other running line, and no one had told him.[1471]

The Helensburgh express had consisted of an engine and tender, seven carriages and a rear van. Its driver stated that the distant signal had come off when hew had whistled for it and had not been reversed before he passed it, this being confirmed by his guard. As he had come around the corner, he had seen the crane, whistled, applied the tender brake and reversed his engine. He thought he had been travelling at 20 to 25 miles an hour when his engine struck the crane.

Colonel W. Yolland, the inspecting officer, thought it likely that the up distant signal had not been reversed until after the engine of the express had passed it, but was not satisfied that the driver of the express had been keeping a good look out, or that he had begun to reduce speed as soon as he stated that he had. Nor was it likely that the signalman had looked along the up line before operating the lever of his up distant signal, situated outside on the Glasgow side of his box so that he could not see down the line while operating it. It was only after returning to his box that he had seen the danger and attempted to remedy matters.

Perhaps surprisingly, Colonel Yolland did not attach any blame to Mayberry, the crane operator, whom he said had not been supplied with the company's timetable or book of regulations. As to whether a guard should have accompanied the travelling crane the inspecting officer thought he should but the officers of the North British Company

---

[1471] Board of Trade report

had varying opinions; Mr McLaren, the superintendent of the line, thought that he should but Mr Wheatley, the locomotive superintendent, disagreed, stating that the stationmaster would still be responsible for the safety of the travelling crane while it was working. The inspecting officer felt that there was no excuse for the stationmaster absenting himself while the crane was working, especially as it would have given him his first opportunity in 17 years of seeing one in action. Had he been present he would have seen at once that the rear of the crane was fouling the up line and taken measures to protect it or stopped the crane from working until the express had passed. The inspecting officer concluded his report by criticising the lack of brake power of the express, with only one brake, exclusive of the tender, to seven carriages, and recommending the at least partial introduction of a continuous brake.[1472 1473]

Stationmaster Kirkman, Crane Operator Mayberry and Signalman Stewart were all apprehended after the accident to face possible criminal charges. The signalman and stationmaster subsequently stood trial on a charge of culpable neglect at the Glasgow Circuit Court at the end of December. The jury found the case against Signalman Stewart 'Not proven' but Stationmaster Kirkman they found 'Guilty' and he was sentenced to four months' imprisonment by Lord Deas.[1474]

The passengers on the 4.25 p.m. mixed train from Toowoomba to Warwick, Queensland, on 1st February 1875 had a narrow escape. The train arrived at Cambooya at 5.25p.m. It consisted of an engine and tender, a brake van, three waggons, two sheep trucks, a first-class and two composite carriages. As the train departed from Cambooya a very severe storm broke out, with torrential rain, and although all the windows and guards were closed the rain penetrated the crevices until there were at least two inches of water in the passenger carriages. The train had travelled three miles towards Warwick when a sudden gust of wind hurled the carriages over onto their sides. Only the engine remained on the track and its tender and a truck were blown into a culvert. It was very fortunate that the only casualty was a Mr Wilson, 'of the Warwick Reserve' who suffered a sprained ankle. The passengers were taken back to Cambooya on two lengthsmen's trolleys and a special train was later sent from Toowoomba, news of the accident having been taken there by a man on horseback. The passengers spoke of having been well cared for after the accident, their only complaint being the lack of any telegraph facilities at Cambooya.[1475]

Collisions between trains on adjoining lines did not always involve derailed vehicles or open doors. Goods which had not been properly loaded onto a waggon and had become displaced during the journey could also cause accidents. Early on the morning of 3rd July 1877 the 12.35 a.m. London and North Western Railway up passenger train from Carlisle, England, left Crewe for the south at 5.12 a.m., two minutes late. It was in the charge of Guard Anthony Johnston with Driver John Pugh and Fireman Joshua Taylor and consisted of an engine and tender, a brake van, five carriages and a second brake van.

1472 Ibid.

1473 The Times, London, England, 18th October 1871

1474 Lloyds Weekly Newspaper, London, England, 31 December 1871

1475 The Maryborough Chronicle, Wide Bay and Burnett Advertiser, Queensland, 4th February 1875

Meanwhile, in the opposite direction a down goods train, in the charge of Brakesman John Newnes, had left Birmingham at 2 a.m. with Driver Joseph Hanchett and Fireman William Stone. It picked up 19 waggons at Bescot Junction, seven of which were loaded with channel bar iron. Brakesman Newnes told the subsequent Board of Trade inquiry that the channel bar iron waggons had been loaded in the usual manner, were about the usual weight and did not appear longer than usual. Prior to the accident the train last did work at Stafford, and left there with 41 loaded waggons, six empty waggons and a brake van. Before leaving Stafford, the waggons had been checked by Waggon Examiner William Gough, who stated that the bars had been loaded as they always were, straight, without chains and projecting five or six feet over the waggons, as they often did.

The goods train passed Signalman William Bailey at Wrine Hill box at 5.20 a.m. His post was on the down side, and hence he could not see the six-foot side of the train (the side next to the line on which the up passenger train would pass). He had noticed nothing unusual with the train.

Betley Road box stood on the up side at the north end of the station but Signalman James Jervis there saw nothing amiss with the down goods train either as it passed him. The Carlisle train was already in the section on the up line and passed his box shortly after the goods. Seeing damage to the passenger train and alarm signals from its occupants he sent the 'Stop and Examine Train' signal to Wrine Hill but then saw that the train had come to a stand 100 yards beyond his box anyway. He did not send any signal forward to stop the goods train, not being aware at that time that there was anything wrong with it.[1476]

As the passenger and goods trains were passing one another a displaced bar of channel iron struck the front brake van and the first three carriages of the passenger train, having somehow missed the engine and tender. Mrs Crofts, of Caton-Green, Lancashire, travelling with her husband and son from Lancaster to Market Harborough, was struck on the head and killed instantly, and one other person in the same compartment had his head cut by flying glass. Ironically, Mrs Crofts had just swopped seats with her husband because of a draught.[1477]

Major-General C.S. Hutchinson, the Board of Trade inspecting officer, noted that the accident had been the result of a method of loading channel iron bars which though it was said had been used for 20 years without an accident, had, no doubt, always been liable to cause one. The bar must have become displaced during shunting operations and gradually shifted its position, projecting further and further into the 6 ft space, and although it had cleared the preceding up mail train at Norton Bridge it had shifted far enough to catch the passenger train at Betley Road. It was quite possible that the bar had suddenly shifted its position and hence not been seen by the Betley Road signalman or by the goods train fireman when he last looked back at his train. To prevent a recurrence of the accident it was most desirable that bars projecting from the ends of waggons should be chained together or otherwise secured, and he considered that the railway company should refuse to accept loads of this description until they were so secured.[1478]

The inquest on Mrs Crofts returned a verdict of 'Accidental death,' the jury adding a

[1476] Board of Trade report

[1477] The Times, London, England, 5th July 1877

[1478] Board of Trade report

presentment that greater care should be taken in loading trucks with great bars of iron.[1479]

The railway between Wellington and Masterton in the New Zealand had to cross the Rimutaka mountain range and to avoid a long and expensive tunnel it included the very steep Rimutaka Incline, with three miles of 1 in 15, worked on the Fell system, with an extra centre rail for extra traction and braking.

At 8.30 a.m. on 11th September 1880 a mixed train left Greytown for Wellington, hauled from there by an "L" class engine, for which a special Fell engine was substituted at Cross Creek. From the latter station the train consisted of two passenger carriages, a brake van with Guard Osborne, the Fell locomotive, two goods waggons loaded with timber, produce and so on and finally the Fell brake van, with Fell Guard Berry. A moderate breeze was blowing as the train began its ascent of the mountain, but the wind strength gradually increased as the train steamed up the incline. About two-thirds of the way up the incline was a point known as Siberia Curve, where the line ran on a high embankment between a cutting and a tunnel, a place described as being bleak and desolate even in midsummer, and a point where especially strong winds blew down the mountainside. As the train emerged from the cutting onto the embankment, travelling only at about five miles an hour, it was struck by an especially fierce gust, and both carriages and the brake van were blown off the line. The leading carriage was blown against some lineside boulders and shattered, and its passengers were blown down the embankment with fragments of the carriage. Fortunately, the couplings held, and the four horizontal steel wheels of the Fell engine prevented it from being blown over and the carriages and van remained hanging over the embankment attached to the engine.

Seeing what had occurred Guard Berry detached the Fell brake van at the rear of the train and allowed it to descend by gravity to Cross Creek, where he could obtain assistance. As soon as the Fell van was detached the two goods waggons were also blown off the line by the gale, but they too remained attached to the engine. Three children, sisters Ida and Ellen Pharazyn, and a boy named Francis Nicholas were killed almost at once and a number of passengers were injured, some very badly, including a third sister, Ella Phararzyn, and Stanley Nicholas, brother of Francis, who later succumbed to his injuries to become the fourth fatality in New Zealand's first serious railway accident.[1480]

At the inquest on the victims of the accident the jury returned a verdict of 'Accidental death' and recommended that 'the Government should take immediate action, by the construction of wind sheds or other protective means, to prevent as far as possible the recurrence of similar accidents on the dangerous part of the incline.'[1481] Such 'breakwinds' were subsequently erected, and two locomotives were subsequently employed when there were strong winds.[1482]

[1479] The Times, London, England, 5th July 1877

[1480] Wellington Evening Post, Wellington, New Zealand, via Sydney Daily Telegraph, Sydney, New South Wales, 24th September 1880

[1481] New Zealand Times, Wellington, 13th September 1880, via Australasian Sketcher, Adelaide, 16th October 1880

[1482] Wikipedia – Rimutaka Incline railway accident

Date	Location	Railway Company	Description	Passenger Fatalities
26 Jun 1840	Portici, Naples, Kingdom of the Two Sicilies		Runaway train fell from uncompleted bridge	20
7 Aug 1840	Near Howden, England	Hull and Selby	Derailment, obstruction on line (faulty loading}	5
23 Aug 1840	B St Petersburg and Czarskojeselo, Russia		Head-on collision, single line.	?
24 Dec 1841	Sonning, England	Great Western	Derailment, Train ran into cutting slip	9
8 May 1842	B Belleville and Meudon (Versailles)	Chemin de fer de Paris à Versailles rive gauche	Derailment, Broken axle, fire	57
17 May 1844	Vieux-Dieux, Belgium		Derailment, broken axle? track defects?	8
1 Mar 1846	Near Pierre Bénite Tunnel, France	Chemin de fer de Saint Étienne à Lyon	Head-on collision	9
8 Jul 1846	Fampoux, France	Chemin de fer du Nord	Derailment, Excessive speed	15
24 Jul 1846	Near Monroe, New York, USA	New York and Erie	Derailment, broken wheel	6
5 Jun 1847	Wolverton, England	London and North Western	Train diverted onto wrong line, Collision	7
6 Nov 1847	Brookline, Massachusetts, USA	Boston and Worcester	Derailment, Defective vehicle	6
10 May 1848	Shrivenham, England	Great Western	Collision with Waggons	6
3 Nov 1848	Salem, Massachusetts, USA	Eastern	Train diverted onto wrong line, head-on collision	6
10 Feb 1849	Near Rockcliffe, England	Caledonian	Derailment, Engine defect	5
1 Aug 1850	Cowlairs, Scotland	Edinburgh and Glasgow	Rear collision	5
30 Apr 1851	Sutton Tunnel, England	Birkenhead Lancashire and Cheshire Junction	Rear collision in tunnel	6

15 Sep 1852	Almond, New York, USA	Erie	Rear collision	5
7 Oct 1852	Weirs, New Hampshire, USA	Boston Concord and Montreal	Rear collision	5
4 Mar 1853	Mount Union, PA USA	Pennsylvania	Rear collision	7
27 Mar 1853	Cheat River, West Virginia, USA	Baltimore and Ohio	Derailment, Track defects	9
25 Apr 1853	Chicago, Illinois, USA	Michigan Central/Michigan Southern	Side collision on rail/rail level crossing	21
6 May 1853	Norwalk, Connecticut, USA	New York and New Haven	Train fell from open drawbridge	46
12 Aug 1853	Valley Falls, RI, USA	Providence and Worcester	Head-on collision, single line	12
5 Oct 1853	Near Straffan, Ireland	Great Southern and Western	Rear collision	16
2 Jun 1854	Near Lobo, Canada	Great Western of Canada	Collision with cow, derailment	6
4 July 1854	Riderwood, Maryland, USA	Baltimore and Susquehanna	Head-on collision, single line	34
6 Jul 1854	Near Thorold, Canada	Great Western of Canada	Collision with horse, derailment	7
27 Oct 1854	Baptiste Creek, Canada	Great Western of Canada	Head-on collision, single line	55
Nov 1854	Near Minoka (Near Joliet), Illinois, USA	Chicago and Rock Island	Derailment, Broken axle	16
18 Nov 1854	Near Dowagiac, Michigan, USA	Michigan Central	Collision	7
29 Aug 1855	Burlington, New Jersey, USA	Camden and Amboy	Collision with horse and carriage on open level crossing, derailment	23
9 Sep 1855	Vaugirard. France	Chemins de fer de l'Ouest	Sidelong collision?	9
21 Oct 1855	B Thomery and Moret, France	Chemin de fer de Paris à Lyon	Rear collision	16
1 Nov 1855	Gasconade River, Missouri, USA	Pacific of Montana	New bridge collapsed under train	29
30 Apr 1856	Near Wilton, Iowa, USA	Mississippi and Missouri	Derailment	12
6 May 1856	'About 9 miles from Aspinwall,' New Granada	Panama	Derailment, Excessive speed over poor track	43
17 Jul 1856	Camp Hill, Pennsylvania, USA	North Pennsylvania	Head-on collision, single line, Boiler explosion, Fire	66
22 Aug 1856	Laufach, Bavaria		Runaway coaches, Derailment	6

16 Sep 1856	Near Villasequilla, Spain	Aranjuez	Fire	5
8 Dec 1856	Alliance, Ohio, USA	Cleveland and Pittsburg/Pittsburg Fort Wayne and Chicago	Side collision on rail/rail level crossing	8
26 Feb 1857	Near Long Mountain Tunnel, Pennsylvania, USA	Pennsylvania	Rear collision	6
12 Mar 1857	Hamilton, Canada	Great Western of Canada	Derailment, Broken axle. Bridge collapsed under train	58
28 Jun 1857	Lewisham, England	South Eastern	Rear collision	12
11 May 1858	Sauquoit Bridge, New York, USA	New York Central	Bridge collapsed under train	9
14 May 1858	Kafir Lais, Egypt		Waggons ran off end of line	?
31 May 1858	Near La Louvière, Belgium		Runaway waggons, Collision	21
15 Jul 1858	Shin Hollow, New York, USA	Erie	Derailment, Broken rail	6
23 Aug 1858	B Round Oak and Brettel Lane, England	Oxford Worcester and Wolverhampton	Runaway coaches, collision	14
31 Dec 1858	Adam's Mills, Georgia, USA	Columbus and Macon	Bridge collapsed under train	13
4 May 1859	San Michele, near Porta Vescovo, Venetia		Head-on collision? explosion	23
15 Jun 1859	Near Wooster, Ohio, USA	Pittsburg Fort Wayne and Chicago	Collision with horse, derailment	5
27 Jun 1859	Springbrook Bridge, Mishawaka, Indiana, USA	Michigan Southern and Northern Indiana	Derailment, Track subsidence	39
25 Jul 1859	San Paolo, (Kingdom of) Sardinia		Head-on collision, single line	11
1 Nov 1859	Johnson's Creek, Wisconsin, USA	Chicago and North Western	Collision with ox, Derailment	11
20 Feb 1860	Tottenham, England	Eastern Counties	Derailment, Broken tyre	5
4 Sep 1860	Helmshore, England	Lancashire and Yorkshire	Runaway coaches, Collision	11
14 Sep 1860	Delavan, Wisconsin, USA	Racine and Mississippi	Rear collision	5
28 Sep 1860	B Riga and Dünaburg, Russia		Collision	5
16 Nov 1860	Atherstone, England	London and North Western	Collision with shunting train	9

25 Aug 1861	Clayton Tunnel, England	London Brighton and South Coast	Rear collision in tunnel	23
1 Sep 1861	Near Cockeysville, Maryland, USA	Northern Central	Derailment, excessive speed, sabotage	7
2 Sep 1861	Kentish Town, England	Hampstead Junction	Collision with shunting train	15
3 Sep 1861	Platte River Bridge, Missouri, USA	Hannibal and St Joseph	Bridge collapsed under train, Sabotage	17
17 Sep 1861	Huron, Indiana, USA	Ohio and Mississippi	Bridge collapsed under train, Probable sabotage	28
26 Sep 1861	Near Paris Nord, France	Chemin de fer du Nord	Sidelong collision, Boiler explosion	5
15 Jan 1862	Near St George's, Delaware, USA	Delaware	Train fell from open drawbridge	5
27 Feb 1862	Near Ponchatoula, Louisiana, USA	New Orleans Jackson and Great Northern	Head-on collision, single line in fog	34
15 Mar 1862	Near Tivoli, New York, USA		Derailment, Broken rail	5
26 March 1862	Near Chicago, Illinois	Chicago and North Western	Derailment, Broken axle	10
13 Oct 1862	Near Winchburgh, Scotland	Edinburgh and Glasgow	Head-on collision during temporary single line working	16
19 Oct 1862	Near Duck Hill, Mississippi. USA	Mississippi Central	Head-on collision, single line	34
2 Nov 1862	Near Cleveland, Tennessee, USA	East Tennessee and Virginia	Derailment, Broken axle	17
7 Nov 1862	Near Lafourche, Louisiana, USA	Algiers	Explosion	11
19 Feb 1863	Chunky Creek Bridge, Mississippi, USA	Southern	Bridge collapsed under train	40
5 Jun 1863	Malkinia, Russia		Derailment, Sabotage	153
3 Aug 1863	Near Wootton, England	Great Eastern	Collision with bullock, Derailment	7
7 Oct 1863	Alabern Bridge, Hostalric, Spain	Gerona	Bridge collapsed under train	21
5 Jan 1864	Pierrefitte, France	Chemin de fer du Nord	Engine suffered broken connecting rod, rear collision, fire	5
7 Jun 1864	Egham, England	London and South Western	Rear collision	6
27 June 1864	B Titwalla and Wassind, India	Great Indian Peninsular Railway	Derailment, Track washed away	15

29 Jun 1864	Beloeil (St Hilaire), Canada	Grand Trunk	Train fell from open drawbridge	97
15 Jul 1864	Near Shohola, Pennsylvania, USA	Erie	Head-on collision, single line	70
1 Sep 1864	Barnesville, Georgia, USA	Macon and Western	Head-on collision, single line	54
21 Sep 1864	Thompsontown, Pennsylvania, USA	Pennsylvania	Rear collision, Fire	5
15 Oct 1864	Near Old Saybrook, Connecticut, USA	Shore Line	Derailment, broken rail	11
31 Oct 1864	Near Culver's, Indiana, USA	Lafayette and Indianapolis	Head-on collision, single line	28
23 Dec 1864	Sandsjö, Sweden		Collision with level crossing gates, Collision	7
7 Jan 1865	Gongitota, Ceylon		Collision with platelayer's trolley	?
3 Mar 1865	Near New Orleans, Louisiana, USA	New Orleans Opelousas and Great Western	Collision with mule, Derailment	10
7 Mar 1865	Bristol, Pennsylvania, USA	Pennsylvania	Rear collision, Fire	6
7 Jun 1865	Near Rednal, England	Great Western	Derailment, Inadequate protection of PW work.	11
9 Jun 1865	Staplehurst, England	South Eastern	Bridge collapsed under train, Inadequate protection of bridgework	11
22 Jun 1865	Loogootee, Indiana, USA	Ohio and Mississippi	Double collision	15
c18 Jul 1865	Near Fort Deposit, Alabama, USA	Mobile and Montgomery	Head-on collision?	5
21 Jul 1865	B Summit and Pataskala, Ohio, USA	Central Ohio	Derailment, Engine defect	9
30 Jul 1865	B Titwalla and Wassind, India	Great Indian Peninsular	Derailment, Cause unknown	8
2 Aug 1865	Cawnpore, India	East Indian	Train diverted onto wrong line, Collision	7
15 Aug 1865	Near Bridgeport, Connecticut, USA	Housatonic River	Rear collision, Boiler explosion	12
24 Aug 1865	Titusville, Pennsylvania, USA	Oil Creek	Head-on collision, single line	6
25 Aug 1865	Near Reynolds, Tennessee, USA	Nashville and Decatur	Trestle collapsed under train	35
28 Aug 1865	Near Jamaica, New York, USA	Long Island	Head-on collision, single line	5

7 Sep 1865	Nashville, Tennessee, USA	Nashville and North Western	Explosion	6
14 Oct 1865	Lancaster, Pennsylvania, USA	Pennsylvania	Derailment, Broken axle	8
1 Dec 1865	White House, New Jersey, USA	Central of New Jersey	Runaway waggons, Collision	7
1 Dec 1865	White House, New Jersey, USA	Central of New Jersey	Runaway waggons, Collision	7
5 Aug 1866	B Brünn and Prerau, Bohemia		Derailment, High speed with unstable rolling stock	5
2 Sep 1866	Kafres-Kayat, Egypt		Boiler Explosion	5
4 Sep 1866	Near St Maixent, France	Chemin de fer de Paris à Orléans	Derailment, Broken axle	6
6 Sep 1866	Brynkir, Wales	Carnarvonshire	Derailment on points	6
21 Sep 1866	Kolontaievska, Russia		Collision	14
13 Nov 1866	Wesleyville, Pennsylvania, USA		Derailment, Inadequate protection of PW work	5
23 Dec 1866	B Franois and Besançon, France	Chemins de fer de Paris à Lyon et à la Méditerranée	Head-on collision, single line	12
28 Dec 1866	B Zwickau and Werdau, Saxony		Train diverted onto wrong line, Head-on collision in blizzard	8
20 Feb 1867	B Koslow and Riäsan, Russia		Derailment, Excessive speed? Broken axle?	10
23 May 1867	Whitby, Canada	Grand Trunk	Rear collision	5
26 Jun 1867	B Bhosawul and Khundwah, India	Great Indian Perninsular	Bridge washed away, Train fell into river	60
29 Jun 1867	Walton Junction, Warrington, England	London and North Western	Train diverted onto wrong line, Collision	8
1 Aug 1867	St Albain. France	Chemins de fer de Paris à Lyon et à la Méditerranée	Derailment, Inadequate protection of PW work	6
5 Sep 1867	Thomaston, Connecticut, USA	Naugatuck	Bridge collapsed under train	10
9 Sep 1867	B Dove Holes Tunnel and	Midland	Rear collision in tunnel, Runaway	5

	Romiley, England		waggons, Collision, Runaway train	
21 Nov 1867	Lockland, Ohio, USA	Cincinati Hamilton & Dayton	Rear collision, Fire	5
18 Dec 1867	Angola, New York, USA	Buffalo and Erie	Derailment, Defective vehicle, Part of train fell from bridge, Fire	42
14 Apr 1868	Carr's Rock, N Y, USA	Erie	Derailment, Sabotage, Fire	6
7 May 1868	Near Samnuggur, India	Eastern Bengal	Derailment, Points moved under train	17
8 Jul 1868	Union Mills, Pennsylvania, USA	Philadelphia and Erie	Bridge collapsed under train	5
20 Aug 1868	Near Abergele, Wales	London and North Western	Runaway waggons, Collision, Fire	31
10 Nov 1868	Near Horowitz, Bohemia		Rear collision in blizzard	29
25 Jan 1869	Bhor Ghat Incline, India	Great Indian Peninsular	Runaway train, Derailment	18
23 Apr 1869	Willow, near Jamaica, New York, USA	Long Island	Derailment, Broken rail	5
28 Jun 1869	Near Etola, India	Bombay Baroda and Central India	Collision with bullock, Derailment	11
14 Jul 1869	Mast Hope, Pennsylvania, USA	Erie	Sidelong collision, Fire	8
10 Oct 1869	Long Eaton Junction, England	Midland	Rear collision in fog	7
14 Nov 1869	Near Brooklyn Park (Near San Leandro), California, USA	Western Pacific	Head-on collision, single line in fog	9
25 Feb 1870	Oxford, Mississippi, USA	Mississippi Central	Derailment, Track defects, Train fell from trestle	19
8 May 1870	Dommeldange, Luxembourg	William Luxemburg	Runaway train? Collision	12
12 May 1870	Near Eureka, Missouri, USA	Missouri Pacific	Head-on collision, single line	19
21 Jun 1870	B Newark and Claypole, England	Great Northern	Derailment, Broken axle, Collision	16
10 Jul 1870	St Nicholas Crossing, Carlisle, England	London and North Western/North Eastern	Side collision on rail/rail level crossing	5
6 Aug 1870	Jerry's Run, near Covington, Virginia, USA	Chesapeake and Ohio	Derailment, broken wheel	12

9 Aug 1870	Near Salina, Kansas, USA	Kansas Pacific	Runaway train, Collision	6
20 Sep 1870	Plessis, France		Rear collision	11
4 Oct 1870	Critot, France		'Buffer stop' collision at end of trap siding	14
26 Nov 1870	Harrow, England	London and North Western	Collision with shunting goods	7
12 Dec 1870	Stairfoot, England	Manchester Sheffield and Lincolnshire	Runaway waggons, Collision	15
26 Dec 1870	Near Bell Bar, Hatfield, England	Great Northern	Derailment, Broken tyre, Collision	6
1 Jan 1871	Saint-Sornin, France		Derailment	11
2 Jan 1871	B Jerez de la Frontera and El Cuervo, Spain		Derailment, Sabotage?	7
3 Jan 1871	Nonconnah Bottom, Tennessee, USA	Mississippi and Tennessee	Derailment, Broken axle, Fire	7
5 Feb 1871	B Bandoz and St Nazaire, France	Chemins de fer de Paris à Lyon et à la Méditerranée	Explosion	80
6 Feb 1871	New Hamburg, New York, USA	New York Central and Hudson River	Derailment, Broken axle, Collision, Fire, Trestle collapsed under train	18
26 Feb 1871	Near Fougeray-Langon, France	Chemins de fer de l'Ouest	Head-on collision, single line	9
10 Mar 1871	Puteaux, France	Chemins de fer de l'Ouest	Rear collision	10
18 Mar 1871	Colombier, Switzerland		Collision	24
12 May 1871	Griswold, New York, Erie	Erie	Rear collision	5
4 Jun 1871	Ablon, France	Chemins de fer de Paris à Orléans	Rear collision	24
21 June 1871	B Rackwitz and Zschortau Germany		Derailment, Defective vehicle	18
2 Jul 1871	Pont-Mort, France	Chemin de fer de Paris à Lyon et à la Méditerranée	Derailment, Track defects?	6
3 Jul 1871	Kingston Springs, Tennessee, USA	Nashville and Northwestern	Bridge collapsed under train	18

25 Jul 1871	Near Edwardsville, Illinois, USA	Toledo Wabash and Western	Head-on collision, single line	6
29 Jul 1871	Near Navasota, Texas, USA		Bridge collapsed under train	7
26 Aug 1871	Revere, Massachusetts, USA	Eastern	Rear collision, Fire	29
3 Sep 1871	Seclin, France	Chemin de fer du Nord	Collision with shunting train	9
16 Sep 1871	Near Pont-sur-Yonne, France	Chemin de fer de Paris à Lyon et à la Méditerranée	Derailment, Broken axle,	11
1 Feb 1872	Wild Run, Pennsylvania, USA	Lehigh Valley	Derailment, Broken rail, Fire	8
22 Jun 1872	Near Shannonville, Ontario, Canada	Grand Trunk	Derailment, Broken wheel	31
3 Aug 1872	Agecroft Sidings, England	Lancashire and Yorkshire	Sidelong Collision	6
5 Sep 1872	Near Independence, Ohio, USA	Baltimore and Ohio	Head-on collision, single line	7
9 Sep 1872	B Hospitalet and L'Ametlla, Spain		Bridge collapsed under train	14
2 Oct 1872	Kirtlebridge, Scotland	Caledonian	Train diverted onto wrong line, Collision	11
24 Dec 1872	Prospect, Pennsylvania, USA	Buffalo Corry and Pittsburg	Derailment., Broken wheel, Part of train fell from bridge, Fire	25
2 Jan 1873	B Bear Creek and Lovejoy's, Georgia, USA	Macon and Western	Head-on collision, single line	7
19 Apr 1873	Richmond Switch, Rhode Island, USA	New York Providence and Boston	Bridge washed away, Train fell into river, Fire	5
4 May 1873	Near Pesth, Hungary	Austrian State	Derailment	21
8 Jun 1873	Yuquery Incline, Paraguay		Runaway train, Collision	5
2 Aug 1873	Wigan, England	London and North Western	Derailment on points	13
16 Aug 1873	B Willow Springs and Lemont, Illinois, USA	Chicago and Alton	Head-on collision, single line in fog	17

28 Aug 1873	Near Muir, Michigan, USA	Detroit and Milwaukee	Rear collision	6
11 Sep 1873	Near Viana de Cega, Spain		Derailment, Sabotage? Part of train fell from bridge	17
27 Jan 1874	Bo'ness Junction, Scotland	North British	Collision with shunting goods	16
28 Feb 1874	Komoka, Ontario, Canada	Great Western of Canada	Fire	10
11 Mar 1874	Montabliz, Spain		Runaway waggons, Collision	9
10 Sep 1874	B Norwich Thorpe and Brundall, England	Great Eastern	Head-on collision, single line	22
24 Dec 1874	Near Shipton-on-Cherwell, England	Great Western	Derailment, Broken tyre	34
5 Jul 1875	Near Far Rockaway, New York, USA	Southern of Long Island	Head-on collision, single line	8
6 Jul 1875	Near Limache, Chile		Bridge collapsed under train, Fire	5
14 Aug 1875	Near Jubbulpore, India	East India	Derailment, engine defect	9
28 Aug 1875	Kildwick, England	Midland	Rear collision	7
28 Sep 1875	Yamaska, Quebec, Canada	South Eastern	Derailment, Vandalism	10
4 Nov 1875	Near Schanzenau, Austria		Derailment, Sabotage	7
15 Nov 1875	Lagerlunda, Sweden		Head-on collision, single line	5
8 Jan 1876	Filigut, Near Odessa, Russia		Derailment, Inadequate protection of PW work, Fire	68
21 Jan 1876	Abbot's Ripton, England	Great Northern	Double collision in blizzard	14
6 Mar 1876	Near Bowman, Virginia, USA		Bridge collapsed under train	8
24 Jun 1876	B Tàrrega and Cervera de Segarra, Spain		Derailment, Defective vehicle? Excessive speed?	17
7 Jul 1876	Near Palézieux, Switzerland		Head-on collision	10

7 Aug 1876	Foxcote, England	Somerset and Dorset Joint	Head-on collision, single line	12
25 Dec 1876	Near Amboor, India	Madras	Head-on collision, single line	12
29 Dec 1876	Ashtabula, Ohio, USA	Lake Shore and Michigan Southern	Bridge collapsed under train, Fire	92
25 Mar 1877	Morpeth, England	North Eastern	Derailment, track defects	5
29 Aug 1877	Four Mile Creek, Near Des Moines, Iowa, USA	Chicago and Rock Island	Bridge washed away, Train fell into stream	17
4 Oct 1877	Phoenixville, Near Kimberton, Pennsylvania, USA	Pickering Valley	Derailment, Track washed away	8
15 Jan 1878	Tariffville, Connecticut, USA	Western Connecticut	Derailment, cause unknown, Bridge collapsed under train	12
3 Feb 1878	Lübben, Germany		Sidelong collision	5
8 May 1878	B Klein Poort and Barkley Junction, Cape Colony	Cape Government	Derailment	7
7 Aug 1878	Near Mingo Junction, Ohio, USA		Head-on collision, single line	18
31 Aug 1878	Sittingbourne, England	London Chatham and Dover	Collision with waggons	5
8 Oct 1878	Wollaston, Massachusetts, USA	Old Colony	Derailment, Points moved under train	21
12 Oct 1878	Portugal	Minho	Derailment	7
19 Oct 1878	Near Pontypridd, Wales	Taff Vale	Sidelong collision	13
19 Dec 1878	B Rostov and Vladikavkas, Russia		Derailment	18
10 Jan 1879	Suide, Russia		Collision with shunting goods	15
11 Jan 1879	B Philippopolis and Adrianople, Turkey		Bridge collapsed under train	200
9 May 1879	Near Cádiz, Spain		Derailment	5
3 Aug 1879	Xeuilley, France	Chemins de fer de l'Est	Train diverted onto wrong line. Buffer stop collision. Sabotage	12

15 Aug 1879	B Flers and De Vassy Montsecret, France	Chemins de fer de l'Ouest	Head-on collision, single line, Fire	5
10 Oct 1879	Jackson, Michigan, USA	Michigan Central	Collision with light engine	14
31 Oct 1879	Near Arconum, India	Madras	Derailment, Track subsidence	19
28 Dec 1879	Tay Bridge, Scotland	North British	Bridge collapsed under train during gale	56
15 Jan 1880	Brickfield Sidings, Near Burscough, England	Lancashire and Yorkshire	Head-on collision	5
3 Feb 1880	Near Clichy Junction, France		Rear collision in fog	11
17 Feb 1880	Near Mittri, India		Collision with waggons	7
20 Mar 1880	Halle, Germany		Collision	7
24 May 1880	Near Santa Cruz, California, USA	South Pacific Coast	Derailment, Excessive speed	16
11 Aug 1880	Wennington Junction, England	Midland	Derailment, Excessive speed, Slight track defects	8
11 Aug 1880	Mays Landing, New Jersey, USA	West Jersey and Atlantic	Rear collision	26
11 Sep 1880	Loco Junction, Nine Elms, England	London and South Western	Collision with light engine	5
9 Oct 1880	28th Street Crossing, Pittsburg, Pennsylvania, USA	Pennsylvania	Rear collision	32
23 Dec 1880	Indian Creek Trestle, Near Lincolnton, North Carolina, USA	Carolina Central	Trestle collapsed under train, Fire	6

INDEX

ABOUT THE AUTHOR

The author has been interested in railways for many years, ever since he was invited up a signal box at the age of 14 back in 1966. Not long afterwards he 'cabbed' the engine of a goods train which was shunted out of the way of a faster train and its driver showed him a copy of 'Red for Danger,' the late L.T.C. Rolt's book about British railway accidents. This was the first railway book that he ever bought, several month's saved up pocket money going towards a paperback copy of it. Later he came across a copy of 'A History of Railroad Accidents, Safety Precautions and Operating Practices' by Robert B. Shaw, dealing with American railroad accidents, and his interest in the subject grew. When the local signal boxes were all done away with in a modernization scheme in the mid-1980's he began a more detailed study of world railway accidents. There was no internet back then of course and many hours were spent in the local reference library slogging through bound volumes of The Times, writing out longhand in large notebooks details of every accident found. Later it was possible to view the paper on microfilm and print pages, which made things a little easier, and with the internet the amount of information available vastly increased. Some years ago, the author contributed to 'Railway Disasters of the World' by P.W.B. Semmens (which deals only with 20th century accidents) and more recently the idea came of writing his own book or books on world accidents, starting at the beginning. This book was initially going to cover the whole of the 19th century but there proved to be too much to cover so it stops at 1880, with hopefully further volumes to follow.